A COMPANION VOLUME TO

Collegiate Dictionary of Botany

BY DELBERT SWARTZ

ROBERT W. PENNAK, Ph.D., University of Wisconsin, is Professor of Biology and Chairman of the Department at the University of Colorado. He has also served as a visiting investigator at the Oceanographic Institute at Woods Hole, Massachusetts. Dr. Pennak is member of a number of professional societies and is Past President of the American Society of Limnology and Oceanography, the Society of Systematic Zoology, the Invertebrate Section of the American Society of Zoologists, and the American Microscopical Society. He has contributed a great many articles on lake and stream biology to professional journals and is the author of *Fresh-Water Invertebrates of the United States*, published by The Ronald Press Company.

COLLEGIATE DICTIONARY OF ZOOLOGY

ROBERT W. PENNAK
UNIVERSITY OF COLORADO

ROBERT E. KRIEGER PUBLISHING COMPANY
MALABAR, FLORIDA
1988

Original Edition 1964
Reprint Edition 1988

Printed and Published by
ROBERT E. KRIEGER PUBLISHING COMPANY, INC.
KRIEGER DRIVE
MALABAR, FLORIDA 32950

Printed in the United States of America

Library of Congress Cataloging-in-Publication Data

Pennak, Robert W. (Robert William)
Collegiate dictionary of zoology.

Reprint. Originally published: New York : Ronald Press, 1964.
1. Zoology—Dictionaries. I. Title.
QL9.P4 1988 591'.03'21 85-23983
ISBN 0-89874-921-2

10 9 8 7 6 5 4 3 2

Dedicated to my students

PREFACE

The total number of technical terms and proper names in the literature of zoology is truly enormous, and the average non-zoologist does not realize that many of the former and most of the latter are not to be found in the largest encyclopedias and unabridged dictionaries.

Even for the trained zoologist, it is sometimes difficult to locate and ascertain the precise meaning of an unfamiliar zoological term, especially since many terms are used with two or more distinct meanings. Most terms can eventually be found in various references and textbooks, but sometimes such words are merely used in context, without an attempt at a definition. In general, textbook "glossaries" are quite incomplete and superficial. A few special fields of zoology are fortunate in having excellent dictionaries or extensive glossaries, e.g. entomology and genetics; certainly medicine is well fortified with dictionaries (one well-known medical dictionary has more than 100,000 entries).

Aside from the "common" (uncapitalized) terms in zoology, an even greater problem is the frequent use of generic and specific names, as well as the proper names of higher taxa. Although a zoologist of long experience is able to recognize or associate a surprisingly large number of such proper names, most of these are apt to be widely used or in groups of animals which are his specialties. For example, nearly every zoologist is familiar with *Paramecium, Unio, Salmo, Drosophila,* and *Microtus*. Some will recognize *Bodo, Nucula, Nemoura, Myxocephalus,* and *Dasypus*. Few, however, are able to identify *Haplosporidium, Hydridella, Atilax, Stenosmylus,* and *Plecodus*. The magnitude of this problem is emphasized by the fact that the *Nomenclator Zoologicus* lists more than 200,000 different generic names!

Theoretically, a "complete" dictionary of zoology is hopelessly and inextricably scattered through a large number of texts, reference works, and zoological periodicals.

On a much simpler scale, most of us who teach zoology courses are convinced that a major task of both the undergraduate and graduate student is the acquisition of a working vocabulary. Such a vocabulary is necessary before the student can gain an understanding of some of the ideas, principles, and countless facts that make up the many ramifications of modern zoology. Indeed, "vocabulary frustration" is one of the most common hurdles in zoology.

This conviction, plus the inadequacy of many textbooks and the lack of a suitable one-volume dictionary, were collectively responsible for my decision to compile the present book. It contains about 19,000 selected entries, of which about 8,600 are proper names. The entries are "zoological" and rather specific. Most entries are necessarily quite brief. All zoological disciplines are included, e.g. taxonomy, morphology, anatomy, ecology, genetics, physiology, embryology, histology, cytology, etc.

Although the great majority of entries reflect my own preferences and interpretations, there are many with which I am not in agreement but which are nevertheless worded in accordance with prevailing textbook usage. The emphasis is on items and terminology as they are used in the United States, and the geographical distributions of many taxa are elaborated only with reference to the United States (or North America). The new *International Code of Zoological Nomenclature* specifies the elimination of diacritic marks but many are nevertheless retained in this volume in keeping with common textbook usage in the past.

Many zoological terms, especially the names of various taxonomic categories, are frequently used in the more generalized (and uncapitalized) sense, e.g.: Ascidiacea—ascidian, Brachiopoda—brachiopod, Distomata—distome, Hydrophilidae—hydrophilid, Hydrozoa—hydrozoan, Metazoa—metazoan, Passeriformes—passerine or passeriform, Tricladida—triclad. The inclusion of such terms would have resulted in unreasonable repetition, lengthy cross-referencing, and a volume much larger than desirable. They are, however, almost invariably understood by referring to the obvious capitalized parent term.

It is hoped that the taxonomic *Appendix* will serve as a useful reference and adjunct to the main body of this volume.

Our intentions are broad, and this book is written for students, research zoologists, libraries, and high school, college, and university teachers of biology and zoology.

Several of my colleagues at the University of Colorado have been most helpful in critically reading certain of the entries in preliminary drafts of the manuscripts; I am especially indebted to Dr. Robert E. Gregg, Dr. E. R. Helwig, Dr. T. Paul Maslin, Dr. Charles H. Norris, and Dr. Olwen Williams. I am also indebted to those anonymous referees who were appointed by The Ronald Press Company to criticize entries for certain of the zoological fields represented in this volume. I must, however, be completely candid and confess that it is difficult to get two zoologists to agree on the precise wording of the definitions of many common zoological terms, including such familiar examples as: meiosis, mitosis, Mendelism, osmosis, niche, vitamin, evolution, Krebs cycle, and cell membrane. All lapses and errors are strictly my own responsibility, especially since I have personally typed the final copy from which the plates have been made.

Robert W. Pennak

Boulder, Colorado
January, 1964

Å. Abbreviation for angstrom unit.
aardvark. See TUBULIDENTATA.
aardwolf. See PROTELES.
abacterial. Free of bacteria.
abalone. See HALIOTIS.
Abastor. See RAINBOW SNAKE.
abaxial. Out of or directed away from the central axis of a part or body.
Abbé condenser. Substage system of wide-angled lenses used in concentrating light of a compound microscope.
abdomen. 1. In vertebrates, that portion of the trunk containing the chief visceral organs except for the heart and lungs. 2. In arthropods, the posterior portion of the body composed of a group of similar segments and containing especially the reproductive organs and posterior portion of the digestive system.
abdominal breathing. Diaphragmatic breathing; that fraction of breathing movements produced by the up and down movements of the diaphragm combined with in and out movements of the abdominal wall.
abdominal cavity. Chief body cavity in the abdominal region of vertebrates and arthropods; in mammals, it is separated from the thoracic cavity by the diaphragm.
abdominal rami. Caudal rami.
abdominal ribs. Riblike ossifications in the fibrous connective tissue of the abdominal wall of certain reptiles.
abdominal vein. Large vein draining the legs and lower abdominal region of the lower vertebrates.
abducens nerves. One of the pairs of cranial nerves of vertebrates; innervate the externus rectus muscles of the eye; motor nerves.
abduction. Drawing away from the body axis, e.g. raising the arm away from the side of the trunk.
abductor. Muscle that draws a part away from the axis of the body or limb.
Abedus. Common genus of giant water bugs in the Family Belostomatidae.
Abietinaria. Genus of marine hydrozoan coelenterates.
abiocoen. Collectively, the non-living factors or aspects of an environment.
abiogenesis. See SPONTANEOUS GENERATION.
abioseston. Tripton.
Ablepharus. Genus of Old World lizards; lower eyelid a transparent membrane fused to the upper lid.
abomasum. Fourth and last chamber of stomach of ruminants; receives food from the omasum and passes it on to the small intestine; often called the "true" stomach.
aboral. Of or pertaining to a region away from or most distant from the mouth.
aboral stomach. Pyloric stomach of a sea star.
abortion. Premature expulsion of a fetus, especially before it is independently viable; miscarriage.
Abothrium. Genus of pseudophyllidean tapeworms; in intestine of fishes.
Abra. Genus of small orb-shaped marine bivalves; Atlantic and Gulf coasts.
Abramis. See BREAM.
abranchiate. Without gills.
Abrocomidae. Family of rat-sized Bolivian rodents; rat-chinchillas.
abscess. Localized mass of pus formed from inflammation and tissue disintegration in any part of the body, especially by bacterial action.
absolute zero. Theoretical lowest attainable temperature, -273°C.; all reactions and atomic motion cease at this temperature.
Abyla. Genus of coelenterates in the Order Siphonophora; with two unlike nectophores.
abysmal. Abyssal; see ABYSS.
abyss. That portion of a body of water below a depth of 1000 m.; 6000 ft. or 4000 m. are also sometimes used as upper limits of the abyss. Adj. abyssal.
abyssobenthos. Collectively, those organisms living on the sea bottom at

depths of more than 1000 m.; term sometimes restricted to depths exceeding 4000 m.

Acadian epoch. Middle of three subdivisions of the Cambrian geological period.

Acadian flycatcher. Small flycatcher of eastern and central states.

acaleph. Obsolete taxon formerly used to include the medusoid coelenterates.

Acalyptratae. Large group of families of Diptera having the second antennal segment without a dorsal seam.

Acanthella normani. Bush coral, an anthozoan in the Order Alcyonacea.

acantha. 1. Any spine. 2. Spinous process of a vertebra.

Acantharchus pomatis. Mud sunfish, a small centrarchid of the coastal plain, N. Y. to Fla.

acanthella. Early developmental stage of an acanthocephalan; develops from a newly hatched egg and forms by growth and differentiation of the acanthor.

Acanthella. Genus of marine sponges in the Subclass Monaxonida; folded leaflike expansions form cuplike spaces; warm coastal waters.

acanthin. Organic internal skeletal material of radiolarians.

Acanthina. Genus of knobbed and ridged marine snails; drills.

Acanthion. Genus of forest-dwelling porcupines; quills only on posterior part of body; southeastern Asia.

Acanthis. Genus which includes the redpolls; small finches with streaked, gray-brown coloration and bright red cap.

Acanthisittidae. Small family of tiny insectivorous wrenlike New Zealand passerine birds.

Acanthobdella. See ACANTHOBDELLIDA.

Acanthobdellida. One of the three orders of leeches; anterior sucker, jaws, and proboscis absent; coelom divided by transverse septa; two pairs of setae on each of segments II to IV; one sp. of Acanthobdella on fish in Lake Baikal.

Acanthocephala. Phylum which includes the spiny-headed worms; adults live in the intestine of vertebrates and larvae (acanthors) in the tissues of crustaceans and insects; an anterior retractile proboscis bears rows of hooks; short neck region and a long trunk; flame bulb system present; body cavity a pseudocoel with no sign of a digestive tract; body wall composed of a cuticle, thick syncitial hypodermis, circular muscle fibers, and longitudinal muscle fibers; adult male has a full complement of reproductive organs, but the adult female contains only a large mass of eggs; about 300 spp.; Macracanthochynchus and Neoechinorhynchus typical.

Acanthocephalus. Genus of Acanthocephala; adults parasitic in fishes and amphibians.

Acanthocheilonema. Dipetalonema.

Acanthochondria. Genus of bizarre copepods parasitic on marine flatfishes.

Acanthocotyle. Genus of external trematode parasites of fishes.

Acanthocybium solandri. Wahoo; a solitary mackerel of the Gulf Stream and tropical reefs.

Acanthocystis. Genus of spherical heliozoan protozoans; cell covered with siliceous scales.

Acanthodes. Genus of extinct placoderm fishes having a single dorsal fin; Carboniferous and Permian.

Acanthodii. Order of extinct placoderm fishes; body to several in. long, covered with small diamond-shaped bony scales; some spp. with a row of extra paired fins between pectorals and pelvics; fresh water; Climatius and Acanthodes best known.

Acanthodoris. Genus of nudibranchs.

Acanthogorgia. Genus of marine coelenterates in the Order Alcyonacea; colony profusely branched; deep waters.

acanthoid. Spine-shaped, spiny, or spinous.

Acantholeberis curvirostris. Uncommon littoral cladoceran; mostly in bog lakes.

Acanthometra. Genus of radiolarians.

Acanthometron. Genus of radiolarians.

Acanthonosotoma. Genus of marine amphipods.

Acanthophacetus reticulatus. Millions; small Barbados fish which feeds on mosquito larvae and pupae.

Acanthophis. See DEATH ADDER.

Acanthopleura. Genus of Caribbean chitons.

Acanthopterygii. Superorder of fishes; includes most bony fishes; term seldom used in modern taxonomy.

Acanthoptilum. Genus of elongated, slender sea pens.

acanthor. Microscopic spiny larva of Acanthocephala, contained within the egg shell.

acanthosoma. Larval stage of certain marine decapods.

acanthostyle. Sponge spicule which is rounded at one end and pointed at the other; covered with small thorny processes.

Acari. Acarina.

acariasis. Mite infestation.

Acarina. Large order of small to microscopic arachnids; includes ticks and mites; cephalothorax and abdomen fused into a single ovoid mass; segmentation lacking; chelicerae and pedipalps variously developed; respiration by tracheae or general body surface; life history usually with egg, larva (three pairs of legs), nymph, and adult stages; herbivorous, carnivorous, omnivorous, or parasitic; many families; probably more than 20, 000 described spp.

Acarnus. Genus of red, encrusting sponges to 1 in. or more thick.

Acaroidea. Large superfamily of terrestrial mites; parasitic and free-living.

acarologist. Specialist in mites.

acarology. Study of Acarina.

acarophytium. Symbiosis between a plant and mites.

Acartia. Common genus of marine calanoid Copepoda.

Acarus siro. Common harmless mite living on cheese.

Acaulis. Genus of solitary coelenterates in the Suborder Anthomedusae.

acceleration center. Cardioaccelerator center.

accelerator nerves. Nerve fibers of the sympathetic system; originate in a pair of accelerator centers of the medulla oblongata and in one or more pairs of accelerator centers of the spinal cord; innervate the heart, increasing the rate of beat by liberating small amounts of sympathin.

accentor. 1. Any bird in the Eurasian genus Prunella, especially the European hedge sparrow; see PRUNELLIDAE. 2. Asiatic ovenbird or spinetail.

accessory gland. Any of a variety of glands associated with the male or female reproductive system in many invertebrates.

accessory nerve. Spinal accessory nerve.

accessory nidamental gland. One of two spheroidal glands in the anterior part of the mantle cavity of the female squid; secretes material contributing to the gelatinous substance surrounding extruded eggs; partly hidden by the much larger nidamental glands.

accessory sex organs. All glands, ducts, and other organs, aside from ovaries or testes, associated with reproduction.

accessory shell gland. Large mucus-secreting gland just inside the female genital pore in many gastropods.

Accipiter cooperi. Cooper's hawk; a short-winged, long-tailed sp., slightly smaller than a crow; common around wooded areas in N. A.; usually called chicken hawk from its attacks on fowl.

Accipiter velox velox. Sharp-shinned hawk; feeds chiefly on birds; widely distributed.

Accipitridae. Family of birds in the Order Falconiformes; includes kites, hawks, eagles, and Old World vultures.

accipitrine. Pertaining to hawks or birds of prey.

acclimation. Physiological and behavioral adjustments of an organism to a new habitat or to marked changes in its normal habitat.

acclimatization. Physiological and behavioral adjustments of a particular sp. over several generations to a new habitat or to marked changes in the normal habitat.

accommodation. Automatic adjustment of the eye to focus on objects at different distances; effected in man and a few other mammals by changing the curvature of the lens, chiefly through contraction and relaxation of the ciliary muscle by its association with the suspensory ligament; in typical birds and reptiles the ciliary muscle squeezes on the periphery of the lens; in fish and amphibians the lens moves toward or away from the retina.

accretion. Growth by simple external addition, as opposed to incorporation.

Acella. Outdated generic name now inincluded in Lymnaea.

acenaphthene. Chemical agent which induces polyploidy; $C_{10}H_6(CH_2)_2$.

acentric chromosome. 1. Chromatid fragment lacking a centromere; a meiotic aberration. 2. Chromosome fragment in mitosis.

acentrous. Having a persistent notochord and no vertebral centra, e.g. certain primitive fishes.

Acephala. 1. Pelecypoda. 2. Old taxon which included tunicates, brachiopods, bryozoans, and a few other small groups.

acephalous. Lacking a distinct head.

Acerata. Old taxon which included the Merostomata and Arachnida.

acerate. Needle-shaped.

Aceratherium. Genus of hornless extinct rhinoceros-like mammals; Oligocene to Pliocene.

Acercus. Common genus of fresh-water Hydracarina.

Acerentulus. Genus in the insect Order Protura; A. barbari common.

acerous. 1. Without horns. 2. Without tentacles. 3. Without antennae.

acetabulum. 1. True sucker; generally circular, with raised muscular rim and central depression; separated from underlying tissues by a distinct septum; used for adhering by suction rather than sticky secretions; especially in Trematoda. 2. Cup-shaped socket at the hip joint in the pelvic girdle of tetrapods. 3. Cavity in the thorax of insects for insertion of a leg. 4. Cavity in the proximal end of an echinoid spine. 5. Cephalopod sucker. 6. Placental cotyledon in ruminants. Pl. acetabula.

acetaldehyde. Transitory substance in the cellular oxidation of organic protoplasmic materials; derived from pyruvic acid; CH_3CHO.

acetic acid. 1. Transitory substance in cellular metabolism; commonly derived from pyruvic acid and converted into carbon dioxide and water. 2. Common commercially available acid having many uses in biology, medicine, and industry; CH_3COOH.

acetoacetic acid. Ketone accumulating in blood during extreme diabetes and starvation.

acetone body. Ketone body.

acetylcholine. Neurohumor generally present in vertebrates and many invertebrates in minute quantities; secreted in vertebrates by motor fibers, parasympathetic fibers, and fibers connecting the central nervous system with the sympathetic ganglia; prepared commercially from ergot; used medicinally to decrease blood pressure and stimulate peristalsis; $C_7H_{17}O_3N$.

ACh. Abbreviation for acetylcholine.

Achaenodon. N. A. genus of piglike Eocene mammals.

Achatina fulica. Giant land snail, native to Mauritius, now introduced into southern and eastern Asia, Philippines, East Indies, and Hawaii; several invasions into Calif. have been exterminated; feeds voraciously on vegetation and is a serious pest in many areas.

Achatinella. Large genus of Hawaiian land snails.

acheiropodia. Absence of hands and feet, the limbs being short stumps; a developmental abnormality.

Achelia. See PYCNOGONIDA.

Acherontia. See DEATH'S HEAD MOTH.

Acheta. Common genus of crickets.

Achilles tendon. Large tendon connecting the calf muscles to the heel bone.

Achirus. See HETEROSOMATA.

achlamydate. Without a mantle, as in a few mollusks.

achlorhydria. Lack of acid secretion by the gastric glands.

Acholoë. Genus of polychaetes commonly found on gills of certain chitons.

achondroplastic dwarf. Human dwarf having a trunk of normal size but short limbs, large head, stumpy nose, and deformed hands.

Achorutes. Common genus of insects in the Order Collembola; A. armatum is sometimes a pest in mushroom beds; A. nivicola, called the "snowflea," is sometimes found in enormous numbers on snowbanks in late winter and early spring.

Achromadora. Common genus of microscopic, free-living nematodes.

achromatic. 1. Pertaining to the usual type of microscope optical system; objectives are spherically corrected for one color and chromatically for two colors. 2. Pertaining to cellular elements which do not take a chromatin stain.

achromatic apparatus. That portion of the mitotic spindle exclusive of chromatin material, namely poles and fibrils.

achromatic spindle. Achromatic apparatus.

achroödextrin. One of several polysaccharide digestion products intermediate between starch and maltose; follows erythrodextrin, precedes maltose.

Achtheres. Genus of highly modified copepods parasitic on fresh-water fishes.

aciculate. Needle-like.

aciculum. Long, especially stout chitinous seta imbedded in and stiffening a polychaete parapodium. Pl. acicula.

acid-base balance. Normal physiological and chemical balance between all salts, acids, and bases in a tissue or organ; always basic except for stomach and urine contents.

acidemia. Decrease in normal pH value of blood or other body fluid when excessive acids neutralize the normal alkali reserve.

acid gland. 1. One of the complex of

honeybee poison glands. 2. One of the lateral glands of the trunk segments of a millipede; secretes hydrogen cyanide and/or other substances.

acidity. Relative excess of hydrogen ions over hydroxyl ions in a solution; the more acid a solution, the farther its pH reading below pH 7.0 (neutrality).

acidophil. Histological element readily colored by acid stains.

acidosis. Condition of blood or other body fluids due to abnormally low concentrations of sodium and potassium salts, which neutralize acids in the blood.

acid stain. Biological dye consisting of an acid organic portion combined with a metal; especially useful for cytoplasmic staining.

Acila. See NUT CLAM.

Acilius. Common genus of dytiscid beetles.

Acineta. Genus of fresh-water and marine suctorian protozoans; stalked, loricate, with one to three clusters of tentacles.

Acinonyx. Genus of cats in the Family Felidae; includes the cheetahs; long-legged, swift-running; coat short, woolly, and spotted; sometimes trained to hunt; Africa and southwestern Asia.

acinus. 1. One of the small lobes of a compound gland; cf. ALVEOLUS. 2. Small saclike cavity at the end of a passage, e.g. air sacs of the lungs. Pl. acini.

Acipenser. Genus of sturgeons; A. sturio (sea sturgeon) common along Atlantic Coast, ascends rivers to spawn; A. transmontanus (white sturgeon) a West Coast sp. of similar habits, to 1200 lbs.; A. fulvescens (lake sturgeon, rock sturgeon) in large lakes and rivers of southcentral Canada and St. Lawrence and Mississippi drainages; A. huro (beluga) in Black and Caspian seas. Also see SCAPHIRHYNCHUS.

Acipenseridae. Family of bony fishes which includes the sturgeons; about 23 large spp. with five longitudinal rows of keeled bony plates; to 10 ft. long; tail heterocercal, snout long; mouth ventral, without teeth, and barbels anterior to mouth; feed on invertebrates and bottom vegetation; marine and large rivers and lakes; desirable food fish; eggs made into caviar; Acipenser and Scaphirhynchus typical American genera.

Acipenseriformes. Taxon similar to the fish Order Chondrostei; includes sturgeons and paddlefishes.

Acmaea. Common genus of marine limpets.

Acnidosporidia. Subclass of the protozoan Class Sporozoa; same as Subclass Sarcosporidia, although some protozoologists combine both the Subclass Sarcosporidia and Subclass Haplosporidia into the Subclass Acnidosporidia.

Acoela. Order of free-living marine flatworms; 1 to 4 mm. long; proctostome anterior or ventral, but gastrovascular cavity absent (pharynx sometimes present); flame bulb system absent; Polychoerus a common East Coast genus.

Acoelomata. Group of lower phyla lacking a coelom; includes such phyla as Coelenterata, Platyhelminthes, Nematoda, etc. Adj. acoelomate.

acoelous vertebra. Amphiplatyan vertebra.

acone. 1. Lacking cones in the ommatidia of a compound eye. 2. An acone compound eye.

acontium. One of many fine tangled threads in the basal region of the anthozoan gastrovascular cavity; each originates as a septal filament at the edge of a mesentery. Pl. acontia.

acorn barnacle. Rock barnacle; see BALANUS.

acorn moth. Moth whose larva feeds on acorns; Valentina typical.

acorn weevils. See BALANINUS.

acorn worm. Worm having an acorn-like appearance of the proboscis and collar; see ENTEROPNEUSTA.

Acotylea. Suborder of polyclad turbellarians; sucker absent; genital pores posterior.

acoustic. Pertaining to hearing.

acoustic nerve. Auditory nerve.

acoustico-lateralis system. Lateral line system.

acquired character. Peculiarity in structure, physiology, or behavior arising during the life of an individual as a response to environmental factors or from a functional cause; such characters are not inheritable; see LAMARCKISM.

acquired characteristics, inheritance of. Lamarckian theory that parental responses to environmental influences (variations) may be transmitted to the offspring; e.g. exceptionally well developed muscles in children of an athlete or otherwise muscular, and skin darker than normal in the children of parent(s) who have sun-tanned skin.

acquired immunity. Immunity possessed by an organism through the presence of antibodies induced by a previous infection.

acraein. Substance secreted by certain African butterflies (Acraeinae); presumably distasteful to birds and therefore a protective mechanism.

Acrania. Cephalochordata.

Acraniata. Taxon which includes groups of chordates having no cranium, e.g. Tunicata and Cephalochordata.

acraspedote. Pertaining to coelenterates having no velum.

acrea moth. See ESTIGMENE.

Acrididae. Family of insects in the Order Orthoptera; same as Locustidae; grasshoppers and locusts; body deeper than wide, antennae shorter than body, hind legs long and modified for jumping; males produce sounds; many spp. are plant pests; common genera: Melanoplus, Dissosteira, Schistocerca, Romalea, and Locusta.

Acris gryllus. Common American cricket frog (Family Hylidae); central Canada to Tex. and eastward; to 25 mm. long; in grass and marshy places.

Acrobates. Genus of small marsupials resembling flying squirrels in habits; pygmy flying possums, gliding feathertails; body only 3 in. long; Australia and New Guinea.

Acrobeles. Common genus of microscopic, free-living nematodes.

Acrocera. Widely distributed genus of flies having very small heads; larvae feed on spiders and spider eggs; in the Family Acroceratidae.

Acrocheilus alutaceum. Chiselmouth, a large minnow of the Columbia River system and Malheur Lake drainage; to 12 in. long.

acrocyst. Chitinoid sac which protrudes from the gonangium in certain Hydrozoa; contains the planulae in the final stages of development.

acrodont. Having teeth attached to the edge of the jaws instead of being inserted in sockets on the surface.

Acrodus. Genus of Mesozoic sharks.

acrodynia. Pyridoxine deficiency disease of rats, dogs, and pigs; marked by swelling and necrosis of paws, ears, and snout; also, a disease of human infants marked by swollen hands and feet, digestive upsets, arthritis, and weakness.

acromegaly. Pathological condition produced by excessive secretion of the growth-stimulating hormone of the anterior lobe of the pituitary during adult life; forehead, nose, and lower jaw become massive, and the facial skin becomes thick and coarse.

acromion. Flat upper lateral projection of the scapula.

acron. 1. Prostomial region of certain mollusk trochophore larvae. 2. Anterior nonsegmented portion of an embryonic arthropod; produces eyes and first antennae.

Acroneuria. Common genus of Plecoptera.

Acropora. Genus of reef-building corals of tropical Pacific and Indian oceans, especially elk-horn or staghorn coral.

Acroporus. Genus of cladocerans common in aquatic vegetation.

acrorhagi. Hollow spherical or oval bodies covered with nematocysts and situated on the periphery of the oral disc in many sea anemones. Sing. acrorhagus.

acrosome. Apical body or mass of a spermatozoan; derived from Golgi materials of the spermatid.

acrostichial bristles. Minute bristles arranged in one to several median longitudinal rows on the dorsal surface of the mesonotum of certain flies.

Acrothoracica. Small order of barnacles (Subclass Cirripedia); only three pairs of thoracic appendages; shell absent but mantle voluminous; bores into the shells of mollusks and other barnacles; sexes separate, male minute; Alcippe common.

Acrotreta. Common genus of small fossil brachiopods; Cambrian to Ordovician.

acrotrophic. Having apical vitelline cells, as in certain insect ovarioles.

Actaeon. Genus of small opisthobranch snails having an ovate shell.

Acteocina. Genus of small West Coast marine snails.

ACTH. Adrenocorticotrophic hormone secreted by the anterior lobe of the pituitary; stimulates the adrenal cortex to produce hormones.

Actias. Genus of large moths in the Family Saturniidae.

actin. Essential contractile protein of muscle.

actinal. Pertaining to that part of a radially symmetrical animal which contains the mouth (or anus) and from which tentacles or arms radiate.

Actinaria. Suborder of coelenterates in the Order Actiniaria; mesenteric filaments with ciliated areas; Metridium, Gonactinia, Anthopleura, Adamsia, Actininia, and Edwardsia typical.

Actinauge. Genus of sea anemones (Order Actiniaria); to 6 in. long.

Actiniaria. Order of zoantharian anthozoans; true sea anemones; skeleton absent; polyp more or less columnar, with large spp. to 12 in. in diameter; attached to substrate by a pedal disc; septa paired; sessile but capable of slow creeping movements across the substrate; solitary but often gregarious; examples are Metridium, Adamsia, and Gonactinia.

Actininia. Genus of sea anemones usually found growing on the shells of hermit crabs.

Actinistia. Coelacanthini.

Actinobdella. Uncommon genus of fresh-water leeches; to 15 mm. long; northeastern states.

actinoblast. Spicule-forming cell in the Porifera.

Actinobolina. Genus of ovate to spherical fresh-water holotrich ciliates; tentacles scattered among cilia.

actinobranch. Gill-like appendage in certain Anthozoa.

Actinocoma. Genus of spherical fresh-water amoeboid protozoans; radiating and branching pseudopodia.

Actinolaimus. Common genus of microscopic, free-living nematodes.

actinomere. Radial section of a radially symmetrical animal.

Actinometra. Genus of feather stars; usually in shallows.

Actinomma. Genus of radiolarians.

Actinomonas. Genus of holozoic fresh-water protozoans; single flagellum and numerous radiating pseudopodia.

Actinomyxidia. One of the three orders in the protozoan Subclass Cnidosporidia; each spore with three valves and three polar capsules; parasitic in the coelom and intestinal epithelium of marine and fresh-water annelids and sipunculids; example is Triactinomyxon.

Actinonaias. Genus of U.S. bivalve mollusks in the Family Unionidae.

actinophore. Bony or cartilaginous supporting structure in a fish fin ray.

Actinophrys. Genus of spheroidal fresh-water amoeboid protozoans with numerous axopodia.

Actinopoda. Subclass of the protozoan Class Sarcodina; characterized by axopodia.

Actinopterygii. Taxon of fishes sometimes used to include the subclasses Palaeopterygii and Neopterygii.

Actinosphaerium. Genus of spherical fresh-water Protozoa in the Order Heliozoa; numerous nuclei and vacuolar ectoplasm.

actinost. One of several small bones which articulate between the girdles and rays of the paired fins in a typical fish.

Actinostola. Genus of large sea anemones (Order Actiniaria); oral disc to 10 in. across.

actinotrichia. Horny rays forming the anterior and posterior portions of the pectoral and pelvic fins in elasmobranchs. Sing. actinotrichium.

actinotrocha. Modified trochophore larva in the Phylum Phoronidea; numerous finger-like lobes originating near the middle of the body and projecting radially or posteriorly.

Actinozoa. Anthozoa.

actinula. Young unattached polyp stage in certain hydrozoan coelenterates.

action current. Rapid flow of ions resulting from a localized potential difference in a tissue as it transmits a stimulus or is subjected to a stimulus; especially evident in nerves, heart muscle, skeletal muscle, and glands but also present in many other tissues.

action potential. Magnitude of the localized and momentary change in the electrical potential between the inside and outside of a nerve fiber as it carries an impulse; it marks the position of a nervous impulse (action current) as it travels along the fiber; in a state of rest the interior protoplasm is negative to the outer surface, but during passage of a stimulus the outside is momentarily negative to the inside; similar action potentials occur in stimulated muscle tissues, the heart, and glands. Cf. ACTION CURRENT.

Actipylea. Suborder of radiolarians; skeleton composed of strontium sulfate.

Actitis macularia. Common spotted sandpiper (Family Scolopacidae) of North and South America; shorebird of lakes and streams; ashy olive above; under parts whitish, with round black spots; white line over eye; constant teetering habits.

activated sludge. Thick or thin watery suspension of sewage and associated protozoans, micrometazoans, and bacteria in certain sewage treatment processes; the suspension is constantly stirred and usually air is blown through it; the organisms convert much of the organic fraction into gases, salts, and inert organic compounds.

active immunity. Acquired immunity.

active transport. Movements of materials across membranes against osmotic gradients.

actomyosin. Type of protein forming fine threads (micellae) in muscle fibers; such threads can be seen only with an eletron microscope; fundamental contractile substance, formed of actin and myosin.

Acuaria. Cheilospiuria.

acuchi. See MYOPROCTA.

aculea. One of the many minute spines on the wing membrane of some Lepidoptera. Pl. aculeae.

aculeate. 1. Bearing a stinger. 2. Bearing aculeae.

acumen. Spine at tip of the rostrum in many decapod crustaceans.

acuminate. Ending in a point, or tapering to a sharp point.

adambulacral spines. Rows of spines on the oral surface of a sea star; around the mouth and along edges of the ambulacral grooves.

Adam's apple. Projection of the thyroid cartilage on the ventral part of the neck in the human male.

Adamsia. Genus of sea anemones usually found on the shells of hermit crabs.

adaptation. Peculiarity of structure, physiology, or behavior of an organism which especially aids in fitting the organism to its particular niche or habitat; a fundamental adjustment property of living matter.

adaptive coloration. Coloration and color patterns of animals which tend to make them more effectively fitted to their habitats; such features may have the function of concealing or making conspicuous, depending on the species.

adaptive convergence. Superficial similarity of physiology, behavior, or body form and structure in widely different groups inhabiting the same environment; an example is the streamlined bodies of ocean vertebrates, including seals, whales, and fishes.

adaptive radiation. Marked evolutionary changes in structure, physiology, and behavior in the members of a single phyletic line by which relatively primitive and unspecialized spp. become specialized to fit numerous distinctive and greatly diverse ecological niches; the wide variety of marsupials of Australia and the characin fishes of South America are examples.

adaxial. Located near or toward the axis.

Addax. Genus of ruminants in the Family Bovidae; addaxes; large, light-colored antelopes with long spiral horns; northern Africa, Arabia, Syria, etc.

adder. 1. Any of a wide variety of venomous snakes such as the puff adder, European viper, krait, etc. 2. Any of several harmless N.A. snakes such as the hognose snake and milksnake.

Addison's disease. Human disease resulting from an insufficiency of the adrenal cortex hormones; marked by bronzing of the skin, weakness, low blood pressure, and digestive disturbances.

additive gene effects. Gradations in a character caused by the relative numbers of dominant and recessive multiple factors; e.g. black, white, and intermediate skin shades in man.

Addra. Genus of slender whitish gazelles (addras) of the Sahara.

adductor. 1. Muscle that draws a part toward the axis of the body or limb. 2. Muscle in bivalve mollusks which runs transversely from one valve to another.

adductor brevis. Short thigh muscle in certain tetrapods; adducts, rotates, and flexes the thigh.

adductor longus. Muscle of the thigh in certain tetrapods; adducts, flexes, and rotates thigh.

adductor magnus. Large adductor muscle of the thigh in certain tetrapods.

Adelea. Genus of coccidian sporozoans parasitic in arthropods.

Adelgidae. Family of very small apterous or winged insects in the Order Homoptera; body soft, fragile, and often covered with threads and masses of white cotton-like wax; on twigs, needles, or in galls on coniferous trees; Pineus common.

Adelie penguin. Common penguin of the periphery of Antarctica.

Adelphocoris. Genus of plant-infesting bugs in the Family Miridae.

adenine. Component of nucleic acids involved in cellular metabolism; present in ADP and ATP; white crystalline base; $C_5H_5N_5$.

adenohypophysis. That portion of the pituitary which is derived from Rathke's pouch; it partly envelopes the neurohypophysis and consists of the pars distalis, pars tuberalis, and pars intermedia.

adenoid. Mass of lymphoid tissue at the back of the nasal chamber, especially in children; pharyngeal tonsil.

adenoma. Tumor of glandular tissue.

adenosine diphosphate. ADP; intermediate organic coenzyme present in all living cells; composed of adenine, ribose, and two phosphate radicals; by virtue of its high-energy phosphate bond, it functions chiefly as an energy transfer mechanism, especially in muscle contraction, chemical synthesis, and active transport phenomena; converted to adenosine triphosphate by the incorporation of one additional phosphate radical and energy; may also be reduced to adenosine monophosphate.

adenosine monophosphate. See ADENOSINE DIPHOSPHATE.

adenosine triphosphate. ATP; a complex coenzyme found in all cells; composed of adenine, ribose, and three phosphate radicals; by virtue of its two high-energy phosphate bonds, it functions chiefly as an energy transfer mechanism, especially in muscle contraction, chemical synthesis, and active transport phenomena; by losing one phosphate group to a monosaccharide it is temporarily converted into adenosine diphosphate, while the monosaccharide then becomes a phosphorylated sugar which may be oxidized to pyruvic acid, thereby releasing hydrogen and phosphate, the latter converting the adenosine diphosphate back to triphosphate.

Adenota. Genus of African antelopes; includes the kob and puku; sleek, orange-red spp. of meadow lands.

adenylic acid. Decomposition product of nucleic acid and adenosine triphosphate; a factor in the vitamin B complex.

Adephaga. One of the two suborders in the insect Order Coleoptera; antennae threadlike, hind wings with only one or two cross veins. Cf. POLYPHAGA.

ADH. Antidiuretic hormone.

adhesive capsule. Type of nematocyst found in many coelenterates; used chiefly in attaching to objects temporarily.

adhesive cells. 1. Various cells which secrete a sticky adhesive substance, as in the pedal disc of Hydra and the epidermis of turbellarians. 2. See COLLOBLASTS.

adhesive pad. Sucker of adhesive mass near the tip of the tentacles of some hydrozoan medusae; used for clinging to seaweeds.

Adineta. Genus of bdelloid rotifers.

Adinia xenica. Small killifish (Cyprinodontidae) in fresh waters from Fla. to Tex. along Gulf Coast.

adipose fin. Small fleshy median fin behind the main dorsal fin in certain fishes, such as the salmon, whitefish, and grayling.

adipose tissue. Fat tissue; consists of spherical or polygonal cells, each containing a large fat droplet which displaces the cytoplasm and nucleus to the cell periphery.

A disc. Anisotropic disc.

adjustment. Temporary change in the physiology or behavior of an organism in response to transient variations in environmental factors; fundamental property of living matter.

adjustor neurons. Neurons which variously interconnect between sensory and motor neurons in the brain and spinal cord.

adjutant stork. See LEPTOPTILUS.

Admete. Genus of small marine gastropods.

adnate. Pertaining to unlike parts which have grown together.

adobe bug. See HAEMATOSIPHON.

adobe tick. Argas persicus.

Adocia. Genus of encrusting marine sponges.

adolescaria. Cercaria or metacercaria stage in the life history of a fluke. Pl. adolescariae.

adoral membrane. Double lamella of fused cilia in the region of the cytostome in most ciliate Protozoa.

adoral wreath. Adoral zone.

adoral zone. Row of membranelles on margin of the peristome in many ciliate Protozoa; brings food particles to the cytostome.

ADP. Adenosine diphosphate.

adradial canal. One of several types of gastrovascular canals extending radially and joining the circumferential canal in certain jellyfish.

adrenal cortex. See CORTEX.

adrenal gland. Vertebrate endocrine gland; one is associated with each kidney in mammals but there are multiple adrenal glands in certain other vertebrates; in tetrapods each adrenal has two distinct portions having different functions; the inner medulla secretes adrenalin, and the outer cortex secretes hormones controlling certain aspects of metabolism; in fishes the cortex and medulla are distinct and separate organs.

Adrenalin. Trade-mark for synthetic epinephrine.

adrenaline. Epinephrine.

adrenergic. Pertaining to those vertebrate post-ganglionic sympathetic nerve endings which, when stimulated, secrete minute quantities of an adrenalin-like substance; this substance produces action by the effector which is so innervated.

adrenal medulla. See MEDULLA.

adrenin. Adrenine; hormone(s) of the adrenal medulla.

adrenocorticotrophic hormone. See ACTH.

adrenotoxin. Substance toxic to adrenal glands.

adrenotrophic hormone. Hormone secreted by the anterior lobe of the pituitary; stimulates secretory activity of the adrenals.

adrostral. Near or toward the rostrum or beak.

adsorption. Adhesion of molecules to solid surfaces, often resulting in monomolecular layers.

ADT. Adenosine triphosphate.

adventitia. Outermost connective tissue coat of an artery, digestive organ, or reproductive tube.

adventitious. 1. Pertaining to something accidental or acquired; not hereditary. 2. Occurring in an unusual or abnormal place.

adventitious plankton. Bottom or epiphytic organisms found accidentally or fortuitously in the plankton; e.g. Tardigrada.

Aechmophorus occidentalis. Western grebe; black and white bird in the Family Podicipitidae; neck long and slender, bill yellow.

aedeagus. Copulatory organ in certain male insects.

Aedes aegypti. Widely distributed tropical mosquito which is especially important as a transmitter of human yellow fever and dengué.

Aega psora. Salve bug; marine isopod parasitic on skate, cod, halibut, etc.; used as a salve by fishermen.

Aegeria. Genus of moths in the Family Aegeriidae; larvae of many spp. are pests, such as A. exitiosa (peach tree borer) and A. tipuliformis (currant borer).

Aegeriidae. Small family of sluggish moths; clear-wing moths; day-flying; wings scale-less and transparent; Aegeria a widespread and important plant pest genus.

Aegidae. Family of marine isopods; body broad and domed, head small, compound eyes large, pleopods adapted for both swimming and respiration.

Aegina. Genus of marine coelenterates in the Suborder Narcomedusae; open ocean.

Aeginella. Genus of caprellid amphipods.

Aeginopsis laurenti. Circumpolar marine hydrozoan; medusa pelagic, as deep as 1000 m.; hydroid stage lacking.

Aegires. Genus of nudibranchs; dorsal surface tuberculous.

Aegolius. Genus of small owls; A. acadica is the saw-whet owl, widely distributed over much of U.S. forests; A. funerea richardsoni is Richardson's owl, a tame sp. of Canadian forests; sometimes Cryptoglaux is the preferred genus name.

Aegoryx. Genus of whitish antelopes (oryx) of the southern Sahara.

Aegothelidae. Small family of birds in the Order Caprimulgiformes; owlet frogmouths; chunky body to 12 in.; nocturnal forest dwellers; feed on ground and flying insects; eight spp. in Australia and New Guinea.

Aeolidia. Small genus of nudibranchs having simple unbranched cerata arranged in transverse lows on either side of the midline; tentacles of unequal length.

Aeolidiidae. Family of nudibranch mollusks; tips of cerata rich in defensive nematocysts obtained from eating hydroids and sea anemones.

Aeolis. Large genus of sea slugs or nudibranchs; body ovate; gills cylindrical and numerous, extending along entire dorsal surface.

Aeolosoma. Common genus of translucent fresh-water oligochaetes; body often contains red, yellow, or greenish globules in integument; reproduction by transverse fission, often resulting in chains of two or more developing zooids; total length to 10 mm.

Aepyceros. Genus of ruminants (Family Bovidae) which includes the impala or pala; large antelopes of southern and eastern Africa; forest spp., never far from water.

Aepyornis. See AEPYORNITHIFORMES.

Aepyornithiformes. Order of flightless birds including the elephant birds (Aepyornis) which have been extinct for several hundred years; four toes; wings vestigial; to 10 ft. tall; eggs to 13 in. long; Madagascar.

Aequipecten. See SCALLOP.

Aequorea. Genus of marine hydrozoan coelenterates; medusae well developed; hydroids minute and poorly known.

Aerobacter. Genus of flagellated bacteria normally found in the intestine.

aerobe. Any organism able to live in the presence of free oxygen.

aerobic respiration. Metabolic breakdown or oxidation of materials in the presence of molecular oxygen; the vast majority of organisms are aerobic.

aerobiology. Study of air-borne organisms.

aeroembolism. Caisson disease; condition resulting from formation of nitrogen bubbles in blood and spinal fluid; occurs in pilots at high altitudes if precautions are not taken; also occurs when divers return to the surface too rapidly.

Aeronautes saxatalis. White-throated swift in the Family Caprimulgidae; a bird of the western mountain states, south to El Salvador.

aeroplankton. Collectively, those microorganisms carried about in air.

Aeschna. Common genus of dragonflies (Order Odonata).

aesthetask. Long, slim, blunt, minute sensory process on the antennae of certain crustaceans, e.g. copepods; (sometimes written aesthetasc).

aesthetes. 1. Microscopic light sensory structures on the surface of the calcareous plates of certain Amphineura. 2. Aesthetask.

aestivation. Type of inactivity and dormancy during times of drouth, especially in temperate and tropical regions; in lungfish, some snails, and many reptiles and insects.

aestivo-autumnal malaria. Malignant tertian malaria; see MALARIA.

Aetea. Genus of marine Bryozoa; tubular, erect zooecia arise from a stolon.

Aethus. Widely distributed genus of burrowing bugs, especially in sand dunes.

Aetobatus. Genus of sting rays.

afferent. Conducting, conveying, or leading inward or toward, such as a blood vessel or nerve.

afferent branchial artery. One of the eight arteries supplying blood to the gills in typical fishes.

afferent neuron. Sensory neuron or one leading to the central nervous system or central ganglion.

afterbirth. Placenta and decidual membranes as they are discharged from the uterus and vaginal canal after the birth of the mammalian young.

after-image. Visual image or other sensation which persists after the object has ceased to be visible; especially true for bright objects.

aftershaft. In many birds, a small, delicate, accessory, plumelike feather arising from the main feather at the point where the quill merges into the shaft of the feather.

Agabetes. Genus of dytiscid beetles.

Agabus. Common genus of dytiscid beetles.

Agama. Genus of Old World brightly-colored lizards in the Family Agamidae.

agamete. 1. Any small amoebula which develops into a normal amoeba. 2. Any single germ cell which develops into an adult individual. 3. See MESOZOA.

agamic. 1. Asexual. 2. Pertaining to reproduction without fertilization by a spermatozoan.

Agamidae. Family of lizards found in Asia, Europe, Africa, and Australia; more than 200 spp.; examples are Agama, Draco, and Moloch.

Agaontidae. Small family of minute wasps inhabiting the fruits of certain figs; fig wasps; female is the sole means of pollinating these plants; Blastophaga the most important genus.

Agapetus. Genus of Trichoptera.

Agapornis. See LOVEBIRD.

Agaporus. Laccornis.

agar. Agar agar; gelatinous hydrophilic substance for bacteriology culture media; prepared commercially as powder, granules, or thin sheets.

Agarica. Genus of leaf corals; growth form often flat and leaflike.

Agassiz, Alexander. American naturalist (1835-1910); son of Louis Agassiz; made several important biological oceanographic trips; contributed much to our knowledge of coral reefs and marine invertebrates.

Agassiz, (Jean) Louis (Rodolphe). Swiss-American zoologist and geologist (1807-1873); wrote Fishes of Brazil, a five-volume Recherches sur les Poissons Fossiles, Nomenclator Zoologicus, Fresh Water Fishes of Central Europe, and a four-volume Contributions to the Natural History of the United States; American career began in 1846; established Harvard Museum of Comparative Zoology; famous for his doctrine of "Study nature, not books"; gave great impetus to American zoology, although he was an oponent of Darwinism.

agastric. Without a digestive cavity, e.g. a tapeworm.

Agathon. Genus of net-winged midges in the Family Blepharoceridae.

Agaue. Genus of marine mites in the Family Halacaridae.

Agauopsis. Genus of marine midges in the Family Halacaridae.

age and area. Correlation between the relative area of the earth occupied by a sp. (or other taxon) and the age of that sp.; in general, the older the sp. the greater the area occupied.

age immunity. Resistance which individuals of a certain age group may have against a particular infection.

Agelaius. See BLACKBIRD.

Agelena. Common genus of grass spiders in the Family Agelenidae; flat webs in grass.

Agelenidae. Family of funnel-web spiders; spin sheet or platform webs with a tube or funnel at one edge; Agelenopsis and Tegenaria typical.

Agelenopsis naevia. Grass spider; builds large flat web with funnel at one edge; in grass and houses.

Age of Amphibians. Collectively, the Permian and Carboniferous periods, extending from about 250 million to 120 million years ago; characterized by an abundance and dominance of amphibians.

Age of Fishes. Devonian period, extending from about 350 million to 250 million years ago; characterized by an abundance and dominance of sharks and bony fishes.

Age of Invertebrates. Collectively, the Silurian, Ordovician, and Cambrian periods, extending from about 550 million to 350 million years ago; marked by the evolution and dominance of invertebrate phyla.

Age of Mammals. Tertiary geological period, lasting for about 70 million years and characterized by the appearance, dominance, and rapid development of placental mammals, as well as modern plants.

Age of Man. Present time and past 1 million years, including the Quaternary and Pleistocene periods; characterized by the appearance and evolution of man and manlike apes.

Age of Reptiles. Mesozoic era, extending from about 190 million to 70 million years ago; especially marked by the evolution and appearance of a wide variety of reptiles.

agglutination. Sticking together of organisms or particles, as the clumping of bacteria in the presence of antibodies or the clumping of erythrocytes when incompatible bloods are mixed.

agglutinin. Type of antibody which will agglutinate or clump the antigen, protozoans, bacteria, blood corpuscles, etc. for which it is specific.

agglutinogen. Type of antigen which induces the formation of an agglutinin specific for it.

Aggregata. Large genus of coccidian sporozoans; parasitic in crustaceans and cephalopods.

aggregation. General term for any group of organisms in a habitat; usually all of the same sp.; e.g. school of fish, flock of geese.

aggression. Unprovoked attack of an animal against members of its own sp.; may be associated with breeding, nesting, feeding, protection of young, territory maintenance, or obscure behavior mechanisms; sometimes such attacks are complicated and ritualistic.

aggressive mimicry. A type of mimicry in which the mimic is unrecognized as a predator by its prey; e.g. a bee fly, by resembling a bee, is enabled to enter the nest of bees and oviposit there so the larvae which hatch out will feed on the bee larvae; a questionable theory.

Agkistrodon. Genus of venomous snakes in the Family Crotalidae inhabiting N.A., South America, and Asia; A. piscivorus is the American water moccasin or cottonmouth found from Va. to Ill. and south to Fla. and Tex.; a heavy-bodied sp. to 4 ft. long; omnivorous and semi-aquatic; A. mokasen is the copperhead, which is slightly smaller and more common in rocky places; ranges from Mass. to Ill. and south to Fla. and Tex.

Aglais. Nymphalis.

Aglaisma. Genus of coelenterates in the Order Siphonophora; with two unlike nectophores.

Aglaja diomedea. Small opisthobranch found burrowing in sand flats; shell internal; no conspicuous flaps or tentacles; West Coast.

Aglantha. Genus of marine hydrozoan coelenterates; medusa reaches 30 mm. in diameter; no polyp stage in the life history.

Aglaophenia. Genus of marine hydrozoans forming large colonies shaped like ostrich plumes; often cast up on beaches.

Aglaspis. Genus of Cambrian king crabs.

Aglaura. Genus of marine coelenterates in the Suborder Trachymedusae.

aglomerular. Pertaining to the kidneys of certain fishes which have no glomeruli.

Aglossa. Group of tongueless toads which have a common eustachian opening in the pharynx, e.g. Pipa.

Aglypha. Group of genera in the snake Family Colubridae; teeth solid; mostly harmless spp., including such familiar American forms as garter snakes, water, black snakes, bull snakes, and gopher snakes.

Aglyphodonta. Aglypha.

Agnatha. One of the eight classes of vertebrates; ostracoderms and cyclostomes; no true jaws or paired appendages (one pair in ostracoderms).

Agonidae. Poachers; a family of small bizarre marine fishes; elongated body covered with bony armor; cold waters of the Northern Hemisphere; Hypsagonus typical.

Agonostomus. Genus of fresh-water mullets valuable for food; East Indies, West Indies, and Mexico.

Agosia chrysogaster. Longfin dace, a minnow found in the Lower Colorado River system.

agouti. 1. Tropical American nocturnal rodent in the Family Dasyproctidae; to 18 in. long; not unlike small pigs but more like a mouse-deer in general appearance; swift-moving forest and plains spp.; coat speckled and chestnut colored. 2. Alternate light and dark grizzled color banding, as in many rodents.

agouty. Agouti.

agranulocyte. Non-granular leucocyte; nongranulocyte.

agranulocytosis. Acute human disease marked by an almost complete absence of agranulocytes from the blood; the body therefore cannot combat bacterial infections.

agraphia. Mental condition in which a person cannot express himself in writing, or write from dictation.

Agraylea. Genus of Trichoptera; larval case purselike.

agricultural ant. See POGONOMYRMEX.

Agrilus. Genus of small beetles in the Family Buprestidae; larvae girdle trees and shrubs.

Agriocharis ocellata. Ocellated turkey; small, brightly colored wild sp. of Yucatan, Honduras, and Guatemala.

Agriochoerus. Genus of piglike Oligocene ungulates.

Agriolimax. Deroceras.

Agrion. Common genus of damselflies.

Agrionidae. Large and common family of damselflies.

Agriotes. Large, cosmopolitan genus of beetles in the Family Elateridae; larvae (wireworms) often destructive to truck and field crops.

Agromyzidae. Family of leaf- and stem-mining flies; small to minute insects; larvae burrow in leaf tissues.

Agrotis. Euxoa.

aguara. Guara; South American wild dog or maned wolf (Chrysocyon).

A horizon. Uppermost layer of soil, especially in areas where rainfall exceeds evaporation; includes surface organic debris and the underlying dark-colored material; stratum of leaching and accumulation or organic materials; variously subdivided. Cf. B HORIZON, C HORIZON.

ahuatle. Brine fly eggs (Ephydra), occasionally used as human food in rural Mexico.

ai. Three-toed sloth (Bradypus tridactylus) of Tropical America; feeds almost exclusively on leaves, buds, and stems of cecropia.

aigrette. Nuptial plumes of the crown and back in the snowy egret (Egretta).

ailanthus silk moth. See PHILOSAMIA.

Ailuropoda melanoleuca. Giant panda, a rare mammal in the Family Procyonidae; large, short-tailed, black and white; feeds mostly on bamboo; mountains of western China and Tibet.

Ailurus fulgens. Lesser panda; a mammal in the Family Procyonidae; stocky, about the size of a badger; red-chestnut above, tail ringed and long; eastern Himalayas.

aimara. Carnivorous South American river fish (Hoplias) in the Family Characinidae.

Aimophila. Genus which includes 13 American spp. of sparrows, four in the U.S.

Ainu. Race of stocky caucasoids living on a few islands northeast of Asia; skin light brown, very hairy; head long, nose concave; language unique and not related to Japanese or Russian.

air bladder. Swim bladder.

air sac. 1. In birds, one of numerous thin-walled air-filled extensions of the lungs; located between the visceral organs, around the neck vertebrae, and in cavities in the larger bones; supplied by bronchioles and serve to increase respiratory efficiency and dissipate body heat. 2. In insects, thin-walled diverticula of some of the larger tracheae which serve to increase efficiency of respiration. 3. Swim bladder of fishes. 4. Lung alveolus.

air-sac mite. See CYTOLEICHUS.

Aistopoda. Group of late Paleozoic amphibians sometimes included under the Subclass Stegocephalia; limbs absent.

Aix galericulata. Mandarin duck, a crested Asiatic sp. often domesticated.

Aix sponsa. Common wood duck of temperate N. A.; a richly colored sp. which nests in hollow trees; becoming scarce.

Ajaia ajaja. See SPOONBILL.

ajaja. Roseate spoonbill.

Akeley, Carl E. American naturalist and sculptor (1864-1926); developed methods in taxidermy and museum habitat groups; variously affiliated with the Chicago Field Museum and the American Museum of Natural History; made several important African expeditions.

ala. 1. Wing or winglike process. 2. Thin projecting longitudinal ridge of fin; common on the cuticle of nematodes. Pl. alae.

Alaimus. Common genus of microscopic free-living nematodes.

alanine. Amino acid; $C_3H_7O_2N$.

alar. Alary.

Alaria. Genus of holostome trematodes; stomach and intestine of fox and dog.

alarm bird. 1. Australian wattled lapwing (Lobibyx novae-hollandiae); flutters about rapidly and screams when approached. 2. African touraco.

alary. 1. Of or pertaining to wings. 2. winglike or wing-shaped.

Alasmidonta. Genus of fresh-water bivalve mollusks (Unionidae); sandy and muddy rivers; Mississippi River and eastward.

Alassostoma. Genus of amphistome trematode parasites of turtle intestines.

alate. 1. Having wings, such as a winged insect. 2. Furnished with winglike expansions or extensions.

Alauda. See ALAUDIDAE.

Alaudidae. Family of passerine birds which includes the larks; chiefly an Old World group of small somber-colored spp.; usually in open country; the American horned lark is a brownish, terrestrial sp. with two small erectile blackish horns; in open country over much of U. S. and Canada; Alauda and Eremophila typical.

Alaus. Genus of large click beetles (Elateridae); larvae predaceous.

albacore. 1. Tuna. 2. General term covering a group of tuna-like fishes.

albatross. See DIOMEDEIDAE.

Albertia. Genus of rotifers living as internal parasites of fresh-water oligochaetes.

Albia. Widely distributed genus of Hydracarina.

albinism. Absence of pigmentation, especially the absence of melanins, in an animal; typically a Mendelian recessive.

albino. Any animal lacking pigmentation.

Albula. See ALBULIDAE.

Albulidae. Family of tropical marine fishes; bonefish and ladyfish; silvery game fishes with ventral mouth used for feeding on bottom animals in shallow inlets; Albula common.

albumen. White of an egg, especially in birds; comtains a high percentage of protein.

albumen gland. Type of gland found in a wide variety of invertebrates, especially in certain mollusks; usually secretes an albuminous mass around the ova.

albumin. Common type of protein found in most animals and many plants.

albuminoid. 1. Protein. 2. Scleroprotein.

albuminuria. Presence of albumin in the urine.

Albunea. Genus of decapod crustaceans (Suborder Anomura); abdomen small; in sand flats near low tide line.

Alcae. Suborder of birds in the Order Charadriiformes; includes the auks (Family Alcidae).

Alcaligenes. Genus of bacteria found in the normal animal intestine and in milk.

alcapton. Alkapton.

Alca torda. Razor-billed auk (Family Alcidae) of North Atlantic coasts and islands; able to fly well, but superb swimmers and divers by the use of both wings and feet; feed on fish and marine invertebrates.

Alcedinidae. Cosmopolitan family of birds (84 spp.) which includes the kingfishers and kookaburra; bill long and straight; forests and woodland, especially near water; Megaceryle alcyon the common belted kingfisher; Chloroceryle americana the green kingfisher, from southern Tex. to Argentina.

Alcelaphus. Genus of ruminants (Family Bovidae) which includes the hartebeest, bubal, and tora; large antelopes of plains and thin forests; restricted to Africa although the bubal occurs in Arabia.

Alces. Genus of ruminants in the Family Cervidae; includes the European elk (A. alces) and the American moose (A. americana); the latter weighs up to 1400 lbs., favors marshy lakes, and occurs in Canada, Maine, N.D., Wyo., Mont., and Ida.; the Alaskan moose (A. gigas) weighs up to 1800 lbs.; female without antlers males with broad, heavy, palmate antlers.

Alcidae. Family of birds in the Order Charadriiformes; includes the auks, murres, puffins, etc.; swimming and diving sea birds of the Arctic and Subarctic; usually nest in dense colonies on sea cliffs; Plautus, Uria, Alle, Alca, and Fratercula typical.

Alciopidae. Family of Polychaeta.

Alcippe. See ACROTHORACICA.

alcohol. 1. Any of a group of organic compounds formed from hydrocarbons by the substitution of one or more hydroxyl groups for an equal number of hydrogen atoms. 2. Common ethyl alcohol or ethanol; intoxicating substance derived from the fermentation of various plant materials; has a wide variety of uses in biology; C_2H_5OH. Cf. METHYL ALCOHOL.

Alcyonacea. Order of alcyonarian corals; the soft corals, characterized by a gelatinous mass from which only the oral ends of the polyps protrude; skeleton of calcareous spicules.

Alcyonaria. One of the two subclasses of the anthozoan coelenterates; colonial polyps with eight pinnately branched tentacles, eight complete septa, one siphonoglyph, and an endoskeleton; soft corals, horn corals, and sea pens.

Alcyonidium. Genus of marine bryozoans in the Order Ctenostomata.

Alcyonium. Genus of soft corals in the Order Alcyonacea.

alderfly. One of a group of flies in the Order Megaloptera; weak flying adults, usually brownish, live only a few days; larva carnivorous and with segmented lateral abdominal tracheal gills; in streams, lakes, and ponds; a highly desirable fish bait; Sialis the most common genus.

Alderia. Genus of nudibranchs.

Aldisa. Genus of nudibranchs.

aldosterone. Potent mineralocorticoid discovered in 1953.

Alebion. Genus of marine copepods (Suborder Caligoida); body greatly depressed and broad; parasitic on fishes.

alecithal egg. Any ovum containing little or no yolk.

Alectis crinitus. Cobblerfish, threadfin, threadfish; a marine fish in the Family Carangidae; body very deep and compressed; first few rays of dorsal and anal fins extremely long and threadlike; coasts of Tropical America.

Alectoris. Genus of Old World partridges; chukar partridge introduced in U.S.

Alectrion. Nassa.

Aleochara. Widely distributed genus of rove beetles (Staphylinidae).

Alepas. Genus of stalked barnacles.

alepidote. Without scales.

Alepisauridae. Lancetfishes; family of elongated carnivorous deepsea fishes; scales absent, dorsal fin high and long, body short.

Aletes. Genus of Pacific marine snails; shells having the form of long twisted wormlike tubes, commonly found in masses; food becomes entangled in mucus secreted at the tube entrance.

alewife. See ALOSA PSEUDOHARENGUS.

Alexia. Common genus of small pulmonate snails found along the high tide line.

alexin. Nonspecific substance present in plasma and serum; able to destroy bacteria by lysis.

Aleyrodidae. Family of insects in the Order Homoptera; the white flies; minute spp. covered with a white waxy powder; nymphs scale-like, sessile; Trialeurodes and Asterochiton important genera.

alfalfa caterpillar. 1. In the broad sense, any caterpillar attacking alfalfa. 2. In the narrow sense, the larva of a small yellow butterfly, Colias eurytheme, an alfalfa pest in the southwestern states.

alfalfa hopper. Treehopper (Stictocephala festina) which feeds on alfalfa.

alfalfa weevil. See PHYTONOMUS.

aliform. Wing-shaped.

Aligena. Genus of minute marine bivalves.

alima. First larval stage of certain stomatopod crustaceans.

alimentary canal. Alimentary tract; a tube or cavity in any animal concerned with digestion and absorption of food; terminal openings are usually mouth and anus; about 30 ft. long in man.

alimentation. Collectively, the processes of ingestion, digestion, and egestion.

alisphenoid. Small bone on the anterior lateral wall of the brain case.

alitrunk. Last two (wing-bearing) segments of a typical insect thorax; includes first abdominal segment of Hymenoptera also.

alizarin. Red biological dye.

alkalemia. Increase in the normal pH value of blood or other body fluid; produced by an unusual increase in bicarbonate.

Alkaligenes. Alcaligenes.

alkaline gland. One of the glands making up the complex of the poison glands in the honeybee; thought to secrete an alkaline material.

alkalinity. Relative excess of hydroxyl ions over hydrogen ions in a solution; the more alkaline a solution is, the further its reading above pH 7.0.

alkali reserve. 1. Relative quantity of buffer substances in the blood which are capable of neutralizing acids; e.g. sodium bicarbonate, dipotassium phosphate, and proteins. 2. Relative quantity of bicarbonates in the blood. 3. Collectively, those dissolved materials in natural waters which tend to maintain a constant pH.

alkalitrophic. Pertaining to excessively alkaline lakes and ponds.

alkaloid. Any of a large group of nitrogen-containing basic organic substances found in certain plants; usually bitter and sometimes physiologically important; e.g. atropine, caffeine, morphine, nicotine, quinine, and strychnine.

alkapton. Homogenistic acid; intermediate breakdown product of phenylalanine; excreted as a black pigment in the urine of persons having an inadequate enzyme system for the complete breakdown of phenylalanine.

alkaptonuria. Presence of a black pigment, alkapton, in the urine.

allantois. Saclike outgrowth from the ventral surface of the posterior part of the embryonic gut of reptiles, birds, and mammals; sometimes it appears before the gut; it becomes so elongated as to lie imbedded in the wall of the yolk sac of birds and reptiles, and between the chorion and amnion of mammals; rich in blood vessels and functions in respiration, excretion, and nutrition.

Allantonema. Genus of nematodes parasitic in bark beetles.

Allantosoma. Genus of suctorian Protozoa found attached to ciliates in the colon of the horse.

Allee, Warder C. American ecologist (1885-1955); pioneer investigator in animal aggregations, peck order, social dominance, and biologically conditioned water; professor at University of Chicago.

Allee's principle. Each population of animals has an optimum density which affords optimum population growth and survival in a particular habitat.

Alleghanian zone. See TRANSITION ZONE.

Allegheny subregion. Zoogeographic subdivision of the Nearctic region; includes most of U.S. east of meridian 100 and small portions of southeastern Canada.

allele. 1. One of two or more alternative states of a gene, or of a character associated therewith, inherited in such a way that a gamete may have either but not both of them. 2. Allelomorph; one of two or more alternative expressions of genes having the same locus on homologous chromosomes.

allelomorph. See ALLELE.

Allenopithecus. A poorly-known genus of greenish guenon-like African monkeys; Allen's swamp monkey.

Allen's rule. In a given sp. or group of related spp., the more northern populations have proportionately smaller extremities, e.g. ear size in rabbits and hares; applies only to warm-blooded animals.

Allen's swamp monkey. See ALLENOPITHECUS.

allergy. Excessive sensitivity to a common substance such as pollen, many foods, natural oils, hair, dust, feathers, medicines, etc.; the sensitivity is manifested in many ways, including hives, rash, asthmatic attacks, higher blood pressure, headaches, and nasal congestion; such symptoms are external manifestations of antigen-antibody reactions, the antigens being the substances mentioned above, and the antibodies being their specific antagonistic materials in the tissues and blood plasma.

allied reflexes. Combination of coordinated reflexes which produces certain complex activities such as walking or crawling.

Alligator. Small genus of reptiles in the Family Alligatoridae; the alligators; entire head broad, snout bluntly rounded, teeth not protruding at sides of jaw; do not attack man except in self defense; American alligator, A. mississippiensis, found in Atlantic drainages from N.C. to Rio Grande; 16 ft. maximum length; only other sp. is A. sinensis of the Yangtze-Kiang River, China, which attains 6 ft.

alligator fish. Northwestern American marine fish (Podothecus acipenserinus); sturgeon sea poacher.

alligator gar. See ATRACTOSTEUS.

Alligatoridae. Family of reptiles which includes the alligators; See ALLIGATOR.

alligator lizard. See ANGUIDAE and GERRHONOTUS.

alligator snapper. See MACROCLEMYS.

Allocapnia. Common genus of stoneflies (Order Plecoptera).

allochthonous. 1. Of or pertaining to material generated outside a particular habitat but brought into that habitat, such as debris brought into a lake by a river. 2. Of or pertaining to an animal or taxon which has immigrated from another area. Cf. AUTOCHTHONOUS.

Allocrangonyx pellucidus. Blind amphipod found in springs, caves, and seeps in Mo. and Okla.

allocryptic. Habit exhibited by some animals which cover themselves with various objects; method of concealment.

Alloeocoela. Order of small turbellarian worms; mostly marine, few in fresh water; gastrovascular cavity sometimes with short diverticula; nervous system with three or four pairs of longitudinal trunks; Plagiostomum and Prorhynchus common.

allogene. Recessive gene of a pair of allelic genes.

Allogromia. Genus of fresh-water foraminiferans; shell simple and membranous.

Alloioplana. Genus of polyclad turbellarians.

Allolobophora. Common genus of earthworms; A. caliginosa is probably the most common N.A. earthworm.

allometry. Heterogony; disproportionate size of a structure in relation to the rest of the body; applies to phylogenetic development, speciation, and life history growth. Adj. allometric.

Alloniscus. Genus of terrestrial pillbugs (Isopoda) often found on marine beaches.

allopatric. Pertaining to two or more related populations which occupy mutually exclusive (but usually adjacent) geographical areas and do not interbreed; this situation is often called contiguous allopatry; where two related populations are separated by a wide gap in which neither occurs, the situation is known as disjunct allopatry. Noun allopatry. Cf. SYMPATRIC.

allopelagic. Relating to open-water marine or fresh-water organisms found at various depths which are not determined by temperature.

Alloperla. Genus of Plecoptera; mostly east of Rockies.

allophore. Cell containing red pigment, especially in lower vertebrates.

allopolyploidy. Chromosome condition in an organism in which one or more complete diploid sets of chromosomes come from each of two widely different strains or spp.

Allopora. Genus of hydrozoan coelenterates which have a coral-like calcareous growth form.

Allorchestes. Genus of marine amphipods.

all-or-none law. Any irritable tissue, especially muscle or nerve tissue, will respond in only two ways to stimuli: if the stimulus is below the threshold intensity, no reaction will occur; if the stimulus is above threshold, a reaction will occur, and this reaction cannot be increased with any increase in the intensity of the stimulus.

Allosaurus. Genus of large, carnivorous Jurassic reptiles.

alloscutum. In certain ticks, a dorsal sclerite just behind the scutum.

allosematic. Having protective coloration similar to that of another dangerous or distasteful sp.

Allosmerus. Whitebait or whitebait smelt of inshore Pacific waters; small, slender, silvery sp.

allosome. Unpaired chromosome present in one half of the gametes produced by an organism.

allotetraploidy. Amphiploidy; chromosome conditions in an organism in which one complete diploid set of chromosomes has been derived from each of two different strains or spp.; many natural and artificially-induced hybrids have appeared in this fashion.

Allotheria. One of the three subclasses of mammals; contains the single fossil Order Multituberculata.

Allothrombium. Small genus of large mites (Trombidiidae).

Allotriognathi. Order of marine bony fishes which includes the peculiar and greatly differing ribbonfish, oarfish, and opah; eyes large; teeth feeble or absent; Lampris, Trachypterus, and Regalecus examples.

Allotriognathida. Allotriognathi.

allotype. Supplementary type specimen of an organism, where the original description was of the opposite sex.

allozooid. Any zooid which differs from its parent.

Alluaudomyia. Genus of ceratopogonid Diptera.

Alona. Common genus in the crustacean Order Cladocera.

Alonella. Uncommon genus of littoral Cladocera.

Alonopsis. Genus of littoral Cladocera.

Aloplex. Genus of Arctic foxes; coloration white in winter and either gray-brown or parti-colored in summer.

Alopias. Genus of large sharks characterized by an upper tail lobe which is curved and as long or longer than combined head and trunk; thresher sharks; warm and temperate seas; to 15 ft. long.

Alosa pseudoharengus. Sawbelly, alewife; a herring-like fish in Atlantic coastal waters; moves up rivers to spawn; young return to sea when 2 to 4 in. long.

Alosa sapidissima. American shad; a herring-like fish found off the Atlantic Coast of U.S. and Canada and also introduced along Pacific Coast in 1870; ascends streams to spawn; important food fish.

Alouatta. Common genus of Tropical American howler monkeys (Family Cebidae); tail very long and prehensile; make remarkably loud booming sounds.

alpaca. Partially domesticated South American herbivorous mammal in the genus Lama; related to the camel, guanaco, and vicuna, and presumably descended from the guanaco; somewhat sheeplike but larger; bred chiefly for its long, lustrous wool in the mountains of Peru, Chile, and Bolivia.

alpha globulins. Group of plasma proteins similar to albumins.

alpha tocopherol. See TOCOPHEROL.

Alpheidae. Snapping shrimps, pistol shrimps; family of marine shrimps with two pairs of chelate legs, the first pair being greatly enlarged and capable of producing a snapping noise; burrowers or commensals; Crangon common.

Alpine. 1. Race of medium to tall caucasoids native to a large area extending from France through Russia; skin pale or light tan, head short. 2. Pertaining to high-altitude habitats or high-altitude animals, especially above timber line.

alternation of generations. Same as heterogenesis and heterogony; the alternation of bisexual and unisexual (or asexual) portions of the life cycle; the phenomenon is best known among certain Coelenterata and Platyhelminthes, but occurs also in some rotifers, cladocerans, and aphids.

Altica. Holarctic genus of blue or green chrysomelid beetles.

Alticamelus. Genus of giraffe-like camels of late Miocene and early Pliocene times.

altricial. 1. Pertaining to the young of songbirds, etc. which are naked and helpless at hatching, and which must be cared for in the nest by the parents. 2. Pertaining to the young of rodents, carnivores, etc. which are born blind and helpless and require further development before they are able to move about freely.

alula. 1. That portion of a bird wing which corresponds to the thumb; it bears a few short quills. 2. Scale-like structure between the base of the wing and halter in certain Diptera; squama; calypter. 3. Small basal posterior wing lobe in certain Diptera, and the elytron of certain Coleoptera. Pl. alulae.

alum cochineal. Bright scarlet biological dye made from the secretion of cochineal insects.

Alutera. See MONACANTHIDAE.

Alvania. Genus of minute marine snails.

alveolar ducts. Short ducts which open

into the individual air sacs of the mammalian lung.

alveolar gland. Gland, such as the salivary gland, having numerous small branching ducts, each of which opens into a hollow terminal cavity.

alveolus. 1. Minute air-filled sac at the end of a bronchiole in the lungs of mammals and a very few other tetrapods; it has a very thin wall and is invested in a capillary network; the exchange of carbon dioxide and oxygen occurs between the alveolus and the surrounding capillaries. 2. The socket holding a tooth. 3. One of the terminal saclike cavities of a compound gland. Pl. alveoli.

Alytes obstetricans. European midwife toad; after deposition by the female, the eggs are gathered around the hind legs of the male and carried thus until they hatch.

amadavat. Small Indian songbird, sometimes kept and used for fighting.

Amaea. Genus of polychaete worms.

Amage. Widely distributed genus of small tube-building polychaetes.

Amaroucium. Common genus of colonial tunicates; individuals greatly elongated; one sp. called sea pork because of its superficial resemblance to a piece of salt pork.

Amathia. Genus of marine Bryozoa.

Amaurobius. Common genus of spiders which build an unorganized, tangled, silken net.

Amauropsis. Genus of marine snails.

amaurotic idiocy. Fatal degeneration of retina and nervous system during first two years of life.

Amazilia. Genus of Central and South American hummingbirds; one sp. occurs in Tex. Gulf Coast lowlands.

amazon. Any of a large group of green parrots having a short, broad, rounded tail; Tropical America; Amazona common.

Amazona. See AMAZON.

Amazon ant. Obligatory slave-making ant in the genus Polyergus; it rears various spp. of Formica to care for its own eggs and young.

Amazon dolphin. See INIA.

amberfish. Any of certain tropical fishes in the genus Seriola; some edible.

amber fossils. Perfect and opaque fossil molds of insects and spiders trapped in the gummy resin of trees as much as 30 million years ago in the Baltic Sea area; such resins were subsequently converted to hard transparent amber.

ambergris. See SPERM WHALE.

amberjack. See SERIOLA.

ambidextrous. Able to use both hands with equal ease, rather than being only right-handed or only left-handed.

ambiens. Special muscle found in birds and reptiles; originates on the pelvis, its tendon inserting on the toes; facilitates grasping and perching.

Amblema. Common genus of unionid mollusks; in small to large rivers of Mississippi River system.

Amblonyx. Genus of otters found in India and southeastern Asia.

Ambloplites rupestris. Rock bass or redeye, a common robust centrarchid fish originally of the Mississippi basin.

Amblycephalidae. Small family of nonpoisonous tropical snakes; head broad, eyes large, teeth long.

Amblyomma. Large genus of ticks; one sp. attains a length of 30 mm.

Amblyopsidae. Family of small fishes in the Order Cyprinodontes; cave fishes, blindfishes; a few spp. have well developed eyes and live in surface waters, but most are blind, whitish, and restricted to caves and underground rivers, especially in Ind. to Ala.; Typhlichthys and Amblyopsis common.

Amblyopsis spelaeus. Small white fish with rudimentary eyes; especially in streams of Mammoth Cave, Ken.; see AMBLYOPSIDAE.

Amblyopsoidei. Suborder of fishes sometimes used to include the Amblyopsidae.

Amblyosyllis. Genus of small polychaetes; dorsal lobes of parapodia elongated to form long, coiled cirri.

Amblypoda. Pantodonta.

Amblypygi. Order of arachnid Subclass Arachnida; abdomen without a terminal filament; pedipalps raptorial, first legs long and sensory; humid tropics and subtropics; Tarantula typical.

Amblyrhynchus cristatus. Large crested iguana of the Galapagos Islands; gregarious sp. which feeds on algae in the sea.

Amblystoma. Ambystoma.

ambon. Ring of fibrous cartilage surrounding an articular socket.

ambrosia beetles. Those spp. of beetles in the Family Scolytidae which live in dead wood and cultivate fungi to feed their grubs; Xyleborus is cosmopolitan.

Ambrysus. Common genus of creeping

water bugs (Family Naucoridae).

ambulacral areas. Five radially arranged areas or series of plates, which bear rows of tube feet, especially in the Echinoidea; alternate with interambulacral areas.

ambulacral groove. Groove running the length of each arm on the oral surface of sea stars; tube feet originate in the groove.

ambulacral ridge. Internal ridge along the oral body wall of each arm of a sea star; formed of calcareous ossicles; externally the ambulacral ridge is an ambulacral groove.

ambush bug. See PHYMATIDAE.

Ambystoma. Common and widely distributed genus of N.A. salamanders; about 20 spp.; A. tigrinum, the tiger salamander, is found over much of U. S. and northern Mexico; to 12 in. long; blackish or sooty ground color with blotchy markings of white, cream, or yellow; breeds in lakes, ponds, and slow streams; especially in the western states, many individuals remain in the gilled aquatic larval stage to 12 in. long and exhibit neotenic reproduction; such larvae are called axolotls; A. opacum is the eastern marbled salamander which deposits eggs in moist places.

Ambystomidae. Family of N.A. salamanders; vomerine teeth in a transverse row across roof of mouth; eyelids present; usually terrestrial except during breeding season; adults with lungs and without gills; Ambystoma, Dicamptodon, and Rhyacotriton typical.

Ambystomoidea. Suborder of amphibians which includes the Family Ambystomidae.

amebiasis. Amoebiasis.

Ameiuridae. North American catfish family; catfish, bullheads, madtoms, etc.; scales lacking; whisker-like barbels on anterior part of head; omnivorous; anterior ray of dorsal and pectoral fins a stout spine; widely distributed, usually in quiet waters; larger spp. excellent food fish; Ictalurus, Ameiurus, and Schilbeodes typical.

Ameiurus. N.A. genus which includes small catfish and bullheads; body stout, head broad; mostly in warm sluggish waters; A. nebulosus (common bullhead, brown bullhead, horned pout) dark yellowish brown, common from Maine to N.D. and south to Ken. and Tex. but introduced elsewhere; A. natalis (yellow bullhead, yellow cat) yellowish with darker cloudiness, Great Lakes southward to Va. and Tex., introduced elsewhere; A. melas (black bullhead) common in Great Lakes and Mississippi drainage west to Rocky Mountains.

Ameiva. Genus of small, active lizards in the Family Teiidae; Tropical America and West Indies.

Ameletus. Common genus of Ephemeroptera.

ameloblasts. Cells which secrete enamel during the formation of a tooth.

amensalism. Association of two different spp. whereby one sp. is not affected and the other is harmed or inhibited; e.g. the association of an antibiotic-producing mold and bacteria.

American leishmaniasis. Espundia; disease of man caused by Leishmania braziliensis which forms dermal and epidermal ulcers; mostly in Tropical America; transmitted by Phlebotomus, a small sandfly.

Amerind. Race of mongoloids including the American Indian; skin yellow to red; North, Central, and South America; occasionally incorrectly used to include the Eskimo.

Ametabola. Group of orders of insects having primitive metamorphosis and in which no distinct external changes, except in size, are evident in the life history; characteristic of such orders as Protura, Thysanura, and Collembola. Adj. ametabolous.

ametamorphic. Having no metamorphosis, e.g. silverfish insects.

Ametor. Old World genus of hydrophilid geetles.

Amia. See PROTOSPONDYLI.

amictic. Pertaining to a female rotifer which produces only diploid eggs incapable of being fertilized by a male. 2. Pertaining to a diploid, unfertilized egg released by an amictic female rotifer.

Amicula. Genus of giant Pacific chitons.

Amiidae. Small family of bony fishes in the Order Protospondyli; includes the fresh-water dogfish or bowfin.

Amiiformes. Taxon which includes the single fish Family Amiidae; similar to the Order Protospondyli.

amino acid. See PROTEIN.

aminopeptidase. Digestive enzyme secreted by the wall of the small intestine; acts on amino acid end of a peptide chain, splitting off amino acids until dipeptides remain.

Aminopterin. Proprietary name for an

antimetabolite of folic acid.

Amitermes. Genus of termites (Order Isoptera) living in southwestern desert soils.

amitosis. Uncommon method of nuclear division involving simple constriction and separation into two halves without the formation of a spingle and chromosomes; occurs in plant endosperm tissue, mammalian cartilage, ciliate and suctorian macronuclei, and pathological tissue; often there is no accompanying division of the cytoplasm and the result is a multinucleate cell.

ammochaeta. Stiff bristle occurring in clusters on the heads of certain ants; used to clean the protibial spurs.

ammocoetes. Larval stage of lampreys found in streams; blind and toothless; usually in U-shaped tubes; stage lasts for three to seven years.

Ammocrypta. Small genus of sand darters (Family Percidae); eastern half of U.S.

Ammodorcas. Genus of gazelles found in Ethiopia and Somaliland; dibitangs; tail long and tasseled; female without horns.

Ammodramus. See SPARROW.

Ammodytes. Genus of fishes which includes the widely distributed sand eels or sand launces; small, eel-like marine spp. which often burrow in sandy shoals.

ammonia. Important transitory substance in intermediate metabolism; NH_3.

Ammonidea. Suborder of the Order Tetrabranchia in the Class Cephalopoda; ammonites, extinct cephalopod mollusks with a straight or more or less coiled chambered shell; related to the living pearly nautilus; Silurian to Cretaceous times.

ammonification. Complex of processes involved in the bacterial conversion of proteins to ammonia and other compounds.

ammonite. Any member of the mollusk Suborder Ammonidea.

ammonotelic. Pertaining to most aquatic non-chordates which excrete ammonia, especially polychaetes, many arthropods, echinoderms, cephalopods, and gastropods. Noun ammonotelism.

Ammophila. Sphex.

Ammophilactis. Genus of sea anemones (Order Actiniaria); commonly almost buried in sand.

Ammospiza. Genus which includes four N.A. sparrows.

Ammothea. Genus of Pycnogonida.

Ammothella. Genus of Pycnogonida.

Ammotragus. Genus which includes the aoudad.

Ammotrecha. Genus of Sulpugida; southern states.

Ammotrypane. Genus of burrowing polychaetes; head without appendages, parapodia rudimentary.

Amnia. Amniota.

Amnicola. Common genus of freshwater snails; rare west of Continental Divide.

amnion. 1. Inner, fluid-filled sac enclosing the embryo of reptiles, birds, and mammals. 2. In insects, an inner membrane which covers the germ band and eventually the entire embryo.

Amniota. Those vertebrates whose embryos have an amnion and allantois, i.e. reptiles, birds, and mammals.

amniote. Having an amnion.

amniotic cavity. Space between embryo and its enveloping amnion.

amniotic fluid. Lymphlike fluid filling the amniotic cavity of higher vertebrates; protects embryo from shock.

amniotic head fold. Crescent-shaped blastodermic fold immediately in front of the embryonic head, as in a chick embryo; grows posteriorly and fuses with the amniotic tail fold.

amniotic tail fold. Blastodermic fold posterior to the tail bud, as in a chick embryo; grows forward and fuses with the amniotic head fold.

Amoeba. Genus of naked Protozoa which move about on the substrate by means of pseudopodia; in the Class Sarcodina; usually 20 to 600 microns in diameter; common in fresh and salt waters; Amoeba proteus is a common sp. in the debris of stagnant ponds and puddles.

amoebiasis. Infection with a parasitic amoeboid protozoan, especially an intestinal sp.

amoebic dysentery. See ENTAMOEBA.

Amoebina. Order of the Class Sarcodina; includes all naked amoeboid protozoans having little cortical differentiation, lobopodia, and no pellicle or test; Entamoeba and Amoeba typical.

amoebocyte. Any body cell capable of independent movements by pseudopodia; several types of blood cells and certain tissue cells are amoebocytic; an important element of sponge tissues.

amoeboid. 1. Pertaining to an Amoeba

or related protozoan. 2. moving by means of pseudopodia. 3. Having movement resembling that of an Amoeba.

amoeboid cell. Any cell capable of amoeboid movement or capable of forming pseudopodia; in the general body tissues of many lower phyla and in the circulatory system of higher phyla.

amoeboid movement. Complicated interrelated physical and chemical changes in protoplasm which collectively result in movement of an amoeboid protozoan or other amoeboid cell.

Amoebotaenia sphenoides. Cestode parasite of fowl; transmitted by earthworms.

amoebula. Miniature amoeboid form in the life history of certain Protozoa.

amorph. Mutant allele which has little or no effect on the development of a character.

AMP. Adenosine monophosphate.

Ampelisca. Genus of marine Amphipoda.

Ampharete. Genus of small tube-building Polychaeta.

Amphiagrion saucium. Common damselfly; widely distributed in U. S.

Amphiascus. Genus of marine harpacticoid Copepoda.

amphiaster. The spindle-shaped achromatic figure in a dividing cell; consists of two centrosomes and their radiating and connecting fibrils. 2. Type of spicule found in certain sponges; stellate at each end.

Amphibdella. Genus of external monogenetic trematode parasites of marine fishes.

Amphibia. One of the eight classes of vertebrates; includes salamanders, frogs, and toads; skin moist, without scales; two pairs of legs used in swimming and terrestrial locomotion; up to five toes; hind feet often webbed; two nostrils connected to mouth cavity; mouth often with fine teeth, tongue usually protrusible; skeleton bony; ribs, if present, not attached to sternum; heart with two atria and one ventricle; respiration through lungs, skin, gills, and lining of mouth in varying degrees; eggs with gelatinous covering; usually an aquatic immature stage; fresh-water and terrestrial, but never in dry habitats; more than 2600 living spp.

amphibiotic. Aquatic during one part of the life history and terrestrial during the rest of the life history.

amphibious. Pertaining to an organism capable of living in both water or moist places, and on drier land.

amphiblastic. Pertaining to a telolecithal ovum with complete but unequal cleavage.

amphiblastula. Unique, free-swimming larval stage in certain spp. of marine sponges; essentially a blastula larva with a small blastocoel; cells of the animal pole are small and flagellated, cells of the vegetal pole are large and not flagellated.

Amphiblestrum. Genus of marine Bryozoa.

Amphibolella. Genus of fresh-water rhabdocoel turbellarians.

amphibolic. Capable of being turned forward or backward; e. g. the outer toe of an owl.

Amphicoela. One of the five orders of Amphibia in the Subclass Salientia; primitive toads with amphicoelous vertebrae and two tail-wagging muscles, even though no tail is present; includes a single family, Ascaphidae.

amphicoelous. Pertaining to the centrum of a vertebra which is concave at both ends, as in certain fishes.

amphicondylous. Provided with two occipital condyles.

Amphicora. Genus of small Polychaeta.

Amphicteis. Widely distributed genus of small polychaetes; tube of hardened mucus covered with mud.

amphid. One of a pair of anterior chemoreceptor sense organs of nematodes; cuticular excavation having a pocket, spiral, or circular shape.

Amphidelus. Genus of microscopic, free-living nematodes.

Amphidinium. Genus of marine and fresh-water dinoflagellates; annulus anterior; two flagella.

amphidiploid. Hybrid allopolyploid having two genomes of each of the two spp. which gave rise to it.

Amphidiscophora. Order of sponges in the Class Hexactinellida; amphidisks present; star-shaped spicules absent; e. g. Hyalonema.

amphidiscs. Amphidisks; birotulate spicules forming the outermost layer of a gemmule in certain fresh-water sponges.

amphigastrula. In certain sponges, the gastrula which develops from an amphiblastula larva.

amphigenesis. Fusion of egg and sperm (or any two unlike gametes) to form a zygote.

Amphigyra alabamensis. Snail found in the Coosa River, Ala.

Amphileptus. Genus of flask-shaped holotrich ciliates; fresh and salt waters; often commensal on other colonial Protozoa.

Amphilimna. Genus of brittle stars (Order Ophiurae).

Amphilocus. Genus of marine Amphipoda.

Amphimerus. Genus of distome trematode parasites of the liver in mammals and birds.

amphimixis. True sexual reproduction involving union of sperm and egg nuclei.

Amphimonas. Genus of small, colorless, marine and fresh-water biflagellate Protozoa; sometimes attached by a fine stalk.

amphimorula. Morula stage derived from the cleavage of an amphiblastic egg.

Amphineura. Class of mollusks which includes the chitons; elongated body covered with mantle in which are imbedded eight dorsal calcareous plates and/or numerous calcareous spicules; head small and hidden under mantle; foot broad and flat and also completely hidden by mantle; radula present; 6 to 80 pairs of ctenidia; nervous system primitive; 10 mm. to 25 cm. long; exclusively marine; larva a trochophore; about 700 spp.

Amphiodia. Genus of brittle stars (Order Ophiurae).

Amphioplus. Genus of small brittle stars (Order Amphiurae).

Amphioxus. See CEPHALOCHORDATA.

Amphipholis. See SNAKE STAR.

amphiplatyan vertebra. 1. Vertebra having both anterior and posterior surfaces of the centrum more or less flat. 2. Acoelous vertebra.

amphiploidy. Allotetraploidy.

amphipneustic. 1. Pertaining to certain Diptera larvae which have functional spiracles on the prothorax and last abdominal segment. 2. Pertaining to a few amphibians which have both lungs and gills throughout life.

Amphipoda. Large order of malacostracan Crustacea; includes the sideswimmers, sand hoppers, etc.; body compressed; first thoracic segment fused with head; no true carapace; mostly scavengers; most spp. marine, burrowing or moving about on the bottom and debris; some parasitic, a few in fresh waters, and several terrestrial; 4300 spp.

Amphiporus. Genus of nemertines in the Order Hoplonemertea.

Amphisbaenidae. Family of cylindrical, wormlike, burrowing lizards; worm lizards; limbs rudimentary or absent; skin and scales in the form of segmental rings; Tropical America, West Indies, and Africa; _Bipes_ and _Rhineura_ typical N. A. genera.

Amphiscolops. Genus of turbellarians in the Order Acoela; found especially on sargassum weed.

Amphispiza. Genus which includes two western N. A. spp. of sparrows, i.e. the black-throated sparrow and Bell's sparrow (sage sparrow).

Amphissa. Genus of small marine snails found on rocky and gravelly shores.

Amphistomata. Suborder of digenetic Trematoda; includes the Family Paramphistomidae; oral sucker anterior; acetabulum at posterior end.

amphistylic jaw suspension. Jaw suspension involving support from both the hyomandibular and brain case; present in a few sharks.

Amphithoë. Genus of marine Amphipoda.

Amphitrite. Common genus of marine polychaete annelids; head with a large cluster of long threadlike tentacular gills; parapodia reduced; in tubes in the sea bottom.

Amphiuma. See AMPHIUMIDAE.

Amphiumidae. Family of sluggish salamanders having an elongated cylindrical body, small weak legs, and short compressed tail; adult with lungs, gills, and a pair of small gill slits; _Amphiuma means_ (congo eel) in swamps, ditches and rice fields from Va. to Fla., west to La., and up the Mississippi Valley to Mo.; to 30 in. long.

Amphiura. Genus of brittle stars (Order Ophiurae).

Amphizoa. See AMPHIZOIDAE.

Amphizoidae. Small family of aquatic beetles; three spp. of _Amphizoa_ in swift, clear streams of western states.

ampholyte. Substance which can function as both an acid and a base, e.g. amino acid.

Amphoriscus. Genus of small cylindrical calcareous marine sponges.

Amplexopora. Genus of fossil Bryozoa in the Order Treptostomata.

amplexus. An embrace, especially the copulatory embrace, as in many amphibians.

ampulla. 1. Small saclike swelling at

one end of each of the three semicircular canals of vertebrates; within each ampulla is a cluster of hair-cells, or crista, which is innervated by a branch of the vestibular nerve; functions in equilibration. 2. Any membranous vesicle. 3. Muscular bulblike internal end of a tube foot in the echinoderms; see WATER VASCULAR SYSTEM. Pl. ampullae.

ampullae of Lorenzini. System of minute jelly-filled canals which open on the rostral and head regions of elasmobranch fishes; presumably a temperature receptor mechanism.

ampulla of Vater. Eminence on the inner surface of the duodenum marking the entrance of the bile duct and pancreatic duct.

Ampullaria. Genus of tropical and subtropical fresh-water snails; apple shells.

Amyda. Trionyx.

Amydetes. Genus of fireflies (Family Lampyridae).

amylase. Any of a group of digestive enzymes which convert starches and glycogen to maltose.

amylolysis. Enzymatic conversion of starches to dextrins and sugars.

amylopsin. Enzyme in pancreatic juice of vertebrates; converts starches to disaccharides.

Amynodon. Genus of Eocene hornless rhinoceros-like mammals; canines in the form of curved tusks.

Anaata. Genus of red marine sponges.

Anabas. See CLIMBING PERCH.

anabiosis. State of suspended animation, especially as found in the Tardigrada under unfavorable environmental conditions. Adj. anabiotic.

Anablepidae. Small family of bony fishes in the Order Cyprinodontae; four-eyed fishes; eyes divided, with upper half for aerial vision and lower half for aquatic vision; small spp., chiefly in rivers of Central America and northern South America; Anableps common.

Anableps. See ANABLEPIDAE.

anabolism. Sum of all metabolic processes involved in building up new protoplasm.

Anabrus simplex. Mormon cricket; an insect in the orthopteran Family Tettigoniidae; clumsy, wingless dark brown or black, and to 45 mm. long; important pest of farm and truck garden crops in the western U.S., especially the Great Basin.

Anacaena. Genus of hydrophilid beetles.

Anacanthida. Anacanthini.

Anacanthini. Order of bony fishes; includes the cods, pollacks, hakes, and burbots; mostly marine bottom dwellers of great economic importance.

Anachis. Genus of small marine gastropods.

anaconda. See EUNECTES.

Anadara. See ARK SHELL.

anadromous. 1. Pertaining to fishes which spend most of their lives in salt water but migrate into fresh waters to spawn; e.g. salmon, shad, striped bass, etc. 2. Term is sometimes used in the broader sense to include upstream migrations and offshore to inshore migrations.

anaerobic respiration. Liberation of energy by the breakdown of substances without the use of molecular oxygen; e.g. the breakdown (fermentation) of glycogen to lactic acid and carbon dioxide in vertebrate muscle and the conversion of glucose to ethyl alcohol and carbon dioxide by yeasts; a few bacteria and perhaps certain Protozoa are permanently anaerobic; many small metazoans can maintain an anaerobic existence for weeks or months.

Anaitides. Genus of polychaete worms.

anal. Pertaining to anus.

anal angle. Posterior distal angle of a typical insect wing.

anal cirrus. One of a pair of filamentous projections on the last segment of many polychaete annelids.

analeptic. 1. Drug or pertaining to a drug which stimulates the central nervous system, e.g. caffeine. 2. Restorative medicine or measure.

anal fin. Median posteroventral fin in a typical fish; usually just posterior to the anus.

Analgesidae. Large family of feather mites; minute, elongated spp. which feed on bird feathers and dead epidermis as commensals; Analgopsis common.

anal gill. Gill closely associated with anus in many fresh-water insect larvae and nymphs; many of the so-called "gills" are of great importance in osmoregulation.

anal gland. Gland associated with the rectum or anus in a wide variety of vertebrates and invertebrates; functions various, usually secretory or excretory.

Analgopsis. See ANALGESIDAE.

analogous. 1. Similar in function and superficial structure but not in fundamental origin; e.g. the wing of a butterfly and the wing of a bird are analogous. 2. The term is also occasionally used with reference to activity, e.g. the flight of a butterfly and the flight of a bat. Noun analogy.

anal pit. Slight invagination on the surface of a vertebrate embryo (e.g. frog) which later breaks through to form the anus and join the gut.

anal plate. 1. Membrane which closes the anal aperture in early embryos of higher vertebrates. 2. Posterior plate of a turtle plastron. 3. Preanal plate in many snakes.

anal pore. Cytopyge.

anal vein. One of the posterior veins in an insect wing; often a series of two or more are present and they are then numbered I, II, etc.

Anamnia. Anamniota.

Anamniota. Group of vertebrates having no amnion during early development, e.g. fishes and amphibians.

Anamygdon. Genus of painted bats; coloration and patterns variable; Africa and southern Asia.

Anaperus. Genus of minute, free-living flatworms in the Order Acoela; red to yellow coloration.

anaphase. That stage in mitosis characterized by the separation of the two chromatids of each chromosome and the migration of each member of each pair toward opposite asters; after complete separation, each chromatid becomes a daughter chromosome.

anaphylactic shock. An anaphylaxis reaction which is so pronounced as to cause violent circulatory, nervous, and respiratory symptoms; unconsciousness and death may result.

anaphylaxis. Essentially the same as allergy, but more commonly used to designate a serious hypersensitivity to certain proteins with which a person has been previously inoculated and for which he has developed specific antibodies; the term is used especially to indicate sensitivity to drugs and protein components in horse serum which is often used for administering antitoxins, etc.

Anaplasma. Genus of minute sporozoan parasites of the erythrocytes, especially in cattle; transmitted by several ticks and flies.

Anaplectus. Common genus of microscopic, free-living nematodes.

anapophysis. Small dorsal projection from the transverse process of a lumbar vertebra in many mammals.

Anapsida. One of the five subclasses of reptiles; turtles and many primitive extinct reptiles; no temporal openings in the skull behind the eye.

Anarcestes. Genus of fossil ammonites.

Anarhichadidae. Family of fishes which includes the wolf fishes, especially the genus Anarhichas; long, tapered spp. with a formidable set of teeth and a spiny dorsal fin extending from head to tail; bottom forms, to five ft. long.

Anarhichas. See ANARHICHADIDAE.

anarthrous. Without joints.

Anas. Genus of surface-feeding ducks in the Family Anatidae; A. platyrhynchus is the common mallard duck; domesticated and widely distributed over the Northern Hemisphere; to 24 in. long; chief duck shot by sportmen; A. rubripes is the black duck of the eastern states; A. carolinensis is the small green-winged teal of N.A.; A. discors the blue-winged teal; A. cyanoptera the uncommon teal of the West; A. strepera is the gray duck, common over much of N.A.; A. acuta is the common pintail of the Northern Hemisphere; two tail feathers 5 to 9 in. long.

Anasa. See COREIDAE.

Anaspida. Order of ostracoderms; active swimmers with the tail bent downward; dorsal nostril between the eyes; round gill openings on the anterior flank; Pterolepis typical.

Anaspidacea. See SYNCARIDA.

Anaspides. See SYNCARIDA.

anastomosis. 1. Union by means of a branching network, as in blood vessels or nerves. 2. Artificial or natural communication between two vessels or other hollow organs.

Anatanais. Genus of marine Isopoda.

Anathana. Genus of Indian tree shrews.

Anatidae. Large family of birds in the Order Anseriformes; contains many spp. of ducks, geese, and swans.

Anatina canaliculata. Hat shell, channeled duck shell; a whitish, delicate bivalve to 2.5 in. long; N.J. to Brazil.

anatomy. Structure of an organism or of any of its parts, usually as revealed by dissection.

Anatonchus. Genus of microscopic, free-living nematodes.

Anatopynia. Genus of tendipedid midges.

Anatosaurus. See TRACHODON.

anatoxin. Type of toxoid made by adding a small amount of formaldehyde to a bacterial toxin.

Anax. Common genus of dragonflies

(Order Odonata).

Anaximander. Greek philosopher (611-547 B.C.); anticipated the theory of organic evolution; believed that man received his physical state by adaptation to the environment, that man developed from fish, and that living things first arose from wet primordial mud.

ancestrula. First-formed polyp of a bryozoan colony.

Anchistropus minor. Littoral cladoceran.

Anchitherium. Genus of horselike Miocene mammals; three toes on each foot.

Anchoa. See ENGRAULIDAE.

anchor ice. Layer or mass of ice attached to the bottom of a stream; forms in very cold weather, especially at night; highly destructive to bottom invertebrates.

anchovy. See ENGRAULIDAE.

Ancistrodon. Agkistrodon.

anconeus. One of the muscles of the forearm in certain tetrapods.

Ancon sheep. Sheep mutation which first appeared in 1791 in Mass.; short, crooked legs and a long body.

Ancula. Genus of nudibranchs.

Anculosa. Genus of fresh-water snails; from Ohio River southward.

Ancyloceras. Genus of fossil ammonites.

Ancylostoma caninum. Hookworm which lives in the dog intestine.

Ancylostoma duodenale. "Old World" hookworm; a widely distributed nematode in the tropics and subtropics, and the dominant human hookworm of southern Europe, northern Africa, and eastern Asia; rare in U.S.; feeds on mucosa of the small intestine; has a free-living stage in soils that have been infected with feces containing eggs; man obtains the disease by penetration of the strongyliform larva through the bare feet or some other part of the body that touches infected soil.

Ancylus. Genus of small fresh-water limpets.

Ancyromonas. Genus of very small marine and fresh-water Protozoa; ovate to triangular; with one trailing flagellum.

Ancyronyx. Genus of elmid beetles.

Andalusian poultry. Breed of poultry; black Andalusians crossed with whites produce blue Andalusians; the latter do not breed true.

Andrena. See ANDRENIDAE.

Andrenidae. Cosmopolitan family of mining bees; each female constructs a burrow in the ground, but many burrows are often associated together; valuable as flower pollinators; Andrena common.

andrenergic nerve fiber. Nerve fiber producing sympathin or sympathin-like substances at its synapses with other nerve fibers.

Andrias scheuchzeri. Fossil cryptobranchid salamander; several hundred years ago said to be the remains of a human victim of the Deluge.

Andricus. Genus of gall wasps; larvae produce abnormal growths of plant tissues.

androconia. Modified scales on the wings of certain butterflies; produce a sexually attractive scent. Sing. androconium.

androgen. Any substance having male sex hormone activity in vertebrates, especially one governing the development and maintenance of male sex features.

androsterone. Androgen found in the urine of both men and women; $C_{19}H_{32}O_2$.

androtype. In taxonomy, a male type specimen.

Aneides. See TREE SALAMANDER and GREEN SALAMANDER.

anelectrotonus. Relative insensitivity and polarization of a nerve or muscle at the anode when an electric current is passed through the tissue.

anellus. In male Lepidoptera genitalia, a sclerotized support for the aedeagus.

anemia. 1. Deficiency of hemoglobin, reduced number of erythrocytes, or both; produces pallor, weakness, and breathlessness; the result of many different pathological situations. 2. Hypemia. Adj. anemic.

anemone. Sea anemone; sedentary, cylindrical marine coelenterate with oral disc bearing tentacles; see ANTHOZOA.

Anemonia sulcata. Sea anemone with especially long tentacles.

aner. Male insect, especially in ants.

anestrus. Interval of sexual inactivity between two estrus periods.

aneuploidy. Condition of genetic polyploidy where at least one haploid set of chromosomes has one or a few chromosomes more or less than a euploid (whole set). Cf. euploidy.

aneurin. Thiamine.

aneurysm. Localized dilation of an artery or vein, with associated weakened wall.

angelfish. 1. Any of several brightly colored compressed tropical fishes in the Family Chaetodontidae. 2. Small South American aquarium fish with silvery black-barred body and long, tapered fins; Pterophyllum scalare, the common scalare. 3. Either of two raylike sharks of North Atlantic coastal areas.

Angelichthys. Genus of tropical marine fishes; includes the isabelita and some of the angelfishes; colorful, compressed spp.

angel shark. See SQUATINA.

angel wings. Any of numerous whitish bivalve marine mollusks which burrow in mud, clay, or peat; wing shells; valves elongated, brittle, sculptured, and toothed; when laid out flat in pairs, the valves have the appearance of outspread wings; Pholas typical.

angina. 1. Sore throat. 2. Severe pain from any cause.

angina pectoris. Attack marked by excessive thoracic pain with a feeling of suffocation; usually the result of an insufficient blood supply to the myocardium brought on by physical effort or excitement.

angioblast. Early embryonic mesenchyme tissue giving rise to blood vessels and blood cells.

angiogenesis. Embryonic blood vessel development.

angiotensin. Blood protein which causes an increase in blood pressure when there is a decrease in pressure in a kidney; formed by union of renin and a plasma globulin component.

angiotonin. Angiotensin.

Anglaspis. Genus of ostracoderms; nostrils paired and ventral; gills with a common slit behind the armor plate on each side.

angler. See CERATIIDAE and MELANOCETIDAE.

angler fish. See LOPHIIDAE.

angle-wing. Any of the spp. of Polygonia, a common U.S. genus of butterflies whose wings are variously notched and pointed along the margins; tawny, with dark markings; includes the "comma" and the "question mark."

angleworm. Any earthworm, especially a large specimen.

angoumois grain moth. See SITOTROGA.

angstrom unit. 0.0001 micron; unit of length used especially for expressing the wave length of electromagnetic waves; usually written Å.

Anguidae. Family containing the alligator lizards; scales heavy, tail long, body slender and with a lateral skin fold on each side; limbs sometimes small or absent; tail lost and regenerated easily; southern and western U.S. to northern South America, and Europe, northern Africa, and India; typical genera are Gerrhonotus, Ophisaurus, and Anguis.

Anguilla. Common genus of fresh-water eels; American eel, A. bostoniensis, breeds in Sargasso Sea, and young eels migrate to N.A. and the females thence up Gulf and Atlantic river systems where most of the life history is spent inland; males remain in brackish estuaries; upon maturity, after three to seven years, both sexes migrate back to the Sargasso Sea where they die after spawning; A. anguilla is the European eel which has similar habits; a few investigators believe these two are actually one and the same sp.

Anguillidae. Family of bony fishes which includes the fresh-water eels; probably one European and one American sp., and about 15 western Pacific spp.; Anguilla typical. See APODES.

Anguilliformes. Same as Apodes; order of fishes which includes the eels.

Anguillula aceti. Vinegar worm, vinegar eel; a small free-living nematode often abundant in unsterilized vinegar; sometimes put in the genus Turbatrix.

Anguillulina petrefaciens. Nematode parasite of onion bulbs; sometimes found in man.

Anguina. Genus of tylenchid nematodes; several spp. are important plant parasites which produce galls on the inflorescences of wheat and grasses.

Anguinella. Genus of marine Bryozoa.

Anguis fragilis. Limbless, snakelike lizard in the Family Anguidae; blindworm, slowworm; Europe, western Asia, and Algeria.

Anguispira. Genus of land snails.

angulare. Dermal bone of the lower jaw in certain vertebrates; absent in amphibians and mammals.

angulosplenial. Bone which forms the posterior half of the lower jaw on each side in the frog and certain other vertebrates.

angwantibo. See ARCTOCEBUS.

Anhima. See ANHIMIDAE.

Anhimidae. Small family of South American gooselike wading and swimming birds in the Order Anseriformes; screamers; bill slender, feet not webbed; a primitive family, possibly ancestral to birds of prey on one hand and ducks and geese on the other hand; Anhima typical.

Anhinga anhinga. Water turkey or snake bird (Family Anhingidae); a large blackish sp. with silvery patches on the wings, long white-tipped tail, and a long serpentine neck; often perches above water; excellent swimmer; obtains food by pursuing prey under water; southern states and Tropical America.

Anhingidae. See ANHINGA.

anhydrous. Having no water, especially as applied to certain chemical salts which usually have one or more loosely associated water molecules per molecule of salt.

ani. See CROTAPHAGA.

Anilidae. Small family of burrowing snakes of tropical South America and southeastern Asia; eyes small, tail short; Anilius common in South America.

Anilius. See ANILIDAE.

animal community. Collectively, all of the animals living and interacting in a community; see COMMUNITY.

animalcule. Obsolete term used to include all microscopic animals.

animal hemisphere. That half of an early embryo which contains the animal pole.

animal pole. 1. One of two general regions in the early embryo of many animals; consists of small, dark cells containing little yolk and is opposite the vegetal pole. 2. That portion of a zygote which contains little yolk.

anion. Negatively charged ion.

aniridia. Absence of iris in the eye; Mendelian dominant in man.

Anisembia. Genus of insects in the Order Embioptera; southwestern states.

anisodactyl. Having feet with three toes extending forward and one backward.

Anisodoris. Genus of Pacific nudibranchs.

Anisogammarus. West Coast genus of Amphipoda; marine and coastal streams.

anisogamy. Heterogamy; condition in which two uniting gametes are unlike as to size or structure.

Anisolabis maritima. Dark brown earwig insect to 24 mm. long; on marine beaches just above the high water mark.

Anisomorpha. Genus of large wingless phasmid insects; common in N.A.

Anisomyaria. Pelecypod order designation used especially by paleontologists; includes shells having various tooth arrangements; posterior adductor large, anterior small or absent; mantle open, siphons absent, gill lamellae variable; e.g. Ostrea, Pecten, Mytilus.

Anisonema. Genus of marine and freshwater biflagellate Protozoa; cell asymmetrical, ovoid, flattened; one flagellum directed forward, one trailing.

Anisops. Old World genus of backswimmers (Family Notonectidae).

Anisoptera. One of the two suborders in the insect Order Odonata; dragonflies, snake doctors, or mosquito hawks; wings extended laterally when at rest; hind wings wide near base; naiads robust and with the rectum specialized as a muscular respiratory organ by taking in and expelling water; common genera are Gomphus, Aeschna, Macromia, Anax, Libellula, and Sympetrum.

Anisotremus. See GRUNT.

anisotropic discs. Microscopic, dark-colored, transverse stripes or discs as seen in surface view of striated muscle fibers; alternate with light-colored isotropic discs; same as A discs and Q discs.

Ankylosauria. Suborder of Cretaceous dinosaurs in the Order Ornithischia; heavy, quadrupedal, herbivorous spp.; strongly armored with thick bony plates completely encasing dorsal and lateral surfaces of body and tail; Ankylosaurus to 20 ft. long.

Ankylosaurus. See ANKYLOSAURIA.

ankylosis. Abnormal growing together and consolidation of a joint.

anlage. Primordium.

Anna's hummingbird. See CALYPTE.

annectent. Pertaining to intermediate linking spp. or genera.

Annelida. Phylum consisting of elongated, segmented worms; includes earthworms, leeches, and many kinds of marine forms; body commonly covered with thin, moist cuticle over a glandular epithelium; with an extensive coelom (usually), closed circulatory system, paired nephridia (usually metanephridia), brain, and ventral nerve cord; marine, fresh-water, terrestrial, and parasitic; 7000 spp.

annelid theory. Phylogenetic theory put forth independently by Semper and Dohrn (1875) which sought to derive chordates from modified annelids; another, related theory was advanced by Delsman (1913).

Anniella. See ANNIELLIDAE.

Anniellidae. Family of legless, snake-like lizards; includes only two spp. of Anniella in coastal areas of Calif. and Lower Calif.

annual production. Total amount of new protoplasm formed during a year by an organism or group of organisms.

annular ligaments. Ligaments in man which run transversely over the top of the ankle and wrist areas.

Annulata. Annelida.

annulated. Having a serially ringed appearance.

annulus. 1. Ring, ringlike mark, or part, as in a leech. 2. Circular groove. 3. One year's growth, as shown by the arrangement of circuli on the scale of a typical fish. Pl. annuli.

annulus ventralis. Seminal receptacle; a grooved, elliptical, calcified area in the midline between the bases of the fourth and fifth pereiopods in female crayfish.

Anoa. Small genus of buffalos in the Family Bovidae; includes the tamarao, a small, black buffalo of the Philippines, and the anoa, a still smaller forest buffalo of the Celebes.

Anobiidae. Family of small somber-colored beetles; death watch beetles; larvae and adults commonly in burrows in old and well-seasoned hardwoods; other spp. in many kinds of dried animal and vegetable products; Stegobium and Xestobium typical.

Anodonta. Genus of American fresh-water mussels; shell thin and without teeth; the paper shells.

Anodontia. Genus of marine bivalves.

Anodontoides. Genus of fresh-water bivalve mollusks (Unionidae); streams and wave-swept shores; Great Lakes, St. Lawrence, and Mississippi drainages.

anole. Lizard in the genus Anolis.

Anolis. Large tropical American genus of small, slender, arboreal lizards in the Family Iguanidae; A. carolinensis is the familiar "American chameleon" found from N.C. to Tex.; changes its color to match that of its background.

Anomalagrion hastatum. Damselfly of eastern U.S.

Anomalocardia. Genus of small Caribbean marine bivalves; sandy shores.

Anomalocera. Genus of marine plankton Copepoda (Suborder Calanoida).

Anomalodesmacea. In some classifications, one of the three orders in the molluscan Class Pelecypoda; mantle lobes united except for openings for the siphons and foot; valves usually unequal; mostly burrowers; all marine.

Anomalops. Genus of luminescent deep-water tropical marine fishes.

Anomaluridae. Scale-tails; a family of squirrel-like rodents; see ANOMALURUS.

Anomalurus. Genus of African rodents resembling flying squirrels and having scales on the ventral side of the tail; scale-tails, scaly-tailed squirrels.

Anomia. Genus of marine bivalves which includes the jingle shells; thin, bright, pearly or yellowish shells of warm waters; one valve flat and permanently anchored to substrate with byssus threads, the other valve being more convex.

Anomocoela. One of the five orders in the amphibian Subclass Salientia; pelobatid toads, spadefoot toads; sacral vertebrae procoelous; free ribs absent; toadlike or froglike in appearance; single Family Pelobatidae; Scaphiopus the only American genus.

Anomopoda. Tribe of cladocerans in the Suborder Calyptomera; five or six pairs of legs; first and second pairs prehensile; includes most Cladocera.

Anomura. One of the three suborders in the crustacean Section Reptantia; hermit crabs, sand crabs, and related forms; abdomen bent ventrally, without lateral plates; third pair of legs never chelate; all marine.

Anonchus. Common genus of microscopic, free-living nematodes.

Anonyx. Genus of marine Amphipoda.

Anopheles. Genus of mosquitoes having world-wide tropical and subtropical distribution; important because of their role in transmitting malaria.

anophthalmos. 1. Congenital absence of one or both eyes; genetic recessive. 2. Person born without eyes.

Anopla. One of the two classes in the Phylum Nemertea; mouth posterior to brain, central nervous system sub epidermal, and proboscis unarmed.

Anoplocephala. Genus of cyclophyllidean tapeworms found in horses and rodents.

Anoplodactylus. Genus of Pycnogonida.

Anoplophrya. Genus of holotrich ciliates parasitic in annelids and crustaceans.

Anoplura. Order of insects which includes the true lice, or sucking lice;

parasitic on mammals; to 6 mm. long; body depressed, tough, and wingless; simple metamorphosis; mouth parts piercing, sucking, and retractile, used for taking blood from the host; legs short, stout, and with strong terminal claw; eyes reduced or absent; important genera are Pediculus, Phthirus, and Haematopinus.

Anostraca. Order of Crustacea which includes the fairy shrimps; elongated, without a carapace; eyes stalked; 11 pairs of swimming legs; usually 6 to 25 mm. long; common in temporary pools and salt lakes; Artemia, Branchinecta, and Eubranchipus familiar genera.

Anotomastus. Genus of polychaete worms.

Anourosorex. Genus of mole-shrews of eastern and southeastern Asiatic forests; tail and limbs very short; eyes minute, ears invisible.

Anous. Noddies; a genus of tropical oceanic terns having dark plumage and short tail.

anovulatory cycle. Menstruation without ovulation.

anoxia. Hypoxia.

Anser. Genus which includes several Old and New World geese, such as the lesser snow goose, blue goose, and graylag; A. albifrons is the Holarctic white-fronted goose which breeds in the Arctic and in the U.S.; winters on West and Gulf coasts.

Anseriformes. Large, cosmopolitan order of birds which includes the ducks, geese, and swans; bill broad, covered with a thick, soft, cornified layer containing numerous small sense pits; edge of bill toothed or fluted; legs short, feet webbed; tail short, bearing many feathers.

Anserinae. See GOOSE.

ant. See FORMICIDAE.

antagonism. 1. Inhibition or interference with the growth of an organism owing to unfavorable conditions created by the presence of another sp., such as the production of inhibitory excretory materials (antibiotics) or utilization of available food. 2. Opposite or neutralizing effects of two different drugs or hormones. 3. Opposing action of two different muscles or sets of muscles, such as the action of flexors and extensors at a movable joint.

antagonistic muscles. Any group of two or more muscles whose contractions have opposite effects on the movement of a joint.

ant bear. Large South American anteater; see MYRMECOPHAGIDAE.

ant bird. See THAMNOPHILUS and FORMICARIIDAE.

anteater. Any of various mammals which feed chiefly on ants; see MYRMECOPHAGIDAE, PHOLIDOTA, TACHYGLOSSUS, and ZAGLOSSUS.

antebrachium. Forearm, between elbow and wrist.

Antechinomys. Genus of Australian marsupials; pouched jerboas, phascogales; adapted for desert life; hind legs stiltlike, feet furred, tail long and tufted.

Antechinus. Genus of carnivorous, mouselike Australian marsupials.

antecosta. Anterior marginal ridge on the internal surface of a tergum or sternum of many insects; serves as an area for attachment of longitudinal muscles.

Antedon dentata. Common feather star of Atlantic Coast.

antefurca. Internal forked prosternal process in many insects; attachment area for muscles.

antelope. Any of a wide variety of light, graceful ruminants in the Family Bovidae; horns unbranched, with a bony core; Africa and Asia; see ANTILOPE. 2. See ANTILOCAPRIDAE.

antenna. 1. Long, slender, sensory (occasionally locomotor)paired appendage on the head of all arthropods except Arachnoidea. 2. In the specific sense, the first pair of the two pairs of such structures in Crustacea. Pl. antennae.

antenna comb. 1. Structure at the proximal end of the anterior tarsus in the honeybee and its relatives; consists of a bristle-lined notch and an enclosing spur from the distal end of the tibia; the antennae may be drawn through the device which removes pollen and other foreign material from them. 2. Strigil.

Antennariidae. Family of bony fishes in the Order Pediculati; frogfishes, sargassum fishes; pectoral and ventral fins with fleshy bases; body scaleless and bearing numerous warts, fingerlike processes, or prickles; in or among floating seaweeds in tropics; able to distend stomach with air and float at surface; Histrio pictus (the mousefish) common.

antennary artery. One of two anterior arteries in the crayfish and certain other Malacostraca; supply blood to antennae, green glands, stomach, and anterior musculature.

Antennata. Mandibulata.

Antennularia. Genus of sessile, colonial, marine Hydrozoa.

antennules. Second, usually smaller, antennae in Crustacea.

antepenultimate. Third last, such as the third last segment of an antenna.

anterior. Of, pertaining to, or toward the front end; in human anatomy, the head end of the body is "superior," and the ventral surface "anterior."

anterior adductor. Large, short mussle in the anterior portion of many bivalve mollusks; runs transversely from one valve to the other.

anterior chamber. Anterior cavity of the eyeball, between the cornea and iris; filled with aqueous humor.

anterior choroid plexus. Thin, membranous, vascular area covering the roof of the diencephalon in certain vertebrates.

anterior lacerate foramen. Opening in front of the alisphenoid bone of mammals; carries the eye muscle nerves and part of the trigeminal nerve.

anterior pituitary. Anterior lobe of the pituitary gland; embryologically, it is derived from Rathke's pocket; see PITUITARY.

anterior root. In man, same as ventral root of a spinal nerve.

anterior vena cava. Large vein in tetrapods bringing blood to the right auricle (or sinus venosus) from the forelimbs, head, and anterior trunk region; in many mammals there is a single anterior vena cava, but in other tetrapods there are two.

anteroposterior. From front to back; from head to tail.

anthelminthic. Any drug used to stupefy or kill internal parasitic worms in a host.

Anthicidae. Cosmopolitan family of small beetles having the head greatly narrowed behind the eyes; base of prothorax also narrow.

Anthidium. Genus of bees in the Family Megachilidae; mason bees; construct nests of mud lined with plant material.

Anthobothrium. Genus of tetraphyllidean tapeworms found especially in the spiral valves of sharks.

Anthocoridae. Family of small hemipteran insects which includes the predaceous flower bugs.

anthogenesis. In certain plant lice, the production of both males and females by a parthenogenetic individual. Adj. anthogenetic.

Anthomastus. Genus of soft corals in the Order Alcyonacea.

Anthomedusae. Gymnoblastea; a suborder of the coelenterate Order Hydroida; hydrotheca absent, gonophores naked; medusae without statocysts; Hydra, Pelmatohydra, Chlorohydra, Corymorpha, Tubularia, Bougainvillia, and Eudendrium typical.

Anthomyiidae. Family of small to medium-sized flies; root maggot flies; larvae of many spp. are important pests on the roots of cultivated plants; Hylemyia an important genus; larvae of some spp. of Anthomyia sometimes occur in the human intestine.

Anthonomus grandis. Cotton boll weevil; one of the most important of all insect pests; larvae feed on cotton bolls.

anthophilous. Living upon or feeding upon flowers, e.g. bees.

Anthophora. Cosmopolitan genus of stout, solitary bees; usually nest in sand or clay banks.

Anthophoridae. Large family of Hymenoptera; includes the digger bees or hairy flower bees; stout, solitary (but sometimes colonial) bees with greatly elongated tongue; nest in sand or clay banks.

Anthophysis. Genus of minute biflagellate Protozoa; in clusters at the tips of branching stalks; stagnant fresh waters.

Anthopleura. Genus of very large Pacific Coast sea anemones, often colored green.

Anthoptilum. Genus of sea pens in the Order Pennatulacea; to 2 ft. tall; deep sea.

Anthostoma. Genus of brightly colored and stout but fragile Polychaeta.

Anthothela. Genus of anthozoans (Order Alcyonacea) found attached to stones, shells, etc.; polyps elongated.

Anthozoa. One of the three classes in the Phylum Coelenterata; corals, sea anemones, sea pens, etc.; attached cylindrical polyps without a medusa stage in the life history; with an oral disc bearing hollow tentacles; with a stomodaeum and usually one or two siphonoglyphs; gastrovascular cavity divided by partial and complete radially arranged septa; skeleton present or absent; marine, colonial or solitary; 6000 spp.

Anthracosauria. Leptospondyli.

anthrax. 1. Infectious and often fatal disease of cattle, sheep, and goats;

may be transmitted to man, especially when unsterilized hides are handled; produced by a bacterium (Bacillus anthracis) and characterized by subcutaneous swellings. 2. Carbuncle or other infection caused by the anthrax bacillus.

Anthrenus. Cosmopolitan genus of small beetles in the Family Dermestidae; larvae are serious museum and household pests.

anthropoid. Pertaining to the great apes in the primate Family Pongidae.

Anthropoidea. Suborder of mammals in the Order Primates; includes man as well as the various apes and monkeys; digits with flattened or curved nails; large, forward-looking eyes; arboreal or terrestrial; active during daytime.

anthropomorphism. Assignment of human characters and abilities to nonhuman organisms.

Anthus. See MOTACILLIDAE.

antiae. Feathers at base of bill in some birds.

Antiarchii. Order of extinct placoderm fishes; head and anterior trunk covwith boxlike bony armor; fresh water; Pterichthyodes typical.

Antias. Genus of marine Isopoda.

antibactericidin. Constituent of certain snake venoms which inhibits bactericidal action of phagocytes.

antibiotic. Substance produced by a living organism which diffuses into its surroundings and is sufficiently toxic to prevent the growth of certain other spp. in that area, such as the production of penicillin by a mold and the inhibition of bacterial growth by the penicillin; several antibiotics are produced commercially and are of great importance in treating many kinds of virus, bacterial, and amoebic infections in man; streptomycin, chloromycetin, and aureomycin are further examples.

antibody. Specific serum globulin produced in the body fluids in response to the introduction of a foreign protein or polysaccharide (antigen) below the epithelial surfaces; has a neutralizing or antagonistic action toward the antigen which is responsible for its formation; the antigen may be a parasite, any microorganism, or any foreign protein or polysaccharide; some antibodies remain effective for short periods, but others last a life time.

anticoagulant. 1. Any substance which prevents the coagulation of a colloid or other type of suspension or solution. 2. Substance which prevents clotting of blood, e.g. heparin.

antidiuretic hormone. ADH; hormone elaborated by the posterior pituitary; increases water resorption from the urinary tubules and thereby decreases amount of urine voided.

Antidorcas euchore. South African springbuck or springbok (Family Bovidae); a gazelle noted for its grace and jumping ability.

anti-enzyme. Substance inhibiting enzyme action.

antifertilizin. Glycoprotein presumed to occur in sperm cells; during fertilization it neutralizes the fertilizin surrounding an ovum.

antigen. Foreign substance (usually a protein or polysaccharide) capable of inducing the formation of antibodies when introduced beneath epithelial surfaces; may be a parasite, any microorganism, or any foreign protein or polysaccharide.

antigen-antibody reaction. Any reaction between an antigen and its corresponding antibody; may involve agglutination, precipitation, complement-fixation, etc.; the "sensitivity" of "allergic" reactions of many individuals toward certain kinds of foods, pollen, dust, feathers, hair, drugs, etc. are all antigen-antibody reactions.

antihistamine. Any drug which counteracts the action of histamine, e.g. benadryl and pyribenzamine.

antiketogenic. Pertaining to certain foods, e.g. glucose, which antagonize ketogenesis.

Antilocapra. See ANTILOCAPRIDAE.

Antilocapridae. Family of ruminants containing a single sp. (Antilocapra americana), the prongbuck, pronghorn, or American antelope; horns in both sexes grow over permanent bony cores and shed annually; each horn slightly curved, with one lateral prong; upper parts rich reddish brown or tan, with a large white rump patch; males to 120 lbs.; a graceful sp. of the western plains.

Antilope. Genus of ruminants (Family Bovidae) which includes the "true" antelope or black buck; a sp. of open, cultivated plains of India; males with long, spiral, annulated horns.

antimeres. Left and right halves of a bilaterally symmetrical animal; also, symmetrically corresponding parts of a radially symmetrical animal, e.g.

the arms of a sea star.

antimetabolite. Any inactive organic compound which is similar to a particular metabolic substance or enzyme and may replace or displace it.

antimorph. Mutant allele which inhibits the effects of the normal ancestral allele.

Antinoë. Genus of small to minute polychaetes; one unusual sp. is a parasite on other polychaetes.

Antipatharia. Order of tropical zoantharian Anthozoa; black or thorny corals; slender, branching, shrub-like colonies; skeleton composed of a horny material; polyps small; Antipathes is a common West Indian genus.

Antipathes. See ANTIPATHARIA.

antiperistalsis. Peristalsis occurring in a direction opposite to the usual direction.

antiprothrombin. Substance (heparin) combined with prothrombin in the circulating blood plasma; when clotting occurs, the antiprothrombin unites with thromboplastin, thus releasing prothrombin; see CLOTTING.

antirachitic. Pertaining to vitamin D, which prevents rickets.

antiscorbutic. Pertaining to vitamin C, which prevents scurvy.

antiseptic. 1. Any agent which inhibits growth of microorganisms without necessarily causing their destruction; e.g. alcohol, carbolic acid, iodine, sunlight, and hot water. 2. Pertaining to an antiseptic agent.

antiserum. Blood serum or body fluid containing antibodies specific for a particular antigen.

antitoxin. Type of antibody induced by the presence of a disease-causing organism or its toxin; diphtheria antitoxin is produced in the blood of horses which have been intentionally infected with the toxin of diphtheria bacteria; serum of such horse blood, when injected into human beings, confers an acquired immunity to diphtheria.

antivenin. Immunized blood serum of a horse which is injected into the victim of snakebite.

antivitamin. Any substance which inactivates a vitamin.

antivivisection. Opposition to the use of living animals for experimentation.

antixerophthalmic vitamin. Vitamin A.

ant king. Any of several South American birds (Grallaria) which feed on ants.

antler. The entire horn of one side in the deer family.

antlia. 1. Tubular proboscis of Lepidoptera. 2. Dilated post-pharynx of Diptera.

ant lion. Predaceous larva of certain insects in the Order Neuroptera; the doodle bug; some spp. hide in sand and debris and seize passing insects and other small arthropods; other spp. lie concealed at the bottom of a sandy funnel-shaped pit into which their prey falls and is seized with the large sickle-shaped jaws.

Antocha. Genus of crane flies (Tipulidae).

ant pipit. See CONOPOPHAGIDAE.

antrorse. Turned forward.

Antrostomus carolinensis. Chuck-will's-widow, a bird of the southern states resembling the whip-poor-will; a type of goatsucker.

Antrostomus vociferus. Common whip-por-will of N.A. wooded areas east of the Rockies; a subspecies also occurs in the southwest and Mexico; general grayish coloration; feet and legs small and weak; wide mouth margined with long bristles to aid in catching insects on the wing; eggs laid on ground; no nest; a type of goatsucker.

Antrozous. Genus which includes the pallid or pale bats; to 5 in. long, pale coloration; western and southwestern states.

antrum. 1. Cavity or sinus, especially in vertebrates. 2. Cavity of a Graafian follicle. Pl. antra.

antrum of Highmore. Maxillary sinus.

Anura. See SALIENTIA.

Anuraea. Keratella.

Anuraeopsis fissa. Uncommon fresh-water plankton rotifer; lorica of two convex plates; foot absent.

Anuraphis maidi-radicis. Small aphid which feeds on roots of certain cultivated plants, especially corn; corn root aphid.

anuria. Failure of renal function.

Anurida maritima. Common collembolan of the intertidal zone.

anus. Posterior opening of the digestive tract; in vertebrates a true anus is used only for the elimination of feces. Cf. CLOACAL APERTURE and VENT.

anvil. Incus.

Anystidae. Family of predatory mites.

Aonyx. Genus of clawless otters found along African rivers south of the Sahara; feed on crabs.

aorta. 1. In mammals, the thick artery which leaves the left ventricle and supplies blood to essentially all of the body except the lungs, via a series of smaller branched arteries; among vertebrates in general this is commonly called the aortic trunk. 2. A large vessel taking blood from the heart in certain invertebrates.

aortic arches. Six paired arteries in vertebrate embryos which connect the ventral aorta with the dorsal aorta; located in the visceral arches between the embryonic gill slits; the last four or five are present in most adult fishes; in most adult tetrapods the first, second, and fifth arches disappear; the third arches become the internal carotid arteries, the fourth become(s) the systemic arch(es), and the sixth forms a part of the pulmonary arteries; in addition, the fifth arch persists in most adult urodele amphibians.

aortic trunk. 1. Ventral aorta, a large vertebrate artery, especially close to the heart just before it branches to the gills or aortic arches. 2. In mammals the same as aorta.

Aotes. Genus of nocturnal, fruit-eating, South American monkeys; body about a foot long, fur woolly, eyes enormous.

aoudad. Large North African wild sheep; Barbary sheep; genus Ammotragus.

Apanteles congregatus. Braconid wasp; important parasite of pest caterpillars.

Apantesis. Large genus of tiger moths; forewings usually dark with branching light stripes; hind wings red, pink, or yellow, with black spots; larvae sometimes pests on truck crops.

apar. South American three-banded armadillo, Tolypeutes tricinctus.

apatetic. Pertaining to misleading and protective coloration pattern of an animal being preyed upon.

Apathy, Istvan. Hungarian zoologist (1863-1922); best known for work on nerve physiology and Die Mikrotechnik der Thierischen Morphologie.

Apatosaurus. Genus of giant American dinosaurs; herbivorous, to 65 ft. long.

ape. 1. In the broad sense, any monkey. 2. In the narrow sense, any large Old World tail-less monkey in the primate Family Pongidae, such as the gorilla, chimpanzee, orang utang, and gibbon.

Apeltes quadracus. Four-spined stickleback (Gasterosteidae); salt water but sometimes entering streams; Labrador to Md.

aperture. Opening into the first whorl in the shell of a gastropod; variously shaped but usually more or less round or ovoid.

Aphanapteryx. Genus of flightless, long-billed rails of Mauritius; recently exterminated.

Aphaniptera. Siphonaptera.

Aphanolaimus. Common genus of microscopic, free-living nematodes.

Aphanostoma. Genus of small, free-living flatworms in the Order Acoela; usually found crawling about on algae.

aphasia. Mental condition brought about by injury to association areas in the brain; marked by loss of powers of expression and comprehension.

Aphasmidea. One of the two classes of the Phylum Nematoda; phasmids absent; amphids various, not porelike; excretory system a single ventral gland; caudal alae absent.

Aphelenchoides. Genus of microscopic, free-living and parasitic nematodes.

Aphelenchus. Genus of herbivorous and and plant-parasitic nematodes.

Aphelinidae. Sun flies, scale parasites; a family of Hymenoptera in which the immature forms are minute parasites within aphids and coccids; adults feed on honeydew and tissue exudates of aphids and coccids; Aphelinus mali is an important parasite of the woolly apple aphid.

Aphelinus. See APHELINIDAE.

Aphelocoma. Small genus of North and Central American jays; includes the scrub jay and ultramarine jay of southeastern U.S.

Aphetohyoidea. Placodermi.

aphid. See APHIDIDAE.

Aphididae. Family of minute to small aphids or plant lice in the insect Order Homoptera; usually wingless; parthenogenesis common; life history often complex; feed exclusively on the juices of living plants; many excrete honeydew via the anus; abdominal wax glands often present; many spp. are serious economic pests.

Aphis. Large genus in the homopteran insect Family Aphididae; important pests on many cultivated plants; A. gossypii is the cotton or melon aphid, also found on certain vegetables.

aphis lion. Predaceous larva of neuropteran insects in the Family Chrysopidae; feeds on aphids and other small insects.

aphis wolf. Larva of neuropterans in the Family Hemerobiidae; predaceous on aphids, scale insects, and other small spp.; Hemerobius common.

Aphodius. Large genus of small beetles in the Family Scarabaeidae; larvae develop in feces of farm animals.

aphodus. Narrow canal between a flagellated chamber and excurrent canal in leucon sponges. Pl. aphodi.

Aphonopelma. Eurypelma.

Aphoruridae. Family of Collembola without a furcula.

aphotic zone. That portion of a body of water which receives no light of biological significance from the surface; usually below a depth of 800 m.

Aphredoderidae. Family of fishes in the Order Salmopercae; includes Aphredoderus sayanus, the pirateperch, a small sp. found from Minn. to Ontario, south to the Gulf, and in N.Y. to Fla.

Aphredoderus. See APHREDODERIDAE

Aphriza virgata. Surf-bird; a stocky, sandpiper-like marine shorebird of the Pacific Coast of the Western Hemisphere; Family Aphrizidae.

Aphrizidae. Small N.A. family of birds in the Order Charadriiformes; surfbird and turnstones; inhabitants of rocky and wave-swept marine shores; robust, short-legged spp.; Aphriza and Arenaria typical.

aphrodisiac. Any drug which arouses the sexual impulses, e.g. cantharidin.

Aphrodita. Genus of marine polychaete annelids; sea mice; body ovate or oblong, convex dorsally; dorsal surface covered with a thick mat of hairlike setae.

Aphrophora. See CERCOPIDAE.

Aphrosylus. Genus of small flies in the Family Dolichopodidae.

aphytal zone. That portion of a lake or ocean which is too deep to support photosynthesis.

apiary. Group of beehives or a place where bees are kept for their honey.

apical. Pertaining to the tip, apex, or top.

apiculture. Commercial bee-keeping.

Apidae. Family in the insect Order Hymenoptera; includes genus Apis, the honeybees; build extensive combs of wax cells; adults in a colony (hive) of three castes; a queen (functional female), drones (functional males), and workers (sterile females); colonies of 10,000 to 40,000 individuals; valuable as pollinators as well as honey producers.

Apis mellifera. Common honeybee.

APL, Anterior pituitary-like hormone.

Aplacophora. One of the two orders in the mollusk Class Amphineura; wormlike, shell absent, but calcareous spicules present in mantle; Neomenia and Chaetoderma are North Atlantic genera.

aplanatic. Pertaining to a microscope ocular which has been specifically corcorected for spherical aberration.

Aplexa hypnorum. Common fresh-water snail of northern states and Canada; swales, intermittent streams, and stagnant pools.

Aplodinotus grunniens. Fresh-water drum, in the Family Sciaenidae; Canada, Mississippi River system, and south to Guatemala; a large food fish, to 60 lbs.

Aplodontia. See APLODONTIDAE.

Aplodontidae. Small family of rodents which includes only one living sp., the mountain beaver or sewellel (Aplodontia rufa) of the Pacific Coast states; tail greatly reduced, ears small; body to 14 in. long; fossorial and nocturnal, usually near streams.

Aplysia. Genus of Gastropoda which includes the sea hares; shell rudimentary and internal; anterior angles of head extended into tentacle-like folds; large, massive, and sluglike; to 18 in. long.

Aplysilla. Genus of marine sponges in the Subclass Keratosa; encrusting, often yellowish.

apnea. Temporary cessation of respiratory movements following forced breathing.

Apneumona. Group of sea cucumbers lacking a respiratory tree.

apneustic. Pertaining to certain aquatic Diptera larvae which have no functional spiracles.

Apocheiridium. Cosmopolitan genus of pseudoscorpions.

apochromatic. Pertaining to an optical system (especially in a microscope) exceptionally free of both spherical and chromatic aberration; such objectives are spherically corrected for two colors and chromatically for three colors; intended for use only with compensating oculars.

Apocope. Genus of minnows commonly included in Rhinichthys.

Apocrangonyx. Genus of eyeless Amphipoda; wells, seeps, and caves in Ill.

apocrine gland. Type of gland in which the superficial part of the cell disintegrates as the secretion is released; such cells are quickly reconstituted.

Apocrita. Clistrogastra.

Apoda. 1. Echinoderm order in the Class Holothuroidea; body greatly elongated, often wormlike; 10 to 25 branched tentacles; tube feet and respiratory tree absent; Synapta and Chiridota are familiar genera. 2. See GYMNOPHIONA.

apodal. Without feet.

apodema. Apodeme.

apodeme. Internal ridge, process, or other type of inward projection in an arthropod skeleton.

Apodemus. Genus of Old World field mice.

Apodes. Order of bony fishes which includes various families of eels; body elongated, slender, and snakelike; dorsal, anal, and tail fins continuous; scales minute or lacking; chiefly marine.

Apodida. Apodes.

Apodidae. Micropodidae.

Apodiformes. Micropodiformes.

apodous. Without legs or feet.

apoenzyme. Protein portion of a holoenzyme; closely associated with a coenzyme.

Apogon. Cardinal fishes; a genus of tropical and subtropical marine spp., some of which incubate the eggs in the mouth cavity.

Apoidea. Superfamily of Hymenoptera which includes the social and solitary bees.

Apolymetis. Genus of tropical and subtropical marine bivalves.

aponeurosis. Fascia; broad fibrous sheet of tissue investing muscle or connecting muscle to a movable part.

apophysis. Any bony process, outgrowth, or swelling.

apoplexy. Condition produced by vascular lesions in the brain, especially a hemorrhage, thrombosis, or embolism; results in coma, paralysis, and often death.

apopyle. One of many small internal openings in sponges; located between radial canals or flagellated chambers and the main spongocoel cavity.

Aporidea. Small order of tapeworms in the Subclass Cestoda; proglottids absent; scolex with four acetabula and an armed rostellum; a few spp. in the intestine of swans; Gastrotaenia typical.

Aporrhais occidentalis. Duck's foot shell; common Atlantic Coast snail with a tapered yellowish spire; outer lip greatly expanded, with a blunt process at the upper angle; to 3 in. long.

aposematic coloration. Warning coloration.

Apostomea. Suborder of Ciliata in the Order Holotricha; cytostome obscure or a small rosette-like aperture; endoparasites of marine invertebrates.

apotype. Any specimen used as the basis for supplementary published taxonomic information.

Appalachian Revolution. Extensive crustal upheavals which resulted in the formation of the Appalachian Mountains during the close of Permian times.

appendage. Protruding structural part used for locomotion, sensory reception, feeding, or other purposes; e.g. legs, mouth parts or arthropods, tentacles, etc.

appendices epiploicae. Small tabs of fat covered with mesothelium along the inner surface of the colon.

Appendicularia. Common genus of pelagic tunicates in the Class Larvacea; same as LARVACEA.

appendicular skeleton. Collectively, the limbs, pectoral girdle, and pelvic girdle of vertebrates.

appendix. 1. Small, finger-like diverticulum of the caecum in certain mammals; contains much lymphoidal tissue. 2. In the broad sense, any appendage or supplementary or dependent part attached to a main part or organ. Pl. appendices.

apple shell. Fresh-water snail in the genus Ampullaria.

apple snail. See POMACEA.

apposable. Ability to touch the tip of the thumb to the tip of any other finger on the same hand.

apposition image. See COMPOUND EYE.

apricot shell. See LORIPINUS.

Apseudes. Genus in the malacostracan Order Tanaidacea.

Apsidospondyli. Amphibian taxon (subclass) sometimes used to include the Labyrinthodonti and Salientia; these large groups both have centra primitively formed by pleurocentra and intercentra.

Aptenodytes. See SPHENISCIFORMES.

Aptera. Small order of primitive insects; small, pale, blind, slender, and apterous; primitive metamorphosis; biting mouth parts; in damp places in vegetable debris; Campodea and Japyx typical.

apterium. Bare space between feathered areas of a bird skin. Pl. apteria.

apterous. Without wings.

Apterygiformes. Order of flightless terrestrial birds which includes the three spp. of kiwi (Apteryx) of New Zealand; bill long, slender; wings degenerate; feathers hairlike and fluffy; to 30 in. long; feed mostly on insects and worms; nocturnal, live in burrows.

Apterygota. Subclass of the Class Insecta which includes those orders of insects which are primitively wingless and have little or no metamorphosis.

Apteryx. See APTERYGIFORMES.

Apus. 1. Common genus of tadpole shrimps. 2. Genus of Old World swifts (birds).

aquatic caterpillars. Caterpillars of a few genera in the lepidopteran Family Pyralididae; modified for an aquatic habitat; Nymphula and Elophila in U.S. streams, ponds, and lakes.

aqueduct. Cerebral aqueduct, aqueduct of Sylvius; narrow tube connecting the third and fourth ventricles in the vertebrate brain.

aqueous humor. Watery fluid filling the space between the lens and cornea in the vertebrate eye.

Aquila chrysaëtos. Golden eagle; found in the northern part of the Northern Hemisphere; mountainous regions of western U.S.; rare east of the Mississippi; wingspread to 7.5 ft.; nest usually on inaccessible cliffs, sometimes large trees; feeds mostly on small mammals.

ara. Large blue and yellow Brazilian macaw.

Arab. Semite; race of caucasoids living in the Arabian area; skin tawny, hair black.

Arabella. Genus of Polychaeta; to 18 in. long; A. opalina is the common opal worm of the Atlantic Coast.

arachidonic acid. Essential unsaturated fatty acid found in lecithin; $C_{20}H_{32}O_2$.

Arachnida. One of the three subclasses of the arthropod Class Arachnoidea; spiders, scorpions, ticks, mites, etc.; abdomen without appendages, eyes simple, pedipalps usually sensory; eat fluids only; without gills; metamorphosis only in the Acarina; mostly terrestrial.

arachnid theory. Phylogenetic theory of Gaskell (1908) which sought to derive chordates from modified eurypterids; Patten (1912) proposed a further refinement of this theory.

Arachnoidea. One of the eight classes in the Phylum Arthropoda; spiders, scorpions, ticks, mites, etc.; except in the Acarina, body of two main divisions; cephalothorax and abdomen; cephalothorax bearing one pair chelicerae, one pair pedipalps, and four pairs of legs; respiration by gills, book lungs, tracheae, a combination of these, or respiratory structures absent; excretion by Malpighian tubules or coxal glands; mostly predaceous and terrestrial; more than 70,000 described spp.

arachnoid membrane. Delicate vascular tissue surrounding the vertebrate brain and spinal cord; covers pia mater, and is covered by dura mater in tetrapods.

Aramidae. Family of birds which includes rail-like birds with the habits of herons; one sp., the limpkin (Aramus vociferus), found along Fla. swamps and streams.

Aramus. See ARAMIDAE.

Araneae. Order of arachnids which includes the true spiders; cephalothorax and abdomen unsegmented and constricted at their union; chelicerae short, with a poison duct in terminal claw; pedipalps short but leglike; two or four book lungs; tracheae sometimes present; females of many spp. have ventral abdominal spinnerets for the extrusion of silk; chiefly terrestrial; many families, only a few of which are included in this volume; probably nearly 45,000 described spp.

Araneida. Araneae.

Araneidae. Family of orb-weaving spiders; web usually large and carefully constructed of radial and spiral threads; Araneus and Argiope common.

Araneomorphae. Labidognatha.

Araneus marmoreus. Marbled spider or false shamrock; common throughout N.A. and Europe; webs in trees and shrubs.

Arapaima. Genus of northern South American large-scaled river fishes; to 15 ft. long and 500 lbs.

Arbacia. Genus to which the common Atlantic Coast purple sea urchin belongs; Order Camarodonta.

arboreal. 1. Living in trees, e.g. monkeys. 2. Treelike or pertaining to trees.

arborizations. Finely branched terminations of certain neurons.

arbor vitae. Treelike outline of white matter as seen in a median section of the cerebellum.

Arca. Large genus of bivalves in all seas; ark shells, in the Order Filibranchia; shell heavy, boxlike, and usually with prominent parallel ribs radiating out from the umbo.

Arcella. Genus of amoeboid fresh-water Protozoa having most of the cell covered with a secreted chitinoid test.

Archaeoceti. Extinct suborder of whale-like mammals in the Order Cetacea; zeuglodonts; Middle Eocene to Upper Oligocene.

Archaeomysis. Genus of small crustaceans in the Order Mysidacea; in sand of open beaches.

Archaeopterygiformes. Order of extinct birds in the Subclass Archaeornithes.

Archaeopteryx. See ARCHAEORNITHES.

Archaeornis. See ARCHAEORNITHES.

Archaeornithes. One of the two subclasses of birds; "lizard birds" of Upper Jurassic deposits in Germany; only two good known specimens (Archaeornis and Archaeopteryx) and a few miscellaneous fragments; about as large as a crow; both jaws with teeth in sockets; three separate fingers and metacarpals; tail with more than 13 vertebrae.

archaeostoma. Mouth which is a direct derivative of the blastopore.

Archelon. Genus of extinct giant Cretaceous marine turtles.

archenteric pouches. Two pouches projecting forward from the anterior end of the gut, as in an Amphioxus embryo; eventually become the preoral cavity and Hatschek's groove.

archenteron. Main cavity of an embryo in the gastrula stage of development; lined with endoderm, opens to the outside by means of a pore (blastopore), and eventually becomes the digestive tract.

Archeocidaris. Genus of extinct echinoids in the Order Cidaroida.

archeocytes. Unspecialized mesenchyme cells of sponges which digest and circulate food and probably form reproductive cells.

Archeozoic era. Earliest of the five major subdivisions of geological time; lasted from about 1,800 million to 1,000 million years ago; a time of igneous activity, sedimentation, and erosion; unicellular organisms were probably in existence, but no fossils are known.

archer fish. Toxotes jaculator, a small yellow and black fresh-water fish of Thailand and East Indian ponds and streams; shoots a stream of water from the mouth onto insects flying or perched just above the water, and in this way gets its food.

Archeria. Genus of primitive labyrinthodonts.

archetype. Prototype.

Archiacanthocephala. Class in the Phylum Acanthocephala; main lacunar channels median; proboscis spines concentric; protonephridia present; male with eight cement glands; in terrestrial hosts; Macracanthorhynchus common.

Archiannelida. One of the five classes in the Phylum Annelida; small, simple, perhaps primitive, marine worms; composed of only a few segments, or numerous poorly-defined segments; with or without external cilia; nervous system imbedded in hypodermis; setae present or absent; parapodia simple or absent; in marine plankton or on marine substrates (one fresh-water sp. and one rare sp. of damp forest soils); larva a trochophore; about 35 spp., including Dinophilus and Polygordius.

archibenthos. Community of organisms living on the bottom of aquatic environments between depths of 200 and 1000 m. Adj. archibenthic.

archicephalon. Embryonic preoral segmented region of arthropods.

archicerebrum. 1. The ganglionic mass of the annelid prostomium. 2. Median ganglionic mass anterior to the first true somite in arthropods. 3. Procerebrum; collectively, the ganglia of the first true somite plus the more anterior ganglia in Arthropoda.

Archidoris. Genus of Pacific nudibranchs.

Archigetes. Genus of small tapeworms parasitic in annelids.

Archilestes. Genus of damselflies; West Coast.

Archilocus colubris. Common ruby-throated hummingbird (Family Trochilidae); from Alberta to Tex. and eastward.

archipallium. Dorsal median mass of nerve tissue in a cerebral hemisphere; becomes the hippocampus in mammals.

archipterygium. Primitive fin having a long segmented central skeletal axis with jointed rays on each side; e.g. in Neoceratodus.

Archistia. Order of extinct bony fishes; head covered with bones, body usually covered with ganoid scales, tail heterocercal, teeth conical, no centra in vertebrae; Cheirolepis and Platysomus typical.

Architectonica. Genus of tropical marine snails with a greatly depressed toplike shell; sundial shells.

Architeuthis. Genus to which the giant squid belongs; largest known invertebrate; including the arms, a specimen attaining a length of 55 ft. has been recorded; offshore waters.

Archoplites interruptus. Sacramento perch, the only native western centrarchid fish; to 18 in. long; Sacramento and San Joaquin river basins.

Archosargus. See SPARIDAE.

Archosauria. One of the six subclasses of reptiles; includes the crocodiles and their relatives and a host of extinct forms such as dinosaurs, flying reptiles, etc.; the "Ruling Reptiles."

Arcidens confragosus. Heavy-shelled fresh-water bivalve mollusk (Unionidae); common in central U.S.

Arcopsis. Genus of very small ark shells.

Arctica. Genus of thick-shelled marine bivalves; valves equal, oval; foot thick; some spp. called quahogs.

Arctic-Alpine zone. Extreme northern latitudes and high mountain regions; vegetation is mostly lichens, herbs, and grasses; one of Merriam's life zones in N.A.

Arctic fox. See ALOPLEX.

Arctic grayling. See THYMALLIDAE.

Arctictis. Binturongs; genus of omnivorous nocturnal mammals found in forests of southeastern Asia; tail prehensile, hairy, and strongly tapered; hair shaggy and greenish; total length to 6 ft.

Arctiidae. Family of thick-bodied, hairy moths containing many pest spp.; tiger moths; caterpillars known as woolly bears because of their dense coat of hairs; Hyphantria and Isia common.

Arctocebus. Genus of small lemurs of West African forests; the angwantibos; face foxlike, ears large.

Arctocephalus. Genus of eared seals found in Antarctic regions and along the coasts of the Southern Hemisphere.

Arctocorisa. Genus of water boatmen (Family Corixidae); Mont., Wyo., and Colo.

Arctocorixa. Old genus name for certain water boatmen (Corixidae).

Arctogaea. One of three primary zoogeographic areas of the world; includes N.A., northern Mexico, Europe, Africa, Asia, and certain islands southeast of Asia. Cf. NEOGAEA and NOTOGAEA.

Arctogalidia. Genus of weasel-like mammals of southeastern Asia.

Arctonoë. Genus of Polychaeta commonly found as commensals in the ambulacral grooves of sea stars.

Arctonyx. Genus of sand-badgers; insectivorous nocturnal spp. of rocky and desert areas of the eastern Himalayas; legs and tail long, snout flat and naked.

Arctopsyche. Genus of Trichoptera.

arcualia. 1. Small rodlike structures above the notochord of the lamprey; similar to neural arches of higher vertebrates. 2. Arch components of all vertebrae. Sing. arcualium.

arcuate. Arched or curved.

arcules. Basal crossvein between the radius and cubitus of Odonata wings.

Arcynopteryx. Genus of Plecoptera; Rocky Mountain area.

Ardea herodias. Great blue heron of the Western Hemisphere (Family Ardeidae); largest American heron, to 50 in. long; head crested, with long plumes; feeds in shallows of fresh waters; nest a rickety collection of twigs, grass, and branches in tall trees.

Ardeidae. Widely distributed family of wading birds in the Order Ciconiiformes; herons, bitterns, and egrets; usually live along edges of fresh waters; neck folded when in flight; Ardea, Butorides, Nycticorax, Casmerodius, and Egretta typical.

area opaca. Area opaca vitellina; that outer opaque part of the blastoderm of an egg such as the chick's in which the peripheral cells are not completely delimited by membranes and are actually continuous with the yolk.

area pellucida. That area of the blastoderm of an egg such as the chick's in which the cells are complete, with inner, outer, and lateral membranes; surrounded by area opaca.

Arenaeus. Common genus of marine crabs (Brachyura).

Arenaria. Genus of stocky, marine shore birds in the Family Aphrizidae; turnstones; found along stony beaches of North and South America; noted for their habit of turning over stones in search of invertebrate food.

Arenicola. Genus of large, stout, polychaete annelids which burrow deeply into the sandy sea bottom; lugworms;

to 40 cm. long; with branched segmental gills and rudimentary parapodia.

arenicolous. Living in sand or burrowing in sand.

areola. 1. Portion of the iris which borders the pupil of the eye. 2. Small closed cell of certain Lepidoptera forewings. 3. Narrow, more or less hourglass-shaped median area of the carapace just posterior to the cervical groove in crayfishes. 4. Pigmented area surrounding a nipple.

areolar tissue. Spongy vertebrate connective tissue made up chiefly of interlacing collagen fibers.

areoles. Irregular, rounded, or polygonal thickened surface areas of the cuticle in most Nematomorpha.

Argasidae. Family of soft ticks; mouth parts ventral, body without dorsal shield; Argas and Ornithodorus common genera.

Argas persicus. Common fowl tick of poultry, pigeons, quail, etc.

Argentina. Genus of small, silvery, smeltlike, marine fishes in the Family Argentinidae; argentines; Atlantic and Pacific.

Argentine ant. See IRIDOMYRMEX.

Argentinidae. See ARGENTINA.

Argia. Common genus of damselflies (Order Odonata).

Argiallagma. Genus of damselflies (Order Odonata).

arginase. Enzyme which splits arginine into urea and ornithine in the liver.

arginine. Amino acid; $C_6H_{14}O_2N_4$.

Argiope. Genus of large garden spiders, usually prominently marked with glack and yellow; webs in gardens and tall grasses; A. aurantia (black and yellow garden spider) and A. trifasciata (banded garden spider) common throughout U.S.

Argiopidae. Formerly a family of spiders but now usually considered a subfamily (Argiopinae) in the Family Araneidae.

Argobuccinum. Genus of large intertidal snails; shell brown, "hairy," to 6 in. long; Pacific Coast.

Argonauta. Genus in the cephalopod Suborder Octopoda; paper nautilus; generally resembling the octopus, but the female (not the male) inhabits a thin, white, spiral shell without septa; she leaves and returns to the shell at will and uses it as an egg case; worldwide in warm waters; shell commonly to 20 cm. long; male small and without a shell.

Argulus. See BRANCHIURA.

Argyroneta. Genus of Eurasian spiders; water spiders; the female builds an underwater silken canopy in ponds which assumes a dome shape when she places air bubbles under it; her eggs are then laid under the canopy.

Argyropelecus sladeni. Silvery hatchetfish; a small, deepsea fish having a short compressed body, large head, large eyes, and almost vertical jaws; coloration silvery and blackish, with large photophores; to 60 cm. long; widely distributed.

Arhynchobdellida. One of the three orders of leeches; mouth medium to large, without a proboscis; jaws present or absent; usually five annuli per segment in middle of body; blood red; Hirudo and Haemopis common.

Aricia. Deep-water genus of Polychaeta.

Ariidae. Family of bony fishes which includes the sea catfishes; shallow coastal seas; Galeichthys and Arius typical, especially in warm waters.

Arilus cristatus. Wheel bug, a large brown hemipteran having a semicircular cogwheel-like crest on the pronotum.

Ariolimax. Genus of Pacific Coast terrestrial slugs; to 7 in. long.

Arion. Genus of Old World terrestrial slugs; several spp. introduced into U.S.

arista. Bristle-like process at or near the tip of the antenna of certain Diptera.

aristate. Having a slender or spinelike tip.

aristopedia. Developmental anomaly, especially in Drosophila; the arista is replaced by a more or less perfect leg.

Aristotle. Greek philosopher, first noted zoologist, and one of the greatest thinkers of all time (384-322 B.C.); wrote Historia Animalium which was equivalent to about 500 pages of modern printing; it included many miscellaneous observations on the structure, behavior, and habits of animals of the eastern Mediterranean region; many personal observations were amazingly accurate, but some items from other sources were erroneous and fantastic; portions of two other works, On the Reproduction of Animals and On the Parts of Animals, also survive.

Aristotle's lantern. Complicated pentagonal lantern-shaped apparatus surrounding the esophagus just inside the mouth of most echinoderms in the Class Echinoidea; consists of a symmetrical framework of ossicles and associated bands of muscle and connective tissue; five movable teeth project from the oral end; the whole lantern may be retracted or partially thrust downward through the peristomial area; used in feeding and in moving the animal along the substrate in a clumsy fashion; originally, the "lantern" described by Aristotle was the whole scraped sea urchin test with its five small projecting teeth.

Arius. See ARIIDAE.

Arizona. See GLOSSY SNAKE.

ark shell. Any of numerous widely distributed marine bivalves; hinge line long and toothed; surface of shell usually ribbed; Arca, Anaddra, and Barbatia typical.

Armadillidium vulgare. Common terrestrial pillbug; able to roll up into a ball; generally distributed in damp places.

armadillo. See DASYPODIDAE.

Armadilloniscus. Genus of marine Isopoda.

Armandia. Genus of marine beach Polychaeta.

Armenoid. Assyroid; race of caucasoids living northeast of the Mediterteranean; skin swarthy, hair brown and wavy, nose hooked.

Armiger. Genus of fresh-water snails; northern states and Canada.

armored catfish. Any of certain small South American catfishes having a covering of bony plates.

armored scales. See DIASPIDIDAE.

army ants. Driver ants; see ECITON.

army worm. See LEUCANIA.

Arneth's formula. Percentage of polymorphonuclear leucocytes having nuclei with 1, 2, 3, 4, or 5 lobes.

arolium. Terminal pad between the tarsal claws in certain insects and arachnids.

arrector pili. Small muscle capable of erecting a hair of the mammalian skin. Pl. arrectores pilorum.

Arrenurus. Large and common genus of fresh-water Hydracarina; usually in vegetation.

Arrhenius equation. Theoretical expression of the rate of a biological or physiological phenomenon with respect to temperature:

$$\frac{Y_2}{Y_1} = e^{.5\mu}\left(\frac{1}{T_1} - \frac{1}{T_2}\right),$$

where Y is the speed of a phenomenon, T is absolute temperature, μ is a coefficient, and e is natural logarithm base.

Arrhenius' law. Only solutions containing ionized substances of high osmotic pressure are electrically conductive.

arrhenotoky. Male production by a parthenogenetic female.

arrow worm. Member of the Phylum Chaetognatha.

arsphenamine. Salvarsan; an organic arsenic compound used in the treatment of syphilis, yaws, and other spirillum infections; cf. NEOARSPHENAMINE.

Artamidae. Small family of Australian and East Indian passerine birds; wood swallows; small, chunky spp. that catch insects on the wing.

Artedius. Northern Pacific genus of sculpins.

Artemesina. Genus of small marine sponges (Subclass Monaxonida); flat, cushion-shaped; cold seas.

Artemia salina. Common brine shrimp (Order Anostraca) found in Great Salt Lake and similar saline lakes and ponds scattered throughout the world; 4 to 10 mm. long.

artenkreis. Superspecies.

arteriole. Small artery; in man, less than 0.3 mm. in diameter.

arteriosclerosis. Loss of elasticity, hardening, and thickening of arteries, especially beyond middle age.

artery. Vessel carrying blood away from the heart to the tissues; usually relatively thick-walled, elastic, and muscular.

arthrobranch. Malacostracan gill which is attached to the joint area between the body and the first segment of a leg.

arthrodia. Joint that permits only a sliding or gliding movement.

Arthrodira. Order of extinct placoderm fishes; head and shoulder region armed with heavy bony plates; no spiracular openings; to 2 ft. long; Dinichthys and Coccosteus best known.

arthromere. See SEGMENT.

Arthroplea. Genus of Ephemeroptera; uncommon in northeastern U.S.

Arthropoda. Largest phylum; characterized by a segmented body, segmented appendages, chitinous exoskeleton, and and an extensive hemocoel; coelom reduced, circulatory system open, digestive tract complete, brain dorsal, nerve cord ventral, and typically having paired, more or less fused ganglia in each segment; includes crustaceans, insects, spiders and their relatives, centipedes, millipedes, etc.; in all types of habitats; about 1,150,000 spp.

Arthrostraca. Outmoded taxon designation which includes the Amphipoda, Isopoda, and usually the Tanaidacea.

Artibeus. Genus of American fruit bats.

article. Segment of any jointed structure, such as a segment of an insect leg.

articulamentum. Lower layer of the calcareous plates of certain Amphineura.

articular process. 1. Zygapophysis; a small vertebral projection with a smooth, flat, dorsal or ventral face; a typical vertebra has two anterior articular processes and two posterior articular processes used in articulating with the preceding and subsequent vertebrae. 2. In the broad sense, any process functioning in an articulation.

Articulata. 1. One of the two classes of Brachiopoda; valves unlike, anus absent; includes Terebratulina and Magellania. 2. Old taxon which included the Annelida and Arthropoda.

articulation. Joint (usually movable) between two bones, cartilages, or body or appendage segments.

artifact. Abnormal structure or appearance of a cell or tissue produced by treatment with some chemical or physical agent, e.g. the abnormal microscopic appearance of certain protoplasmic details after fixation.

artificial classification. Classification of plants or animals based on arbitrary characters of convenience without reference to phylogenetic relationships; classifying all animals according to whether they live on land, in fresh water, in salt water, or as parasites is a simple example.

artificial immunity. Type of immunity derived from the inoculation of a serum, vaccine, or antitoxin.

artificial insemination. Removal of semen from the male and its artificial introduction into the female in order to initiate pregnancy; a common practice in breeding desirable strains of livestock.

artificial parthenogenesis. Laboratory phenomenon involving the maturation of an ovum and its development into an individual without fertilization; the essential stimuli include shaking, pricking, heating, and treating with dilute organic acids; the phenomenon can be initiated in frogs and many marine invertebrates; in a very few instances animals produced by artificial parthenogenesis have been reared to adults.

artificial selection. Selective breeding and the perpetuation of desirable mutations; most modern domestic animals and cultivated plants are the result of long-continued artificial selection.

Artiodactyla. Large order of even-toed hoofed mammals which includes a wide variety of spp. such as the antelope, pig, hippopotamus, camel, deer, sheep, goat, giraffe, bison, and cattle; two toes (rarely four) on each foot; many spp. with antlers or horns; stomach usually of four compartments; many are cud-chewers; native to all continents except Australia; composed of three suborders: Suiformes, Tylopoda, and Ruminantia.

Aruga. Genus of marine Amphipoda.

Arvicola. Genus of Old World water rats.

arytenoid. Jug- or pitcher-shaped.

arytenoid cartilage. One of the two small cartilages at the back of the larynx.

Asbestopluma. Genus of long, slender, stalked sponges (Subclass Monaxonida); cold seas.

Ascalabota. Division of lizards which includes the geckos, iguanas, agamids, and chameleons.

Ascalaphidae. Owl flies; cosmopolitan family of Neuroptera having a superficial dragonfly-like appearance.

Ascaphidae. Liopelmidae; small family of primitive frogs (bell toads) found in New Zealand and northwestern U.S.; examples are Liopelma and Ascaphus.

Ascaphus truei. Primitive frog found in or near mountain streams in western Mont., Wash., Ore., and northern Calif.; male has a tail-like copulatory organ; tadpole with a ventral sucking disc used for adhering to rocky substrates in swift currents; head and body of adult to 2 in. long.

ascariasis. Infection with Ascaris.

Ascaridia lineata. Nematode parasite of the chicken small intestine.

Ascaridina. Suborder of rhabditid nematodes parasitic in insects, vertebrates, and land snails; mouth surrounded by three large lips, six small lips, or no lips; caudal alae with papillae; common genera are Ascaris, Enterobius, and Heterakis.

Ascaris. Genus of rhabditid nematodes parasitic in mammals; A. lumbricoides is a large and familiar parasite in the small intestine of man, pigs, and other domestic and wild mammals; man becomes infected by ingesting embryonated eggs in contaminated food and water.

Ascaris megalocephala. Maw worm; large roundworm parasite occurring in the horse intestine; same as Parascaris equorum.

Ascaroidea. Order of Nematoda equivalent to the Suborder Ascaridina.

ascending colon. First portion of the large intestine of man; runs anteriorly (superiorly) from the ileocaecal valve.

Ascheim-Zondek test. Human pregnancy test; subcutaneous injection of urine from a pregnant woman into an immature female mouse produces premature maturation of the follicles in the mouse ovaries.

Aschelminthes. Phylum category used by some zoologists; contains the following wormlike pseudocoelomate groups (classes), each of which is given separate phylum status by other zoologists: Rotatoria, Gastrotricha, Kinorhyncha, Priapulida, Nematoda, and Nematomorpha.

Aschiza. One of the three tribes in the dipteran Suborder Cyclorrhapha; with a suture around base of antenna.

Ascidia. Cosmopolitan genus of sessile tunicates in the Order Enterogona; test translucent.

Ascidiacea. Class of Tunicata; sac-shaped, mostly sessile, simple or colonial, and covered by a common tunic; common genera are Styela, Amaroucium, Botryllus, Ciona, and Molgula.

Ascidicola. Genus of marine copepods; male free-living; female wormlike and commensal in digestive tract of ascidians and echinoderms.

ascites. Abnormal accumulation of fluid in the body cavity.

Ascomorpha. Genus of sac-shaped plankton rotifers.

ascon. Type of canal system in simplest sponges; consists of openings leading from the ostia directly to the spongocoel, as in Leucosolenia.

Asconosa. Homocoela.

ascorbic acid. Vitamin C; essential to proper structure and function of capillary walls and the formation of intercellular cement; promotes oxidation of fatty acids; increases phagocytosis; deficiency produces scurvy (subcutaneous and mucous membrane bleeding); abundant in citrus fruits and tomatoes; $C_6H_8O_6$.

Ascothoracica. Order of barnacles; six pairs of thoracic appendages, but adults highly modified and parasitic; especially common on anthozoans; shell lacking.

Aselli, Gasparo. Italian anatomist (1581-1626); first to realize the importance of lymphatics in fat absorption and transport.

Asellota. Suborder of Isopoda; aquatic spp. in which the pleon consists of two segments and the uropods are terminal.

Asellus. Genus of fresh-water Isopoda; A. communis is common in streams and lakes in Europe; A. militaris is possibly the most common American sp.

asepsis. 1. Prevention of the access of microorganisms. 2. Freedom from infection.

asexual reproduction. Any type of reproduction not involving gamete(s), e.g. budding and proliferation.

Asiatic cholera. See CHOLERA.

Asilidae. Large family of small to large Diptera, to 2 in. long; robber flies, assassin flies; body hairy; predaceous adults capture smaller insects on the wing and suck out body fluids; Erax and Asilus common.

Asilus. See ASILIDAE.

Asio. Genus which includes the widely distributed long-eared owl and short-eared owl.

asity. See PHILEPITTIDAE.

asker. Old name for a salamander.

asp. See CERASTES.

asparagus beetle. One of several small beetles whose larvae feed on asparagus shoots; introduced into U.S. from Europe.

aspartic acid. Amino acid; $C_4H_7O_4N$.

aspect. Seasonal appearance of a community.

aspection. Seasonal succession of ecological phenomena in a habitat.

Aspelta. Genus of free-swimming rotifers usually found in bogs and acid waters.

asphyxia. 1. Suspended animation owing to suffocation or anoxia. 2. Suffocation.

asphyxy. 1. Asphyxia. 2. Inactive condition common in Tardigrada, usually when there is insufficient food or oxygen; body swollen, rigid, and turgid. 3. Coma resulting from oxygen deprivation.

Aspidiophorus. Uncommon genus of fresh-water Gastrotricha; body covered with complex stalked plates.

Aspidiotus. Quadraspidiotus.

Aspidisca. Genus of small marine and fresh-water hypotrich ciliate Protozoa; ventral surface flat; macronucleus horseshoe-shaped; cirri large.

Aspidobranchia. Suborder of Gastropoda in the Order Prosobranchiata; nervous system not especially concentrated; with one, two, or no plumelike ctenidia; two auricles, two nephridia; limpets, periwinkles, and conches.

Aspidochirota. Echinoderm order in the Class Holothuroidea; shield-shaped tentacles; respiratory tree present; body cylindrical or flattened; tube feet reduced; 15 to 30 tentacles; some are deepsea forms; Holothuria and Stichopus common.

Aspidocotylea. One of the three orders of the Class Trematoda; oral sucker absent; ventral surface provided with an enormous sucker which is subdivided into compartments by a series of ridges, or with a ventral row of suckers; without intermediate hosts; endoparasites in clams, snails, fishes, and turtles.

Aspidogaster. Genus of trematodes in the Order Aspidocotylea; parasites in the renal and pericardial cavities of fresh-water clams.

Aspidogastraea. Aspidocotylea.

Aspidosiphon. Widely distributed genus in the Phylum Sipunculoidea; proboscis covered with minute hooks.

Asplanchna. Genus of large, saclike, plankton rotifers; foot, intestine, and anus absent; very common.

Asplanchnopus. Genus of large, saclike, plankton rotifers; intestine and anus absent; foot small; uncommon.

ass. See EQUIDAE.

assassin bugs. Common name of a group of predaceous hemipterans in the Family Reduviidae.

assassin flies. See ASILIDAE.

assimilation. Absorption and building up of complex organic protoplasmic constituents from simpler food materials.

association. 1. One of the major climax community types forming a large, distinctive segment of a biome; thus, the oak-hickory forest and the beech-maple forest are two different associations of the deciduous forest biome; similarly, the wet prairie of the Midwest and the Argentine pampas are two different associations but are parts of the same world-wide grassland biome; some ecologists use the term to apply to very much smaller natural units of vegetation. 2. Collectively, all those animals living together in any given combination of environmental conditions.

association areas. Portions of the cerebral cortex functioning in memory, learning, reasoning, imagination, and personality.

association neurons. Neurons which extend from one level of the spinal cord or brain to another and serve to modify, inhibit, or coordinate composite reflex actions.

associative neuron. Adjustor neuron.

associes. Temporary, non-climax association.

assortive mating. Non-random sexual reproduction in a population; tendency of certain types of males to breed with certain types of females, resulting in parents that are mentally and physically more similar than would be expected by chance.

Assyroid. Armenoid.

astacene. Astacin; red pigment of certain crustaceans, echinoderms, and fishes; $C_{40}H_{48}O_4$.

Astacidae. Common family of crayfishes; includes all U.S. fresh-water crayfishes.

Astacura. Group of Reptantia which includes the lobsters and crayfishes, as distinct from the spiny lobsters (Palinura).

Astacus. Genus of fresh-water crayfishes of the Pacific northwestern states and Europe and Asia; in general morphology, they resemble the American lobster (Homarus) but body usually only 7 to 20 cm. long; rivers and lakes; A. nigrescens occurs in Wash., Ore., and Calif.; A. fluviatilis is a European sp.

Astarte. Genus of small, brown, bivalve mollusks having thick shells and conspicuous growth ridges; astarte shells, chestnut shells; roughly triangular in outline; usually in temperate to cold seas.

astasia. Unsteady walking or standing.

Astasia. Genus of colorless monoflagellate Protozoa; body elongated but plastic; stagnant fresh waters.

Astenophylax argus. Caddis fly of northeastern U.S.

aster. System of minute radiating fibrils appearing with or without a centriole during mitosis, meiosis, and fusion of gametic nuclei; located at the ends of the spindle when the spindle is differentiated; not present in higher plants.

Asterias. Most common genus of American sea stars; A. forbesi (common eastern sea star) from Maine to Gulf of Mexico; A. vulgaris (purple star) from Labrador to Cape Hatteras.

Asterina. Genus of sea stars found along Atlantic Coast; arms short.

Asterocheres. Genus of copepods parasitic on sea stars.

Asterochiton. See ALEYRODIDAE.

Asteroidea. Class of the Phylum Echinodermata which includes the true starfishes or sea stars; body pentagonal, with five to 50 arms which are not set off from disc; surface with short spines, dermal branchiae, and pedicellariae; madreporite aboral; oral surface directed downward; two to four rows of tube feet in the ambulacral grooves; larva a bipinnaria; 1100 spp.; Asterias the common Atlantic Coast genus.

Asteromeyenia. Uncommon genus of fresh-water sponges; central and southern states.

Asteronyx. Cosmopolitan genus of basket stars (Order Euryalae); five arms very long and unequal.

Asterope. Genus of marine Polychaeta.

Asterophora. Genus of gregarine Protozoans found in Neuroptera and Coleoptera guts.

Asterozoa. Echinoderm subphylum; includes Asteroidea and Ophiuroidea.

asthenia. Weakness of muscle contraction.

asthenobiosis. Inactive period immediately preceding pupation in certain insect larvae.

astigmatism. Defective curvature of a refractive surface of the eye; results in a diffuse focus of light rays on the retina.

Astomata. Suborder of holotrich ciliates; cytostome absent; parasitic in invertebrates.

Astraea. 1. Genus of corals; includes the star corals. 2. Genus of turban shells (marine Gastropoda); on all coasts of U.S. except north Atlantic.

astragalus. Talus; foot bone between the ends of the tibia and fibula and the calcaneum.

Astrangia. Genus of corals; A. danae occurs in temperate and north Atlantic, with colonies usually not more than 12 cm. in diameter; A. insignifica occurs along the southern Calif. coast.

Astrapotheria. Order of extinct South American mammals; large spp. with protruding lower jaw; Eocene to Miocene.

astrocytes. Star-shaped neuroglia cells of the brain.

Astrometis. Genus of active West Coast sea stars.

Astropecten. Common genus of sea stars.

Astrophyton. Gorgonocephalus.

Astur gentilis. Goshawk; a large, long-tailed, short-winged hawk, considerably larger than a crow; common over much of N.A. but nests mostly in Canada; the most destructive of hawks to game and song birds.

Astyanax fasciata. Banded tetra; a small fish in the Family Characinidae; introduced into southern U.S., especially Tex. and N.M. streams.

Astylinifer. Genus of marine sponges.

Asymmetron. Common genus of cephalochordates, especially along coast of Fla. and West Indies.

asymmetry. Lack of any symmetry; condition obtaining in animals in which no plane can be drawn through the body which will separate the animal into two similar halves, e.g. some Protozoa and sponges. Adj. asymmetrical.

Asyndesmus lewis. Lewis's woodpecker; glossy greenish black on back, belly rose red, neck with a wide gray collar; a large sp. of western pine forests.

atabrine. Proprietary synthetic acridine dye used extensively as an antimalarial; usually taken by mouth as hydrochloride tablets; acts on the asexual reproductive cycle of Plasmodium in the erythrocytes; also effective for certain tapeworm and roundworm infections.

Atalopedes campestris. Common field skipper butterfly; larvae feed on grasses.

Atamosia. Genus of robber flies.

atavism. Apparent inheritance of a feature from remote rather than immediate ancestors; result of random recombination of genes which produce the ancestral character.

Atax. Unionicola.

ataxia. Irregularity or failure of muscular coordination.

ateleiosis. Type of dwarfism featured

by retention of infantile characters; caused by underactivity of the anterior pituitary.

Ateleopida. Ateleopii.

Ateleopii. Order of deep-sea fishes.

Ateles. Common genus of American spider monkeys (Family Cebidae); limbs and prehensile tail very long; thumb absent or rudimentary.

Ateuchus. Scarabaeus.

Atheca. One of the two suborders of the reptile Order Chelonia; large tropical marine turtles; ribs and vertebrae separate from carapace; limbs paddle-like, forelimbs especially long; Dermochelys coriacea typical.

Athecata. Anthomedusae or Gymnoblastea.

athecate. In certain Hydrozoa, the absence of a chitinoid covering around the hydranths.

Atherinidae. Family of omnivorous percomorph fishes of shallows; silversides; small and smelt-like, with a lateral silvery stripe; mostly marine but the fresh-water brook silversides (Labidesthes sicculus) is common in streams of the eastern half of the U.S., and several other spp. of Menidia occur from Mass. south and west to Tex. and Mexico.

Atherix. See RHAGIONIDAE.

atherosclerosis. Form of arteriosclerosis involving the intima of the arteries and producing fatty masses in the arterial cavities.

Atherura. Genus of brush-tailed porcupines; west and central Africa and southeastern Asia.

athlete's foot. Widespread minor fungus infection of the skin, especially between the toes; usually transmitted in shower rooms and other damp places.

Athripsodes. Genus of Trichoptera.

Atilax. Genus of African marsh mongooses.

Atlanta. Genus of translucent, pelagic, free-swimming, marine gastropods; shell large but translucent and delicate; foot modified into a ventral fin and a posterior tail-like portion used in locomotion.

atlas. First vertebra of vertebrates, just behind the skull.

atlas beetle. Giant beetle (Chalcosoma atlas) of tropical Asia; to 4 in. long; two long curved pronotal horns.

Atlas moth. See ATTACUS.

atoke. Anterior part of a marine polychaete as distinct from the posterior part (epitoke) during the breeding season.

atokous. Without offspring.

atoll. Large, thick, coral mass encircling a lagoon in tropical oceans; the lagoon may be 0.5 to 50 mi. in diameter and of varying depth; channels through the reef connect the lagoon with the open sea; the upper surface of the reef is more or less exposed at low tide; sometimes portions of the reef become built up with sand, silt, soil, and vegetation to become a terrestrial environment. Cf. BARRIER REEF and FRINGING REEF.

Atolla. Common genus of disc-shaped jellyfishes in the Order Coronatae.

Atoperla ephyre. Stonefly with wide distribution east of Rockies.

ATP. Adenosine triphosphate.

Atractaspis. Genus of African burrowing poisonous snakes.

Atractides. Common genus of Hydracarina; usually in cold streams.

Atractonema. Genus of nematodes parasitic in cecidomyiid midges.

Atractosteus. Genus of bony fishes in the Order Ginglymodi; alligator gars; large specimens to 13 ft. long; voracious; similar to gar pikes but larger and with shorter and broader snout; several spp. in warm lakes and rivers from lower Mississippi Valley to Cuba and Central America.

Atremata. Order of brachiopods in the Class Inarticulata; Lingula and Glottidia common.

atresia. 1. Degeneration of primary follicles in the mammalian ovary; process involving the great majority of follicles, since (in man) only about 400 of the half million primary follicles ever reach maturity; degenerate (ing) follicles are known as corpora atritica. 2. Congenital absence of a normal body opening, or the pathological closure of such an opening.

Atretia. Genus of very small Brachiopoda.

atretic follicle. Mammalian ovarian follicle which develops partially, then stops and degenerates.

atrial cavity. Peribranchial cavity.

atrial chamber. 1. Peribranchial chamber. 2. Atrium.

Atrichopogon. Common genus of ceratopogonid Diptera.

Atrichornithidae. Family of one sp. of passerine bird; scrub bird; a poorly-known wrenlike sp. of eastern tropical Australian scrubland.

atrichous isorhiza. Isorhiza nematocyst

which is small and has a smooth tubule.

Atrina. Genus of large pen shells; see PINNA.

atriopore. 1. Small ventral pore about two-thirds of the way back in cephalochordates; the opening of the peribranchial chamber or atrium. 2. Tadpole spiracle.

atrioventricular bundle. Bundle of His; special strand of heart tissue located in the wall between the two ventricles in higher vertebrates; an extension of the pacemaker system.

atrioventricular node. Small concentration of special tissue in the heart of higher vertebrates; located between the auricles just above the ventricles; an extension of the pacemaker; gives rise to the atrioventricular bundle.

atrium. 1. One of the chambers of the heart; see AURICLE. 2. Small chamber or cavity (genital atrium) just below the body surface in many invertebrates and into which a genital pore opens, especially in the Platyhelminthes. 3. Large anterior cavity in which the pharynx hangs, as in *Amphioxus*; also, the large chamber containing the pharynx in tunicates; same as peribranchial chamber. Pl. atria.

atrocha. Type of polychaete larva in which most of the body is uniformly ciliated and the preoral ciliary band is absent.

Atrochus tentaculatus. Uncommon spindle-shaped rotifer; foot absent; corona without cilia but with many small peripheral tentacles.

atrophy. Decrease in size or wasting away of an organ or tissue; caused by disuse or arrested development.

Atropidae. Family of insects in the Order Corrodentia; small spp., wings absent or vestigial; includes most book lice.

atropine. Drug which stimulates the sympathetic nerve endings and depresses the cerebrospinal nerves; chief ingredient of belladonna.

Atropos pulsatorium. Minute, pale, wingless insect in the Order Corrodentia; book louse; 1.5 to 2.0 mm. long; common in homes, museums, and libraries, as well as outside.

Atta. Genus of fungus ants which feed on fungi cultured in underground galleries.

attachment constriction. Spindle attachment.

Attacus atlas. Giant tropical moth with wingspread attaining 10 in.

attenuate. Long, slender, drawn out.

Attheyella. Genus of marine, brackish, and fresh-water copepods.

Atthis. Hummingbird genus; one sp. in mountains of southwestern states to central Mexico.

Attidae. Salticidae.

Atylopsis. Genus of marine Amphipoda; usually associated with brown algae.

Atypidae. Family of tube-web spiders or purse-web spiders; build silk-lined burrows in the ground, with the silk extending above the ground as a silken tube; *Atypus* common.

Atypus. See ATYPIDAE.

Auchenorrhyncha. One of the two suborders in the insect Order Homoptera; beak arising plainly from base of head; antennae short; active, free-living spp.; cicadas, leafhoppers, etc.

audiogenic seizures. Epileptic seizures evoked in some strains of mice by such sounds as whistles and bells.

auditory capsule. That part of the vertebrate skull which encloses an auditory organ.

auditory nerves. Pair of cranial nerves of vertebrates; originate on the medulla oblongata and innervate the inner ears; sensory nerves.

auditory organ. Any sense organ for detecting vibrations of the surrounding medium; the vertebrate auditory organ also houses the receptors for equilibrium, gravity, and acceleration.

auditory ossicles. One or three small bones of the middle ear; transmit vibrations from the tympanic membrane to the endolymph of the inner ear in tetrapods; these bones are connected end to end in a lever system which decreases the amplitude of vibrations; see COLUMELLA, MALLEUS, INCUS, and STAPES.

auditory pit. One of two saucer-like depressions in the ectoderm of an early vertebrate embryo marking the location of the future auditory vesicles.

auditory placode. Thickened ectodermal plate in early vertebrate embryos; gives rise to auditory pit.

auditory vesicle. Hollow capsule which buds off internally from the ectoderm in the region of the hindbrain of embryonic higher vertebrates; develops into the inner ear.

Audouinia. Genus of Polychaeta.

Audubon, John James. American ornithologist and bird artist (1780?-

1851); famous for his paintings in The Birds of America which was printed in Britain in elephant folio size and accompanied by five volumes of his Ornithological Biography; also wrote most of The Viviparous Quadrupeds of North America.

Auerbach's plexus. Plexus of autonomic fibers located between the muscle layers of the intestine.

Aufwuchs. See PERIPHYTON.

auger shell. See TEREBRIDAE.

augmentor nerves. Accelerator nerves.

auk. See PINGUINIS and ALCA.

auklet. Any of several small auks of the North Pacific; see PTYCHORAMPHUS and CERORHINCA.

Aulacantha scolymantha. Common marine radiolarian.

Aulactinia. Genus of long, slender sea anemones with many distal tubercles; Atlantic Coast.

Aulodonta. Order of Echinoidea; test rigid, gills present, teeth grooved; Diadema typical.

Aulophorus. Genus of fresh-water Oligochaeta; in tubes; with posterior gills and finger-like terminal processes.

Aulostomidae. See AULOSTOMUS.

Aulostomus. Genus of long, slim, tropical, marine fishes; flute mouths, trumpet fishes; Family Aulostomidae; body slender and head enormously elongated.

Aurelia. Aurellia; a genus of scyphozoan jellyfish in the Order Discomedusae; the white sea jelly; to 10 in. in diameter; common North Atlantic.

aureomycin. Trade-mark for antibiotic derived from a soil bacillus, Streptomyces aureofaciens; effective against a wide variety of soil bacteria and certain human bacterial infections.

auricle. 1. One of the less muscular chambers of a heart; receives blood from blood vessels or a blood chamber and passes it to a ventricle; present in certain higher invertebrates as well as vertebrates; fishes have a single auricle but tetrapods all have two; ("atrium" and "auricle" are generally used interchangably, but the former is preferred). 2. The external ear; pinna. 3. Any earlike lobe or process.

auricularia. Free-swimming bilaterally symmetrical ciliated larva of the echinoderm Class Holothuroidea.

auriculoventricular valve. One of the valves which prevents backflow of blood when the ventricle contracts in the vertebrate heart; present also in the heart of certain invertebrates.

Auriparus. See VERDIN.

aurochs. 1. See URUS. 2. Name sometimes applied to the wisent or European bison (Bison bonasus), a sp. now nearly extinct. (The spelling aurochs is both sing. and pl.)

auscultation. Type of examination in which the examiner listens for sounds within the chest or elsewhere, usually with a stethoscope.

Austral. See UPPER AUSTRAL and LOWER AUSTRAL.

Australian. Race of Australian caucasoids having dark brown skin, long head, prominent jaw, and flat nose.

Australian realm. One of the six major zoogeographic realms of the world; consists of Australia, Tasmania, New Zealand, New Guinea, Celebes, and many small associated islands; characterized by monotremes, emu, birds of paradise, a lungfish, sphenodon, kiwi, etc.

Australian water rat. See HYDROMYS.

Australopithecus. Genus of South African fossil primates; some human and some apelike features.

Austroriparian zone. See LOWER AUSTRAL ZONE.

autacoid. Active principle secreted by an endocrine gland.

Autarchoglossa. One of the two divisions of the reptilian Suborder Sauria (lizards): with less than four transverse rows of ventral scales per trunk segment.

autecology. 1. Ecology of a single individual. 2. Ecology of all individuals in a single sp. Cf. SYNECOLOGY.

autocatalysis. Catalytic reaction which gradually accelerates, presumably because some of the products of the catalysis act as catalytic agents also.

autochthonous. Native; generated within a particular habitat, such as fish in a lake; indigenous. Cf ALLOCHTHONOUS.

autoclave. Chamber for sterilizing objects and biological materials, usually with steam under pressure.

autogamy. Paedogamy; nuclear reorganization resembling conjugation but occurring within a single ciliate protozoan.

autogenous. Pertaining to anything produced within the same organism.

autointoxication. 1. Discarded term used to denote the headache and other symptoms of constipation; actually produced by rectal reflexes and has nothing to do with absorption of "toxic" materials from the large intestine. 2. In the broad sense, resorption of toxic substances produced by the body.

autolysis. Self-dissolution or self-digestion of certain tissues after death, due largely to their own enzymes.

Autolytus. Genus of marine Polychaeta in which reproduction commonly occurs by posterior buds which become separated.

Automeris. Large pinkish-brown or yellowish moths; io moths; genus found chiefly east of the Rockies in hardwood forests.

autonomic nervous system. Special system of motor nerves and ganglia in vertebrates; not under voluntary control; innervates the smooth muscles of the digestive, respiratory, urogenital, and circulatory systems, as well as blood vessels, many glands, and the skin; most evident as a chain of sympathetic ganglia in the body cavity on either side of the spinal column in the thoracic and lumbar region; each of these ganglia has connections with its associated spinal nerve, and each ganglion has one or more autonomic nerves which innervate certain of the organs included in the categories listed above; that portion of the autonomic nervous system in the thoracic and lumbar area is the sympathetic nervous system, and stimuli reaching tissues from this system have a general excitatory effect (but slow digestion); the parasympathetic system, on the other hand, has inhibiting effects on tissues and organs (but stimulates peristalsis); it consists of certain fibers in some of the cranial nerves (especially the vagus) along with certain fibers from the ventral roots of some of the sacral spinal nerves; many internal organs have double autonomic innervations, that is, from both the parasympathetic and sympathetic systems, and the stimuli may therefore act in an antagonistic manner; preganglionic fibers leaving the central nervous system stop short of effectors and form synapses with postganglionic fibers either in the sympathetic ganglia or in the organ at the site of the effector.

autoplastic grafting. Transplantation of a graft from one place to another on the same individual.

autoploid. Autopolyploid.

autopolyploid. Autoploid; polyploid in which all the chromosomes of a hybrid have come from the same parent sp.

autopsy. Internal and external examination of a body after death, especially to determine the cause of death.

autoradiograph Radioautograph; image on a photographic emulsion produced by the variable distribution of radioactive substances in a section, organ, or whole animal.

autosomal linkage. Linkage of groups of genes on the same autosome.

autosome. One of the chromosomes other than the sex chromosomes; man has 22 pairs of autosomes and one pair of sex chromosomes.

autostylic jaw suspension. Articulation of primitive upper jaw of vertebrates directly to the neurocranium; in amphibians, reptiles, and some fishes.

autosynthesis. Ability of self-reproduction, as in protoplasm or a gene.

autotetraploidy. Tetraploid in which the four genoms have been derived by duplication within one sp. and are cytologically identical; usually originates by fusion of two diploid gametes.

autotomy. Self-amputation, especially with respect to the five pairs of walking legs of decapod crustaceans; if such an animal is roughly handled and seized by a leg, reflex action will break the leg off across a special breaking joint near its base; during subsequent molts a normal leg is regenerated; tails of salamanders and lizards may be similarly shed, and sea stars may sometimes break off an arm if roughly handled; temporarily winged ants and termites may break off their wings by autotomy.

autotriploid. Triploid in which the three genoms have been derived by triplication within one sp. and are cytologically identical; usually originates by fusion of a haploid gamete and a diploid gamete.

autotrophic nutrition. Type of nutrition occurring in green plants and in some bacteria and true fungi; characterized by the ability to utilize simple inorganic substances for the synthesis of more complex organic compounds.

autozooids. Fully-formed feeding polyps as found in hydrozoans in the Order Pennatulacea. Cf. SIPHONOZOOIDS.

avahi. See LICHANOTUS.

availability. Ease with which a sp. can secure a particular food item under a given set of ecological conditions.

Aves. Class of vertebrates which includes all birds; warm blooded; with feathers and wings; hind limbs variously adapted for perching, swimming, walking, scratching, etc.; shanks and toes usually without feathers, covered with cornified scales; beak or bill without teeth and covered with a horny sheath; neck flexible and often long; sternum large, usually with a prominent median keel; heart four-chambered, right aortic arch persistent; red blood corpuscles with nuclei; lungs with associated thin-walled air sacs ramifying between visceral organs; syrinx at base of trachea; bladder absent, excretions semisolid; female with only one ovary and oviduct; fertilization internal; eggs incubated externally, with large yolk and limy shell; embryonic membranes present; upon hatching, young variously developed and dependent upon parents; 8600 living spp.

Avicenna. Ibn ben Sienna, Arabian physician (980-1037); famous for five-volume Canon of Medicine.

Avicula. Genus of bivalves which includes the wing shells; hinge line greatly drawn out to form an anterior and a posterior ear or wing; warm seas.

Avicularia. Genus of bird spiders or tarantulas.

avicularium. Curiously modified zooid found scattered about on the surface of certain marine bryozoans; has the shape of a miniature bird's head, with two jaws that snap shut on foreign materials and animals settling on the colony. Pl. avicularia.

Aviculidae. Pteriidae.

avidin. Antivitamin found in raw egg white; inactivates biotin.

avifauna. Collectively, the birds characteristic of a particular region.

avitaminosis. Abnormal condition produced by the lack of sufficient quantities of vitamin(s).

avocet. See RECURVIROSTRA.

awning clam. See SOLEMYA.

axenic culture. Culture consisting of one or more individuals of a single sp. in a non-living culture medium.

axial filament. Fine central rod in long, permanent pseudopodia of certain amoeboid Protozoa.

axial gradient. Graded change in general metabolic activity or some unknown metabolic property along the three main body axes (anterior-posterior, dorsoventral, median-lateral), especially in certain invertebrates; presumed to have controlling influence over embryonic development and regeneration.

axial skeleton. Collectively, the skull, vertebral column, sternum, and ribs.

axilla. 1. Armpit. 2. Triangular sclerite on each side of the scutellum in some Hymenoptera. Pl. axillae.

axillar. One of the innermost feathers on the undersurface of a wing; grows from the armpit.

Axinella. Genus of marine sponges (Subclass Monaxonida); stalked and lobed; Atlantic Coast.

axioplasm. Cytoplasm of the axis cylinder of a neuron.

Axiothella. Genus of Arctic and Subarctic Polychaeta.

axis. Second vertebra of tetrapods; just behind the atlas.

Axis. Genus of small deer (Family Cervidae) of India and nearby parts of Asia; forest spp. with white spotted coat; the axis deer or chital.

axis cylinder. Axon.

axis deer. See AXIS.

axolotl. See AMBYSTOMA.

axon. Process of a neuron which transmits impulses away from the nerve cell body; unbranched and short to long.

axone. Axon.

axoneme. 1. Central thread of the stalk in certain stalked ciliate Protozoa. 2. Central bundle of 11 fibrils in a cilium or flagellum. 3. Central thread of a chromosome. 4. Fibril extending from the base of a flagellum to its blepharoplast.

Axonopsis complanata. Brightly colored water mite; in cold lakes.

axopodium. More or less permanent pseudopodium composed of an axial rod and a cytoplasmic envelope; present in most Heliozoa and Radiolaria.

axostyle. Stiffening longitudinal rod in certain complex flagellate Protozoa.

aye-aye. See DAUBENTONIOIDEA.

Aysheaia. Genus of Peripatus-like animals of Cambrian marine deposits.

Aythya. Same as Nyroca; genus of ducks (Family Anatidae); A. valisneria is the common N.A. canvasback, highly esteemed for its flavor; A. americana is the N.A. redhead; A. marila is the greater scaup, or broadbill, found over the northern part of the Northern Hemisphere; A. affinis is the N.A. lesser scaup duck; A. collaris is the ring-necked duck; all are excellent divers and occur on shallow marine waters as well as fresh.

azoic. Without life, especially with reference to earliest geological history.

Azotobacter. Genus of nitrogen-fixing bacteria; important in the nitrogen cycle.

Azygia. Genus of distome trematode parasites of the intestine of fishes.

azygos vein. Mammalian vein running along the right side of the vertebral column; enters the right precaval vein near the heart.

azygous. Unpaired.

B

B_1. Thiamine.
B_2. Riboflavin.
B_6. Pyridoxine.
B_8. Discarded vitamin designation for adenylic acid.
B_{10}. Discarded vitamin designation for folic acid.
B_{12}. Cobalamin.
babbler. See TIMALIIDAE.
Babesia. Genus of Sporozoa occurring in the erythrocytes of cattle and other mammals. B. bigemina produces Texas cattle fever, transmitted by the cattle tick and now largely exterminated.
Babirussa. Genus of wild hogs in the Family Suidae; the babirussa of Celebes; upper tusks grow through the upper jaw, curve backward, and often touch the forehead.
baboon. See CYNOCEPHALUS.
baby's ear. See SINUM.
bacciform. Berry-shaped.
baccivorous. Berry-eating.
bacillary. 1. Rod-shaped or composed of small rodlike structures. 2. Pertaining to bacilli.
bacillus. 1. Any rod-shaped bacterium. 2. General term for any schizomycete fungus. Pl. bacilli.
Bacillus. Large genus of rod-shaped bacteria; aerobic spore-formers; B. anthracis causes anthrax.
backbone. Vertebral column.
backcross. Genetic mating of an animal with a parent or parent stock.
backswimmer. Any member of the hemipteran insect Family Notonectidae; swim with the ventral surface upward.
bacteria. Large group of unicellular fungi found wherever conditions are not sterile; usually without an organized nucleus; rod-shaped (bacillus), spherical (coccus), or spiral (spirillum); largest dimension 0.1 to 100 microns; reproduction by fission; aerobic, anaerobic, or facultative anaerobic respiration; nutrition parasitic or saprophytic; some spp. disease-producing, some valuable in industry and agriculture, others of no special importance; many of fundamental importance in conversion of dead organic matter into simple inorganic compounds. Sing. bacterium.
bacteriochlorophyll. One of the green photosynthetic pigments; present in sulfur and non-sulfur purple and brown bacteria; associated with various proteins; $C_{55}H_{74}O_6N_4Mg$.
bacteriolysis. Type of lysis involving bacterial disintegration.
bacteriophage. Ultramicroscopic virus or enzyme-like substance which disintegrates bacteria; grows only in the presence of living bacteria.
bacteriostasis. Inhibition of bacterial growth without actual destruction of the bacteria.
bacterioviridin. One of the green photosynthetic pigments; present in green sulfur bacteria and closely related to chlorophyll A; $C_{55}H_{72}O_6N_4Mg$.
Bacterium. Large genus of rod-shaped bacteria; not spore-forming.
Bactrites. Genus of fossil ammonites.
Bactrurus. Genus of eyeless Amphipoda found in caves, seeps, and springs in central U.S.
Baculites. Genus of ammonites in which only the earliest chambers exhibit coiling.
baculum. Penis bone of certain mammals.
badger. Any of several kinds of carnivorous burrowing mammals, especially in the Family Mustelidae; body robust, legs short and thick, claws of forefeet long; N.A., Europe, Asia, and East Indies; common American sp. is Taxidea taxus; European sp. is Meles meles.
Baeolophus. Genus of passerine birds in the Family Paridae; titmice; B. bicolor is the common tufted titmouse of eastern U.S.; upper parts slate-gray, under parts white and reddish; crest prominent and pointed.

Baermann apparatus. Device for concontrating infective larval nematodes from soil; a small wire basket lined with cheesecloth is placed in a funnel supplied with a petcock; the basket is filled with soil and enough warm water is then placed in the funnel to come in contact with the lower part of the soil; thermopositive nematodes creep downward and fall into the funnel where they may be drawn off by means of the petcock.

Baetis. Common American genus of Ephemeroptera.

Baetisca. Genus of Ephemeroptera; eastern U. S.

Bagous. Large and common genus of aquatic weevils; feed on aquatic vegetation.

Bagre marina. Gafftopsail catfish; marine catfish of the southeastern coast of U. S.; long ribbon-like dorsal and pectoral fins; male incubates the eggs in the mouth.

bagworm. See PSYCHIDAE.

bailer. Scaphognathite.

Bainbridge reflex. Reflex which increases the rate of heart beat when the amount of blood entering the right auricle increases.

Baiomys taylori. Pygmy mouse; body 59 to 66 mm. long; gray coloration; prairies and brushlands of Tex. and eastern Ariz.

Baird, Spencer F. American zoologist (1823-1887); secretary of the Smithsonian Institution for many years, influential in founding the U. S. Bureau of Fisheries, and commissioner of Fish and Fisheries; organized Albatross expeditions; made many important contributions in ichthyology, herpetology, and mammalogy.

Baird beaked whale. See BERARDIUS.

Balaena. Genus of whalebone whales which includes the right whales or bowhead whales of circumpolar seas; to 65 ft. long, head very large.

Balaeniceps. See SHOEBILL.

Balaenicipitidae. See SHOEBILL.

Balaenidae. Family which includes the right whales, e. g. Balaena and Eubalaena; large spp. with exceptionally large head; throat without longitudinal grooves.

Balaenoptera. Genus of large and widely distributed whalebone whales; rorquals, finback whales, sulfur-bottom and blue whales; to 105 ft. long; largest animal of all time; longitudinal grooves on throat and breast.

Balaenopteridae. Family which includes a group of large whales with relatively small head; throat with numerous longitudinal grooves; in all oceans; rorquals, finback whales, humpback whales; Balaenoptera, Megaptera, and Sibbaldus typical.

balancers. 1. Halteres. 2 In certain immature salamanders, a pair of rodlike lateral head appendages which function as props.

balancing organ. Any special organ for detecting the nature of body movements and maintaining balance, e. g. statocysts of invertebrates.

Balaninus. Widely distributed genus of acorn weevils; long snouts and legs.

Balanoglossus. Common and cosmopolitan genus in the Phylum Enteropneusta.

Balanomorpha. Group of genera which includes the rock barnacles.

Balanophyllia. Genus of Pacific Coast solitary cup corals; usually orange-red.

Balanosphyra formicivora. California woodpecker or acorn woodpecker of the far western states; red-crowned, white-rumped, black-backed sp. of oaks and pines; with a red crown only; sometimes placed in the genus Melanerpes.

Balantidium. Genus of oval, ellipsoid, or subcylindrical ciliate Protozoa occurring as parasites in the digestive tract of vertebrates and invertebrates; B. coli causes ciliate dysentery in man.

Balanus. Genus of rock barnacles or acorn barnacles; body enclosed in a short cylindrical or conical shell made up of a number of plates; stalk absent; extremely abundant on rocks of the intertidal zone.

Balbiani, Édouard-Gérard. Italian histologist (1825-1889); first saw what were later identified as giant chromosomes in the salivary glands of certain flies.

Balcis. Genus of small, white, marine snails.

bald eagle. See HALIAEETUS.

baldhead. 1. Blue goose. 2. Type of domestic pigeon. 3. Type of fruit crow.

baldpate. See MARECA.

baleen. See MYSTICETI.

Balfour, Francis. Scottish embryologist (1851-1882); famous for his two-volume Treatise on Comparative Embryology and A Monograph on the Development of Elasmobranch Fishes.

Balistes. See BALISTIDAE.

Balistidae. Family of bony fishes; includes the triggerfishes; body deep and compressed; projecting incisor teeth; first dorsal fin spine hornlike; Balistes common.

Balladyna. Genus of fresh-water hypotrich ciliates.

ball-and-socket joint. Type of joint at the hip and shoulder of tetrapods permitting considerable movement and rotation of the leg and arm.

ballistocardiograph. Apparatus used for calculating cardiac output; heartbeat of the subject causes movements of a special supporting table, and these movements are recorded.

ballooning. Transport of small spiders through the air, sometimes for long distances, by means of wind action on a long secreted silken thread.

ball python. Small, burrowing, central African snake; twists into a spherical mass when alarmed.

Balticina. Genus of tall, wandlike sea pens (Order Pennatulacea); to 3 ft. long.

Baltimore. See MELITAEA.

baluchitherium. Primitive Miocene rhinoceros; probably the largest land mammal of all time, with a shoulder height of nearly 18 ft.

bamboo rat. See RHIZOMYIDAE.

bananaquit. Small nectar-feeding warbler of West Indies and South America.

banana spider. See HETEROPODA.

Bancroft's law. Organisms and communities tend to reach and maintain a state of equilibrium with their environment.

banded krait. Poisonous snake of southeastern Asia; bright bands of black and yellow or tan; Bungarus fasciatus.

banded mongoose. See MUNGOS.

banded purple. Butterfly (Basilarchia arthemis) of northeastern states; blue-black wings with a broad white band; white admiral.

banded tetra. See ASTYANAX.

bandicoot. One of several spp. of marsupials found in Australia, Tasmania, and New Guinea; see PERAMELIDAE.

bandicoot-rat. See BANDICOTA.

Bandicota. Genus of bandicoot-rats; southern India.

band shell. See FASCIOLARIA.

band-tailed pigeon. See COLUMBA.

Bankia. Genus of shipworms; long wormlike pelecypods similar to Teredo; one large sp. makes burrows up to a meter long and 12 mm. in diameter.

bank swallow. See HIRUNDINIDAE.

banteng. Large wild and domesticated ox (Bibos) of southeastern Asia and East Indies.

Banting, Sir Frederick Grant. Canadian physician (1891-1941); along with Macleod, first isolated insulin in 1922 for which he shared the 1923 Nobel prize in physiology and medicine.

Bantu. Race of negroids having yellow, chocolate, or black skins; central and southern Africa.

barasingha. Swamp deer of India and Assam (Cervus).

barb. One of the numerous parallel filamentous extensions protruding from either side of the central shaft in a typical feather or from the calamus in a down feather; collectively form the vane. See BARBULE.

Barbary ape. Tail-less ape of northern Africa and Gibraltar.

Barbary sheep. Aoudad.

Barbatia. See ARK SHELL.

barbel. 1. Slender, external, tactile, whisker-like process, fleshy protuberance, or filament on the chin or elsewhere near the mouth of certain fishes, as in the Family Ameiuridae; usually two or more present. 2. One of several European fresh-water fishes in the genus Barbus (Cyprinidae).

barbet. Any of numerous small gaudy tropical birds in the Family Capitonidae (Order Piciformes) having a stout bill beset with bristles at the base.

barbicel. See BARBULE.

Barbour, Thomas. American zoologist (1884-1946); director of the Harvard Museum of Comparative Zoology for many years and the author of numerous important papers in herpetology, especially of the Central American and West Indian area.

barbule. One of many parallel microscopic linear extensions protruding from both the proximal and distal sides of a barb of a typical feather; each barbule is supplied with numerous hooklets, or barbicels, which hold opposing rows of barbules together; not present in down feathers. See BARB.

Barcroft apparatus. System of flasks and calibrated tubes used in manometric determination of respiration.

Barentsia. See PEDICELLINIDA,

bar eye. Mutation from the wild type of Drosophila in which the number of eye facets is greatly reduced.

bark beetles. Those spp. in the beetle

Family Scolytidae which form burrows just beneath the bark of trees.

barking deer. See MUNTIACUS.

barking frog. See ROBBER FROG.

barking squirrel. Prairie dog (Cynomys).

bark lice. See CORRODENTIA.

Barleeia. Genus of minute marine snails.

barnacle. Any member of the crustacean Subclass Cirripedia.

barnacle goose. Northern European goose, occasionally found in N. A.; Branta leucopsis.

barndoor skate. North Atlantic skate, to 6 ft. long; feeds on bottom animals; Raja laevis.

Barnea truncata. Angel wing; common East Coast rock- and clay-boring marine clam; to 3 in. long.

barn owl. See TYTO.

barn swallow. See HIRUNDINIDAE.

barracuda. See SPHYRAENIDAE.

barramunda. Native name for Australian lungfish (Neoceratodus); see NEOCERATODONTIDAE.

barred owl. See STRIX.

barrel-shell snail. Actaeon punctocoelatus; a West Coast mud-flat sp.; shell spindle-shaped, black and white.

barren grounds. Open lands of Canada north of the tree line; featured by thin soil and much exposed rock, with many swamps, ponds, and lakes; a special kind of tundra; inhabited by caribou and musk ox.

barrier. Any factor which limits further geographic spread of a sp. or other taxon; such barriers may be physical, chemical, climatic, or biological; e.g. a river, desert, lack of proper food, salinity, predators, etc.

barrier reef. Large, thick, coral mass more or less surrounding an island or parallelling the mainland shore in tropical areas; separated from shore by a lagoon 30 to 350 ft. deep and .5 to 10 or more miles wide; the reef itself may be 15 to 400 ft. wide, and its upper surface may be more or less exposed at low tide; channels through the reef connect the lagoon with the open sea. Cf. FRINGING REEF and ATOLL.

Bartholin's glands. Vestibular glands.

Bartonella bacilliformis. Bacterium which produces Oroya fever in man on the Andean slopes of northern South America; transmitted by the bite of Phlebotomus, a sand fly.

Bartramia longicauda. Upland plover; a large streaked buff-brown shore bird found over much of North and South America; grassy plains, pastures.

basal disc. More or less flattened basal surface by which Anthozoa and other coelenterate polyps are attached to the substrate.

basal field. Anterior portion of a fish scale; usually covered by the scale just anterior to it.

basal granule. Kinetosome; small intracellular granule at the base of a flagellum or cilium in Protozoa; in flagellates it usually has a closely associated blepharoplast, with which it is sometimes fused.

basalia. Collectively, the two or three cartilaginous rods supporting the base of the pectoral and pelvic fins of elasmobranch fishes.

basal membrane. Basement membrane.

basal metabolic rate. Basal metabolism.

basal metabolism. Amount of heat given off by an animal at rest (but not asleep), measured at least 14 hours after eating; expressed as calories per square meter of body surface (or per unit weight) per hour; determined by calorimetry or by measurement of oxygen consumed or carbon dioxide given off; also called basal metabolic rate or BMR.

basal pigment cells. See OMMATIDIUM.

basement membrane. Thin, intercellular, fibrous connective tissue at the base of most epithelial tissues; often difficult to distinguish.

Baseodiscus. Genus of large marine nemertine worms.

Basiaeschna janata. Dragonfly of eastern and central U. S.

basibranchial. Median ventral cartilage or bony element in a branchial arch.

basic stain. Biological dye consisting of a basic organic portion combined with an inorganic acid; especially useful for nuclear staining.

basidopodite. Second segment of pereiopod in most malacostracan Crustacea.

basihyal. Median ventral cartilage or bony element of the hyoid arch.

Basilarchia. Common genus of butterflies in the Family Nymphalidae; color pattern variously purple, brown, and greenish blue; the purples; includes B. archippus, the common viceroy butterfly, whose coloration resembles that of the monarch which it is believed to mimic.

basilar membrane. Thin membrane in the cochlea lying just beneath the receptor hair cells.

Basilochiton. Genus of chitons.

Basiliscus. See BASILISK.

basilisk. 1. Any of several Tropical American lizards of the genus Basiliscus; related to the iguanas; with an air-filled membranous bag on the head and an erectile crest along the dorsal midline. 2. Fabulous winged African lizard or dragon whose very glance was said to be fatal.

Basilona imperialis. Large moth of eastern U.S.; imperial moth; female yellow, marked with pinkish purple; male darker; larva feeds on tree foliage.

basipod. Basipodite.

basipodite. 1. Basipod; distal of the two segments making up the protopodite of a typical biramous appendage, the basal segment being the coxopodite. 2. Occasionally, a synonym of protopodite.

basipterygium. Pubic bone or cartilage to which a pelvic fin is attached in a typical fish. Pl. basipterygia.

basis. Second segment of a typical walking leg in many malacostracan Crustacea.

basisphenoid. Bone in fetus of higher vertebrates; becomes posterior part of the sphenoid; a distinct bone in the skull of bony fishes.

basisternum. Principal sclerite of an insect sternum.

basket cockle. Cardium corbis, a common West Coast gastropod which burrows in sand flats.

basket star. See EURYALAE.

basking shark. See CETORHINUS.

Basommatophora. Suborder of pulmomate snails; one pair of tentacles; chiefly fresh-water spp., a few marine.

basonym. Original genus and sp. names of an animal.

basophil. Type of leucocyte in vertebrate blood; characterized by basophilic granules; only about 0.5 per cent of human leucocytes are basophils; function unknown.

basophilic. Having an affinity for basic dyes; especially characteristic of nucleoproteins.

bass. 1. Any of several edible spiny-finned American fresh-water fishes; see especially MICROPTERUS. 2. Any of a wide variety of marine spiny-finned fishes; many edible.

Bassaricyon. The cuataquils; genus of nocturnal and arboreal mammals of Central America; sluggish, fruit-eating, with a gingery, squirrel-shaped body and long, bushy tail.

Bassariscus astutus. Small nocturnal mammal in the carnivore Family Procyonidae; cacomistle, ring-tail cat, bassarisk; related to the raccoon but much smaller and more slender; southwestern U.S.

bassarisk. See BASSARISCUS.

bastard wing. Alula; tuft of about three feathers growing from the thumb in birds.

Bastiania. Genus of microscopic, free-living nematodes.

bat. See CHIROPTERA.

bat bug. See POLYCTENIDAE.

Batea. Genus of marine Amphipoda.

Bates, Henry Walter. English explorer and naturalist (1825-1892); noted for his biological explorations of the upper Amazon basin and an explanation of mimicry; wrote The Naturalist on the River Amazon.

Batesian mimicry. Form of animal mimicry; a relatively scarce, palatable sp. resembles an abundant, unpalatable sp. so that the former is protected by a "deception" mechanism; e.g. the great similarity in color pattern between the Viceroy butterfly (mimic, palatable to birds) and the monarch butterfly (model, unpalatable); other examples of model and mimic, respectively, are: bees and flies, wasps and crickets, tiger beetles and crickets; zoologists are not agreed on the reality and effectiveness of such supposed mimicry.

Bateson, William. English geneticist (1861-1926); especially well known for work on Mendelism, sex determination, and linkage.

batfish. Any of several misshapen marine fishes, especially the California batfish (Aetobatus), flying gurnard (Dactylopterus), and Ogcocephalus (Caribbean and East Coast).

Bathophilus. See MELANOSTOMIATIDAE.

bathyal. Pertaining to the deep sea, especially the bottom between depths of 200 and 4000 m.

bathybius. Gelatinous material in mud from the sea bottom; Huxley assumed it was protoplasm, but later it was shown to be inorganic.

Bathyclupei. Order of bony fishes which includes the deep-sea herrings of the tropics and subtropics.

Bathycrinus. Genus of deep-sea stalked crinoids.

Bathyergidae. Blesmols, a family of fossorial rodents; eastern Africa south of the Sahara; see BATHYERGUS and HETEROCEPHALUS.

Bathyergus. Strand-rats; a genus of burrowers in the Family Bathyergidae; coastal flats of southeastern Africa; dig with the front teeth and use the short limbs for scraping the dirt backward.

Bathynella. See SYNCARIDA.

Bathynellacea. See SYNCARIDA.

Bathynomus giganteus. Giant marine isopod; to 14 in. long.

bathypelagic. Pertaining to the community of organisms suspended in water at depths of 200 to 1000 m. (or 1000 to 4000 m., depending on the authority).

Bathyplotes. Genus of sea cucumbers in the Order Aspidochirota.

Bathystoma. See GRUNT.

bathythermograph. Instrument which makes a visible record of temperature as a function of depth, especially in the ocean; towed behind a boat or hauled up from appropriate depths.

Batoidea. Suborder of selachian fishes which includes the sawfishes, skates, and rays; body depressed; pectoral fins greatly enlarged and used as chief organs of locomotion; five pairs of ventral gill slits; upper eyelids not free; familiar genera are Pristis, Rhinobatos, Raja, Torpedo, and Manta.

Batrachia. Amphibia.

Batrachoidida. Haplodoci.

Batrachoseps. See WORM SALAMANDER.

batracin. See PHYLLOBATES.

battery cell. Epitheliomuscular cell of a coelenterate containing one or more nematocysts.

bat tick. See NYCTERIBIIDAE.

bay lynx. Common wildcat of eastern U.S.

BCG. Weakened strain of the tuberculosis bacillus used, especially in Europe, to produce active immunity.

Bdella. Snout mites in the Family Bdellidae; a large genus of small predaceous mites found in moss, rotten wood, and duff.

Bdellocephala. Genus of white planarians found in Europe and Asia.

Bdellodrilus. Genus of small, leechlike Oligochaeta commensal on the gills and other structures of crayfish.

Bdelloidea. Order of rotifers in the Class Digononta; males unknown; body highly contractile in a telescoping manner; free swimming or creeping in a leechlike fashion.

Bdellonemertea. Order in the Phylum Nemertea; proboscis unarmed; intestine sinuous, without diverticula; parasitic.

Bdellostoma. Genus of hagfishes; west coast of South America.

Bdelloura. Genus of large triclad Turbellaria; ectocommensal on Limulus.

Bdeogale. Dog-mongooses, a genus of mongooses found in forested parts of Africa; live in base of hollow tree and bark like dogs.

beach flea. Sand hopper.

beaded lizard. See HELODERMATIDAE.

Beadle, George Wells. American biologist (1903-); made important contributions to the genetics of enzyme formation and discovered (with Tatum) that the formation of an enzyme is often dependent upon the action of a single gene; shared a Nobel prize in 1958.

beagle. Small, houndlike, short-legged, smooth-haired dog with pendulous ears.

Beagle. Small British naval vessel in which Darwin cruised around the world (1831-1836); during the voyage he made many important observations essential to the formulation of his concept of organic evolution.

beaked whale. Any of a group of small toothed whales having the snout prolonged into a beak and the teeth reduced to one or two in the lower jaw; Family Ziphiidae.

bean clam. Donax gouldi, a small wedge-shaped clam of Pacific Coast beaches.

bean ladybird. See EPILACHNA.

bean lycaenid. See STYMON.

bean weevil. See LARIIDAE.

bear. See URSIDAE.

bear animalcule. See TARDIGRADA.

bearded seal. See ERIGNATHUS.

Beatragus. Genus of large African hartebeests or damalisks.

Beaumont, William. American physician and army surgeon (1785-1853); for ten years he studied the physiology of the stomach of Alexis St. Martin, who had been badly hurt by a gunshot and whose stomach was more or less exposed and perforated ventrally during that period.

beaver. See CASTORIDAE.

beaver rat. Aquatic web-footed rodent-like marsupial of Australia and Tasmania.

beche de mer. Trepang.

bedbug. See CIMICIDAE.

bee. Any one of many insects in the

Order Hymenoptera which have hairy bodies, gather pollen, and have biting as well as sucking mouth parts; most spp., such as the honeybee and bumblebee, are social insects and live in highly organized colonies, but others are solitary (carpenter bees); see APIDAE.

bee bread. Pollen gathered by bees and used as a protein source for the larvae; stored in the honeycomb in brown masses.

bee eater. Any member of the Family Meropidae; brightly-colored insectivorous birds; bill long and slender; Old World tropics; arboreal but nest in river bank burrows.

bee flies. See BOMBYLIIDAE.

beef tapeworm. See TAENIA SAGINATA.

bee jelly. Royal jelly.

bee killer. Any large robber fly (Asilidae).

bee moth. See GALLERIA.

beeswax. Wax secreted by glands on the ventral surface of the abdomen of worker bees; used in construction of the comb.

beetle. Any member of the insect Order Coleoptera (except for weevils).

beetle mite. See ORIBATIDAE.

Beggiatoa. Genus of bacteria found growing in stagnant water; contain sulfur granules.

behavioral isolation. Prevention of interbreeding between two or more populations because of differences in behavior or modes of life; the fish crow vs. the common crow, and the head louse vs. the body louse are examples.

behemoth. 1. Any large and powerful animal (colloquial). 2. As mentioned in the Bible, probably the hippopotamus.

beira. See DORCATRAGUS.

beisa. See ORYX.

belemnite. Conical internal calcareous skeleton of many extinct cephalopods; posterior end solid, anterior end hollow and often with transverse laminae.

Belemnites. Genus of fossil mollusks in the Suborder Decapoda; shell (belemnite) internal and chambered.

Bell, Charles. Scottish anatomist (1774-1842); made major contributions to anatomy of the human nervous system, including a differentiation between motor and sensory nerves.

belladonna. Extract of leaves of the nightshade plant; contains atropine and used chiefly as an antidote for spasma of coughing, asthma, and cramps, and to suppress glandular secretion.

bellbird. Any of several birds whose notes are similar to the sound of a bell.

bell magpie. See CRACTICIDAE.

bellows fish. Any of several fishes having a deep, compressed body and long, tubular snout.

Bell's sparrow. See AMPHISPIZA.

bell toad. See ASCAPHIDAE.

belly. 1. Abdominal area or cavity; ventral part of a vertebrate body. 2. Thickest fleshy portion of a typical skeletal muscle.

Belon, Pierre. French naturalist (1517-1564); wrote two volumes on the natural history of aquatic animals (especially vertebrates) and a third volume on the natural history of birds.

Belonidae. Family of marine fishes which includes the bellfish, salt-water gar, houndfish, and needlefish; predaceous, both jaws elongated to form a slim beak armed with many teeth; body very slender; Strongylura common.

Beloniformes. Order of fishes which includes the Belonidae and related families; equivalent to the Synentognathi.

Belonorhynchii. Order of extinct bony fishes; four series of bony scutes; jaws very long; Belonorhynchus typical.

Belonorhynchus. See BELONORHYNCHII.

Belostoma. Common genus of water bugs in the hemipteran Family Belostomatidae; B. grandis attains 4.5 in.

Belostomatidae. Family of insects in the Order Hemiptera; giant water bugs; American spp. to 4.5 in. long; in ponds and shallows of lakes; often fly to lights at night; vicious biters; common genera are Lethocerus, Abedus, and Belostoma.

Belostomidae. Belostomatidae.

beluga. See DELPHINAPTERUS and ACIPENSER.

Bembex. Bembix; see BEMBICIDAE.

Bembicidae. Cosmopolitan family of large wasps; sand wasps; adults build open burrows in sandy places; Bembix common.

Bembix. See BEMBICIDAE.

Benacus griseus. Large aquatic bug of the southern and southeastern states.

benadryl. Proprietary antihistamine useful in hay fever and urticaria.

bends. See CAISSON DISEASE.

benign neoplasm. Any neoplasm which remains localized and does not invade adjacent tissues.

benthic. Pertaining to the benthos, or to the bottom in a pelagic area.

Benthodytes. Deep-sea genus of sea cucumbers (Order Elasipoda).

benthon. Benthos.

Benthopecten. Widely distributed genus of sea stars (Order Phanerozonia).

benthos. Collectively, all those animals and plants living on the bottom of a lake or sea, from the water's edge to greatest depths; the term is often used with reference to animals only. Adj. benthic. Cf. ABYSSOBENTHOS, ARCHIBENTHOS, and NERITIC.

bent mussel. Hooked mussel, Mytilus recurvus; R.I. to Tex.

bent-nosed clam. Macoma nasuta, a very common sp. of West Coast mud flats; lies on its side in the mud at a depth of 6 to 8 in.; posterior edge of shell turned upward where the siphons extend up to the surface.

benzamine hydrochloride. Anesthetic for certain small aquatic Metazoa; 2 per cent solution.

Beraea. Genus of Trichoptera.

Berardius bairdi. Baird beaked whale; a North Pacific sp. to 40 ft. long; upper parts black, underparts pale.

Berber. Race of caucasoids living in northwestern Africa; skin whitish, head long, hair dark and curly.

Bergmann's principle. Within a single sp. or in closely related spp. of birds and mammals, individuals at high latitudes are larger than those at lower latitudes.

beri beri. See THIAMINE.

Berlese funnel. Any apparatus in which small animals, primarily arthropods, are driven down and out of soil by the application of heat, drying, and/or light from above; a sample is placed on a screen at the top of a steep-sided funnel and the animals fall through and down into jars which are usually filled with preservative.

Berlese's organ. Reproductive gland in the abdomen of certain female Hemiptera.

Berlese's theory. Relative form and complexity of an early insect larva is determined by the particular moment of eclosion.

Bernard, Claude. French physiologist (1813-1878); founder of experimental medicine; made major contributions to the physiology of the liver, pancreas, and vasomotor system; emphasized constancy of the internal environment.

Beroë. See NUDA.

Berosus. Common genus of hydrophilid beetles.

Bertiella. Genus of cyclophyllidean tapeworms found especially in apes, monkeys, and rodents.

Berycomorphi. Order of marine fishes; squirrel fishes; eye large, fins with some spiny rays, and general body surface with numerous short, sharp spines; principally in warm waters; Holocentrus common.

Berycomorphida. Berycomorphi.

bestiary. Popular type of medieval book describing animals and their "human traits"; a mixture of folklore, fable, imagination, and fanciful art.

Betaeus. Genus of littoral marine shrimps.

beta globulins. Group of plasma proteins associated with transportation of thrombin and prothrombin.

beta oxidation. Primary oxidation of a fatty acid chain at the beta carbon atom (the second carbon atom from the carboxyl group).

bettong. See BETTONGIA.

Bettongia. Genus of ratlike Australian kangaroos; boodies, squeakers, bettongs; dusty gray, hopping animals commonly found as scavengers around human habitations.

betweenbrain. Diencephalon.

Bewick's wren. See THRYOMANES.

Bewuchs. Complex assemblage of microorganisms living on the surface of dead and inanimate objects in aquatic habitats. Cf. PERIPHYTON.

bezoar. 1. Concretion in the alimentary tract of certain ruminants; usually composed of calcium phosphate or fibers and hairs in a resinous matrix. 2. Similar concretion in the human digestive tract. 3. Coprolite.

Bezzia. Genus of ceratopogonid Diptera.

bharal. See PSEUDOIS.

B horizon. Soil stratum where soil materials are deposited as the result of movements of the ground water; underlies the A horizon; variously subdivided. Cf. A HORIZON and B HORIZON.

Bianium. Genus of distome trematode parasites of marine fishes.

Bibio. See BIBIONIDAE.

Bibiocephala. Common genus of net-winged midges (Blepharoceridae).

Bibionidae. Family of small to medium-sized hairy flies; March flies; larvae primitive and apodous; some larvae injure economic plants; Bibio typical.

Bibos. Genus of large ruminants in the

Family Bovidae; includes the gaur, gayal, and banteng.

Bicellariella. Genus of marine Bryozoa.

biceps. Large muscle of the upper arm in most tetrapods; flexes the arm at the elbow.

biceps brachii. Biceps.

biceps femoris. Large thigh muscle in certain vertebrates.

Bichat, Marie Francois. French anatomist and physiologist (1771-1802); founder of modern animal histology.

bichir. See POLYPTERUS and CLADISTIA.

Bicidiopsis. Genus of elongated burrowing sea anemones (Order Actiniaria).

bicipital. Pertaining to ribs which have a double articulation with vertebrae.

Bicosoeca. Genus of fresh-water monoflagellate Protozoa; cell on a fine stalk, inside a vaselike lorica.

bicuspid. 1. Two-pointed. 2. In man, one of the two double-pointed (premolar) teeth on each side of each jaw; see PREMOLAR.

bicuspid valve. Mitral valve; valve separating the left auricle and left ventricle in the mammalian heart; consists of two membranous flaps and prevents backflow when the ventricle contracts.

Bidder's organ. Small organ at the anterior end of each testis in a male toad; remnant of the gonad cortex (ovary) of a previous sexually undifferentiated stage.

Bidessus. Common genus of small dytiscid beetles.

Biemma. Common and widely distributed genus of marine sponges (Subclass Monaxonida); irregularly encrusting or massive.

biflagellate. Bearing two flagella.

bifurcate. Forked; divided into two branches.

big brown bat. See NYCTICEIUS.

bigeneric. Relating to a hybrid resulting from a mating between representatives of two different genera.

bigeye. See PRIACANTHIDAE.

big-headed flies. See PIPUNCULIDAE.

bighorn. Rocky Mountain sheep or bighorn sheep (Ovis canadensis); large, wild sp. with hairy rather than woolly pelage; excellent climbers; females with small horns, males with massive curled horns; upper coloration brown to grayish brown; to 300 lbs.; mountainous areas from Canada to Mexico; a related sp., O. dalli, is found in Alaska and Canada.

bilateral cleavage. Early cleavage and arrangement of blastomeres in certain embryos; results in a bilaterally symmetrical embryo.

bilateral symmetry. Disposition of parts in an animal so that there are right and left sides, dorsal and ventral surfaces, and anterior and posterior ends; only one plane can be passed through the body so that there are two mirror-image halves.

bilbies. See MACROTIS.

bile. Yellow secretion of the vertebrate liver; contains organic salts, acids, pigments derived from disintegrated erythrocytes, cholesterol, lecithin, and inorganic salts; stored temporarily in the gall bladder.

bile duct. Duct which carries bile from the liver and to the small intestine in vertebrates.

bile salts. Sodium taurocholate and sodium glycocholate, normally found in bile.

Bilharzia. Schistosoma.

bilharziasis. Schistosomiasis; a blood fluke infection produced by Schistosoma.

bilirubin. Pigment derived from the disintegration of old erythrocytes; formed in the liver and excreted as a component of bile.

biliverdin. Green bile pigment formed from the bilirubin.

billfish. See BELONIDAE and SCOMBERESOCIDAE.

Bimeria. Genus of marine hydrozoan Coelenterata.

binary fission. Asexual reproduction in animals by a simple division of the body into two parts; the two new individuals may be the same or different sizes; common in Protozoa, Turbellaria, some Oligochaeta, and a few other groups.

binary nomenclature. Binomial nomenclature.

binocular microscope. 1. Microscope with two eyepieces and one or two objectives for attaining binocular vision of an object; those with two objectives are usually called binocular dissecting microscopes. 2. Stereoscopic microscope.

binocular vision. Type of vision whereby each eye obtains an image of the same object; owing to the slightly different angle of vision for each eye, the animal is able to perceive relative distance and depth of vision; occurs in primates and many other higher vertebrates, especially predators.

binomial nomenclature. System of plant and animal nomenclature first standardized by Linnaeus; each sp. is designated by a double scientific name consisting of the genus name and the sp. name.

binominal nomenclature. Binomial nomenclature.

binturong. See ARCTICTIS.

bio-assay. Establishment of the character and strength of a drug or other substance by measuring its effect upon a living laboratory animal.

biocenosis. Collectively, the plants and animals of a community; biocenose.

biochemical genetics. Study of the chemistry and physics of genes as well as the manner in which genes control development and physiological coordination.

biochemistry. Study of the chemistry of living organisms and their processes.

biochore. Fundamental geographic terrestrial environment, characterized by certain vegetation types; the four such biochores are: forest, savanna, grassland, and desert.

biocycle. One of the three fundamental subdivisions of the biosphere: marine, fresh-water, and terrestrial; sometimes a fourth (parasitic) category is also used.

bioecology. Ecology with special reference to the interrelationships of plants and animals.

bio-electricity. In the broad sense, the production of electricity by living animals, whether it be the faint currents produced by nerve metabolism or the production of high voltages by electric eels and catfishes.

bioenergetics. Energy transformations in biological processes and the application of the laws of thermodynamics to living systems.

biogenesis. Reproduction. See PRINCIPLE OF BIOGENESIS.

biogenetic law. See RECAPITULATION.

biogeochemistry. Distribution, utilization, and flow of chemical materials between living organisms and their inanimate surroundings.

biogeography. Study of the geographic distribution of organisms.

biological clock. Innate physiological or behavior rhythms which are synchronized with daily, lunar, or annual cycles of tide, light, temperature, etc.; such rhythms may include such phenomena as cell division, bioluminescence, breeding, migrations, color changes, feeding, metabolic rate.

biological control. Reduction or elimination of a plant or animal pest by the introduction of a predator or parasite that normally feeds on it, such as the use of killifish to feed on mosquito larvae.

biological productivity. Amount of protoplasm produced per unit time; a sp., habitat, or community character.

biological warfare. BW; Use of living microorganisms and their toxic products for reducing the military effectiveness of man.

biology. 1. Science dealing with all matters relating to living plants and animals. 2. A treatise on the subject. 3. Animal and plant life, as of a particular region.

bioluminescence. Production of light by living organisms, a phenomenon common to many groups of marine and terrestrial animals, as well as certain bacteria and other fungi; the most familiar biochemical mechanism operates as follows: in the presence of oxygen and an enzyme (luciferase), certain proteins (luciferins) are converted to oxyluciferins with the liberation of light; luminescence in animals usually occurs in definite organs, in the secretions of mucous glands, or in masses of bacteria which are on or in tissues; examples are the firefly, marine fishes, Crustacea, Coelenterata, and Annelida.

biomass. Total weight of all organisms in a particular habitat or area; the term is also used to designate the total weight of a particular sp. or group of spp.

biome. World-wide complex of communities, characterized by the prevailing climatic and soil conditions; thus all grassland areas collectively form the grassland biome, and tropical rain forests everywhere collectively form the tropical rain forest biome; the deciduous forest biome is another example. Cf BIOCHORE.

biometry. Application of statistics to the study of organisms; biometrics.

Biomyxa. Genus of fresh-water amoeboid Protozoa; pseudopodia anastomosing.

bionomics. Outmoded term for ecology.

biophysics. Physics of living systems.

biopsy. Removal of a small bit of living tissue from the body for examination and diagnosis.

bios I. Inositol.

bios II. Obsolete term for biotin.

bioseston. Collectively, the living organisms of the seston.

biosphere. 1. Long period of earth's history characterized by great floral and faunal climaxes; cf. LITHOSPHERE and NOÖSPHERE. 2. That portion of the upper crust of the earth and the lower portion of the atmosphere which is inhabited by plants and animals.

biosynthesis. Formation of organic compounds from simpler compounds by living organisms.

biota. Collective flora and fauna of any particular area.

biotic community. See COMMUNITY.

biotic factors. In ecology, those environmental factors or influences which are the result of living organisms and their activities; distinct from physical, chemical, climatic, and edaphic factors.

biotic formation. Biome.

biotic potential. Theoretical measure of the relative ability of a sp. to survive in the face of environmental adversities.

biotic provinces. Divisions of an area (continent, country, etc.) into characteristic regions on the basis of their natural vegetation and fauna.

biotin. Vitamin H; vitamin of the B complex; essential for growth and for carbon dioxide fixation and oleic acid synthesis by bacteria; deficiency produces dermatitis in rats and chicks; abundant in yeast, cereal grains, cane molasses, eggs, vegetables, and fresh fruits; obsolete name is bios II; $C_{10}H_{16}O_3N_2S$.

biotope. Niche, especially in the geographic or spatial sense.

biotype. 1. Population or group of individuals having the same genotype. 2. A sp. made up of two or more distinct races.

Bipalium. Genus of elongated terrestrial Turbellaria; to 12 in. long; common in tropics and subtropics, occasional in greenhouses in temperate areas.

bipennate muscle. Any muscle in which the fibers pass obliquely from either side to a central tendon.

Bipes biporus. Sp. of worm lizard in the Family Amphisbaenidae; forelimbs present; Lower Calif.

bipinnaria. Free-swimming bilaterally symmetrical ciliated larva of Asteroidea; later develops into a brachiolaria larva.

bipolar neuron. Neuron having one main axon.

biradial cleavage. Type of holoblastic early cleavage in Ctenophora; the eight-cell stage has the form of a curved plate, the short axis being oral-aboral and the long axis being the tentacular plane of the adult.

biradial symmetry. Bilateral symmetry superimposed on fundamental radial symmetry, especially where the number of certain repeated parts is only two; e.g. certain sea anemones and comb jellies.

biramous appendage. Typical crustacean appendage; consists of a short basal two-segmented portion (protopodite), a median branch (endopodite), and a lateral branch (exopodite); each of the two branches (rami) is segmented and of varying size; the fundamental structure of a biramous appendage is often greatly modified in accordance with a variety of functions; frequently one of the two rami is lacking.

bird. See AVES.

bird lice. Common name of most insects in the Order Mallophaga.

bird lime. Sticky or gummy substance used for catching small birds; composition variable.

bird mite. Liponyssus sylviarum, a small mite (Parasitidae) found on domestic and wild birds.

bird of paradise. Any of 43 spp. of brilliantly colored birds in the Family Paradisaeidae; males with long tails and fantastic plumes; elaborate display patterns; forests of New Guinea and adjacent islands.

bird spider. See THERAPHOSIDAE.

bird tick. See HIPPOBOSCIDAE.

Birge, Edward A. American limnologist (1851-1950); made many fundamental contributions to limnology, especially with reference to Wisconsin lakes; professor, Dean, and President of the University of Wisconsin at various times.

Birgus latro. Purse crab or palm crab; large terrestrial crab of tropical islands of the Pacific and Indian oceans; feeds on coconuts and other vegetation; to 20 lbs.; returns to the sea at breeding time.

Birkenia. See OSTRACODERMI.

birotulate spicules. Amphidisks; type of sponge spicule having a disc or series of radial spokes at each end.

birth canal. Collectively, the cervix,

vagina, and vulva through which the young passes during birth in mammals.

birthmark. Some peculiar skin blemish or mark or congenital origin; usually pigmented, sometimes hairy.

bisexual. 1. Pertaining to an individual possessing both male and female reproductive organs; hermaphroditic. 2. Pertaining to a population consisting of individuals of both sexes.

bisexual reproduction. Usual type of reproduction involving both male and female gametes for the formation of zygotes.

Bismarck brown. Intra-vitam stain useful for certain Protozoa.

bison. Large, shaggy-maned oxlike ruminant in the Family Bovidae; short, stout horns, massive head, heavy forequarters, and a dorsal hump; includes the European bison (Bison bonasus) and the American bison or buffalo (B. bison); the eastern subspecies of the American bison became extinct in about 1800; a small population of wood bison, a large subspecies, occurs in northwestern Canada.

Bispira. Genus of fan worms.

bisulcate. 1. cloven. 2. With two grooves.

Bithinia. Genus of small fresh-water snails; B. tentaculata common in the Great Lakes area, introduced from Europe.

biting housefly. See STOMOXYS.

Bitis. Small African genus of venomous snakes in the Family Viperidae; sluggish, stubby-bodied spp. with short tail, heart-shaped head, and bright coloration; B. arietans is the African puff adder, to 4.5 ft. long; it hisses violently.

Bittacomorpha. Common genus of phantom craneflies (Ptychopteridae).

Bittacus. Genus of cosmopolitan insects in the Order Mecoptera.

bitterling. See RHODEUS.

bittern. Any of several small to large migratory heron-like birds, common in N. A.; seldom perch but remain in marsh vegetation on the ground.

bittersweet clam. See GLYCYMERIS.

Bittium. Genus of small brownish marine gastropods; horn shells.

Biuret reaction. Biochemical test for proteins; a strong potassium hydroxide solution and a few drops of copper sulfate are added to the unknown suspension; pinkish-violet color (positive test) then indicates the presence of protein.

bivalent. Collectively, each two associated homologous chromosomes during the prophase of the first division of meiosis.

bivalve. 1. Any member of the molluscan Class Pelecypoda. 2 Having a shell of two parts which are joined by hinge, as in pelecypods and lamp shells.

bivium. Collectively, those two arms of a sea star situated on either side of the madreporite.

bivoltine. Having two generations per year.

black abalone. Haliotis cracherodi, a dark-colored abalone of the Pacific Coast.

black-banded snake. Coniophanes imperialis, a small, smooth, night-snake of the southern tip of Tex. and adjacent Mexico; dorsal surface brown, with three dark stripes; 12 to 20 in. long.

black-bellied plover. See SQUATAROLA.

blackbird. Any of a wide variety of birds in which the males are chiefly black; most spp. are in the Family Icteridae; common American examples are the yellow-headed blackbird (Xanthocephalus xanthocephalus), redwing blackbird (Agelaius phoeniceus), and Brewer's blackbird (Euphagus cyanocephalus).

black buck. See ANTILOPE.

black chiton. Katharina tunicata, a common dark-colored chiton of the Pacific Coast.

black clam. Arctica islandica, a heavy-shelled quahog of the North Atlantic.

black crab. Edible land crab of Fla. and West Indies.

Black Death. Bubonic plague.

black duck. See ANAS.

black eye. Bruise formed by a clot produced by a break in a blood vessel under the skin around the eye; disappears as the clot is absorbed.

blackfin. Common name for a sp. of whitefish (Coregonus) found from Lake Superior to Hudson Bay.

blackfish. 1. Small toothed whale (Glocephala) of blackish coloration; to 20 ft. long; gregarious. 2. Common name of several marine fishes, especially the tautog and black sea bass (Centropristes). 3. Small oily food fish (Dallia pectoralis) of ponds and streams in Alaska and northern Siberia; reputed to be able to survive freezing.

black flies. See SIMULIIDAE.

Blackfordia. Genus of marine coelenterates in the Suborder Leptomedusae.

blackhead. 1. Encysted glochidium in superficial tissues of a fish. 2. Comedo; a plug of dried and discolored sebum in a sebaceous gland duct. 3. Any of certain birds with a blackish coloration. 4. Protozoa disease of the digestive system of turkeys, chickens, and a few other birds.

black-headed snakes. Group of small, slender snakes in the genus Tantilla; head flat, dorsal surface brown; southern half of U. S.

blackout. Temporary blindness and loss of consciousness owing to an insufficient blood supply to the brain; occurs when airmen make a sharp turn at high velocity with the head directed toward the center of the circular movement.

black petrel. See LOOMELANIA.

black scale. See SAISSETIA,

blacksnake. See COLUBER.

black swamp snake. Seminatrix pygaea pygaea, a thick-bodied, red-bellied, swamp-loving snake of N.C. to Fla.; 9 to 20 in. long.

black swift. See CYPSELOIDES.

black-tailed deer. See ODOCOILEUS.

black tern. See CHLIDONIAS.

black-throated sparrow. See AMPHISPIZA.

black tongue. Nicotinic acid deficiency disease in dogs.

black widow spider. See LATRODECTUS.

bladder. 1. Sac for the storage of a fluid, e.g. gall bladder and urinary bladder. 2. Gas- or air-filled sac, such as the air bladder of fishes.

bladderworm. Cysticercus.

Blainville, Henri. French zoologist (1777-1850); exponent of Lamarckism and author of Faune Francaise.

Blanding's turtle. See EMYS.

Blarina brevicauda. Short-tailed shrew or American mole shrew; body to 4 in. long; forests and grasslands of eastern half of U.S.

blastaea. Hypothetical ancestral organism from which the gastraea, planulaea, and the Phylum Porifera are assumed to have arisen; it had the structure of a blastula larva but was an abundant adult organism.

blastema. Mass of undifferentiated cells which later develops into an organ; also functions in regeneration in lower invertebrates.

Blastocerus. Genus of marsh deer of the Amazon basin; ears enormous, coloration red above and white below.

blastocoel. Large central cavity of a blastula or blastodermic vesicle.

blastocyst. Blastodermic vesicle; blastula stage in early mammalian development; a mass of special cells on one area of a thin-walled hollow sphere becomes the embryo proper.

Blastocystis hominis. Harmless vegetable microorganism found in human stools; has general appearance of a protozoan cyst.

blastoderm. 1. Blastodisc. 2. Thin layer of cells on the surface of a developing insect egg. See SUPERFICIAL CLEAVAGE.

blastodermic vesicle. Early embryonic stage, especially in mammals; consists of an outermost layer of cells (trophoblast), an inner cell mass (embryo proper), and a cavity (blastocoel); by further embryonic development the primitive blastocoel is obliterated and replaced by other cavities.

Blastodinium. Genus of dinoflagellates in the digestive tract of certain marine copepods.

blastodisc. Disclike area of protoplasm on the surface of a large, yolky egg; by cell division it produces the embryo; especially in fertilized eggs of birds, reptiles, and many fishes.

Blastoidea. Class of extinct echinoderms; calyx budlike and composed of 13 major plates; Ordovician to Permian; Pentremites typical.

blastokinesis. Movements of an embryo by which it changes its orientation within the egg, especially in insects.

blastomeres. Individual cells making up an early embryo during the process of cleavage, especially before the attainment of a distinct gastrula stage.

blastomycosis. Chronic fungus infection marked by suppurating skin lesions or by lesions of the liver, lungs, bones, spleen, kidneys, and subcutaneous tissues.

Blastophaga psenses. Fig wasp which pollinates edible figs (Ficus carica var. smyrnica) by transferring pollen from the inedible caprifig (F. carica var. sylvestris) to the flowers of the edible variety.

blastophore. That residual protoplasm resulting from the transformation of spermatid(s) into functional spermatozoa.

blastopore. Single external opening of the archenteron (primitive digestive tract) in the gastrula of most animals; depending on the phylogenetic group, the blastopore becomes the mouth or anus, or it may close with the subsequent formation of an anus in the same general area.

blastostyle. Living, axial portion of a gonangium; modified hydrozoan polyp from which medusae are budded.

blastozooid. Tunicate zooid which arises by budding. Cf. OOZOOID.

blastula. Early embryological stage in many animals; a hollow mass of cells. Pl. blastulae.

Blasturus. Genus of Ephemeroptera.

Blatta orientalis. Dark brown or blackish cockroach; Oriental roach; to 1 in. long; common domestic pest in U.S.

Blattaria. Order of omnivorous insects; the cockroaches; body wide and depressed; often foul-smelling; antennae very long; hidden in crevices during the day but active at night; several pest spp. in homes and buildings, especially where food is prepared or processed; numerous families; common genera are Periplaneta, Blatta, and Blatella.

Blattella germanica. Pale brownish cockroach; German roach or croton bug; to .5 in. long; the most common cockroach in homes, restaurants, and places where food is processed.

Blattidae. Family of insects in the Order Orthoptera; includes the cockroaches; usually these insects are placed in a separate order, the Blattaria.

bleeder. 1. Person afflicted with hemophilia. 2. Person who bleeds easily. 3. Person from whom blood is taken as a supposed cure for various ills.

bleeding tooth. Nerita peleronta, a common marine snail of Fla. and West Indies; with a blood red stain on the central part of the columella.

blending characters. Paired or multiple characters which merge and do not show distinct Mendelian segregation.

blending inheritance. Inheritance produced by crossing two parents widely different in some quantitative character and resulting in offspring which are intermediate between the parents; occurs when variation of the character is produced by a number of paired genes, each with a small and additive effect.

Blennidae. Large family of small carnivorous percomorph fishes (blennies) found chiefly along rocky marine shores, especially in tide pools, inlets, and estuaries; body elongated, with long dorsal and anal fins.

blenny. See BLENNIDAE. Pl. blennies.

Blepharipoda occidentalis. Spiny sand crab, a burrowing sp. of Calif. intertidal sandy beaches.

Blepharisma. Large genus of marine and fresh-water heterotrich ciliates; spindle-shaped, pyriform, or ellipsoid.

Blepharocera. Genus of net-winged midges (Blepharoceridae).

Blepharoceridae. Net-winged midges; family of elongated, mosquito-like insects of world-wide distribution; larvae and pupae tightly attached to rocks in rushing brooks; larvae flattened, composed of seven divisions, and with six median ventral suckers; Bibiocephala and Blepharocera common.

blepharoplast. Intracellular granule associated with a basal granule and base of a flagellum; sometimes the blepharoplast and basal granule are fused.

blesbok. See DAMALISCUS.

blesmol. See BATHYERGIDAE.

blindcat. See SATAN and TROGLOGLANIS.

blind fish. See AMBLYOPSIDAE.

blind goby. See PINKFISH.

blind salamander. See TYPHLOMOLGE and TYPHLOTRITON.

blind shell. See CAECUM.

blind snake. See LEPTOTYPHLOPIDAE.

blind spot. Small area of the vertebrate retina where the optic nerve leaves the eye and hence there are no rods and cones, and no vision.

blindworm. See ANGUIS.

Blissus. See LYGAEIDAE.

blister beetles. See MELOIDAE.

blister mite. See ERIOPHYIDAE.

blister pearl. Pearly growth on the inner surface of a bivalve shell; sometimes blister-shaped but often asymmetrical; see PEARL.

bloater. See LEUCICHTHYS.

blood. Chief circulatory fluid of animals; may or may not be entirely confined to a system of definite vessels; functions chiefly in the transport of materials throughout the body; also has important respiratory, temperature-regulating, and disease prevention roles.

blood bank. Any storage place (usually a hospital or clinic) where liquid blood, blood plasma, or dried blood plasma are stored ready for use in transfusions.

blood clam. Marine clam in the genus

Arca; reddish coloration.

blood corpuscles. Collectively, the various types of cells suspended in blood.

blood film. Thin, dry smear of blood on a glass slide; when properly fixed and stained, it may be used for examining the various types of corpuscles and for detecting blood parasites.

bloodfin. Small silvery South American aquarium fish with bright red fins.

blood gill. 1. Thin-walled gill in many immature fresh-water insects; cavity continuous with the hemocoel and filled with blood. 2. Projecting filamentous gill in the neck region of certain immature aquatic salamanders.

blood glucose. Blood sugar.

blood group. Group of human beings or other primates whose blood may be mixed without a resulting clumping of the corpuscles; four general blood groups are recognized: A, B, AB, and O; agglutination occurs when blood from two different groups are mixed, and such agglutination is produced by an antigen-antibody reaction between an agglutinogen (antigen) in the erythrocytes of one blood and an antagonistic agglutinin (antibody) in the plasma of the other blood; erythrocytes of blood groups A, B, AB, and O contain antigens A, B, A and B, and no antigens, respectively; the plasmas of these same blood groups contain the following corresponding antibodies: b, a, none, and a and b; blood from persons of the same group may therefore be mixed without any agglutination; for purposes of blood transfusions, the following donor blood may also be used without danger: O donor for A, B, and AB patients; A donor for AB patient; B donor for AB patient; antigens called M, N, or MN are also contained in the corpuscles of each human being, but their corresponding agglutinins are not present in the plasma, hence they have no significance in transfusions, except that a person may acquire agglutinins as the result of repeated transfusions. See RH FACTOR.

blood islands. Embryonic patches of mesoderm in which primitive erythroblasts first appear.

blood plasma. See PLASMA.

blood platelets. Minute granular bodies, 2 to 3 microns across, suspended in mammalian blood and functioning in blood clotting; about 250,000 per cubic mm. of human blood.

blood pressure. Pressure exerted by the blood as the result of the pumping action of the heart and the resilient artery walls; according to the stage of heart contraction and other factors, normal human arterial pressure ranges from 80 to 120 mm. of mercury; pressure falls progressively as blood passes from arteries to capillaries and veins.

blood protein. One of several typical proteins found in blood, e.g. serum albumin and fibrinogen.

blood sea star. Henricia sanguinolenta, a red circumboreal sea star in the Order Spinulosa.

blood serum. Fluid remaining after a blood clot forms; similar to blood plasma but lacks certain proteins and some calcium.

blood sinus. Open, irregular channels and spaces through which blood passes; found especially in the tissues of animals having an open circulatory system, e.g. mollusks.

bloodsucker. 1. Leech. 2. Any of several Indian and Australian lizards having reddish throats.

blood sugar. Blood glucose; normal quantity of glucose circulating in the blood of vertebrates; usually 80 to 180 mg. per 100 ml. of blood in man; derived from glycogen in the liver and other tissues, and utilized for metabolism by all cells.

blood transfusion. Transfer of blood from the venous system of one animal into that of another.

blood type. Blood group.

blood vessel. Any tube or canal through which blood flows; in the higher animals they are usually recognized as arteries, veins, and capillaries.

bloodworm. 1. Fresh-water larva of many spp. in the dipteran Family Tendipedidae; red coloration produced by invertebrate hemoglobin dissolved in the circulating body fluid. 2. Any of numerous bright red marine Polychaeta.

bloody clam. Combed ark shell; Arca campechiensis, a blood-red marine bivalve of the East Coast.

blowfish. Fish able to inflate its body with water or air.

blowflies. See SARCOPHAGIDAE.

blowhole. Nostril on the top of the head of a whale.

blubber. Thick subdermal layer of insulating fat in many marine mammals.

blue abalone. Haliotis fulgens, a thin-shelled iridescent bluish abalone of the Calif. coast.

blue baby. Condition of a new-born baby in which the ductus arteriosus fails to close or there is some heart lesion and the blood is not properly oxygenated; corrected by surgery.

blueback salmon. Sockeye salmon; see ONCORHYNCHUS.

bluebird. See SIALIA.

bluebottle fly. See CALLIPHORA.

blue crab. See CALLINECTES.

blue dog. See LAMNA.

bluefish. Any of a wide variety of fishes having a blue coloration. see especially GADIDAE and POMATOMUS.

blue-foot. Quadrula costata, a common unionid clam of the Mississippi and St. Lawrence drainages; used for button manufacture.

bluegill. See LEPOMIS.

blue goose. See ANSER.

blue grosbeak. See GUIRACA.

blue jay. See CYANOCITTA.

blue mud shrimp. Upogebia pugettensis; a bluish-white West Coast shrimp which burrows in mud flats.

blue racer. See COLUBER.

blues. See LYCAENIDAE.

blue sucker. See CYCLEPTUS.

blue-tailed skink. See EUMECES.

blue-top snail. Calliostoma costatum, a common blue and yellow snail of the Pacific Coast.

blue whale. See BALAENOPTERA and SIBBALDUS.

bluntnose minnow. See PIMEPHALES.

BMR. Basal metabolic rate.

boa. Any large constrictor snake; see BOIDAE.

Boa. Tropical American genus of boa constrictors.

boa constrictors. See BOIDAE.

boarfish. Any of several fishes having a piglike snout.

boat-billed heron. See COCHLEARIIDAE.

boat shell. See CREPIDULA.

boat-tailed grackle. See CASSIDIX.

bobcat. Any of several cats in the genus Felis, especially F. rufus, the American bobcat, wildcat, or bay lynx; much like the American lynx but smaller and restricted chiefly to the U.S. and Mexico; sometimes placed in the genus Lynx.

bobolink. See DOLICHONYX.

bob-white. See COLINUS.

bocaccio. See SEBASTODES.

Boccardia. Genus of polychaete worms.

Boderia. Genus of delicate marine amoeboid Protozoa; pseudopods long and finely branched; thin test covers the main cell mass.

Bodo. Genus of small colorless flagellates found chiefly in stagnant waters; occasional in human feces; with one trailing and one anterior flagella.

body cavity. General term applied to any extensive fluid-filled cavity occurring between body wall and digestive tract; much of the cavity customarily contains visceral organs.

body whorl. First (large) coil of a gastropod shell.

Bohr effect. Action of carbon dioxide in reducing the affinity of hemoglobin for oxygen.

Boidae. Family of snakes which includes many large pythons, constrictors, etc.; vestigial internal hind legs project as a pair of spurs on each side of the anus; kill prey by constriction; habits variable; world-wide distribution except for New Zealand; Python, Eunectes, and Charina typical.

Boiga. Genus of tree snakes in the Family Colubridae; usually long, slender spp.

boil. Furuncle; circumscribed bacterial inflammation of the corium and subcutaneous tissue, enclosing a central core; formation favored by local irritation and nutritional disturbances.

Boleosoma. Genus of darters now usually included in Etheostoma.

Bolinopsis. Genus of ctenophores; laterally compressed and with two oral lobes and four large auricles at the oral end; North Atlantic.

boll weevil. See ANTHONOMUS.

bollworm. Larva of a moth (Heliothis armigera) which feeds on cotton bolls, corn ears, and beans.

Bolocera. Genus of large sea anemones (Order Actiniaria); small commensal crustaceans are often present in the gastrovascular cavity.

Boloceroides. Genus of small, primitive sea anemones; swim by tentacular lashing.

Boltenia. Genus of large, sessile ascidians.

bolus. Mass of food about to be swallowed, or a mass passing along the esophagus or intestine.

bombardier beetles. Beetles in the Family Carabidae which discharge an irritating fluid or vapor when disturbed; Brachinus common in southwestern states.

bomb calorimeter. Calorimeter for measuring the potential energy of food;

a known amount is placed in a platinum dish inside a steel container (bomb) filled with oxygen and surrounded by a water jacket; the food is ignited and the heat taken up by the water jacket is easily measured.

Bombidae. Family of medium to large social bees with three castes; bumblebees, tongue long; vicious stingers; each queen hibernates and starts a new colony in spring, usually in a mouse nest or rodent burrow; valuable as flower pollinators; Bombus common.

Bombina. Genus of Eurasion toads, one sp. being the fire-bellied toad owing to the black and bright red or yellowish ventral surface.

Bombus. See BOMBIDAE.

Bombycidae. Small family of moths which includes the silkworm moth, Bombyx mori.

Bombycilla. See BOMBYCILLIDAE.

Bombycillidae. Small family of fruit-eating passerine birds which includes the waxwings and palm chat; coloration chiefly rich brown or fawn, with a showy crest and velvety plumage; Northern Hemisphere; Bohemian waxwing (Bombycilla garrula) and cedar waxwing (B. cedrorum) common American spp.; palm chat of Hispaniola builds a communal nest.

Bombyliidae. Family of Diptera superficially resembling bees; bee flies; body hairy, legs and proboscis long and slender; adults hover about flowers and feed on nectar and pollen; larvae parasitic or predaceous on other immature insects; Bombylius common.

Bombylius. Common genus of bee flies (Bombyliidae); larvae parasitic on larvae of certain bees.

Bombyx mori. Silkworm moth; larva spins (from salivary secretions) a silken cocoon from which silk thread is produced commercially, chiefly in the Orient; larva feeds exclusively on leaves of white mulberry tree; one cocoon yields about 1000 ft. of fiber, and 25,000 cocoons are necessary for one pound of silk thread.

Bomolochus. Genus of marine copepods (Suborder Cyclopoida); parasitic on fish gills.

Bonasa umbellus. Common American ruffed grouse, a large red-brown or gray-brown chicken-like upland game bird of wooded areas; in the Family Tetraonidae; often called partridge in northern states.

bone. 1. skeletal connective tissue peculiar to most vertebrates; consists of scattered cells located in spaces in the surrounding matrix; the latter is hard and contains about 60 or 65% calcium phosphate and calcium carbonate, as well as organic connective tissue fibrils; bone cells are connected with each other by extremely fine channels permeating the matrix everywhere. 2. one of the discrete hard pieces or parts of the endoskeleton of most vertebrates.

bonefish. See ALBULIDAE.

Bonellia. Genus in the Phylum Echiuroidea; proboscis very long and branched at the tip; notable because of the degenerate, microscopic male, which becomes a parasite in the nephridium of the adult female.

bone marrow. Highly vascular spongy tissue found in the central cavity of bones; source of all erythrocytes and most leucocytes.

bongo. See BOOCERCUS.

bonito. Any of several marine fishes, mostly related to the mackerels.

Bonnet, Charles. Swiss philosopher and naturalist (1720-1793); discovered parthenogenesis in aphids, studied insect metamorphosis, and elaborated the preformation theory.

bontebok. See DAMALISCUS.

bony fishes. Generally, the fishes in the vertebrate Class Osteichthyes; skeleton more or less bony.

bony labyrinth. Small mass of bone surrounding the membranous labyrinth of the inner ear and having a complicated cavity corresponding to the shape of the enclosed membranous labyrinth; the whole bony labyrinth is imbedded in the temporal bone.

booby. See SULA.

Boöcercus. Genus of bongos; large, heavy-bodied ruminants of African forests; coat orange-red with white stripes.

boody. See BETTONGIA.

book gills. Flat respiratory plates on the abdomen of Xiphosura; each of ten abdominal appendages bears 150 to 200 such plates arranged as the leaves of a book; each plate is actually a double lamella which encloses a flat extension of the hemocoel.

book lice. See CORRODENTIA and TROCTES.

book lungs. Paired respiratory structures in many arachnids; each consists of a cavity containing numerous double-walled leaves arranged as the pages in a book; blood circulates between the two lamellae of each leaf; each book lung opens externally by a ventral pore or slit.

book scorpion. See CHEIRIDIUM.

bookworm. Common name of the larvae of many insects which feed on the binding and paste of books.

boomslang. See DISPHOLIDUS.

Boophilus annulatus. Cattle tick which transmits the causative organism of Texas cattle fever; see Babesia bigemina.

booster shot. Second injection of an antigen which markedly raises the available antibody level in the tissues.

Bopyridae. Family of highly modified marine Isopoda; parasitic on decapod crustaceans.

Bopyrina. Genus of Isopoda parasitic on marine shrimps.

Bopyroides. Genus of Isopoda parasitic on marine decapods.

Bopyrus. Genus of marine Isopoda; highly modified and parasitic in the gill chambers of prawns.

Boreal zone. Collectively, the Arctic-Alpine, Hudsonian, and Canadian life zones.

Boreus. Genus of wingless insects in the Order Mecoptera; snow scorpion flies, snow fleas; adults superficially resemble nymphal grasshoppers; often found on snow on warm winter days.

Borhyaena. South American genus of puma-like Miocene marsupials.

boring clam. Any of numerous spp. of marine clams which burrow into soft rock, e.g. Pholadidea.

boring sponge. Common type of sponge found in marine shallows; genus Cliona; irregular yellow masses; when attached to mollusks they may dissolve the shell in spots and thus kill the mollusk.

Borrelia. Genus of bacteria causing several types of relapsing fever; mostly transmitted by the bite of ticks and body lice.

Bos. Genus of ruminants in the Family Bovidae; includes the urus (B. primigenius), yak (B. grunniens), Asiatic domestic cattle (B. indicus), and European domestic cattle (B. taurus).

Boselaphus tragocamelus. Large Indian antelope in the Family Bovidae; the nilgai; male has short, straight horns.

Bosmina. Very common genus of limnetic Cladocera.

Bostrichobranchus. Common genus of globular, sessile ascidians.

Bostrychidae. Small family of elongated, cylindrical beetles; larvae feed on dead wood; adults of certain spp. are pests by their habit of boring into wine casks, corks, and lead telephone cables.

bot. Larva of a botfly.

botflies. See OESTRIDAE, CUTEREBRIDAE, and GASTEROPHILIDAE.

Bothidae. Common family of flatfishes; sand dabs; eyes and coloration on left side of body.

bothridium. One of four, broad, leaf-like muscular structures on the scolex of certain tapeworms; often armed with hooks. Pl. bothridia.

Bothriocephalus. Genus of pseudophyllidean tapeworms found especially in fishes.

Bothriocidaris. See BOTHRIOCIDAROIDA.

Bothriocidaroida. Subclass of Ordovician Echinoidea; madroporite ambulacral; Bothriocidaris typical.

Bothriocyrtum. Common trapdoor spider of Calif.; cylindrical burrow in the ground closed with a wafer-like lid.

Bothriolepis. Genus of placoderm fishes having a single pair of jointed flippers.

Bothrioplana. Genus of large alloeocoel Turbellaria.

Bothriotaenia. Genus of tapeworms; mature form parasitic in birds.

bothrium. One of two or four elongated suction grooves on the scolex of certain tapeworms; sometimes the margins of a bothrium may project or are ruffled or fused so as to forma a tubular structure. Pl. bothria.

Bothromesostomum. Genus of large rhabdocoel Turbellaria.

Bothrops. Small genus of venomous snakes in the Family Crotalidae; Tropical America and West Indies; B. atrox, to 6 ft. long, is the fer-de-lance.

Botrylloides. Genus of compound tunicates.

Botryllus. Common genus of tunicates; soft, fleshy colonies with individuals arranged in groups discharging by a common cloacal aperture.

botryoidal tissue. Loosely arranged mesenchyme tissue found in the coelomic spaces of certain leeches; pigmented and traversed by intracellular capillaries.

bottlenose whale. See HYPEROODON.

bottom fauna. Benthos; collectively, the animals living in or on the bottom of a body of water.

Botula. Genus of elongated inflated marine bivalves.

botulism. 1. In man, a type of food poisoning caused by a toxin (botulin) produced by bacteria in improperly preserved food. 2. In the broad sense, a toxic poisoning caused by such bacteria (Clostridium botulinum) in soil, alkaline muds, carrion, and wounds; these poisons may affect fowl, ducks, stock animals, and fish.

Bougainvillia. Genus of sessile, colonial, marine hydrozoan Coelenterata.

Bouin's fluid. Common killing and fixing fluid for cytological and histological work; 75 parts of saturated picric acid solution, 25 parts formalin, and 5 parts glacial acetic acid.

Bourletiella spinata. Common springtail; often found on the surface film of ponds and lakes.

bouto. See INIA.

bouton. Terminal lappet of the ligula in certain Hymenoptera.

Boveri, Theodor. German zoologist (1862-1915); made important contributions to the cytology of fertilization in ascarids and sea urchins; demonstrated random distribution of chromosomes and realized that somatic cells are diploid and gametes haploid.

Boveria. Genus of holotrich ciliates often found as commensals on gills of many marine invertebrates.

Bovidae. Large family of mammals in the Suborder Ruminantia; horns paired, unbranched, and composed of keratin which grows out slowly and continuously from the base of projecting bony corelike prolongations of the frontal bones; horns usually present in both sexes; includes many kinds of antelope, buffalo, cattle, gazelle, goat, sheep, etc.

bowel. Intestine, especially the large and small intestine of man.

Bowerbankia. Genus of marine Bryozoa.

bowerbird. Any of 19 spp. of birds in the Family Ptilonorhynchidae; Australia and New Guinea; males build large elaborate bowers or playhouses to attract females.

bowfin. See PROTOSPONDYLI.

bowhead whale. See BALAENA.

Bowman's capsule. See MALPIGHIAN BODY.

box crab. See CALEPPA.

box shell. Ark shell.

box turtle. See TERRAPENE.

Boyeria. Genus of dragonflies.

brachial. Pertaining to an arm or armlike process.

brachial artery. Large artery in the forelimb of many tetrapods.

brachial enlargement. Slightly thickened portion of the spinal cord in the region of the second and third spinal nerves of e.g. the frog.

brachialis. Tetrapod muscle originating on the lower half of the humerus and inserting on the coronoid process of the ulna; flexes forearm.

brachial nerve. Large nerve supplying the shoulder and arm region in tetrapods.

brachial plexus. Network formed by interconnections of several spinal nerves in the shoulder region of tetrapods.

brachial vein. Large vein in the forelimb of many tetrapods.

Brachiata. See POGONOPHORA.

brachiation. Progression by swinging from one hand-hold to another, as the gibbon.

Brachidontes. Genus of small marine mussels; all U.S. coasts.

Brachinus. See BOMBARDIER BEETLES.

brachiolaria. Free-swimming bilaterally symmetrical ciliated larva of the Class Asteroidea; develops from the bipinnaria and is characterized by three preoral holdfast processes. Pl. brachiolariae.

Brachionidae. Large family of loricate rotifers; chiefly fresh water.

Brachionus. Large and common genus of plankton rotifers; lorica stout and rigid; foot long, annulated, and retractile; alkaline waters, occasionally marine.

Brachiopoda. Marine phylum which includes the lamp shells; most of the soft parts are enclosed within dorsal and ventral calcareous or hornlike shells which are slightly different in size but which nevertheless are superficially similar to the two valves of a clam shell; a fleshy peduncle or pedicle extends from the posterior end, and the valves can be gaped somewhat at the anterior end to allow partial extrusion of a lophophore and the free circulation of sea water into the mantle cavity; the lophophore occupies much of the mantle cavity; typically, it has two, long, rigid, lateral lobes and often also has a ventral flat strip which may be coiled or extruded; all three parts of the lophophore have a peripheral row of ciliated tentacles used in gathering minute food particles; digestive tract short and often with a blind end; it is situated in the coelom which occupies the posterior half of the animal; small heart, nephridia (nephromixia), nerve ring, and free-swimming ciliated larva (a modified trochophore); attached to the substrate by means of the peduncle; to 4 in. long; about 260 living spp. and 30,000 fossil spp.

brachioradialis. Mammalian muscle which originates on the humerus and inserts on the radius; flexes forearm and aids in supination.

Brachiosaurus. Genus of giant Jurassic quadrupedal herbivorous dinosaurs.

brachium. 1. Upper arm between shoulder and elbow. 2. Any arm or armlike process in a wide variety of animals

Brachycentrus. Common genus of caddis flies; larval case either round or square in cross section.

Brachycephalidae. Large family of Neotropical toads; two halves of pectoral girdle more or less fused along midline; Brachycephalus and Dendrobates common.

Brachycephalus. Genus of small Neotropical toads.

brachycephaly. Broad-headed condition. See CEPHALIC INDEX.

Brachycera. One of the two insect tribes in the dipteran Suborder Orthorrhapha; antennae short, usually three-segmented; larvae with an incomplete head.

brachycerous. With short antennae.

Brachycerus. Common genus of Ephemeroptera; eastern U.S.

brachydactyly. Occurrence of abnormally short fingers and toes (or other appendages in arthropods); a Mendelian dominant in man.

brachydont. Pertaining to teeth having low crowns, long roots, and narrow root canals, e.g. in man.

brachyphalangy. Short-fingered condition in human beings; a dominant lethal character when homozygous.

Brachyptera. Common genus of Plecoptera.

brachypterous. With short or reduced wings, as in certain insects.

Brachystola magna. Lubber grasshopper; large, heavy-bodied, short-winged sp. of southern and southwestern states.

Brachyura. One of the three suborders in the crustacean Section Reptantia; true crabs; cephalothorax short and very broad; abdomen bent sharply under thorax and fitting into a ventral thoracic groove; mostly marine.

brachyury. Short-tailed condition; mutation in mice.

brackish. Pertaining to sea water which is diluted with fresh water, or vice versa; characteristic of bays, estuaries, etc.

Braconidae. Family of small to minute insects in the Order Hymenoptera; braconid wasps; adults frequent flowers; larvae parasitic in larvae and pupae of other insects; pupation usually outside of host; Apanteles and Lysiphlebus are examples.

Brada. Genus of small burrowing Polychaeta.

bradyauxesis. Relative growth of a part more slowly than the growth of the whole organism.

bradycardia. Abnormally slow heart rate, usually less than 60 per minute in man.

Bradyodonti. Order of extinct cartilaginous fishes; teeth modified as crushing plates; Devonian to Permian; examples are Cochliodus and Psammodus.

Bradypodidae. Family of mammals in the Order Edentata; includes the sloths; sluggish Tropical American spp.; hang upside down from branches by means of long, curved claws; hair long and shaggy; mostly herbivorous; Bradypus includes the three-toed sloth (the ai), and Choloepus includes the two-toed sloth (unau).

Bradypus. See AI and BRADYPODIDAE.

Bradysaurus. Genus of extinct primitive cotylosaur reptiles.

Bradysia. Genus of fossil fungus gnats.

bradytelic. Evolving at an unusually slow rate, with slow speciation.

brain. Enlarged anterior part of the central nervous system in most bilaterally symmetrical animals.

brain coral. See DIPLORIA and MEANDRINA.

brain sand. Corpora arenacea.

brain stem. Vertebrate brain exclusive of the cerebral hemispheres and cerebellum.

brain waves. See ELECTROENCEPHALOGRAPH.

bramble shark. See ECHINORHINUS.

brambling. Brightly colored Eurasian finch.

Bramidae. Pomfrets; a family of pelagic marine fishes; body compressed, head rounded, caudal peduncle slender; Brama typical.

Branchellion. Remarkable genus of leeches parasitic on gills of skates and rays; with lateral leaflike gills.

branchia. Gill. Pl. branchiae.

branchial. Of or pertaining to gills.

branchial arch. See VISCERAL SKELETON.

branchial chamber. Pharyngeal cavity of tunicates and cephalochordates.

branchial cleft. Gill slit.

branchial heart. One of two hearts in cephalopod mollusks; pump blood to the gills.

branchial vessel. Aortic arch.

branchial sac. Pharynx of a tunicate.

Branchinecta. Common genus of fairy shrimps (Order Anostraca); in temporary ponds.

Branchiobdella. Genus of small leechlike oligochaetes; crayfish commensals, especially on the gills.

Branchiobdellidae. Aberrant leechlike family of Oligochaeta; parasitic or commensal on fresh-water crayfish; with a posterior sucker, annulated segments, and jaws; setae absent.

branchiocardiac grooves. External furrows on the dorsal surface of the carapace of many decapod crustaceans; mark the position of the underlying gill chambers and viscera.

branchiocardiac sinuses. Series of narrow vessel-like sinuses in the crayfish and many other Malacostraca; transfer blood from the gills, upward along the median surface of the two branchial chambers, and into the pericardial sinus.

Branchiocerianthus imperator. Giant deep-sea hydrozoan coelenterate.

Branchiocoetes. Genus of fresh-water holotrich Ciliata; commensal on Isopoda and Amphipoda.

branchiomere. Early embryonic mass from which a visceral arch develops.

Branchiopoda. One of the subclasses in the arthropod Class Crustacea; water fleas, fairy shrimps, etc.; flat, leafy, thoracic appendages used for respiration; parthenogenesis common; chiefly in fresh waters.

branchiosaur. See PHYLLOSPONDYLII.

Branchiosaurus. Genus of extinct stegocephalian Amphibia.

branchiostegals. Bony rays supporting the under surface of the head of a fish; attached to the hyoid arch below the opercular bones behind the lower jaw.

Branchiostegidae. Family of marine fishes; includes the tilefishes; bottom-living spp. with the general structure of sea basses but with the soft portion of the dorsal fin much longer than the spiny part; Lopholatilus chamaeleonticeps is the common Atlantic Coast tilefish; it is brilliantly colored and has a fleshy process at the nape of the neck; important food fish; to 50 lbs.

branchiostegite. Expanded lateral portion of a carapace forming a gill cover and gill chamber of certain malacostracan Crustacea.

Branchiostoma. Common genus of cephalochordates found along U. S. coasts; see CEPHALOCHORDATA.

Branchiura. One of two orders in the crustacean Subclass Copepoda; fish lice; body greatly modified, circular, and depressed; eyes compound; fixed parasites in the branchial chamber or on the general body surface of many fresh-water and marine fishes; Argulus the only American genus.

Branchiura sowerbyi. Uncommon fresh-water oligochaete; with dorsal and ventral gills on posterior segments; to 8 in. long.

brandling. 1. Small yellowish earthworm found especially in dung and rich organic soils. 2. Young salmon; parr.

brant. See BRANTA.

Branta. Genus of geese (Family Anatidae); B. canadensis is the common and widely distributed Canada goose, represented by several subspecies; B. bernicla is the American brant, found chiefly in the eastern states and northern Canada; B. nigricans, the black brant, occurs in western N. A.

Braula. See BRAULIDAE.

Braulidae. Family of minute apterous insects in the Order Diptera; parasitic on the honeybee; adults cling to the body of the host and claw food from the mouth parts of the bee; larvae feed on honey and food of the bee larvae; originally European but now widely distributed; only one common sp., Braula caeca, plus another rare sp. from the Congo.

breaking joint. Double crease on the ischium of a walking leg in most decapod Crustacea; marks the point at which the legs may be broken off by reflex action when the decapod is seized or irritated.

bream. 1. European fresh-water fish (Abramis brama) in the minnow family. 2. Common name of several fresh-water centrarchid fishes, especially in the southern states.

breastbone. Sternum.

breeding potential. Reproductive potential.

Brehm, Alfred Edmund. German naturalist (1829-1884); author of the classical Tierleben.

Bresslaua. Genus of broadly reniform fresh-water holotrich Ciliata.

Brevicipitidae. Microhylidae.

Brevoortia tyrannus. Menhaden or mossbunker; a fish of Atlantic coastal areas; travels in enormous schools; feeds on plankton; to 18 in. long; used for bait, fish meal, fertilizer, and fish oil.

Bridges, Calvin B. American geneticist (1889-1938); in collaboration with other investigators, did much to develop modern concepts of genetics through his work with Drosophila; worked on proof of the chromosome theory of heredity, theory of genic balance, and position of genes on giant salivary chromosomes.

Bright's disease. Any one of a group of kidney diseases characterized by albumin in the urine and edema.

Brillia. Genus of tendipedid midges.

brine fly. Any of certain flies, especially in the Family Ephydridae, which breeds in highly saline or alkaline waters.

brine shrimp. Artemia salina.

Brisinga. Genus of deep-sea sea stars; arms long, slender, and sharply set off from disc.

bristle. 1. Any short, stiff hair or hairlike projection, either macroscopic or microscopic. 2. In certain birds, such as the whip-poor-will, one of many hairlike feathers around the mouth; the slender shaft has a basal tuft of small barbs.

bristletail. See THYSANURA.

bristly porcupine. See CHAETOMYS.

bristly rabbit. See CAPROLAGUS.

brit. Common name applied by fishermen to dense swarms of the marine copepod Calanus finmarchicus.

brittle star. Common name of members of the echinoderm Class Ophiuroidea.

broadbill. 1. See NYROCA and SPATULA. 2. See EURYLAIMIDAE.

broad-shouldered water strider. See VELIIDAE.

broad tapeworm. Fish tapeworm of man; see DIBOTHRIOCEPHALUS LATUS.

brock. European badger, Meles meles.

brocket. 1. Second-year male European red deer. 2. Any of a group (Mazama) of small deer of Central and South America; horns spikelike.

bronchiole. One of the many fine branches of a bronchus; each, after considerable rebranching, eventually terminates in many alveoli of a lung.

bronchus. Large respiratory tube of a tetrapod, a main one supplying each lung; the two bronchi branch and rebranch considerably before becoming bronchioles; each bronchus originates at the lower end of the trachea; a bronchus wall contains cartilage and smooth muscle; the epithelium consists of mucous-secreting ciliated cells. Pl. bronchi.

Brontosaurus. Genus of giant Jurassic quadrupedal, herbivorous, semiaquatic dinosaurs; neck and tail long; total length to 75 ft.

brood capsule. In a hydatid cyst, a bit of infolded germinal epithelium which contains internal proliferated tapeworm scolices.

brood chamber. 1. Dorsal space between the body and carapace in Cladocera; it retains developing eggs and newly-hatched young. 2. Space between the ventral surface of the thorax and the oostegites of crustaceans in the Subdivision Peracarida; the eggs hatch and the young undergo early development in the chamber.

brood pouch. 1. Brood chamber. 2. In Bryozoa, an internal coelomic pouch which holds a larva until its release to the outside.

Brooks, William Keith. American teacher and zoologist (1848-1908); did much to clarify the biology and propagation of the oyster; founded Chesapeake Zoological Laboratory and wrote The Foundations of Zoology.

brook trout. See SALVELINUS.

Brosme brosme. Cusk; a large edible codlike fish of the Atlantic Coast; on the bottom in moderately deep water. See GADIDAE.

Brotulidae. Large family of elongate, compressed fishes; scales, when present, minute and embedded; chiefly marine, but habits variable; several spp. blind, fresh-water, and cavernicolous.

Brown, Robert. Scottish botanist and and explorer (1773-1858); authority on the flora of Australia and curator at British Museum; observed Brownian movement and discovered the cell nucleus in plants.

brown bat. See EPTESICUS.

brown body formation. Phenomenon occurring in a few spp. of fresh-water

Bryozoa; a polypide degenerates and forms a compact, dark-colored mass; a new polypide is then regenerated from the body wall and the dark-colored mass comes to lie in the new stomach and may or may not be voided through the anus.

Brownian movement. Random movement of small particles suspended in a fluid; the result of kinetic molecular energy; easily seen in colloids under the microscope.

brown mite. See BRYOBIA.

brown shark. Sandbar shark, Carcharhinus milberti; a widely distributed sp., to 8 ft. long.

brown-tail moth. See EUPROCTIS.

brown trout. See SALMO.

browse line. Height to which animals (usually ungulates) have removed browse from a stand of trees or shrubs.

Brucella. See BRUCELLOSIS.

brucellosis. 1. Undulant fever; generalized infection caused by a bacterium in the genus Brucella; usually obtained through milk from goats and cattle; marked by varying fever, malaise, sweating, constipation, weakness, and anemia. 2. Any infection caused by Brucella; common in stock and farm animals.

Bruchophagus funebris. Wasp in the Family Eurytomidae; larva a pest in clover seed.

Brunner's glands. Tubular glands extending into the submucosa in the mammalian duodenum; secrete mucus.

brush border. Brushlike appearance on the exposed surface of certain epithelial cells, especially gland and absorptive cells.

brush rabbit. Small brownish rabbit (Sylvilagus) of the Pacific Coast.

brush-tailed porcupine. See ATHERURA.

Brychius. Genus of crawling water beetles (Family Haliplidae); western states.

Bryobia praetiosa. Brown mite, clover mite; a small mite (Tetranychidae) injurious to clover and fruit trees.

Bryocamptus. Common genus of freshwater copepods; usually found in debris and damp moss.

Bryophrya. Genus of ovoid to ellipsoid holotrich Ciliata; fresh waters.

Bryophyllum. Genus of elongated holotrich Ciliata; usually in wet mosses.

Bryozoa. Ectoprocta; phylum which includes the "moss animalcules"; small tufted or branched marine and freshwater organisms a few mm. high; attached to substrates in shallows; single individuals often connected with each other by a common coelom, and the whole branched or massive colony is covered with a continuous zooecium secreted by the thin body wall and composed of calcium carbonate, chitin, or a chitinoid or gelatinous material; the distal end of each polypide projects from an opening in the zooecium; it consists mainly of a horse-shoe shaped retractile lophophore bearing numerous ciliated tentacles as well as the mouth; digestive tract U-shaped, anus just below the lophophore; gonads develop from peritoneum; young larva retained in a special brood pouch until its release to the outside; reproduction and growth also by asexual budding and colony proliferation; about 4000 living spp., nearly all marine.

bubal. See ALCELAPHUS.

Bubalus bubalis. Common buffalo or water buffalo of southern Asia; a semidomesticated oxlike draft animal with large curved horns and little hair; carabao.

bubble shells. Group of small tropical marine gastropods; shell thin, globose, oval or oblong-cylindrical, with sunken spire; usually less than 2 in. long; Hydatina and Bulla common.

bubo. Inflammation of a lymph gland, especially in the groin or armpit.

bubonic plague. Acute disease of man and rats caused by the bacterium Pasteurella pestis which is transmitted chiefly by the bite of the rat flea, Xenopsylla cheopis; infections occur in the lymph nodes, lungs, and general circulation; death rate very high; great epidemics occurred during the Middle Ages, but today it is chiefly confined to crowded populations of rats, ground squirrels, marmots, etc. and is seldom transmitted to man; local and temporary foci occur in most large seaports and in central and southern Asia.

Bubo virginianus. Great horned owl of North and South America; to 2 ft. long; with prominent ear tufts.

Bubulcus ibis. Cattle egret, a white bird with buff plumes, yellow bill, and black feet; originally in Spain, Africa, and tropical Asia, but now also in northern South America and eastern and southeastern U.S.

buccal cavity. Mouth cavity, just inside the mouth opening; term used especially with reference to certain invertebrate groups.

buccal cirri. Series of short tentacles at the anterior end of cephalochordates; surround opening to vestibule.

buccal mass. Thick, dense mass of tissue in the head of gastropod, amphineuran, scaphopod, and cephalopod mollusks; surrounds the anterior part of the digestive tract and contains the radula and much supporting musculature and connective tissue.

Buccinum. Genus of carnivorous gastropods of temperate and cold seas; whelks; shell thick, fusiform or pear-shaped, and with few whorls; one sp. used as human food in northern Europe.

Bucconidae. See PUFFBIRD.

buccopharyngeal respiration. Respiratory exchange occurring through the mucous membrane of the mouth cavity in amphibians.

Bucephala. Common genus of ducks (Family Anatidae) found in Europe and Asia; to some extent also in N.A.; B. clangula americana is the American goldeneye, primarily associated with fresh waters; an excellent diver; often called the whistler because of its swift whistling flight; B. albeola is the small bufflehead which winters over most of U.S.

Bucephalus. Genus of digenetic trematodes parasitic in the digestive canal of marine and fresh-water fishes; sporocysts are in bivalves (first intermediate host), and cercariae encyst on small fishes (second intermediate host).

Buceratodidae. See HORNBILL.

buck. Male deer, antelope, goat, rat, rabbit, or hare.

buckeye. See JUNONIA.

bud. 1. Asexually-produced protuberance which developes into a new individual. 2. To develop or put forth buds.

budding. Type of asexual reproduction involving the appearance of new animals as outgrowths from an older individual and their subsequent growth and maturation, the end result being a colony, although buds often detach and become independent; the process is common in sponges, coelenterates, bryozoans, tunicates, and some polychaetes.

budgerigar. See MELOPSITTACUS.

Budorcas. Takins; a genus of wild ox-like mammals of forested slopes of the Himalayas.

Buenoa. New World genus of back swimmers (Family Notonectidae).

buffalo. See BUBALUS, ANOA, SYNCERUS, and BISON.

buffalo fish. See ICTIOBIUS and CATOSTOMIDAE.

buffalo gnats. See SIMULIIDAE.

buffalo treehopper. See MEMBRACIDAE.

buff-breasted sandpiper. See TRYNGITES.

buffer. 1. Any substance in a solution which dampens or lessens the change in hydrogen-ion concentration when an acid or base is added to the solution; usually a protein or the salt of a weak acid and strong base. 2. Any substance which moderates or prevents a reaction that a therapeutic agent would produce if administered alone.

buffer species. Any sp. which is not the normal prey of a predatory sp. and which thereby decreases the predatory pressure exerted on the normal prey; e.g. cattle may be the buffer sp. between the mountain lion and deer; term mostly used by game management biologists.

buffer zone. Zone surrounding a specific ecological area and having such a composition that it protects that ecological area from encroachment and disturbance.

bufflehead. See BUCEPHALA.

Buffon, Georges. French zoologist (1707-1788); the first to discard the theory of special creation; devoted much of his life to a 44-volume work, Histoire naturelle, a popular natural history interspersed with his own ideas and theories.

Bufo. Common genus of toads; more than 100 spp., of which about 16 occur in N.A.; terrestrial and nocturnal, in burrows or under objects during daytime; enter water only to spawn; B. marinus, the giant toad of Central and South America, has a head and body length to 8.5 in.; B. terrestris generally in N.A. east of Rockies and south of Hudson Bay; B. cognatus (Great Plains toad); B. boreas western states to Alaska; B. punctatus (spotted toad) from Tex. to eastern Calif.; B. woodhousi (Woodhouse's toad) over most of U.S.

Bufonidae. Family of amphibians in the Order Anomocoela; includes many of the true toads; maxillary teeth usually absent; parotid gland conspicuous; world-wide distribution except for Madagascar and the Australian region; Bufo the only New World genus.

bug. In the strict sense, any member of the insect Order Hemiptera; sometimes also used to include the Order Homoptera; the layman often incorrectly uses "bug" to apply to almost any invertebrate.

Bugula. Very common genus of Bryozoa found in shallow seas; in tufted, mosslike colonies.

bulb mite. See RHIZOGLYPHUS.

bulbo-urethral glands. Cowper's glands; pair of small glands whose mucous secretion is released into the base of the urethra in male mammals at the time of sperm discharge.

bulbul. Any of a large family (Pycnonotidae) of southern Asiatic and African songbirds; usually somber colored, medium-sized spp.

Bulbus. Genus of North Atlantic Gastropoda.

bulbus arteriosus. 1. Indistinct tip of the embryonic vertebrate ventricle (left) just before the truncus arteriosus. 2. Expanded muscular base of the ventral aorta of some vertebrates, chiefly teleost fishes.

Bulimnaea. Discarded gastropod generic name now included in Lymnaea.

Bulimus. Genus of small fresh-water snails; not in U.S.

bulla. 1. Small projection of the mammalian skull which encloses the middle ear. 2. Blister-like structure. 3. Shieldlike sclerite closing a spiracle in certain insects. 4. Weak spot in a wing vein of an insect.

Bulla. Large and common genus of tropical marine Gastropoda; shell globose, oval, or oblong, and with a sunken spire; bubble shells.

Bullaria gouldiana. Bubble-shell snail; a very common mud flat snail of the Pacific Coast; brown, paper-thin shell; body large and yellow.

bullfinch. Any of several European songbirds in the finch family; often caged.

bullfrog. See RANA CATESBEIANA.

bullhead. See AMEIURUS.

Bullock's oriole. Common yellow, orange, and black oriole of western U. S. and Mexico; Icterus bullockii.

bullsnake. See PITUOPHIS.

bumblebee. Any of a wide variety of large bees, especially Bombus; see BOMBIDAE.

bundle of His. Atrioventricular bundle; muscular band containing nerve fibers and connecting the auricles with the ventricles of the heart; imbedded in wall between the auricles and spreads out through the ventricular walls; conveys stimuli and causes the heart to beat in a coordinated fashion.

Bungarus. Southeastern Asiatic genus of venomous snakes; kraits; to 4 ft. long; no "hood" present.

Bunodera. Genus of distome trematode parasites of fresh-water fishes.

Bunodes. Genus of Paleozoic arachnids, somewhat similar to Limulus.

bunodont dentition. Having rounded cusps on the molar teeth, as in man and swine; crushing molars.

Bunodontia. Suiformes.

Bunonema. Genus of nematodes which feed mostly on moss and other vegetation.

Bunostomum trigonocephalum. Nematode parasite of the sheep small intestine.

Bunsen-Roscoe law. For very short light flashes, the photochemical effect on the retina equals the product of intensity and duration.

bunting. Any of various brightly colored plump birds in the Family Fringillidae; common American spp. are the lazuli bunting (Passerina amaena), varied bunting (P. versicolor), painted bunting (P. ciris), indigo bunting (P. cyanea), and the snow bunting (Plectrophenax nivalis).

Buprestidae. Large family of medium to large metallic colored beetles; metallic wood borers; larvae and adults of many spp. are pests of forest trees; Chrysobothris femorata, a flat-headed apple tree borer, is an important American sp.

burbot. See GADIDAE.

Burhinidae. Family of birds in the Order Charadriiformes; thick-knees, stone curlews; gray-brown shorebirds not found in N. A.

Burmeisteria. Genus of South American armadillos.

burro. See EQUIDAE.

burrowing owl. See SPEOTYTO.

bursa. 1. Pouch, sac, or saclike cavity. 2. In higher animals, a cavity filled with viscid fluid in a place where friction between tissues would otherwise occur.

Bursa. Genus of bluntly spinous marine gastropods; frog shells; rocky coasts, Calif. to Mexico.

bursa copulatrix. 1. Specialized pouch or sac forming part of the female reproductive system of certain invertebrates; receives spermatozoa and stores them temporarily; e. g. present in many insects and Turbellaria. 2. Specialized part of the posterior end of the male in many invertebrates; used during copulation for holding the female body and genital pore in proper position; an eversible or permanent hemispherical structure in Acanthocephala and a few nematodes; a set of two posterior lateral flaps or finlike structures in other nematodes.

bursa of Fabricius. Cloacal pocket in birds.

Bursaria. Genus of large, fresh-water Ciliata; cell ovoid, with a truncate anterior end.

Bursaridium. Genus of fresh-water heterotrich Ciliata.

Bursella. Genus of oval, fresh-water holotrich Ciliata.

bursitis. Inflammation of the bursa of a joint.

burying beetle. Carrion beetle or sexton beetle; see SILPHIDAE.

bush coral. See ACANELLA.

bush baby. See GALAGO.

bushbuck. See STREPSICEROS.

bush dog. See SPEOTHOS.

Bushman. Race of negroids including the natives of the Kalahari Desert area; skin yellow, buttocks large in females; to 5 ft. tall.

bushmaster. See LACHESIS.

Bushnell's wasp. See ODYNERUS.

bush pig. See POTAMOCHOERUS.

bush squirrel. See PARAXERUS.

bushtit. See PSALTRIPARUS.

bushy-backed slug. Dendronotus frondosus; a brown to reddish nudibranch with seven pairs of branching, tree-like cerata; tidal zone to 360 ft.; Labrador to R.I.

Buskia. Genus of marine Bryozoa.

bustard. Any of several large game birds of the Eastern Hemisphere; related to cranes and plovers; see OTIDIDAE.

Busycon. Genus of large marine snails; the whelks or conchs; shell to 10 in. long, with a short spire and very large body whorl; egg capsules lens-shaped, attached in a row to a common cord.

butcher bird. See LANIIDAE and CRACTICIDAE.

Buteo. Nearly cosmopolitan genus of large soaring hawks with broad wings and broad, rounded tails; common American spp. are B. borealis (red-tailed hawk), B. swainsoni (Swainson's hawk), B. regalis (ferruginous rough-leg hawk), B. lineatus (red-shouldered hawk), and B. lagopus (rough-legged hawk).

Buteogallus. Small genus of Tropical American hawks; includes the crab hawk of southwestern U.S.

Buthus. Widely distributed genus of scorpions; not in U.S.

Butorides virescens. Green heron of eastern N.A. (Family Ardeidae); to 18 in. long; around bodies of fresh water, where it feeds on fish and invertebrates.

Bütschli, Otto. German zoologist (1848-1920); made important contributions to protozoology and to the anatomy and embryology of mollusks, insects, and nematodes; emphasized the alveolar nature of protoplasm.

butter clam. See SAXIDOMUS.

buttercup shell. See LORIPINUS.

butterfish. Any of several fishes covered with a heavy coat of slippery mucus, as well as certain other unrelated spp. See PORONOTUS.

butterfly. See RHOPALOCERA.

butterfly fish. Any of a wide variety of tropical fishes characterized by large fins and/or bright color patterns.

butterfly ray. A very broad sting ray with a short tail; body almost triangular; to 4 ft. wide; warm and temperate seas.

butterfly shell. See DONAX.

butterfly squid. Stoloteuthis leucoptera, a small uncommon squid with short, globose body.

button. Terminal conical portion of the rattle of a rattlesnake.

button quail. See TURNIX.

button-shell snail. Trimusculus reticulatus, a small West Coast pulmonate limpet.

butyn. Anesthetic for small aquatic metazoans; used in 0.1 to 2.0% solution.

buzzard. 1. General term applied to any of several heavy, slow-flying hawks, especially in the genus Buteo. 2. Any of several large carrion-feeding birds, such as Cathartes in the U.S.

BW. Biological warfare.

Byblis. Genus of marine Amphipoda.

byssus. Bundle of silken anchorage fibers produced by many marine bivalves; a single sensory byssus thread is found in some fresh-water glochidia.

byssus pit. Concavity on the ventral surface of the posterior part of the foot in certain gastropods, especially mussels; secretes fluid which flows anteriorly in a groove and hardens into a thread; new threads are formed as the mussel shifts about, the result being a cluster of byssus threads all originating in the pit.

C

caaing whale. North Atlantic blackfish (Globicephala).
Cabassou. Genus of very large armadillos of the Amazon drainage; 11 to 13 bands of scales; dig deep, extensive burrows.
cabbage butterfly. See PIERIS.
cabbage worm. Any insect larva that feeds on cabbage.
Caberea. Genus of marine Bryozoa.
cabezon. See SCORPAENICHTHYS.
Caborius. Genus of Trichoptera.
cabrilla. Any of several edible fishes in the Family Serranidae.
Cacajao. Genus of northern South American monkeys; uacaris; hair long, scraggly, and red; live in tree tops.
cachalot. See SPERM WHALE.
cachexia. Emaciation in adults resulting from insufficient somatotrophic hormone or some serious debilitating disease.
cacogenics. Racial deterioration resulting from repeated mating of inferior or defective stocks.
cacomistle. See BASSARISCUS.
cacomitl cat. See JAGUARUNDI.
Cactosoma. Genus of sea anemones; frequently among kelp holdfasts.
cactus moth. South American moth (Cactoblastis cactorum) whose larva feeds chiefly on prickly pear cactus; introduced into Australia where it is extremely important in cactus control.
cactus wren. See CAMPYLORHYNCHUS.
cadaver. Human corpse after death, especially one intended or being used for dissection and study.
caddis fly. See TRICHOPTERA.
caddis worm. Larva of a caddis fly; see TRICHOPTERA.
Caddo. Genus of daddy longlegs in the Order Phalangida.
Cadlina. Genus of nudibranchs.
cadophore. Elongated dorsal process on certain pelagic tunicates; site of asexual bud formation.
caducous. Falling off easily or early, e.g. insect hairs.
Cadulus. Genus of Scaphopoda; shell cucumber-shaped.
caecilian. See GYMNOPHIONA and CAECILIIDAE.
Caeciliidae. Family of stegocephalian Amphibia; caecilians; limbless spp. of damp tropics; Ichthyophis, Gymnopis, and Typhlonectes representative.
caecum. Any blind, hollow outpouching of a digestive or other organ; it may be small and insignificant or very large; in man it is cup-shaped and occurs just below the junction of the small and large intestines.
Caecum pulchellum. Tube shell, blind shell; a curious minute marine gastropod of the Atlantic Coast; shell short, cylindrical, and almost straight in the adult.
Caenagnithiformes. Order of fossil birds.
Caenestheriella. Common genus of clam shrimps.
Caenis. Common genus of Ephemeroptera.
caenogenesis. Cenogenesis.
Caenolestes. See CAENOLESTIDAE.
Caenolestidae. Small family of uncommon, rat-sized marsupials restricted to South America; two spp. of Caenolestes in Andes.
Caenomorpha. Genus of bell-shaped heterotrich Ciliata; sapropelic in fresh and salt waters.
Caesalpinus, Andreas. Italian botanist and physiologist (1519-1603); showed that the blood in the veins flows toward the heart.
Caesar. See GRUNT.
caffeine. Alkaloid occurring in tea, coffee, and maté; diuretic and nerve and heart stimulant; $C_8H_{10}N_4O_2$.
Caiman. Genus of reptiles in the Family Alligatoridae; caimans; about five spp. in tropical Central and South American rivers; general structure and habits similar to those of crocodiles and alligators; one sp. attains 20 ft.
caique. Any of numerous small, stout, aggressive parrots with unusual color

patterns; tropical America.

Cairina moschata. Common domestic muscovy duck, supposedly derived from wild stock in Central or South America.

caisson disease. The bends; a disease produced in men working at high atmospheric pressures in tunnel construction and in deep-sea diving; upon returning to normal pressures too rapidly, the excess nitrogen dissolved in the blood is released as fine bubbles in the blood and tissues; characterized by dizziness, joint pains, difficult breathing, and paralysis; death sometimes follows; patients are placed in a compression chamber and subjected to slow decompression.

cake urchins. Spp. of echinoderms in the Class Echinoidea; test depressed but not discoidal, often concave orally; usually on sandy bottoms.

calabar swellings. Swellings the shape of half an egg and occurring on various parts of the surface of the body in Loa loa infections.

calamistrum. One or two rows of curved spines on the metatarsus of some spiders.

Calamoichthys. Genus of eel-like African fresh-water fishes in the Order Cladistia.

Calamospiza melanocorys. Lark bunting (Family Fringillidae); a small blackbird-like sp. with large white wing patches in the male; female brownish with striped breast; Great Plains.

calamus. Quill of a feather.

Calanoida. Suborder of free-living marine and fresh-water copepods; movable articulation between sixth and seventh thoracic segments; first antennae long, 22 to 25 segments, right or left modified as a grasping device in the male; metasome much wider than urosome.

Calanus finmarchicus. Brit; limnetic copepod which is exceptionally abundant in northern seas; to 4 mm. long; basic food organism for herring, mackerel, some whales, etc.

Calappa. Genus of sluggish box crabs.

Calathura. Genus of marine Isopoda.

Calathus. Genus of herbivorous beetles in the Family Carabidae.

calcaneum. Calcaneus.

calcaneus. Large, irregular bone forming the back of the heel.

calcar. 1. Spurlike outgrowth. 2. Small elevation on the floor of the lateral ventricles of the brain. 3. Small, spurlike, accessory digit bone on the median edge of the posterior limb at the base of the first metatarsal; e.g. in the frog.

Calcarea. One of the three classes of Porifera; skeleton composed of separate calcium carbonate spicules which are one-, three-, or four-rayed; examples are Leucosolenia and Sycon.

calcareous. Containing or impregnated with calcium carbonate.

Calcarius. Holarctic genus in the Family Fringillidae; longspurs; small, sparrow-like spp. of the tundra and northern plains; three spp. winter in U.S.

calciferol. Crystalline vitamin D_2 produced by ultraviolet radiation of ergosterol.

calciferous glands. Three pairs of small glands associated with the anterior part of the digestive tract in the earthworm; presumably their secretion neutralizes acids in the digestive cavity.

Calcispongiae. Calcarea.

calculus. Stonelike deposit formed in a hollow organ or duct, e.g. in the gall bladder or salivary duct. Pl. calculi.

Caleppa flammea. Box crab; a large crab (Brachyura) of tropical and temperate seas.

calico bass. Strawberry bass or black crappie; see POMOXIS.

calico clam. See CALLISTA.

calico crab. See OVALIPES.

calico shell. See CALLISTA.

Calidris canutus. Knot; chunky brownish or grayish shore bird found mostly along East and West coasts.

California herring. See SARDINOPS.

California mussel. Mytilus californicus.

Californian subregion. Zoogeographic subdivision of the Nearctic region; includes most of Calif., Ore., Wash., and extreme southwestern Canada.

Caligoida. Suborder of marine copepods; adults parasitic on marine and fresh-water fishes and a few marine mammals and invertebrates; body greatly depressed; metasome and urosome distinct but most segments obscure.

Caligus. Common genus of marine copepods parasitic on fishes; highly modified.

Caliroa limacina. Hymenopteran in the Family Tenthredinidae; pear slug, cherry slug; larvae are pests on these fruit trees.

Callaeidae. Small family of New Zealand passerine birds; wattle birds; three crowlike spp. with fleshy wattles at the corners of the jaws.

Callianassa. Common genus of elongated marine burrowing decapod crustaceans (Suborder Anomura).

Callibaetis. Common genus of insects in the Order Ephemeroptera.

Callicorixa. Genus of water boatmen (Family Corixidae); mostly in cold waters.

Callilepis imbecilla. Common orange-brown spider (Gnaphosidae).

Callinectes. Genus of marine decapod Crustacea; includes the well-known blue or edible crab of the East Coast; carapace to 7 in. wide; common on mud bottoms in shallow water or swimming about among seaweed or near the surface; ranks with lobster and shrimp as a human food.

Calliobothrium. Genus of tetraphyllidean tapeworms found especially in the spiral valve of sharks.

Calliopius. Genus of marine Amphipoda.

Calliostoma. See TOP SHELL.

Callipallene. Genus of Pycnogonida.

Callipepla squamata. Scaled quail, a small bird of southwestern U.S. and the Mexican plateau; coloration mostly gray and brown.

Calliphora. Genus of flies in the Family Calliphoridae; bluebottle flies; larval food habits variable; some spp. feed on flesh of domestic animals, especially open wounds.

Calliphoridae. Family of medium to large flies; blowflies, greenbottle flies, bluebottle flies; variously colored, often with a metallic iridescence; many pest spp.; Phormia, Cochliomyia, Lucilia, and Calliphora are important genera.

Callipus. Genus of small millipedes.

Callisaurus. Genus of lizards of arid southwestern U.S. and Mexico; gridiron-tailed lizard, zebra-tailed lizard; trunk length to 7 in.

Callispongia. See VASE SPONGE.

Callista. Genus of marine clams of south Atlantic and Gulf coasts; includes the calico or checkerboard clams which have a brownish plaid pattern; C. maculata is the calico shell of the Atlantic Coast.

Callithricidae. Family of anthropoid mammals which includes the marmosets; tropical New World spp. about as large as squirrels; head often with lateral tufts of hair; tail not prehensile; Callithrix typical.

Callithrix. See CALLITHRICIDAE.

Callitroga. Important genus of flies in the Family Calliphoridae; larvae (screwworms) of several pest spp. penetrate wounds, nose, and navel of domestic animals (occasionally man).

Callocardia. Genus of whitish marine clams; North Atlantic.

Callorhinus alascanus. Common Alaskan fur seal; congregate in large rookeries on shores of Pribilof Islands during summer; total population about 4,000,000; each mature male controls a harem of 5 to 60 females, thus leaving many surplus and bachelor bulls; each pregnant female gives birth to a single pup; mating occurs shortly thereafter, and the herds take to the open sea for the ensuing nine months; males to 500 lbs.; sealskins of commerce mostly derived from the annual harvest of bachelor bulls.

Callorhynchus. Genus of southern temperate marine fishes; chimaera-like, with a pendant tactile process on the elongated snout.

Callosamia promethea. Large moth of eastern half of U.S.; promethea moth; female brownish, male dark maroon, and both with light-colored markings.

Callosciurus. Genus of Oriental tree-squirrels; richly colored spp. which make no nests; southeastern Asia.

callosity. Callus.

callous. Hardened, or having a callus.

callow. 1. Worker ant which is newly-emerged from the pupa. 2. Teneral; newly-emerged adult insect which is soft and not yet fully colored.

callus. 1. Callosity; any hard, thickened, epidermal area; usually the result of friction or other irritation. 2. Exudate formed around the fractured area of a broken bone; involved in tissue repair and eventually replaced by bone tissue. 3. Thickening which more or less fills the umbilicus in many snail shells.

Calocalanus. Genus of pelagic marine copepods (Suborder Calanoida).

Calocaris. Common genus of burrowing decapod crustaceans (Suborder Anomura).

Calonympha. Genus of symbiotic flagellates living in termite guts; cell spherical, with many nuclei and many flagella.

Calonyx. Small genus of cold-water Hydracarina.

Calopsectra. Genus of tendipedid midges.

calorie. Amount of heat necessary to raise the temperature of one gram of water from 15° to 16° C.; this unit is so small that it is seldom used in making metabolic determinations; instead the kilocalorie (Calorie) is usually used.

Calorie. Kilocalorie; amount of heat necessary to raise the temperature of one kilogram of water from 15° to 16° C.; used as a measure of metabolism in animals; the average daily energy output of an active man (expressed as heat, work, or energy content of excretions) is equivalent to 3000 Calories.

calorimeter. Any instrument used to measure heat exchange in a system; in physiology, it is usually used to measure heat produced by a living animal or by a bit of food which is ignited.

Calosoma. Common genus of ground beetles; caterpillar hunters; some spp. with bright metallic colors.

Calotermes. Kalotermes.

Calothorax. Small genus of hummingbirds (Family Trochilidae); includes C. lucifer, the Lucifer hummingbird in mountains of southwestern U.S. and Mexican plateau.

calotte. Anterior cap of four cells in Dicyemidae (Phylum Mesozoa).

calvarium. Dome of the skull; calvaria.

Calycella. Genus of marine coelenterates in the Suborder Leptomedusae.

Calycophora. Suborder of the Siphonophora; polyps arranged on a long stemlike growth; one or more swimming bells present, but air sac absent.

Calycopsis. Genus of marine coelenterates in the Suborder Anthomedusae.

calyculus. Small bud- or cup-shaped structure.

calymma. Outer protoplasmic layer of certain radiolarians; consists largely of mucilage-filled alveoli.

Calypte. Genus of hummingbirds; Anna's and Costa's hummingbirds; southwestern states.

calypters. Tegulae, squamulae, squames, alulae; a pair of membranous lobes at the posterior angle of each wing base in certain Diptera.

Calyptoblastea. Same as Leptomedusae, but often used with reference to the hydroid stage in the life history.

Calyptomera. Suborder of Cladocera in which the body is covered with a bivalve shell and the legs are foliaceous; includes almost all Cladocera.

calyptopis. Free-swimming larval stage of a euphausid.

Calyptraea. Genus of limpet-like marine snails; shell subconical, apex inclined to one side.

Calyptratae. Group of families of Diptera in which there is a longitudinal seam on the second antennal segment.

calyx. 1. Visceral mass of an entoproct, as distinct from the slender supporting stalk. 2. Cup-shaped central portion of the body of crinoid echinoderms. 3. Cuplike part of the kidney pelvis. 4. Cuplike cavity in many animal structures.

Camallanina. Suborder of spirurid nematodes; immature stages in copepods; mature stages in vertebrates.

Camallanus. Genus of spirurid nematodes parasitic in the digestive tract of cold-blooded vertebrates.

Camarodonta. Large order of Echinoidea with rigid test, external gills, and keeled teeth; includes most littoral spp.; Arbacia and Strongylocentrotus common.

Cambala. Common genus of millipedes.

Cambarellus. Genus of small, fresh-water crayfishes, to 30 mm. long; southern U.S.

Cambarincola. Genus of small, leechlike oligochaetes commensal or parasitic on crayfishes; in the Family Branchiobdellidae.

Cambarus. Genus of fresh-water crayfishes; ponds, streams, and lakes in eastern and central U.S.; body morphology closely resembles American lobster (Homarus); body usually 2 to 6 in. long; C. bartoni mostly in springs, marshy areas, and small streams; C. diogenes burrows in marshy meadows; C. virilis especially common, more properly Orconectes virilis.

Cambrian period. Earliest of the six geological subdivisions of the Paleozoic era; lasted from about 550 million to 480 million years ago; age of lowlands and mild climates; fauna characterized by abundance of lower invertebrates; trilobites dominant.

camel. See CAMELUS.

camel cricket. See STENOPELMATIDAE.

Camelidae. See TYLOPODA.

camelopard. Giraffe.

Camelus. Genus of herbivorous domesticated mammals in the Order Tylopoda; stomach three-chambered; includes the Arabian or dromedary camel (C. dromedarius) and the Asiatic or Bactrian camel (C. bactrianus); the former is one-humped, swifter, and used chiefly in Arabia and North Africa; the latter is two-humped and occurs chiefly in Central Asia; humps are mostly stored fat; flat cloven hoofs and the ability to go without water for several days are desert adaptations; in addition to being used as a beast of burden, the camel is valued for its hair, flesh, and milk; paleontologically, camels originated in N.A.

cameo shell. Any of several tropical helmet shells used for cameo manufacture, e.g. Cassis.

camera lucida. Simple optical device which is attached to a microscope ocular and which allows the object to be viewed through the microscope at the same time that a virtual image is projected upon a plane surface beside the microscope; used in drawing an accurate image of the object being studied.

Camerina. Nummulites.

camerostome. Opening through which the beak is protruded in certain Acarina.

Camisia. Genus of mites (Oribatidae) found on trees, rocks, or moss.

Camnula pellucida. Clear-winged grasshopper; occasionally migratory.

camouflage. Coloration and color patterns of animals which tend to blend with the background and render the animals less visible to their predators or prey.

Campanella. Genus of fresh-water peritrich ciliates; inverted bell-shaped; colonial, with long dichotomously-branched non-contractile stalks.

Campanularia. Common and widely distributed genus of marine Hydrozoa; colony branched or simple.

Campanulina. Genus of marine hydrozoan Coelenterata.

Campascus. Genus of fresh-water amoeboid protozoans; most of cell enclosed by a retort-shaped test.

Campeche wing. Pholas campechiensis, a white wing shell to 3 in. long; N.C. to Central America.

Campeloma. Common genus of fresh-water snails of the eastern half of U.S.

Campephagidae. Large family of passerine birds; minivets are small, slender, tree-top birds of eastern and southeastern Asia.

Campephilus principalis. Ivory-billed woodpecker, a glossy blue-black sp. with white markings, now probably extinct; formerly in bottom land forests of southeastern states.

Campodea. Genus of insects in the Order Aptera; usually in ground debris of meadows and woods.

campodeiform larva. 1. Type of caddis larva with compressed abdomen, no lateral line, and case not portable, or case absent. 2. Type of active insect larva with well developed legs.

Campodes. Common genus of small millipedes.

Camponotus. Genus of large black carpenter ants; nest in dead wood or earth.

Campostoma anomalum. Stone roller, a small stream fish of the U.S.

Campsurus. Uncommon but generally distributed genus of Ephemeroptera.

Camptocercus. Common genus of Cladocera; in aquatic vegetation.

Camptocladius. Genus of small non-swarming midges in the Family Tendipedidae.

Camptonema. Genus of Radiolaria; inner ends of axopodia attached to nuclear membranes.

Camptorhynchus labradorius. Labrador duck, a sp. of the Atlantic seaboard; white, black, and brown coloration.

Camptosaurus. Genus of Cretaceous dinosaurs in the Order Ornithischia; camptosaurs; large, heavy-bodied, bipedal herbivores.

Campylorhynchus brunneicapillus. Cactus wren; a large, heavily-spotted wren of the cactus country.

Canachites canadensis. Spruce grouse, fool hen; a gray, black, and brown grouse of Canadian and northern U.S. spruce forests.

Canada balsam. Oleoresin of American balsam fir; when purified and dissolved in xylene, it is used to mount tissue sections between slide and coverslip; after drying, it is clear and has a favorable refractive index; rapidly being replaced by synthetic resin.

Canada goose. See BRANTA.

Canada jay. See PERISOREUS.

Canadian subregion. Zoogeographic subdivision of the Nearctic region; includes most of Canada, all of Alaska and Greenland, and small portions of the northern U.S.

Canadian zone. One of Merriam's life zones, located between the Hudsonian and Transition zones; an irregular strip along the Canadian border in eastern U.S., the higher elevations of the Appalachians, and large areas in the Rocky Mountain west; characterized by evergreen forest.

canaliculate. Having small channels, grooves, or canals.

canaliculus. See HAVERSIAN SYSTEM. Pl. canaliculi.

canal of Volkmann. Canal in bone which carries blood vessels from the surface or marrow to the Haversian systems.

canary. See SERINUS.

Cancellaria. Genus of marine gastropods; shell robust and cross-ribbed; warm waters.

cancellous. Having a reticular, spongy, or porous structure, especially as applied to bone.

cancer. Cellular malignant tumor which is eventually fatal if not properly removed or treated.

Cancer. Genus of marine decapod Crustacea; includes the rock crabs; carapace flattened, more or less elliptical; many large and edible spp.

Cancridae. Common family of typical marine crabs; all legs adapted for walking; carapace broadly oval.

cancroid. Crablike.

Candelabrum. Genus of marine hydrozoan coelenterates.

candlefish. Smeltlike North Pacific coastal marine fish (Thaleichthys pacificus) which is so oily that it can be supplied with a wick and burned as a candle when dried; often called eulachon.

Candona. Common genus of fresh-water ostracods.

Canidae. Family of carnivorous mammals which includes the dogs, wolves, coyotes, and foxes; feet digitigrade, with four or five digits; claws nonretractile; examples are Canis, Urocyon, and Vulpes.

canine. 1. Pertaining to one of the four pointed teeth in the jaws of many mammals; one such tooth is located on each side of each jaw just behind an incisor; variously modified, especially large in the wild boar. 2. Eye tooth. 3. Pertaining to dogs or the Family Canidae.

caninus. Small, deep, facial muscle which originates on the maxilla and inserts at the angle of the mouth; raises corner of mouth.

Canis. Genus in the Family Canidae; includes all varieties of the domestic dog (C. familiaris), wolves, jackals, and coyotes; N.A. has about nine spp. of wolves, the most familiar being the gray or timber wolf (C. nubilus); there are about seven spp. and additional varieties of American coyotes, the most common and largest being the northern coyote, prairie coyote, or brush wolf (C. latrans).

cankerworm. See GEOMETRIDAE.

Cannacria gravida. Dragonfly of southern U.S.

cannibalism. Eating of any sp. by its own kind.

cannon bone. In hoofed mammals, the shank bone.

cannula. Small tube for insertion into a vessel or cavity; used for drainage or the addition of fluid to tissues.

canoe shell. Any of several small marine gastropods with an egg-shaped shell having the narrow aperture widely open below; e.g. Scaphander.

Cantharidae. Family of predaceous beetles often found around showy flower clusters; soldier beetles.

cantharidin. Crystalline solid extracted from certain beetles in the Family Meloidae; used as an aphrodisiac, vesicant, or diuretic; Spanish fly, Lytta vesicatoria, is an important source.

Canthocamptus. Fresh-water genus of harpacticoid copepods; in wet moss and on the bottom of ponds and littoral areas.

canthus. Angle where the lids meet on each side of an eyeball.

Canthydrus. Suphisellus.

canula. Cannula.

canvasback. See NYROCA.

canyon wren. See CATHERPES.

capelin. See MALLOTUS VILLOSUS.

Capella gallinago. Wilson's snipe (Family Scolopacidae) of N.A. and northern South America; common striped brown game bird of wet meadows, bogs, and streamside; bill long, used for probing.

capercaillie. Large European grouse (Tetrao urogallus); capercailzie.

Capillaria. Genus of enoplid nematodes parasitic in birds, mammals, salamanders, and fishes; C. columbae common in chicken small intestine.

capillary. Fine blood or lymph vessel; the wall is usually only a single cell in thickness; collectively form a network between arterioles and venules in vertebrates and some invertebrates; most respiratory, nutrient, and excretory exchanges occur through the capillary walls.

capillitium. Mass of anastomosing bands or fibrils within a slime mold sporangium. Pl. capillitia.

Capitella. Common genus in the polychaete Family Capitellidae.

Capitellidae. Family of marine Polychaeta; head without tentacles or palps but with a pair of retractile, ciliated, tentacle-like processes; parapodia rudimentary.

capitellum. Capitulum (def. 1).

Capito. Genus of Central and South American birds in the Family Capitonidae; barbets.

Capitonidae. See BARBET.

Capitosaurus. Genus of extinct stegocephalian Amphibia.

capitulum. 1. Small knoblike bony protuberance. 2. That part of a barnacle which is enclosed in the mantle. 3. Swollen end of a hair, antenna, or proboscis. 4. Anterior headlike portion of a tick; bears the hypostome, chelicerae, and pedipalps. 5. In sea anemones, the distal thin-walled portion of the column.

Capnia. Common genus of small stoneflies.

capon. Castrated cock.

Capra. Genus of ruminants (Family Bovidae) which includes the many varieties of domesticated goats, markhor, tur, and ibex.

Caprella. See CAPRELLID.

caprellid. Any of certain marine Amphipoda; skeleton shrimps; body long and thin, abdomen reduced to a terminal button, and some of the head and thoracic appendages are lacking; clamber about on hydroids, algae, etc.; Caprella common.

Capreolus capreolus. Small roe deer of northern Europe and Asia.

Capricornis. Genus of goat antelopes or serows (Family Bovidae) of eastern Asia; dark-colored, heavy mammals, sometimes with manes.

Caprimulgidae. Family of birds in the Order Caprimulgiformes; nightjars, nighthawks; nocturnal spp.; stippled gray and brown coloration.

Caprimulgiformes. World-wide order of birds which includes nighthawks, goatsuckers, frogmouths, potoos, and whip-poor-wills; bill small and weak; mouth very wide, used for capturing dusk- and night-flying insects on the wing; legs and feet small, weak; plumage soft; build no nests; Antrostomus, Phalaenoptilus, and Chordeiles typical.

Caprimulgus. Antrostomus.

caproic acid. A fatty acid.

Caprolagus. Bristly rabbits, a genus of burrowing Indian rabbits; small ears and minute eyes.

Capromyidae. Family of rodents which includes the hutias and the nutria or coypu (Myocastor), a large South American rodent, introduced into southern and western U.S. marshlands; hind feet partially webbed; tail long and naked; fur of some economic importance; feral populations established in Wash., Ore., Calif., N.M., and La.; see CAPROMYS and GEOCAPROMYS.

Capromys. Genus of arboreal ratlike long-tailed hutias; includes the hutiacougas and hutiacarabalis; West Indies.

caprylic acid. Common fatty acid.

capsid bugs. Common name of hemipteran insects in the Family Miridae.

capsular filtrate. Collectively, all the material which filters through a glomerulus and passes into a Bowman's capsule in the kidney.

capsular ligament. Ligament which surrounds and encloses the joint area between two long bones.

capsule. Investing sheath of connective tissue, such as the capsule of the kidney and the capsule of a joint.

captacula. Filamentous processes with sucker-like tips; arise from the dorsal surface of the head of some scaphopods; sensory and food-getting. Sing. captaculum.

capuchin. See CEBUS.

capybara. Sp. of rodent in the Family Hydrochoeridae; generally similar to a giant guinea pig; amphibious in rivers of northern and central South America; largest of all rodents, weighing to 100 lbs.; feet partially webbed, hair brownish.

carabao. See BUBALUS.

Carabidae. Very large family of long-legged, terrestrial, nocturnal beetles; ground beetles; most larvae and adults carnivorous and feed on pest insects, although a few spp. are herbivorous; Harpalus and Calosoma common.

caracal. Reddish-brown catlike animal of Asia and Africa; total length 3 ft.

caracara. Any of several large South American vulture-like hawks; one sp. in southern U.S.

Caranx hippos. Common jack or crevalle; a greenish to golden fish with deeply forked tail (Family Carangidae); famous game sp., to 25 lbs.; common in warm waters of East and West coasts.

carapace. 1. Dorsal and lateral shield-like plate covering the cephalothorax of decapods and certain other crustaceans. 2. Dorsal part of the shell of turtles; consists of exoskeletal plates to which the ribs and vertebral column are usually fused. 3. Rigid cuticle of many rotifers.

Carapus. Genus of small fishes living as commensals in the cloaca of certain sea cucumbers.

Carassius auratus. Common goldfish; native of Asia and domesticated for thousands of years; many unusual breeds; to 12 in. long; widely distributed and common in warm waters of eastern U. S. and in warm and temperate areas elsewhere.

carbamino-hemoglobin. Carbhemoglobin.

carbhemoglobin. Carbamino-hemoglobin; loose chemical combination of hemoglobin and carbon dioxide; a normal but minor method of carbon dioxide transport by the blood.

carbohydrase. Hydrolytic enzyme catalyzing the splitting of carbonydrates.

carbohydrate. Compound of carbon, hydrogen, and oxygen, with the latter two in a ratio of two to one, as in water; common examples are glucose - $C_6H_{12}O_6$, starches, cellulose, and glycogen - $(C_6H_{10}O_5)_n$, and lactose - $C_{12}H_{22}O_{11}$.

carbon cycle. World wide circulation and reutilization of carbon atoms, chiefly due to metabolic processes; the air and oceans are the main reservoirs for carbon, in the form of carbon dioxide in the former, and carbon dioxide, carbonates, and bicarbonates in the latter; this inorganic carbon is utilized by plants to form organic carbon compounds through photosynthetic and other anabolic processes; such carbon atoms may then be assimilated into the bodies of one or more successive animals; excretion, respiration, burning, and bacterial action on dead plants and animals return the carbon atoms to an inorganic state.

carbon dating. Determination of the age of a biological material by establishing the relative proportion of carbon with the normal atomic weight of 12 to that having an atomic weight of 14 (radioactive isotope); based on the fact that the atmosphere contains a constant small percentage of carbon 14 which has a half-life decay rate of 5568 years even when incorporated into organisms; practical analytical limit about 40, 000 years.

carbonic acid. Unstable acid formed by the union of carbon dioxide and water; H_2CO_3.

carbonic anhydrase. Widely distributed respiratory enzyme which speeds the conversion of carbon dioxide into carbonic acid in the tissue capillaries and the conversion of carbonic acid back into carbon dioxide in the lung capillaries; also concerned with secretion of hydrochloric acid by gastric mucosa.

Carboniferous period. One of the six geological subdivisions of the Paleozoic era; between the Permian and Devonian periods; lasted from about 250 million to 215 million years ago; age of lowlands and swamps, mild climates, rich vegetation, seed ferns, and lycopods; fauna characterized by dominant amphibians and insects, first reptiles, and a decline of brachiopods; subdivided into Pennsylvanian and Mississippian epochs.

carbon monoxide poisoning. Type of suffocation produced by the irreversible combination of hemoglobin and carbon monoxide (carboxyhemoglobin).

carboxyhemoglobin. See CARBON MONOXIDE POISONING.

carboxylase. Enzyme which splits the carboxyl group from pyruvic acid and other organic acids.

carboxypeptidase. Digestive enzyme in pancreatic juice; converts polypeptides to dipeptides.

carbuncle. Localized subcutaneous inflammation larger than a boil; usually perforates and discharges pus through more than one opening.

carcajou. See GULO.

Carcharias. Common and widely distributed genus of sharks; gray sharks, dogfish sharks, sand sharks.

Carcharinus. See BROWN SHARK.

Carcharodon carcharias. White shark; a large, viviparous, tropical sp. reaching 36 ft. ; occasionally attacks man.

Carchesium. Genus of colonial ciliate Protozoa of fresh and salt water; body bell-shaped, with a long basal stalk, each of which contracts independently of other individuals in the colony.

Carcinides maenas. Green crab, shore crab; marine decapod of Europe and U. S. ; along rocky shores.

carcinogen. Any substance capable of producing cancer as the result of prolonged contact or repeated applications to the tissues; e. g. certain hydrocarbons.

carcinology. Study of Crustacea.

carcinoma. Malignant cancerous growth originating in epithelial cells but proliferating and becoming deeper; many varieties recognized.

Carcinonemertes. Genus of nemertines in the Order Hoplonemertea; commensal on gills and egg masses of certain marine crabs.

Carcinoscorpius. See XIPHOSURA.

Carcinus. Genus of marine decapod crabs; the genus name Carcinides is preferred.

cardate mastax. Rotifer mastax specialized for suction and oscillatory grinding.

cardiac. Pertaining to the heart.

cardiac center. Nerve center in the medulla oblongata; stimulated by carbon dioxide in the blood and sends impulses to the heart by way of the autonomic nerves; such impulses increase the heart rate; another type of cardiac center has an inhibitory function.

cardiac chamber. Swollen anterior chamber of the stomach in decapod crustaceans and most other Malacostraca. Cf. PYLORIC CHAMBER.

cardiac gland. Type of gastric gland in the stomach wall; secretes mucin.

cardiac muscle. Type of involuntary muscle tissue peculiar to the vertebrate heart; consists of long, cylindrical, branching and interconnecting fibers containing delicate cross striations; nuclei are scattered in the interior of the fibers instead of being arranged at the periphery as in voluntary muscle; electron micrographs show that the cells are not syncitial, as was formerly believed.

cardiac sphincter. Sphincter muscle at the junction of esophagus and stomach.

cardiac stomach. Large, loosely folded stomach just inside the mouth of a typical sea star; capable of being extruded during feeding.

cardinal. See RICHMONDENA.

cardinal teeth. Groups of ridges, protuberances, and grooves on the inner surface of the two valves near the anterior end of the hinge in many Pelecypoda; hold the valves in juxtaposition.

cardinal vein. One of two short veins returning blood from most of body to a duct of Cuvier on each side and thence to the heart in fishes and tetrapod embryos; each cardinal vein is actually divided so that there are two anterior and two posterior cardinal veins.

cardioaccelerator center. One of a pair of medullary centers which accelerate heart action through the sympathetic system.

cardioblast. Embryonic cell destined to become a part of the heart.

Cardiocladius. Genus of tendipedid midges; larvae in running waters.

cardiogram. Any tracing or written record indicated by a cardiograph.

cardiograph. Any instrument used for indicating the nature and intensity of heart movements.

cardioinhibitor center. One of a pair of medullary centers which inhibit heart beats through the vagus nerve.

cardiotoxin. Any substance producing low blood pressure and heart failure; a component of certain snake venoms.

Cardisoma. Genus of terrestrial crabs; fields and woods of West Indies.

Cardita. Genus of common cockle-like bivalves; especially on the West Coast.

Cardium. Large genus of Pelecypoda; includes many cockles; valves of equal size and heart-shaped when the two are viewed from either end; valves with saw-toothed margins; to 6 in. in diameter; some spp. edible.

cardo. 1. Small basal piece of a typical insect maxilla. 2. Basal ring in the genitalia of Hymenoptera. 3. Proximal segment in the protopodite of certain Crustacea appendages. 4. Bivalve shell hinge. Pl. cardines.

Caretta. Genus of loggerhead turtles in the Family Chelonidae; to 500 lbs.; Atlantic Coast.

Cariamidae. Family (two spp.) in the Order Gruiformes; the seriemas; grayish brown running birds of South America.

caribe. Piranha; voracious South American fish; often attacks man and animals which enter the water.

caribou. See RANGIFER.

Caridea. Suborder of crustaceans in the Section Natantia; many spp. of shrimps and prawns in which the third pair of legs is not chelate; pleura of second abdominal segment overlapping those of the first; Crangon, Spirontocaris, and Palaemonetes typical.

Caridion. Genus of small, deep-water, marine shrimps.

caridoid facies. Morphological type which includes many malacostracan crustaceans; prawnlike and adapted for swimming and a pelagic habitat.

caries. Molecular decay or death of bony materials; applied especially to tooth decay.

carina. 1. Long, median, posterior

calcium carbonate plate on the surface of the mantle of many barnacles. 2. platelike projection on the sternum of birds and bats; wing muscles are attached on each side. 3. Elevated ridge or keel. Pl. carinae.

Carinaria. Genus of translucent, pelagic, free-swimming, marine gastropods; shell diminutive and transparent; foot modified into a ventral fin and a posterior tail-like portion which functions in locomotion.

carinate. Keeled; bearing a carina.

Carinatae. Discarded taxonomic category which included those birds with well-developed wings and a keel or carina on the sternum for the attachment of wing muscles. Cf. RATITAE.

Carinella. Tubulanus.

Carinifex. Genus of western fresh-water snails.

Carinina. Genus of marine littoral Nemertea.

Carinogammarus. Genus of marine Amphipoda.

Carinoma. Genus of marine nemertine worms; body thick, cylindrical anteriorly, and flattened posteriorly.

Carmarina. Geryonia.

Carmia. Genus of red marine sponges.

carmine. Red biological dye.

carnassial teeth. Molar or premolar teeth specialized for shearing in a bladelike manner, as in the dog and cat.

Carnivora. Large order of mammals which includes many families of flesh-eating spp.; small to large, usually with five clawed toes on all feet (sometimes only four); canine teeth more or less modified as fangs; carnassial teeth usually present; dogs, cats, bears, weasels, skunks, seals, sea lions, walruses, etc.

carnivore. 1. Any animal which is solely or chiefly dependent upon catching other animals for its food. 2. See CARNIVOROUS PLANT.

carnivorous plant. Any plant specialized for trapping insects and other small arthropods, e.g. Venus flytrap, sundew, pitcher plant, and bladderwort; such plants are presumed to digest their prey and utilize them as a nutrient source.

Carnosa. Order of sponges in the Subclass Tetractinellida; skeleton principally jelly-like.

Carnosauria. Group of carnivorous dinosaurs; teeth daggerlike and recurved.

carnosine. Dipeptide of histidine and alanine in vertebrate muscle tissue.

Carolina mantid. See MANTODEA.

Carolina paroquet. See CONUROPSIS.

Carolina wren. See THRYOTHORUS.

Carolinian zone. See UPPER AUSTRAL ZONE.

Carollia. Genus of leaf-nosed bats; New World tropics.

carotene. Red or orange crystalline hydrocarbon, $C_{40}H_{56}$, synthesized by green plants; occurs in the chloroplasts and also in other plastids where chlorophyll is absent, as in carrots and certain flowers, as well as in egg yolk; when ingested by vertebrates it is converted to vitamin A in the intestinal mucosa.

carotenoid. Any of a large number of organic compounds found in every major group of plants and animals; except for green, white, and black, these substances are responsible for most of the colors of living organisms.

carotid artery. Carotid arch; one of the two main arteries supplying blood to the head region in vertebrates.

carotid gland. Spongy thickening on the internal carotid artery of certain Amphibia; presumed to facilitate a uniform distribution of blood to the head.

carotin. Carotene.

carp. See CYPRINUS.

carpal. One of the bones of the carpus.

carpal ligament. Ligament running transversely around much of the wrist region in man; holds the muscles, tendons, and ligaments in place.

carpenter ant. See CAMPONOTUS.

carpenter bees. See XYLOCOPIDAE.

carpenter moth. See COSSIDAE.

carpet beetle. Any of certain small beetles in the Family Dermestidae whose larvae feed on rugs, woolens, fur, leather, and hair.

Carphophis amoenis. Common ground snake or cone-nosed snake; a small non-poisonous sp. with a pointed snout; brown above and pink below; southeastern U.S.

Carpiodes. Genus of medium to large fresh-water fishes in the Family Catostomidae; quillbacks or carpsuckers; anterior margin of dorsal fin elongated and quill-like; not edible; several spp. in Mississippi and Great Lakes drainages, and on the Atlantic Slope.

Carpocapsa pomonella. Codling moth, an extremely important sp. whose larva feeds on apples in all parts of the world; often two or three generations per year.

Carpodacus. See FINCH.

Carpoidea. Class of extinct echinoderms; calyx laterally flattened and attached with or without a horizontal stalk; Cambrian to Devonian; Trochocystites typical.

carpometacarpus. Bone in birds representing the fusion of carpal and metacarpal bones.

carpopodite. Fifth segment of pereiopod in most malacostracan Crustacea.

carpsucker. See CARPIODES.

carpus. 1. Region of the wrist in the foreleg of tetrapods. 2. One of the bones of the wrist. 3. Carpopodite.

Carrel, Alexis. American surgeon and experimental physiologist (1873-1944); noted for tissue culturing, organ transplantation, blood vessel suturing, transfusions, and a "mechanical heart"; received the Nobel prize in physiology and medicine in 1912.

carrier. 1. Person who harbors disease organisms in his body and is capable of transmitting it to someone else; he may or may not have clinical symptoms. 2. Any substance in a cell which can accept hydrogen (thus a hydrogen carrier) and become reduced, and subsequently can lose its hydrogen and become oxidized. 3. Heterozygous individual carrying a hidden recessive allele.

carrier shell. See XENOPHORA.

carrion beetles. See SILPHIDAE.

carrion fly. 1. Any fly that oviposits in decaying animal matter. 2. The common blowfly, Phormia terrae-novae.

carrying capacity. Theoretical maximum number or weight of individuals in a sp. which can be maintained in a habitat without depletion of the food supply or without being depleted by any other ecological factor.

Carteria. Large genus of small green fresh-water Protozoa; four anterior flagella.

Carterius. Discarded name for certain fresh-water sponges; usually incorporated in Heteromeyenia.

Cartesian diver. Device used in highly sensitive manometric respiration measurements; the diver is a small glass vessel containing a gas bubble and having the neck sealed with an oil droplet; it is immersed in fluid in a larger vessel and will sink or rise as the pressure on the surrounding fluid rises or falls, or as the volume of the gas within the diver increases or decreases; measurements are made by adjustment of pressure, so as to maintain constant volume.

cartilage. Gristle; skeletal connective tissue of vertebrates; consists of spherical cells scattered in a homogeneous matrix which may contain connective tissue fibers; cartilage forms much of the skeleton of adult lower vertebrates and much of the skeleton of all immature vertebrates; in adult human beings it is largely replaced by bone; invertebrates do not have true cartilage.

cartilage bone. Any bone which has replaced a corresponding embryonic cartilage mass; the vertebral column and limb bones of man are examples.

cartilaginous fishes. Generally, all members of the vertebrate Class Chondrichthyes.

caruncle. 1. Any fleshy outgrowth, either normal or abnormal, such as the comb and wattles of fowl. 2. Small protuberance at the inner corner of upper and lower eyelids. 3. Disc on the tarsus of a mite. 4. Horny spine on the upper jaw of a hatchling turtle.

Carunculina. Genus of unionid mollusks; eastern half of U.S.

Carybdeida. Cubomedusae.

Carychium. Genus of long, cylindrical land snails; shell thin; Europe and N.A.

Caryocrinites. See CYSTOIDEA.

Caryocrinus. Genus of extinct pelmatozoan echinoderms.

Caryophyllaeus. Genus of tapeworms in the Order Pseudophyllidea; adults in fresh-water fishes, larvae in copepods.

Caryophyllia. Genus of solitary corals.

casebearer. See COLEOPHORIDAE.

casein. Important protein in milk; basis of curd and cheese.

caseinogen. Soluble milk protein which is converted to whey protein and paracasein by rennin in the stomach.

case-making clothes moth. See TINEA.

cask shell. See TONNA.

Casmerodius albus. Egret of temperate and tropical Americas (Family Ardeidae); breeds along both U.S. coasts; total length sometimes exceeds 40 in.; plumage entirely white; along shores of fresh and brackish waters; nest a platform of sticks.

casque. 1. Horny process or shield on the bill of the hornbill and a few other birds. 2. Collectively, the bony plates covering the head of certain extinct fishes.

Cassida. Genus of tortoise beetles (Chrysomelidae).

Cassididae. Family of large marine gastropods; shell thick and heavy; mostly tropical; Cassis often used for making cameos.

Cassidisca. Genus of marine Isopoda.

Cassidix mexicanus. Boat-tailed grackle; a large bird, over a foot long, with a keel-shaped tail; male black, head and neck with a purple cast; Gulf and Atlantic Coast states north to Va., south to northern South America.

Cassiduloida. Order of subtropical and tropical echinoderms in the Class Echinoidea; test round, somewhat elongated, and flattened or disclike; spines small; Aristotle's lantern absent; most spp. extinct; Cassidulus and Echinolampas modern, Galeropygus extinct.

Cassidulus. See CASSIDULOIDA.

Cassiopeia. Genus of scyphozoan jellyfishes in the Order Discomedusae; habitually lie upside down on the bottom of shallow lagoons.

Cassis. See HELMET SHELL.

cassowary. See CASUARIUS.

Castalia. Genus of Polychaeta; some of setae scale-like.

caste. One of several forms or kinds of mature individuals among social insects, such as the worker, queen, and drone forms of the honeybee.

Castianeira descripta. Common black hunting spider in the Family Clubionidae.

castings. Mass of excreta, especially of the earthworm and its relatives.

Castle, William E. American zoologist and geneticist (1867-1962); noted for his work on mammalian ovarian transplantation, mammalian genetics, and eugenics.

Castor. See CASTORIDAE.

castorbean tick. See IXODES.

Castoridae. Family of mammals which includes the several spp. of N.A. beavers (Castor); largest N.A. rodents; robust, amphibious, hind feet webbed, and with a broad, flat, scaly tail; Alaska and Labrador to the Rio Grande; feed on bark and cambium of deciduous trees; adept at cutting down trees; build dams and large houses of logs, sticks, stones, mud, and rubbish; sometimes live in burrows in the banks of ponds.

Castrada. Genus of rhabdocoel Turbellaria.

castration. 1. In the broad sense, removal of the testes or ovaries, or rendering an animal incapable of reproduction by various other agencies, e.g. hormone imbalance and infections. 2. In the narrow sense, the removal of the testes.

Castrella. Genus of rhabdocoel Turbellaria.

Casuariidae. Family in the bird Order Casuariiformes; includes three spp. of cassowaries.

Casuariiformes. Order of birds; includes the cassowaries (Casuarius) and emu (Dromiceius); terrestrial, three toes on each foot.

Casuarius. Genus of flightless birds in the Order Casuariiformes; cassowaries; terrestrial, pugnacious, and to 5 ft. or more tall; plumage dark and glossy; head and neck unfeathered, brightly colored, and wattled; forehead with a horny protrusion; Australia and New Guinea.

catabolism. Katabolism.

catadromous. 1. Pertaining to fishes which spend most of their lives in fresh water but migrate into salt water to spawn, such as the eel (Anguilla). 2. Sometimes used in the broader sense to include fish which migrate downstream, or inshore to offshore, for breeding.

catalase. Enzyme found in essentially all living cells except anaerobic bacteria; catalyzes decomposition of metabolic hydrogen peroxide to water and oxygen.

catalo. Offspring of mating between cattle and buffalo.

Catalina perch. See GIRELLA.

catalufa. See PRIACANTHIDAE.

catalysis. Marked increase in the speed of a chemical reaction due to the presence of small quantities of a catalytic substance which does not form a part of the end product of the reaction.

catamount. Local name of a lynx or cougar.

cataract. Any of a wide variety of opacities developing in the lens of the eye or in its capsule.

Catarrhina. Cercopithecoidea.

catastrophism. Obsolete geological doctrine which maintains that the earth has had periodic cataclysms each of which destroyed all life existing at that time.

catbird. See DUMETELLA.

Catenula. Uncommon genus of rhabdocoel Turbellaria; each zooid composed of a cephalic lobe and body.

caterpillar. Larva of Lepidoptera and

sawflies (Hymenoptera).

caterpillar hunter. Any of numerous beneficial carabid beetles, especially in the genus Calosoma, which feeds chiefly on caterpillars and grubs.

Catesby, Mark. English naturalist (1679?-1749); wrote the first North American natural history (Natural History of Carolina, Florida, and the Bahama Islands).

cat-eyed snake. See LEPTODEIRA.

catfish. See ICTALURUS, AMEIURUS, BAGRE, and GALEICHTHYS.

Catharacta skua. Large, brownish, maritime bird of the North Atlantic; nests in grassy and tundra areas; in the Family Stercorariidae.

Cathartes aura. Common turkey vulture or turkey buzzard (Order Falconiformes); body to 30 in. long, wingspread to 6 ft., coloration blackish; head and neck red, without feathers; soars, sometimes for hours, searching for carrion; eyesight exceptionally keen; builds no nest; eggs laid on ground or in cavity; tropical and temperate America.

cathartic. Any medicine which quickens bowel movements and produces purgation.

Cathartidae. Family of large soaring birds in the Order Falconiformes; includes the New World vultures; see GYMNOGYPS, CATHARTES, and CORAGYPS.

cathepsin. 1. One of a complex of intracellular enzymes functioning in autolysis of proteins; an endopeptidase. 2. One of a complex of intracellular enzymes associated with the synthesis of complex organic compounds from simpler compounds, e.g. synthesis of proteins from amino acids.

Catherpes mexicanus. Canyon wren; a reddish-brown sp. of canyons and rocky slopes in western N.A.

catheter. Tubular instrument (metal, rubber, etc.) used for withdrawing fluid from a body cavity, especially the urinary bladder.

cation. Any ion having a positive charge.

Catocala. Large genus of forest-inhabiting moths (Family Plusiidae); underwing moths; forewings cryptic, hind wings brightly colored.

Catonotus. Genus of darters now usually included in Etheostoma.

Catophragmus. Primitive genus of barnacles.

Catoptrophorus semipalmatus. Willet; a large gray and white shore bird of North and South America; fresh and salt-water marshes, shores, and beaches.

Catostomidae. Family of fresh-water bony fishes; includes the suckers, buffalofish, and redhorse; fins soft-rayed, scales cycloid, and mouth more or less ventral, with fleshy lips; bottom dwellers of lakes, ponds, and streams; feed on bottom by suction; omnivorous; sometimes used for human food, but generally considered rough fish; nearly 100 N.A. spp., and several elsewhere; Carpiodes, Ictiobus, Hypentelium, Moxostoma, and Catostomus typical.

Catostomus. Genus of fresh-water bony fishes which includes the common suckers; numerous widely distributed N.A. spp.; often obnoxious through overpopulation and driving out game spp.; bony but edible.

cat's paw. See PLICATULA.

cattalo. Hybrid offspring of bison and cattle.

cat tapeworm. See TAENIA PISIFORMIS and DIPYLIDIUM CANINUM.

cattle egret. See BUBULCUS.

Caucasoid. Any of a general group of races of modern man characterized by smooth to wavy hair and white to dark brown skin; includes the following races: Ainu, Alpine, Arab, Armenoid, Australian, Berber, Dravidian, Ethiopian, Ibero-insular, Indo-Afghan, Littoral, Nordic, and Mediterranean.

cauda. 1. Tail or tail-like appendage. 2. Tube at the tip of the abdomen in certain insects.

caudad. Toward the tail or posterior end.

cauda equina. Group of posterior spinal nerves of higher vertebrates; pass longitudinally beyond the tip of the spinal cord.

caudal. Pertaining to the tail or posterior end.

caudal aorta. Large posterior artery, especially in many invertebrates.

caudal fin. Dorsoventral tail fin in a fish.

caudal furca. Caudal rami; collectively, the two terminal processes on the last abdominal segment in certain Crustacea, especially copepods and ostracods.

caudal peduncle. In a fish, the slender posterior part of the body which bears the tail fin.

caudal rami. Caudal furca.

Caudata. One of three subclasses of Amphibia; salamanders, newts, etc.;

body long, with head, trunk, and tail; limbs weak, of similar size; eggs laid in ponds and brooks; larvae with gills.

caudifemoralis. Group of two ventral hind leg muscles of reptiles and urodeles.

caudilioflexorius. Ventral muscle of the bird leg.

Caudina. Genus of sea cucumbers; tube feet absent; body tapered to a tail-like posterior end; Atlantic and Pacific coasts.

caudofemoralis. Mammalian muscle which arises on the second and third caudal vertebrae and inserts on the patella; abducts the thigh and extends the shank; not in man.

caul. 1. Portion of the amnion which sometimes envelops the head of a child at birth. 2. Great omentum.

Caularchus. Genus of clingfishes; small elongated fishes which slither about over smooth damp surfaces or cling to rocks of the intertidal zone.

Cauleriella. Genus of Polychaeta.

Caulibugula. Genus of marine Bryozoa.

Cauligastra. Taxon which includes those orders of arachnids in which there is a constriction between mesosoma and prosoma; e.g. spiders and whip scorpions.

Cauloramphus. Genus of marine Bryozoa.

cavalla. Any of several edible marine fishes of Tropical America, especially Caranx.

cavallo. See SCOMBEROMORUS.

cave cricket. See STENOPELMATIDAE.

cave fishes. See AMBLYOPSIDAE.

cavernicole. Cave-inhabiting animal. Adj. cavernicolous.

cavernous bodies. Three longitudinal masses of erectile tissue in the mammalian penis.

cave salamander. Yellow to orange salamander averaging 5 in. long; found in or near limestone areas of east central states; Eurycea lucifuga. See TYPHLOTRITON.

Cavia. See GUINEA PIG.

caviar. Commercially salted and prepared eggs of the sturgeon and certain other large fishes.

Caviidae. Family of rodents which includes the common guinea pig (Cavia) and cavies.

Cavolina. Genus of sea butterflies (Gastropoda); shell symmetrically globular, with three posterior spines.

cavy. 1. Any of several South American short-tailed, rough-haired, herbivorous rodents in the Family Caviidae; the Patagonian cavy is long-legged, harelike, and to 30 in. long. 2. Guinea pig. Pl. cavies.

cayman. See CAIMAN.

cc. Cubic centimeter(s); also written cm^3.

Cebidae. Family of anthropoids which includes the spider monkeys, howlers, and capuchins; New World tropics.

Ceboidea. Superfamily in the Suborder Anthropoidea; includes the New World monkeys of Central and South America; wide, flat space between nostrils.

Cebuella. Genus of pygmy marmosets; tropical slopes of the Andes in eastern Ecuador; breed and live in tree holes in treetops.

Cebus. Common genus of tropical American capuchin monkeys (Family Cebidae); very small spp., often with a crown of hair resembling a monk's cowl.

ceca. Pl. of cecum; preferred sing. spelling is caecum, with pl. caeca.

Cecidomyiidae. Family of small to minute, mosquito-like insects in the Order Diptera; gall midges, gall gnats; wide variety of food habits, many pest spp.; larvae and eggs often produce plant galls; important genera are Phytophaga, Diarthronomyia, and Miastor.

cecropia moth. See SAMIA.

Cecrops. Genus of large marine copepods in the Suborder Caligoida; enormously modified, depressed, and parasitic on fishes.

celiac. Coeliac.

Celithemis. Genus of dragonflies; eastern U.S.

cell. 1. Microscopic unit of protoplasmic structure in all living organisms; in animals the cell is bounded by a living membrane; in plants the cell is bounded by a dead secreted cellulose wall outside the membrane; each cell characteristically contains one nucleus. 2. In an insect, a membranous portion of a wing separated from similar adjacent areas by one or more veins.

Cellaria. Genus of marine Bryozoa.

cell body. Term occasionally used to designate the main cell mass containing the nucleus as distinct from any branches, filaments, or processes extending out from the main cell mass.

cell constancy. Constant number and arrangement of cells for mature animals in a given sp., e.g. rotifers and tardigrades.

cell division. Division of cytoplasm and nucleus of a cell; the nucleus usually undergoes mitosis, rarely amitosis; in animals the cytoplasm divides by constriction, but in plants a rigid middle lamella and cell wall are laid down across the center of the cell.

cell lineage. Ancestral pattern of a tissue or organ in terms of the succession of cells, beginning with the zygote.

cell membrane. 1. Thin, outermost membrane in most animals cells. 2. Plasma membrane.

cell theory. Theory elaborated by Schleiden and Schwann in 1838 and 1839; all plants and animals are made up of cells and the products of cells, and the cells are the units of structure and function.

cellulase. Cellulose-hydrolyzing enzyme.

cellulitis. Connective tissue inflammation.

cellulose. Chief polysaccharide constituent of the cell wall of green plants and a few fungi; it has the approximate generalized formula $(C_6H_{10}O_5)_{2000}$ and when hydrolyzed produces the corresponding number of glucose molecules.

cell wall. Outermost rigid layer of a plant cell; closely invests the plasma membrane (cell membrane); not present in animals.

Celsus, Aulus Cornelius. Latin encyclopedist; wrote eight books on medicine around 30 A.D.

cement. 1. Layer of bonelike material which covers the roots, neck, and sometimes part of the crown of a mammalian tooth; fastens the tooth to the jawbone. 2. Substance present in small amounts between the individual cells of an epithelium; presumably adhesive of cementing in nature.

cement gland. 1. One of a cluster of small glands, or a single large gland associated with the reproductive system of male Acanthocephala; gives off a secretion which functions during copulation. 2. Small gland in each of the two toes of most rotifers and gastrotrichs; their secretion is used for temporary attachment to the substrate.

cementum. Cement.

Cemophora coccinea. Scarlet snake; a small, burrowing sp. with incomplete red, black, and yellow rings; N.J. to Okla. and south to the Gulf.

Cenocorixa. Genus of water boatmen (Family Corixidae); western U.S.

Cenocrinus asteria. Sessile crinoid with a long stalk and cirri; West Indies in 500 to 2000 ft. of water.

cenogenesis. All or part of the development of an individual which does not repeat the phylogenetic development of its taxon. Cf. PALINGENESIS.

cenosis. Community having two or more spp. of dominant animals; e.g. a carnivorous bird and a carnivorous mammal.

cenospecies. 1. A group of two or more spp. separated only by extrinsic factors. 2. An interbreeding population whose individuals are completely unable to interbreed with any other population; lions and tigers can freely interbreed and are therefore technically in the same cenospecies; the same is true for dogs and wolves. Cf. PLASTOSPECIES.

Cenozoic era. Present geological division of time; began about 70 million years ago; subdivided into Quaternary and Tertiary periods; characterized by man, modern plants, and the rapid evolution and dominance of mammals.

centaur. Mythical animal, half man and half horse.

center of origin. Area in which a sp. first appeared and from which it subsequently spread.

centers. Certain aggregations of nerve cell bodies in the brain of vertebrates.

centipede. See CHILOPODA.

centralia of manus. Series of one to four small bones in the tetrapod wrist; sometimes absent.

centralia of pes. One to four ankle bones in tetrapods; sometimes absent.

central nervous system. Collectively, the brain and spinal cord; the chief coordinating mechanism of most animals; in invertebrates it usually consists of one or more solid cords of nervous tissue plus their associated ganglia.

central vein. Small vein in the center of a liver lobule; empties into a sublobular vein.

Centrarchidae. Large family of N.A. fresh-water percomorph fishes; sunfish family; basses, crappies, sunfishes, and rock basses; spiny and soft-rayed portions united into a single dorsal fin; examples are Micropterus, Lepomis, Pomoxis, and Archoplites.

Centrarchus macropterus. Flier; a small brightly colored centrarchid fish found from Ill. and Ind. south to Va. and Tex.

Centrechinoidea. Diadematoidea.

Centrechinus. Genus of dark-colored tropical sea urchins.

centriole. Small dark-staining dense granule in the centrosome area of animal cells and certain lower plant cells; it doubles before mitosis, and each portion forms a pole and aster of the spindle in the mitotic figure.

centroacinar cells. Small non-secretory cells of the pancreas.

Centrocercus urophasianus. Common sage hen of western sagebrush areas (Family Tetraontidae); large, gray-brown sp., the largest American grouse; males have spectacular premating habits, including drumming, dancing, and great, pale, inflated air sacs at the sides of the throat.

centrolecithal egg. Type of egg characteristic of insects, the yolk being concentrated in the center of the egg and most of the cytoplasm around the periphery.

centromere. 1. Kinetochore, kinomere; a clear, constricted area at the bend or angle of many chromosomes; it delimits the arms of such chromosomes and usually is the point of origin of spindle fibers during mitosis. 2. Neck region of a sperm cell.

Centropages. Common genus of pelagic marine copepods (Suborder Calanoida).

Centropomidae. Family of tropical marine fishes related to the basses but more slender; some spp. enter rivers or are confined to fresh waters; snook and robalos (Centropomus); fine game and food spp.

Centropomus. See CENTROPOMIDAE.

Centropristes. See BLACKFISH.

Centroptilum. Common genus of Ephemeroptera.

Centropyxis. Large genus of freshwater amoeboid Protozoa; most of cell enclosed by a test composed of sand grains or diatoms.

centrosome. Region of differentiated cytoplasm containing the centriole in cells; usually near the nucleus.

centrosphere. Structure found during animal mitosis at each end of the mitotic spindle; consists of centriole and aster.

centrum. Ventral heavy part of a vertebra; each is attached to preceding and following centra by fibrous and cartilaginous connective tissue; body of a vertebra. Pl. centra.

Centruroides. Common southern genus of scorpions.

Centurio. Genus of centurion bats; face composed of a series of grotesque folds; New World tropics.

Centurus. Genus of American woodpeckers; often included in Melanerpes.

Ceophloeus. Dryocopus.

Cepedea. Genus of ciliate Protozoa parasitic in the intestine of Amphibia; in the Subclass Protociliata.

Cephalaspis. See OSTRACODERMI.

cephalic. Pertaining to the head or anterior end.

cephalic flexure. Slight ventral bending of the brain in the region of the midbrain, especially during early embryology.

cephalic index. Number obtained by dividing the maximum width of the primate cranium by its maximum distance from back to front and multiplying by 100; an index of 80 or more indicates brachycephaly; less than 80 indicates dolichocephaly.

cephalin. Common phospholipid in many tissues and foods.

cephalization. Tendency to concentrate, centralize, or move forward into or toward the head, especially with reference to sense organs, nervous system, and appendages.

Cephalobaena. See CEPHALOBAENIDA.

Cephalobaenida. Class of Linguatulida; mouth hooks lacking a basal arm; genital pore anterior; no intermediate host; Cephalobaena in South American snakes and Reighardia in gulls and terns.

Cephalobus. Genus of microscopic, free-living nematodes.

Cephalocanthida. Dactylopteridae.

Cephalocarida. Subclass in the Class Crustacea; rare, primitive, wormlike crustaceans reported only from the bottom of Long Island Sound and San Francisco Bay; head horseshoe-shaped; ten pairs of thoracic appendages; eight abdominal segments; Hutchinsonella macracantha and Lightiella serendipita the only known spp.

Cephalochordata. Subphylum in the Phylum Chordata; includes the marine lancelets; body slender, elongated, compressed, and pointed at each end; to 4 in. long; no distinct head; pharynx, with numerous diagonal slits, hangs suspended in a cavity (atrium) which opens to the outside by an atriopore about two-thirds of the way from the anterior end; anus subterminal; with dorsal fin, tail fin, preanal fin, and two metapleural folds anterior to the anal fin; with notochord, hollow dorsal nerve cord, myotomes, and complete digestive tract; circulatory system without heart; paired nephridia (protonephridia) and gonads; usually buried in sandy bottom of shallows with anterior end projecting into the water; tropical and temperate sea coasts; about 20 spp.; Branchiostoma the most common U.S. genus; Amphioxus and Asymmetron also typical.

cephalocone. One of several cephalic tentacle-like projections in certain pteropods.

Cephalodasys. See MACRODASYOIDEA.

Cephalodella. Large and common genus of rotifers; lorica thin; chiefly in fresh-water littoral.

Cephalodiscidea. Class in the Phylum Pterobranchia; collar with several branched arms; solitary; two gonads, two gill slits; Cephalodiscus typical.

Cephalodiscus. See CEPHALODISCIDEA.

cephalon. 1. Head or head region. 2. Head shield of a trilobite.

Cephalophus. See DUIKER.

Cephalopoda. One of six modern classes in the Phylum Mollusca; includes the carnivorous squid, nautilus, octopus, etc.; shell external, internal, or absent; mouth with two horny jaws; radula present; head surrounded by eight or ten arms, or by many tentacles; siphon present; eyes well developed; all except Nautilus have an ink gland that secretes a dark fluid which is stored in an ink sac; locomotion by creeping and by the forcible ejection of water from the mantle cavity out through the flexible siphon.

Cephalorhynchus. Genus of piebald dolphins; coloration black and white; southern seas.

cephalosome. Head or anterior region of an arthropod.

Cephalothamnium. Genus of small biflagellate Protozoa; clusters attached to fresh-water plankton organisms by a common stalk.

cephalotheca. Head covering of an insect pupa.

cephalothorax. Body division in many Crustacea and Arachnoidea representing a fusion of one or more thoracic segments with the head; in such forms as the crayfish and spider the head and all of the thorax are fused into a single body region.

Cephalothrix. Genus of intertidal nemertines; body slender and coiled, pointed anteriorly.

cephalotrocha. Turbellarian larva with eight processes surrounding the mouth.

Cephidae. Family of Hymenoptera which includes the stem sawflies; fragile, wasplike spp.; Cephus is sometimes an important pest of grasses and cereals.

Cephus. See CEPHIDAE.

Cepphus. Guillemots; a genus of northern auklike sea birds; nest on rocky cliffs.

Ceramaster. Genus of sea stars.

Cerambycidae. Large family of slender beetles to 3 in. long; long-horned beetles; often brightly colored; antennae longer than body; adults eat foliage or bark; grubs bore in wood; many pest spp.; Saperda and Ergates typical.

Ceramica. See ZEBRA CATERPILLAR.

Cerapus. Genus of marine amphipods which build tubes out of foreign material.

Cerastes. Small genus of venomous North African desert snakes in the Family Viperidae; C. cornutus is the horned viper, or asp, which has a sharp upright spine over each eye.

cerata. External gills of certain nudibranchs; variously shaped, may contain gut diverticula. Sing. ceras.

Cerataspis. Genus of shrimps.

Ceratias. See CERATIIDAE.

Ceraticelus. Genus of minute spiders; webs irregular.

Ceratiidae. Family of bizarre bony fishes in the Order Pediculati; deep-sea anglers; black spp., seldom more than 6 in. long, usually live at great depths; first dorsal spine modified as a bioluminescent lure; body flaccid; eyes small, blindness common; males often dwarfed, attached to females as parasites; Ceratias typical.

ceratine. Hornlike substance secreted by certain anthozoan coelenterates.

Ceratiomyxia. Genus of Mycetozoa forming masses on damp wood.

Ceratites. Genus of fossil ammonites.

Ceratitis capitata. Mediterranean fruit fly (Family Trypetidae); serious pest of fruits in tropics and subtropics.

Ceratium. Genus of fresh-water and marine dinoflagellates; cell covered with a cellulose test, with one anterior and one to four posterior hornlike processes.

ceratobranchials. Bones of the branchial arches of a fish, just below their lateral angles.

Ceratodontidae. Family of lungfishes;

scales large and cycloid; paired fins flat and leaflike; Ceratodus extinct; Neoceratodus forsteri the only living sp., in stagnant waters of Queensland, Australia; uses modified swim bladder for much of respiratory exchange.

Ceratodus. See CERATODONTIDAE.

ceratohyal. 1. One of the bones of the hyoid or second visceral arch. 2. Pertaining to a projection of the hyoid bone in mammals.

Ceratonereis. Genus of Polychaeta.

Ceratophyllus. Genus of fleas (Order Siphonaptera); parasitic on wild birds and mammals.

Ceratopogon. Helea.

Ceratopogonidae. Family of tiny biting flies in the Order Diptera; no-see-ums, punkies; larvae and pupae usually in fresh water or damp places; Leptoconops, Culicoides, and Helea common.

Ceratopsia. Suborder of Upper Cretaceous dinosaurs in the Order Ornithischia; horned dinosaurs; massive quadrupedal herbivores having skulls adorned with a large posterior bony frill, a parrot-like beak, and horns near the nose and eyes; Triceratops typical.

Ceratostomella. Genus of fungi which produce Dutch elm disease; transmitted by Scolytus, a bark beetle.

ceratotheca. That part of an insect pupal case which encloses the antennae.

Ceratotherium. Genus of rhinoceroses of South Africa; the "white" rhinoceros; large gray sp. with two horns on snout.

ceratotrichia. Long horny rays which support the fins of a shark.

Ceratotrocha. Genus of bdelloid rotifers.

Cercaertus. Dormouse-phalangers, a genus of very small insectivorous arboreal phalangers of Australia; hibernators.

cercaria. Miniature immature tailed fluke produced by a redia, daughter redia, sporocyst, or daughter sporocyst in the tissues of a snail; such cercariae break out into the surrounding water and must penetrate or encyst on a further host within a short time.

Cercocebus. Genus of large equatorial African monkeys; the mangabeys; coat gray and silky, tail long and straight.

Cercomonas. Genus of colorless freshwater flagellates; cell highly plastic; one flagellum projects anteriorly, the other trails posteriorly.

Cercopidae. Family of small insects in the Order Homoptera; froghoppers, spittle bugs; feed on plant sap; nymphs of many spp. live within a mass of whitish froth expelled from the anus; Aphrophora a common N.A. genus.

Cercopithecidae. Family of Old World monkeys in the Superfamily Cercopithecoidea.

Cercopithecoidea. Catarrhina, a superfamily of primates; includes the Old World monkeys; langur, baboons, mandrills, macaques, etc.; differ from New World monkeys in having the nostrils pointing downward, with a narrow septum; tail never prehensile.

Cercopithecus. Large and common genus of African monkeys; the guenons; mostly with soft, grizzled, olive or yellowish coat.

cercopods. 1. Two long filamentous projections at the posterior end of crustaceans in the Order Notostraca. 2. Cerci; segmented appendages of the last abdominal segment, especially in insects.

Cercops. Genus of marine Amphipoda.

cercus. One of a pair of sensory appendages of the tenth (sometimes called the eleventh) abdominal segment of typical insects; usually filamentous and segmented but may be short and blunt. Pl. cerci.

cere. Soft, swollen mass of skin at base of the beak in birds of prey and parrots; bears nostrils.

cereal psocid. See LIPOSCELIS.

cerebellum. Large dorsal swollen portion of the vertebrate hindbrain; it overlaps the anterior end of the medulla oblongata and is the chief muscle coordination center.

cerebral aqueduct. Small canal between third and fourth ventricles of the brain.

cerebral cortex. Layer of gray matter investing the cerebral hemispheres; well developed only in mammals; seat of conscious sensation and action, memory, and intelligence.

cerebral ganglion. In invertebrates, one of the two or more ganglia making up the brain.

cerebral hemisphere. One of the two large anterior dorsal lobes of the vertebrate brain; concerned chiefly with the sense of smell in lower vertebrates, but with coordination, voluntary movements, and consciousness in higher vertebrates; form much the largest portions of the mammalian brain.

cerebral peduncle. See CRURA CERE-

BRI.

cerebral vesicle. Slightly enlarged anterior end of the dorsal nerve cord of cephalochordates.

Cerebratulus. Genus of nemertines in the Order Heteronemertea; often in burrows on sandy or muddy sea bottom; well known because of much experimental embryological work in this genus.

cerebropleural ganglion. In bivalve mollusks, a ganglion just above the mouth.

cerebrosides. Group of lipids probably common to all living cells.

cerebrospinal. Pertaining to the vertebrate brain and spinal cord.

cerebrospinal fluid. Colorless watery fluid filling the ventricles of the brain, the cavity of the spinal cord, and the pia-arachnoid spaces around them; secreted continuously by the choroid plexuses and absorbed by the blood vessels on the surface of the brain; contains little protein and few cells.

cerebrum. Collectively, the two cerebral hemispheres of the vertebrate brain.

Ceresa. See MEMBRACIDAE.

Cereus. Genus of sea anemones.

Ceriantharia. Order of solitary zoantharian Anthozoa; skeleton lacking; polyp slender, tapering, and cylindrical; tentacles numerous and arranged in two circles; most of body buried in tube in muddy or sandy bottom with only the oral disc and tentacles exposed; Cerianthus a common genus.

Cerianthus. See CERIANTHARIA.

Ceriodaphnia. Genus of fresh-water Cladocera; usually in ponds and littoral areas.

Cerithidea. Genus of tropical and subtropical marine snails; on mud flats.

Cerithiopsis. Genus of small marine gastropods with many whorls forming a long spire.

Cerithium. Genus of small marine snails with long spires; usually in warm seas.

Cermatia forceps. Common house centipede; active sp. with long, delicate legs.

cero. See SCOMBEROMORUS.

Cerorhinca monocerata. Rhinoceros auklet, a small sea bird of the North Pacific.

Certhia. See CERTHIIDAE.

Certhiidae. Family of brownish passerine birds which includes the creepers; small spp. which creep about on trees and bushes searching for insects; bill decurved; long, stiff tail used as a prop in climbing; Certhia familiaris is the common European and American brown creeper, but the family is chiefly Old World in distribution.

cerumen. 1. Wax secreted by skin glands in the ear passage of mammals. 2. Mixture of wax, earth, resins, etc.; used by certain stingless bees for nest construction.

cervical. Pertaining to the neck.

cervical flexure. Slight forward bending of the myelencephalon where it joins the spinal cord, especially in early embryology.

cervical groove. Transverse groove marking the general separation of head and thoracic areas on the carapace of many decapod Crustacea.

cervical vertebrae. Vertebrae of the neck and upper trunk region; seven such vertebrae in the great majority of mammals.

Cervidae. Large family of ruminant mammals in the Suborder Ruminantia; includes deer, elk, moose, caribou, etc.; male with a pair of solid calcareous antlers which are shed and grown anew each year; typical genera are Moschus, Muntiacus, Dama, Cervus, Elaphurus, Odocoileus, Mazama, Ozotoceras, Pudu, Alce, Rangifer (both sexes with antlers), and Capreolus.

cervix. 1. Cylindrical posterior tip of mammalian uterus which projects into the vagina. 2. Narrow necklike part of an organ.

Cervus. Genus of deer in the Family Cervidae; includes the American elk (wapiti), red deer, maral, sambar, etc.

cesarean operation. Delivery of a child (or other mammal) by means of a ventral incision of the abdominal and uterine walls.

Cestida. One of the orders of the ctenophore Class Tentaculata; greatly compressed so that the body is ribbon-like and up to 1 m. wide and 5 cm. high; travel by sinuous movements; tropical spp.; Venus' girdle is Cestum veneris.

Cestoda. One of the two subclasses of the Class Cestoidea; usually with few to many proglottids and a scolex which bears suckers and sometimes hooks; larva with six hooks; often with an intermediate host; includes the great majority of tapeworms.

Cestodaria. One of the two subclasses of the Class Cestoidea (tapeworms);

scolex absent and undivided body with a single set of reproductive organs; anterior protrusible proboscis and a posterior frilled attachment organ (rosette); adult parasitic in the coelom and intestine of lower fishes; Gyrocotyle common in the intestine of chimaeroids.

Cestoidea. One of the three classes in the Phylum Platyhelminthes; the tapeworms; body composed of (1) a scolex bearing suckers and often attachment hooks and usually (2) few to many serially arranged proglottids each of which has a complete set of reproductive organs; adult unpigmented, covered by a thick cuticle, and without a gastrovascular cavity; parasitic in the intestinal tract of vertebrates; life cycle usually involving intermediate as well as definitive hosts; 3000 spp.; some spp. of Taenia, Hymenolepis, and Dibothriocephalus parasitic in man.

Cestracion. Genus of sharks.

Cestum. See CESTIDA.

Cetacea. Order of oceanic mammals which includes the whales, dolphins, porpoises, etc.; medium to very large spindle-shaped spp.; forelimbs (flippers) paddle-like; functional hind limbs absent; tail long and ending in two transverse flukes; one or two nostrils on top of head; hair lacking except on muzzle; thick layer of blubber under skin.

Cetonia aurata. Goldsmith beetle, a brightly colored golden beetle (Family Cetoniidae) of Europe and Asia.

Cetoniidae. Family of medium-sized to very large tropical and subtropical beetles; larvae feed chiefly on roots, adults often on flowers and fruits; includes Cetonia, Cotinis, and Goliathus.

Cetorhinus maximus. Large open-water shark, to 35 ft. long; basking shark; sluggish, feeds on plankton.

Ceuthophilus. Genus of wingless cave crickets.

Ceuthorrhynchus. Genus of weevils; includes the turnip gall weevil.

chachalaca. See ORTALIS.

Chaenea. Genus of long, spindle-shaped holotrich Ciliata; marine and fresh waters.

Chaenobryttus coronarius. Warmouth bass, a common centrarchid fish in the Mississippi basin; often confused with the rock bass.

chaeta. Essentially the same as seta but usually restricted to such structures in the Annelida; largely British usage. Pl. chaetae.

Chaetobranchus. Genus of large, tube-building Polychaeta.

Chaetoderma. Genus of Amphineura of the North Atlantic; body covered with spicules; plates absent.

Chaetogaster. Common genus of small, translucent, fresh-water oligochaetes; on substrates, and commensal on other invertebrates; reproduction by transverse fission, resulting in chains of individuals, or zooids.

Chaetognatha. Small phylum which includes torpedo-shaped translucent planktonic marine worms 10 to 70 mm. long; arrow worms; body roughly quadrangular in cross section and supplied with lateral and caudal fins; digestive tract complete, and mouth supplied with strong sickle-shaped hooks on each side; coelom present; no special larval stages; about 30 spp.; Sagitta common.

Chaetogordius. Genus of annelids in the Class Archiannelida; to 30 mm. long; on substrates in marine littoral.

Chaetonotoidea. Class in the Phylum Gastrotricha; adhesive tubules absent or posterior only; two flame bulbs; parthenogenetic; fresh water and marine; Chaetonotus, Lepidodermella, and Neodasys common.

Chaetonotus. Common genus of fresh-water Gastrotricha.

Chaetophractus. Genus of South American armadillos; peludos; with six bands of armor and dense long hair.

Chaetopleura. Common East and West coast genus of chitons; small spp., less than 1 in. long.

Chaetopoda. Taxon sometimes used to include both the Oligochaeta and Polychaeta.

Chaetopterus. Genus of bizarre luminescent polychaete Annelida; parchment worms; in parchment-like U-shaped tubes in the sea bottom; worm to 10 in. long.

chaetosema. Series of short sensory bristles on the head of some Lepidoptera.

chaetotaxy. Pattern of bristles on part of an animal.

Chaetura. See MICROPODIDAE.

chaffinch. European finch, often kept in a cage as a songbird pet.

Chagas' disease. Human disease in Central and South America; produced by Trypanosoma cruzi which especially infects cells of glands, muscles, and the nervous system; transmitted by bugs in the genera Rhodnius and Triatoma.

chain pickerel. See ESOCIDAE.

chain reflexes. Series of reflex actions, with the response of one often serving as the stimulus for the next.

chain snake. See LAMPROPELTIS.

Chalastogastra. One of the two suborders in the insect Order Hymenoptera; sawflies, horntails, etc.; base of abdomen broadly joined to thorax; larva with thoracic and usually abdominal prolegs; phytophagous.

chalaza. One of two spiral cords of dense albumin extending from opposite ends of the yellow yolk to the membrane inside the shell in a bird's egg. Pl. chalazae.

Chalcididae. Family of minute wasps; chalcid wasps; parasitize larvae and pupae of other insects.

Chalcosoma. See ATLAS BEETLE.

Chalina. Genus of marine sponges; finger sponges.

chalone. Any internal secretion having an inhibitory action, e.g. enterogastrone.

Chama. Genus of marine bivalves; jewel boxes; shell with spinelike foliations.

Chamaea. See CHAMAEIDAE.

Chamaeidae. Family of passerine birds which includes only the wren-tit, Chamaea fasciata; several subspecies along coastal belt and interior valleys of Calif. and the coastal belt of Ore.; small, sparrow-sized birds with long tails; characters intermediate between those of wrens and titmice.

Chamaeleo. See CHAMAELEONTIDAE.

Chamaeleontidae. Family of lizards occurring in Africa, Madagascar, and India; true chamaeleons; body compressed, head angular, tail prehensile, toes opposed; tip of tongue large and coated with mucus, can be shot out several inches beyond head to capture prey; large, indpendent eyes; color changes rapidly to match that of background; Chamaeleo a common genus.

chambered nautilus. See NAUTILOIDEA.

chameleon. Chamaeleon; see CHAMAELEONTIDAE and ANOLIS.

chamois. See RUPICAPRA.

chancre. Syphilis lesion which develops at the primary site of infection; a reddish ulcer.

chancroid. Soft chancre; venereal bacterial disease of the external genitals; usually in the form of a localized virulent ulcer.

channel catfish. See ICTALURUS.

channeled duck shell. See ANATINA.

channeled lathe shell. See TORNATINA.

channeled whelk. Busycon caniculatum; a large and common whelk of the Atlantic Coast; to 9 in. long.

channel shell. Conch.

Chaoborus. Very common genus of phantom midges in the Family Culicidae; larvae abundant in the plankton of lakes and ponds.

Chaos. See PELOMYXA.

chaparral. Type of plant community in which shrubs are dominant and often quite dense; usually in areas of 10 to 20 in. of precipitation per year, too dry for forests and too wet for deserts; common in western and southwestern U.S.; typical plants are scrub oak, mountain mahogany, mesquite, and manzanita.

chapparal cock. See GEOCOCCYX.

char. See SALVELINUS.

Characidae. Characinidae.

Characinidae. Characins; very large family of small fresh-water fishes of Central and South America and Africa; exhibit many modifications and modes of life.

Characoidei. Suborder of fishes which includes the Family Characinidae.

character gradient. Cline.

Charadriidae. Cosmopolitan family of birds in the Order Charadriiformes; plovers and killdeers; wading and shore birds; compactly built, and with shorter, stouter bills than sandpipers; some spp. migrate very long distances.

Charadriiformes. Large order which includes shore birds such as the gulls, terns, auks, etc.; most spp. longlegged; toes more or less webbed.

Charadrius. Oxyechus.

Charina bottae. Rubber snake, a heavy-bodied, grayish constrictor in the Family Boidae; to 24 in. long; crepuscular and nocturnal, usually in damp woods; feeds on lizards and small mammals; five far western states.

Charisea. Genus of sea anemones; cold waters.

Charitonetta. Bucephala.

charr. See SALVELINUS.

Chasmistes. Genus of large suckers in deep lakes of the Great Basin; uncommon.

chat. Largest American warbler; upper parts olive green, throat and breast yellow, and belly white; bill large, eyes encircled with white; yellow-breasted chat (Icteria virens) in eastern U.S., Mexico, and Central America; long-tailed chat is a western subspecies.

Chaudhurei. Order of fresh-water fishes which includes a small eel-like fish from northern Burma; to 2 in. long.

Chaudhurida. Chaudhurei.

Chauliodes. Common genus of Megaloptera; larvae aquatic, with paired lateral abdominal appendages.

Chauliodontidae. Viperfishes; family of deepsea spp. having a large head, long fangs, and very long slender body.

checkerboard clam. See CALLISTA.

checkerspot. See EUPHYDRYAS.

cheese mite. See TYROGLYPHUS.

cheese skipper. See PIOPHILA.

cheetah. See ACINONYX.

Cheilonereis. Genus of Polychaeta.

cheilosis. See RIBOFLAVIN.

Cheilospiura hamulosa. Nematode parasite of the chicken gizzard.

Cheilostomata. Order of Bryozoa in the Class Gymnolaemata; zooecium chitinoid or calcareous; aperture with operculum; Bugula common.

Cheimatobia. Genus of geometrid moths; female wingless.

Cheiridium. European genus of book scorpions (Pseudoscorpionida).

Cheirogaleus. Genus of catlike lemurs of Madagascar; dwarf lemurs.

Cheirolepis. Genus of extinct bony fishes in the Order Archistia.

chela. 1. Short, broad sponge spicule having recurved hooks, plates, or flukes at each end. 2. Arthropod appendage modified to form a pincer, such as a pincer claw of the lobster and crayfish. Pl. chelae.

chelate. Pertaining to an arthropod appendage modified to form a pincer.

Cheleutoptera. Phasmida.

chelicerae. First pair of appendages of Arachnoidea; usually modified for seizing and crushing. Sing. chelicera.

Chelicerata. Taxon (subphylum) often used to include the Merostomata, Pycnogonida, and Arachnida.

Chelifer. Common genus of pseudoscorpions often found in homes.

Chelifera. Same as Tanaidacea; taxon (order) sometimes split off from the Isopoda to include a group of aberrant spp. of isopods in which the first two thoracic segments are fused with the head, and the first legs have heavy pincers; includes Tanais.

chelipeds. First pair of legs in most decapod crustaceans; specialized for seizing and crushing; not used in locomotion.

Chelodina. Genus of turtles in the Family Chelydidae; Australia and New Guinea; spp. with an exceptionally long neck.

Chelonethida. Pseudoscorpionida.

Chelonia. 1. Order of reptiles which includes the turtles, terrapins, and tortoises; broad body typically encased in a rigid shell which consists of an arched dorsal carapace and a flat ventral plastron, joined laterally, and covered with polygonal scutes or leathery skin; teeth absent but jawbones often sharp and powerful; thoracic vertebrae and ribs usually fused to shell; marine, fresh-water, and terrestrial; cosmopolitan except for New Zealand and western South America. 2. Genus of green turtles found in warm oceans; to 400 lbs.; legs modified into flippers; prized as food.

Chelonidae. Family of giant tropical and subtropical marine turtles; limbs flipper-like; carapace covered with smooth, horny shields; eggs laid on shore; examples are Caretta, Eretmochelys, and Chelonia.

Chelura. Genus of marine Amphipoda; wood borers.

Chelydidae. Small family of snake-necked turtles; head and neck very long and not completely retractile beneath edge of carapace; South America, Australia, and New Guinea; Chelys and Chelodina typical.

Chelydra serpentina. Common American snapping turtle; southern Canada to Gulf Coast; to 30 in. long and 85 lbs.

Chelydridae. Family of reptiles which includes the snapping turtles; small cross-shaped plastron; head, neck, and limbs stout and incapable of being retracted into shell; voracious; tip of upper jaw hooked; sluggish fresh waters of North and Central America; Chelydra and Macrochelys common.

Chelys. Genus of South American turtles in the Family Chelydidae; the matamatas; each shield of the flattened shell rising to a prominent protuberance; head with a tubular snout and scalloped fleshy filaments.

Chemical components of Protoplasm. Living protoplasm regularly contains about 20 elements as follows: oxygen, carbon, hydrogen, nitrogen, sulfur, phosphorus, potassium, iron, magnesium, calcium, sodium, chlorine, silicon, copper, aluminum, manganese, boron, cobalt, iodine, fluorine, and bromine; of these, carbon, hydrogen, and oxygen commonly constitute more than 95%; water and dissolved salts make up 50 to 97% of most plants and animals, while carbohydrates, lipids, and proteins constitute the remainder.

chemoautotrophic. Chemosynthetic.

chemoreceptor. Any receptor which detects and distinguishes substances in suspension or solution in air or water; smell and taste receptors.

chemosynthetic. Chemoautotrophic; type of metabolism found in certain bacteria which obtain energy from the oxidation of simple compounds, e.g. the oxidation of hydrogen sulfide to elemental sulfur.

chemotaxis. Taxis movement in response to a chemical gradient; shown by protozoans swimming away from a harmful substance in a culture.

chemotherapy. Treatment of disease by chemicals which affect the causative organism unfavorably.

chemotrophic. 1. In the broad sense, a type of nutrition characterized by the acquisition of energy through chemical processes independently of light. 2. In the usual sense, bacterial nutrition which is independent of light.

Chen. Genus of geese (Family Anatidae); C. hyperboreus hyperboreus is the N. A. snow goose, found especially in western areas; C. hyperboreus nivalis is an eastern form, the greater snow goose; C. caerulescens is the uncommon blue goose, winters in La.

cherry slug. See CALIROA.

cherrystone clam. Young Venus mercenaria clam.

chest breathing. That fraction of breathing movements produced by in and out movements of the chest wall.

chestnut shell. See ASTARTE.

chestnut top shell. See KNOBBY TOP SHELL.

Cheumatopsyche. Genus of Trichoptera.

chevron bone. Hemal arch.

chevrotain. See TRAGULIDAE.

chewink. See TOWHEE.

Cheyne-Stokes breathing. Series of cyclic respiratory movements apparent during nervous coma, deep sleep of some individuals, and at high altitudes; the breathing movements gradually decrease in depth, followed by cessation for 5 to 30 seconds, and then by gradually increasing intensity or exceptionally deep breaths.

chiasma. See MEIOSIS and OPTIC CHIASMA. Pl. chiasmata.

chickadee. See PENTHESTES.

chickaree. See SCIURUS.

chicken mite. See DERMANYSSIDAE.

chicken snake. See ELAPHE.

chicken turtle. See DEIROCHELYS.

chigger. See TUNGA and TROMBIDIIDAE.

chigoe. See TUNGA.

chilarium. One of a pair of rudimentary appendages at the posterior end of the cephalothorax in Limulus. Pl. chilaria.

Child, Charles Manning. American zoologist (1869-1954); noted for his work on axial gradients and the physiology of the invertebrate nervous system and invertebrate behavior.

Childia. Genus of marine turbellarians (Order Acoela); 1 mm. long, yellowish coloration; low tide zone.

chilipepper. See SEBASTODES.

Chilo. Important genus of moths; often pests on cereals and grasses.

Chilodon. Chilodonella.

Chilodonella. Large genus of marine and fresh-water ciliate Protozoa; ovoid, with anteriorly flattened dorsal surface and a transverse row of heavy bristle-like cilia near the cytostome.

Chilodontopsis. Genus of fresh-water holotrich Ciliata; elongate ellipsoid.

Chilognatha. Large subclass of Diplopoda; millipedes which have a heavily sclerotized exoskeleton containing large quantities of calcium carbonate.

Chilomastix. Genus of pyriform flagellate Protozoa with four flagella; parasitic in the vertebrate intestine.

Chilomeniscus cinctus. Banded sand snake, a burrowing colubrid of southwestern Ariz. and adjacent Mexico; snout shovel-like.

Chilomonas. Genus of fresh-water biflagellates; no chromatophores; body elliptical, with a firm pellicle.

Chilomycterus. See DIODONTIDAE.

Chilonycteris. Genus of leaf-lipped bats; Tropical America.

Chilophrya. Genus of ovoid to ellipsoid marine and fresh-water holotrich Ciliata.

Chilopoda. Class of terrestrial arthropods which includes the centipedes; elongated, depressed body; head with one pair of long antennae, labrum, mandibles, and two pairs of maxillae; poison claws (maxillipeds) on first trunk segment; 15 to 173 trunk segments; one pair of long legs per segment except last two; ocelli present; genital pore posterior; tracheae and Malpighian tubules; nocturnal, active, and carnivorous; 2000 spp.; common genera are Geophilus, Lithobius, Scolopendra, and Cermatia.

Chilostomata. Cheilostomata.

chimaera. 1. Any member of the Order Holocephali, a group of cartilaginous marine fishes having no scales, large pectoral fins, a long whiplike tail, and a large tooth plate in each jaw; in cold oceans, often in deep waters; some spp. known as ratfishes. 2. Organisms whose tissues are of two or more genetic origins; can occur as the result of abnormal distribution of chromosomes in one cell (and the resulting descendant cells) during early cleavage; can also occur as the result of grafting two different plants whose cells may become partially associated, producing a mixing of characters. 3. A mythical two-headed animal, usually having the head and body of a lion plus the head of a goat.

Chimaera. Common genus of fishes in the Order Holocephali.

Chimarra. Genus of Trichoptera.

Chimarrogale. Genus of Asiatic water-shrews; large, stream-inhabiting forms of southeastern Asia.

chimpanzee. See PAN.

chinch bug. See LYGAEIDAE.

Chinchilla. See CHINCHILLIDAE.

Chinchillidae. Small family of rodents which includes the chinchilla (Chinchilla), a small Andean sp. with soft gray fur; now raised in captivity in South America and the U.S.

Chinese liver fluke. Clonorchis sinensis.

chink shell. Any of several conical, thin-shelled marine snails; aperture half-moon-shaped; Lacuna typical.

chinook salmon. See ONCORHYNCHUS.

Chione. Genus of small, blotched or striped, grooved marine bivalves; usually in sand flats.

Chionactis. See SHOVEL-NOSED SNAKE.

Chionididae. Small family (two spp.) of birds in the Order Charadriiformes; sheathbills; white, dovelike scavengers; subantarctic coasts and islands.

chipmunk. See TAMIAS and EUTAMIAS.

Chiridota. Genus of long, wormlike, translucent sea cucumbers; usually buried in mud or sand with tentacles projecting into the water.

Chiridotea. Common genus of marine Isopoda.

chiro. Large tropical marine fish (Elops) related to the tarpon; not edible but a good game fish.

Chirocephalopsis bundyi. Common and widely distributed U.S. fairy shrimp.

Chirocephalus. Common and widely distributed genus of fairy shrimps; not in U.S.

Chiromeles. Genus of ugly hairless bats of the Malay region.

Chironectes. Genus of water opossums or yapoks; semi-aquatic marsupials having certain features of rats and otters; feet large and webbed; pelt short, dense, and gray; tail ratlike; exude a vile odor; Guatemala to southern Brazil.

Chironomidae. Tendipedidae.

Chironomus. Tendipes.

Chiropotes. Genus of tropical South American fruit-eating monkeys; sakis; dense woolly coats.

Chiropsalmus. Genus of jellyfishes in the Order Cubomedusae.

Chiroptera. Large world-wide order of mammals which includes the bats, the only true flying mammals; wings consist of greatly elongated forelimbs and second to fifth digits supporting a thin integumental membrane which includes the hind legs; teeth sharp; mostly nocturnal; commonly divided into two suborders, Megachiroptera and Microchiroptera.

chiru. See PANTHALOPS.

chiselmouth. See ACROCHEILUS.

chi-square. Statistical procedure used to determine the probability of obtaining merely by chance a deviation from the expected ratio greater than that obtained with the observed data; written χ^2 or X^2.

chital. 1. Venomous Indian sea serpent (Hydrophis). 2. See AXIS.

chitin. Tough and resistant nitrogenous polysaccharide having the formula $(C_8H_{13}O_5N)_n$; insoluble in water, alcohol, alkalis, dilute acids, and digestive juices of most animals; an important constituent of the cuticle of arthropods; found sparingly in certain structures in some sponges, hydroids, bryozoans, brachiopods, mollusks, nematodes, acanthocephalans, and annelids.

chitinase. Enzyme which hydrolyzes chitin; found in certain invertebrates.

chitinoid. Chitin-like in general structure and rigidity but not necessarily containing the specific chemical substance, chitin.

chitinous. Pertaining to chitin.

chiton. Common name of mollusks belonging to the Class Amphineura.

Chiton. Common genus of chitons, especially in the West Indies and Pacific area.

Chlamydobotrys. Genus of green, fresh-water, biflagellate Protozoa; in colonies of 8 or 16 cells.

Chlamydodon. Genus of holotrich ciliate Protozoa; elongate to ellipsoid or triangular; fresh and salt water.

Chlamydomonas. Genus of small, spherical, ovoid, or elongated green biflagellate Protozoa; fresh water.

Chlamydophrys. Genus of fresh-water amoeboid Protozoa; only the pseudopodia project from the enveloping vase-shaped test.

Chlamydoselachus. Primitive genus of frilled sharks; body elongated; gill slit covers frilled.

Chlamydotheca. Uncommon genus of fresh-water Ostracoda.

Chlamyphorus. Genus of fairy armadillos; drier parts of western Argentina and Bolivia.

Chlamys. See SCALLOP.

Chlidonias nigra. Black tern; a Holarctic sp. which breeds as far south as Penn., Tenn., Nev., and Calif.; winters in Southern Hemisphere.

chloragen. Chlorogogue.

chloral hydrate. Deliquescent crystalline substance used as a general anesthetic for small aquatic metazoans; usually in 10% solution.

Chloraster. Genus of marine and fresh-water Protozoa; four lateral lobes and five anterior flagella.

chloretone. Proprietary name for chlorobutanol; a white crystalline substance used medicinally as a local anesthetic and as a general anesthetic for aquatic invertebrates; 0.1 to 0.6% solution.

chloride cells. Certain epithelial cells of gill filaments in some fishes; excrete chlorides.

Chloridella. See HOPLOCARIDA.

Chloroceryle americana. Green kingfisher, Texas kingfisher; a very small sp. found from southern Tex. to Argentina.

chlorocruorin. Green iron-containing respiratory pigment dissolved in the blood plasma of certain polychaetes; chemically related to hemoglobin.

chlorogog. Chlorogogue.

chlorogogue. Special type of spongy tissue which envelopes the stomach-intestine of the earthworm and some other annelids; modified peritoneum; cells probably aid in fat distribution and excretion.

Chlorogonium. Genus of spindle-shaped green biflagellate Protozoa; fresh waters.

Chlorohydra viridissima. Common sp. of fresh-water hydrozoan; the "green" hydra; contains symbiotic zoochlorellae in the gastrodermis.

Chloromonadina. Order of green biflagellate Protozoa; reserve food as oil droplets; fresh waters; Gonyostomum typical.

chloromycetin. Trade-mark for a mold antibiotic originally extracted from Streptomyces but now produced synthetically; effective against various bacterial, virus, and rickettsia infections.

Chloroperla. Genus of green Plecoptera.

chlorophyll. One of a group of green pigments found in green plants, a few Protozoa, and certain bacteria; in the presence of light and chlorophyll, such organisms are able to synthesize carbohydrates from carbon dioxide and water; the seven known types of chlorophyll are: chlorophyll a, b, c, d, and and e; bacteriochlorophyll, and bacterioviridin; their presence and abundance vary from one group of organisms to another.

Chloropidae. Frit flies; a family of small to minute pale-colored Diptera; maggots bore into stems of cereals and grasses; Oscinosoma common.

chloroplast. Type of plastid found in certain plant cells and some Protozoa; contains chlorophyll, imparts a green color to the cells containing them.

Chlorops. Large and cosmopolitan genus of small flies; some spp. pests on grasses and grains.

Chloropseidae. Small family of south-eastern Asiatic passerine birds; leafbirds; brilliantly colored forest spp.; mostly fruit and nectar eaters.

chloroquine. Antimalarial drug, especially useful against stages in the circulating blood.

Chlorura chlorura. Green-tailed towhee (Family Fringillidae); a ground-dwelling finch, slightly larger than a house sparrow; back olive-green, crown rufous, throat white, breast gray; western states.

choanae. Internal nares. Sing. choana.

Choanichthyes. One of the three subclasses of fishes in the Class Osteichthyes; nostrils connected to mouth cavity; each of the paired fins with a central linear fleshy mass; includes lungfishes, coelacanths, etc.

Choanites. Genus of marine sponges; northern seas.

choanocyte. Collar cell; a type of cell peculiar to sponges; they form an epithelium in certain chambers and passages through the sponge; each cell has a funnel-shaped, mucus-covered, gelatinous collar and a single flagellum originating from the cell in the center of the collar; the beating of the flagella of these many cells creates currents of water through the sponge; a few Protozoa have the general structure of a choanocyte.

Choanoflagellata. Group of flagellate Protozoa having much the same structure as sponge choanocytes.

Choanophrya. Genus of stalked freshwater Suctoria; tentacles greatly thickened.

choanosome. Endosome; inner region of leuconid sponges; includes the clusters of flagellated chambers.

Choeropsis. Genus of mammals in the Family Hippopotamidae; includes the pygmy hippopotamus of Liberia; about 400 lbs.; lives in pairs in marshes and shady forests.

Choeropus. Genus of diminutive deer-like mammals of Australia; pig-footed bandicoots; nocturnal and uncommon; make nests of grass.

cholecystokinin. Hormone of the intestinal mucosa; causes the gall bladder to release bile.

choledochal duct. Bile duct.

cholera. Asiatic cholera; acute infection produced by Vibrio cholerae; transmitted in polluted water or food which has come in contact with the excreta of infected persons; symptoms are severe diarrhea, vomiting, and collapse; high death rate; sulfa drugs and vaccination are essential; now chiefly restricted to the Orient.

cholesterol. White, fatty, crystalline alcohol (sterol) in the cells of most animals; especially abundant in bile and nerve tissue.

choline. Constituent of certain important natural fats; essential in the diet of some laboratory animals and has a wide variety of physiological roles; sometimes considered one of the vitamin B complex.

cholinergic nerve fiber. Any nerve fiber which secretes acetylcholine as the essential neurohumor facilitating transmission of impulses to adjacent nerve fibers; vertebrate somatic motor fibers, some parasympathetic fibers, and fibers connecting the central nervous system and sympathetic ganglia are cholinergic.

cholinesterase. Enzyme which functions in the destruction of acetylcholine shortly after the secretion of the latter by nerve fibers.

Choloepus. See UNAU and BRADYPODIDAE.

Chologaster. Small genus of pigmented blind fishes in the Family Amblyopsidae; swampfish, springfish; swamps and streams from Va. to Ga., and springs and caves of Tenn., Ken., and Ill.

Chonchoecia. Genus of marine Ostracoda in the Order Myodocopa.

Chondestes grammacus. Lark sparrow; breeds in plains areas of western U.S. and Mexico but also found as far east as W. Va. and Ala.

Chondracanthus. Genus of copepods parasitic on the gills of marine fishes.

Chondractinia. Genus of sea anemones (Order Actiniaria).

Chondrichthyes. Class of vertebrates which includes the cartilaginous fishes (sharks, rays, skates, and chimaeras); skin usually with minute placoid scales; skeleton cartilaginous only; vertebrae numerous and complete; movable jaws; both median and paired fins; mouth ventral, with many vitrodentine-covered teeth; one or two olfactory sacs not opening into mouth cavity; intestine with spiral valve; heart with sinus venosus, one auricle, one ventricle, and conus arteriosus; five to seven pairs of gills (three in chimaeras), all in separate clefts about 290 spp.; almost exclusively marine.

chondrification. Conversion into cartilage.

chondrin. Gelatinous matrix of cartilage.

chondriosomes. Mitochondria.

chondroblast. Embryonic cell destined to produce cartilage.

chondroclast. Large cell concerned with the destruction of cartilage.

chondrocranium. Part of skull first formed in the vertebrate embryo; cartilaginous capsules surrounding part of brain, olfactory organs, and inner ear; in most vertebrates some or all of this cartilage is later replaced by true bone, and the skull is further formed by the appearance of superficial membrane bones external to the chondrocranium.

chondrocyst. Especially large type of rhabdoid.

chondrocyte. Cartilage cell.

chondrodystrophy. Abnormal growth of cartilage at the ends of long bones and the formation of cartilaginous and bony tumors on the shafts of long bones; a Mendelian dominant in man.

Chondronema. Genus of tylenchid nematodes parasitic in the hemocoel of certain insects.

chondrophore. Small calcareous shelf, ridge, or tooth to which the ligament or resilium is attached in a bivalve mollusk.

Chondrosia. Genus of sponges in the Subclass Tetractinellida; stiff mesoglea but spicules absent.

chondroskeleton. Cartilaginous skeleton

Chondrostei. Order of bony fishes which includes the sturgeon and paddlefish; body of living spp. scaleless or with longitudinal rows of bony scutes; snout pronounced; teeth absent; tail heterocercal; skeleton mostly cartilaginous; vertebrae acentrous.

Chondrosteica. Chondrostei.

Chone. Genus of Polychaeta.

Choniostoma. Genus of marine copepods parasitic on gills of Crustacea.

Chonotricha. Order of marine ciliate Protozoa; vase-shaped, with a complicated cytoplasmic collar; on objects or other invertebrates; Spirochona common.

chorda. In anatomy, a cord or chord.

chordae tendinae. Tendinous strings connecting the auriculo-ventricular valves with the wall of the ventricles; prevent these valves from turning inside out.

chorda-mesoderm. In early vertebrate embryos, the mass of embryonic cells which later gives rise to the notochord and mesoderm tissue.

Chordata. Phylum characterized by a hollow dorsal nerve cord, notochord, and usually pharyngeal gill slits at some stage in the life history; segmentation (often obscure), large coelom, and complete digestive tract are other characters; includes certain primitive wormlike marine forms as well as fishes, amphibians, reptiles, birds, and mammals; 55,000 spp.

chordate. 1. Having a notochord. 2. Any member of the Phylum Chordata.

Chordeiles minor. Common nighthawk of North and South America; upper parts sooty-black, irregularly mottled; feet and legs small and weak; wide mouth used in catching insects on the wing; especially active at dusk; flight erratic and often high; eggs laid on ground, no nest.

Chordeuma. Widely distributed genus of millipedes.

Chordodes. Genus of Nematomorpha.

chordotonal receptors. Rodlike or bristle-like receptors for vibrations; located on various parts of the insect body.

chorio-allantoic graft. Small piece of living tissue which is placed on the outer surface of the chorion of a bird (chick) embryo in the area where the allantois is fused with the chorion; after the egg shell is resealed, the embryo continues developing, and the foreign tissue is invaded by blood vessels and thus maintained.

chorio-allantoic placenta. Placenta consisting of a large chorion fused with a large allantois.

chorioid plexus. Choroid plexus.

chorion. 1. Outermost embryonic membrane of reptiles, birds, and mammals; encloses the embryo and its other associated membranes, and is an important device for the nourishment, excretion, and respiration of the embryo. 2. Outer shell of an insect egg.

chorionic gonadotrophin. Gonad-stimulating substance elaborated by the placenta in pregnant women and mares; excreted in urine and used as a pregnancy test; induces follicular growth and corpus luteum formation in the female, and stimulates interstitial cells in the male; formerly called prolan.

chorionic villi. Abundant finger-like projections of the chorion in the mammalian placenta; become imbedded in uterine tissue and greatly increase the absorptive area between embryonic and maternal tissues. Sing. villus.

Chorioptes. Genus of mites which infest mammals and sometimes cause mange.

Choristida. Order of sponges in the Subclass Tetractinellida; skeleton principally four-rayed spicules.

C horizon. Deep stratum of a soil profile which consists of unconsolidated more or less weathered parent material. Cf. A, B, and D HORIZON.

choroid coat. Thin, vascular, pigmented tissue lining the sclerotic coat of the vertebrate eyeball and forming the iris anteriorly.

choroid fissure. Fissure formed by the optic vesicle and its stalk in early vertebrate embryology; permits ingrowth of tissue which later becomes vitreous humor.

choroid plexus. Non-nervous epithelial tissue on the roof of some or all of the ventricles in the vertebrate brain; nourishes the brain and secretes cerebrospinal fluid.

chorology. Study of migrations, range, and geographic distribution; term seldon used.

Choroterpes. Genus of Ephemeroptera.

chorus frogs. See HYLIDAE.

chough. Old World bird in the crow family (Corvidae); legs red, plumage black.

chouka. Indian four-horned antelope.

chousingha. See TETRACEROS.

Chriopeops goodei. Redfin killifish of Fla.

Chromadora. Common genus of chromadorid nematodes; fresh-water and marine soils.

Chromadorida. Order of nematodes in the Class Aphasmidea; esophagus divided into three regions and not elongated.

Chromadorina. Suborder of chromadorid nematodes; mouth cavity with one or three large teeth or six inwardly directed teeth; fresh-water and marine soils; common genera are Chromadora, Cyatholaimus, and Desmoscolex.

chromaffin system. Collectively, clumps of brown-staining endocrine cells in the suprarenal medulla, autonomic ganglia, carotid bodies, etc.

Chromagrion conditum. Damselfly of northeastern U. S.

chromatic aberration. Inherent defect in an optical system (including microscope objectives) whereby the images formed by the different colors of the spectrum do not all lie in the same plane and are therefore not equally sharp; largely corrected by combining lenses of different refractive character in an objective; achromatic objectives are corrected chromatically for two colors, and apochromatic objectives for three colors.

chromatid. One of the two (or multiples of two) strands of a chromosome which lie closely side by side during prophase and metaphase of mitosis; after separation in early anaphase they are known as daughter chromosomes; chromatids are also formed in the first and second meiotic divisions.

chromatin. Nucleoprotein material of chromosomes, characterized by its affinity for basic dyes; usually present in the form of more or less densely arranged granules in the interphase nucleus and in the chromosomes during mitosis and meiosis.

Chromatium. Genus of reddish or or brownish water bacteria.

chromatocyte. Pigment cell.

chromatoid bodies. Elongated, refractile bodies in certain Protozoa, especially during precystic and early cystic stages.

chromatolysis. Neuron damage (especially asphyxiation) marked by the disappearance of Nissl bodies.

chromatophore. 1. Special cell, usually in the dermis; contains an abundance of pigment granules which are usually capable of being dispersed or concentrated; some such cells are simple in structure, others are large and complicated. 2. The term is also occasionally used as a synonym for chloroplast or chromoplast.

chromidium. Extranuclear granule (possibly chromatin) in the cytoplasm of many Protozoa. Pl. chromidia.

chromocenter. Heterochromatic region of a chromosome on both sides of a centromere; transverse bands of chromatin less discrete than in the rest of the chromosome.

chromocyte. Pigmented cell.

Chromodoris. Genus of West Coast nudibranchs.

Chromogaster. Genus of plankton rotifers; lorica of two convex plates fused laterally; foot absent.

chromogen. 1. Chemical precursor of a biological pigment. 2. Strongly pigmented or pigment-producing microorganism.

chromomere. Dark-staining beadlike chromatin granules arranged in a linear fashion in a chromosome.

chromonema. The primary axial structure of each chromosome in non-dividing cells; a series of twisted, interlaced, and coiled filaments containing quantities of chromatin; by shortening and condensation, they are contained in the chromatids at the beginning of mitosis, as well as in the resulting chromosomes; giant salivary gland chromosomes of Diptera are relatively straight chromonemata.

chromophil substance. Cytoplasmic granules which take a nuclear stain, as in neurons.

chromophobe cells. Cells which do not

readily take up biological stains.

chromosomal aberrations. Abnormalities in chromosome structure, including deficiencies, duplications, translocations, inversions, polyploidy, and aneuploidy.

chromosome. One of the small chromatin-containing bodies carrying the genes; a permanent part of the nucleus which becomes coiled, condensed, and definitive during mitosis; the great majority of chromosomes are paired, identical in appearance, and homologous, but the different pairs are often visibly distinct from each other in size and shape; each chromosome is composed largely of nucleoprotein and consists of hundreds or thousands of genes arranged in a linear fashion. See CHROMATID.

chromosome map. Schematic map showing the relative linear distance between genes on the various chromosomes in an organism; determined chiefly by the relative frequency of crossing-over between any two genes located on one chromosome; first worked out for Drosophila by T. H. Morgan and his associates.

Chromulina. Genus of fresh-water monoflagellate Protozoa; slightly amoeboid; one or two golden brown chromatophores.

chronaxie. 1. Latent interval between electrical stimulus and muscle response. 2. Minimal latent interval in milliseconds required with an electric current having an intensity twice that necessary for excitation, especially when the stimulus is prolonged.

Chrosomus. Genus of small dace (Cyprinidae); brilliant red or yellow in spring; widely distributed in eastern Canada and eastern half of U.S.

Chrotogale. Genus of hemigales; small mammals of the Tonkin area; related to the civets and genets; muzzle elongated, teeth small.

chrysalid. Chrysalis.

chrysalis. Pupal stage of butterflies. Pl. chrysalides.

Chrysamoeba. Genus of fresh-water Protozoa; flagellate stage transient, usually amoeboid, with chromatophores.

chrysanthemum shell. See THORNY OYSTER.

Chrysaora. Genus of oceanic jellyfishes.

Chrysapsis. Genus of small fresh-water monoflagellate Protozoa; chromatophore diffuse or branching.

Chrysemys. Genus of painted turtles, perhaps the most common of all American turtles; ponds and sluggish waters everywhere; omnivorous; carapace smooth, margined with red; plastron yellow, sometimes tinted with red; head and limbs streaked with yellow; carapace to 5 in. long.

Chrysidella. Genus of small yellow biflagellate Protozoa; symbionts in foraminiferans and radiolarians.

Chrysididae. Family of jewel wasps or cuckoo wasps; brilliant blue, green, red, or purple spp.; eggs laid in nests of other wasps and bees; Chrysis common.

Chrysis. Genus of small to medium-sized wasps; cuckoo wasps; females deposit their eggs in the egg cells of various bees; larvae then parasitize the bee larvae.

Chrysobothris. See BUPRESTIDAE.

Chrysochloridae. Golden moles; a family of moles with metallic golden fur; limbs imbedded in body wall; feet greatly modified for burrowing; eyes non-functional; external ear absent; feed chiefly on earthworms; Africa south of the Sahara.

Chrysochus. Common genus of leaf beetles (Chrysomelidae).

Chrysococcus. Genus of fresh-water monoflagellate Protozoa; covered with a smooth or sculptured test; chromatophores present.

Chrysocyon. See AGUARA.

Chrysomelidae. Very large family of leaf beetles; often brilliantly colored and metallic; larvae and adults feed mostly on living plants; many pest spp.; important genera are Leptinotarsa, Diabrotica, and Donacia.

Chrysomonadina. Order of flagellate Protozoa; one or two flagella; cells small, solitary or colonial; yellow, brown, or no special pigmentation; holophytic or holozoic; common examples are Dinobryon, Synura, and Uroglena.

Chrysopa. Large cosmopolitan genus of insects in the Order Neuroptera; larvae called "aphis lions" because of their habit of feeding on aphids as well as many other small insects.

Chrysopetalum. Genus of Polychaeta.

Chrysopidae. Family of insects in the Order Neuroptera; green lacewings, stinkflies; larvae called lions because of their habit of feeding on aphids as well as other small insects.

Chrysops. Cosmopolitan genus of deerflies (Family Tabanidae); female a vicious biter on warm-blooded animals.

Chrysopyxis. Genus of fresh-water monoflagellate Protozoa; lorica and chromatophores present; attached to algae.

Chrysosphaerella. Genus of colonial monoflagellate fresh-water Protozoa; each cell with two, long siliceous rods; chromatophores present.

Chthamalus. Genus of small barnacles.

Chthonerpeton. Genus of gymnophionids.

Chthonius. Genus of pseudoscorpions.

chub. 1. In general, any one of a large group of small fresh-water spp. in the minnow family (Cyprinidae); usually relatively stout-bodied; often used as bait. 2. Any of a large group of small-mouthed marine fishes resembling porgies; scavengers, found along reefs, around wrecks, and following ships. 3. See LEUCICHTHYS.

chubsucker. See ERIMYZON.

chuckwalla. See SAUROMALUS.

chuck-will's-widow. See ANTROSTROMUS CAROLINENSIS.

chukar partridge. Alectoris graeca; a small Old World partridge, native to Asia and southern Europe; introduced into western states.

chum salmon. See ONCORHYNCHUS.

churrworm. Local name for mole cricket.

Chydorus. Common genus of littoral Cladocera.

chyle. Milky fluid taken up by the lacteals of the small intestine; consists of lymph containing emulsified fats; passes into the blood stream by way of the thoracic lymph duct.

chylomicrons. Minute fat globules found in the blood during the digestion of fat; about 1 micron in diameter.

chyme. Grayish semiliquid mass of partially digested food as it passes from the stomach into the duodenum.

chymotrypsin. Digestive enzyme in pancreatic juice; converts protein molecules into proteoses, peptones, and polypeptides.

chymotrypsinogen. Inactive enzyme secreted in pancreatic juice; activated to chymotrypsin in the presence of trypsin.

ci. Cubitus interruptus.

cibarium. Food pocket of the preoral cavity just under the clypeus in certain insects.

cicada. See CICADIDAE.

cicada killer. Large solitary wasp which stings cicadas to immobility, places tham in a burrow, and oviposits on them; Sphecius speciosus.

Cicadellidae. Large family of small, herbivorous insects in the Order Homoptera; leafhoppers; forewings often thickened and brightly colored to match head and thorax; feed on plant sap and transmit many virus and bacterial diseases of plants; exude "honeydew"; Circulifer an important genus.

Cicadidae. Family of medium to large, robust insects in the Order Homoptera; cicadas or harvest flies; males usually with a sound-producing device on each side at the base of the abdomen, and drumming or buzzing occurs on hot, sunny days; nymphs subterranean and feed on root sap; series of nymph stages lasts from two to 17 years, depending on climate and the sp.; Magicicada septendecem (17-year locust) the most familiar American sp.; its nymph instars live for 13 to 17 years before the adult emerges from the ground; females may damage trees and shrubs by ovipositing in small twigs.

cicatrix. Scar or scarlike mark.

Cichlasoma cyanoguttata. Rio Grande perch; streams of Tex. and Mexico; Family Cichlidae.

Cichlidae. Very large family of fresh-water fishes in Tropical America, Africa, and Asia; wide variety of adaptations and habits; superficially resemble the sunfishes.

Cicindela. Large cosmopolitan genus of tiger beetles (Cicindelidae).

Cicindelidae. Family of active, long-legged beetles, often brightly colored; tiger beetles; predaceous larvae in burrows in ground; Cicindela and Omus familiar genera.

Ciconia ciconia. Common white stork of continental Europe and northwestern Africa; see STORK.

Ciconiidae. Family of birds in the Order Ciconiiformes; storks; cosmopolitan except for Oceania, New Zealand, northern N. A., and much of Australia.

Ciconiiformes. Large order of birds which includes the herons, storks, ibises, flamingos, etc.; wading birds with long neck and legs; mostly tropical and subtropical; feed chiefly on fish and aquatic invertebrates; usually nest in trees.

Cidaridae. Family of primitive sea urchins in the Order Cidaroida.

Cidaris. See CIDAROIDA.

Cidaroida. Order of primitive echinoderms in the Class Echinoidea; anus aboral, central; without peristomial gills; more or less spherical sea urchins; in tropical seas; Cidaris common.

cigar fish. Small cigar-shaped West Indian marine fish.

ciliary body. Thickened rim of the choroid coat of the vertebrate eye; bears the ciliary muscle.

ciliary feeding. Type of feeding by means of a current of water drawn towards or through an animal by ciliary action, as in rotifers, bryozoans, and clams.

ciliary glands. Glands of Moll.

ciliary muscle. Small circular ring of vascular and muscle tissue, wedge-shaped in cross section, which projects into the eyeball cavity just behind the iris; by relaxing and contracting, it pulls or relaxes the suspensory ligament, which, in turn, flattens or thickens the lens.

Ciliata. One of the five classes of the Phylum Protozoa; cilia for locomotion and feeding; surface of cell is covered with a thin, secreted pellicle; most spp. have a single large macronucleus and one or more small micronuclei; from the standpoint of organization, the most complex protozoans; free living, parasitic, marine, and freshwater; probably 6000 described spp.

ciliated bands. Linear ribbon-like tracts of cilia having a bilateral arrangement, especially in echinoderms and tornaria larvae.

ciliated epithelium. Any epithelium having cilia on the exposed surfaces of the cells.

ciliated groove. Endostyle of prochordates.

ciliate dysentery. Type of dysentery in man; produced by Balantidium coli infections in the large intestine.

Cilicacea. Genus of marine Isopoda.

Ciliophora. One of the two subphyla of the Phylum Protozoa; includes all Protozoa having cilia for locomotion in some stage of the life cycle (Ciliata and Suctoria); macro- and micronuclei.

cilium. 1. Short, hairlike living process in some types of animal cells; when present, cilia are usually numerous on a single cell; capable of beating in unison and moving particles along the surface of the cell or producing locomotion in ciliate Protozoa; the action of each cilium is governed by its basal granule in the peripheral cytoplasm of the cell; each cilium consists of two central fibrils surrounded by a ring of nine fibrils which, in turn, is enveloped by a thin membrane corresponding to the cell membrane); a rhizoplast, if present, does not connect with the nucleus or a parabasal body; cf. FLAGELLUM. 2. Eyelash. Pl. cilia.

Cimbex. See CIMBICIDAE.

Cimbicidae. Small family of sawflies in the Order Hymenoptera; large, robust spp. with caterpillar-like larvae which feed on many deciduous trees, shrubs, and perennials; Cimbex is a serious pest on trees in Europe and N. A.

Cimex. See CIMICIDAE.

Cimicidae. Small family of hemipteran insects which includes the bedbugs; temporary ectoparasites of man, bats, birds, and a few other warm-blooded animals; broad, depressed, and with vestigial forewings only; Cimex lectularius is the common bedbug in temperate and subtropical regions; it takes a blood meal only at night when the host is asleep, and hides in cracks, furniture, and under rugs in the daytime; C. rotundatus is the common tropical sp.

Cinclidae. Small family of passerine birds; includes the unique water ouzels, or dippers; excellent underwater swimmers; mountainous or hilly districts of Europe, Asia, western N. A., and South America; the common American dipper, Cinclus mexicanus, is a dark, slate-colored bird with a short tail; it is found bobbing around mountain streams where it feeds on bottom insects.

cinclides. Pores in the body wall of certain sea anemones; serve as a release for internal water when the animal contracts suddenly; sometimes acontia are forced out through these pores. Sing. cinclis.

Cinclus. See CINCLIDAE.

cinera. Gray matter of the nervous system.

Cinetochilum. Genus of marine and fresh-water holotrich Ciliata; oval to ellipsoid, flattened, and with several long caudal cilia.

cingulum. 1. Any girdle-like structure or band of color. 2. Bundle of association fibers in the brain. 3. Outer ciliary zone on the coronal disc of a rotifer. 4. Clitellum. 5. Ridge around the base of a tooth crown.

Cinygma. Genus of Ephemeroptera;

western U.S.

Cinygmula. Genus of Ephemeroptera; mostly western.

Ciona. Common and cosmopolitan genus of sessile tunicates; body usually elongated.

Circaëtus. Old World genus of large hawks.

Circinalium. Genus of colonial ascidians.

circular muscle. Layer of muscle tissue in the digestive tract, respiratory tract, urogenital ducts, etc. of vertebrates and many invertebrates; the muscle cells are arranged in a circular fashion in the walls of such organs.

circulation. 1. Movement of circulatory fluids through the animal body, especially lymph and blood. 2. The movement of cytoplasm within individual cells.

circulatory system. Collectively, all organs and tissues associated with the distribution of materials throughout the animal body; in addition to blood, heart, veins, arteries, and capillaries in higher animals, the circulatory system also usually includes lymph, lymph vessels, and lymph nodes.

Circulifer tenella. Sugar beet leafhopper which transmits a virus disease, sugar beet curly top.

circulus. 1. One of many small ringlike markings on a fish scale. 2. Any ringlike arrangement, especially of small blood vessels. Pl. circuli.

circumboreal. Pertaining to geographical distribution which encompasses all land areas or all marine areas in the colder parts of the Northern Hemisphere, especially north of parallel 45 or 50.

circumcision. 1. Removal of all or a part of the prepuce or foreskin in human males. 2. In the broad sense, sometimes used to include also the removal or cutting of the internal labia in human females.

circumenteric nerves. Circumesophageal commissures.

circumesophageal commissures. Two large nerves which originate in the brain of typical arthropods, pass around the esophagus, and unite ventrally at the subesophageal ganglion; also present in certain other phyla, e.g. Annelida, Chaetognatha, etc.

circumferential canal. Ring canal.

circumpharyngeal connectives. Two nerve strands, one on each side of the esophagus, in annelids and arthropods; connect the brain and subesophageal ganglia; same as circumesophageal commissures.

circumvallation. Pseudopodial entrapment of food particles without actual contact of pseudopodia and particles.

circumvallate papillae. Mushroom-shaped papillae in a broad arrangement toward the rear of the tongue; they do not project above the general surface of the tongue but have a deep surrounding groove whose epithelium contains taste buds in abundance; there are only seven to 11 circumvallate papillae, each of which is 1.5 to 3.5 mm. across.

Circus hudsonicus. Marsh hawk or harrier; a gray, low-perching hawk of open country.

Cirolana. Genus of marine Isopoda; sometimes pests on bathing beaches.

Cirphis. Leucania.

cirrate. Bearing cirri.

Cirratulus. Common genus of marine polychaetes; slender, with parapodial cirri greatly elongated and threadlike; worm to 6 in. long; in burrows in sea bottom; often called fringed worms.

cirral ossicles. Small ossicles of a crinoid cirrus.

Cirriformia. Genus of Polychaeta.

Cirripathes. Genus of horny, slender, unbranched, colonial anthozoan coelenterates; black corals; tropical seas.

Cirripedia. Subclass in the Class Crustacea; includes the barnacles; adults sessile and attached or parasitic; mostly with an external calcareous shell composed of several to many parts imbedded in a soft mantle which surrounds the animal; attached to the substrate along dorsal surface; up to six pairs of slender thoracic appendages, used in straining food particles from the water; abdomen vestigial; hermaphroditic; 600 spp.; exclusively marine.

Cirrodrilus thysanosomus. Small leechlike oligochaete commensal and parasitic on western U.S. crayfishes; in the Family Branchiobdellidae.

cirrus. 1. In hypotrich Protozoa, a long, tapered, conical locomotor structure composed of cilia which are completely fused; may beat in any direction. 2. Hairlike tuft on an insect appendage. 3. Small, slender projection on the surface of certain crinoid echinoderms; used for temporary attachment. 4. Filamentous thoracic appendages of barnacles. 5. Muscular male copulatory organ in many invertebrate phyla; used to transmit sperm into the female reproductive tract. 6. Tactile barbel in certain fishes. Pl. cirri.

cirrus sac. Sac or channel which contains the cirrus (copulatory organ) in a variety of invertebrates, e.g. tapeworm.

cisco. See LEUCICHTHYS. Pl. ciscos or ciscoes.

Cistenides gouldi. Atlantic Coast polychaete found commonly at low water line and below; tube trumpet-shaped, constructed of sand grains.

cisterna. Among vertebrates, a closed reservoir-like space, especially as applied to the subarachnoid spaces.

cisterna chyli. Large, thin-walled expansion of the thoracic duct in certain vertebrates.

cisterna magna. Subarachnoid reservoir between the ventral posterior surface of the cerebellum and the medulla oblongata.

Cistothorus. Genus of marsh wrens in the Family Troglodytidae; two widely distributed U.S. spp.

Citellus. Common genus of American rodents in the Family Sciuridae; ground squirrels, spermophiles; includes many spp. which burrow in open lands, feed on vegetation, and sometimes damage crops; Mississippi Valley and westward.

Citheroniidae. Family of insects which includes the regal moths (Order Lepidoptera).

citric acid cycle. Krebs cycle, tricarboxylic acid cycle; series of intracellular respiratory processes during which an acetyl group, derived from carbohydrates, fatty acids, or amino acids, is combined with four-carbon oxaloacetic acid to form the six-carbon citric acid, which is subsequently dehydrogenated and decarboxylated through a series of enzyme-catalyzed steps, thereby releasing carbon dioxide and water, again forming oxaloacetic acid, and providing energy for the synthesis of ATP molecules.

citrin. Vitamin P.

citrulline. Amino acid, first found in watermelons; $C_6H_{13}O_3N_3$.

citrus mealybug. See PSEUDOCOCCUS.

citrus nematode. Microscopic nematode (Tylenchulus semipenetrans); parasitic in roots of citrus plants.

Cittotaenia. Genus of cyclophyllidean tapeworms found especially in rodents.

civet. 1. Any of numerous spp. of small, catlike, nocturnal mammals in the Family Viverridae; body elongated, legs short, claws only partly retractile, coat often handsomely marked; certain spp. have glands which secrete civet; Africa and southern Asia; Viverra the most common genus. 2. Soft yellow or brown substance found in a pouch near the external genitalia of civet cats; used as a perfume fixative.

Civettictis. Genus of African civets.

Claassenia arctica. Uncommon western stonefly.

Cladistia. Order of bony fishes; elongated body covered with thick scales; base of pectoral fins fleshy and covered with scales; dorsal fin subdivided into eight or more finlets; tail diphycercal; vertebrae amphicoelous; air bladder lunglike; Polypterus and Calamoichthys examples.

Cladocarpus. Genus of marine coelenterates in the Suborder Leptomedusae.

Cladocera. Order of crustaceans which includes the water fleas; a folded carapace covers the body but not the head; true abdomen suppressed; segmentation indistinct; with five or six pairs of thoracic appendages; second antennae enlarged and used for swimming; usually 0.5 to 3.0 mm. long; 300 spp.; found everywhere in fresh waters; several marine spp.; common genera are Daphnia, Alona, and Chydorus.

Cladocopa. One of the four orders of Ostracoda; only one pair of trunk appendages; both pairs of antennae used in locomotion; heart absent; with a bent flat process at the posterior end of the body; all spp. marine; Polycope typical.

Cladodus. Genus of primitive Carboniferous sharks.

Cladonema. Genus of marine hydrozoan coelenterates with well-developed medusae.

Cladorhiza. Genus of long, slender marine sponges (Subclass Monaxonida).

Cladoselache. See CLADOSELACHII.

Cladoselachii. Order of extinct Devonian cartilaginous sharklike fishes; fins with broad base; Cladoselache common.

clam. Common name of typical members of the bivalve mollusk Class Pelecypoda.

clam shrimp. Common name of members of the crustacean Order Conchostraca.

clam worm. See NEREIS.

Clangula hyemalis. Old squaw; common sea duck of the northern part of the Northern Hemisphere; male mostly dark with white flanks, belly, and patch around the eye; long pointed tail.

Clarke-Bumpus plankton sampler. Quantitative plankton sampling device consisting of a silk net and metal frame; towed at the proper depth, the amount of water passing through the net is metered by a propeller and revolution counter; the device is opened and closed by means of metal messengers sent down the line.

Clark's nutcracker. See NUCIFRAGA.

claspers. Paired external copulatory processes, especially in male sharks and insects.

class. Major subdivision of a phylum; e.g. the Phylum Protozoa includes the following classes: Mastigophora, Sarcodina, Sporozoa, Ciliata, and Suctoria; in a few phyla some zoologists recognize certain superclasses and/or subclasses.

classification. Taxonomy.

Clathrina. Genus of marine sponges in the Class Calcarea.

Clathriopsamma. Genus of marine sponges.

Clathrostoma. Genus of ovoid, freshwater, holotrich Ciliata.

Clathrulina. Genus of stalked, spheroidal amoeboid Protozoa; main body mass contained in a latticed sphere of tectin.

Claudisium. Genus of marine cyclopoid copepods; commensal on Callianassa.

clava. Knoblike tip of the antenna in certain insects.

Clava. Common genus of marine hydromedusae; hydranths long and tapered, with scattered threadlike tentacles.

clavate. Club-shaped.

Clavelina. Genus of colonial pelagic tunicates in the Order Enterogona.

clavicle. Ventral bone of the pectoral girdle on each side; the collarbone of man.

Clavicornia. Taxon which includes numerous families of beetles having pronounced club-shaped antennae.

clavicularium. Epiplastron of turtles.

claviform. Club-shaped.

clavobrachialis. Sheet of muscle on the side of the lower neck and upper part of the chest in certain vertebrates.

Clavodoce. Genus of Polychaeta.

clavotrapezius. Wide sheet of muscle on the back and side of the neck in certain mammals.

Clavularia. Genus of anthozoan coelenterates in the Order Stolonifera; polyps grow upright from a stolon-like base; skeleton of warty spicules.

clawed toad. See XENOPIDAE.

clawless otter. See AONYX.

clearing. One of the stages involved in preparing sections, tissues, or small whole organisms for permanent mounts used for microscopical examination; soaking in clearing agents such as cedar oil, xylene, or toluol renders the tissues translucent.

clear-winged moths. See AEGERIIDAE.

cleavage. Repeated process of nuclear and cell divisions by mitosis in an animal ovum; corresponding process in plants usually termed segmentation.

cleg. Common name of various large biting flies.

cleidomastoid. Muscle between the clavicle and mastoid region of the skull.

Clemmys. Genus of spotted turtles and wood tortoises or wood turtles found in southeastern Canada, New England to Fla., and westward to Minn., also on Pacific Slope; heavy, keeled carapace with deep concentric grooves; moist woods, swamps, ponds, and sluggish waters.

cleptobiosis. Form of symbiosis in ants; small ants feed on refuse food or steal it from workers of a larger sp.

Cleridae. Family of predaceous beetles; larvae feed on eggs and larvae of nesting and burrowing insects.

Clethrionomys. Small genus of red-backed voles; two spp. in damp, cool forests of N. A.

Cleveland, L. R. American protozoologist, parasitologist, and cytologist (1892-); especially noted for his work on the physiological relationships between termites and the symbiotic flagellates in their guts.

Clibanarius. Genus of hermit crabs.

click beetle. See ELATERIDAE.

click mechanism. In insect flight, a pulling inward of the thoracic pleuron which produces a rapid conclusion of a wing beat upward or downward.

cliff swallow. See HIRUNDINIDAE.

Climacia. See SISYRIDAE.

Climacostomum. Genus of oval, flattened, holotrich Ciliata; brackish and fresh waters.

climacteric. Period of endocrine change and decreasing gamete production in men and women, culminating in the menopause for women.

climatic isolation. Prevention of interbreeding between two or more populations because of differential preferences in temperature, humidity, etc., often in adjacent areas; a special type of geographic isolation; the long-tailed vs. the mountain weasel is an example in the western U. S.

Climatius. Genus of placoderm fishes with large fin spines and five pairs of accesory ventral fins.

climax community. Climax formation; final or stable community in a successional series (sere); it is self-perpetuating, in equilibrium with the physical and chemical factors of the environment, and composed of a definite group of plant and animal spp.; for convenience, climax communities are usually characterized in terms of their dominant vegetation, e.g. prairie, taiga, tundra, beech-maple forest, oak-hickory forest, tropical rain forest, spruce-fir forest, etc.; some aquatic biologists are convinced that certain deep and/or cold lakes also represent a type of climax.

climax formation. See CLIMAX COMMUNITY.

climax species. Particular spp. of plants and animals characteristic of a climax community.

climbing perch. Perchlike "walking fish" of southeastern Asia in the genus Anabas; dark brown fresh-water spp. to 10 in. long; with thick skin and scales; able to move over the ground and even climb trees by means of the fins, tail, and opercula; suffocates if it has no opportunity to come to water frequently.

cline. Gradual and nearly continuous change of a character through a series of geographically continuous populations; character gradient.

clingfish. See XENOPTERYGII.

Clinidae. Kelpfishes; temperate and tropical family of small colorful tidepool fishes; body compressed, dorsal fin very long.

Clinocardium. See COCKLE.

Clinocottus. Genus of small tide-pool cottid fishes.

Clinostomum. Genus of distome trematodes found in the mouth of fish-eating birds.

Clinotanypus. Genus of tendipedid midges.

Clio. Genus of sea butterflies.

Cliona. Genus of sponges (Order Monaxonida); includes the boring sponges which bore through pieces of limestone to form anastomosing galleries; often destroy mollusks by erosion, boring, and envelopment.

Clione. Genus of sea butterflies; shell and mantle absent; body tapered to a posterior point; sometimes present in immense schools in the North Atlantic.

Clistogastra. One of the two suborders in the insect Order Hymenoptera; ants, bees, and wasps; base of abdomen constricted; larvae apodous; parasitic, omnivorous, and phytophagous.

Clitellio. Genus of earthworms found especially along the high-tide line.

clitellum. Thickened saddle-like portion of certain mid-body segments in many oligochaetes and leeches; in some spp. it is prominent only in sexually mature individuals; forms rings of epithelial tissue and mucus as cocoons in which eggs are enclosed; during copulation it also gives off mucus which envelopes the anterior ends of the two individuals.

clitoris. Small erectile body at the anterior angle of the female vulva; homologous to the penis in the male mammal.

cloaca. 1. Posteriormost chamber of the digestive tract in monotremes, birds, reptiles, amphibians, and many fishes; receives the feces, excretory products, and reproductive cells. 2. Terminal portion of the digestive tract in certain invertebrates in which it also functions as a respiratory, excretory, or reproductive duct.

cloacal aperture. 1. External opening of the cloaca. 2. Vent.

Cloeon. Genus of Ephemeroptera.

clone. All descendents derived from a single individual by asexual reproduction or parthenogenesis; members of a particular clone have the same genetic constitution.

Clonorchis. Genus of trematodes in the Order Digenea; C. sinensis is the common human liver fluke occurring over large areas of the Orient; first intermediate host is any of a variety of aquatic snails, and the second intermediate host is any of numerous fresh-water fishes (mostly Cyprinidae); man and fish-eating mammals become infected by handling or ingesting raw or poorly-cooked fish.

clonus. Irregular reflex contractions and partial relaxations of skeletal muscles; pathological convulsive spasma induced by abnormal motor neuron stimuli.

closed circulatory system. Any circulatory system in which the circulatory fluid is everywhere confined to definite vessels and does not circulate freely among the tissues or in tissue spaces; e.g. the vertebrate circulatory system.

closed community. 1. Any community in which all available ecological niches are presumably occupied. 2. Any community in which a plant canopy completely shades the soil surface.

closed society. Aggregation of individuals of the same sp. into which additional individuals will not be accepted; e.g. a pack of wolves or family of otters.

Clostridium. Large genus of anaerobic, spore-forming, gram-positive, rod-shaped bacteria; widely distributed in soils and intestinal tracts; includes spp. causing tetanus, gas gangrene, and botulism; others are important in commercial fermentation processes.

clot. Mass of coagulated blood, blood serum, lymph, etc. both in vertebrates and in many invertebrates.

clothes moth. See TINEA.

clotting. Series of processes involved in the formation of a blood clot; in man the injured tissues and disintegrating blood platelets release thromboplastin (thrombokinase) which combines with antiprothrombin (heparin), thus releasing prothrombin, which, in turn, combines with calcium ions from the plasma to form thrombin; the thrombin converts fibrinogen (a soluble blood protein) into fibrin which becomes a mass of fine fibrils in which the corpuscles become entangled, the resulting mass being a clot.

cloud-rat. See PHLOEOMYS.

cloven hoof. Hoof which is divided into two or more parts, e.g. cattle.

clover mite. See BRYOBIA.

Clubiona. Common genus of hunting spiders (Clubionidae).

Clubionidae. Family of two-clawed hunting spiders; usually make tubular retreats in litter; sexes similar; Micaria a large and common genus.

Clupea harengus. Common herring of North Atlantic; usually 6 to 8 in. long; used extensively as human food, either fresh, salted, smoked, or pickled; young canned as sardines.

Clupeidae. Family of bony fishes; shad and herring; mostly small marine spp. which breed in inshore waters; body long, scales cycloid; commonly in enormous schools; examples are Clupea, Sardinops, Dorosoma, Sardinia, and Alosa.

Clupeiformes. Order of fishes; includes the salmons, chars, trouts, pikes, herrings, shad, tarpon, and anchovies.

Clupeoidei. Taxon of fishes sometimes used to include the Elopidae, Clupeidae, Engraulidae, etc.

clutch. Nest of eggs; brood of young birds.

Clymenella. Genus of tube-building Polychaeta.

Clypeaster. Common genus of cake urchins.

Clypeastroida. Order of echinoderms in the Class Echinoidea; cake urchins and sand dollars; test depressed or flattened; anus on oral, aboral, or peripheral areas; peristomial gills absent; spines small; Clypeaster and Dendraster common.

clypeate. Shield-shaped.

clypeus. Median anterior plate on the surface of an insect head.

Clytemnestra. Common genus of marine copepods (Suborder Harpacticoida).

Clytia. Genus of marine coelenterates in the Suborder Leptomedusae; hydranths on a long stalk.

cnemial crest. Strong ridge on the front margin of the head of the tetrapod tibia.

cnemidium. Lower part of a bird's leg, generally scaly.

Cnemidocarpa. Genus of elliptical sessile ascidians.

Cnemidocoptes laevis. Itch mite of fowl; attacks skin.

Cnemidocoptes mutans. Small, flat mite which burrows under the scales on the legs of poultry; scaly-leg mite.

Cnemidophorus. Genus of whiptail lizards or racerunners in the Family Teiidae; common over much of southern U.S. but extending as far north as Minn.; also in Mexico, Central, and South America; extremely active spp. with very long tails.

Cnephia. Genus of black flies (Simuliidae).

cnida. Nematocyst.

Cnidaria. Coelenterata.

cnidoblast. See NEMATOCYST.

cnidocil. See NEMATOCYST.

cnidoglandular band. Septal filament.

cnidophore. Modified polyp or other specialized part of a coelenterate bearing abundant nematocysts.

Cnidosporidia. One of the four subclasses of Sporozoa; trophozoite amoeboid; each spore with one to four polar capsules.

C.N.S. Central nervous system.

coachwhip. Masticophis flagellum, a long slender colubrid snake of the southern states and Mexico; to 7 ft. long.

coaction. Range of complex ecological interactions among plants, plants and animals, and animals alone.

coagulation time. Time interval between collection of a blood sample and the first sign of clotting; for human blood about 3 to 4 min.

coagulin. Precipitin.

coagulum. Any coagulated mass.

coalfish. The pollack; see GADIDAE.

coarctate. With a narrow base and enlarged tip.

coarctate pupa. Type of dipteran pupa enclosed by the hardened last larval exoskeleton.

coati. See NASUA.

coat-of-mail shell. See CHITON.

cobalamin. Vitamin B_{12}; $C_{63}H_{90}N_{14}O_{14}PCo$; a growth vitamin which catalyzes formation of purine and pyrimidine deoxyribosides and functions in transmethylation; deficiency produces pernicious anemia; abundant in liver, fish, meat, milk, and eggs.

Cobb, N. A. Nematologist (1859-1932); associated with the U.S. Dept. of Agriculture; made many important contributions to nematode biology.

cobblerfish. 1. Common killifish (Fundulus heteroclitus). 2. Pompano (Trachinotus). 3. Threadfish (Alectis).

cobia. Sergeant fish (Rachycentron).

cobra. See NAJA.

cobweb. Web spun by a spider, especially a small, irregular web by a small spider.

co-carboxylase. Thiamine pyrophosphate.

Coccidae. Large family of insects in the Order Homoptera; coccids, soft scales, tortoise scales; adults flattened to globular, with obscure segmentation; many spp. of great economic importance as pests (e.g. Saissetia) but others secrete wax which is of commercial importance.

Coccidia. Order of sporozoan Protozoa; parasites of the digestive epithelium; both schizogony and sexual reproduction by anisogamy occur in the same host body; Eimeria common.

coccidiosis. Any of a large group of sporozoan infections of the digestive epithelium in many invertebrates and vertebrates.

Coccinella. Large and cosmopolitan genus of coccinellid beetles; especially beneficial since they feed on aphids and other small pest insects.

Coccinellidae. Family of minute to small beetles whose larvae and adults feed chiefly on plant exudates and soft-bodied insects, especially aphids and coccids; ladybird beetles, ladybugs; often brightly colored and spotted; Coccinella, Hippodamia, and Epilachna are familiar genera; Novius cardinalis an important Australian sp.

Coccolithophoridae. Family of marine plankton biflagellates in the Order Chrysomonadina; cell with bits of calcium carbonate (coccoliths) imbedded in a membrane investing the cell.

Coccomonas. Genus of fresh-water biflagellate Protozoa; cell enveloped in a large capsule.

Coccosteus. Genus of extinct placoderm fishes.

coccus. Spherical bacterium.

coccygeal glomus. Small vascular mass near the tip of the coccyx in man.

coccygeal vertebra. Small vertebra at the end of the spinal column; four present in man.

coccygeofemoralis. Ventral muscle of the bird leg.

coccygeoiliacus. Longitudinal muscle on either side of the midline in the posterior part of the body wall in certain vertebrates.

coccygeosacralis. Small, flat muscle in the dorsal body wall of certain vertebrates.

coccyx. End of the vertebral column beyond the sacrum in primates; in man, consists of three to five fused vestigial vertebrae.

Coccyzus. Genus of birds in the Order Cuculiformes; includes the yellow-billed and black-billed cuckoos of North and South America.

cochineal. Crimson dye commercially extracted from the dried bodies of the cochineal insect Dactylopius coccus; this sp. lives especially in Mexico and feeds on the opuntia cactus.

cochlea. That tubular part of the membranous labyrinth which is concerned with the reception of sound vibrations; it is an outgrowth from the sacculus and occurs only in crocodiles, birds, and mammals; in mammals it is spirally coiled, tapered, and divided into three chambers as seen in cross section.

cochlear canal. Scala media.

cochlear duct. Scala media.

Cochleariidae. Family of birds in the Order Ciconiiformes; boat-billed heron (Cochlearis cochlearis); a broad-billed sp. of Central American mangrove swamps.

Cochliocopa. Circumpolar genus of land snails.

Cochliodus. See BRADODONTI.

Cochliomyia. Callitroga.

Cochliopa. Small genus of fresh-water snails; southern Tex.

Cochliopodium. Genus of fresh-water amoeboid Protozoa; most of cell in a hemispherical test.

Cochlodesma. Common genus of marine bivalve mollusks; shell thin, oval.

cockateel. Small Australian parrot; see PSITTACIFORMES.

cockatoo. Any of many brightly-colored gregarious parrot-like birds of the Australian region; usually have pronounced crests; feed on seeds and insects; Kakatoe a common genus. See PSITTACIFORMES.

cockchafer. See MELOLONTHIDAE and SCARABAEIDAE.

cockle. Any of certain marine heart-shaped bivalve mollusks with strong radiating ribs and sawtooth margins; to 6 in. across; in mud and sand of shallows; Cardium edule the common edible European sp.; other typical genera are Clinocardium, Trachycardium, and Dinocardium.

cock-of-the-rock. One of several brightly colored crested birds of Tropical America; see RUPICOLA.

cockroach. Common name of insects in the Order Blattaria.

cocoon. 1. Specialized capsule or other structure in which many aquatic invertebrates deposit their eggs, as in certain Turbellaria, leeches, and earthworms. 2. Silk or other fibrous material which is spun or constructed by many insect larvae as a protective covering for the subsequent pupa. 3. Silk or other fibrous material which is spun by certain terrestrial arthropods, especially spiders, and is used as a covering for egg masses.

Codakia. See LUCINA SHELLS.

codehydrogenase. Name sometimes applied to DPN and TPN.

codfish. See GADIDAE.

codling moth. See CARPOCAPSA.

codominant species. Two or more equally dominant spp. in a habitat; one of several plants or animals dominating a community, no one to the exclusion of the others; e.g. the hawk and coyote are codominants of a prairie community.

Codonella. Genus of fresh-water oligotrich Ciliata; in urn- to pot-shaped loricas.

Codonoeca. Genus of marine and fresh-water uniflagellates; cell in a stalked, vase-shaped lorica.

Codosiga. Fresh-water genus of small colorless flagellates; cell on a stalk and with a thin conical collar surrounding the base of the single flagellum.

Coelacanthini. Order of lobe-finned marine fishes; head with both bone and cartilage; pineal eye absent; tail with three groups of fin rays; Macropoma from Upper Cretaceous; Latimeria modern, a few specimens having been caught recently off the east coast of Africa.

Coelambus. Hygrotus.

Coelenterata. Cnidaria; phylum which includes the hydroids, jellyfishes, corals, and sea anemones; body with a fundamental radial saclike construction; a single opening (proctostome) to the gastrovascular cavity serves as both mouth and anus; nematocysts present; body wall composed of epidermis, mesoglea, and gastrodermis; mesoglea with or without tissues of mesodermal origin; with a diffuse nerve net; many spp. are polymorphic and exhibit metagenesis; mostly marine but a few in fresh waters; 10,000 spp.

coelenteron. Gastrovascular cavity.

coeliac. Pertaining to the abdominal cavity.

coeliac artery. Artery in vertebrates usually supplying the esophagus, stomach, duodenum, spleen, pancreas, liver, and gall bladder.

coeliaco-mesenteric artery. Short artery supplying the abdominal viscera in many vertebrates.

coeloblastula. Simple blastula whose cavity (blastocoel) is formed by the drawing apart and further division of the cells, leaving a single layer enclosing the blastocoel.

Coeloceras. Genus of extinct ammonites.

Coelocormus. Genus of compound ascidians.

coelogastrula. Typical gastrula derived from a coeloblastula.

coelom. Main or secondary body cavity of higher phyla; it is formed between layers of mesoderm and is everywhere lined with epithelium of mesodermal origin. Adj. coelomic and coelomate.

Coelomata. Taxon including all phyla having a true coelom.

coelomic cavity. See COELOM.

coelomic epithelium. Layer of mesodermal epithelium which covers all organs and lines all surfaces in a coelom.

coelomic pouch. Mesodermal pouch; one of the paired embryonic dorsolateral outpocketings which grow outward from the endoderm, especially in echinoderms and certain chordates; forms a double layer of mesoderm on each side, the enclosed cavities being the embryonic coelom.

coelomocytes. Corpuscles (usually amoebocytes) in the fluid of the coelom or pseudocoel of invertebrates.

coelomoduct. In the broad sense, a type of nephridium; a paired ciliated duct of mesodermal origin having one opening on the external surface of the body and the other internally in the coelom (coelomic sac); in onychophorans and mollusks; excretory and/or reproductive function(s).

coelomostome. Internal opening of a coelomoduct.

Coelomyaria. Taxon sometimes used to include those nematodes in which the protoplasmic portion of the musculoepithelial cells bulges into the pseudocoel and in which the fibrillar portion of these cells extends up its sides.

Coeloplana. Aberrant genus of Ctenophora; flat, oval, creeping on the sea bottom and on coelenterate colonies.

Coelotanypus. Genus of tendipedid midges.

coelozoic. Inhabiting a cavity within the body of an animal.

Coenagrion resolutum. Danselfly of northern states and Canada.

Coendou. Genus of tree-dwelling Central and South American porcupines; coendous; tail prehensile.

coenenchyme. Common tissue connecting adjacent polyps in alcyonarian coelenterates; consists of mesoglea, imbedded spicules, and epidermis.

Coenobita. Genus of tropical land hermit crabs.

coenobium. Colony of Protozoa having a constant shape, arrangement, and number of cells. Pl. coenobia.

coenoblast. Embryonic germ layer giving rise to both endoderm and mesoderm.

coenoecium. Common secreted investment of a bryozoan colony; may be gelatinous, chitinoid, or calcareous.

coenosarc. Axial living portion of the upright branched parts of certain hydrozoan coelenterate colonies; often surrounded by a dead secreted perisarc.

coenosite. Commensal organism which is also able to live independently.

coenospecies. Collectively, those related spp. which may hybridize.

coenosteum. Calcareous skeleton of a colonial coral.

Coenothecalia. Order of alcyonarian anthozoans; skeleton a bluish mass of fused crystalline calcareous fibers; polyps brownish; includes only the blue coral (Heliopora) of Pacific Ocean coral reefs.

coenurus. Hollow, fluid-filled, bladderlike structure containing many larval tapeworm scoleces attached to its inner surface; imbedded in host tissues, especially the brain; formed in the life history of certain tapeworms.

coenzyme. Any of several organic cellular substances which play accessory roles in enzyme-catalyzed processes; often acts as a carrier for some transient compound; some are not so specific in their function as the enzymes themselves and may be involved in reactions catalyzed by two or more enzymes; examples are adenosine triphosphate, riboflavin, and thiamin.

coenzyme I. Diphosphopyridine nucleotide.

coenzyme II. Triphosphopyridine nucleotide.

coenzyme A. Coenzyme derived from the union of pantothenic acid and betathioethylamine plus adenosine 3'-5'-diphosphate; functions chiefly in oxidative decarboxylation in cell metabolism, and in breakdown and synthesis of fatty acids.

Coerebidae. Drepanididae.

coffee-bean shell. See TRIVIA.

Cogia. Kogia.

Cohnheim's fields. Discrete longitudinal groups of contractile fibrillae (sarcostyles) visible in a transversely cut voluntary muscle fibril; sarcoplasm fills the spaces between adjacent Cohnheim's fields.

Cohnilembus. Genus of slender, spindle-shaped, holotrich Ciliata; marine, fresh-water, and endozoic.

coitus. Copulation; coition.

Colacium. Genus of stalked green colonial Protozoa; often attached to small fresh-water invertebrates; free-swimming stage flagellated.

Colanthura. Genus of marine Isopoda.

Colaptes. Common U.S. genus of flickers in the Family Picidae; large, brightly colored, woodpecker-like spp. which drill into trees for insects but also feed on animal and vegetable matter on the ground.

cold-blooded. Poikilothermous.

cold light. Light given off by bioluminescent organisms; involves relatively little heat.

Coleonyx. Genus of geckos; southwestern U.S. to Panama.

Coleophoridae. Family of small grayish or brownish moths whose larvae construct portable silken cases; casebearers; larvae are minor pests of fruit trees.

Coleoptera. Largest insect order; over 250,000 spp.; beetles and weevils; minute to large insects with tough integument, complete metamorphosis, biting or chewing mouth parts; forewings (elytra or wing covers) useless in flight, thick, leathery, veinless, and meeting along midline; hind wings membranous, with few veins, and folded under elytra when at rest; prothorax large, movable; meso- and metathorax united to abdomen; active larva (grub) with legs; pupa exarate; over 150 families.

Coleps hirtus. Common barrel-shaped ciliate protozoan of fresh and salt waters; ectoplasm with small regularly-arranged plates, often with small posterior spines.

Colias philodice. Clouded sulfur butterfly; caterpillar a pest on alfalfa, clover, etc.

Colidotea. Genus of Isopoda usually found clinging to the spines of sea urchins.

coliform. Pertaining to the bacterium Escherichia coli and other spp. normally found in the human large intestine.

Coliidae. Colies or mouse birds; only family of long-tailed African birds in the Order Coliiformes; creep about on tree branches; six spp. of Colius.

Coliiformes. See COLIIDAE.

Colinus virginianus. American bobwhite or quail in the Family Phasianidae; formerly common only in eastern N.A. but widely introduced in western states; small, chicken-like bird to 10.5 in. long; nests on ground; remarkably successful under farming and ranching conditions.

colitis. Inflammation of the colon.

Colius. See COLIIDAE.

collagen. Fibrous protein material in bone, tendons, and other connective tissues; in small amounts it binds together many cells and tissues; relatively inelastic.

collarbone. Clavicle.

collar cell. Choanocyte.

collared lizard. See CROTAPHYTUS.

collateral ganglion. An autonomic ganglion lying outside the synpathetic chain, e.g. coeliac ganglion.

collecting tubule. One of the larger tubules in the medullary portion of the kidney of higher vertebrates; formed by the confluence of numerous small urinary tubules.

Collembola. Order of small insects (springtails) of marshy and damp places, especially in the debris of woodland and forest floor; to 5 mm. long; whitish or colored; no wings, compound eyes, or Malpighian tubules; no metamorphosis; chewing mouth parts; tracheae usually absent; able to leap by the action of a ventral abdominal springing organ (furcula) which is held in place by a trigger-like hook (hamula); common genera are Sminthurus, Achorutes, and Entomobrya.

collenchyma. Undifferentiated mesenchyme cells loosely scattered through gelatinous or other non-cellular material, especially in the lower metazoans.

collencytes. Stellate amoebocyte connective tissue cells of sponges.

colleterial glands. Paired tubular reproductive glands in certain female insects; provide a sticky fluid which forms an egg case or fastens the eggs to surfaces.

Colletes. Large and common genus of bees; mostly blackish, solitary spp.; burrow in sand or clay soil.

colliculus. Corpus quadrigeminum.

Collinia. Anoplophrya.

colloblasts. Adhesive cells; special sticky cells covering much of the surface of the two tentacles of ctenophores. aid in food-getting.

Collodictyon. Genus of colorless, spherical to heart-shaped, fresh-water Protozoa; four anterior flagella.

colloid. See COLLOIDAL SYSTEM.

colloidal system. Permanent suspension of finely divided particles; in living organisms the particles (dispersed phase) are commonly molecules or aggregates of molecules of proteins, lipids, and polysaccharides suspended in the watery fluid (liquid phase) of cells; most particles suspended in a typical colloidal system are between 0.000001 and 0.0005 mm. in diameter; colloidal particles will not pass through the plasmalemma of cells, and they remain in suspension largely by virtue of their size and the electrical charge on the individual particles; death of cells brings about destruction of the colloidal system and coagulation of the protoplasm; milk is a colloidal suspension of fat and protein particles in water; a syspension of a liquid in another liquid is an emulsion type of colloidal system, and many types of protoplasm appear to have this physical structure; protoplasmic colloids may undergo reversible sol-gel transformations; when water is "withdrawn" from the usual sol (fluid) stage of cytoplasm, the original dispersed phase becomes a continuous spongelike network which holds water droplets in its meshes, and the result is the more solid gel state of the colloid (a liquid suspended in a solid). Cf. PROTOPLASM and CRYSTALLOID.

collophore. Ventral tubular projection on the first abdominal segment of Collembola.

Collosphaera. Genus of Radiolaria.

Collotheca. Genus of sessile, case-building, fresh-water rotifers.

Collothecacea. Order of sessile rotifers in the Class Monogononta; no lorica, but often with a secreted tube; corona very large, with a single ciliary wreath.

Collozoum. Common genus of Radiolaria.

collum. 1. Neck. 2. Collar or necklike part of a structure, e.g. dorsal plate of first body segment of a millipede.

Collyriclum. Genus of monostome trematodes parasitic in cysts on passerine birds.

Colobidae. Family of leaf-eating monkeys in the Superfamily Cercopithecoidea; langurs, sakis, and titis.

Colobognatha. Order of millipedes; 30 to 60 body segments; male with one pair of gonopods each on seventh and eighth segments; Platydesmus in southern U.S.

Colobus. Guerazas; genus of large African monkeys in the Superfamily Cercopithecoidea; hair long, black and white; thumbs absent or vestigial.

Colocephali. Order of bony fishes which includes marine moray eels in the Family Muraenidae; brightly colored, voracious, and mostly tropical and subtropical; Muraena typical.

colon. Large intestine of vertebrates, exclusive of the terminal rectum or cloaca.

colonial animal. Any sp. living in groups of separate, attached, or incompletely separated individuals, as seals, barnacles, Hydrozoa, Bryozoa, etc.

Colorado potato beetle. See LEPTINOTARSA.

Colorado tick fever. Non-fatal virus disease of man transmitted by the tick Dermacentor; infections characterized by malaise and high fever, but no rash.

color blindness. Human sex-linked recessive character marked by inability to distinguish red from green; much more common in males; some color-blind individuals have difficulty only with certain pastel shades; a very few are totally blind to all color.

Colossendeis. See PYCNOGONIDA.

colostrum. Special type of protein-rich mammalian milk secreted during the first few days before and after birth of young.

colpeo. See DUSICYON.

Colpidium. Common genus of marine and fresh-water holotrich Ciliata; elongated, slightly flattened, with preoral region curved to right.

Colpius. Genus of dytiscid beetles.

Colpocephalum. Genus of bird lice in the Order Mallophaga.

Colpoda. Common genus of flat, kidney-shaped, holotrich Ciliata; stagnant fresh waters.

Coluber. Genus of long, slim, non-poisonous snakes; N.A. and northeastern Asia; coachwhips, whipsnakes, blacksnakes, and racers; active spp. which may climb trees and bushes; feed on small mammals, reptiles, and frogs; do not constrict prey but may kill it by pressing it against the ground; C. constrictor is the common blacksnake, blueracer, or hoop snake found over most of U.S.; to 6 ft. long.

Colubridae. Very large cosmopolitan family of snakes; contains about 75% of known snakes; both jaws with solid or grooved teeth; habits vary widely; most spp. harmless, a few poisonous; representative genera: Natrix, Thamnophis, Coluber, Masticophis, Pituophis, Lampropeltis, Boiga, and Tantilla.

colulus. Slender conical organ lying just anterior to the spinnerets of certain spiders.

Columba. 1. Genus of birds in the Family Columbidae; pigeons; C. fasciata is the band-tailed pigeon of forested western U.S. and south to Central America; distinguished from the mourning dove by its broad rounded tail; domesticated pigeons derive from the wild rock pigeon (C. livia) of Europe. 2. See SALMOPERCAE and PERCOPSIDAE.

Columbella. See DOVE SHELL.

Columbidae. Cosmopolitan family of birds (289 spp.) in the Order Columbiformes; pigeons and doves; Columba, Ectopistes, and Zenaidura typical.

Columbiformes. Order of herbivorous land birds which includes pigeons and doves; bill usually short, slender, and with a cere at base; crop large, produces "pigeon's milk" used in feeding young; cosmopolitan distribution.

Columbigallina passerina. Ground dove in the Family Columbidae; small dove of southern U.S. to Ecuador and Brazil.

columella. 1. Rod of bone or cartilage connecting and transmitting vibrations from the tympanum to the inner ear in reptiles, birds, and Anura; homologous with the hyomandibula of fishes and the stapes of mammals. 2. Axial portion of a snail shell around which the coil is arranged. 3. In certain corals, a central column in the cuplike skeleton formed by an individual polyp. 4. Central bony axis of the cochlea.

columella muscle. Large muscle in a snail; attached in the upper portion of the spire and used to draw the soft parts into the shell.

columellar lip. Internal lip of the aperture of a gastropod shell.

columnar epithelium. Epithelium composed of cells that are very tall, usually three or more times their width, e.g. the lining of the vertebrate stomach and intestine; sometimes such tissue is ciliated, as the lining of the human trachea.

Columnaria. Genus of fossil corals.

column chromatography. Analysis and separation of dissolved materials from each other by percolating the solution through a proper absorptive and porous material in a cylindrical container; each substance forms a transverse (colored) band at a particular level in the column.

Colurella. Genus of small rotifers; strongly compressed; lorica of two lateral plates open along anterior, ventral, and posterior margins; chiefly in fresh water.

Colus. Common genus of small to large carnivorous marine snails.

coly. See COLIIDAE.

Colymbetes. Common genus of dytiscid beetles.

Colymbiformes. Discarded name for Podicipitiformes.

Colymbus auritus. Horned grebe, a sp. in the bird Order Colymbiformes; ponds and lakes in northern part of Northern Hemisphere; excellent swimmer and diver; this sp. sometimes placed in Podiceps. See EARED GREBE.

Comactinia. Genus of feather stars occasionally found in American waters.

comatulid. Feather star; a free-swimming stalkless echinoderm.

comb. Fleshy median crest on top of the head in the chicken, pheasant, and a few other birds.

combed ark. See BLOODY CLAM.

comb jelly. Any member of the Phylum Ctenophora.

comb plate. Short ridge bearing very large cilia which are more or less fused and have a superficial resemblance to a short, ragged comb; arranged in eight meridional rows in the Phylum Ctenophora, and the animal moves by waves of beats of the comb plates along each row.

comedo. Blackhead.

comma. See ANGLE-WING.

commensal. One of the two members of a commensalism association.

commensalism. Close association between two different spp. whereby one member of the association derives an advantage and the other member has neither an advantage nor a disadvantage; examples are the harmless Protozoa in the human large intestine and the special barnacles attached to the surface of whales.

commiscuum. Group of individuals all of which can potentially interbreed.

commissure. 1. Band of nerve fibers connecting two distinct areas of the brain or spinal cord, especially in vertebrates. 2. Bundle of nerve fibers connecting paired ventral ganglia of certain Arthropoda, Annelida, etc.

common carotid artery. Carotid arch.

communalism. Association of three or more insect castes in a permanent colony.

community. Collectively, all of the organisms inhabiting a common environment and interacting with each other; this term is used with a wide variety of implications by different ecologists; some include only the living organisms within the concept; others maintain that the living organisms cannot be separated from their inanimate surcoundings and that "community" must include both of these major aspects; "community" is also variously used with reference to very small to very large units; thus, a culture jar of pond water and its contained organisms is a community; at the other extreme, all organisms inhabiting the Great Plains area of the U.S. form a community. Cf. ECOSYSTEM.

community succession. See SUCCESSION.

comparative morphology. Study of groups of animals with respect to their fundamental structural differences and similarities.

compass. One of the five slender radial ossicles on the aboral surface of Aristotle's lantern.

compensation point. 1. In aquatic habitats, that depth at which algae in culture bottles strike a balance between photosynthesis and respiration. 2. That depth at which algal photosynthesis in the natural plankton balances respiration and decomposition processes for all the plankton. 3. That light intensity at which the release of photosynthetic oxygen equals the utilization of respiratory oxygen.

compensatory hypertrophy. Increase in size of remaining tissue when a part or counterpart has been removed, is functioning improperly, or is nonfunctional; e.g. the enlargement of the remaining kidney when one is removed or the enlargement of the thyroid in cases of simple goiter.

competitive inhibition. When two processes are competing for a common source of raw material, one process will become dominant by diversion of the raw material, thus inhibiting the second process; especially true in biochemistry and enzyme systems.

complement. Nonspecific protein material in blood plasma and cytoplasm of cells; in combination with specific antibodies in the blood it causes the destruction or inactivation of the corresponding antigen; the antigen may be protein, polysaccharide, bacteria, blood corpuscles, etc.

complemental air. Amount of air which can be taken into the lungs in excess of the amount normally inspired; about 2500 ml. in man.

complemental male. 1. Small, degenerate male living attached to a female; functions only in reproduction; e.g. angler fish and Bonellia. 2. Small male form associated with the usual hermaphroditic form of a sp., e.g. some barnacles.

complementary genes. Interaction of two or more dominant alleles of corresponding genes, all of which are necessary for the production of a character.

complement fixation. Binding or fixation of complement along with the union of an antibody and its corresponding antigen; the basis of certain diagnostic tests for syphilis and gonorrhea.

complete metamorphosis. See HOLOMETABOLA.

complete protein. Any protein containing the amino acids essential to the requirements of a particular animal.

complex metamorphosis. See HOLOMETABOLA.

compound eye. Complex eye of typical insects and some Crustacea; usually consists of many long tapering elements (ommatidia), each with light-sensitive cells and a refractive system which can form an image; the ommatidia are separated from each other by varying degrees of pigmentation; some compound eyes form a simple apposition or mosaic image in which each ommatidium focuses only a narrow pencil of light parallel to its long axis, and constitutes only a very small part of the whole field of view; superposition images are formed by the overlapping of visual fields of adjacent fields of adjacent ommatidia; the outer surface of the compound eye is covered with a transparent cuticle (cornea) which is divided into hundreds or thousands of square or (usually) hexagonal facets or lenses, each of which represents the outermost portion of an ommatidium.

compound gland. Any gland having two or more branches.

compound microscope. 1. In the broad sense, any microscope with two or more lenses. 2. In the usual sense, the familiar student microscope having a set of lenses forming the ocular and another set of lenses built into an objective; most such microscopes are equipped with two or three objectives and have magnifications of 50 to 440 or 1000.

compressed. Flattened from side to side.

compressorium. Simple device for examining objects microscopically while they are being subjected to slight and graduated mechanical pressure.

Compsilura. See TACHINIDAE.

Compsognathus. Genus of small Jurassic dinosaurs; total length to 30 in.; hind legs exceptionally long.

Compsothlypidae. Mniotiltidae.

Comstock, John Henry. American entomologist (1844-1919); associated with Cornell University for many years and the author of An Introduction to Entomology.

concealing coloration. Any color or color pattern which more or less blends with an animal's background and thus presumably affords some protection from predators or concealment from prey.

conch. Any of a wide variety of large, heavy, marine gastropods of several genera; the shell is sometimes made into cameos, ornaments, or a kind of horn; tropical and subtropical; some spp. are called cameo or helmet shells; Strombus the common American genus; see also FASCIOLARIA.

concha. 1. Anatomical term used to apply to various shell-shaped structures. 2. That part of the outer ear which projects from the head. 3. Turbinated bone, as in the nasal passages of higher vertebrates.

conchiolin. Albuminoid material forming much of the organic matrix of a mollusk shell.

Conchoderma. Genus of barnacles common on whales and ships; plates vestigial.

Conchoecia. Genus of Ostracoda of northern oceans; one of the most common of all Myodocopa.

conchology. Study of mollusk shells.

Conchophthirus. Genus of oval, flattened, holotrich Ciliata; in mantle cavity and gills of marine and fresh-water mussels.

Conchostraca. Order of Crustacea which includes the clam shrimps or pod shrimps; body compressed and enclosed in a carapace consisting of two lateral shell-like valves; usually 3 to 10 mm. long; on the mud bottom of temporary ponds, especially in the Great Plains area; Lynceus and Leptestheria common.

concrescence. Modified type of epiboly occurring during gastrulation of the chick embryo.

conditioned reflex. Reflex action which has been modified by experience, particularly by the substitution of an abnormal stimulus for the normal one; most familiar example involves ringing a bell when food is placed in a dog's mouth; after sufficient repetitions the mere ringing of the bell alone is enough to cause the dog to salivate by reflex action.

condor. See GYMNOGYPS and VULTUR.

Condylactis. Genus of large tropical sea anemones.

Condylarthra. Order of extinct early ungulates; Paleocene to Eocene.

condyle. 1. Prominence on a bone, especially on the end of a bone; usually serves to articulate with an adjoining bone, often fitting into a socket. 2. Similar process on an arthropod segment.

Condylostoma. Genus of marine and fresh-water spirotrich Ciliata; cell ellipsoid, anterior end truncate; wide anterior peristome.

Condylura cristata. Star-nosed mole; body to 5 in. long; brownish black coloration; uses surface runways and tunnels in wet meadows and marshes; southeastern Canada and northeastern U. S.

conenose. Any of certain hemipteran insects (Family Reduviidae) having the base of the beak cone-shaped; some (Triatoma) are blood suckers which transmit protozoan parasites; mostly southern and western U. S.

cone-nosed snake. See CARPHOPHIS.

Conepatus mesoleucus. Common hog-nosed skunk or rooter skunk of the Southwest and Mexico; as large as the familiar striped skunk but with a single median white stripe from crown to end of tail; muzzle naked and hoglike; feeds on insects, small vertebrates, and cactus fruits; habits similar to those of other skunks but more of a digger.

cones. Elongated cone-shaped receptor cells in the retina of the vertebrate eye; in man a cone is about 30 microns long and 7 microns wide; especially concerned with bright light and color vision; much less numerous than rods; each cone is associated with a neuron.

cone shells. Large group of tropical marine carnivorous gastropods (Family Conidae); shell cone-shaped and often brightly or delicately colored with yellows and browns; to 4 in. long; some spp. are expensive collectors' items; Conus common.

coney. 1. See HYRACOIDEA. 2. Any of several unrelated marine fishes.

confused flour beetle. See TRIBOLIUM.

congeneric. Pertaining to a group of spp. in the same genus.

congenital. Present at or before birth; not necessarily inherited.

Conger. Genus of marine eels attaining 7 ft.; dorsal fin begins just behind the head; upper jaw projects past the lower; do not enter fresh water.

conger eel. See CONGRIDAE and CONGER.

Congeria leucophaeata. Small mussel-like marine bivalve; to 0.8 in. long; attached to substrate in clumps; N.Y. to Mexico.

congo eel. See AMPHIUMIDAE.

congo snake. Congo eel; see AMPHIUMIDAE.

Congridae. Widely distributed family of conger eels; marine spp. of the littoral and continental slope; thick-bodied, heavy eels to 9 ft. long and much larger than Anguilla; upper jaw projects beyond the lower; Conger common.

Conidae. See CONE SHELLS.

coniferous forest. Forest consisting chiefly of gymnosperm evergreen trees such as pines, spruces, firs, etc.

Conilurus. Genus of Australian jumping jerboa rats.

Coniophanes. See BLACK-BANDED SNAKE.

Coniopterygidae. Small neuropteran family of very small to minute insects more or less covered with powdery wax; dustywings; larvae predatory.

conjoined twins. Siamese twins.

conjugant. One of the two temporarily attached individuals during the process of conjugation in Paramecium and other Ciliata. Cf. EX-CONJUGANT.

conjugated protein. Protein molecule containing some radical(s) other than amino acids, e.g. hemoglobin and nucleoproteins. Cf. SIMPLE PROTEIN.

conjugation. 1. In many Ciliata and Suctoria, a temporary union of individuals in pairs during which there is an exchange of nuclear material and a complex series of micro- and macronuclear divisions and reorganization; after separation, each of the two conjugants has several further nuclear and cell divisions before normal metabolism and binary fission are resumed; the process of conjugation effects genetic reshuffling and perhaps some "rejuvenation." 2. Union of male and female gametes to form a zygote. 3. Syngamy. 4. Synapsis; see MEIOSIS.

conjunctiva. 1. Outermost thin epithelial layer covering the exposed surface of the vertebrate eyeball. 2. Flexible portion of exoskeleton between adjacent segments in arthropods, especially insects.

Conklin, Edwin G. American biologist (1863-1952); professor at Princeton University; principal contributions in cytology, embryology, and genetics.

connective tissue. Any animal tissue containing an abundance of dead, secreted, intercellular material; the latter may be fibrous or a homogeneous matrix; blood, bone, fat, ligaments, fibrous tissue, cartilage, etc. are all examples.

connexivum. Flattened peripheral portion of the abdomen of certain Hemiptera.

Connochaetes. Genus which includes a small South African gnu; cf. GORGON.

Conochiloides. Genus of limnetic rotifers; case gelatinous; solitary or colonial.

Conochilus. Genus of fresh-water colonial rotifers in which the individuals in a colony radiate out from a central point by means of their elongated feet.

conodont. Minute toothlike Paleozoic fossil having one or more cusps; presumed to be cyclostome tooth but may be some type of invertebrate remains.

Conopeum. Genus of marine Bryozoa.

Conopia. Genus of N.A. moths; blue wings marked with yellow; larvae of two spp. are important borers in peach trees.

Conopidae. Small family of thick-headed flies; small to medium-sized elongated wasplike spp.; larvae parasitic in bees and wasps.

Conopophagidae. Small family of birds in the Order Passeriformes; ant pipits, gnateaters; somber brown and green; to 5.5 in. long; Amazon forests.

Conotrachelus nenuphar. Important weevil pest; plum curculio; larvae feed on apples, plums, cherries, and peaches.

Conradilla caelata. Unionid mollusk in Tennessee, Elk, and Flint rivers.

consanguineous marriage. Marriage between persons of near kin.

conscutum. In certain ticks, a large

dorsal shield formed by the fusion of scutum and alloscutum.

consociation. Climax animal community dominated by a single sp.

consocies. 1. Portion of an association in which not all of the dominant spp. are present. 2. Developmental community with a single dominant.

consortes. Associated organisms of different spp., other than symbionts, commensals, parasites, and hosts. Sing. consors.

conspecies. Subspecies; variety.

conspecific. Pertaining to individuals or populations in the same sp.

constipation. Retention or infrequent difficult evacuation of the feces; opposite of diarrhea.

constrictor. 1. Muscle which compresses an organ or decreases the size of a cavity or opening. 2. See BOIDAE.

consumers. Collectively, those organisms in an ecosystem which feed upon other organisms; often divided into primary consumers (plant eaters), secondary consumers (carnivores which eat primary consumers), etc.

contact poison. Chemical substance, especially an insecticide, which kills the animal without being ingested; may plug the spiracles or penetrate the body surface.

contagious. Transmissible from one individual to another, as a disease.

contagium. Any organism capable of producing a communicable disease.

Contarinia pyrivora. Pear midge; a small dipteran whose larvae are sometimes pests in pears.

Contia. See SHARP-TAILED SNAKE.

continental drift. Theory which suggests that all continents were once united into a single mass and have since separated and moved apart.

continental island. Island in shallow water close to a main continental land mass, e.g. the British Isles, Long Island, and Vancouver Island; the flora and fauna of such islands are similar to those of the adjacent mainland.

continental shelf. That portion of the sea and the bottom between the tidal zone and a depth of 100 fathoms.

continental slope. That portion of the sea and the bottom between depths of 100 and 1000 fathoms.

continuous phase. In a colloidal system, the medium in which particles are suspended, e.g. water in the case of milk and most protoplasm.

continuous spindle fibers. Mantle fibers.

continuous variations. Small gradations in genetic characteristics owing to the quantitative effects of multiple pairs of genes having the same general effect, e.g. the many possible gradations between Negro and white skin color.

continuum. Pattern of gradually overlapping populations of different spp. along a gradient.

Contopus. Common genus of U.S. pewees in the Family Tyrannidae; small forest birds.

contour feather. One of the many outermost feathers of a bird; collectively, they establish the general body contours.

contraception. Prevention of conception or impregnation.

contractile vacuole. One or more clear, fluid-filled cell vacuoles in fresh-water Protozoa and a few lower metazoans; functions in a cyclic manner chiefly by taking up water from the surrounding protoplasm, growing, bursting, and releasing its contents to the outside, and then reforming; although the contractile vacuole may function incidentally in the elimination of nitrogenous wastes, its main role is osmoregulation.

Conulata. Group of fossil Scyphozoa; with a pyramidal sclerotized and calcium phosphate skeleton; middle Cambrian to early Triassic.

conure. Any of numerous small parrot-like tropical American birds with pointed tails.

Conuropsis carolinensis. Carolina parakeet; a true American parrot, once common from Va. to Wis. and Colo. and southward to the Gulf, but extinct since about 1925; formerly trapped and shot as a fruit pest, and for its brightly colored plumage.

Conus. See CONE SHELLS.

conus arteriosus. 1. Muscular valvular region of the vertebrate ventral aorta. 2. Conical portion of the right ventricle where the pulmonary artery originates.

convergence. Morphological similarity among distantly related forms, e.g. the similar shape of whales and fishes.

convergent evolution. Evolutionary processes resulting in superficial morphological similarity among distantly related forms, e.g. the wings of bats, flying reptiles, and birds.

Convoluta. Genus of small, free-living

flatworms in the Order Acoela; sandy intertidal zone; green color produced by symbiotic algae.

convoluted tubule. That portion of a urinary tubule of the higher vertebrate kidney which is especially tightly coiled and is situated near its corresponding Bowman's capsule.

cony. 1. See OCHOTONIDAE. 2. Any of several unrelated marine fishes.

coon oysters. Small oysters, especially those occurring in clusters on mangroves or other marine and brackish vegetation.

coon-striped shrimp. Pandalus danae, common marine shrimp; Calif. to Alaska.

Cooper's hawk. See ACCIPITER COOPERI.

coot. See FULICA.

cooter. 1. See PSEUDEMYS. 2. Box tortoise. 3. Snapping turtle.

cootie. Body louse.

Cope, Edward D. American naturalist (1840-1897); made many contributions in the field of fossil vertebrates; supporter of Lamarck.

Copelatus. Common genus of dytiscid beetles.

Copepoda. Subclass in the Class Crustacea; copepods; body more or less cylindrical; head fused with one or more thoracic segments to form a cephalothorax; with five or six free thoracic segments, each with a pair of swimming legs; three to five abdominal segments and a pair of caudal rami; usually less than 10 mm. long, many spp. microscopic; common in fresh and salt waters; free-living, commensal, or parasitic; 6300 spp.; Calanus, Diaptomus, and Cyclops typical.

copepodid. One of the several instars preceding the sexually mature adult copepod.

Copidognathus. Genus of marine and fresh-water mites (Halacaridae).

Copidosoma. Genus of parasitic Hymenoptera; larvae parasitize caterpillars; noted for exceptional polyembryony, where 100 or more embryos may result from a single zygote.

copperhead. See AGKISTRODON.

coppers. See LYCAENIDAE.

coprodeum. 1. Dorsal portion of the monotreme cloaca into which the intestine opens. 2. Innermost portion of the cloaca of birds, lizards, and snakes.

coprolite. Fossil feces, droppings, or excrement.

Copromonas. Genus of colorless monoflagellate ovoid Protozoa; coprozoic in amphibians and man.

coprophagous. Feeding on feces.

coprophilous. Coprozoic.

coprozoic. Living in feces.

Coptotomus. Genus of dytiscid beetles.

copula. Basibranchial.

copulation. Temporary special physical contact of the two sexes to facilitate reception of sperm by the female.

copulatory bursa. Wings, flaps, or other extensions in certain invertebrate worms; used to hold the male and female genital pores in juxtaposition during copulation.

copulatory spicule. One of two minute needle-like sclerotized spicules in special pouches in most male nematodes; also in Kinorhyncha; facilitate the transfer of sperm into the vagina during copulation.

coquina shell. See DONAX.

Coracias. Genus of birds which includes the common roller; in an Old World family (Coraciidae).

coracidium. Early larval stage of certain tapeworms; essentially an oncosphere enclosed in a layer of ciliated or unciliated epithelium; often motile and aquatic; e.g. forms when the egg of Dibothriocephalus latus is deposited in water. Pl. coracidia.

Coraciidae. Family of birds in the Order Coraciiformes; rollers; 17 spp. of Old World bluish tropical birds; legs short, feet weak; hop clumsily and do not walk on ground.

Coraciiformes. Large order of birds which includes the motmots, bee eaters, hornbills, kingfishers, etc.; temperate and tropical; bill large and strong; third and fourth toes fused basally.

coracoarcuales. Collectively, the ventral mass of musculature in the throat region of fishes.

coracobrachialis. Tetrapod muscle originating on the coracoid and intermuscular septum, and inserting on the shaft of the humerus; flexes and adducts arm.

coracoid. Ventral bone or cartilage of the pectoral girdle in the region of the glenoid cavity in some tetrapods; also in the shark and some bony fishes.

Coragyps atratus. Common black vulture in the Order Falconiformes; southern U.S. to southern South America; smaller and stockier than the turkey vulture; head naked and black.

coral. 1. Hard exoskeleton, chiefly calcium carbonate, secreted by certain solitary or colonial polyps in the Class Anthozoa of the Phylum Coelenterata; colonial coral organisms may form extensive submerged masses associated with islands and mainland in tropical areas; such coral reefs are of three general types: fringing reefs, barrier reefs, and atolls; true corals do not exhibit metagenesis; a few hydrozoans (millipores) also form massive calcareous skeletons, but these are not "true" corals. See SOFT CORAL and HORNY CORAL. 2. Ovaries of a lobster.

coral fish. Any of numerous brightly-colored fishes living in coral reefs.

Corallimorpharia. Suborder of coelenterates in the Order Actiniaria; mesenteric filaments without cilia; Corynactis typical.

coralline. Any non-coral organism which bears a superficial resemblance to a mass of coral, e.g. some Hydrozoa, Bryozoa, and algae.

Corallium. Genus of corals in the Order Gorgonacea; the red corals of commerce, used for jewelry; skeleton composed of calcareous spicules cemented together with calcium carbonate.

coral reef. Large thick coral mass of various size and extent occurring in tropical seas near shore or in other shallow areas; three intergrading types are recognized; see ATOLL, FRINGING REEF, and BARRIER REEF.

coral snake. See MICRURUS and MICRUROIDES.

Corambe. Genus of nudibranchs; usually on floating kelp.

Coras medicinalis. Common U.S. spider in the Family Agelenidae; yellowish brown; to 12 mm. long.

corbiculum. Pollen basket of a bee.

Corbicula fluminea. Asiatic bivalve mollusk now also found in sloughs and rivers of Wash., Ore., Calif., Tenn., and probably other states.

corbina. See MENTICIRRHUS.

corbula. Phylactocarp bearing leaflike protective outgrowths which arch over the enclosed gonangia in certain hydroids.

Corbula. Genus of small marine clams; valves unequal.

Cordulegaster. Common genus of dragonflies; eastern U.S.

Cordulia shurtleffi. Dragonfly of northeastern U.S.

Cordylophora. Genus of colonial hydrozoan coelenterates; in brackish waters and a few inland rivers; branching colony 10 to 100 mm. high and composed of feeding polyps and gonophores.

Coregonidae. Family of bony fishes which includes the various whitefishes, ciscoes, and chubs; mouth small, with no teeth or weak ones; adipose fin present; glaciated areas of N.A., Europe, and Asia; all but a few spp. confined to fresh waters throughout their lives; examples are Prosopium, Coregonus, and Leucichthys.

Coregonus clupeaformis. Common whitefish of cold waters of the Great Lakes area and southeastern Canada; delicious food fish, now greatly depleted because of overfishing and the depredations of the sea lamprey.

Coreidae. Family of herbivorous insects in the Order Hemiptera; squash bugs; many pest spp.; Anasa tristis an important pest of squash, melon, and pumpkin.

corella. Any of several parakeets and cockatoos.

Corella. Genus of simple tunicates.

Corethra. Chaoborus.

cor frontale. Small accessory muscular structure on the ophthalmic artery of the crayfish; aids in pumping blood to the anterior end of the body.

coriaceous. Tough and leathery.

Corisella. Genus of water boatmen in the Family Corixidae.

corium. 1. Dermis; derma. 2. Elongated middle part of the forewing of Hemiptera.

Corixa. Common Old World genus of hemipteran insects in the Family Corixidae.

Corixidae. Family of insects in the Order Hemiptera; includes the water boatmen; swim about in ponds and lakes by means of fringed oarlike hind legs; feed on algae and detritus; Sigara and Trichocorixa common.

corm. In certain crustacean appendages, the pronounced linear arrangement of the protopod and large endopod, with the exopod extending out laterally.

cormidium. Assemblage or cluster of various zooids which occur at intervals along the long, dangling stemlike structure in pelagic siphonophores; ordinarily includes a hydrophyllium, and one or more gastrozooids, dactylozooids, and gonozooids. Pl. cormidia.

cormorant. See PHALACROCORACIDAE.

cornborer. See PYRAUSTA.

cornea. 1. Transparent anterior portion of the sclerotic coat of the eyeball in vertebrates. 2. In invertebrates, the transparent epidermis or cuticle forming the surface of the eye.

corneagen cell. See OMMATIDIUM.

corn earworm. See HELIOTHIS.

corneous. Horny or hornlike.

cornet fish. See FISTULARIA.

corneum. Stratum corneum.

cornification. Deposition of keratin in a cell or tissue, especially in epidermis.

cornified layer. Outermost dead epidermal cells, especially in the vertebrate integument.

corn root aphid. See ANURAPHIS.

corn snake. See ELAPHE.

cornu. Horn or hornlike projection. Pl. cornua.

cornua of hyoid. Two to four paired processes extending out from the hyoid apparatus in tetrapods.

Cornularia. Common genus of tropical and temperate anthozoans; with a horny investment on the basal portion; shallow waters.

Cornulariella. Genus of sea anemones in the Order Alcyonacea; to 0.5 in. long.

Corolla. Uncommon genus of Pteropoda; shell a transparent, slipper-like pouch.

corona. 1. Main ciliary wreath surround- the anterior end of a rotifer. 2. Central mass and arms of a crinoid. 3. test of an echinoid.

coronal disc. In a rotifer, the anterior part of the head region; bears the ciliary wreath and ciliary tufts.

corona radiata. Layer of protective cells surrounding a ripe human ovum.

coronary blood vessels. Those vertebrate arteries and veins which circulate blood through the wall of the heart.

coronary thrombosis. Stoppage of blood flow by a clot in the coronary artery or one of its branches.

Coronaster. Genus of sea stars (Order Forcipulata).

Coronatae. Order of scyphozoan jellyfish; upper surface of the bell has a circular furrow located inward from the scalloped margin; Periphylla common.

Coronula. Genus of barnacles typically found attached to Cetacea.

Corophium. Common genus of marine Amphipoda.

corpora. Pl. of corpus.

corpora allata. Pair of small, ovoid, glandular bodies closely associated with the cerebral ganglia of larval insects; secrete a "juvenile hormone" which inhibits growth and metamorphosis but permits molts to occur; not produced during the last larval instar, and its action is counteracted by the molting and pupation hormones secreted by cells of the cerebral ganglia. Sing. corpus allatum. Cf. PARS INTERCEREBRALIS.

corpora arenacea. Brain sand; laminated and lobed salt concretions in the pineal body of older people.

corpora atretica. See ATRESIA.

corpora bigemina. Two oval masses behind the third ventricle of the brain; centers of optic reflexes in fishes, amphibians, reptiles, and birds.

corpora quadrigemina. Four oval masses behind the third ventricle of the brain, forming the dorsal part of the mesencephalon; centers of optic and auditory reflexes in mammals.

corpus. 1. Whole body of an animal, especially man. 2. Mass of special tissue. 3. Main part of an organ. Pl. corpora.

corpus albicans. Mass of fibrous tissue which replaces a regressing corpus luteum.

corpus callosum. Transverse band of fibers connecting the two cerebral hemispheres in mammals.

corpus cavernosum. One of two cylindrical spongy erectile bodies forming the upper portion of the penis and clitoris.

corpuscle. 1. See BLOOD CORPUSCLE. 2. Any small mass or organ.

corpuscle of Ruffini. See ORGAN OF RUFFINI.

corpuscles of Hassal. Spherical masses of concentrically arranged degenerated cells in the thymus gland.

corpus haemorrhagicum. Blood clot formed within the empty Graafian follicle immediately after ovulation.

corpus luteum. Mass of yellowish tissue which rapidly fills the cavity of a Graafian follicle after rupture and release of the mature ovum from the follicle; this tissue secretes progesterone and persists if the egg is fertilized; if the egg is not fertilized, the corpus luteum progressively decreases in size and secretes less and less progesterone, until so little is secreted that there is insufficient to maintain the thickened condition of the uterine wall, and the lining of the latter disintegrates in the process of menstruation; if the ovum is fertilized, the corpus luteum persists and secretes progesterone throughout pregnancy; in addition to mammals, a structure similar to the corpus luteum occurs in some other viviparous vertebrates. Pl. corpora lutea.

corpus spongiosum. 1. Longitudinal spongy erectile mass making up the lower median portion of the penis; contains the urethra. 2. Spongy mass of connective tissue.

corpus striatum. One of the two basal ganglia of the cerebral hemispheres. Pl. corpora striata.

Correns, Karl Erich. German botanist (1864-1933); one of several scientists who independently rediscovered and confirmed Mendel's 1866 paper; author of researches on heredity and sex determination in plants.

Corrodentia. Order of insects; includes psocids, book lice, bark lice, and dust lice; minute to small spp. with simple metamorphosis and biting mouth parts; membranous wings present or absent; many spp. are pests on cereal grains, cereal products, insect collections, and the paste and glue of books; others are common in vegetable debris, under bark, and on shrubbery and herbaceous vegetation; common genera are Leposcelis, Troctes, and Psocus.

corsairs. See RASAHUS.

cortex. 1. Any outer layer or rind. 2. Outer portion of the vertebrate adrenal gland; more than 100 different substances are produced by this tissue; in general, such compounds: promote water retention in cells, promote sodium and chloride retention, promote excretion of potassium ions, promote the formation of glucose from glycogen, influence gonad activities, exert some control over carbohydrate metabolism, etc.; cortisone is one specific secretion of the adrenal cortex; it adapts the organism to severe injury and stimulates healing. 3. The outer layer of the kidney of higher vertebrates. 4. See CEREBRAL CORTEX.

Corthylio calendula. Ruby-crowned kinglet, a very small warbler-like bird; upper parts olive gray with a white ring around the eye; male with a scarlet patch on top of head; usually associated with evergreen trees; widely distributed in N. A.

corticalization. Increasing degree of control by cerebral cortex in the vertebrate phylogenetic series.

corticosteroids. Group of several steroid endocrines elaborated by the adrenal cortex; increase metabolic protein breakdown and increase the amount of glycogen in the tissues.

corticotrophin. ACTH.

cortin. Mixture of hormones of the adrenal cortex.

cortisone. See CORTEX.

Corvidae. Family of passerine birds (100 spp.) which includes the ravens, crows, magpies, and jays; habits variable; world-wide distribution.

corvina. Any of several edible marine fishes, especially Micropogon.

Corvus. See CROW.

Corydalus cornutus. Insect in the Order Megaloptera; the common eastern dobsonfly; adult with greatly elongated hornlike mandibles; larva the common aquatic hellgrammite.

Corymbites. Large genus of click beetles in the Family Elateridae; many important agricultural pests.

Corymorpha. Genus of solitary marine hydrozoan coelenterates.

Corynactis. Genus of Californian sea anemones.

Coryne. Genus of marine Hydrozoa; commonly on algae in tide pools.

Corynebacterium. Large genus of slender, rod-shaped, gram positive bacteria which often develop some club-shaped or pointed individuals; one sp. (C. diphtheriae) causes diphtheria.

Corynitis. Genus of marine coelenterates in the Suborder Anthomedusae; cylindrical, unbranched hydranths.

Corynoneura. Genus of tendipedid midges.

Corynorhinus rafinesquei. Lump-nosed bat; yellowish to dark brown; widely distributed in U. S. except for northeast and southeast.

Coryphaena. Genus of dolphins (fishes); the dorados; see CORYPHAENIDAE.

Coryphaenidae. Family of fishes which includes the two spp. of dolphins; moderately slender oceanic fishes with compressed bodies, a massive blunt head, and a long, high, dorsal fin lacking spines; cosmopolitan in warm seas; to 6 ft. long; common dolphin is Coryphaena hippurus.

Coryphaeschna ingens. Dragonfly of southeastern U. S.

Coryphella. Genus of nudibranchs; cerata very numerous.

Coryphodon. Genus of fossil mammals; five toes, plantigrade; up to size of rhinoceros; Europe and N. A., Upper Paleocene to Lower Eocene.

Corystes. Genus of marine crabs; antennae form an inhalent tube when the animal is buried in the substrate.

Corythosaurus. Genus of Cretaceous dinosaurs in the Order Ornithischia; duckbills; massive, bipedal, herbivorous spp.; jaws broad and flat.

cosmine. Hard, dentine-like material forming the cosmoid scales of primitive dipnoan and crossopterygian fishes.

cosmogony. 1. Origin of the earth or universe. 2. Theory regarding such an origin.

cosmoid scales. See COSMINE.

cosmopolitan. Having essentially a world-wide distribution, wherever a suitable habitat occurs; the term refers to the geographical range of a taxon; e.g. the housefly, the common water flea (Daphnia pulex), and a great many spp. of Protozoa are cosmopolitan.

cosmozoic theory. Theory that life on earth arrived here from some planet or other source in the universe.

Cossidae. Small family of medium to large moths including the carpenter or goat moths; larvae burrow in wood of temperate deciduous forests; Prionoxystus robiniae an important N. A. sp.; Cossus cossus important in Europe.

Cossus. See COSSIDAE.

costa. 1. Vein forming anterior thickened edge of a typical insect wing. 2. Elevated, curved, ridgelike structure.

costal. Of or pertaining to a rib.

costal angle. Tip of an insect wing.

costal plate. One of a series of paired dorsolateral plates in a typical tortoise.

Costa's hummingbird. See CALYPTE.

costate. Bearing raised riblike structures.

Costia. Genus of Protozoa ectoparasitic on fresh-water fishes; four flagella.

coterie. Closed social group of animals which defend their common territory against encroachment by other coteries.

Cothurnia. Genus of marine and fresh-water ciliate Protozoa; elongated, cylindrical, sessile, and surrounded by a lorica.

cotinga. See COTINGIDAE.

Cotingidae. Large family of brightly-colored passerine birds which includes the umbrella birds, fruit-crows, cock-of-the-rock, tityras, and cotingas; structure diverse, but typically large-headed, thick-billed birds resembling flycatchers; often crested; forest spp. of tropical America, occasionally in extreme southwestern U. S.

Cotinis nitida. Metallic green and brownish beetle almost 1 in. long; fig eater; common in eastern half of U. S.; larva feeds on many kinds of roots, stems, and leaves; adults on fleshy fruits.

Cottidae. Family of carnivorous mail-cheeked fishes having large, spiny heads, broad mouths, and large fan-like pectorals; sculpins; predominantly in cold seas, but Cottus (often called muddlers or miller's thumbs) includes small N. A. fresh-water spp.; Hemitripterus, Myoxocephalus, and Scorpaenichthys common marine genera.

Cottogaster. Genus of Percidae now included in the fish genus Percina.

Cottoidei. Suborder of fishes which includes the Cottidae.

cotton aphid. See APHIS.

cottonmouth. See AGKISTRODON.

cotton rat. See SIGMODON.

cotton stainer. See PYRRHOCORIDAE.

cottontail. See SYLVILAGUS.

cottony cushion scale. Pest sp. of scale insect (Icerya purchasi); feeds on citrus trees, sometimes in massive infestations; secretes cotton-like fibers at the body periphery; introduced into U. S. from Australia.

Cottus. See COTTIDAE.

Coturnix coturnix. European quail.

Cotylaspis. Genus of trematodes in the Order Aspidocotylea; sucking disc broad; in mantle cavity of mussels and intestines of turtles.

Cotylea. Suborder of polyclad Turbellaria; ventral sucker present; genital pores anterior.

cotyledonary placenta. Placenta whose chorionic tissue has more or less evenly spaced dense clusters of villi (cattle, sheep, deer).

Cotylogaster. Genus of trematodes in the Order Aspidocotylea; sucking disc long and narrow; intestine of marine and fresh-water fishes.

Cotylosauria. Primitive order of extinct (Permian) reptiles showing certain amphibian characters; Seymouria typical.

cotype. In taxonomy, a specimen accompanying the holotype.

cougar. American puma, mountain lion, panther, or painter; largest New World unspotted cat, to 200 lbs.; several spp. (Felis couguar most common) and numerous subspecies; originally over most of U.S. and ranging from parts of southern Canada to Patagonia but now uncommon in U.S.; kill small mammals, deer, and sometimes cattle but normally avoid man.

countershading. Type of protective coloration characterized by light ventral colors and darker dorsal colors.

couprey. Rare sp. of wild cattle (Bos) discovered in Cambodia in 1936; both sexes with a large dewlap; horns of male with a ring of shredded fibers on the last turn.

courlan. See ARAMIDAE.

Couthouyella. Genus of small, long-spired, marine Gastropoda.

cover. Collectively, the vegetation, debris, and irregularities of the substrate; variously used by animals for concealment, sleeping, feeding, breeding, etc.; the term is used especially with reference to game animals.

coverslip. Coverglass; very thin piece of glass used to cover specimens on a microscope slide; usually 0.08 to 0.30 mm. thick; thickness indicated by numbers: number 0 thinnest, number 2 thickest.

covert. One of the special feathers covering the bases of the quills of the wing and tail feathers of a bird.

covey. Small flock, brood, or hatch of birds, e.g. quail.

cowbird. See MOLOTHRUS.

Cowdry, Edmund Vincent. American zoologist (1888-); author of several important books on cytology, histology, and microtechnique; many important research contributions in pathology, anatomy, and parasitology.

cowfish. 1. Common northern Pacific dolphin, Tursiops gilli. 2. Any of certain trunkfishes having two hornlike projections on the top of the head. 3. Certain beaked whales in the genus Mesoplodon. 4. Manatee (poor usage).

cow killer. Large, hairy wasp in the Family Mutillidae; coloration whitish, yellow, and red; female wingless and a powerful stinger.

cow-nosed ray. See RHINOPTERA.

Cowper's glands. Bulbo-urethral glands.

cowrie. Cowry; see CYPRAEA.

cowry. See CYPRAEA.

cowsucker. Any of several N.A. snakes alleged to milk cows.

coxa. Short basal segment of an insect leg by which the leg is articulated to the thorax; also applies to the basal segment of the legs of certain other arthropods.

coxal gills. Flattened respiratory sacs extending downward from the inner surface of some of the coxal plates in certain Amphipoda.

coxal gland. Gland which opens at the base of the leg in some insects and many arachnids; presumably excretory.

coxal plate. 1. Epimera. 2. Platelike expansion of the hind coxa in certain aquatic beetles.

coxite. 1. Basal segment of any leglike appendage among insects. 2. Lateral abdominal plate in thysanuran insects; a rudimentary abdominal appendage.

coxopodite. 1. Basal segment of pereiopod in most malacostracans. 2. Basal of the two segments making up the protopodite of a typical biramous appendage, the distal segment being the basipodite.

coyote. See CANIS.

coypu. See CAPROMYIDAE.

cozymase. Diphosphopyridine mucleotide.

crab. 1. Any of numerous marine or shore crustaceans in the Order Decapoda, especially those characterized by broad cephalothorax and a small abdomen bent up against the ventral surface of the cephalothorax; walking legs well developed, antennae small; see CALLINECTES, OVALIPES, CANCER, CARCINIDES, MENIPPE, OCYPODE, UCA, PINNOTHERES, LIBINIA. 2. Any of numerous fresh-water crayfish and a few marine crayfish; see ASTACUS and CAMBARUS. 3. Vernacular name for any large decapod crustacean. 4. Vernacular name for Phthirus pubis.

crab catcher. Any of numerous large shore birds which eat crabs.

crab eater. Any of several fishes, birds, and seals which feed on crabs; see RACHYCENTRIDAE.

crab-eating dog. See DUSICYON.

crab hawk. See BUTEOGALLUS.

crab louse. See PHTHIRUS PUBIS.

crab plover. See DROMADIDAE.

Crabro. Cosmopolitan genus of square-headed wasps; nest in wood and store flies for larval food.

crab spiders. See THOMISIDAE.

Cracidae. Family of birds in the Order Galliformes; includes chachalacas, guans, and curassows; tropical and subtropical America.

Cracticidae. Small passerine family of Australian and Papuan birds; Australian butcher birds impale prey on thorns; bell magpies are large crowlike spp. noted for their large nests of sticks and (often) pieces of wire.

Crago. Common genus of marine shrimps; only the first two pairs of legs chelate; cephalothorax somewhat depressed; usually less than 3 in. long.

Cragonidae. Family of marine shrimps with a short rostrum; Sabinea typical.

crake. 1. Local name of several U. S. rails. 2. Common name for several Old World rails.

Crambidae. Family of small silvery or brownish moths; larvae of some are borers in grasses; Diatraea saccharalis is the sugarcane borer.

Crampton's muscle. Radial muscle in the eye of a bird; functions in bringing close objects into focus.

crane. See GRUIDAE.

crane flies. See TIPULIDAE.

Crangon. See ALPHEIDAE.

Crangonidae. Alpheidae.

Crangonyx. Common genus of fresh-water Amphipoda; especially in eastern and central U. S.

Crania. Common genus of Brachiopoda.

craniad. Toward the cranium, head, or anterior end.

cranial. Of or pertaining to the vertebrate skull.

cranial nerve. One of the 10 to 13 pairs of peripheral nerves arising from the brain of vertebrates; several arise as both dorsal and ventral roots which do not fuse but remain separate; beginning at the anterior end, these commonly are: terminalis, olfactory, optic, oculomotor, trochlearis, trigeminal, abducens, facialis, auditory, glossopharyngeal, vagus, spinal accessory, and hypoglossal.

Craniata. Taxonomic category often used to include those chordates having a cranium, i. e. all Vertebrata.

Craniella. Genus of bristly, egg-shaped, marine sponges in the Class Demospongiae.

cranium. Collectively, those bones enclosing the brain and sense organs of the head.

crappie. See POMOXIS.

Craspedacusta sowerbyi. "Fresh-water jellyfish"; a hydrozoan coelenterate occurring sporadically in small lakes, ponds, and old quarries; bell-shaped medusa with numerous peripheral tentacles, to 22 mm. in diameter; hydroid stage simple, to 8 mm. long.

Craspedochilus. Genus of intertidal, elongated, wormlike Amphineura.

Craspedosoma. Genus of small millipedes.

craspedote. Of or pertaining to a coelenterate medusa stage which possesses a velum

Crassinella. Common genus of small marine bivalve mollusks.

Crassostrea. Large genus of marine bivalves; includes certain oviparous oysters, some of economic importance; shell irregular and rough, with the left valve larger; hinge teeth absent; single large adductor muscle; animal fixed to the substrate; C. virginica, the familiar edible American oyster, usually 2 to 4 in. long, although reported to 12 in.; C. gigas is the Japanese edible oyster introduced on the Pacific Coast.

Craterolophus. Genus of sessile scyphomedusae.

Cratogeomys castanops. Plains pocket gopher in the Family Heteromyidae; yellowish-brown; high plains and mountain basins, Colo. to central Mexico.

crawdad. Crayfish; see CAMBARUS and ASTACUS.

crawfish. Crayfish; see CAMBARUS and ASTACUS.

crawling water beetle. See HALIPLIDAE.

crayfish. Crawdad, crawfish; see CAMBARUS and ASTACUS.

creatine. Nitrogenous substance in muscle and brain fluids of vertebrates; $C_4H_9O_2N_3$; the phosphate form is energy-rich.

creek chub. See SEMOTILUS.

creel census. Data relating to number, kind, and weight of fish caught by anglers in a particular body of water.

creeper. See CERTHIIDAE.

Creeper. Short-legged, short-winged variety of chickens having the heterozygous Creeper gene condition; individuals homozygous dominant for the Creeper character die in the shell; a lethal gene.

creeping water bugs. See NAUCORIDAE.

cremaster. 1. Special muscle for drawing testes into the body cavity through the inguinal canals; e. g. in squirrels, mice, and rats. 2. Stout terminal abdominal spine in certain terrestrial insect pupae. 3. Anal hook for suspending an insect pupa.

crenate. Having a scalloped or notched margin.

crenation. Abnormally shrunken or notched appearance of erythrocytes in hypertonic solution.

Crenella. Genus of small mussels often found among algae holdfasts along Atlantic Coast.

Crenichthys. Springfishes; small genus of cyprinodontid fishes; eastern Nev.

Crenitis. Uncommon genus of hydrophilid beetles; eastern U.S.

Crenobia alpina. Common northern European planarian; restricted to cold brooks.

Crenothrix polyspora. Filamentous bacterium having iron in the sheath; grows abundantly in reservoirs and water pipes; may sometimes impart a noxious taste to water and may clog pipes.

Creodonta. Extinct suborder of the mammal Order Carnivora; small, slender spp. with long skulls and no specialized carnassial teeth; Paleocene to Pleistocene.

Creophilus. Common genus of rove beetles; feed chiefly on carrion insects.

Crepidula. Marine gastropod genus; includes slipper shells or boat shells; shell oval and flat, with a horizontal internal shelflike plate in the posterior half; spire inconspicuous; to 2 in. long; common and widely distributed.

Crepipatella. Genus of boat shells; internal shelf not attached on left side.

crepuscular. Active in dim illumination; e.g. some snails, earthworms, bats, etc.

Creseis. Widely distributed genus of pelagic Pteropoda; shell long, slender, and tapered.

crested flycatcher. Any of several flycatchers having a prominent crest; e.g. Myiarchus.

crested hamster. See LOPHIOMYS.

crested swift. See HEMIPROCNIDAE.

Cretaceous period. Last of three geological subdivisions of the Mesozoic era; extended from about 120 million to 70 million years ago; Rocky Mountains and Andes were being formed; many extensive swamps, cool climate; flowering plants were developing; also characterized by modern fishes and invertebrates, disappearance of bizarre reptiles, and appearance of snakes and small marsupials.

cretin. Individual suffering from cretinism.

cretinism. Pathological condition produced by marked undersecretion of thyroxin during childhood or adolescence; featured by dwarfism, coarse facial features, low metabolic rate, retarded mental development, deafness, undeveloped gonads, and obesity; more or less alleviated by administration of thyroxin.

crevalle. See CARANX.

cribellum. 1. Transverse platelike spinning organ just anterior to spinnerets in some spiders. 2. Sievelike plate on the mandibles of certain insects.

cribriform organs. Rows of thin integumental folds between certain marginal plates in some sea stars.

cribriform plate. Plate of the ethmoid bone separating cranial and olfactory cavities in mammals; contains openings for the passage of olfactory nerves.

Cribrilina. Genus of European marine Bryozoa.

Cribrina. Anthopleura.

Cricetidae. Large family of rodents sometimes combined in the Muridae; includes the native American rats and mice and a few Old World forms; mostly small nocturnal spp. with no premolars; tail generally nearly naked or scaly; typical genera: Onychomys, Reithrodontomys, Peromyscus, Oryzomys, Sigmodon, Neotoma, Synaptomys, Lemmus, Dicrostonyx, Phenacomys, Evotomys, Microtus, Neofiber, and Ondatra.

Cricetinae. Subfamily of the Cricetidae; includes the New World rodents.

Cricetomys. Genus of West African hamster-rats; large ratlike spp.; cunning, strong, and with packrat habits.

Cricetus cricetus. Hamster; small rodent native to temperate Europe and western Asia; thickset body 6 in. long; gray of brown; head broad, ears round, and cheek pouches well developed; widespread as a pet, and an especially useful laboratory animal owing to ease of maintenance.

cricket. See GRYLLIDAE.

cricket frog. See ACRIS.

cricoid cartilage. Small ring-shaped cartilage of the larynx.

Criconema. Common genus of microscopic, free-living nematodes.

Criconemoides. Genus of microscopic, free-living nematodes.

Cricotopus. Genus of tendipedid midges.

Crinoidea. Class of Echinodermata; sea lilies and feather stars; body a disc or cup-shaped cluster of calcareous plates to which are attached 10 flexible arms bearing many slender lateral branches, or pinnules, arranged like barbs on a feather; in some spp. a long, jointed aboral stalk attaches the animal to the sea bottom by rootlike outgrowths; other spp. are motile with a cluster of aboral leglike cirri; anus on upward-facing oral surface; tube feet tactile and respiratory; madreporite and pedicellariae absent; ciliated larva ovoid; deep-sea forms; 800 living and 2100 fossil spp.

Crisia. Genus of marine Bryozoa in the Order Cyclostomata.

crissum. 1. Area of the cloacal aperture of birds. 2. Collectively, the feathers of that region; under tail coverts.

crista ampullaris. Prominent thickened area of the membranous lining of an ampulla of a semicircular canal; its epithelium contains auditory cells.

cristate. Having a crest.

Cristatella mucedo. Fresh-water bryozoan in the Class Phylactolaemata; each colony a creeping gelatinous mass to 8 in. long.

Cristigera. Genus of ovoid holotrich Ciliata; fresh and salt water.

Cristivomer namaycush. Lake or Mackinaw trout in the Family Salmonidae; a very large trout of deep, cold lakes of Canada and northern U.S.; widely introduced into western states; spawns on rocky reefs in autumn; placed in the genus Salvelinus by some authors.

crithidia. Temporary or permanent morphological type in certain hemoflagellates; flagellum originates at mid-length of cell; undulating membrane correspondingly short.

Crithidia. Genus of Protozoa in the Family Trypanosomidae; parasitic in arthropods and other invertebrates.

croaker. See SCIAENIDAE.

Crocethia alba. Sanderling; a small, plump, sandpiper-like bird with gray and white plumage; widely distributed, especially along marine beaches and sand flats.

crochets. 1. Minute hooks or curved spines on the apical surface of a caterpillar proleg and on the cremaster of pupae. 2. Setae in certain Oligochaeta; tip bifurcate. 3. Balancers in certain larval salamanders. 4. Projections on certain lophodont molar teeth.

Crocidura. Common genus of musk-shrews; large, nocturnal African spp.

crocodile. See CROCODYLUS.

crocodile bird. Plover-like bird which alights on African crocodiles and feeds on insects and parasites.

Crocodilia. Loricata.

Crocodylidae. See LORICATA.

Crocodylus. Genus of reptiles in the Family Crocodylidae; includes about 12 spp. of crocodiles; head triangular, very narrow toward snout; some teeth protruding at sides of jaw; tropics of Old and New worlds; more active and vicious than alligators; several spp. notoriously dangerous to man; American crocodile, C. acutus, in brackish waters in southern Fla., West Indies, Mexico, Central America, and northern South America; maximum length 14 ft.; the Nile crocodile C. niloticus, occurs over much of Africa.

Crocuta. Genus of mammals; includes the large spotted hyena of Africa.

Cro-Magnon man. Fossil man dating from 40,000 years ago onward; abundant remains in European caves, associated with stone implements and paintings; closely related to modern man and usually placed in a subspecies, Homo sapiens fossilis.

crop. 1. Muscular enlargement of the esophagus of grain-eating birds for storing and moistening ingested food. 2. In certain invertebrates, e.g. earthworms and some insects, an expanded part of the digestive tract near the anterior end, for storage and/or digestion.

crop milk. Pigeon's milk; secretion formed by the crop epithelium in both sexes; for nourishing nestlings.

croppie. Crappie; see POMOXIS.

Crossarchus. Cusimanses; a genus of West African semifossorial omnivorous mongooses.

Crossaster. Common genus of sunstars in the Order Spinulosa; 8 to 14 arms; circumboreal.

crossbill. See LOXIA.

cross fertilization. 1. Mutual exchange of sperm between two hermaphroditic animals and subsequent union of eggs and sperm, e.g. many snails. 2. Fusion of male and female gametes originating from two different varieties or spp.

crossing over. See MEIOSIS.

Crossobothrium. Genus of tetraphyllidean tapeworms found especially in the spiral valve of sharks.

Crossomys. Genus of water-rats; large, muskrat-like spp. with webbed, paddle-like feet, small eyes, and no external ears; New Guinea.

Crossopterygii. Superorder which includes primitive lobe-finned fishes; chiefly fresh-water. See RHIPIDISTIA and COELACANTHINI.

Crossopterygiida. Crossopterygii.

crossover. The product of separation of alleles originally linked on the same chromosome; see MEIOSIS.

crossover value. Frequency of crossing over between any two gene loci on a chromosome; the percentage of gametes in which one of the genes has been exchanged for its allelomorph on the homologous chromosome; the greater the crossover value, the farther apart two genes are on the same chromosome.

Crotalidae. Family of venomous snakes; includes pit vipers and rattlesnakes; paired erectile fangs in front part of upper jaw are folded back when not in use; pitlike depression on each side between nostril and eye; widely distributed but not in Africa or Australia; common genera: Agkistrodon, Bothrops, Lachesis, Sistrurus, and Crotalus.

Crotalus. Genus of North and South American rattlesnakes in the Family Crotalidae; tail with series of cornified button-like terminal structures forming the rattle; in most of U.S.; feed chiefly on small mammals; C. horridus, timber rattler, to 6 ft., generally east of Mississippi R. but extending into Iowa and Tex.; C. viridis, prairie rattler, to 5 ft., in most of the western half of the U.S. and into extreme southwestern Canada; C. atrox, western diamondback, to 7.5 ft., from Ark. and Okla. west and south; C. adamanteus, the eastern diamondback, to 8 ft., in swamps and wet woods from N.C. to La.; C. cerastes, sidewinder or horned rattlesnake, to 30 in., in southwestern deserts, has prominent erect scales over eyes and a curious angular locomotion.

Crotaphytus. Genus of large, brightly-colored colored lizards with very long tail and prominent neck-fold; collared lizards; mostly in southwestern states and Mexico but extending into Ore. and Mo.

Croton bug. Blattella germanica.

Crotophaga sulcirostris. Groove-billed ani; Tex. and Ariz. south to northern South America; coal-black cuckoo-like bird with short wings and a high, grooved ridge on the upper mandible; usually gregarious.

crow. Any of various large black birds in the Family Corvidae; Corvus brachyrhynchos is the common American crow; C. ossifragus is the fish crow of the Atlantic and Gulf coasts; C. corax is the common raven of northern Europe, Asia, and N.A.

crown. 1. Top of a tooth. 2. That part of a tooth extending above the gum.

Crucibulum striatum. Cup-and-saucer limpet; a small Atlantic Coast sp. with a cup-shaped process on the inner surface of the shell.

cruciform. Cross-shaped.

Crucigera. Genus of Polychaeta.

crumb-of-bread sponge. See HALICHONDRIA.

crumen. Pouch or concavity; may be secretory and/or a receptacle for various structures.

Crumenula. Genus of green, fresh-water, monoflagellate, ovoid Protozoa; pellicle striated.

crura. 1. Legs, leglike parts or any of a wide variety of elongated supporting structures in anatomy. 2. Legs from knee to ankle. Sing. crus.

crura cerebri. Two fiber tracts on the ventral side of the mesencephalon containing most of the fibers connecting the cerebral hemispheres with the lower part of the body.

crural glands. Saclike glands in the lateral ventral body cavity of male Onychophora.

Cruregens. Genus of marine Isopoda.

crureus gluteus. Dorsal muscle of the hind leg of the frog.

crurococcygeus. Ventral muscle of the mammalian hind leg.

crus. See CRURA.

crus commune. Common stalk for the adjacent origins of the two vertical canals of the membranous labyrinth.

Crustacea. One of the eight classes in the Phylum Arthropoda; includes water fleas, crayfish, lobster, crabs, barnacles, etc.; body composed of head, thorax, and abdomen; cephalothorax also present, owing to fusion of one or more thoracic segments with the head; head composed of five fused segments and bearing two pairs of antennae, a pair of mandibles, and typically two pairs of maxillae; thoracic segments with appendages; abdominal segments with or without appendages; appendages fundamentally biramous but often greatly modified; mostly aquatic spp.; gills present or absent; 30,000 spp.

cryoplankton. Kryoplankton.
cryoscopy. Determination of the freezing point of a liquid, especially as modified by dissolved materials.
Cryphocricos. Tropical and subtropical genus of creeping water bugs in the Family Naucoridae.
cryptic coloration. Resemblance of an animal's coloration to that of its surroundings; presumably a concealment device.
Cryptobia. Genus of flagellate Protozoa parasitic in reproductive organs of invertebrates and fishes.
Cryptobranchidae. Family of large salamanders; body depressed; skin soft, flabby, and folded laterally; eyelids absent; vomerine teeth in an arched series parallel with jaws; permanently aquatic; Cryptobranchus (eastern U.S.) and Megalobatrachus (China and Japan) typical.
Cryptobranchoidea. Suborder of primitive amphibians; includes Family Cryptobranchidae; aquatic, adults without gills; large trunk; eyes small, lidless.
Cryptobranchus alleganiensis. Common salamander or hellbender of rocky streams from N.Y. to Ohio to La.; spotted yellowish to red and brown; to 20 in. long.
Cryptocellus. Genus of arachnids in the Order Ricinulei.
Cryptocerata. One of the two suborders in the insect Order Hemiptera; includes spp. which are aquatic and mostly predaceous; antennae short.
Cryptocerus. Genus which includes the common brown wingless cockroach; 23 to 30 mm. long; in rotting wood and under bark of dead trees.
Cryptochironomus. Common genus of tendipedid midges.
Cryptochirus. Genus of tropical marine crabs.
Cryptochiton. Chiton genus of southern U.S. Pacific Coast; occasionally used for human food.
Cryptocotyle. Genus of flukes found in the intestine of gulls.
Cryptodira. One of the superfamilies of the reptilian Order Chelonia; includes most turtles, tortoises, and terrapins; carapace covered with horny shields; if the head is retractile, the neck bends in a vertical S; pelvis not fused to shell.
Cryptodon. Genus of small marine bivalve mollusks.
Cryptoglaux. Aegolius.
Cryptoglena. Genus of fresh-water monoflagellate Protozoa; cell rigid, flattened.
Cryptolithodes. Genus of marine crabs in the Suborder Anomura; abdomen completely bent under cephalothorax.
cryptomitosis. Aberrant mitotic phenomenon in certain Protozoa; chromatin assembles at the equatorial plate without the formation of discrete chromosomes.
Cryptomonadina. Order of flagellate Protozoa; small, oval cells with two flagella; chromatophores various or absent; holozoic or holophytic but usually with a gullet; Chilomonas and Cryptomonas typical.
Cryptomonas. Genus of fresh-water biflagellate Protozoa; green chromatophores; body elliptical, with a firm pellicle.
Cryptomya. Genus of marine bivalves; usually buried in substrate with the siphons protruding into the burrows of other marine invertebrates.
Cryptonchus. Common genus of microscopic, free-living nematodes.
Cryptophagidae. Small and widely distributed family of beetles; silken fungus beetles; feed on fungi and decaying organic matter, sometimes as scavengers in the nests of ants and wasps.
Cryptoprocta. Genus of bright brown mammals of Madagascar; fossas; although true cats, the general build is foxlike.
Cryptops. Genus of small centipedes.
cryptorchid. 1. Male mammal in which the testes abnormally fail to descent into the scrotum but remain in the main body cavity; such individuals are sterile. 2. Male mammal in which retention of the testes in the main body cavity is the normal condition (elephant).
Cryptorhynchus. Large and widely distributed genus of weevils; includes many economic spp.
Cryptostemma. Genus of arachnids in the Order Ricinulei.
Cryptostomata. Order of fossil Bryozoa.
Cryptosula. Genus of marine Bryozoa.
Cryptothir. Genus of Isopoda; parasitic on barnacles.
Cryptothrips. Important genus of thrips (Order Thysanoptera).
Cryptotis parva. Least shrew; a small sp. with body less than 3 in. long; grassy areas of southeastern U.S. as far north as N.Y. and Iowa.
cryptozoic. Pertaining to animals inhabiting crevices or living under

stones, leaves, etc.

crypts of Lieberkühn. Deep cylindrical glands which open at the base of the villi of the small intestine; secrete mucus and intestinal juice.

Cryptus. Genus of ichneumon flies.

crystal cells. Vitrellae.

Crystallaria asprella. Crystal darter, a small fish in the Family Percidae; Minn. to Ohio and south to Okla., La., and Miss.

crystalline cone. See OMMATIDIUM.

crystalline style. Translucent cylindrical proteinaceous mass in the stomach of many bivalve mollusks and some gastropods; formed in a small caecum and whirled on its long axis by ciliary action, the projecting end being worn away by abrasion on the gastric shield of the stomach wall; such abrasion releases carbohydrate-digesting enzymes into the stomach cavity.

crystalloid. Substance such as a salt or sugar which forms a true solution.

c.s.f. Cerebrospinal fluid.

Ctenacodon. Genus of small Jurassic mammals.

Ctenidae. Family of wandering, hunting spiders, especially in warmer parts of the U.S.

ctenidium. 1. Gill, especially an invertebrate gill, such as those in many mollusks. 2. Comblike row of spines. Pl. ctenidia.

Ctenizidae. Family of trapdoor spiders; build silk-lined burrows closed by a trapdoor lid; southern and western states; Pachylomerides, Ummidia, and Bothriocyrtum common.

Ctenocephalides. Genus of fleas in the Order Siphonaptera having a cosmopolitan distribution; C. felis is the common cat flea which also feeds on dogs, rats, and man; C. canis is the common dog flea which also feeds on cats.

Ctenodactylidae. Family of African hystricomorph rodents; pale-colored spp. of rocky scrub and semidesert areas; mostly nocturnal and communal; Ctenodactylus and Pectinator typical; gundis.

Ctenodactylus. See CTENODACTYLIDAE.

Ctenodipterini. Order of extinct lungfishes.

Ctenodiscus. Small, widely distributed genus of sea stars with pentagonal to stellate body; mud stars.

Ctenodrilus. Genus of minute Polychaeta.

ctenoid. Having a margin of small teeth, as in a comb.

ctenoid scales. Thin, rounded, overlapping dermal scales covering most of the body in advanced bony fishes; posterior portion more or less roughened; posterior margin serrate or comblike.

Ctenomyidae. Family of burrowing hystricomorph rodents of South American plains; tucotucos; tunnel just below surface of sandy soils in a mole-like fashion; body 6 to 12 in. long.

Ctenophora. Phylum which includes the comb jellies and sea walnuts; free-swimming, more or less spheroidal (occasionally flat), biradially symmetrical, hermaphroditic, and marine; with eight meridional rows of comb plates, a gastrovascular cavity, and a nerve net; nematocysts and polymorphism absent; translucent, gelatinous, delicately colored, and often bioluminescent; about 90 spp.; Pleurobrachia and Beroë common.

Ctenoplana. Rare aberrant genus of Ctenophora; swims as a pelagic organism or creeps on the bottom.

Ctenopoda. Tribe of cladocerans in the Suborder Calyptomera; with six pairs of foliaceous legs; first and second pairs not prehensile; e.g. Holopedium and Diaphanosoma.

Ctenostomata. 1. Order of bryozoans in the Class Cymnolaemata; zooecium chitinoid or gelatinous, with toothlike processes for closing aperture; Alcyonidium and Paludicella are examples. 2. Suborder of the ciliate Order Spirotricha; ciliation reduced or absent; cell compressed, with a carapace; e.g. Epalxis.

cuataquil. See BASSARICYON.

Cubaris. Genus of terrestrial Isopoda; Calif.

Cubitermes. Common genus of tropical termites.

cubitus. 1. One of the veins in an insect wing; sometimes two or more are present. 2. Forearm. 3. Ulna.

cubitus interruptus. A gene (ci) which determines the length of the cubitus vein in the wing of Drosophila.

cuboidal epithelium. Epithelium consisting of cells that are roughly cuboidal in shape, e.g. salivary gland cells; sometimes such cells are ciliated, as in the reproductive tract of many invertebrates.

Cubomedusae. Order of scyphozoan jellyfish; with four tentacles or groups of tentacles at the edge of the bell; Tamoya is a common Atlantic Coast genus.

cuckold. 1. Trunkfish in the genus Lactophrys. 2. Cowbird.

cuckoo fly. Cuckoo wasp; see CHRYSIDIDAE.

cuckoos. Group of insectivorous arboreal birds in the Order Cuculiformes; European sp., Cuculus canorus, lays its eggs in the nests of other birds for them to hatch, but American cuckoos (Coccyzus) do not have this habit.

cuckoo-shrike. See CAMPEPHAGIDAE.

cuckoo spit. Frothy mass found on plants; secreted by nymphs of spittle insects (Cercopidae).

cuckoo wasps. See CHRYSIDIDAE.

Cucujidae. Family of greatly depressed beetles; mostly under bark of trees; some pest spp.; widely distributed.

Cuculidae. Family of birds in the Order Cuculiformes; includes the cuckoos, roadrunners, and coucals; body slender, wings long; mostly forest spp.; world wide.

Cuculiformes. Order of birds which includes the touracos, cuckoos, roadrunners, and anis; long-tailed spp., either arboreal or terrestrial and ground scratching; foot with two toes pointed forward and two backward; nests absent or of crude construction; widely distributed in temperate and tropical areas; Crotophaga, Geococcyx, and Coccyzus typical in U.S.

Cuculus. See CUCKOO.

Cucumaria. Common genus of thick-bodied sea cucumbers.

cucumber beetles. See DIABROTICA.

cucumber wilt. Disease of cucumbers caused by a bacterium, Bacillus tracheiphilus; transmitted by cucumber beetles (Diabrotica).

cud. See RUMEN.

cui-ui. Sp. of Chasmistes (a sucker) of Pyramid Lake and the Truckee River, Nev.

Culcita. Genus of sea stars having vestigial arms.

Culex. Large and common genus of "nuisance" mosquitoes; although they regularly feed on human blood, they transmit no diseases except filariasis.

Culicidae. Large family of insects in the Order Diptera; mosquitoes; slender, delicate insects with a humped body; females with long, piercing mouth parts; feed chiefly on blood of birds and mammals; males may feed on plant juices; larvae and pupae aquatic; adults of some spp. transmit human dengue, malaria, filariasis, and yellow fever; important genera are Aedes, Culex, and Anopheles.

Culicoides. Genus of minute biting flies in the Family Ceratopogonidae; larvae and pupae in fresh waters and damp places.

Culiseta. Common genus of mosquitoes.

culmen. Dorsal ridge along the bill of birds.

cultch. Mass of broken shells, pebbles, and debris dumped into the sea and serving as a substrate for oyster growth.

cultellus. One of a set of bladelike lancet mouth parts in some flies.

culture medium. Any natural or artificial preparation used for growing organisms, especially microorganisms.

culture pearl. See PEARL.

Cumacea. Order of malacostracan Crustacea; small, shrimplike spp., 3 to 15 mm. long; carapace short, abdomen slender; burrow in sand or mud of sea bottom; Diastylis common.

Cumingia. Genus of small bivalves of the Atlantic Coast.

Cumopsis. Genus of Cumacea.

cumulative factors. See CUMULATIVE GENES.

cumulative genes. Multiple genes; independent pairs of genes, each of which acting alone may produce a character, and when acting together cause the character to be accentuated in a quantitative manner; thus, a pure-blooded Negro has two pairs of genes, BB and B'B', for black pigmentation, while a pure-blooded white has the recessive bb and b'b' condition; mating of two such individuals produces BbB'b' mulatto offspring of intermediate pigmentation; matings of such mulattoes result in offspring ranging from pure black to pure white skin color in a ratio of 1:4:6:4:1.

cumulus oophorus. Follicle cells surrounding an ovum in a Graafian follicle.

cuneate. Wedge-shaped.

cuneiform. Wedge-shaped.

Cuniculus. Common genus of pacas, rodents which grow to be more than 2 ft. long; eyes bulbous; powerful spp. which rummage about on the forest floor with the four feet and large upper front teeth; forests of Mexico to Argentina.

Cunina. Genus of marine coelenterates in the Suborder Narcomedusae; life history complicated.

cunner. Small fish of the Atlantic Coast;

Tautogolabrus adspersus; similar to the basses and wrasses; also called sea perch, blue perch, bergall, and chogset.

Cunoctantha. Genus of tropical marine coelenterates in the Suborder Narcomedusae.

Cuon. See DHOLE.

cup-and-saucer limpet. See CRUCIBULUM.

cup coral. Any of numerous spp. of corals in which the polyp is in a cup-like depression in the calcium carbonate skeleton.

Cupelopagis vorax. Uncommon saclike rotifer usually found attached to aquatic plants by a ventral disc; corona a large concave chamber.

cupola. Cupula.

cupula. Mass of gelatinous material at the tip of each sense organ of the fish lateral line system; also present at the tip of each crista in an ampulla of the semicircular canals; secreted by the underlying neuromast cells.

Cupulita. Genus of coelenterates in the Order Siphonophora; with four to six nectophores in each of two rows.

curare. Powerful toxic alkaloid which blocks the junctions between motor nerve fibers and striated muscle fibers.

curassow. Any of several large chicken-like forest birds of Central and South America; blackish or brownish, with an erectile crest; Mitu and Crax typical.

Curicta. Uncommon genus of water scorpions (Family Nepidae); extreme southwestern U.S.

Curculionidae. Very large family of Coleoptera; weevils, snout beetles; exoskeleton hard, mouth parts at the end of a prolonged snout, larvae apodous; adults and larvae feed on a wide variety of plants; many pest spp.; the largest family in the animal kingdom, more than 40,000 spp.; typical genera are Anthonomus, Conotrachelus, Sitophilus, and Phytonomus.

curlew. Any of several large brown shore birds with long down-curved bills; in the Family Scolopacidae; characteristically breed in far north.

currant borer. See AEGERIA.

Cursoria. Group of families of walking or running insects in the Order Orthoptera, e.g. walking sticks and mantids.

cursorial. Specialized for running.

Curtisia foremani. Planarian reported from several eastern states.

cururo. See SPALOCOPUS.

cuscus. See PHALANGER.

cushion star. See PTERASTER.

cusimanse. See CROSSARCHUS.

cusk. See BROSME.

cusk eel. See OPHIDIIDAE.

cusp. 1. Pointed tip or apex. 2. One or more pointed projections on a larger structure. 2. Projection on the biting surface of a mammalian molar tooth.

cuspid. Tooth having one point or cusp; canine tooth.

Cuspidaria. Genus of marine clams in the Order Septibranchia; mostly in deep water.

Cuspidella. Genus of marine coelenterates in the Suborder Leptomedusae.

cutaneous. Of or pertaining to the skin, especially in the higher animals.

cutaneous abdominis. Narrow dorsoventral muscle in the posterior portion of the body wall of certain vertebrates.

cutaneous artery. Artery carrying blood to the skin in Amphibia.

Cuterebra. Genus of flies in the Family Cuterebridae; larvae parasitic under skin of rodents.

Cuterebridae. Family of large hairy flies; botflies; larvae parasitic under the skin of mammals; Cuterebra and Dermatobia are important pests.

cuticle. 1. Cuticula; general term for a dead non-cellular organic layer secreted by the external epithelium of many types of invertebrates, including arthropods, nematodes, earthworm, etc.; the chief functions are support and protection; the epithelium which secretes cuticle is more properly called hypodermis rather than epidermis. 2. Epidermis, especially in man.

cutis. Vertebrate skin, composed of dermis and epidermis.

cutis plate. Dermatome.

cutlass fish. Any of several very long and exceedingly thin marine fishes, especially Trichiurus (Trichiuridae); widely distributed.

cutlip chub. See EXOGLOSSUM.

cutthroat trout. See SALMO.

cuttlebone. See SEPIA.

cuttlefish. See SEPIA.

cutworms. Caterpillars in a group of genera in the moth Family Noctuidae; many spp. are pests on cultivated and wild plants because of their habit of cutting the stems of young plants.

Cuvier, Baron Georges. French zoologist and geologist (1769-1832); founder of comparative anatomy and paleontology; especially noted for studies on the osteology of mammals and anatomy of fishes; proponent of the theory of catastrophism.

Cuvierian duct. See DUCT OF CUVIER.

Cuvierian organs. Cluster of whitish sticky tubes attached to the cloaca in certain sea cucumbers; when disturbed, the animals extrude them over the bodies of enemies.

Cuvierina. Genus of pelagic pteropods; shell cylindrical vase-shaped.

Cyamus. Genus of Amphipoda parasitic on whales; whale lice.

Cyanea. Genus of scyphozoan jellyfish in the Order Discomedusae; the pink jellyfish or "lion's mane"; disc to 6 ft. in diameter; tentacles dangling as much as 150 ft. downward; the largest of all jellyfishes; common in North Atlantic; occasionally fatal to swimmers.

Cyanocephalus. Gymnorhinus.

Cyanocitta cristata. Common bluejay of eastern N.A.; blue above, whitish below, and crested.

Cyanoplax. Genus of chitons.

cyanopsin. Bluish pigment in the cones of the vertebrate retina.

cyanosis. Blueness of skin and mucous membranes produced by hypoxia, especially in mountain sickness and in abnormally slow capillary circulation as in some types of heart disease.

Cyanthus. Genus of hummingbirds (Family Trochilidae); mountains of southwestern states and Mexican Plateau.

Cyathocephalus. Genus of pseudophyllidean tapeworms found especially in fishes.

Cyathocormus. Genus of compound ascidians; colony vase-shaped.

Cyatholaimus. Common genus of chromadorid nematodes; terrestrial and aquatic.

Cyathomonas. Genus of small, oval, flattened, biflagellate Protozoa; holozoic and fresh-water.

cyathozooid. Primary or stem zooid of certain compound tunicates; structure imperfect; perfect secondary zooids bud off from it.

Cyathura. Genus of marine Isopoda.

cybernetics. 1. Science of communication and control of the body by means of the nervous system. 2. Science of recording, communication, and problem-solving by means of electonic-mechanical devices.

Cybister. Common genus of dytiscid beetles.

Cyclas. Genus of weevils (Family Curculionidae).

Cyclaspis. Genus of crustaceans in the Order Cumacea.

Cycleptus elongatus. Blue sucker; a southern sp. found from the Mississippi to the Rio Grande.

cycles of abundance. See POPULATION CYCLE.

Cyclidium. Common genus of marine and fresh-water holotrich ciliates.

Cyclochaeta. Genus of curious bell- or saucer-shaped peritrich Ciliata; ectocommensal on marine and fresh-water sponges and fishes.

Cyclocoelum. Genus of monostome trematodes; in air passages of water birds.

Cyclocypria. Common genus of fresh-water Ostracoda.

Cyclocypris. Common genus of fresh-water Ostracoda.

cycloid scales. Thin, rounded, overlapping dermal scales covering most of the body in the more primitive families of bony fishes; posterior margin smooth.

cyclomorphosis. Series of gradual changes in gross morphology occurring during successive seasonal generations, especially in plankton spp. of Cladocera and Rotatoria.

Cyclomyaria. Suborder of tunicates having a barrel-shaped body; Doliolum common.

Cyclonaias tuberculata. Common thick-shelled unionid mollusk; Mississippi River drainages.

cyclophorase. Enzyme system involved in the citric acid cycle and presumed to be present in mitochondria.

Cyclophylla. Genus of dragonflies.

Cyclophyllidea. Large order of tapeworms in the Subclass Cestoda; scolex with four acetabula (suckers) and sometimes with an apical rostellum armed with hooks; adults of most spp. parasitic in the intestine of birds and mammals; most of the common tapeworms of man are members of this order, especially Taenia, Echinococcus, and Hymenolepis.

cyclopia. Abnormal condition in vertebrates characterized by a single median eye.

Cyclopoida. Suborder of marine and fresh-water copepods; metasome much wider than urosome; movable articulation between fifth and sixth thoracic segments; first antennae shorter than in the Calanoida and longer than in the Harpacticoida, both modified for clasping in male; free-living, commensal, and parasitic.

Cycloporus. Genus of marine polyclad Turbellaria; unusual because of its numerous "anal pores."

Cycloposthium. Genus of oligotrich Ciliata commonly found in the caecum and colon of the horse.

Cyclops. Very common genus of fresh-water copepods.

Cyclopteridae. Family of marine bottom fishes in the Order Scleroparei; lumpfishes; short, thick, high-arched body, and a ventral bony sucking disc; skin with tubercles; Cyclopterus common.

Cyclopterus lumpus. Lumpsucker, lumpfish; a soft sluggish fish found on both sides of the North Atlantic; body with seven longitudinal ridges marked by pointed tubercles; on the bottom or in floating masses of rockweed; to 18 in. long.

Cyclorrhapha. One of the two suborders of insects in the Order Diptera; without a crescent-shaped sclerite over the base of the antennae; head of larva poorly developed; puparium formed from the last larval skin, opens by an anterior circular lid.

Cyclosa. Common genus of spiders (Araneidae).

Cyclosalpa. Genus of pelagic salps found mostly in tropical and subtropical oceans.

cyclosis. Streaming movement of protoplasm in certain cells.

Cyclospondyli. Taxon which includes certain sharks; with a single calcified sheath around the notochord in each vertebral centrum.

Cyclostomata. 1. Subclass of the vertebrate Class Agnatha; lampreys and hagfishes; body cylindrical, long, and scale-less; suctorial mouth with horny teeth; single median nostril; median fins; skull, visceral arches, and imperfect neural arches cartilaginous; notochord persistent; two-chambered heart; seven to 14 pairs of gills in expanded chambers of tubes leading from pharynx to outside; eight or ten pairs of cranial nerves; marine and fresh-water; about 50 modern spp. 2. Order of marine bryozoans in the Class Stenolaemata; zooecium calcareous and tubular; no operculum; Tubulipora common.

Cyclothrix. Limnochares.

Cycloxanthops. West Coast genus of crabs; intertidal zone.

Cyclura. Genus of lizards or iguanas; Mexico and southward.

cydippid. Free-swimming larval type characteristic of the development of most ctenophores; superficially similar to the adult.

Cydippida. One of the orders of the ctenophore Class Tentaculata; tentacles branched and retractile into special sheaths; Pleurobrachia common.

cyesis. Pregnancy.

cygnet. Young swan.

Cygnus. Genus which includes the swans (Family Anatidae); widely distributed aquatic birds with heavy bodies, very long necks, white coloration, and vegetarian food habits; C. columbianus, the whistling swan, winters in U.S.; C. olor is the domesticated mute swan; C. buccinator, the trumpeter swan, is fast disappearing; its range is interior and western N.A.

Cylichna. Genus of small opisthobranch snails having a solid subcylindrical shell.

Cylindrolaimus. Common genus of chromadorid nematodes; fresh-water and terrestrial.

Cylindroleberis. Genus of marine Ostracoda.

Cylisticus. Terrestrial genus of Isopoda.

Cyllene robiniae. Common black beetle with yellow cross bands; Family Cerambycidae; locust borer, a pest of honey locust trees.

Cylloepus. Genus of elmid beetles.

Cymatia americana. Water boatman of northern states and Canada; in the Family Corixidae.

Cymatium. Genus of very large tropical marine gastropods; shell to 20 in. long; formerly used as trumpets.

Cymbiodyta. Common genus of hydrophilid beetles.

Cynarioides. Genus of primitive therapsids.

Cynictis. Genus of South African mongooses.

Cynidiognathus. Genus of primitive therapsid reptiles.

Cynipidae. Family of small to minute wasps; gall wasps; abdomen compressed; eggs usually deposited in tissues of oak trees and occasionally roses and composites; activities of larva in tissues produce a characteristic abnormal growth, or gall; males rare, parthenogenesis common; Andricus and Diplolepis common.

Cynocephalus. Papio.

Cynodictis. Genus of primitive weasel-like mammals of Paleocene and Eocene times.

Cynogale. Genus of otter-civets; fairly large mammals of Malaya, Borneo, and Sumatra; semiaquatic spp. which feed on fresh-water vertebrates and invertebrates.

Cynognathus. See THERAPSIDA.

Cynomys. Genus of herbivorous rodents in the Family Sciuridae; prairie dogs; heavy-bodied, short-tailed, terrestrial, squirrel-like rodents about the size of a small woodchuck; upper parts generally cinnamon colored; burrows deep and branching, with a ridgelike mound surrounding the entrance; formerly in large "towns" of millions of individuals, but now greatly reduced in numbers; several spp. and subspecies in the Great Plains.

Cynopithecus. Genus of black apes; large, baboon-like monkeys of mangrove forests of southern Philippines, Celebes, and nearby islands.

Cynopterus. Genus of short-nosed fruit bats of the Oriental region.

Cynoscion. Large genus of marine fishes in the Family Sciaenidae; weakfish and relatives of weakfish; mostly tropical and subtropical.

Cyphoderia. Genus of fresh-water amoeboid Protozoa; most of cell enclosed in a retort-shaped test.

Cyphon. See HELODIDAE.

cyphonautes. Ciliated conical free-swimming larva of certain marine Bryozoa; digestive tract complete.

Cypraea. Marine gastropod genus which includes certain of the cowries; brightly colored ovate shells with concealed spiral, narrowed at both ends, and with a toothed aperture extending the length of the shell; tropical and subtropical; formerly used as money by some native tribes; certain spp. are spectacularly colored, rare, and highly desirable collector's items.

Cypraeolina. Genus of tiny tide pool snails.

Cypretta. Common genus of fresh-water Ostracoda.

Cypria. Common genus of fresh-water Ostracoda.

Cypricercus. Common genus of fresh-water Ostracoda.

Cypriconcha. Genus of fresh-water Ostracoda.

Cypridina. Genus of marine ostracods in the Order Myodocopa; some spp. bioluminescent.

Cypridopsis. Widely distributed and common genus of fresh-water Ostracoda.

Cyprina. Arctica.

Cyprinidae. Large family of fresh-water fishes; the true minnows; with pharyngeal teeth; includes carp, goldfish, dace, chub, shiners, etc.

Cypriniformes. Taxon sometimes used to encompass those families of fishes which include the suckers, minnows, catfishes, characins, etc.

Cyprinodon. Genus of small fishes in the Family Cyprinodontidae; killifish, desert fish, pupfish; springs and streams in southwestern deserts, but extending into Central America.

Cyprinodontes. Order of bony fishes which includes the killifishes, top minnows, and cave fishes; small omnivorous spp., chiefly in fresh waters; pelvic fins abdominal or absent; often viviparous.

Cyprinodontida. Cyprinodontes.

Cyprinodontidae. Family of fresh-water bony fishes in the Order Cyprinodontes; killifishes and desert fishes; small spp.; Cyprinodon and Fundulus common.

Cyprinoidei. Suborder taxon sometimes used to include the fish families Catostomidae and Cyprinidae.

Cyprinotus. Common genus of fresh-water Ostracoda.

Cyprinus carpio. Common carp; a medium to large fish native to Asia but now widely introduced and abundant, especially in warm and temperate waters; commonly used as human food in Europe and Asia but in the U.S. used chiefly in the larger cities, especially during certain religious holidays; a pest sp. which drives out game fish by rooting in the bottom of lakes, ponds, and rivers and thus keeping the water muddy.

cypris. Advanced free-swimming larval stage of barnacles; develops from the nauplius; covered with a bivalve shell; eventually attaches to a substrate and metamorphoses into the adult barnacle.

Cypris. Large genus of fresh-water Ostracoda; many common spp.; 1 to 2.3 mm. long.

Cyprogenia. Genus of unionid mollusks; central U.S.

Cyprois. Common genus of fresh-water Ostracoda; creeks and vernal ponds.

Cypseloides niger. Black swift, a soot-colored bird in the Family Micropodidae; mountainous areas from British Columbia to Costa Rica.

Cypselurus. See EXOCOETIDAE.

Cyrenella floridana. Small bivalve of Fla. brackish waters.

Cyrtidae. Acroceratidae.

Cyrtoceras. Genus of fossil nautiloid mollusks.

Cyrtodaria. Northern genus of blackish marine clams; soft parts extend much beyond valves.

Cyrtolophosis. Genus of ovoid to ellipsoid fresh-water holotrich Ciliata; in a mucilaginous envelope.

Cyrtonyx montezumae. Harlequin quail, Mearns' quail; a small quail of the mountains of the southwestern states and Mexican Plateau.

Cyrtophium. Genus of marine amphipods; build cases out of bits of hollow plant stems.

cyst. 1. Resistant resting stage formed by many different organisms, especially as a response to adverse environmental conditions, such as drouth, low temperature, high temperature, etc.; a typical cyst is spherical, and the outermost layers are thick, tough, and relatively impervious; the cyst stage occurs in many algae, bacteria, Protozoa, and a few small metazoans. 2. Any abnormal pouch or sac without an external opening, usually with a distinct membrane and containing a fluid or semifluid material; sometimes pathological.

cystic duct. Short duct of the gall bladder.

cysticercoid. Larval stage of certain tapeworms, consisting of a scolex with an investing layer of tissue; resembles a cysticercus, but there is no appreciable fluid-filled cavity between the scolex and the investing outer layer; often has a posterior tail-like projection; found encysted in the body cavity or tissues of arthropods, oligochaetes, and other invertebrates, and occasionally in vertebrates; occurs in the life history of Hymenolepis nana.

cysticercus. Bladderworm; a fluid-filled sac containing an invaginated scolex; an inactive larval stage of many tapeworms, imbedded in the muscle or viscera of the intermediate host; upon being ingested by the definitive host, it develops into a mature tapeworm. Pl. cysticerci.

cystid. In Bryozoa, the dead secreted outer parts plus the adherent living underlying layers.

cystine. First amino acid to be discovered; abundant in hair, nails, and and other epidermal derivatives; $C_6H_{12}O_4N_2S_2$.

Cystobranchus. Genus of leeches found occasionally as parasites of fresh-water fishes.

Cystodytes. Genus of ascidians.

Cystoidea. Class of primitive echinoderms; calyx small, rigid; Ordovician to Devonian; Caryocrinites and Pleurocystites typical.

Cystophora cristata. Large, dark-colored seal in the Family Phocidae; hooded seal; male has a large, inflatable muscular bag on top of the head; Newfoundland to Greenland.

Cythere. Genus of marine Ostracoda.

Cythereis. Common genus of marine Ostracoda in the Order Podocopa.

Cytherella. Genus of marine Ostracoda in the Order Platycopa.

Cytheridea. Genus of marine Ostracoda.

cytochrome. Any of a group of related compounds widely distributed in plant and animal cells; protein with an iron-containing group; enters into the long series of chemical processes involved in cell respiration; examples are cytochrome A, B, and C.

cytochrome C oxidase. Enzyme necessary for the oxidation of reduced cytochrome C by oxygen.

cytogenetics. Study of cellular details with special reference to the phenomena of heredity, e.g. meiosis.

cytokinesis. Collectively, the changes taking place in the cytoplasm of a cell during mitosis, meiosis, and fertilization.

Cytoleichus nudus. Air-sac mite of poultry.

cytology. Study of the structure and function of cells and their parts.

cytolysin. Any antibody or other substance which produces breakdown of cells; named according to the type of cell upon which they act, such as hemolysin and neurocytolysin.

cytolysis. Disintegration of cells, especially by destruction of surface membranes.

cytopharynx. Funnel-shaped cavity leading into the cell from the cytostome in ciliates and certain flagellates; usually functions as a passage for particles of food being ingested.

cytoplasm. Collectively, the protoplasmic material making up all of the cell contents inside of the plasma membrane, but not including the nucleus.

cytoplasmic bridges. Strands of cytoplasm connecting adjacent individuals in a colony of Protozoa.

cytoplasmic inheritance. Inheritance of characters by self-propagating cytoplasmic materials; non-Mendelian and distinct from gene inheritance.

cytoproct. Cytopyge.

cytopyge. More or less permanent pore in the posterior portion of certain ciliate Protozoa; a point of discharge for residual indigestible particles remaining in food vacuoles.

Cytorhinus mundulus. Plant bug which feeds on eggs of the sugarcane leafhopper; tropical Pacific.

cytosine. Pyrimidine base; component of nucleic acids involved in cellular metabolism; $C_4H_5ON_3$.

cytosome. Entire cell body inside the plasma membrane.

cytostome. The "mouth" opening of certain holozoic Protozoa; the term also applies to the opening of the canal leading to the reservoir in certain holophytic and saprozoic Protozoa.

cytotrophoblast. Cytotrophoderm.

cytotrophoderm. Cytotrophoblast; thickened inner layer of trophoblast in an early mammalian embryo; forms plasmotrophoderm by proliferation.

D

dab. General name for any small flatfish, e.g. flounders and their relatives; see HETEROSOMATA.

dabchick. Pied-billed grebe (Podilymbus podiceps) of North and South America, or the little grebe (Podiceps ruficollis) of the Eastern Hemisphere.

dace. 1. Small European fish in the minnow family. 2. See RHINICHTHYS and CHROSOMUS.

dactyl. 1. Digit; finger or toe. 2. Dactylopodite. 3. European piddock (Pholas).

Dactylanthus. Genus of anemones in the Suborder Ptychodactiaria; cold seas.

Dactylometra. Widely distributed genus of jellyfishes in the Order Discomedusae; to 8 in. in diameter.

Dactylopiidae. Small family of insects in the Order Homoptera; cochineal insects; red or crimson spp. having the body more or less obscured by white cottony wax secretions; mature cleaned females of Dactylopius coccus are the source of commercial cochineal dye.

Dactylopius. See DACTYLOPIIDAE and COCHINEAL.

dactylopodite. Terminal (seventh) segment of pereiopod in most malacostracan Crustacea.

dactylopores. Small pores on the surface of a millepore coelenterate skeleton through which the dactylozooids protrude.

Dactylopsila. Genus of large striped phalangers; feed mostly on insect larvae in bark and rotting wood; nocturnal and arboreal; Australian region.

Dactylopteridae. Family of fishes which includes the flying gurnards; see DACTYLOPTERUS.

Dactylopterus volitans. Flying gurnard, a marine fish remarkable for its enlarged fanlike pectoral fins; sometimes called batfish.

Dactylopusia. Genus of marine harpacticoid Copepoda.

Dactylosphaera. See PHYLLOXERIDAE.

dactylozooid. One kind of polyp in certain hydrozoan coelenterate colonies; specialized for a tactile role; tentacles and gastrovascular cavity lacking.

dactylus. 1. Dactylopodite. 2. In certain insects, the second tarsal segment which follows an enlarged first tarsal segment. 3. Digit. Pl. dactyli.

Dacus dorsalis. Oriental fruit fly in the Family Trypetidae; serious pest of fruits in tropics and subtropics.

daddy longlegs. See PHALANGIDA.

Dallia. See BLACKFISH.

Dallingeria. Genus of very small Protozoa; one anterior and two lateral flagella; coprozoic and stagnant fresh waters.

Dalmanella. Genus of fossil brachiopods.

Daltonism. Red-green color blindness.

Dalyellia. Large genus in the turbellarian Order Rhabdocoela.

Dama dama. Semidomesticated European fallow deer; coat spotted with white during summer.

Damaliscus. Genus of antelopes in the Family Bovidae; includes the korrigum, tiang, topi, damalisk, bontebok, blesbok, and sassaby; southern Africa.

damalisk. Any of numerous African antelopes, especially Damaliscus and Beatragus.

dammar. Damar; transparent resin used for mounting histological sections, etc. on microscope slides; dissolved in xylol.

damsel bugs. See NABIDAE.

damselfish. See POMACENTRIDAE.

damselfly. Common name of insects in the Suborder Zygoptera of the Order Odonata.

Danaidae. Small family of large, brightly colored butterflies; usually orange-brownish with black markings; Danaus common.

Danaus plexippus. Common monarch or milkweed butterfly; dark orange coloration with black and white markings; shows a true north-south seasonal return migration; presumed to be a "model" for the viceroy butterfly in Batesian mimicry.

dance flies. See EMPIDIDAE.

danio. Any of several brightly colored cyprinid tropical aquarium fishes; native to southeastern Asia.

daphne. Vernacular for Daphnia.

Daphnia. Very common genus of Cladocera; found everywhere in fresh waters; familiar spp. are D. pulex, D. galeata, and D. magna.

daphnid. 1. Of or pertaining to any member of the cladoceran genus Daphnia. 2. Of or pertaining to any cladoceran.

daraprim. Antimalarial drug, apparently effective against most stages of Plasmodium in all parts of the body.

dark adaptation. Relative ability of the human eye to distinguish objects in very dim light; partially determined by an adequate amount of vitamin A in the blood.

dark field microscope. Microscope having a small opaque plate in the center of the condenser field so that only oblique light strikes the object which thus appears bright against a dark background; often called ultra-microscopy.

darkling beetles. See TENEBRIONIDAE.

Darlington, Cyril Dean. British cytologist and geneticist (1903-); worked on chromosome cytology and wrote several important books on genetics and chromosome biology.

darning needle. See ODONATA and ANISOPTERA.

dart. See DART SAC.

darter. Any of a large group of small, brightly colored, fresh-water fishes in the Family Percidae; restricted to clear streams of N. A.; examples are Etheostoma and Percina.

dart sac. Small sac in the basal portion of the reproductive system of certain snails; it forms small sclerotized dartlike structures which are "shot" into the body of another individual as a preliminary to copulation; such a phenomenon is presumed to have a stimulating effect on the recipient.

Darwin, Charles Robert. English naturalist (1809-1882); most famous for his On the Origin of Species by Means of Natural Selection (1859), on which the modern concept of evolution is largely founded; the book was based on an enormous 20 years' accumulation of facts and observations, many of which were gathered during his five-year voyage around the world as naturalist on the Beagle; among his other important works are a monograph on barnacles, the origin and development of coral reefs, Descent of Man and Selection in Relation to Sex (1871), and Variation of Plants and Animals Under Domestication (1868).

Darwin, Erasmus. English physician, poet, and naturalist (1731-1802); grandfather of Charles Darwin; in his Zoonomia (1794-1796) he anticipated later evolutionary theories.

Darwinism. Concept of organic evolution, especially as developed by Charles Darwin and contained in his Origin of Species (1859); his arguments and conclusions may be summarized as follows: all plants and animals show variations in their morphology, physiology, and behavior; because of the efficient reproductive potential of all organisms, the numbers of every sp. would tend to become very large, yet most individuals are soon eliminated by the adversities of the environment; this involves a struggle for existence during which those individuals having slight variations which better fit them to the environment will survive, but those having unfavorable variations are eliminated; this process of natural selection results in the survival of the fittest, and coupled with an environment which changes slowly with geological time, the end result is the formation of new species. See EVOLUTION.

Darwin's finches. Family of finches (Geospizidae) restricted to the Galapagos Islands; exhibit remarkable adaptive radiation; four genera and about 14 spp. variously specialized for seed eating, nectar feeding, vegetation and fruit eating, insect eating, and digging in tree trunks.

Darwin's point. Small lump with a cartilaginous support; points downward and forward from the pinna of the human ear.

Darwinula. Common genus of fresh-water Ostracoda.

Dasineura. Widely distributed genus of midges in the Family Cecidomyiidae; larvae feed on leaves or produce galls on certain economic plants.

dassie. See PROCAVIA.

Dasyatidae. See DASYATIS.

Dasyatis. Common genus of stingrays in the Family Dasyatidae.

Dasybranchus. Genus of Polychaeta.

Dasycercus. Uncommon genus of marsupial mice; deserts of central and southern Australia; size of large mice and habits of lizards; carnivorous and voracious.

Dasydytes. Genus of fresh-water Gastrotricha; body with very long spines.

Dasyhelea. Genus of ceratopogonid Diptera.

Dasypeltis scaber. African egg-eating snake having processes of cervical vertebrae modified into an esophageal rasp used in breaking eggshells.

Dasypodidae. Family of mammals in the Order Edentata; the armadillos; bony platelike armor over head, back, and flanks, and covered externally by horny shell which is divided by transverse furrows permitting the animal to curl up when disturbed; forefeet with powerful digging claws; Tex. to South America; Dasypus novemcinctus is the common Texas armadillo, to 32 in. long.

Dasyproctidae. Small family of large rodents; includes the South and Central American paca and agouti.

Dasypterus. Genus of yellow bats; medium-sized spp. with long, tapering ears and yellow pelage; North and South America.

Dasypus. See DASYPODIDAE.

dasyure. See DASYURIDAE.

Dasyuridae. Dasyures; family of Australian marsupials; carnivorous or insectivorous; Thylacinus, Sarcophilus, Dasyurus, Phascogale, and Sminthopsis typical.

Dasyurus. Genus of carnivorous, arboreal, catlike marsupials in the Family Dasyuridae; tiger cats; Tasmania and Australia.

data. Accumulated facts on a particular subject; (the singular form, datum, is seldom used in the biological sciences).

date mussel. Lithophaga plumula, a Pacific Coast burrowing mussel having the superficial appearance of a date.

Daubentonia. See DAUBENTONIOIDEA.

Daubentonioidea. Suborder of primates which includes the aye-aye, a nocturnal lemur found only in Madagascar (Daubentonia madagascarensis).

Dauermodifikationen. Persisting modifications in generations succeeding one which has been subjected to abnormal environmental conditions, e.g. chemicals in protozoans and heat in Drosophila.

daughter chromosomes. Pair of chromosomes resulting from the separation of each pair of chromatids during mitosis; the members of each pair of daughter chromosomes move toward opposite poles of the spindle.

daughter cyst. In a hydatid cyst, a miniature spherical cyst which is suspended in the hydatid fluid and contains internal proliferated tapeworm scolices.

daughter rediae. Second generation of trematode rediae produced within the snail host by internal proliferation of the parent redia in certain spp.

daughter sporocysts. Second generation of trematode sporocysts produced within the snail host by internal proliferation of the parent sporocyst in certain spp.

Davainea. Genus of cyclophyllidean tapeworms found in mammals and birds.

Davenport, Charles Benedict. American zoologist (1866-1944); noted for his work on human genetics, eugenics, and anthropometry.

Da Vinci, Leonardo. Italian painter, sculptor, architect, engineer, and scientist (1452-1519); as a zoologist, most renowned for his careful drawings and studies of human anatomy.

dayfly. Adult mayfly.

D'Azyr, Felix Vicq. French physician and anatomist (1784-1794); noted for his very extensive studies on vertebrate comparative anatomy.

DDT. Dichloro-diphenyl-trichloroethane; a white, odorless, crystalline organic compound which is extremely effective as an insecticide; acts as both a contact and stomach poison.

dead-leaf butterfly. See KALLIMA.

deadman's fingers. See HALICLONA.

dealate. Among insects, having lost the wings.

dealation. Among insects, the loss of wings by breaking off by the insect itself.

dealfish. Ribbonfish (Trachypterus).

deaminase. See DEAMINATION.

deamination. Partial disintegration of amino acids, especially in the mammalian liver; in the presence of an enzyme (deaminase) and with the use of water and oxygen, amino acids are converted to ammonia and keto acids; as rapidly as it is formed, the ammonia is converted to urea by the liver; the keto acids, having a carbohydrate structure, are converted to either glycogen or fat; in the strict sense, "deamination" refers only to the oxidative removal of NH_2 groups from amino acids to form NH_3 groups.

Dean, Bashford. American zoologist (1867-1928); made many important contributions in ichthyology and hermetrium; curator at the American Museum of Natural History.

death adder. Acanthophis antarcticus, a deadly snake in the Family Elapidae; body stout, tail tipped with a spine; Australian region.

death's head moth. Large blackish European moth, Acherontia atropos; dorsal thoracic markings of brown and yellow have the shape of a human skull.

death watch. See LIPOSCELIS.

death watch beetle. See ANOBIIDAE.

De Blainville, Henri Marie Ducrotay. French zoologist (1777-1850); especially known for his work in comparative anatomy and for some of his ideas about the importance of tissues in the organization of the animal body.

decalcification. 1. Loss of calcium compounds from living tissues. 2. Removal of calcium compounds from bits of tissue in preparation for making thin sections for study with the microscope; such tissues are usually treated with acid.

Decapoda. 1. Order of malacostracan Crustacea which includes the crayfish, crab, lobster, shrimp, etc.; thoracic appendages mostly uniramous; last five thoracic segments with "walking legs"; 8000 spp., mostly marine, some fresh-water, few terrestrial; many edible spp. 2. Suborder of cephalopod mollusks which includes the squids and cuttlefish; with ten arms and tentacles; body more or less elongated.

decarboxylase. Any of several cellular enzymes that facilitate decarboxylation.

decarboxylation. Part of the breakdown process of an organic acid radical; in the presence of a decarboxylase, a carboxyl group is split off and a molecule of carbon dioxide is formed.

decerebrate. 1. Without cerebral hemispheres, as an experimental animal on which such surgery has been done. 2. Without a brain.

decidua. Thick endometrial lining of uterus in pregnant mammals; part or all of it is shed with the placenta at birth or shortly thereafter.

decidua basalis. That part of the endomedtrium of the pregnant uterus which lies between the chorionic sac and the muscular wall of the uterus.

decidua capsularis. That part of the endometrium of the pregnant uterus which covers the chorionic sac and lies between it and the uterine cavity.

decidua parietalis. That part of the enmetrium of the pregnant uterus which forms a general uterine lining except where the embryo is attached.

deciduate placenta. Type of placenta which, upon loss at birth, is accompanied by a layer of maternal uterine tissue; found in primates, insectivores, moles, bats, hedgehogs, shrews, mice, and rabbits.

deciduous forest. Forest consisting chiefly of trees which lose their leaves periodically (in the autumn or dry season); in the U. S. these forests are composed of various pure stands and mixtures of oak, hickory, elm, beech, etc.

deciduous teeth. Milk teeth.

decomposers. Collectively, those organisms (bacteria, fungi) in an ecosystem which convert dead organic materials into inorganic materials.

decompression sickness. Caisson disease.

decorator crabs. Any of numerous brachyuran crabs which camouflage the body with bits of seaweed, bryozoans, sponges, and hydroids.

decurved. Curved downward.

dedifferentiation. Simplification of tissue structure when animals undergo a period of physiological depression or shrinkage in size.

deep-sea angler. See CERATIIDAE and MELANOCETIDAE.

deer. Common name applied to a wide variety of ruminants in the Family Cervidae; usually does not include the elk, moose, and caribou; examples are Moschus, Muntiacus, Dama, Axis, Cervus, Odocoileus, Ozotoceras, and Capreolus.

deer flies. See CHRYSOPS and TABANIDAE.

defassa. Large, heavy-bodied antelope of Africa south of the Sahara; genus Kobus.

defecation. Process of voiding indigestible materials from the digestive tract; used especially with reference to vertebrates; (as distinct from excretion, defecation involves chiefly materials which have never been a part of the living protoplasm). Cf. EGESTION.

defibrinated blood. Blood from which the fibrin has been removed by vigorous stirring or whipping with a small brush or cluster of small straws; such blood remains red and liquid.

deficiency disease. Any disease produced by an insufficient amount of an essential element, mineral, vitamin, amino acid, etc. in the diet.

definitive host. That host which harbors the mature stage in the life cycle of a parasite; thus, man is the definitive host of the pork tapeworm. Cf. INTERMEDIATE HOST.

degeneration. 1. Deterioration of whole or part of an organ or tissue or loss of effective function during the life of an animal. 2. Similar loss of structure or function as an evolutionary process over many generations.

deglutition. Act of swallowing, especially with respect to the interaction of reflexes involved.

De Graaf, Regner. Dutch physician (1641-1673); first ascertained the properties of pancreatic juice.

degrees of freedom. Number of classes upon which a chi-square value is based; thus, in a 3:1 genetic ratio, there is only one degree of freedom; and in a 9:3:3:1 ratio, there are three degrees of freedom.

degu. See OCTODON.

dehydration. 1. Excessive loss of water from a tissue, organ, or whole animal. 2. Extraction of water from fixed tissue in preparation for mounting, sectioning, or imbedding; such tissues are usually placed in dioxan or successively stronger concentrations of ethyl alcohol.

dehydroandrosterone. Androgen synthesized from cholesterol and found in human urine.

dehydrocorticosterone. Hormone of the adrenal cortex; see CORTICOSTERONE.

dehydrogenase. Any of a group of enzymes which oxidize a substrate by removing hydrogen from it; such hydrogen usually combines with a hydrogen acceptor (commonly oxygen), after being passed through a chain of transferring enzymes.

dehydrogenation. In physiology, an enzyme-controlled process involving the removal of hydrogen atoms from an organic molecule by a hydrogen acceptor.

dehydrohydroxycorticosterone. Hormone of the adrenal cortex; see CORTICOSTERONE.

Deima. Genus of deep-sea benthic sea cucumbers; respiratory tree absent.

Deinocerites. Genus of mosquitoes; several rare spp. in Gulf states; breed in burrows of marine crabs.

deirid. One of two lateral cervical sensory papillae in certain nematodes.

Deirochelys reticularia. Chicken turtle; neck especially long; a sp. common in southeastern and southern U.S.

De Kay's snake. See STORERIA.

deletion. Abnormal absence of part of a chromosome as determined by cytological study.

delouse. To remove lice from, especially by means of steam treatment or medication.

Delphinapterus leucas. White whale or beluga of Arctic and Subarctic seas; to 12 ft. long; a toothed whale.

Delphinidae. Family of whales; includes the dolphins and porpoises; small to medium-sized, with numerous teeth; e.g. Pseudorca and Tursiops.

Delphinus. See DOLPHIN.

deltoides clavicularis. Dorsal pectoral girdle muscle of reptiles.

deltoid. Deltoideus; a muscle which raises the arm and carries it forward and backward; originates on the pectoral girdle and inserts on the middle of the humerus; present in certain tetrapods.

delundung. See LINSANG.

deme. 1. Interbreeding population within a sp. 2. Aggregate of single cells.

dementia praecox. Schizophrenia.

demersal. Living on or near the bottom of a lake or sea; e.g. fishes and eggs.

demilunes. Crescentic groups of serous cells in the submandibular and sublingual salivary glands.

Demodex. See DEMODICIDAE.

Demodicidae. Family of elongated follicle mites which live in the skin of mammals; Demodex folliculorum is a common commensal of hair follicles and sebaceous glands of man and domestic mammals.

Demodicoidea. Superfamily in the arachnid Order Acarina; minute, elongated mites with four or eight legs; form galls in plants (Eriophyidae) or parasitic or commensal in skin of mammals (Demodicidae).

demography. Study of human populations with special reference to their physical environment and geographic distribution.

demoiselle. 1. Small Old World crane. 2. Louisiana heron. 3. Any of numerous slender damselfly adults, e.g. Agrion.

Demospongiae. One of the three classes of Porifera; skeleton of silicon dioxide spicules (not triaxon), or horny fibers, or both; commercial sponges belong in this group; examples are Spongia and Spongilla.

denaturation. 1. Any non-proteolytic change in the composition or state of a natural protein, usually as the result of high temperature, low temperature, or various chemicals; boiling an egg is essentially denaturation of its proteins (coagulation). 2. Making ethyl alcohol unfit for drinking by the addition of some toxic or unpalatable material, such as acetone.

Dendragapus obscurus. Common dusky grouse, or blue grouse, of evergreen forests in the Rocky Mountain area; a large, chicken-like, grayish or blackish sp. in the Family Tetraonidae.

Dendraster. Genus of eccentric sand dollars of the Pacific Coast.

dendrites. Processes of a neuron which transmit impulses toward the nerve cell body; commonly short and very much branched and rebranched. Cf. axon.

Dendroa. Genus of small tunicates which grow in berry-like bunches.

Dendroaspis. Genus of snakes which includes the deadly African mamba.

Dendrobates. Genus of small, brightly colored, Neotropical frogs in the Family Ranidae.

Dendrobeania. Genus of marine Bryozoa.

Dendrochirota. Echinoderm order in the Class Holothuroidea; tentacles irregularly arborescent; respiratory tree present; Thyone and Cucumaria common.

Dendrocoelum. Genus of large, pale, triclad Turbellaria; fresh waters; not found in U. S.

Dendrocolaptidae. Family of small brownish creeping birds of the American tropics and subtropics; nest in natural hollows; woodcreepers.

Dendrocometes. Genus of fresh-water Suctoria found especially on gills of Amphipoda; cell spherical, with numerous branched arms.

Dendrocopos. Dryobates.

Dendroctonus. Genus of pine beetles in the Family Scolytidae; burrow extensively in conifers.

Dendrocygna. Genus of birds which includes the tree ducks of the tropics and subtropics; long necks and legs and high whistling voice; combine characters of ducks, geese, and swans.

Dendrodoa. Genus of low, sessile ascidians.

Dendrodoris. Genus of nudibranchs.

Dendrogale. Genus of tree shrews; southeastern Asia and Borneo.

Dendrograptus. See DENDROIDEA.

Dendrohyrax. Genus of tree-hyraxes; omnivorous arboreal spp. of Africa.

Dendroica. See MNIOTILTIDAE.

Dendroidea. Group of bushlike graptozoans; e. g. Dendrograptus.

Dendrolagus. Genus of Australian marsupials in the Family Macropodidae; tree-kangaroos; usually in small trees, although poorly adapted to an arboreal habitat; New Guinea and Queensland.

Dendromonas. Genus of small, stalked, colonial, biflagellate Protozoa; stagnant fresh waters.

Dendronotus frondosus. Bushy-backed sea slug, notable for two dorsal rows of branching treelike cerata; common along north Atlantic Coast; genus found also on Pacific Coast.

Dendrosoma. Genus of suctorian Protozoa which form branching masses of individuals to 2. 5 mm. high.

Dendrostomum. Genus of Sipunculoidea; elongated pear-shaped.

dengué. Non-fatal virus disease of man in the tropics and subtropics; transmitted by Aedes aegypti and other mosquitoes; also called seven days' fever and breakbone fever.

Denisonia. Genus of poisonous snakes in the Family Elapidae.

denitrification. Complex of bacterial conversions of ammonia, nitrites, and nitrates into free nitrogen.

denitrifying bacteria. Certain soil and aquatic bacteria which break down (successively) nitrates, nitrites, and ammonium compounds into gaseous nitrogen during their metabolism.

density-dependent factors. Mortality factors of the environment whose severity and effectiveness is usually dependent upon the density of a population; e. g. competition, food supply, predation, and parasites.

density-independent factors. Mortality factors of the environment whose severity and effectiveness is usually not dependent upon the density of a population; e. g. extreme temperatures and storms.

dental caries. Tooth decay.

dental formula. System of letters and numbers used to indicate the numbers and kinds of teeth present in mammals;

the dental formula for man is $I\frac{2}{2}$, $C\frac{1}{1}$, $P\frac{2}{2}$, $M\frac{3}{3}=32$, indicating two incisors, one canine, two premolars, and three molars on each half of each jaw; for the cottontail rabbit the formula is $I\frac{2}{1}$, $C\frac{0}{0}$, $P\frac{3}{2}$, $M\frac{3}{3}=28$; for the deer $I\frac{0}{4}$, $C\frac{1}{0}$, $P\frac{3}{3}$, $M\frac{3}{3}=34$; white-footed mouse $I\frac{1}{1}$, $C\frac{0}{0}$, $P\frac{0}{0}$, $M\frac{3}{3}=16$.

Dentalium. Common genus of mollusks (Class Scaphopoda).

Dentella. Genus of caprellid Amphipoda.

dental pulp. Soft tissues occupying the pulp chamber and root canals of a tooth (or placoid scale); composed of nerves, blood vessels, and connective tissue.

dentary. 1. Dermal bone forming much of the anterior part of the lower jaw on each side in most vertebrates. 2. The entire lower jaw of mammals. 3. Pertaining to teeth.

dentate. Having a toothed margin.

denticles. 1. Small teeth or toothlike projections. 2. Toothlike scales of elasmobranch fishes.

denticulate. Having small toothlike processes.

dentine. Calcareous material forming the bulk of teeth and placoid scales; usually covered with a layer of enamel; ivory is composed of dentine.

dentition. Number, kinds, and arrangement of teeth in any animal or part of an animal.

Deomys. Genus of small African rats; wading rats; legs long, head long and pointed; wade about in shallow equatorial puddles looking for insects, snails, etc.; improperly referred to as "tree mice."

deoxycorticosterones. Desoxycorticosterones; two closely related steroid endocrines elaborated by the adrenal cortex; control cell permeability through the distribution of sodium and potassium.

deoxycortisol. Hydrodeoxycorticosterone; a steroid endocrine elaborated by the adrenal cortex; controls cell permeability through the distribution of sodium and potassium.

deoxyribonuclease. Enzyme which hydrolyzes deoxyribonucleic acid.

deoxyribonucleic acid. Desoxyribonucleic acid, DNA; important constituent of nucleoproteins of chromatin; composed of long chains of phosphate and sugar molecules (deoxyribose) with several bases arranged as side groups: adenine, guanine, cytosine, and thymine; chief carrier and mediator of genetic information.

deoxyribose. Pentose sugar which is an important part of the deoxyribonucleic acid molecule.

dependent differentiation. Specialization or developmental peculiarity of an embryonic tissue induced by a stimulus from some other tissue.

depolarization. See MEMBRANE THEORY.

depressed. Flattened dorsoventrally.

depression period. Condition found in hydras in old and unsuitable cultures; body shortens, tentacles disappear, and the animal disintegrates into a shapeless brown mass which may or may not revive.

depression slide. Microscope slide that has a concave depression in it; for use with hanging drops or small organisms that are too large to be studied with an ordinary slide and coverslip.

depressor. Muscle that lowers a part of the body.

depressor mandibulae. Muscle that opens the mouth in vertebrates.

Deraeocoris. Important genus of plant bugs in the Family Miridae.

Derby flycatcher. See PITANGUS.

De Réaumur, René Antoine Ferchault. French physicist and naturalist (1683-1757); contributed to regeneration in decapods and to digestion in birds; wrote six volumes on the natural history of insects.

derived protein. 1. Any complex protein-like substance derived from the decomposition of a natural protein. 2. Synthetic protein.

derma. Corium, dermis.

Dermacentor. Genus of wood ticks which are important as transmitters of Rocky Mountain spotted fever; parasitic on many spp. of wild and domestic mammals, as well as man; *D. andersoni* is widely distributed in central and western U.S., *D. variabilis* in the east, and *D. occidentalis* in the Pacific Coast states.

Dermacentroxinus rickettsi. See ROCKY MOUNTAIN SPOTTED FEVER.

dermal bone. Bone which arises near the surface of a vertebrate embryo without a cartilage precursor; e.g. some human skull bones.

dermal branchiae. Papulae; small, hollow, finger-like respiratory projections abundantly scattered over the external surface of the Asteroidea; they are extensions of the coelomic epithelium through pores in the body wall, and each one is covered by a layer of external epithelium; thus each branchia has a thickness of two epithelial cells (plus a very thin cuticle).

dermal denticles. Placoid scales in cartilaginous fishes.

dermal ossicle. Any bone imbedded in the skin, e.g. in certain reptiles.

dermal papilla. Blunt extension of the dermis into the epidermal layer of vertebrate skin.

dermal plica. Ridge along either side of the body of a frog beginning just behind the eyes; a thickened portion of the skin.

dermal scales. Typical scales of fishes; originate by folding and growth of dermal tissues.

Dermanyssidae. Family of mites parasitic on birds, mammals, and reptiles; Dermanyssus gallinae is the common chicken mite.

Dermanyssus. See DERMANYSSIDAE.

Dermaptera. Order of insects which includes the earwigs; small to medium-sized, elongated, heavily sclerotized insects with biting mouth parts and simple metamorphosis; wings short; posterior end of body with a prominent pair of terminal pincers or forceps; nocturnal, usually hidden in crevices; common genera are Labia, Forficula, and Anisolabis.

Dermasterias. Genus of sea stars in the Order Phanerozonia; surface smooth, body thick; leather stars.

Dermatemyidae. Small family of fresh-water turtles; Mexico and Central America; Dermatemys typical.

Dermatemys. See DERMATEMYIDAE.

dermatitis. Inflammation of the skin.

Dermatobia hominis. Human botfly (Family Cuterebridae); larva parasitic in human subcutaneous tissues.

dermatology. Study of the skin and its diseases.

dermatome. One of the paired masses of mesoderm in a vertebrate embryo which is destined to develop into the somatic dermis.

dermatophyte. Fungus parasite of the skin.

Dermestes. Large, cosmopolitan genus of beetles in the Family Dermestidae; some spp. are quite destructive, and is widely used by museums to clean dried tissue from bones.

Dermestidae. Family of small, dark beetles whose larvae and adults feed on furs, woolens, hides, cheese, cereal products, etc.; skin beetles, dermestids; serious pests in museums, warehouses, stores, and homes; Dermestes and Anthrenus common.

dermis. Corium, derma; inner of the two layers of skin in vertebrates, the outer being the epidermis; the dermis is much the thicker and contains fibrous connective tissue, blood vessels, lymph vessels, and the basal portions of sweat glands and hair follicles; it may also contain scales and bone and form a part of feathers.

Dermochelidae. Family of large, tropical, marine turtles in the Suborder Atheca; Dermochelys coriacea (leatherback), to 9 ft. long, is the largest living turtle, found in warm seas.

Dermochelys. See DERMOCHELIDAE.

dermomuscular sac. Body wall of an annelid.

Dermoptera. Order of mammals which includes the flying lemurs; herbivorous spp. the size of a cat; able to leap and glide prodigiously owing to broad web-like skin extensions from the body on either side and between the fore and hind limbs, and between the hind limbs and tail; forested regions of southeastern Asia; Galeopithecus typical.

Dero. Common genus of fresh-water Oligochaeta; with posterior finger-like gills; in tubes on the bottom.

Deroceras. Genus of land slugs; individuals often suspend themselves from vegetation by strands of mucus.

Derocheilocaris. See MYSTACOCARIDA.

Deronectes. Genus of dytiscid beetles.

Derris. Genus of tropical plants whose roots are the source of rotenone, an alkaloid used as an insecticide and for killing fish in lakes and ponds.

Descartes, René. French philosopher and scientist (1596-1650); made contributions to the physiology of the nervous system and reflex action; author of a physiology textbook.

descending colon. That portion of the large intestine just before the sigmoid flexure.

desert dace. See EREMICHTHYS.

desert fish. See CYPRINODON.

desert fox. 1. Desert kit fox of the arid southwest; Vulpes macrotis; a small sp., grayish above, sides and chest buff. 2. In western Asiatic deserts, a similar fox, V. leucopus.

desert shrew. See NOTIOSOREX.

desiccation. Inactive dry state of an organism; occurs under abnormally dry conditions; found in certain rotifers, nematodes, insects, tardigrades, snails, etc.

Desmacidon. Common genus of marine sponges (Subclass Monaxonida); oval or lobed and finger-like.

desman. Small, mole-like, aquatic insectivore (Talpidae) of Russia; also, a similar sp. of the Pyrenees; Desmana.

desmocranium. In early vertebrate embryos, the first indication of a skull; mass of dense expanding mesenchyme appearing at the anterior end of the notochord.

Desmodontidae. Family of true vampire bats; bite various vertebrates and lap up the blood; Mexico to Argentina; Desmodus common.

Desmodus. Genus of vampire bats of the American tropics; large canine teeth used especially in biting stock animals so that it can feed on the streaming blood; rarely bites sleeping human beings; transmits one type of rabies and a trypanosome disease of horses.

Desmognathinae. Subfamily of salamanders in the Family Plethodontidae; lower jaw immovable; larva with four gill slits on each side; Desmognathus and Leurognathus typical.

Desmognathus fuscus. Common dusky salamander; southern Canada to the Gulf, westward to Ill. and La.; chiefly along streams; to 4 in. long.

desmoneme nematocyst. Small nematocyst with a short unarmed tubule; functions mainly in entangling and wrapping around bristles of the prey.

Desmopachria. Common genus of dytiscid beetles.

Desmoscolex. Common genus of chromadorid nematodes; body with serially-arranged thickened rings; fresh-water and marine soils.

Desor larva. Modified pilidium larva; a ciliated postgastrula which remains inside the egg membranes and lacks apical tuft, oral lobes, and oral band.

desoxycorticosterones. Deoxycorticosterones.

desoxyribonuclease. Deoxyribonuclease.

desoxyribonucleic acid. Deoxyribonucleic acid.

desoxyribose. Deoxyribose.

Desulfovibrio. Genus of curved, rod-shaped bacteria essential to the sulfur cycle in soil and water; reduce sulfates to hydrogen sulfide.

determined tissue. Embryonic tissue capable of developing into only one kind of adult tissue or organ.

detorsion. Partial untwisted condition of the original embryonic 180° visceral twisting as seen in many Gastropoda having a reduced shell (not to be confused with coiling of the shell). Cf. torsion.

detoxification. Detoxication; metabolic processes involved in changing toxic substances into non-toxic substances; e.g. may occur in the large intestine, blood, and liver.

detritus. In the ecological sense, any fine particulate debris of organic or inorganic origin.

Deuterophlebia. See DEUTEROPHLEBIIDAE.

Deuterophlebiidae. Mountain midges; small family of poorly known flies taken in mountainous areas of Asia, Japan, and the western U.S.; adults without mouth parts; wings enormous and elaborately folded; bizarre larvae and pupae found only on rocks in swift mountain streams; Deuterophlebia the only genus.

deuteroplasm. Deutoplasm.

Deuterostomia. Group of higher phyla in which cleavage is indeterminate, mesoderm and coelom originate as pouches from the gut, and the mouth is not derived from the blastopore; includes the Echinodermata, Enteropneusta, Chordata, Chaetognatha, and Brachiopoda. Cf. PROTOSTOMIA.

deuteroky. Production of both males and females by a parthenogenetic female.

deuteromerite. In certain gregarine Protozoa, the posterior (usually larger) of the two cells which make up the individual. Cf. PROTOMERITE.

deutonymph. Second immature instar of typical mites.

deutoplasm. Yolk portion of an egg, especially as distinct from the true cytoplasmic fraction.

deutovum. Quiescent stage in certain mites following the rupture of the outer egg shell.

developmental genetics. Study of mechanisms of gene control and modification of embryological processes.

Devescovina. Genus of oblong flagellates in the termite digestive tract; three flagella anterior and one trailing.

devilfish. 1. See DEVIL RAYS. 2. Octopus.

devil rays. Small group of large tropical and subtropical cartilaginous fishes in the Suborder Batoidea; manta, devilfish; body greatly depressed, with enormous, flat, pectoral fins; to 20 ft. across; with two earlike fins extending forward from the head; mouth wide, extending across front of head; harmless, feed on zooplankton; Manta common.

Devonian period. One of the six geological subdivisions of the Paleozoic era, between the Carboniferous and Silurian periods; lasted from 350 million to 250 million years ago; Age of Fishes; extensive seas, and some aridity on land; fauna characterized by many sharks, bony fishes, and mollusks; first amphibians, crabs, and land snails; decline of trilobites and ostracoderms.

De Vries, Hugo. Dutch botanist (1848-1935); developed the experimental method in genetics; coined the term "mutation"; one of several scientists who independently confirmed and rediscovered Mendel's 1866 paper.

dewclaw. 1. Vestigial digit on the foot of a terrestrial mammal. 2. Claw or hoof at the tip of such a digit.

dewlap. 1. Hanging median fold of skin in the neck region; e.g. some reptiles, moose. 2. Wattles under the neck of certain birds. 3. Fold of flesh on the human throat.

dewpoint. Temperature at which an air mass can hold no additional water vapor without condensation.

Dexiospira. Genus of Polychaeta.

dexiotropic cleavage. Common type of spiral cleavage in which the micromeres are displaced in a clockwise direction when viewed from the animal pole. Cf. LAEOTROPIC CLEAVAGE.

dextral. 1. Right-handed. 2. Coiled or turned toward the right. 3. In gastropods, having the aperture to the right of the shell axis when the shell is held with the spire pointing upward and the aperture facing the observer.

dextran. Carbohydrate produced by certain bacteria and also found in milk, beet juice, and molasses; used as a blood plasma substitute for treating severe burns; $(C_6H_{10}O_5)_n$.

dextrin. Intermediate polysaccharide formed during the hydrolysis of starch to glucose; $(C_6H_{10}O_5)_n$.

dextrose. Glucose.

d-glucose. See GLUCOSE.

dhole. Fierce wild "dog" (Cuon); coloration whitish to reddish brown; usually hunt large game in packs; live among rocks or brush in Mongolia, Tibet, and southeastern Asia; not true dogs.

D horizon. Deepest stratum in a soil profile; lies below all weathering. Cf. A, B, and C HORIZON.

diabetes insipidus. Rare pathological condition marked by the voiding of large quantities of watery urine; output may reach 40 liters per day; caused by inadequate secretion of vasopressin by the posterior pituitary.

diabetes mellitus. See INSULIN.

diabetogenic hormone. Hormone secreted by the anterior lobe of the pituitary; raises blood sugar content.

Diabrotica. Genus of cucumber beetles in the Family Chrysomelidae which transmit Bacillus tracheiphilus, a bacterium causing cucumber wilt; larvae feed on roots, adults on foliage of cucumber, squash, etc.; D. vittata is the black and yellow American squash beetle.

diacetin. Glyceryl diacetate; a cellular narcotic.

diad. One of two groups of two chromatids resulting from the first division of a tetrad in meiosis.

Diadectes. Genus of "stem reptiles" (cotylosaurs).

Diadema. Genus of black tropical sea urchins in the Order Aulodonta; usually in coral reefs; poisonous spines to 12 in. long.

Diadematoidea. Order of echinoderms in the Class Echinoidea; anus aboral, central; peristomial gills present; more or less rounded or somewhat flattened sea urchins; Arbacia and Strongylocentrotus common; in modern usage this group is broken into several other orders.

Diadophis. Genus of small colubrid snakes with a flattened head and a yellow ring around the neck; ringnecked snakes; common in moist woods over much of U.S.

diadromous. Pertaining to fishes that migrate freely between fresh and salt waters; e.g. salmon and the sea lamprey.

Diadumene. Genus of pale, slender sea anemones.

diagnosis. In taxonomy, a statement of the important and essential characters which distinguish one taxonomic category from all other related and similar categories.

diakinesis. Pronounced thickening and contraction of chromatids following the formation of chiasmata during prophase of the first meiotic division.

dial bird. Any of several Indian songbirds related to the European robin.

dialysis. Method of separating dissolved salts from colloids in a mixed solution (suspension) by placing the liquid in a suitable collodion bag suspended in water; salts than diffuse through the collodion into the surrounding water, leaving behind an essentially pure suspension of colloids.

diamondback rattler. See CROTALUS.

diamondback terrapin. See MALACLEMYS.

diapause. Period of suspended development or growth, especially in certain insects and mites; characterized by inactivity and decreased metabolism during an egg, larval, nymph, or adult stage; genetically determined but may be eliminated experimentally if such animals are raised under constant and favorable environmental conditions; a similar process occurs in a few crustaceans and snails and perhaps certain other taxa.

diapedesis. Passage of leucocytes through the unruptured wall of a small blood vessel; usually occurs at the membrane separating two adjacent endothelial cells.

Diaphana. Genus of very small, delicate marine gastropods.

diaphane. Essentially the same as euparal.

Diaphanosoma. Common genus of limnetic Cladocera.

Diapheromera. See PHASMIDA.

diaphorase. Cellular enzyme which catalyzes the oxidation of reduced DPN; an outmoded and discarded term.

diaphragm. 1. Transverse arched muscular and tendinous partition separating thoracic and abdominal cavities in mammals; up and down movements are essential for breathing. 2. Delicate transverse dorsal membrane in many Arthropoda; separates the pericardial cavity from the main hemocoel. 3. Well developed dorsoventral septum in certain worms and arthropods.

diaphysis. 1. Shaft of a long bone. 2. The central ventral part of a vertebra. Pl. Diaphyses.

diapophysis. Transverse vertebral process, especially a more dorsal process.

Diaptomus. Common genus of fresh-water copepods; exceptionally long antennae.

diarrhea. Abnormally frequent discharge of watery feces. Cf. DYSENTERY.

Diarthronomyia hypogaea. Chrysanthemum gall fly (Camily Cecidomyiidae); produces small galls in flower heads and leaves.

diarthrosis. Movable joint between two bones.

Diasparactus. Genus of primitive Permian reptiles.

Diaspididae. Very large family of insects in the Order Homoptera; armored scales, scale insects; small, flattened spp., hidden under a secreted wax scale; a smaller ventral scale also sometimes present; feed primarily on perrenial shrubs and trees; many important pest spp., including _Quadraspidiotus perniciosus_, the San Jose scale.

diastase. An amylase.

diastema. Toothless space between two different kinds of teeth, especially the space between incisor and first premolar.

diastole. 1. Dilation or relaxation stage of a heart beat, especially in the ventricles. 2. Period of filling of a contractile vacuole.

diastolic blood pressure. Blood pressure at its weakest point in the heart beat as determined through a radial artery pulse with a sphygmomanometer and stethoscope; about 80 mm. mercury for normal adults at rest.

diastrophism. 1. Collectively, the major processes by which the earth's crust is formed into continents, ocean basins, mountain ranges, etc. 2. In a broad sense, the results of these processes.

Diastylis. Common genus in the malacostracan Order Cumacea.

Diatraea. See CRAMBIDAE.

Diatryma. See DIATRYMIFORMES.

Diatrymiformes. Order of huge Eocene birds; legs large, wings atrophied, bill large; _Diatryma_ typical.

Diaulota. Genus of staphylinid beetles.

Diaulula. Genus of nudibranchs.

dibitang. Dibitag; see AMMODORCAS.

Dibothriocephalus latus. _Diphyllobothrium latum_; large "fish tapeworm" occurring in the small intestine of man and many other fish-eating mammals; larval stages in fresh-water copepods and fishes; especially common in the Baltic area but scattered in some other lake districts of the world.

Dibranchia. One of the two orders in the molluscan Class Cephalopoda; squids, octopus, etc.; shell usually internal and reduced, or absent; body cylindrical or globular, often with fins; 8 or 10 arms with suckers; ink sac present.

Dicaeidae. Family of very small passerine birds of the Oriental and Australian regions; flowerpeckers; bill short, thick; feed on insects, fruits, and berries.

Dicamptodon ensatus. Giant salamander of humid forests of U.S. Pacific Coast drainages and northern Idaho; eggs and larvae in streams.

dicentric chromatid. Single chromatid with two centromeres; the result of certain chromosome aberrations.

Dicerorhinus. Genus of small, two-horned rhinoceroses of Sumatra.

Diceros. Genus of black rhinoceroses of the southern half of Africa; two horns on snout.

Dichilum. Genus of ovoid to ellipsoid holotrich Ciliata; salt and fresh waters.

dichotomous. Dividing into two equivalent or similar branches, or progressively dividing in such a manner.

Dichromanassa rufescens. Red egret, a sp. of the Gulf Coast, West Indies, and Central America.

dichromism. Having two color phases, as the unlike coloration of certain male and female birds.

dickcissel. See SPIZA.

Dick test. Immunological test used to determine whether a person possesses an active immunity to scarlet fever.

Diclidophora. Genus of external monogenetic trematode parasites of marine fishes.

Diclidurus. Genus of whitish tropical American bats in the Family Emballonuridae.

dicondylian. With two occipital condyles.

Dicosmoecus. Genus of Trichoptera; mostly western states.

dicoumarin. Dicumarol.

Dicraneura. Large genus of leafhoppers.

Dicranophorus. Genus of creeping rotifers with ventral corona; trophi forcipate.

Dicrocoelium dendriticum. Cosmopolitan liver fluke of sheep and other herbivores.

Dicrostonyx. Genus of small herbivorous circumpolar rodents in the Family Cricetidae; collared lemmings; to 6 in. long; thickset spp. with small ears, very short tail, and soles of feet hairy; winter coloration white, summer coloration grayish; burrow and make runways in grassy areas; populations cyclic but never as dense as the brown lemming (Lemmus).

dicrotic. Pertaining to a double beat, especially the second arterial expansion following the heart beat.

Dicruridae. Family of Old World aggressive insectivorous perching birds; drongos or king crows; black spp., many crested.

Dictya. See TETANOCERIDAE.

Dictyna. See DICTYNIDAE.

Dictynidae. Hackle-banded weavers; a large family of spiders with a cribellum; construct irregular webs in vegetation or on ground; Dictyna widely distributed.

Dictyocaulus filaria. Nematode parasite of the air passages of sheep; the lung threadworm.

dicumarol. Organic anticoagulant first isolated from spoiled sweet clover and now made synthetically.

Dicyema. 1. One of the two classes in the Phylum Mesozoa; mature nematogen to 8 mm. long; with a single internal cell. 2. A genus of Mesozoa parasitic in the nephridia of squids and octopuses.

Didactylus. Genus of pygmy anteaters; to 12 in. long; snout short, tail furry and prehensile; tropical South America.

Didelphidae. Family of marsupials which includes the opossums; South America, Central America, and U.S.; the only marsupials found outside the Australian region; tail scaly and prehensile; nocturnal and omnivorous; when only about 11 mm. long, the premature young crawl to the marsupium and remain attached to the nipples for about two months; thereafter they become free but frequently return to the pouch for protection; adults often feign death as an escape device; Didelphis virginiana is the common American or Virginia opossum.

Didelphis. See DIDELPHIDAE.

Didemnum. Genus of flat, encrusting, colonial ascidians.

Didinium. Genus of barrel-shaped, ciliate Protozoa which feed on other ciliates, especially Paramecium; fresh and salt waters.

Didymoceras. Genus of Cretaceous ammonites.

Didymops. Genus of dragonflies; eastern

U. S.

diecdysis. Especially in decapod Crustacea, the period between the end of one metecdysis and the beginning of the next proecdysis.

diel. Referring to the 24-hour day (in constrast to that part of a "day" when it is light).

Diemictylus. Triturus.

diencephalon. Posterior part of the vertebrate forebrain; includes the thalamus, posterior pituitary, optic chiasma, pineal, and parapineal.

Dientamoeba fragilis. Minute commensal amoeba of the large intestine of man and monkeys.

diethylstilbestrol. Synthetic estrogen used in treating menopause symptoms, vaginitis, and suppressed lactation.

differentially permeable. Selectively permeable.

differentiation. 1. Progressive changes in structure and function of cells, tissues, or organs during development; usually involves diversification and changes from general to specialized condition. 2. In microtechnique, the proper staining and clearing of tissues in order to make them more easily distinguishable.

Difflugia. Genus of amoeboid fresh-water and soil Protozoa having most of the chitinoid test covered with grains of sand and other debris glued together into an inverted vase-shaped structure.

Difflugiella. Genus of fresh-water amoeboid Protozoa; most of cell in an ovoid test.

diffuse placenta. Placenta whose chorionic tissue has scattered villi, as in the pig, horse, and lemur.

diffusion. Random movement of suspended colloidal or dissolved particles from regions of greater to lesser concentrations.

digastric. Pertaining to a muscle which is fleshy at each end and tendinous in the middle.

digastric muscle. Depressor mandibulae.

Digenea. One of the three orders of the Class Trematoda; usually with two suckers, one encircling the proctostome and the other somewhere on the ventral surface; endoparasitic; complicated life histories, with two or more types of larval forms which are usually parasitic in mollusks, crustaceans, and fishes; e. g. Fasciola, Clonorchis, and Schistosoma.

digenetic. 1. Of or pertaining to the trematode Order Digenea; 2. Of or pertaining to an animal life history involving an alternation of sexual and asexual reproduction, as in most Trematoda.

digestion. Enzymatic processes involved in breaking down complex organic food materials into simpler materials that may be easily absorbed and utilized as a source of energy or for building protoplasm; the processes occur in individual cells of both plants and animals and in the cavity of the digestive system of animals.

digestive enzymes. Large group of enzymes which function in the digestion of foodstuffs; either intracellular or extracellular.

digestive system. Collectively, all of the organs associated with ingestion, digestion, absorption, and defecation; includes the digestive tract and its associated digestive glands.

digestive tract. Collectively, those hollow organs in the digestive system through which food and digestion products actually pass.

digger bees. See ANTHOPHORIDAE.

digger wasp. Any of numerous slender-waisted wasps which dig burrows in mud or sand.

digit. Finger or toe.

digital pads. Swollen padlike structures on the tips of the digits in many arboreal amphibians and lizards.

digitate. 1. Having fingers or toes. 2. Having finger-like or toelike processes.

digitiform. Finger-shaped.

digitigrade. Pertaining to mammals which walk on the toes, with the wrist and heel bones held off the ground, as in dogs and cats.

Digononta. One of the two classes in the Phylum Rotatoria; female with paired ovaries.

dihybrid. With reference to two pairs of alternate characters, an offspring having different alleles at two different gene loci, e. g. AaBb.

dihybrid cross. A mating between parents with reference to two pairs of alternate hereditary characters; one parent may possess two pairs of dominant genes and the other two pairs of recessive genes.

diiodotyrosine. Amino acid derivative found in the thyroid gland and thought to be a precursor of thyroxine.

dik-dik. See MADOQUA.

Dileptus. Genus of marine and fresh-water holotrich Ciliata; with a long snout or necklike prolongation.

Dimastigamoeba. Genus of minute amoeboid Protozoa found in stagnant water, sometimes coprozoic; with a biflagellate stage in the life history.

dimastigate. Having two flagella.

Dimetrodon. Genus of Permian and Carboniferous reptiles in the Order Pelycosauria; heavily-built spp. with enormously elongated vertical dorsal vertebral spines from neck to sacrum.

Dimorpha. Genus of small fresh-water Protozoa; two flagella and slender pseudopodia.

dimorphism. 1. The occurrence of two kinds of individuals in a colonial organism, e.g. feeding and reproductive polyps in Obelia. 2. Having two different forms according to sex, size, coloration, etc.

Dimyaria. Discarded taxon sometimes used to include those bivalve mollusks with two adductor muscles. Cf. MONOMYARIA.

Dina. Common genus of fresh-water leeches; feed mostly on invertebrates.

dinergate. Soldier caste in certain ants; head and mandibles greatly enlarged.

Dineutes. Dineutus.

Dineutus. Common genus of whirligig beetles in the Family Gyrinidae.

dingo. Australian wild dog, perhaps derived from wild dogs of India or Egypt; small, stoutly built, and sandy brown; preys on rabbits, poultry, and sheep, especially at night.

Dinichthys. Genus of extinct placoderm fishes to 30 ft. long; head and thorax covered with bony armor.

Dinobryon. Common genus of yellowish-brown fresh-water flagellates; solitary or colonial; with a hyaline vaselike cellulose test.

Dinocardium. See COCKLE.

Dinocerata. Order of very large extinct ungulate mammals; bones massive, feet broad, skull often grotesquely horned; Paleocene to Eocene.

Dinoflagellata. Order in the Class Mastigophora; most representatives are holophytic and have a cellulose test; predominantly marine and free-swimming; certain spp. become extremely abundant along warm coastal areas and cause some of the "red tides" which are highly toxic to fishes and other animals; some marine spp. bioluminescent; Ceratium and Noctiluca typical.

Dinomonas. Genus of small, colorless, biflagellate Protozoa; in stagnant waters.

Dinomyidae. Family of mammals which includes the rare paracanas; related to the cavies, capybaras, and hutias; small, shaggy spp. with a long, hairy tail; herbivorous burrowers; Peruvian Andes.

Dinophilus. Genus of microscopic annelids in the Class Archiannelida; five or six body segments; external cilia; on substrates along marine coasts.

Dinophysis. Genus of yellow marine dinoflagellates.

Dinornis. See DINORNITHIFORMES.

Dinornithiformes. Order of flightless birds; includes the moas (Dinornis) which have been extinct for several centuries; wings absent, legs large and powerful, with three or four toes; to 8 ft. tall; New Zealand.

dinosaur. Any of a large number of land reptiles of the Mesozoic era; 2 to 90 ft. long; carnivorous or herbivorous; bipedal or quadrupedal; in the reptilian Subclass Archosauria.

dinothere. Any of a group of large extinct proboscideans having no upper tusks and with the lower tusk curved outward and downward from the tip of the lower jaw; Dinotherium.

diocoel. Third ventricle of the brain (diencephalon).

Dioctophymata. Enoplida.

Dioctophymatina. Suborder of enoplid nematodes; intestine with four rows of muscles; male with caudal sucker and one copulatory spicule; common genera are Dioctophyme and Histrychis.

Dioctophyme renale. Giant nematode parasitic in the body cavity of mammals; female nearly 3 ft. long.

Dioctophymoidea. Order of nematodes equivalent to the Suborder Dioctophymatina.

Diodon. See DIODONTIDAE.

Diodontidae. Family of bony fishes whose members are able to swallow water and puff up to globular proportions, thus erecting the many spines which cover the body; burrfish, porcupinefish, globefish; warm-water marine spp.; Diodon and Chilomycterus.

Diodora. Genus of keyhole limpets.

dioecious. Pertaining to an organism in which male and female reproductive organs occur in different individuals.

Diomedea. See DIOMEDEIDAE.

Diomedeidae. Family of birds in the Order Procellariiformes; the albatrosses; large, soaring, oceanic spp.

of the Southern Hemisphere; Diomedea exulans, wandering albatross, has a wingspread to 12 ft.

Dionda. Genus of minnows; Minn. and Wis. to the Ozarks, southwestern states, and Mexico.

Diopatra. Large genus of Polychaeta; in parchment-like tubes; some spp. brightly colored; D. cupraea is the North Atlantic plumed worm, a translucent red sp. which builds tubes to 3 ft. long.

Diophrys. Genus of marine hypotrich Ciliata; cirri large and numerous.

Diopsidae. Stalk-eyed flies; small tropical family in which the eyes are on the tips of long lateral stalks.

diopter. Measure of the refracting power of a lens; reciprocal of the focal length in meters; a one-diopter lens has a focal length of 1 m., and a two-diopter lens has a focal length of 0.5 m.

Dioptidae. Neotropical family of slender, butterfly-like moths.

diorchic. Having two testes.

Diotocardia. Aspidiobranchia.

dipeptidase. Digestive enzyme secreted by the wall of the small intestine; converts dipeptides into amino acids.

dipeptide. Simple intermediate protein digestion substance consisting of only two joined amino acid residues.

Diphasia. Genus of marine coelenterates in the Suborder Leptomedusae.

diphasic response. Response of galvanometer or other suitable instrument as an action current passes over nerve or muscle when similar electrodes are placed in contact with two different regions; the indicator of the instrument moves from zero position, first in one direction, then back to zero, then in the opposite direction, and finally back to zero.

diphosphopyridine nucleotide. Necessary substance for the oxidation of lactic, malic, pyruvic, and triose phosphoric acids; also important in cellular fermentation and glycolysis; same as coenzyme I, DPN, and cozymase.

diphtheria. Contagious epidemic disease of man, produced by the toxin of Corynebacterium diphtheriae; most common in children where it is marked by sore throat, fever, and runny nose; the toxin may damage the heart and nervous system; treated by vaccination and an antitoxin.

diphycercal tail. Tail which tapers to a point, as in lungfishes.

Diphyes. Genus of marine coelenterates; the elongated branching colony of polymorphic hydroids is suspended by the pulsating contractions of two large bell-shaped structures located at the top end of the colony.

diphygenetic. Producing two different types of embryos, e.g. some dicyemids.

Diphyllidea. Order of tapeworms in the Subclass Cestoda; scolex with four bothria fused into pairs and situated at the end of a spiny head stalk; adult in the intestine of elasmobranchs; immature stages in marine mollusks and crustaceans; Echinobothrium the only genus.

Diphyllobothrium latum. Dibothriocephalus latus.

diphyodont. Having two successive sets of teeth, as in man.

Diphyopsis. Genus of coelenterates in the Order Siphonophora; with two pyramidal nectophores.

Diplasiocoela. One of the five orders of amphibians in the Subclass Salientia; pectoral girdle fused to sternum; includes many of the true frogs and toads.

Diplectanum. Genus of small monogenetic external trematode parasites of marine fishes.

Diplectrona. Genus of Trichoptera.

Diplectrum. Marine genus of small perchlike shore fishes in the Family Serranidae.

dipleurula. 1. Hypothetical simple ancestral form of the Phylum Echinodermata; soft, elongated, bilaterally symmetrical, and with three pairs of coelomic sacs. 2. Collective term applied to the various types of ciliated bilaterally symmetrical echinoderm larvae.

diploblastic. Of or pertaining to the fundamental body structure of the Porifera and Coelenterata where only ectoderm and endoderm are well differentiated. (Many zoologists are convinced that such a category is inadmissable, and that all phyla except the Protozoa are triploblastic; the mesoderm of Porifera is said to be mesenchymal and of obscure origin; the mesoderm of the Coelenterata is poorly differentiated in many spp. and sometimes is obscured or replaced by the non-cellular mesoglea.)

Diplocardia. Common genus of earthworms.

diplocardiac. Having the pulmonary and systemic sides of the heart completely separated, as in birds and mammals.

Diplocaulus. Genus of extinct stegocephalian Amphibia.

Diplocentrus. American genus of scorpions.

Diplochlamys. Genus of fresh-water amoeboid Protozoa; cell covered by a hemispherical or cup-shaped test.

Diplococcus. Genus of gram-positive, elongated or lance-shaped bacteria usually growing in pairs; several spp. cause human diseases, especially pneumonia (D. pneumoniae).

Diplodinium. Genus of complex spirotrich Protozoa; commensal in the digestive tract of ruminants.

Diplodiscus. Genus of flukes found in the rectum of frogs.

Diplodocus. Genus of giant Jurassic herbivorous dinosaurs; neck and tail very long, head small; total length to 87 ft.; largest terrestrial animal of all time.

Diplodontus despiciens. Common sp. of fresh-water Hydracarina.

diploë. Layer of cancellous bone between the two layers of compact bone of the cranium.

Diplogaster. Genus of free-living nematodes often found in feces which have been on the ground for some time.

diploglossate. In certain lizards, having the end of the tongue retractile into the base.

Diplograptus. Genus of fossil graptolites.

diploid. Pertaining to cells or organisms having a normal, double (2n) set of chromosomes; usual complement of homologous chromosomes in the nonreproductive tissues of animals and the sporophyte generation of plants.

Diplolepis. Genus of gall wasps; larvae stimulate abnormal growths of plant tissues.

Diplomita. Genus of small biflagellate Protozoa; cell with a lorica and a fine basal stalk; stagnant fresh waters.

diplont. Diploid chromosome condition in any organism.

Diplophrys archeri. Small amoeboid protozoan; most of cell enclosed in a delicate test but pseudopodia projecting from two opposite apertures; on fresh-water vegetation.

diplopia. Double image vision.

Diplopoda. Class of terrestrial arthropods; millipedes; elongated, cylindrical body; head with one pair of short antennae, labrum, one pair of mandibles, and one pair of maxillae (united to form a gnathochilarium); "thorax" usually of four segments, each with one pair of legs; abdomen of 20 to more than 100 (double) segments, each with four short legs, two ganglia, four spiracles, and two pairs of heart ostia; genital pore midventral on second or third body segment; sluggish, feed on decaying animal and vegetable matter; 7000 spp.; common genera are Julus, Narceus, and Polydesmus.

Diplopteraster. Widely distributed genus of sea stars (Order Spinulosa); pentagonal, arms only slightly projecting from disc.

Diploria. Genus of corals whose skeleton has the convoluted appearance of the human cerebral hemispheres.

Diploscapter. Genus of microscopic, free-living nematodes.

Diplosiga. Genus of small, sessile, monoflagellate Protozoa; two distal collars.

diplosis. Diploid chromosome condition.

Diplosoma. Genus of small colonial ascidians.

diplosome. Double centrosome.

Diplostraca. Taxon sometimes used to include the Conchostraca and Cladocera.

diplotene stage. Condition obtaining during the first meiotic division; characterized by the first appearance of chromatids; follows pachytene stage.

Diplovertebron. Genus of primitive Paleozoic Amphibia.

Diplozoon. Genus of monogenetic trematodes; adults on the gills of fishes.

Dipneusti. Dipnoi.

Dipnoi. Superorder of bony fishes; the lungfishes; body long and slender, air bladder lunglike, paired fins linear and partly fleshy; see CERATODONTIDAE and LEPIDOSIRENIDAE.

Dipodidae. Family of small, nocturnal, social rodents; includes the jerboas of arid portions of the Old World; hind legs greatly modified for leaping; forelegs short.

Dipodomys. Genus of mammals in the Family Heteromyidae; includes the many spp. of pocket rats or kangaroo rats; small, jumping rodents with long hind legs, long tail, short forelegs, and external, hair-lined cheek pouches; nocturnal burrowers; warm areas of western N.A., especially the southwestern states.

dipper. See CINCLIDAE.

Diprotodontia. Discarded name for certain families of marsupials, especially the phalangers and opossum rats.

Dipsadomorphus. Boiga.

Dipsosaurus. Small genus of desert lizards confined to southwestern U. S. and northwestern Mexico.

Diptera. Order of insects which includes the true flies; minute to medium size; mouth parts sucking and lapping, piercing, or vestigial; complete metamorphosis; eyes large; forewings transparent, with few veins; hind wings represented by knoblike halteres; larvae without true legs; 90,000 spp. and about 150 families.

Dipterus. Genus of Devonian lungfishes.

Dipylidium caninum. Common tapeworm of the dog and cat, occasional in man; each proglottid with two sets of reproductive organs; larvae in lice and fleas of dogs, cats, and man.

directives. Paired primary mesenteries of many anthozoan coelenterates; their retractor muscle strands face outward.

direct metamorphosis. See PAUROMETABOLA.

dire wolf. Large N. A. wolflike mammal of Pleistocene times.

Dirofilaria immitis. Heart worm; a large, common nematode found in the right ventricle of the dog heart; transmitted by mosquitoes.

Dirona. Genus of nudibranchs.

disaccharide. A double sugar; any of a group of carbohydrates having the general formula $C_{12}H_{22}O_{11}$; upon hydrolysis, they are converted to two molecules of monosaccharide; e. g. sucrose and lactose.

disc. 1. One of several groups of special quiescent cells in an insect larva; during the pupal stage these discs grow and differentiate into adult structures but remain collapsed and folded until emergence of the adult. 2. Area marking entrance of optic nerve into eye. 3. Central body portion of sea stars and brittle stars, as distinct from the arms. 4. See INTERVERTEBRAL DISC.

discal cell. 1. Large cell (section of wing delimited by ridgelike veins) extending from the base of the wing toward the center in Lepidoptera. 2. Cell between branches of the media (vein) in Diptera. 3. A cell in the wings of Trichoptera and Odonata.

Discinisca. Genus of Brachiopoda.

disclimax. Apparent ecological climax resulting from the modification of a previous climax area, especially as the result of introducing new spp. to the area; e. g. the introduction of prickly pear cactus into Australia has produced a disclimax over wide areas.

discoblastic. Pertaining to meroblastic segmentation or a meroblastic embryo.

Discocelis. Genus of polyclad Turbellaria.

Discocephali. Order of bony fishes which includes the remoras or shark suckers; upper surface of head with an oval, furrowed suction disc by which the fish attaches itself to objects or other fishes, usually sharks; tropical seas; Echeneis and Remora common.

Discocephalida. Discocephali.

Discocotyle. Genus of monogenetic trematodes found especially on trout.

Discodoris. Genus of nudibranchs.

Discoglossidae. Family of toads in the Order Opisthocoela; adult with ribs, tongue, and eyelids; Europe, Africa, Asia, and the Philippines; e. g. Discoglossus, Bombina, and Alytes.

Discoglossus pictus. Common grayish aquatic frog of southwestern Europe and northwestern Africa.

discoidal cleavage. Meroblastic cleavage.

discoidal placenta. Placenta whose chorionic tissue has one or two large, disc-shaped patches of villi; e. g. man, insectivores, bats, and rodents.

Discomedusae. Order of scyphozoan jellyfish; manubrium enlarged and prolonged into four grooved oral arms; chiefly coastal; Aurelia, Cyanea, and Cassiopeia common.

Discomorpha. Genus of sapropelic ctenostome Ciliata.

discomorula. Morula derived from a meroblastic egg.

discontinuity layer. Thermocline, mesolimnion, metalimnion.

discontinuous distribution. Occurrence of the same or closely related spp. in two or more widely separated areas of the globe; examples are the tapir in Central America and Malaya, the paddlefish in the Mississippi and certain large Chinese rivers, and the lungfish in South America, Africa, and Australia.

discontinuous phase. In a colloidal system, the dispersed or suspended particles, e. g. fat particles in milk, or granules in protoplasm.

discontinuous spindle fibers. Cytoplasmic fibrils which aid in pulling daughter chromosomes to opposite poles in a cell undergoing mitosis. Cf. MANTLE FIBERS.

discontinuous variations. Sharply defined differences in individuals owing to the differentiating effects of only one or two pairs of genes. Cf. CONTINUOUS VARIATIONS.

Discophrya. Genus of fresh-water suctorians; elongated, with a short, stout pedicel.

Discorbis. Genus of marine Foraminifera.

Discus. Genus of land snails; eastern and central U. S.

disguise. Protective mechanism in certain animals owing to their striking resemblance to a special object or part of their habitat; the Amazon "leaf fish," and the similarity of some insects to twigs and bird excrement are examples.

disinfectant. Chemical or physical agent which destroys infective microorganisms.

dispermy. Penetration of an ovum by two spermatozoa.

dispersion. Spatial distribution pattern of individuals within a population, e. g. random, uniform, or clumped.

Dispholidus. Genus which includes the boomslang, an African tree snake.

disruptive coloration. Bizarre spotting and striped pattern on certain animals; presumed to have a protective function by momentarily confusing predators and drawing their attention to the superimposed pattern rather than to the actual outline and identity of the animal possessing the disruptive coloration; as a protective mechanism, this device depends on psychological confusion; the principle is exemplified in certain brightly patterned coral reef fishes.

disseminule. Any special device which facilitates the geographic spread of a sp.; e. g. spores, resistant eggs, desiccation stages.

dissepiment. 1. Oblique calcareous partition extending from one septum to another in certain corals. 2. Muscular or membranous partition between two adjacent segments in polychaetes and oligochaetes.

dissimilation. Katabolism.

dissoconch. Shell of a veliger larva.

dissogeny. With two different sexually mature periods, one as a larva, the other in the mature stage.

Dissosteira. Common genus of long-winged U. S. grasshoppers; D. longipennis and D. carolina (Carolina "locust") common.

Dissotrocha. Genus of bdelloid rotifers.

distal. Distad; situated away from the base or point of attachment.

distal convoluted tubule. Coiled portion of a kidney tubule just before it empties into a collecting tubule.

distal pigment cell. See OMMATIDIUM.

Distaplia. Uncommon genus of colonial ascidians.

Distesphanus. Genus of marine monoflagellate Protozoa; enveloped by a siliceous skeleton.

Distigma. Genus of elongate, plastic, holozoic Protozoa; one long and one short anterior flagella; stagnant fresh waters.

Distoechurus. Genus of feather-tailed phalangers; nocturnal, arboreal marsupials which glide from branch to branch; New Guinea.

Distomata. Group of digenetic trematode families; oral sucker at anterior end; acetabulum somewhere on the ventral surface but never at the posterior end.

Distyla. Lecane.

Ditmars, Raymond Lee. American herpetologist (1876-1942); for many years curator of reptiles and mammals at the New York Zoological Park; world authority and author of several important books on reptiles.

ditrematous. 1. In hermaphroditic animals, having separate genital pores, e. g. certain snails. 2. Having separate anal and genital openings, as in certain viviparous fishes.

Diurella. Trichocerca.

diuresis. Increased output of urine by kidneys, normally occurring after drinking unusually large amounts of fluids.

diuretic. Any substance promoting the excretion of urine.

diurnal. 1. Activity by daylight; opposite of nocturnal. 2. Occurring every day.

divaricator. 1. Muscle which opens the valves of a brachiopod shell. 2. Avicularium muscle.

Divaricella. Widely distributed genus of small marine bivalves.

diversity index. Ratio between number of spp. and number of individuals in a community.

diverticulum. Any blind sac extending out from a larger cavity. Pl. diverticula.

Dixa. See DIXIDAE.

Dixidae. Family of small, slender flies with U-shaped aquatic larvae; dixa midges; Dixa common.

Dizygocrinus. Genus of extinct crinoids.

dizygous twins. Fraternal twins.

DNA. Deoxyribonucleic acid.

dobsonfly. See CORYDALUS.

Dobsonia. Genus of cave-dwelling fruit bats; New Guinea and west Pacific area.

Dobzhansky, Theodosius. Russian-born American geneticist (1900-); best known for his work on Drosophila genetics, population genetics, and evolution; wrote the book Genetics and the Origin of Species.

Dodecaceria. Genus of small Polychaeta; usually excavate tubes in bivalve shells.

dodo. Gray, flightless, forest-dwelling bird of Mauritius, extinct since late 17th century; large black bill with a horny upper cap; larger than a turkey; unpalatable; Raphus cucullatus, in the Family Raphidae.

doe. Female deer, antelope, hare, etc.

dog. See CANIS.

dogfish. 1. Common name of many spp. of sharks, especially Mustelus, Carcharias, and Squalus. 2. Sp. of bony fish of rivers and lakes of eastern U.S.; see PROTOSPONDYLI.

dog-mongoose. See BDEOGALE.

dog salmon. 1. Oncorhynchus keta of the Pacific coast. 2. Occasionally, O. gorbuscha, the humpbacked salmon.

dog tapeworm. See TAENIA PISIFORMIS and DIPYLIDIUM CANINUM.

dog whelk. Any of a large group of marine snails; shell mostly small and thick; see NASSA and THAIS.

dogwinkle. See THAIS.

Dohrn, Anton. German vertebrate anatomist, embryologist, and marine biologist (1840-1909); established the Naples aquarium and Zoological Station.

Doisy, Edward Adelbert. American biochemist (1893-); contributed much to the knowledge of vitamin K, insulin, blood buffering system, and sex hormones; shared Nobel prize in physiology and medicine in 1943.

dolichocephaly. Long-headed condition; see CEPHALIC INDEX.

Dolichodorus. Genus of microscopic, free-living nematodes.

Dolichoglossus kowalevskii. Common sp. in the Phylum Enteropneusta; to 7 in. long; burrows in sand and mud flats along Atlantic Coast.

Dolichohippus. Genus which includes Grevy's zebra, a shy, crested sp. of Abyssinian scrub deserts.

Dolichonyx oryzivorous. Common American bobolink in the Family Icteridae; known as reedbird or ricebird in southern states.

Dolichopodidae. Cosmopolitan family of minute to small, slender, long-legged flies; predaceous adults usually found near water, especially on flowers, foliage, and bark; larvae terrestrial or aquatic.

Dolichosoma. Genus of Carboniferous Amphibia.

Dolichotis. Genus which includes the maras in the Family Caviidae; long-legged cursorial rodents about as large as a terrier; herbivorous, diurnal burrowers; pampas and other South American treeless deserts.

Dolichovespula maculata. White-faced hornet in the hymenopteran Family Vespidae; large globular nest in trees.

Doliidae. Family of large, thin-shelled gastropods; spire short; tropical seas; shells sometimes used as native utensils.

Doliolida. Order of tunicates in the Class Thaliacea; barrel-shaped; muscle bands complete; few to many slits; Doliolum common in warmer oceans, reproduces alternately by budding and sexual reproduction.

Doliolum. See DOLIOLIDA.

Dolium. Tonna.

dollarfish. See PORONOTUS and VOMER.

Dolly Varden trout. See SALVELINUS.

Dolomedes. Common genus of spiders in the Family Pisauridae; live near water and catch aquatic insects and sometimes very small fish.

Dolophilodes. Genus of Trichoptera.

dolphin. 1. Any of a group of carnivorous cetaceans in the Suborder Odontoceti; head more or less produced into a beak (some true whales are also somewhat beaked); teeth numerous in both jaws; usually to 8 or 10 ft. long; common in most seas, but a few spp. in large rivers; common American sp. is Delphinus delphis; Tursiops truncatus is the bottle-nosed dolphin. 2. See CORYPHAENIDAE.

Domicella. Genus of small Old World parrots.

dominance. See DOMINANT CHARACTER.

dominant character. One of a pair of alternate characters determined by paired genes located on opposite members of a chromosome pair; it exerts its effect so as to exclude manifestations of the alternative (recessive) character.

dominant species. 1. Animal (usually a large carnivore) which "controls" its habitat and food web. 2. Animal which is numerically most abundant in its particular community. 3. Sp. (usually a carnivore) which typifies a particular community.

Donacia. Genus of beetles in the Family Chrysomelidae; adults feed on emergent leaves of aquatic plants; grubs feed on submerged portions.

Donax. Large genus of delicately colored marine bivalves; butterfly, coquina, donax, or wedge shells; shell more or less wedge-shaped or triangular; to 1 in. long; color pattern variable; widely distributed, especially in warm seas.

doncella. Any of several brightly colored West Indian marine fishes.

Dondersia. Genus of small wormlike Amphineura.

donkey. See EQUIDAE.

Donnan equilibrium. When a membrane separates two solutions of electrolytes, one of which contains an ion which cannot pass through the membrane, while the others can, an equilibrium is attained in which there is an unequal distribution of the different ions on the two sides of the membrane.

donor. 1. Animal from which tissue for a graft is taken. 2. Any individual providing blood for direct transfusion or storage for use at a later time.

doodle bug. Ant lion.

dopa. Dihydroxyphenylalanine; intermediate compound in the synthesis of melanin from phenylalanine.

dopa oxidase. Enzyme which catalyzes the oxidation of dopa.

dorado. Dolphin (fish) in the genus Coryphaena.

Doration. Genus of darters now usually included in Etheostoma.

dor beetle. Common European dung beetle, or any other large beetle which flies with a distinct buzzing sound.

Dorcatragus. Genus of small, delicate, African antelopes (beiras) of rocky and mountainous Somaliland and Abyssinia.

Dorippe. Marine genus of small deepwater crabs.

Doris. Large genus of sea slugs; body depressed, often with dorsal tubercles; gills branched or feathery.

dormancy. Any period of inactivity or suspended animation; usually brought about by adverse environmental conditions, although some types are genetically-induced; also occurs in reproductive bodies, spores, cysts, etc. See HIBERNATION, AESTIVATION, INANITION, ENCYSTMENT, and DESICCATION.

Dormitator. See ELEOTRIDAE.

dormouse. See GLIRIDAE.

dormouse-phalanger. See CERCAERTUS.

Dorocordulia. Genus of dragonflies.

Doropygus. Genus of copepods parasitic or commensal in ascidians; in the Suborder Notodelphyoida.

Dorosoma. See GIZZARD SHAD.

dorsal. Of or pertaining to the back or upper surface; in human anatomy the back of the body is said to be posterior.

dorsal abdominal artery. Artery in the dorsal portion of the abdomen of the crayfish and certain other Malacostraca; supplies blood to the intestine, telson, and abdominal musculature.

dorsal aorta. 1. Large dorsal artery of vertebrates located just ventral to the vertebral column; carries blood via numerous branches to all of the body except head, lungs, and (in some spp.) the forelimbs; it originates from one or more of the aortic arches. 2. A large dorsal artery in certain invertebrates.

dorsal fin. Median fin on dorsal surface of any fish, cephalochordate, or amphibian; may be single, double, or variously modified.

dorsalis scapulae. Dorsal muscle of the pectoral girdle in reptiles and amphibians.

dorsalis trunci. Collectively, the epaxial musculature of fishes.

dorsal line. Faint longitudinal line visible on the dorsal surface of some nematodes; produced by a thickening of the hypodermis containing the dorsal nerve strand.

dorsal root. Dorsal (sensory) root of a vertebrate spinal nerve.

dorsal root ganglion. Ganglion located on the dorsal root of a vertebrate spinal nerve.

dorsal vessel. Main dorsal blood vessel in many invertebrates, e.g. earthworm.

dorso-intestinal blood vessel. One of four blood vessels in the gut wall in most segments of typical annelids;

receives blood from the gut wall and delivers it to the dorsal vessel.

dorsolongitudinal muscles. See TERGO-STERNAL MUSCLES.

dorsum. The back or dorsal surface of the body or a part of the body.

Dorvillea. Genus of Polychaeta.

Dorylaimina. Suborder of enoplid nematodes; male with two, one, or no copulatory spicules; without four rows of intestinal muscles; excretory system rudimentary or absent; aquatic, terrestrial, and parasitic; Dorylaimus, Trichinella, and Trichuris common.

Dorylaimus. Very large and common genus of enoplid nematodes; aquatic and terrestrial.

Dosidicus gigas. West Coast squid; body to 3 ft. long.

Dosinia. Genus of Atlantic Coast clams; in sandy shallows; the dosinas; thin, white, circular valves to 3 in. in diameter.

Doto. Genus of nudibranchs; foot very narrow; body with a single row of cerata on each side.

dotterel. Any of several Old World plovers.

double recessive. Individual homozygous recessive at two different loci, e.g. aabb.

double sugar. See DISACCHARIDE.

douc. See PYGATHRIX.

Douglas squirrel. See TAMIASCIURUS.

dourine. Disease of horses, donkeys, etc.; produced by Trypanosoma equiperdum and spread by sexual contact; marked by swollen lymph glands and genital area and by paralysis of the hind legs.

douroucouli. See AOTES.

dove. See COLUMBIDAE.

dovekie. See PLAUTUS ALLE.

dove shell. Any of a very large number of spindle-shaped marine shells; often brightly colored and extremely abundant, especially near the low tide line; usually less than 0.5 in. long; Columbella typical.

dowitcher. See LIMNODROMUS.

down. Special type of soft insulating feather on many young birds; the quill is reduced, and the shaft is short and flexible.

DPN. Diphosphopyridine nucleotide.

Draco. Genus of Malayan lizards having prolonged ribs covered by membranous skin on each side; used in gliding and in a parachute fashion; flying dragons.

Dracunculus medinensis. Guinea worm; a spirurid nematode parasitic in human subcutaneous tissues, especially in the neck, shoulders, arms, and legs; also in many other mammals; female to 3 ft. long; from tropical Africa through southern Asia to the East Indies; the definitive host becomes infected by swallowing fresh-water copepods (e.g. Cyclops) which contain the larvae.

Dracunculus ophidensis. Nematode parasite of the garter snake; immature stages in Cyclops and tadpoles.

dragline. Silken thread paid out by many spiders as they move about.

dragonfish. See HYPOSTOMIDES and MELANOSTOMIATIDAE.

dragonfly. See ODONATA and ANISOPTERA.

dragon lizard. See VARANIDAE.

drake. Male swan, duck, or goose.

dramamine. One of numerous proprietary drugs effective in treating motion sickness.

Drassidae. Gnaphosidae.

Dravidian. Race of short caucasoids living in southern India; skin brownish-black, hair dark and curly.

Dreissensia polymorpha. Small European marine and fresh-water bivalve.

Drepanididae. Family of land birds restricted to the Hawaiian Islands; honey creepers; habits and coloration variable; show considerable adaptive radiation; rapidly disappearing with the introduction of land birds from other parts of the world.

Drepanomonas. Genus of fresh-water holotrich Ciliata; flattened, keeled.

Drepanothrix dentata. Uncommon littoral cladoceran.

Driesch, Hans. German zoologist (1867-1889); father of experimental embryology.

Drifa. Genus of upright colonial anthozoans (Order Alcyonacea).

drift line. Irregular line of debris deposited on a beach by wave action.

drill. 1. Any of numerous small marine gastropods which feed on other mollusks by drilling small holes through the shells; Urosalpinx and Acanthina typical. 2. See MANDRILLUS.

Drilophaga. Genus of illoricate rotifers; usually ectoparasitic on oligochaetes and leeches.

driver ants. See ECITON.

dromedary. See CAMELUS.

Dromadidae. Small family of birds in the Order Charadriiformes; crab plovers; coasts of east Africa and southwestern Asia.

Dromiacea. Taxon which includes certain marine crabs in which the last pair of thoracic appendages are dorsal, chelate, and sometimes used for placing bits of sponge, shells, algae, etc. on the dorsal surface of the cephalothorax.

Dromia erythropus. Sponge crab; a West Indian marine crab which conceals itself beneath a fragment of sponge.

Dromiceius novae-hollandiae. Flightless, swift-running bird in the Order Casuariiformes; emu; to 6 ft. tall; open areas of Australia.

Dromogomphus. Genus of dragonflies; eastern and southern states.

Dromus dromas. Common unionid mollusk of the Tennessee and Cumberland river systems.

drone. Male bee in the insect Order Hymenoptera; a non-worker.

drone fly. Any of a group of flies in the Family Syrphidae; superficially similar to honeybees but breed in carrion and excreta.

drongo. See DICRURIDAE.

droplet infection. Infection by droplets of contaminated coughing or sneezing discharges.

dropsy. Old term for edema or any other excessive accumulation of fluid in tissues or cavities.

Drosophila melanogaster. Small fruit fly extensively used for experimental genetics; easily reared in the laboratory; successive generations are obtained every 10 to 14 days.

Drosophilidae. Family of small flies many of which are attracted to decaying and fermenting vegetation and fruits; fruit flies, vinegar flies, pomace flies; Drosophila melanogaster the most familiar sp.

drugstore beetle. See STEGOBIUM.

drum. See SCIAENIDAE.

drummer. 1. Any of numerous fishes which produce a drumming or vibrating sound when taken from water. 2. Any of several Central and South American cockroaches which make a drumming vibration on woodwork; presumably a mating call.

Drymarchon. Genus which includes the indigo or gopher snake, a heavy blue-black constrictor to 8 ft. long; S.C. to Tex.

Drymobius margaritiferus. Speckled racer in the Family Colubridae; a dark, slender snake to 3 ft. long; southern Tex. and Mexico.

Dryobates. Common U.S. genus of woodpeckers; D. villosus is the hairy woodpecker with black and white upper parts and white under parts; D. pubescens is the downy woodpecker with the same coloration but smaller size; sometimes Dendrocopos is preferred over Dryobates.

Dryocopos pileatus. Pileated woodpecker; black, with a red crest and several white patches and stripes; crow-sized; widely distributed in forests of U.S. and Canada.

Dryomys. Genus of Asiatic tree dormice.

Dryopidae. Small family of aquatic beetles; usually 2 to 6 mm. long; much of body covered with hydrofuge pubescence and a plastron when submerged; feed on algal film of substrate; often leave the water, especially at night; Helichus common.

Dryopithecus. Genus of fossil anthropoids found in Europe and Asia.

Dryops. Genus of dryopid beetles.

Dubois, Eugene. Dutch army officer who discovered the original Pithecanthropus remains in Java in 1891.

Dubois-Reymond, Emil. German physiologist (1818-1896); well known for his studies of nerve and muscle activity; showed that electrical changes accompany muscle contractions.

duck. Any of numerous kinds of wild and domestic swimming fowl smaller than a goose; three major kinds are usually recognized: river ducks (mallard, teal, etc.), sea ducks (canvasback, eider, redhead, etc.), and the fish-eating ducks (merganser); most spp. migrate long distances.

duckbill dinosaur. See TRACHODON and CORYTHOSAURUS.

duck-billed platypus. See ORNITHORHYNCHUS.

duck-bill shell. See THRACIA.

duck hawk. See FALCO.

duck's foot shell. See APORRHAIS.

ductless gland. Endocrine gland.

duct of Bellini. Main urinary collecting emerging from each kidney papilla.

duct of Botallus. Ductus arteriosus.

duct of Cuvier. Common cardinal vein; one of two short blood vessels in fishes and tetrapod embryos; they bring blood from the cardinal veins into the sinus venosus.

duct of Steno. Duct of parotid gland; opens on the inner surface of the cheek near the base of the upper second molar.

duct of Wharton. Duct of submaxillary gland.

duct of Wirsung. Pancreatic duct.

ductuli efferentes. Special tubules which carry sperm from a testis through the kidney to the Wolffian duct, as in many fishes and amphibians.

ductulus aberrans. Small coiled tubule extending from the duct of the mammalian epididymis.

ductus arteriosus. Small remainder of the left sixth aortic arch connecting the pulmonary artery and the dorsal aorta in amphibians and in immature higher vertebrates.

ductus choledocus. Bile duct.

ductus reuniens. Tubule connecting the sacculus and the cochlea in the membranous labyrinth of the ear.

Dugesia. Genus of large, pigmented, triclad, fresh-water Turbellaria; the familiar "planarian"; common in springbrooks, ponds, and marshes; to 30 mm. long; common American spp. are D. dorotocephala (including D. agilis) and D. tigrina (including D. maculata).

Dugesiella. Genus of southwestern tarantulas.

Dugong. Genus of mammals in the Order Sirenia; dugongs; similar to the manatee but the tail is cleft; flippers are without nails, and the male has two large tusks; rivers, estuaries, and coasts of Red Sea, Indian Ocean, New Guinea, and northern Australia.

duiker. Any of numerous kinds of small graceful antelopes of Africa south of the Sahara; range from size of a rabbit to that of a small donkey; Cephalophus, Sylvicapra, and Philantomba the most common genera.

Dujardin, Felix. French protozoologist (1801-1862); first described the physical nature of protoplasm in Protozoa.

Dulichia. Genus of marine Amphipoda.

dulosis. Slave-making among ants.

Dumetella carolinensis. Common American catbird in the Family Mimidae; slaty gray sp. with a black cap and red spot under tail coverts; smaller than a robin; has a catlike mewing note.

dun. Mayfly; see EPHEMEROPTERA.

Dunbaria. Genus of late Paleozoic dragonflies.

dunce-cap limpet. Acmaea mitra, a West Coast sp.

dung beetles. See SCARABAEIDAE.

Dungeness crab. Cancer magister, the most important commercial crab of the N. A. Pacific Coast.

dung fly. See SCATOPHAGIDAE.

Dunhevedia. Genus of littoral Cladocera.

dunlin. Red-backed sandpiper, Erolia alpina; breeds in Holarctic tundra.

duodenum. First portion of the small intestine in certain of the higher vertebrates; contains glands which secrete an alkaline mucus; in man the duodenum is 8 in. long.

duplicate genes. Two or more pairs of genes having the same genetic effect.

Duplicity Theory. Idea fostered by Max Schultze; the vertebrate eye is really a combination of two sense organs, because of the different roles of rods and cones.

dura mater. Outermost layer of vascular connective tissue investing the brain and spinal cord of all vertebrates except fish.

Dusicyon. Genus of South American jackals; grayish spp. with shaggy hair; habits variable; includes the colpeo of the extreme south and the crab-eating dog of coastal, swamp, and forest areas.

dusky grouse. See DENDRAGAPUS.

dusky hawk. See PARABUTEO.

dusky salamander. See DESMOGNATHUS.

dustywings. See CONIOPTERYGIDAE.

Dutch elm disease. Fungus disease of elm trees transmitted by bark beetles (Scolytus).

Dutrochet, René. French biologist (1776-1847); made early contributions to the phenomena of osmosis; announced that plants and animals are composed of cells.

Duva multiflora. Sea cauliflower; a tree-like anthozoan in the Order Alcyonacea; to 5 in. high.

Duvaucelia. Genus of nudibranchs.

Du Vigneaud, Vincent. American biochemist (1901-); made important contributions to the chemistry of insulin, proteins, and biotin; awarded the Nobel prize in physiology and medicine in 1955 for his work on the identity and isolation of oxytocin and vasopressin.

dwarf lemur. See PHANER and CHEIROGALEUS.

dwarf pocket rat. See MICRODIPODOPS.

dwarf tapeworm. See HYMENOLEPIS NANA.

dyad. Diad.

Dynastes. Genus of large tropical and subtropical beetles in the Family Dynastidae; head and pronotum often supplied with horns, crests, or tubercles; *D. hercules* is the Hercules beetle of Tropical America, to 2.5 in. long; *D. tityus* is the unicorn beetle with three pronotal horns.

Dynastidae. Family of medium to large beetles; head and pronotum often armed with long processes or tubercles; mostly tropical; *Dynastes* and *Strataegus* typical.

Dysdera. Common genus of spiders.

Dysdercus. See PYRRHOCORIDAE.

dysentery. Inflammatory intestinal disorder characterized by frequent passage of watery, bloody stools; caused by a variety of chemical irritants, viruses, bacteria, Protozoa, and parasitic worms in the large intestine.

dysgenic. Racially detrimental; contrary to racial or sp. improvement.

Dysnomia. Large and common genus of unionid mollusks; eastern half of U.S.

dyspepsia. Any of many kinds of indigestion.

dysphotic zone. That portion of a body of water in which there is insufficient light for photosynthesis but sufficient light for certain animal responses; usually between depths of 100 and 800 m. in clear ocean water.

dyspnea. Labored breathing in mammals.

Dysteria. Genus of marine and fresh-water holotrich ciliate Protozoa; ovate, ventral surface flat.

dystrophic. Pertaining to a type of lake with brown waters, poor bottom fauna, and pronounced oxygen consumption; many limnologists do not recognize it as a distinct lake type in the developmental series.

dystrophy. 1. Defective nutrition. 2. Dystrophic condition in lakes.

Dythemis. Genus of dragonflies; southern U.S.

Dytiscidae. Large family of fresh-water beetles; predaceous diving beetles; hind legs of adult flat, fringed with hairs; air supply carried under elytra; both adults and larvae (water tigers) predaceous.

Dytiscus. Common genus of aquatic beetles (Dytiscidae).

dzo. Zobo; Himalayan pack and stock animal; cross between a yak bull and the common cow.

E

eagle. Any of several large carnivorous birds in the Order Falconiformes; see AQUILA and HALIAEETUS.

eagle owl. Any of several large owls of Eurasia.

eagle ray. Any of several large sting rays found along both coasts of the Atlantic.

eardrum. See TYMPANIC MEMBRANE.

eared grebe. Sp. of grebe having fan-shaped ear tufts of yellowish feathers; Colymbus caspicus; Europe and western N.A.

eared seal. Any seal in the Family Otariidae; external ears small but well developed; hind legs capable of rotation forward and functioning in terrestrial locomotion.

ear shell. See HALIOTIS.

ear stone. Otolith.

earth snake. See HALDEA.

earthworm. Any terrestrial soil-inhabiting oligochaete; Lumbricus a familiar genus.

earwig. See DERMAPTERA.

Ebner's glands. Small glands on the posterior part of the tongue.

Ebo latithorax. Small gray and white spider in the Family Thomisidae; common in eastern half of U.S.

Ecardines. Group of brachiopods having no shell hinge or internal skeleton; anus present.

ecdysis. Periodic shedding of the exoskeleton to permit an increase in size and/or change of form; the newly exposed exoskeleton quickly hardens a size larger than the old; especially characteristic of nematodes, arthropods, and tardigrades. Pl. ecdyses.

ecesis. Introduction and establishment of an organism in an area previously not occupied by it.

ECG. Electrocardiogram.

Echeneibothrium. Genus of tetraphyllidean tapeworms found especially in the spiral valve of skates.

Echeneis. See DISCOCEPHALI.

Echidna. Tachyglossus; also see ZAGLOSSUS.

Echidnophaga gallinacea. Sticktight flea (Order Siphonaptera) of poultry and other birds; a pest sp.

Echimyidae. Family of porcupine-rats; numerous spp. of small, long-legged, ratlike rodents; usually with small, flattened spines; forests of South America.

Echinanthus. Genus of large cake urchins in the Order Clypeasteroida.

Echinarachnius. Common genus of sand dollars.

Echinaster. Common genus of sea stars; the warty sea stars; coarse, scattered spines.

Echiniscus. Fresh-water and marine genus of tardigrades; common in mosses.

Echinobothrium. See DIPHYLLIDEA.

Echinocardium. See SPATANGOIDEA.

Echinochasmus. Genus of distome trematode parasites of the intestine of various mammals.

Echinococcus granulosus. Minute tapeworm consisting of scolex and only three or four proglottids; adult in intestine of dog, wolf, jackal, etc.; larval stage a complex hydatid cyst in a great many different mammals (including man) but most common in sheep.

Echinocyamus. Genus of small oval sea urchins.

Echinodera. Kinorhyncha.

Echinoderella. Typical genus in the Phylum Kinorhyncha; in muddy sea bottom.

Echinoderes. Common genus in the Phylum Kinorhyncha.

Echinodermata. Phylum consisting of radially symmetrical coelomate marine animals; there is a calcium carbonate skeleton of spicules and/or plates imbedded in the body wall; digestive tract simple; water vascular system used variously for locomotion, respiration, and feeding; circulatory and nervous systems radially arranged and not well developed; sexes separate;

larvae bilaterally symmetrical, motile, and ciliated; includes sea stars, brittle stars, sea urchins, sea lilies, and sea cucumbers; 5200 living spp.

echinoderm theory. Phylogenetic theory which seeks to derive primitive chordates from echinoderms (especially sea cucumbers); serological similarities and the tornaria larva are specifically cited as lines of evidence.

Echinoidea. Class in the Phylum Echinodermata; includes the sand dollars, cake urchins, sea urchins, and heart urchins; body globular, hemispherical, discoidal, or egg-shaped; skeletal plates forming a rigid test which bears many movable spines and three-jawed pedicellariae; tube feet with suckers; arms absent; most spp. have Aristotle's lantern just inside the mouth; common on the sea bottom.

Echinolaelaps echidninus. Rat mite; a common sp. in the Family Parasitidae.

Echinolampas. See CASSIDULOIDA.

Echinomera. Genus of gregarine Protozoa found in centipede guts.

Echinometra. Genus of elliptical sea urchins; includes some rock-boring spp.

Echinoneus. See HOLECTYPOIDA.

Echinophthirius horridus. Louse in the Order Anoplura; parasitic on the harbor seal.

echinopluteus. Pluteus larva of the Echinoidea.

Echinoprocta. Genus of arboreal porcupines; mountain porcupines; equatorial forests north and west of Andes of Ecuador and Colombia.

Echinops. See HEDGEHOG TENRECS.

Echinorhinus brucus. Bramble shark; widely distributed shark, to 9 ft. long; back and sides sparsely strewn with large scales bearing one or two spines.

Echinorhynchus. Genus of acanthocephalans found in fresh-water fishes.

Echinosoma. Genus of very large sea urchins; test to 12 in. in diameter.

Echinosorex. Genus of odorous mammals in the Family Erinaceidae; rat-shaped, with long, scaled, bristly tails; coat of thick woolly fur and an overcoat of coarse bristles; southeastern Asia.

Echinostoma. Large genus of trematodes found in the intestine of birds and mammals; cuticle spiny; large acetabulum near the oral sucker; anterior end with a collar of spines.

Echinus. Common genus of sea urchins.

Echiuroidea. Phylum including about 65 spp. of peculiar worms of mud and sand bottoms in the shallows of warm and temperate seas; some spp. burrow and others make U-shaped tubes; a long, troughlike or threadlike proboscis at the anterior end of the sausage-shaped body is used for feeding; no segmentation and only one to three pairs of setae; coelom very large; circulatory system present; mouth at base of proboscis; one to three pairs of nephridia (metanephridia); ventral nerve cord but no ganglia or brain; to 18 in. long; larva a trochophore; Echiurus, Urechis, and Bonellia are familiar genera.

Echiurus. Common genus in the Phylum Echiuroidea; lives in a burrow and obtains food by extruding its mucus-covered proboscis.

Eciton. Genus of carnivorous ants of American tropics; driver ants, legionary ants; live in temporary nests, and on dull days they migrate in long files and vast armies.

eclampsia. Sudden attack of convulsions associated with hypertension, albuminuria, and edema; occurs during pregnancy or delivery.

Eclipidrilus. Genus of semiaquatic and fresh-water oligochaetes.

eclosion. 1. Hatching of an insect egg. 2. Emergence of an adult insect from the pupal stage.

ecobiotic isolation. Essentially the same as behavioral isolation.

ecoclimatic isolation. Climatic isolation.

ecogeographic isolation. See GEOGRAPHIC ISOLATION.

ecological age. Any of three stages in a life history: prereproductive, reproductive, and postreproductive.

ecological distribution. Ecological range.

ecological efficiency. Relative amount of energy or food used to produce a unit of protoplasm at any particular trophic level.

ecological equivalents. Two distantly related spp. which occupy similar ecological niches in the same or distant communities; e.g. kangaroos of Australia and large wild herbivores of the U.S.

ecological isolation. Prevention of interbreeding between two or more otherwise sympatric populations because of differing ecological requirements; e.g. stream and pond crayfishes may presumably be isolated from each other in this fashion.

ecological niche. 1. Ecological role of a plant or animal with reference to its special place in its inanimate environment and with reference to other spp. associated with it; food and nutrition relationships are of primary importance. 2. A few ecologists use the term in the geographic or spatial sense.

ecological race. Subspecies.

ecological range. Kind of environment(s) or habitat(s) occupied by a particular sp. or other taxon.

ecological succession. The slow, orderly progression of changes in community composition during development of vegetation in any area, from initial colonization to the attainment of the climax typical of a particular geographical area; accompanied by changes in the associated animal communities; beginning with bare ground, and depending on the type of climax attained, the complete sequence may take from 50 years to thousands of years; e.g. a forest (xerarch) succession may traverse the following overlapping stages: algae, lichens, mosses, herbs and grasses, shrubs and bushes, subclimax trees, and climax forest; similarly, events comprising the maturation, filling in, and disappearance of a small lake constitute a hydrarch succession accompanied by aquatic and terrestrial animal community changes; see CLIMAX COMMUNITY.

ecological valence. Theoretical measure of the ability a sp. has to cope with the adversities of its environment.

ecology. 1. Study of the interrelationships between organisms and their animate and inanimate surroundings. 2. Bionomics.

ecophene. Modification of a genotype produced as a response to a local ecological factor.

ecospecies. Two or more populations of a sp. which are able to exchange genes freely without loss of vigor or fertility in the offspring; such populations may be separated by ineffective geographic or ecological barriers.

ecosystem. Collectively, all organisms in a community plus the associated environmental factors.

ecotone. Transition or interdigitated area between two adjacent communities, as the merging zone of adjacent forest and grassland.

ecotopic isolation. Ecological isolation.

ecotype. Group of individuals (constituting a subspecies) which can interbreed with other ecotypes of the same ecospecies but which maintains its individuality through isolation.

ectepicondylar foramen. Small opening for a nerve and blood vessels on the ectepicondyle of many turtles, lizards and a few other vertebrates.

ectepicondyle. Smaller of the two distal humerus expansions; some vertebrates.

Ectinosoma. Very common genus of marine harpacticoid Copepoda.

ectoblast. Epiblast.

ectocommensal. Commensal organism which lives on the external surface of the host and neither benefits nor harms the host.

Ectocotyla. Genus of flatworms in the Order Acoela; commensal on hermit crabs.

ectocrine. Hormone which diffuses into the medium and affects the metabolism of other organisms.

Ectocyclops phaleratus. Uncommon but widely distributed fresh-water copepod.

ectocyst. The outermost layer of the zooecium in Bryozoa.

ectoderm. Outermost layer of cells in an early animal embryo; gives rise to skin, nervous system, sense organs, etc. in vertebrates; the term is most properly applied to this particular germ layer chiefly while it is a distinct region (after gastrulation) and before differentiation of its derived tissues; originally, however, the term was first applied to the outermost cell layer of Coelenterata.

ectoenzyme. Extracellular enzyme.

ectolecithal. Pertaining to zygotes having peripheral yolk granules; early cleavages central.

ectomesoblast. In early embryology, a layer of cells destined to give rise to both ectoblast (epiblast) and mesoblast.

ectomesoderm. Mesoderm derived chiefly from the ectoderm during very early embryology of animals; especially predominant in Porifera, Ctenophora, Coelenterata, and in certain mollusks and annelids.

ectoneural system. Outer (oral) part of the echinoderm nervous system; sensory and motor.

ectoparasite. Any parasite occurring on the external surface of its host, e.g. a louse.

ectopic pregnancy. Abnormal implantation and development of a mammalian embryo in the oviduct or abdominal cavity.

Ectopistes migratorius. Passenger pigeon of forested eastern N.A.; formerly present in enormous numbers but intensively slaughtered for the market and extinct since 1914.

ectoplasm. Peripheral zone of relatively clear cytoplasm, especially in amoeboid Protozoa.

Ectopleura. Genus of marine coelenterates in the Suborder Anthomedusae.

Ectoprocta. Bryozoa.

ectosarc. Outermost relatively nongranular protoplasmic portion of a protozoan.

ectosome. Outer portion of a leuconid sponge; devoid of flagellated chambers but consists of dermal membrane and subdermal spaces.

ectothermic. Poikilothermic.

Ectyodoryx. Genus of marine sponges.

edaphic factor. Ecological factor determined by any one of many physical, chemical, or biological features of the soil or other substrate; distinct from climatic factor.

edaphon. Soil flora and fauna.

Edaphosaurus. Genus of Permian and Carboniferous reptiles in the Order Pelycosauria; heavily-built spp. with enormously elongated vertical dorsal vertebral spines from neck to sacrum.

edema. Swelling of tissues owing to the accumulation of abnormally large amounts of fluid in intercellular spaces.

Edentata. Order of mammals which includes the anteaters, sloths, and armadillos; teeth absent or only molar teeth present, without enamel or roots; same as Xenarthra.

edentulate. Without teeth.

edentulous. Without teeth.

edestin. Simple protein, insoluble in water but soluble in neutral salt solutions; a globulin.

edge effect. Interspersion of types; relative amount of intermingling of habitat types in an environment; in general, the greater the admixture, the greater the ability to support large animal populations, especially game animals; e.g. the intermingling of small patches of cultivated fields, pastures, meadow, brush, and woodland.

Edotea. Genus of marine Isopoda with prehensile legs.

Edrioaster. Genus of extinct pelmatozoan echinoderms.

Edrioasteroidea. Class of extinct echinoderms; calyx globose or disc-shaped, composed of numerous small plates; Cambrian to Carboniferous; Stromatocystites typical.

Edwardsia. Genus of small, slender sea anemones which burrow in soft substrates.

Edwardsiella. Genus of burrowing sea anemones.

E.E.G. Electroencephalogram.

eel. See ANGUILLA, ANGUILLIDAE, COLOCEPHALI, and CONGER.

eelpout. See GADIDAE and MACROZOARCES.

eelworm. Common name of free-living and plant-parasitic nematodes.

effector. Organ, tissue, or cell which is capable of reacting to stimuli; e.g. muscles and glands.

efferent. Discharging, conducting, conveying, or leading outward or away from, such as a blood vessel or nerve.

efferent branchial artery. One of the arteries carrying blood from the gills to the dorsal aorta in typical fishes.

efferent neuron. Motor neuron.

eft. 1. Salamander in the Family Salamandridae. 2. Newt.

egestion. Process of voiding indigestible material, usually by way of the anus; the bulk of the material egested has never been a part of the living protoplasm; this term is usually used with reference to invertebrates, and defecation for vertebrates.

egg. Ovum or zygote; structure ranges from simple and microscopic to large and complex, as the hen's egg.

egg axis. Hypothetical axis extending through animal and vegetal poles of a zygote and early embryo.

egg burster. In insects, a set of special spines, teeth, or ridges on the head; aids in rupturing the shell at time of hatching.

egg capsule. Any gelatinous or mucous mass, or any membranous saclike structure containing developing eggs; usually deposited outside the body by the female.

egg funnel. Short, broad, funnel-shaped portion of an oviduct which receives eggs from the ovary; in the earthworm and its relatives.

egg tooth. Temporary epidermal tooth which develops at the tip of the upper mandible in embryonic birds, reptiles, and monotremes; used to cut the surrounding membranes and egg shell at hatching, then drops off; in lizards the egg tooth is a normal tooth of the premaxillary.

egret. See CASMERODIUS and FLORIDA.

Ehlersia. Genus of Polychaeta.

Ehrenberg, Christian Gottfried. German zoologist (1795-1876); wrote the classical Die Infusionsthierchen als vollkommene Organismen (1838) which contributed greatly to the knowledge of Protozoa and which first separated Protozoa from multicellular animals; unfortunately, he thought he discerned complete organ systems within single protozoans.

Ehrlich, Paul. German bacteriologist (1854-1915); made valuable contributions to hematology, pathology, and biological staining; discovered salvarsan and neosalvarsan for the treatment of syphilis; shared Nobel prize in physiology and medicine in 1908.

eider. See SOMATERIA.

Eigenmann, Carl H. American zoologist (1863-1927); on staff of Indiana University; authority on cave vertebrates and South American freshwater fishes.

Eijkman, Christian. Dutch physician (1858-1930); contributed much to the understanding of beriberi as the result of his studies in the East Indies.

Eimeria. Genus of Sporozoa in the epithelium of the digestive system of arthropods and vertebrates; E. stiedae, an important parasite of rabbits, occurs in cells of the bile duct.

Eisenia foetida. Common and cosmopolitan sp. of pinkish earthworms; found especially in manure and compost heaps; to 5 in. long.

Eiseniella. Genus of small earthworms.

ejaculate. 1. Emitted seminal fluid. 2. To eject a fluid from the body.

ejaculation. Muscular emission of seminal fluid by the male during copulation.

ejaculatory duct. Special muscular portion of the male reproductive system in many invertebrates; usually located between the seminal vesicle and genital pore.

EKG. ECG, electrocardiogram.

Ekman dredge. Relatively light weight metal clamshell dredge used in taking quantitative samples of bottom organisms from lakes and the ocean; useful only on soft bottoms.

eland. See TAUROTRAGUS.

Elanoides. See KITE.

Elanus. See KITE.

élan vital. "Vital force," a supposed unmeasurable feature distinguishing living from non-living.

Elaphe. Genus of large, colorful, nonvenomous snakes with flat, blunt head; N.A., Europe, and Asia; the corn snake occurs in the southern states, the fox snake in the northcentral states, the chicken snake in the southeastern states, and rat snakes generally in the eastern half of the U.S.

elaphocaris. Post-nauplius larval stage in certain decapod Crustacea.

Elaphurus. Genus of ruminants in the Family Cervidae; includes Père David's deer or milu, a long-tailed, marsh-inhabiting sp. of northern China.

Elapidae. Family of terrestrial poisonous snakes; tail cylindrical and tapered; fangs short, stout, and always erect; widely distributed except for northern parts of N.A., Europe, and Asia; typical genera are Naja, Bungarus, Denisonia, and Micrurus.

Elaps. Micrurus.

Elasipoda. Echinoderm order in the Class Holothuroidea; tentacles with shield-shaped ends; respiratory tree absent, tube feet abundant, mouth ventral; pelagic to deep sea; Pelagothuria typical.

elasmobranch. Any fish in the Class Chondrichthyes.

Elasmobranchii. Chondrichthyes.

Elasmopus. Genus of marine Amphipoda.

Elasmosaurus. Genus of giant marine Cretaceous reptiles with paddle-like appendages.

Elassoma. Small genus of pygmy sunfishes; to 40 mm. long; mostly southeastern U.S.

elastic cartilage. Type of cartilage containing some yellow connective tissue fibers; occurs especially in the Eustachian tube and external ears of mammals.

elastic connective tissue. See YELLOW FIBROUS CONNECTIVE TISSUE.

elastin. Yellow elastic fibrous protein connective tissue, especially abundant in walls of large arteries.

Elateridae. Large family of elongated beetles; click beetles; herbivorous adults able to snap a process and socket arrangement on the thorax, producing a distinct clicking sound and a hurtling of the insect into the air; larvae (wireworms) subterranean, feed on plants, and are often pests; Alaus, Melanotus, and Limonius examples.

Electra. Membranipora.

electric catfish. See MELAPTERURUS.

electric eel. See ELECTROPHORIDAE.

electric-light bug. Large aquatic hemipteran (Lethocerus) which often leaves ponds and lakes and flies to lights at night; the "toe-biter."

electric organs. Specialized tissue

which generates high voltages in certain lower vertebrates, e.g. electric ray and electric eel.

electric ray. See TORPEDO.

electrocardiogram. Linear tracing or recording of the electric currents produced by cardiac activity. Abbr. ECG.

electroencephalogram. Record of rhythmic electrical waves in the vertebrate cerebral cortex; such waves can be picked up electronically through the surface of the intact skull; pattern correlated with physiological and neurological condition of the brain.

electroencephalograph. Complicated apparatus used in detecting, amplifying, and recording electrical waves or "brain waves" in the vertebrate cerebral cortex.

electron microscope. Very high power microscope in which a beam of electrons is used instead of a beam of light; magnetic or electrostatic fields focus an image onto a fluorescent screen or photographic plate.

electrophoresis. Migration of colloidal particles in an electric field.

Electrophoridae. Small family of fishes of the Amazon and Orinoco drainages; includes the electric eel, Electrophorus electricus; sometimes more than 6 ft. long; ventral muscle tissue modified as electric organs capable of inflicting severe shocks.

Electrophorus. See ELECTROPHORIDAE.

electrophysiology. Study of electrical phenomena and their correlation with physiological processes in cells.

electrotonus. Change in degree of polarization of tissue at point of contact of stimulating electrodes during the passage of an electic current through the tissue.

Eledonella. Genus of deepsea octopus.

eleidin. See STRATUM LUCIDUM.

eleocytes. Free chlorogogue cells in the coelom of oligochaetes.

Eleodes. Large genus of tenebrionid beetles; common in N.A.; often destructive to important plants.

Eleotridae. Sleepers, a family of percomorph fishes; Eleotris is a marine genus, but Dormitator maculatus penetrates far inland in fresh waters from N.C. to Mexico and Brazil.

elephant. Familiar representative of the mammal Order Proboscidea; the African elephant (Loxodonta africana) has the larger ears, and there are usually tusks in both sexes; the Indian elephant (Elephas maximus) has the smaller ears, and tusks are obvious only in the male; the latter sp. is commonly domesticated and used in circuses.

elephant beetle. Any large beetle in the Family Scarabaeidae having a prominent horn on the front of the head; mostly tropical.

elephant birds. See AEPYORNITHIFORMES.

elephant bug. Vernacular name for a weevil.

elephantiasis. Disease of the human lymphatic system produced by chronic infections of the nematode Wuchereria bancrofti; larvae and adults clog lymph ducts, resulting in the formation of new lymph channels and much connective tissue; the legs, arms, scrotum, and female breast may become swollen to enormous size, and the skin takes on a coarse, cracked texture.

elephant seal. See MIROUNGA.

elephant shrew. See MACROSCELIDIDAE.

elephant's tusk shell. See SCAPHOPODA.

Elephas. See ELEPHANT.

Eleutherodactylus. See ROBBER FROG.

Eleutherozoa. One of the two subphyla in the Phylum Echinodermata; without stalks or cirri; includes all living echinoderms except the Class Crinoidea.

elf owl. See MICRATHENE.

Eliomys. Genus of European carnivorous garden dormice.

elk. 1. Largest European deer (Alce alces); antlers broad and palmate. 2. American elk or wapiti (Cervus canadensis).

Elliptio. Genus of fresh-water mussels; common in eastern half of U.S.

Ellobiidae. Family of small brackish and terrestrial snails; aperture ear-shaped and toothed; tropical and subtropical.

elm bark beetle. See SCOLYTUS.

elm borer. Any of several beetles whose larvae bore into elm trees.

Elmidae. Family of small, blackish aquatic beetles, seldom more than 3.5 mm. long; do not swim but crawl about slowly on the substrate, especially in streams; feed on vegetation and debris; usually covered with a plastron when submerged; Stenelmis the largest U.S. genus.

Elmis. European genus of elmid beetles.

elm scale. See KERMIDAE.

Eloactis. Genus of elongated burrowing

sea anemones in the Order Actiniaria.

Elophila. Genus of moths in the Family Pyralididae; caterpillars of some spp. aquatic and supplied with numerous blood gills; in other spp. the caterpillars are gill-less and live on emergent aquatic plants.

Elopidae. Family of bony fishes closely related to the tarpons and including the ten-pounders; with a bony plate under the throat; usually only 2 to 4 lbs.; in warm waters of marine bays, inlets, and estuaries; Elops common.

Elops. See ELOPIDAE.

Elphidium. Genus of Foraminifera.

Elsianus. Genus of elmid beetles.

Elton, Charles. English ecologist (1900-); especially noted for his many important contributions to cycles of abundance, food webs, pyramids of numbers, and succession.

elver. 1. Young eel. 2. In fresh-water eels, the term refers to eels larger than the leptocephalus stage, especially during their migration up fresh-water drainages.

Elysia. Common genus of sea slugs; cerata absent, respiration through the general ciliated body surface; sides of body with winglike expansions.

elytron. 1. One of the two anterior heavy wings of beetles, serving as a cover for the membranous hind wings and commonly meeting in the mid-dorsal line when at rest. 2. One of many dorsal plates or scales in certain Polychaeta. Pl. elytra.

elytrum. Elytron.

emarginate. Notched, indented, or slightly forked.

Emarginula. Genus of keyhole limpets.

emasculation. Removal of the testes.

Emballonuridae. Family of Old and New World tropical bats; tail lies in a sheath in the membrane stretching between the hind legs; face truncated; insectivorous.

Embata. Genus of bdelloid rotifers.

embed. To infiltrate and encase a bit of tissue in paraffin so that sections may be cut with the microtome and used for study with the microscope.

Embia. Genus of insects in the Order Embioptera.

embiid. Any member of the insect Order Embioptera.

Embioptera. Order of small cylindrical insects of tropics and subtropics; biting mouth parts, incomplete metamorphosis; wings absent in females, present or absent in males; first tarsi enlarged and containing unique spinning glands; colonies in underground nests; feed on decaying vegetation; familiar genera are Embia, Bynembia, Anisembia, and Oligotoma.

Embiotocidae. Family of marine percomorph fishes of the northern Pacific Ocean; surf fishes and viviparous perches.

embolism. Blocking of a blood vessel by a clot or other obstruction brought into place by the blood stream.

embolium. 1. Outer part of the forewing in certain Hemiptera. 2. Enlargement at the base of the forewing in certain Hemiptera.

embolomerus centrum. Double centrum found in some aberrant labyrinthodonts.

embolus. 1. Blood clot, air bubbles, fat globules, or clumps of bacteria which obstruct a blood vessel. 2. Apical portion of a palp in certain spiders.

Embrithopoda. Order of Oligocene rhinoceros-like mammals.

embryo. Organism in the early stages of development, especially previous to hatching from the egg before metamorphosis, and before the organism is active and physiologically independent; in man the term is used only through the seventh week of uterine existence, thereafter being known as a fetus.

embryogeny. Embryology.

embryology. Study of the formation, early growth, and early development of living organisms.

embryonic membrane. One of a series of membranes surrounding or otherwise associated with embryonic reptiles, birds, and mammals; e.g. chorion, amnion, yolk sac, and allantois.

embryophore. Special shell-like or mantle-like covering of the microscopic egg or larva of many tapeworms.

embryotroph. Cellular debris of the uterine wall formed during the early development of the implanted blastodermic vesicle in mammals; absorbed by the embryo for nourishment (embryotrophic or histotrophic nutrition).

embryotrophic nutrition. See EMBRYOTROPH.

Emerita. Genus of marine decapod Crustacea; includes the sand crabs, mole crabs, or sand bugs; abdomen reduced and bent under cephalothorax; first four pairs of legs modified for mole-like digging; found at water's edge along sandy beaches where they burrow rapidly; usually 10 to 30 mm. long.

Emesa. Genus of spider bugs or thread-legged bugs; slender, delicate, long-legged, predaceous spp. in the Family Ploiariidae; usually in vegetation.

emmenin. Estrogen obtained from the human placenta.

emmetropia. Normal eye condition, with neither near- nor far-sightedness.

emperor goose. See PHILACTE.

emperor moth. See SAMIA.

emperor penguin. See SPHENISCIFORMES.

Empetrichthys. Springfishes; a small genus of cyprinodontid fishes of southwestern Nev.

Empididae. Dance flies; a widely distributed family of minute to small slender flies; mate in swarms which move up and down; larvae predaceous and terrestrial or aquatic; Empis common.

Empidonax. Genus in the Family Tyrannidae; includes most U. S. flycatchers; E. minimus is the least flycatcher of Canada and the northern half of the U. S. as far west as Mont. and Wyo.

Empis. See EMPIDIDAE.

Emplectonema. Genus of slender marine nemertine worms; some spp. to 12 ft. long.

Empoasca. See JASSIDAE.

empodium. Small structure of variable form (often padlike) between the terminal claws of the legs of many insects and arachnids. Pl. empodia.

empyema. Presence of pus in the pleural cavity.

emu. See DROMICEIUS.

emulsification. Conversion of suspended fat particles into smaller particles dispersed in an immiscible substance.

emulsion. See COLLOIDAL SYSTEM.

Emydidae. Chelydridae.

Emys blandingi. Blanding's turtle; a fresh-water sp. found from New England westward to Minn., Iowa, and Nebr.; carapace black, with small yellow spots.

enaliosaur. Any of a large group of Mesozoic marine reptiles.

Enallagma. Very common genus of damselflies.

enamel. Hard, glossy outermost covering of teeth; formed by epidermis; chiefly calcium phosphate and carbonate with 3 to 5% organic matrix.

encasement theory. Old fanciful theory that a sperm or egg cell contains a fully formed miniature individual and that its germ cells likewise contain miniature individuals, and so on indefinitely.

Encentrum. Genus of marine and fresh-water rotifers; body cylindrical, corona strongly oblique; closely associated with substrates.

encephalitis. Inflammation of the brain.

encephalomyelitis. Inflammation of the brain and spinal cord.

encephalon. That part of the vertebrate central nervous system enclosed by the cranium.

Enchelydium. Genus of fresh-water holotrich Ciliata; with a swollen oral ring.

Enchelyodon. Genus of ovoid to flask-shaped holotrich Ciliata; marine and fresh-water.

Enchelys. Genus of flask-shaped holotrich Ciliata; marine and fresh-water.

Enchytraeidae. Family of Oligochaeta; slender hermaphroditic spp. on fresh-water and marine substrates and in decaying and living vegetation; four bundles of slender setae per segment; body usually less than 30 mm. long.

Enchytraeus. Genus of whitish oligochaetes in fresh waters, along the seashore, and in decaying and living plants; to 30 mm. long; E. albidus is often cultured to feed aquarium fishes and small laboratory animals.

Encope. Genus of tropical American sand dollars having five prominent marginal notches.

Encyrtidae. Large and cosmopolitan family of minute Hymenoptera; parasitize egg, larval, and pupal stages of many insects.

encyst. To form or become enclosed in a cyst.

encystment. Formation of a resistant cyst by certain microorganisms, especially under unfavorable environmental conditions. See CYST.

Endalus. Genus of aquatic weevils; feed on aquatic vegetation.

Endamoeba. Genus of parasitic amoeboid Protozoa living in the digestive tract of termites and cockroaches. Cf. ENTAMOEBA.

endbrain. Telencephalon.

end bulb of Krause. See ORGAN OF KRAUSE.

endemic. Restricted to a particular geographical area, either large or small; term used with reference to any plant or animal taxon.

Endeodes. Genus of beetles in the Family Melyridae.

end-feet. Synaptic knobs.

endite. Lobe on the basal median margin of many crustacean and arachnid appendages.

endocardium. Heart endothelium.

Endoceras. Genus of giant fossil cephalopod mollusks; to 12 ft. long.

endochondral bone. Bone formed by the replacement of embryonic cartilage as contrasted with membrane bone which is formed directly from mesenchyme.

endochorion. Thick inner layer of an insect egg.

endocoel. That portion of an anthozoan gastrovascular cavity between the two members of an adjacent pair of septa.

endocranium. 1. Collectively, the processes and ridges extending inward from the inner surface of an insect head capsule; tentorium. 2. Cranial dura mater of certain vertebrates.

endocrine. Hormone.

endocrine gland. Ductless gland whose secretions (hormones) are released into the circulatory system and which have important physiological role(s) elsewhere in the body; the pituitary, thyroid, gonads, adrenals, and certain patches of cells in the pancreas are all vertebrate endocrine glands.

endocrine system. Collectively, all of the endocrine glands of an animal.

endocrinology. Study of endocrine glands, hormones, and their production, nature, and effects.

Endocyclia. Taxon which includes echinoid echinoderms in which the mouth is central, the anus apical, and the ambulacra not petaloid.

endocyst. Inner layer of a cyst wall, especially in protozoan cysts.

endoderm. Entoderm; innermost germ layer in an embryo, forming the primitive gut; produces much of the epithelia of the digestive system, thyroid, thymus, epithelium of the respiratory system and urinary bladder, etc. of vertebrates; the term is properly applied to this layer while it is a distinct region (after gastrulation) but before differentiation of its derived tissues; originally the term was first applied to the layer of cells lining the gastrovascular cavity of Coelenterata.

endoenzyme. Intracellular enzyme.

endogenous. Originating within.

endognath. Median ramus or process of an oral appendage in Crustacea.

endolymph. Lymph which fills most of the membranous labyrinth of the inner ear of vertebrates.

endolymphatic duct. Blind, tubular outgrowth of the sacculus of the vertebrate membranous labyrinth.

endolymphatic sac. Small, swollen saccule at the end of the endolymphatic duct of the inner ear.

endolymphatic space. Cavity of the membranous labyrinth of the inner ear; filled with lymph.

endomesoderm. Mesoderm derived chiefly from the endoderm during very early embryology of animals; predominant in most higher phyla.

endometrium. Glandular epithelium of the mammalian uterus; undergoes cyclical thickening and disintegration during the period of sexual maturity of women.

endomitosis. Doubling of chromosome number of a cell without a typical nuclear division; results in polyploidy; occurs chiefly in certain insect, vertebrate, and plant tissues.

endomixis. Periodic nuclear division and reorganization within a single cell in certain ciliates; no fusion of nuclei is involved.

Endomychura. Genus which includes two spp. of murrelets, small sea birds of Calif. and Baja California.

endomysium. Delicate masses of connective tissue found between individual muscle fibers.

endoneurium. Delicate reticulum of connective tissue surrounding individual nerve fibers, as seen in cross section of a nerve.

endoparasite. Any internal parasite, as a tapeworm or malaria parasite.

endophragm. Septum formed by thoracic and cephalic apodemes in Crustacea.

endoplasm. Inner portion of cell protoplasm; in amoeboid Protozoa it is notably more granular than the peripheral ectoplasm.

endoplasmic reticulum. Complicated meandering double membranous network in many cells; often connects with the outermost cell membrane so as to provide minute openings to the outside; may also have similar connections with the nuclear membrane.

endopleurite. In arthropods, a sclerotized infolding between pleurites.

endopod. Endopodite; see BIRAMOUS APPENDAGE.

Endoprocta. Entoprocta.

Endopterygota. Holometabola.

end organ. Especially in higher animals, a group of one or a few cells connected to the central nervous system; usually receptors of stimuli or motor end plates in vertebrate voluntary muscle.

endoskeleton. Any animal skeleton of various composition which is built into the body and gives support and permanent shape to the body and its parts; present in echinoderms and vertebrates.

endosmosis. Movement of materials, especially liquids, inward through a selectively permeable membrane.

endosome. 1. Central concentrated mass of chromatin material in the nucleus of certain organisms. 2. Choanosome.

endosternite. Entosternite; sclerotized ridge or other process on the inner surface of the cephalothoracic exoskeleton of various arthropods; serves for muscle and connective tissue attachment.

endosteum. Connective tissue lining the medullary cavity of bone.

endostyle. Special ciliated ventral groove or pair of grooves in the pharynx (branchial chamber) of tunicates, cephalochordates, and larval cyclostomes; accumulates food particles and passes them to the stomach.

endothelium. Single layer of thin, flat cells forming the lining of blood vessels and lymphatics of vertebrates.

endothermic. 1. Homoiothermic. 2. Featured by or accompanied by heat absorption.

endothermic reaction. Chemical reaction which requires energy input, e.g. organic syntheses.

endothorax. Collectively, the thoracic or cephalothoracic apodemes of arthropods, especially Crustacea.

endotoxin. Any normal internal protein constituent of an organism which causes the production of precipitins, agglutinins, complement-fixation antibodies, etc. (but not antitoxins) when introduced beneath the epithelium of a higher animal (especially man).

endozoic. Living within another animal, either as a commensal or parasite.

end piece. Extreme tip of the flagellum of a sperm cell in many animals; not covered by an enveloping protoplasmic sheath.

end plates. Flattened discoid expansions at the tips of the terminal branches of the axon of a motor neuron; in contact with muscle fibers.

endysis. Process of developing a new exoskeleton, coat of hair, set of feathers, etc.

English sparrow. See PASSER.

Engraulidae. Anchovies; a family of small, marine, herring-like fishes with a large mouth; on Atlantic and Pacific coasts; Anchoa from Me. to Tex.

engraver beetle. See SCOLYTIDAE.

Enhydra lutris. Sea otter of rocky Pacific shores from Alaska to Calif.; to 4 ft. long; fur dark brown to blackish, soft and heavy; hind feet webbed, ears inconspicuous; feeds on fish and marine invertebrates; expert swimmers and divers, seldom leaving the water; not common.

Enneacanthus. Small genus of small centrarchid fishes; includes two common sunfishes of East Coast states.

Enochrus. Common genus of hydrophilid beetles.

Enopla. One of the two classes in the Phylum Nemertea; mouth anterior to brain; central nervous system internal to body wall; proboscis usually armed.

Enoplida. Order of nematodes in the Class Aphasmidea; esophagus long and divided into an anterior muscular portion and a posterior glandular portion.

Enoplina. Suborder of enoplid nematodes; without four rows of intestinal muscles; stylet usually absent; male with two spicules; aquatic soils; common genera are Enoplus and Trilobus.

Enoplus. Common genus of enoplid nematodes; mostly in marine soils.

Ensatina. See PAINTED SALAMANDER.

Enseosteus. Genus of Devonian placoderms.

ensiform cartilage. Lowermost median pointed cartilage in the sternum; xiphoid cartilage.

Ensis. Genus which includes the true razor clams; the common razor clam of the Atlantic Coast is narrow, with gaping ends, and to 6 in. long.

Entamoeba. Genus of parasitic and commensal amoeboid Protozoa living in the digestive tract of terrestrial vertebrates; sometimes written Endamoeba; E. histolytica causes amoebic dysentery in man; E. gingivalis is a commensal living in the mouth, especially along the gum line; E. coli is a common commensal in the large intestine of man and other primates.

entelechy. "Vital principle" presumed to guide the activities of living organisms in the proper direction.

entellus. East Indian long-tailed monkey; hair of the head projects like the peak of a cap.

entepicondylar foramen. Small opening for a nerve and blood vessels on the entepicondyle of some primitive mammals.

entepicondyle. Larger of the two distal expansions of the humerus of many vertebrates.

enteric. Pertaining to the intestines or alimentary tract.

enterobiasis. Infection with Enterobius vermicularis.

Enterobius vermicularis. Common rhabditid nematode parasitic in the caecum, appendix, and upper large intestine of man; the pinworm; gravid females migrate to the perianal region at night and release their eggs; man becomes infected with embryonated eggs through dirty linen, contaminated food and water, and sometimes through eggs momentarily syspended as dust in the air.

Enterocoela. Group of phyla having enterocoelic mesoderm formation.

enterocoelic mesoderm formation. Embryonic formation of mesoderm characterized by the development of pouchlike outfolding from the archenteron which then expands and obliterates the blastocoel, the resulting large cavity completely lined with mesoderm being the coelom; occurs in Chaetognatha, Brachiopoda, Echinodermata, Hemichordata, and Chordata.

enterocoelic pouches. Primordia of mesodermal somites which pouch out on either side of the archenteron in such embryos as those of Amphioxus and other enterocoelic groups.

enterocrinin. Hormone secreted by certain duodenal cells; increases the secretion of intestinal juice.

enterogastrone. Hormone produced by duodenal mucosa and released into the blood stream when fatty materials pass through the duodenum; inhibits secretion of gastric juice and stomach movements.

Enterogona. Order of tunicates in the Class Ascidiacea; body sometimes constricted; gonad single; Clavelina, Ciona, and Ascidia common.

enterokinase. Activator secreted by the wall of the small intestine in higher vertebrates; converts trypsinogen into trypsin.

enteromyiasis. Presence of fly larvae in the intestine.

enteron. 1. Digestive tract, especially in the coelenterates. 2. That portion of the digestive tract which is derived from endoderm.

Enteropneusta. Phylum which includes the acorn worms or tongue worms; soft, wormlike, marine animals found burrowing in sand or mud bottoms; 1 in. to 6 ft. long; segmentation obscure; body composed of long cylindrical to short conical proboscis, a circular collar, and a trunk; digestive tract extends from anterior margin of collar to posterior end of trunk; pharyngeal respiratory slits present; nervous and circulatory systems flimsy; larva a tornaria; an anterior outpocketing of the digestive tract has been likened to a notochord, and the nervous system has a small cavity in some spp.; certain zoologists are therefore inclined to include this group as a subphylum of the Phylum Chordata; about 80 spp.; Dolichoglossus and Balanoglossus common.

Enterozoa. Branch of the Metazoa which includes those phyla having a digestive cavity; includes all phyla except Protozoa and Porifera. Cf. PARAZOA.

Entoconcha. Genus of gastropods which are internal parasites of certain Holothuroidea.

entoconid. Inner posterior cusp of a lower mammalian molar tooth.

Entocythere. Genus of fresh-water ostracods commensal on N.A. crayfish gills.

entoderm. Endoderm.

Entodesma. Genus of marine bivalves.

Entodinium. Genus of complex oligotrich ciliates found in the stomach of cattle and sheep.

entoglossum. Anterior tongue-supporting element of the hyoid apparatus, especially in some reptiles.

Entomobrya laguna. Insect in the Order Collembola.

entomology. In its broadest aspects, the study of all phases of the biology of insects.

entomophagous. Insect-eating.

entomophily. Pollination of a flower by an insect.

entomostraca. General taxonomic category which includes such diverse small crustaceans as the cladocerans, copepods, phyllopods, ostracods, etc.

entoneural system. Part of the echinoderm nervous system; consists of a ring of nervous tissue around the mouth area, five pairs of radiating marginal nerves, and a plexus to the musculature and viscera.

entoplastron. Anterior median plate of a tortoise plastron.

Entoprocta. Phylum which includes minute, sessile, colonial or solitary moss-like organisms; each animal consists of a compact mass at the end of a slender stalk; the former contains the viscera and has a circlet of ciliated tentacles for feeding; digestive tract U-shaped; body cavity a pseudocoel; single pair of flame bulbs; reproduction hermaphroditic, sexual and dioecious, or by asexual budding and proliferation; trochophore type of larva; one genus found rarely in fresh waters, all other forms marine; about 60 spp.; Urnatella and Pedicellina are examples; this phylum is sometimes considered a highly modified class of the Phylum Bryozoa.

Entosiphon. Genus of colorless fresh-water biflagellate Protozoa; cell flattened, rigid; one flagellum trailing.

Entosphenus. Common genus of lampreys; E. tridentatus is the marine lamprey of the Pacific Coast, breeds in coastal streams; E. amotteni, in streams of the upper Mississippi drainage basin, is a harmless sp.

entosternite. Endosternite.

Entotrophi. Aptera.

entotympanic bone. Mammalian bone surrounding the middle ear cavity; sometimes fused with the tympanic to form a compound bulla.

Entovalva. Genus of small bivalves parasitic in the digestive tract of sea cucumbers.

entozoa. Internal parasites.

enucleate. 1. To remove a nucleus. 2. Lacking a nucleus.

environment. 1. Sum of all physical, chemical, and biological factors to which an organism is subjected. 2. One of the major habitat types, such as: marine, terrestrial, rain forest, desert, lake, etc. Cf. HABITAT.

environmental resistance. Restriction imposed upon the numerical increase of a sp. by the rigors of the environmental factors; a theoretical concept.

enzyme. Organic protein catalyst which enormously speeds up the rate of a metabolic chemical reaction; elaborated only by living cells and in minute quantities; there are a great many different kinds of enzymes, and each one activates only a single chemical reaction or a very limited group of reactions; an enzyme produces its effects on its substrate by temporarily combining with the substrate and activating it, so that the substrate undergoes a chemical change, at which time it loses its combination with the enzyme; most enzymes exert their effects within cells, but a few, such as certain digestive enzymes, are normally secreted to the outside of cells. See DIGESTIVE ENZYMES and RESPIRATORY ENZYMES.

Eoacanthocephala. Class of Acanthocephala; main lacunae median; proboscis hooks in alternating rows; protonephridia absent; males with one syncitial cement gland; Neoechinorhynchus typical.

Eoanthropus dawsoni. Formerly called the "Piltdown man," a manlike fossil sp. supposedly found near Sussex, England; fragments of cranium, mandibles, and teeth now known to be forgeries attained by grinding and chemically treating ape material.

Eocene epoch. One of the five geological subdivisions of the Tertiary period, between the Oligocene and Paleocene epochs; lasted about 20 million years; characterized by the disappearance of archaic mammals and the appearance of many modern forms.

Eocyzicus. Uncommon genus of clam shrimps.

Eogaea. Seldom-used zoogeographic term; includes Africa, South America, and Australia.

Eogyrinus. Genus of extinct stegocephalian Amphibia.

Eohippus. Genus of fossil ancestral horses; small forest-dwellers of Eocene times; about 11 in. high at the shoulder; the front feet had four functional toes and the hind feet three; the first of a long series of ancestral horselike creatures.

Eolis. Old generic name for many spp. of nudibranchs.

eosin. Red biological dye.

eosinophil. 1. Type of non-phagocytic leucocyte in vertebrate blood; characterized by acidophilic cytoplasmic granules; the nucleus may be a single mass or sometimes divided into two; about 2 to 5% of human leucocytes are eosinophils. 2. Any histological element readily stained by an eosin dye.

Eosuchia. Order of lizard-like reptiles of Upper Permian to Jurassic times; Proterosuchus typical.

Epactophanes. Genus of fresh-water copepods; found especially in damp moss and sandy beaches.

epaulettes. 1. Branched or knobbed processes on the oral arms of Scyphozoa. 2. In Diptera (sing.), the first haired scale at the base of the costa.

Epalxis. Genus of rounded triangular ctenostome Ciliata; posterior end toothed; sapropelic in fresh and salt waters.

epaxial musculature. Dorsal trunk musculature of most vertebrates.

Epeira. Old generic name for certain orb-weaver spiders; now usually placed in Araneus.

ependymal cells. Special epithelial cells lining the cavities of the brain and spinal cord; often ciliated.

ephedrine. Plant alkaloid now made synthetically; action similar to that of epinephrine; contracts smooth muscle fibers, raises blood pressure, constricts capillaries, and is effective for asthma.

Ephelota. Genus of marine suctorian Protozoa; stalk stout; suctorial and prehensile tentacles numerous and scattered.

Ephemera. Common genus of mayflies.

ephemeral. Pertaining to an animal with a very short life, especially certain insects.

Ephemerella. Common genus of mayflies.

Ephemerida. Ephemeroptera.

Ephemeroptera. Order of insects which includes the mayflies; delicate, soft-bodied insects with vestigial mouth parts in the adult stage; wings membranous and held vertically when at rest; forewings much larger than hind wings; tip of abdomen with two or three long cerci; life history with a series of gilled fresh-water naiad stages; last naiad molts to form an aerial subimago which molts in a few hours to form the sexually mature adult; this stage lives only a few hours or days; it copulates, oviposits, but takes no food; common genera are Ephemera, Leptophlebia, and Callibaetis.

Ephesia. Genus of Polychaeta; head conical.

Ephestia kühniella. Mediterranean flour moth; a dark gray cosmopolitan sp.; larva often a pest in cereals and cereal products.

Ephialtes. Genus of ichneumon flies.

Ephippiorhynchus senegalensis. Saddle-billed stork; one of the jabiru storks; black, saddle-shaped mark on the red bill; body black and white; western tropical Africa.

ephippium. 1. Special postero-dorsal part of the carapace of certain female Cladocera; thick-walled and contains one to several eggs (usually fertilized); the whole device is separated from the body and then it may sink or float, and the contained eggs are highly resistant to adverse environmental conditions; hatching usually occurs during the following spring; a female cladoceran usually produces only a single ephippium but occasionally an additional one or two are formed. 2. Pituitary fossa. Pl. ephippia.

Ephoron. Common genus of Ephemeroptera; nymphs fossorial.

Ephydatia. Old generic name for certain fresh-water sponges; usually incorporated into Heteromeyenia.

Ephydra. See EPHYDRIDAE.

Ephydridae. Holarctic family of Diptera; shore flies; small flies, usually in moist places on the shores of bodies of water; larvae of some spp. are leaf miners; others, especially Ephydra, occur in salt and alkali ponds.

ephyra. Ephyrula; immature, free-swimming scyphozoan jellyfish having arisen from the asexual transverse division of a strobila; with eight more or less distinct radially arranged arms, each with a terminal notch in which a tentaculocyst develops. Pl. ephyrae.

Epiactis. Common genus of sea anemones (Order Actiniaria).

Epiaeschna heros. Eastern U.S. dragonfly.

Epibdella. Genus of broad, flat, external, monogenetic trematode parasites of marine fishes.

epibenthos. Collectively, the organisms living on the sea bottom between low tide line and a depth of 100 fathoms.

epibiont. 1. Relict; an endemic sp. confined to a small area and which is a survivor from a preceding epoch. 2. Animal which lives on the surface of another animal.

epibiotic. Pertaining to a relict endemic sp.

epiblast. 1. Ectoderm, especially in very early embryos. 2. Upper layer of blastodisc.

epiboly. Overgrowth and lengthening process occurring during gastrulation in many animals; involves growth of ectoderm cells downward to cover the cells of the vegetal area.

epibranchial chamber. Suprabranchial chamber.

epibranchial placode. In lower vertebrate embryology, an extodermal thickening at the dorsal end of a branchial cleft.

epibranchials. Bones directly above the angles of the branchial arches in fishes.

epicanthic fold. Inner fold on the upper eyelid of mongoloids; produces the illusion of "slant eyes."

epicardium. Innermost cell layer of the pericardium.

Epicaridea. Suborder of Isopoda; highly modified, bizarre spp., parasitic on decapod crustaceans.

Epicauta. Important pest genus in the beetle Family Meloidae.

Epiceratodus. Ceratodus; see CERATODONTIDAE.

Epiclintes. Genus of marine and freshwater spirotrich ciliate Protozoa; elongated spoon-shaped.

epicoel. 1. Perivisceral cavity formed by the invagination of the main body wall; e.g. atrial cavity of tunicates. 2. Cavity of the midbrain in lower vertebrates.

epicondylocubitalis. Dorsal muscle of the frog forelimb.

epicone. That part of a dinoflagellate which is anterior to the equatorial groove. Cf. HYPOCONE.

epicoracoid. Narrow, longitudinal, median bone of the pectoral girdle of the frog.

Epicordulia. Genus of dragonflies; eastern U.S.

epicranium. Fused upper part of the head capsule in a typical insect, from the frons to the neck.

epicranius muscle. Broad, flat muscle on the top of the scalp.

Epicrates. Genus of tropical American boas.

epicuticle. Outermost hardened layer of an arthropod cuticle.

epicyst. Outer layer of a cyst wall, especially in protozoan cysts.

epidemic. Great but temporary increase in the incidence of an organism, especially a sp. causing a parasitic infection.

epidermal nervous system. Portion of the echinoderm nervous system; consists of an oral ring and five radial epidermal nerves.

epidermis. Outermost layer(s) of cells of the integument; it may be a single layer of epithelial cells or it may be composed of two to many layers of stratified epithelial cells, the outermost of which may be dead and horny, especially in terrestrial vertebrates.

epididymis. 1. Coiled duct which receives sperm from the seminiferous tubules and transmits them to the vas deferens in most male amniotes. 2. A similar duct in certain complex invertebrates.

Epidinium. Genus of complex ciliates occurring in the rumen of cattle, sheep, camels, etc.; E. ecaudatum common.

epifauna. That fauna living on the surface of the bottom deposits in the sea. Cf. INFAUNA.

epigamic character. Reproductive feature of an animal, excluding the organs and tissues of the reproductive system; e.g. mating calls.

epigastric. Pertaining to the upper median part of the abdomen, or to the top (anterior) part of the stomach.

epigastrium. 1. In human anatomy, the epigastric region. 2. In insects, the ventral surface of the meso- and metathorax. 3. In insects, the first entire ventral abdominal sclerite. 4. In spiders, the anterior ventral abdominal surface.

epigean. Pertaining to an animal which lives on the exposed surface of land or in shallow water. Cf. HYPOGEAN.

epigenesis theory. Theory of the middle eighteenth century in opposition to the preformation theory; it emphasized the lack of internal organization of the zygote.

epiglottis. Small flap of tissue at base of tongue in mammals; closes glottis during swallowing.

epignathous. Having the upper mandible long and curved over the tip of the lower mandible, e.g. many birds.

Epigonactis. Genus of sea anemones in the Order Actiniaria; developing young are commonly imbedded in the upper part of the column.

epigynum. Midventral flap covering the genital pore in female spiders.

epihyal. One of the hyoid bones in fishes.

epilabrum. Lateral sclerite on each side of the labrum of certain myriapods.

Epilachna. Very large cosmopolitan genus of coccinellid beetles; feed chiefly on vegetation; a few spp. are pests; E. corrupta (bean ladybird) eats beans and other crops; E. varivestris is the Mexican bean beetle.

epilepsy. Periodic convulsions and loss of consciousness; caused by genetic factors, brain injury, or brain infection.

epilimnion. Uppermost warm layer of a thermally stratified lake during the warm months; wind-circulated and essentially homothermous; usually 1 to 15 m. thick, depending on lake size; occasionally the term is applied to marine waters. Cf. THERMOCLINE and HYPOLIMNION.

epiloia. Tuberous sclerosis; a human dominant mutant marked by brain tumors, visceral tumors, and epilepsy.

Epimartyria. N.A. genus of small moths in the lepidopteran Family Micropterygidae.

epimera. Coxal plate; in many Acarina, a ventral plate from which a leg originates.

epimere. 1. In chordate early development, the dorsal part of the mesodermal mass lying on each side of the neural tube. 2. Dorsal muscle plate of the mesothelium.

epimerite. Anterior anchoring deciduous process of the protomerite in certain gregarine Protozoa.

epimeron. Posterior lateral portion of a pleuron in many Arthropoda.

epimyocardium. Thick layer of tissue forming the bulk of an embryonic vertebrate heart.

epimysium. Fascia.

Epinebalia. Genus of small marine crustaceans in the Order Nebaliacea; usually in algae.

Epinephelus adscensionis. Rock hind; an important polka-dotted food fish from Bermuda and Fla. to Brazil; to 16 in. long; the genus Epinephelus is large and contains many tropical groupers.

epinephrine. Hormone produced by the medulla of the adrenal gland; an "emergency" hormone which raises blood pressure, increases heart rate, liberates liver glycogen, increases muscular power and resistance to fatigue; $C_9H_{13}O_3N$.

epinephros. Adrenal gland.

epineural. Pertaining to the neural arch of a vertebra.

epineural canal. In many echinoderms, a small canal or sinus between each radial nerve and the external dermis.

epineurium. Loose fibrous connective tissue sheath surrounding a peripheral nerve.

epiorganism. Superorganism.

epiopticon. 1. Second ganglion of the optic tract of insects. 2. External mass of the optic segment of the insect brain.

epipelagic. Pertaining to the community of suspended organisms inhabiting an aquatic environment between the surface and a depth of 200 m.

Epiphanes. Small genus of rotifers; E. senta is common in stock ponds that are rich in organic matter.

epipharynx. Organ attached to the inner surface of the labrum in many insects; presumably functions in taste.

epiphragm. Calcareous or parchment-like diaphragm which closes the aperture of the shell of a land snail after the soft parts have been drawn inside; usually formed during dry and cold seasons when the snails are inactive.

epiphragma. Epiphragm.

epiphysis. 1. End part or process of a long bone which undergoes ossification independently of the long diaphysis; after ossification of these two areas is well developed, the separating cartilaginous area also ossifies; 2. One of five small peripheral bars in an Aristotle's lantern. 3. The pineal body.

epiphyton. Periphyton assemblage in which the organisms are scattered on the substrate and not closely and physically associated. Cf. PERIPHYTON, LASION.

epiplastron. One of two anterolateral plates of a tortoise plastron.

epipleura. 1. Turned-down outer margin of the elytra in certain beetles. 2. An uncinate process in birds. 3. In certain fishes, a spine or process attached to a rib and extending toward the skin of the lateral line area.

epipleural bone. One of many small intramuscular bones extending longitudinally between some of the ribs in certain fishes.

epiploic foramen. Foramen of Winslow; opening of the omental bursa.

epipodite. 1. Respiratory process attached near the base of a trunk appendage in trilobites and many crustaceans; usually long and flat but may be branched or folded. 2. Gill separator; narrow process attached at or near the base of certain thoracic appendages in decapod Crustacea; functions in keeping gills separated so that they will not mat together and reduce respiratory efficiency.

epipodium. 1. In certain gastropods, a ridge, lobe, or fold along the lower edge of the foot on each side. 2. Raised ring of an ambulacral plate in Echinoidea.

Epipolasida. Order of sponges in the Subclass Monaxonida; spherical spp. with radiating monaxon spicules.

epipterygoid. Paired bone of the palate in primitive tetrapods.

Epischura. Genus of fresh-water copepods; in plankton of large, cold lakes.

episematic. Pertaining to recognition markings or coloration.

epistasis. Masking effects of one or more genes on the expression of differences usually determined by one or more other gene pairs; e.g. white eye in Drosophila is epistatic over about 40 other eye color loci. Cf. HYPOSTASIS.

episternum. One of the elements of the tetrapod sternum.

epistome. 1. Small flap which covers the mouth of certain Bryozoa. 2. In crustaceans, the area or plate between the mouth and the base of the second antennae. 3. In many insects, the area or plate between labrum and epicranium; clypeus or an intermediate piece. 4. In certain Diptera, the area between the frons and labrum; a portion of the rostrum. 5. In Phoronidea, a flaplike projection which separated mouth and anus.

Epistylis. Large genus of ciliate Protozoa; cell inverted bell-shaped and stalked; colonies on inanimate substrates or on fresh-water and marine animals.

epithalamus. Roof of the diencephalon in the vertebrate brain.

epitheca. 1. External calcareous layer in the basal part of many corals. 2. Covering of the epicone in dinoflagellates.

epithelial tissue. Epithelium.

epitheliomuscular cells. Musculo-epithelium.

epithelium. General type of tissue which covers a body or structure or lines a cavity; the outer surface of the tissue is therefore exposed and the inner surface rests on connective or some other type of tissue. Pl. epithelia.

epitoke. Posterior part of a marine polychaete during the breeding season; swollen with developing gonads and ova or sperm. Cf. ATOKE.

Epitonium. See WENTLETRAP.

epitrochleoanconeus. Muscle which originates on the back of the internal condyle of the humerus and inserts on the inner side of the olecranon; in mammals and reptiles.

epitrochleocubitalis. Ventral arm muscle of the frog.

Epizoanthus. See ZOANTHIDEA.

epizoic. Living on or attached to the surface of another animal but not parasitic upon it.

epizoon. External animal parasite or commensal.

epizootic. 1. Rapidly spreading disease or animals. 2. Epidemic.

epoch. One of the time subdivisions of the various geological periods.

Epomophorus. Genus of African fruit bats; with a tufted sac on each side of the neck.

epoophoron. Parovarium; remnant of a Wolffian body; represented by a portion of the ligament between the ovary and the oviduct, especially in mammals.

Eptesicus. Common genus of bats in N. A., including especially the brown bats; to 4.5 in. long; upper parts brown, under parts lighter.

equatorial plate. Arrangement of metaphase chromosomes in a general plane midway between the two poles during mitosis.

Equidae. Family of mammals in the Order Perissodactyla; one functional digit and hoof on each leg; usually plains and desert spp. which feed chiefly on grasses; Equus the only modern genus; E. caballus includes all varieties of horses from Shetland ponies to the largest percherons; E. asinus is the common African wild ass which has several domesticated forms such as the donkey and burro; the mule is a sterile F_1 hybrid of a male ass and a female horse; E. zebra is the common African zebra; E. kiang is the kiang, a wild ass of Tibet and Mongolia; the onager of India is a very close relative.

equine. Pertaining to a horse or other member of the Family Equidae.

Equus. Genus to which the modern horse belongs; about a dozen fossil Pleistocene spp. are also known; see EQUIDAE.

era. One of the five major divisions of geological time (Cenozoic, Mesozoic, Paleozoic, Proterozoic, Archeozoic).

Erasistratus. Greek physician (third century B.C.); studied and taught human anatomy (by dissection) at Alexandria medical schools; noted for his work on circulation, motor and sensory nerves, and brain convolutions.

Erato. Genus of small, glossy, marine snails; West Coast.

Erax. See ASILIDAE.

erd shrew. European shrew (Sorex).

Erebus agrippina. Sp. of giant tropical moth with a wingspread attaining almost 12 in.

erectile tissue. Tissue containing venous spaces into which small arteries empty directly; such tissue becomes distended when the blood supply is increased, e.g. the penis and clitoris.

erection. 1. Distension and stiffening of the male copulatory organ, especially in mammals; by this means sperm are more readily deposited a favorable distance up the female genital tract during copulation. 2. The distension of any erectile tissue.

erector pili. Small muscles that elevate hairs of mammals.

Eremichthys acros. Desert dace, a small fish known only from desert springs and streams of Humboldt Co., Nev.

Eremobates. Common genus of solpugids in southwestern states.

Eremophila. Otocoris.

erepsin. Mixture of digestive enzymes produced by the wall of the small intestine of vertebrates; chief role is the conversion of peptides to amino acids.

Erethizon. See ERETHIZONTIDAE.

Erethizontidae. Family of mammals which includes the American porcupines; large, clumsy rodents covered with shaggy hair and quills; excellent climbers which feed on buds, bark, and foliage of many trees and shrubs; Erethizon dorsatum is the Canadian porcupine, and E. epixanthum is the Western, or yellow-haired, porcupine.

Eretmochelys. Genus of hawkbill turtles in the Family Chelonidae; upper jaw hooked; to 2 ft. long; overlapping scutes provide commercial tortoise shell; warm seas.

Eretmoptera. Genus of midges in the Family Tendipedidae.

Ereunetes. Small genus of sandpipers which breed in Arctic America; some individuals winter along U.S. coastal areas.

Ergasilus. Genus of marine and freshwater copepods; mature females parasitic on fishes.

ergate. Worker ant.

Ergates. Genus of beetles in the Family Cerambycidae; pests on evergreen trees.

ergosterol. Sterol occurring in certain plant and animal tissues; converted to an antiricketic substance when exposed to ultraviolet light.

ergot. 1. Fungus infection of rye. 2. Preparation from such a fungus for medicinal use; contracts smooth muscle fibers, hastens labor, and checks internal hemorrhaging.

Erichsonella. Genus of marine Isopoda.

Erichthonius. Genus of marine Amphipoda.

erichthus. Free-swimming larval stage of certain stomatopod Crustacea.

Eriociolacerta. Genus of advanced fossil therapsid reptiles.

Ericymba buccata. Silverjaw minnow; Mo. to Mich. to Penn. and south to Fla.

Erignathus. Genus of bearded seals in the Family Phocidae; large, plain-colored seals having a tuft of long bristles on each side of the muzzle; polar seas.

Erimyzon. Chubsuckers; a genus of small suckers found in lakes, ponds, and sluggish streams of the eastern half of the U.S.

Erinaceidae. Small family of mammals in the Order Insectivora; true hedgehogs and gymnures; nocturnal, Old World spp. with dorsal spines in addition to hair; Erinaceus typical; also includes Echinosorex and Gymnura.

Erinaceus. See ERINACEIDAE.

Eriocampoides. Caliroa.

Eriocera. Genus of crane flies in the Family Tipulidae.

Eriococcus. See KERMIDAE.

Eriophyes. See ERIOPHYIDAE.

Eriophyidae. Family of Acarina which includes the gall mites; feed on plant juices and stimulate production of galls; only two pairs of legs; Eriophyes is the blister mite which damages trees and shrubs; Phyllocoptes pyri damages pear trees.

Erioptera. Genus of crane flies in the Family Tipulidae.

Eriosoma. See ERIOSOMATIDAE.

Eriosomatidae. Family of aphid-like insects in the Order Homoptera; wax-secreting spp. which produce galls and other plant malformations; the woolly apple aphid, Eriosoma lanigerum, is an important cosmopolitan sp.

Eriphia. Common genus of marine mud crabs.

Erismatura. Oxyura.

Eristalis. Tubifera.

Erithacus. See ROBIN.

ermine. See MUSTELA.

erne. European gray sea eagle, Haliaeetus albicilla.

Erogala. Common genus of minnows found in clear streams of eastern U.S.

Erolia. One of several genera of sandpipers; E. minutilla is the least sandpiper which breeds in Alaska and Canada.

Erophylla. Genus of West Indian leaf-nosed bats.

Erotylidae. Widely distributed family of beetles; pleasing beetles; feed chiefly on plant stems and fungi; Languria mozardi is a pest on lettuce, alfalfa, and clover in the western U. S.

Erpetogomphus. Genus of dragonflies; southern and western U. S.

Erpetoichthys calabraicus. Small, elongated, West African crossopterygian fish; pelvic fins absent.

Erpobdella. Genus of fresh-water leeches; common and widely distributed; feed on aquatic invertebrates, fish, and frogs, and occasionally attack man; sometimes scavengers.

Errantia. One of the two orders in the annelid Class Polychaeta; all segments similar except for head and anal regions; parapodia all similar; pharynx usually protrusible; free-swimming or in tubes; Nereis and Aphrodite common.

Ersaea. Genus of coelenterates in the Order Siphonophora.

eruciform larva. 1. Type of caddis larva with cylindrical abdomen, lateral line, and portable case. 2. Any caterpillar-like larva.

eruptive population. Population of animals which is normally cyclic or relatively stable in density but which suddenly becomes extremely dense; e. g. grasshopper plagues and mouse plagues.

Eryops. Genus of extinct crocodile-like stegocephalian Amphibia.

Erythemis. Common genus of dragonflies.

erythroblast. Nucleated cell found in bone marrow; develops into an erythrocyte.

erythroblastosis. See RH FACTOR.

Erythrocebus. Genus of military monkeys; reddish-brown coloration; predominantly terrestrial, in savannas and scrubland of east and west Africa; omnivorous.

erythrocruorin. Any of several invertebrate hemoglobins; in some flat-worms, nematodes, Nemertea, Echiuroidea, Phoronidea, and entomostraca; also in a few insects, echinoderms, and mollusks, and in most annelids; some zoologists discourage the use of this term.

erythrocyte. Flat cell found in vertebrate blood, containing hemoglobin; circular in cyclostomes and most mammals, but oval in other vertebrates; formed in bone marrow and carry most of the oxygen transported by blood; in an adult human male at sea level there are about 5, 000, 000 erythrocytes (each 8 microns in diameter) per cubic mm. of blood; the normal life of an erythrocyte is four months in man.

erythrodextrin. One of several intermediate polysaccharide digestion products between starch and maltose; follows soluble starch and precedes achroödextrin.

Erythrodiplax. Genus of dragonflies.

Erythroneura. See GRAPE LEAF HOPPER.

erythrophore. Red pigment cell in the dermis of many vertebrates.

erythroplastid. Non-nucleated red corpuscle of mammals.

erythropoiesis. Production of erythrocytes by bone marrow.

Erythrops. Genus of marine crustaceans in the Order Mysidacea.

Eryx. Small genus of North African boa constrictors.

escargot. Snails cooked for table use.

Escherichia coli. Rod-shaped, gram-negative, motile bacterium found abundantly in the vertebrate intestine.

Eschrichtidae. Family which includes the gray whales; North Pacific spp. to 40 ft. long; Eschrichtius typical.

Eschrichtius. See ESCHRICHTIDAE.

escolar. See RUVETTUS.

escutcheon. 1. Scutellum of beetles and hemipterans. 2. Depression behind or below the beak in certain bivalve mollusks. 3. Area of the skin extending upward and outward from the posterior part of the udder in certain mammals.

Eskimo. Race of mongoloids having yellow-brown skin, long head, and round face; Arctic America and northeastern Asia.

Esocidae. Family of Holarctic bony fishes which includes the pikes and muskellunge; Esox the only genus, all spp. being important game fish; front of head duck-bill shaped when viewed from above; body slender, mouth large, teeth conspicuous; E. vermiculatus (mud pickerel, grass pickerel) a small sp. in quiet waters of eastern N. A.; E. niger (chain pickerel) has similar distribution and habits; E. lucius (northern pike) is a large sp. of northern parts of Northern Hemisphere; E. masquinongy (muskellunge) is the largest American fresh-water game fish (to 100 lbs.), found in Great Lakes drainages, upper Mississippi drainage, southeastern Canada.

Esocoidei. Suborder designation sometimes used to include the fish families Umbridae and Esocidae; equivalent to the Order Haplomi.

esophagus. 1. That tubular portion of the digestive tract immediately posterior to the pharynx; in many invertebrates as well as vertebrates. 2. In a few invertebrates, a tubular part of the digestive tract immediately following the oral cavity; not especially muscular.

Esox. See ESOCIDAE.

Esperella. Genus of marine sponges in the Subclass Monaxonida; slender, flexible, upright, almost cylindrical; other spp. encrusting on Bryozoa.

Esperiopsis. Genus of leaflike or branching marine sponges in the Subclass Monaxonida.

espundia. See AMERICAN LEISHMANIASIS.

essential amino acids. Amino acids necessary in the diet of a particular animal; cannot be synthesized from ingested food.

Estigmene acraea. Common N.A. moth, sometimes called the acrea or saltmarsh moth; abdomen tawny; head, thorax, and wings whitish with dark spots; male with tawny hind wings.

estivation. Aestivation.

estradiol. Female sex hormone produced by the Graafian follicles of the ovary; chiefly responsible for the phenomenon of estrus, or heat, in female mammals.

estrin. Any of several ovarian estrogens.

estrogens. Hormones secreted by the Graafian follicle and promoting estrus; the term is often used to designate one specific compound having the formula $C_{18}H_{22}O_2$ (also known as estrone, estrin, folliculin, and theelin).

estrone. See ESTROGENS.

estrous. Estrus.

estrus. Period of sexual excitement in the female of some mammals during which she will accept mating with the male; a specific period of estrus is not evident in human females.

estrus cycle. Complete reproductive cycle occurring in sexually mature females of certain mammals in the absence of pregnancy; depending on the sp., the cycle may last from five to 60 days; includes a period of heat (estrus), ovulation, growth of the uterine wall, growth and regression of the corpus luteum, menstruation, and subsequent maturation of Graafian follicle(s); the cycle is under complex hormonal control; if pregnancy occurs, the cycle is suspended and there are one or more persistent corpora lutea.

estuary. Area where river water meets and dilutes salt water of the sea. Adj. estuarine; pl. estuaries.

Eteone. Genus of small Polychaeta.

Etheostoma. Large genus of small, brightly colored stream fishes in the Family Percidae; darters; common and widely distributed in N.A.

Ethiopian. Race of caucasoids of northeastern Africa; skin brown, head long, hair dark; same as Hamite.

Ethiopian realm. One of the six major zoogeographic realms of the world; consists of Africa and Arabia south of the Tropic of Cancer, and Madagascar; some characteristic animals are the gorilla, chimpanzee, African elephant, hippopotamus, giraffe, a lungfish, ostrich, and many horned antelopes.

ethmoidal sinuses. Air spaces or cells within the ethmoid bone.

ethmoid bone. Delicate bone composed of thin sievelike plates forming the roof of the nasal chambers and part of orbit walls; highly variable from one vertebrate group to another.

Ethmolaimus. Common genus of microscopic, free-living nematodes.

ethology. Behavior of an organism with respect to its inanimate and living environment.

etiology. Collective knowledge regarding the causes of a disease.

Euapta. Genus of long wormlike sea cucumbers.

Euarctos. Genus of bears in the Family Ursidae; includes the omnivorous American black bears, the most common being E. americanus; found over most of wooded America; to 500 lbs.; cinnamon or brown bears are merely color varieties of the usual black phase.

Eubalaena glacialis. North Atlantic right whale; large, black, whalebone whale to 55 ft. long.

Eubranchiopoda. One of the two divisions in the crustacean Subclass Branchiopoda; includes the fairy shrimps, tadpole shrimps, and clam shrimps; body distinctly segmented, with many thoracic appendages; about 250 spp.; exclusively in inland waters, especially temporary ponds and playas.

Eubranchipus vernalis. Sp. of fairy shrimp in the Order Anostraca; especially common in temporary ponds in the northeastern states.

Eubranchus. Genus of very small nudibranchs.

Eubrianax. See PSEPHENIDAE.

Eucalia. See GASTEROSTEIDAE.

Eucapnopsis. Genus of Plecoptera.

Eucarida. One of the four subdivisions in the crustacean Division Eumalacostraca; carapace large, covering all of thorax; eyes stalked; gills thoracic; includes the orders Euphausiacea and Decapoda.

Eucestoda. Cestoda.

Eucheilota. Genus of marine coelenterates in the Suborder Leptomedusae; hydroid unknown.

Euchlanis. Common genus of fresh-water loricate rotifers.

Euchone. Genus of tube-building Polychaeta.

euchromatin. That portion of a chromosome containing the bulk of the genic material and staining less deeply than the heterochromatin.

Eucidaris. Cidaris; see CIDAROIDA.

Euciliata. Subclass of the Class Ciliata; includes all ciliates having macro- and micronuclei; fresh-water, marine, and parasitic.

Eucoelomata. Group of phyla in which a true coelom is present, e.g. Chaetognatha through Chordata.

Euconulus. Genus of land snails; common over much of U.S.

Eucopella. Genus of marine hydrozoan coelenterates.

Eucopepoda. One of the two orders of copepod crustaceans; with typical copepod characters; eyes simple; mostly free living. Cf. BRANCHIURA.

Eucorethra underwoodi. Phantom midge in the Family Culicidae; northern states and Canada.

Eucratea. Genus of marine Bryozoa; creeping stolon; upright ranks of zooecia.

Eucyclogobius newberryi. A sp. of goby in the Family Gobiidae; creeks and brackish waters of coastal southern Calif.

Eucyclops. Common genus of fresh-water copepods; limnetic and bottom.

Eucypris. Common genus of fresh-water Ostracoda.

Eudendrium. Genus of sessile, colonial, marine hydrozoan coelenterates.

Euderma maculata. Spotted bat; ears very long; white spot at base of tail and on each shoulder; underparts white; Mont. to Calif. and N.M.

Eudistoma. Genus of ascidians.

Eudistylia. See PEACOCK WORM.

Eudocimus albus. White ibis; large bird with most of bill, face and legs red; four outer primaries tipped with black; S.C. south to Peru.

Eudorallopsis. Genus of crustaceans in the Order Cumacea.

Eudorella. Genus of Cumacea found in northern seas.

Eudorina. Fresh-water genus of colonial, free-swimming green flagellates; 16 or 32 cells imbedded in a gelatinous matrix.

Eugenes. Hummingbird genus; one sp. in mountains from Ariz. to Panama.

eugenics. Betterment of the human race by means of the application of the principles of genetics.

Euglandina. Genus of tall, conical land snails; shell to 40 mm. long; tropical and subtropical; several spp. in southern U.S.

Euglena. Large genus of protozoans having a single flagellum and green coloration due to contained chloroplasts; cell more or less spindle-shaped; eyespot reddish; usually 20 to 500 microns long; in fresh waters everywhere, especially stagnant waters; Euglena viridis and E. gracilis common.

Euglenoidina. Order of flagellate Protozoa; cell elongated, with one to three flagella, and green chromatophores, or none; holophytic, holozoic, or saprophytic; mostly fresh-water; Euglena and Astasia typical.

euglenoid movement. Squirming, peristalsis-like movements exhibited by certain green flagellates, especially under abnormal environmental conditions.

Euglypha. Large genus of fresh-water amoeboid Protozoa; most of cell enclosed in a vase-shaped siliceous test.

Euherdmania. Genus of long, wormlike ascidians.

Eukrohnia. Common genus of Chaetognatha.

eulachon. See CANDLEFISH.

Eulalia. Genus of marine Polychaeta; coloration usually greenish; pharynx very large.

Eulamellibranchia. Order of pelecypod mollusks; gills of each side W-shaped in cross section; two well developed adductor muscles; fresh-water clams, cockles, Venus, razor clams, shipworms, and seed clams are examples.

Eulimnadia. Common genus of clam shrimps.

eulittoral zone. 1. In lake biology, that portion of the bottom which begins at the high water mark and which is subjected to wave action. 2. Also in lake biology, that portion of the bottom between the high water mark and the lakeward limit of rooted aquatics. 3. In oceanography, that portion of the bottom extending from the high tide line to a depth of 50 m.

Eumalacostraca. One of the two divisions in the crustacean Subclass Malacostraca; abdomen of six or less segments; includes the great majority of Malacostraca.

Eumastia. Genus of marine sponges in the Subclass Monaxonida; mass of finger-like projections; cold oceans.

Eumeces. Common genus of skinks; about 20 spp. in U. S.; E. fasciatus, the blue-tailed skink, found from Mass., southern Canada, and Mississippi Valley south to the Gulf; E. skiltonianus, an olive green sp. found on West Coast and in Nev. and Utah.

Eumenes. Genus of wasps in the Family Eumenidae; potter wasps; vaselike nests of mud.

Eumenidae. Family of dark-colored wasps with light-colored markings; potter wasps, mason wasps, solitary wasps; nests of mud, or in plant stems or crevices; nest stocked with paralyzed insect prey to feed larvae; Eumenes common.

Eumeros. Genus of flies in the Family Syrphidae; larvae feed on bulbs of narcissus, onion, etc.

Eumetazoa. Collectively, all animal phyla except the Protozoa and Porifera; same as Enterozoa.

Eumetopias jubata. Northern sea lion or Steller's sea lion; West Coast from Bering Straits to Calif.; much larger than the California sea lion, males to 1800 lbs.

Eumida. Genus of Polychaeta.

Eumops. Tropical American genus of mastiff bats; three spp. reach extreme southern U. S.

eumycin. Antibiotic isolated from Bacillus subtilis cultures.

Eunectes murinus. Large, nonvenomous constrictor snake in the Family Boidae; South American anaconda or water boa; partly aquatic and partly arboreal; to 30 ft. long; tropical South America.

Eunereis. Genus of Polychaeta.

Eunice. Genus of marine Polychaeta; E. viridis is the Pacific palolo worm of Samoa and Fiji; for two or three days, beginning on the first day of the last quarter of the October-November moon, the worms swarm and cast off their posterior segments which are crowded with gametes; E. gigantea is a Pacific Coast sp. attaining 10 ft.

Eunicella. Genus of sea fans (hydrozoan coelenterates).

Eunoë. Genus of Polychaeta.

Eunotosaurus. Genus of Permian turtles.

eunuch. Castrated human male.

Euoticus. Genus of African bush-babies in the Family Galagidae.

euparal. Mounting medium used in making microscope slides; valuable for demonstrating delicate cytoplasmic details.

Euparkeria. Genus of primitive reptiles; thought to be a type from which birds, dinosaurs, and crocodilians were derived.

Eupentacta. Genus of bristly, whitish sea cucumbers.

Eupera. Genus of very small freshwater bivalves; Fla. to Tex.

Euphagus. See BLACKBIRD.

Euphausia. Common genus in the crustacean Order Euphausiacea; krill.

Euphausiacea. One of the two orders in the malacostracan Subdivision Eucarida; marine, pelagic, and shrimplike; thoracic appendages all biramous; usually bright red and bioluminescent; to 1 in. long; important food of certain whales; Euphausia most common.

euphotic zone. That uppermost portion of a body of water in which there is sufficient light for photosynthesis; in the sea it may be as much as 300 ft. thick, but in certain lakes it may be only 10 ft. or less thick.

Euphractus. Genus of South American weasel-headed armadillos; about 16 in. long, with very little hair.

Euphrosine. Genus of bright red polychaete worms.

Euphydryas chalcedona. Common checkerspot butterfly of western N. A.; blackish with many yellow and orange-brown spots; larvae sometimes pests on cultivated plants.

Euplanaria. Dugesia.

Euplectella. Genus of hexactinellid sponges which includes Venus' flower basket.

Eupleura. Widely distributed genus of small, rough, fusiform marine gastropods; drills.

Euplexaura. Genus of fan-shaped, colonial anthozoan coelenterates; deep water off Pacific Coast.

euploidy. Condition of polyploidy where all haploid sets of chromosomes are completely represented. Cf. ANEUPLOIDY.

Euplokamis. Genus of oval comb jellies.

Euplotes. Genus of complex ciliate Protozoa; ovoid, flattened body with numerous ventral cirri; fresh-water and marine; E. patella common.

eupnea. Normal and quiet breathing.

Eupoda montana. Mountain plover of the western U.S.

Eupolymnia. Genus of marine Polychaeta.

Eupomatus. Genus of marine Polychaeta.

Euproctis chrysorrhoea. Pest sp. of moth in northeastern states; browntail moth; caterpillar feeds on leaves of cultivated and forest trees.

Euprocyon. Genus of South American crab-eating raccoons; reddish, semiaquatic, long-legged, and crepuscular.

euroky. Euryoky; relatively broad range of ecological conditions tolerated by a particular sp.; theoretical expression of the well developed ability of an animal to maintain itself in the face of environmental variables; e.g. man, carp, mountain lion, and crow. Adj. eurokous or euryokous. Cf. STENOKY.

Eurosta. Genus of goldenrod gall flies.

Euryalae. Order of the echinoderm Class Ophiuroidea; basket stars; arms greatly branched and highly flexible; both disc and arms invested with a soft skin or abundant granulations; Gorgonocephalus is a common carnivorous North Atlantic genus which reaches an over-all diameter of 30 in.; often in tangled masses.

Euryalona occidentalis. Cladoceran found especially in aquatic vegetation in southern states.

eurybath. An animal occurring over a wide depth range in the oceans, e.g. many brachiopods, mollusks, and annelids. Cf. STENOBATH.

eurybenthic. Eurybathic; pertaining to bottom animals which occur over a wide depth range in the sea.

Eurycaelon. Genus of river snails; Tenn. and Ala.

Eurycea. See TWO-LINED SALAMANDER, LONG-TAILED SALAMANDER, and CAVE SALAMANDER.

Eurycercus lamellatus. Common cladoceran found in weedy shallows.

Eurycyde. Genus of Pycnogonida.

Eurygaster. See SCUTELLERIDAE.

euryhaline. Able to withstand a wide range of dissolved salts in the environment; e.g. salmon, a few polychaetes, and certain sharks. Cf. STENOHALINE.

euryhygric. Pertaining to an animal which can maintain itself over a wide range of relative humidity; e.g. man, many common birds, mammals, and insects. Cf. STENOHYGRIC.

Eurylaimidae. Family of bright colored Old World tropical birds in the Order Passeriformes; broadbills; large heads, wide bills; insectivorous.

Eurylepta. Genus of polyclad Turbellaria.

Euryleptoides. Genus of polyclad Turbellaria.

Eurymus. Colias.

euryoky. Euroky.

Eurypanopeus. Genus of marine mud crabs.

Eurypauropus. Genus of Pauropoda.

Eurypelma. Genus of spiders which includes the southwestern tarantulas.

euryphagous. Utilizing a wide variety of foods.

Eurypon. Genus of thin, encrusting marine sponges.

Eurypterida. Sea scorpions; extinct order of large arachnoids in the Subclass Merostomata; tapered abdomen of 12 segments; Cambrian to Carboniferous; Eurypterus and Pterygotus best known.

Eurypterus. See EURYPTERIDA.

Eurypygidae. Bird family which includes only the sun bittern, Eurypyga helias, in the Order Gruiformes; elegant, shy, wading bird of Central and South America.

Eurystheus. Genus of marine Amphipoda.

Eurystomella. Genus of rose-colored marine Bryozoa.

Eurytemora. Genus of calanoid copepods; salt, brackish, and fresh waters, especially in coastal areas.

eurythermal. Pertaining to an animal which is able to maintain itself over a wide range of temperatures, e.g. whales, mountain lion, and oyster. Cf. STENOTHERMAL.

Eurythoë. Genus of Polychaeta.

Eurytomidae. Family of minute to small insects in the Order Hymenoptera; larvae known as straw worms; examples are Bruchophagus and Harmoleta.

eurytopic. Having a wide range of suitable ecological conditions; a sp. characteristic.

Euryurus. Genus of millipedes.

euryzonal. Pertaining to an animal which occurs through a broad range of altitude in mountainous country; e.g.

coyote and robin. Cf. STENOZONAL.

Euscelis. Genus of leafhoppers; pests on many economic plants.

Eusimulium. Common genus of blackflies in the Family Simuliidae.

Euspongia. Spongia.

Eustachi, Bartolomeo. Italian anatomist (?-1574); described many human anatomical features including the Eustachian tube, adrenals, thoracic duct, uterus, and kidneys.

Eustachian tube. Fine canal extending from the middle ear to the pharynx in tetrapod vertebrates; allows equal air pressure on both sides of the ear drum.

Eustala anastera. Common grayish spider in the Family Araneidae.

Eusthenopteron. Genus of Devonian coelacanth fishes.

Eustylochus. Common genus of marine polyclad Turbellaria.

Eusyllis. Genus of tube-building Polychaeta; abundant in kelp.

Eutamias. Genus of western N.A. chipmunks; coloration generally similar to that of the eastern chipmunk (Tamias) but without the reddish rump and hips, and smaller and more slender; several spp. and many subspecies found mostly west of the Great Plains.

Eutardigrada. Class in the Phylum Tardigrada; head without anterior cirri and lateral filaments; e.g. Macrobiotus.

eutely. Constant number of cells in all mature individuals of a single sp., as in rotifers, nematodes, and Acanthocephala.

Euthemisto. Genus of marine Amphipoda; eyes very large.

euthenics. Betterment of environmental conditions for the human race so as to give the best possible expression to the inherent genetic constitution of each individual.

Eutheria. One of the three infraclasses of mammals in the Subclass Theria; true placental mammals; marsupial pouch absent; vagina not double; fetus undergoes complete development within uterus, attached to uterine wall by placenta; includes the great majority of modern mammals.

Euthyneura. Opisthobranchiata.

euthyneury. Symmetry of the nervous system, especially the visceral nervous system in certain Gastropoda.

Euthynnus alleteratus. Little tuna or false albacore of Atlantic Coast waters; to 30 in. long.

Eutima. Genus of marine coelenterates in the Suborder Leptomedusae.

Eutonina. Genus of hydrozoan coelenterates having well developed medusae.

Eutreptia. Genus of marine and freshwater biflagellate Protozoa; somewhat similar to Euglena.

Eutrichomastix. Genus of pyriform flagellates occurring as internal commensals in arthropods and lower invertebrates; four flagella, of which one is trailing.

Eutrombicula. See TROMBIDIIDAE.

eutrophic. Pertaining to a type of lake characterized by partial depletion or absence of oxygen in the deeper waters in midsummer, rich nutrient supply, and rich plankton.

Euxenura galeata. Maguari stork of South America; black and white, with bluish bill, red feet, and red face.

Euxoa. Genus of noctuid moths; larvae are general feeders, often destructive.

Evadne. Marine genus of Cladocera; carapace does not cover the abdomen and legs but serves only as a brood pouch.

Evasterias. Genus of red sea stars; arms slim.

evening grosbeak. See GROSBEAK.

eviscerate. To remove the viscera from an animal.

evisceration. Phenomenon occurring commonly in sea cucumbers when irritated or kept in unsuitable aquarium conditions; the body wall contracts so strongly as to force the respiratory tree and most of the digestive tract out of the body by way of the ruptured anus; if placed in a suitable habitat, the animal can sometimes regenerate lost parts.

evocation. In embryonic tissue, the instigation of a particular developmental phenomenon by a substance (evocator) diffusing from nearby tissue or implant, such as the formation of the first neural tissue from ectoderm as the result of an evocator secreted by the underlying chorda-mesoderm in vertebrate gastrulas.

evolution. Origin, ancestry, and differentiation of organisms; implies the great complex of processes by which organisms have been derived, structurally and functionally, from ancestral forms, including the implication of complex forms being derived from simpler ones by modification and change; descent with modification; includes the slow, long-time continuous adaptation of organisms to their changing environment through the agencies of variation, mutation, and selection. (These remarks refer to organic evolution; there are other kinds of evolution, including geological evolution, social evolution, etc.)

Evotomys. Genus of mouselike herbivorous rodents in the Family Cricetidae; red-backed mice; in cool, damp, mossy places in northern N.A.; distinguished from meadow mice by their reddish dorsal band.

exarate. 1. Grooved or furrowed. 2. Legs and wings free from the body, as in certain insect pupae. Cf. OBTECT.

Excirolana. Genus of marine Isopoda.

exconjugant. An individual Paramecium or other ciliate which has just separated from another individual following the process of conjugation.

excrement. 1. Fecal matter when outside of the body. 2. Also, sometimes a mixture of feces and urine.

excreta. Collectively, the excretory products of the body.

excrete. To separate, concentrate, and eliminate metabolic waste products.

excretion. 1. Any metabolic waste material, especially water, carbon dioxide, alkaloids, organic acids, nitrogenous compounds, etc. 2. The sum total of all processes involved in concentrating and eliminating metabolic waste materials (or other substances present in excess) from the tissues. 3. The act or process of excreting. (As distinct from defecation, excretion involves those waste products which have once been a part of the living protoplasm; as distinct from secretion, an excreted substance usually is of no use to the organism.)

excretory pore. Any pore on the surface of the body through which excretory products are voided, especially in certain invertebrates.

excretory system. Collectively, all tissues and organs of animals associated with the concentration and elimination of metabolic waste materials from the body; the liver, skin, lungs, and kidneys and associated ducts constitute the excretory system of man.

excurrent canal. Radial canal.

excurrent siphon. See SIPHON.

excystment. Emergence of an animal from its cyst stage and the resumption of normal metabolic activities.

exflagellation. Process of microgamete formation from a microgametocyte in certain Sporozoa.

exfoliation. Shedding of superficial layers or sheets of any structure.

exhalation. Expiration.

exhale. To breathe out.

exites. Lobes on the outer margin of a foliaceous appendage in the Eubranchiopoda.

exoccipital. One of a pair of bones at the posterior portion of the skull in the region of the exit of the spinal cord.

exochorion. Very thin outer shell of an insect egg.

exocoel. That portion of an anthozoan gastrovascular cavity between pairs of septa.

Exocoetidae. Family of bony fishes which includes the flying fishes; pectoral fins greatly enlarged and used for planing or skittering through the air just above the water, chiefly in warm seas; Exocoetus and Cypselurus common.

Exocoetus. See EXOCOETIDAE.

exocrine. Secreting outward through a duct or otherwise into a cavity; e.g. digestive glands.

Exocyclica. Taxon which includes the irregular sea urchins; anus acentral and interradial.

exogamy. Fusion of gametes having distinctly different ancestries.

Exoglossum maxillingua. Cutlip chub, a small cyprinid fish of northeastern U.S.; mandible with a strongly trilobed outline.

Exogone. Genus of Polychaeta.

Exogyra. Genus of extinct bivalve mollusks.

exophthalmic goiter. Grave's disease; enlargement and oversecretion of thyroxin by the thyroid gland, resulting in high metabolic rate, excitability, high tension, and protruding eyeballs; may be treated by surgical removal of part of the thyroid.

exophthalmos. Protrusion of the eyeballs.

exopod. Exopodite; see BIRAMOUS APPENDAGE.

exopodite. See BIRAMOUS APPENDAGE.

Exopterygota. Same as Hemimetabola and Paurometabola.

exoskeleton. Any invertebrate skeleton of various composition which forms the outermost covering of the body; it gives more or less support and permanent shape to the body and its parts; present in arthropods, nematodes, mollusks, etc.

exosmosis. Movement of materials, especially water, out of a cell by means of osmosis.

Exosphaeroma. Genus of isopods found on Pacific Coast marine shores, in fresh-water ponds and pools near the coast, and in warm springs in N. M.

exostosis. Abnormal bony spur formed in certain pathological conditions. Pl. exostoses.

exothermic reaction. Chemical reaction which liberates energy, especially a decomposition process.

exotic. Pertaining to any non-native, introduced sp.

exotoxin. See TOXIN.

expectoration. Discharging mucus by spitting.

experiment. Observations and measurements made under controlled conditions in order to discover some facts or suggested truth, to demonstrate some known truth, or to test the validity of a hypothesis, theory, or law.

expiration. Exhalation; the mechanical process of forcing air out of the lungs during respiration; produced by relaxing the rib muscles and diaphragm, thereby producing a decrease in the size of the lung cavity.

explantation. Tissue culture.

expressivity. Relative degree to which a character is expressed in a phenotype; determined by the relative interaction of genotype and environment.

extensor. Muscle that straightens a joint.

extensor carpi radialis. Group of tetrapod muscles which originate on the humerus and insert on the metacarpals; extend and abduct wrist and forearm.

extensor carpi ulnaris. One of the muscles of the lower forelimb of certain tetrapods.

extensor communis digitorum. Muscle which originates on the humerus and inserts on the fingers; extends fingers and forearm.

extensor cruris. Muscle on the front surface of the shank in certain tetrapods.

extensor digiti quinti. Mammalian muscle which originates on the humerus and inserts on the little finger; extends little finger.

extensor digitorum brevis. Tetrapod muscle which extends the first phalanx of fingers and toes; part of extensor digitorum longus.

extensor digitorum communis. Muscle of the leg in certain tetrapods; extends the digits.

extensor digitorum lateralis. Muscle of the forelimb in certain tetrapods; extends the digits.

extensor digitorum longus. Muscle which extends the digits of the hind foot in certain vertebrates.

extensor hallucis longus. Mammalian muscle which originates on the front of the tibia and inserts on the terminal phalanx of the great toe; extends great toe.

external auditory canal. See OUTER EAR.

external auditory meatus. See OUTER EAR.

external carotid artery. Artery supplying the neck, face, and skull in many vertebrates.

external ear. Outer ear.

external fertilization. Union of egg and sperm in the environment outside of the parental body.

external jugular vein. Vein draining a large superficial portion of the head in most vertebrates.

external nares. Nostrils.

external oblique muscle. Superficial sheetlike abdominal muscle in most tetrapods; compresses the abdomen.

external respiration. Exchange of oxygen and carbon dioxide at the general external respiratory surfaces, such as lungs and gills; contrasts with internal, or cellular, respiration.

exteroceptor. Receptor which receives stimuli from the external environment, such as the eye and ear.

extinction threshold. Minimum number or density of individuals in a sp. which are able to survive within a habitat; when the number falls below this minimum, the sp. disappears from that habitat.

extirpation. 1. Extinction of an organism in a given area. 2. Removal of a part.

extracellular. Occurring within an organism but outside of the individual cells, such as in a digestive cavity.

extracellular digestion. Digestion of food in a cavity of the digestive system.

extracolumella. Outermost cartilaginous tip of the ear columella in reptiles and birds.

extraembryonic membrane. Membrane of an embryo which lies outside the body but is concerned with protection and metabolism of the embryo; e. g. yolk sac.

exudate. 1. Substance discharged through pores, lesions, or tissue incisions. 2. Substance accumulated in or on tissues as the result of abnormal metabolism or disease.

exumbrella. Outer, aboral surface of a medusoid coelenterate.

exuviae. Cast exoskeleton or skin of larva, nymph, or other immature or mature stage, especially in arthropods. Pl. exuviae (Both sing. and pl. forms have the same spelling.)

Exuviaella. Genus of marine dinoflagellates; subspherical or ovoid; two anterior flagella.

exuvium. Exuviae. Pl. exuvia.

eye. 1. In the broad sense, any receptor for light, ranging from a simple light-sensitive organelle in certain Protozoa to the complex eyes of insects, cephalopods, and vertebrates. 2. In the narrow sense, a complex receptor for light, usually resulting in the formation of an image.

eye brush. Cluster of stiff hairs on forelegs of many insects; used in cleaning the surface of the compound eye.

eyecup. A cup-shaped outgrowth of the embryonic vertebrate brain; later becomes the greater portion of the eye.

eyed coral. Oculina diffusa, a small branching sp. of deep, cold seas.

eye muscles. Set of six muscles which move a typical vertebrate eyeball; two oblique muscles and four rectus muscles.

eyepiece. The series of two or three lenses arranged in a common mount at the eye end of an optical instrument; ocular.

eyepoint. Point above the ocular of a microscope at which the eye must be placed in order to see the full field of view.

eyespot. Organelle or simple organ specialized for the detection of light; there is no sharp distinction between "eyespot" and "eye"; the former is present in many invertebrates and prochordates, the latter in some invertebrates (arthropods, mollusks, etc.) and essentially all vertebrates; it is probable that an eyespot never produces a true image.

eyestalk. Stalk bearing a terminal compound eye in decapod Crustacea.

eye tooth. See CANINE.

Eylais. Genus of Hydracarina; mostly in quiet waters.

F

F_1. First filial generation; offspring of any mating, especially when the genetic transmission of one or more characters is under consideration.

F_2. Second filial generation; offspring of any mating between F_1 individuals, especially when the genetic transmission of one or more characters is under consideration.

FAA. Formalin-aceto-alcohol, a general purpose fixative; composed of 40 ml. water, 2 ml. glacial acetic acid, 50 ml. 95% alcohol, and 10 ml. formalin.

Fabia. Genus of brachyuran crabs; female often commensal on *Mytilus*.

Fabre, Jean Henri. French entomologist (1823-1915); noted for his careful observations of insect behavior and natural history as well as his essay-like style of writing; many sections of his ten-volume *Souvenirs Entomologiques* have been translated into English.

Fabricius, Hieronymus. Italian anatomist (1537-1619); made important contributions to the comparative embryology of vertebrates and to the structure of the ear, eye, and larynx; discovered venous valves.

facet. See OMMATIDIUM.

facialis nerve. One of the paired cranial nerves of vertebrates; innervates lateral-line structures in lower vertebrates, but superficial facial and scalp muscles, the digastric muscle, and tongue in higher vertebrates; a mixed nerve.

faciation. In ecology, a subdivision of an association determined by spp. composition; thus, the grasslands of western Okla. constitute a faciation which is different from the grasslands faciation occurring in N.D.

facies. 1. General ecological appearance of a habitat. 2. Particular ecological appearance or composition of an association, especially with reference to the relative numerical dominance of certain companion spp. Pl. facies.

facilitation. Effect produced on nerve tissue by the passage of successive impulses; resistance is diminished so that a second stimulus produces a reaction more easily.

factor. 1. Gene. 2. Any element of an environment, e.g. pressure, food supply, and light.

facultative anaerobe. Any organism usually living in the presence of oxygen but which also may live in the absence of oxygen.

facultative parasite. Organism that can live either as a parasite or non-parasitically; e.g. certain leeches.

faded snake. Glossy snake.

faeces. Feces.

fainting. Temporary unconsciousness owing to insufficient blood supply to the brain.

fairy armadillo. See CHLAMYPHORUS.

fairy shrimp. Common name of members of the crustacean Order Anostraca; so-called because of their translucence, delicate colors, and graceful swimming.

fairy tern. Either of two spp. of white terns; tropical seas.

falciform. Falcate; scythe- or sickle-shaped.

falciform ligament. Sickle-shaped ligament which attaches the liver to the diaphragm and separates the right and left liver lobes.

Falco. Genus of falcons; long, pointed wings, long tails, and rapid wing strokes; *F. rusticolus* (gyrfalcon) of Arctic America; *F. mexicanus* (prairie falcon) of arid N.A.; *F. peregrinus* (duck hawk) of North and South America, which is the sp. commonly used in falconry; *F. columbaris* (pigeon hawk) of N.A., chiefly east of the Rockies; *F. tinnunculus*, a small European kestrel.

falcon. See FALCO.

Falconidae. Family of birds in the Order Falconiformes; includes 58 spp.

of true falcons; wings long; body 6 to 24 in. long.

Falconiformes. Very large and cosmopolitan order of birds which includes the vultures, kites, hawks, eagles, etc.; mostly predatory; beak sharp, stout, hooked, with cere at base; feet with sharp, curved claws; strong fliers.

fall armyworm. See LAPHYGMA.

fallfish. See LEUCOSOMUS.

Fallopian tube. One of two reproductive tubes in female mammals; oviduct; conducts egg from ovary to the uterine cavity and sperm from the uterus to that point in the tube where fertilization occurs.

Fallopio, Gabriele. Italian anatomist (1523-1562); discovered the Fallopian tubes and contributed to the anatomy of the ear and throat.

fall overturn. Variable period during autumn when a lake is essentially homothermous and is subject to complete top-to-bottom circulation by wind action; especially true of lakes in the temperate zone.

fallow deer. See DAMA.

fall webworm. See HYPHANTRIA.

false killer whale. See PSEUDORCA.

false labor. Periodic cramps similar to true labor pains but not accompanied by cervical dilation.

false rib. Any rib which does not unite directly with the sternum.

false scorpion. See PSEUDOSCORPIONIDA.

false sunbird. See PHILEPITTIDAE.

false vocal cord. One of two small upper vocal cords in man which are not concerned with sound production.

falx cerebelli. One of the septum-like extensions of the dura mater; extends into the notch between the posterior parts of the cerebellum.

falx cerebri. Sickle-shaped septum extending from the dura mater vertically between the cerebral hemispheres.

family. Taxonomic category including one or more genera which have certain phylogenetic characters in common; e.g. the Family Ursidae (bears) includes genera such as: Tremarctos (spectacled bear), Selenarctos (Himalayan bear), Ursus (black and grizzly bears), Thalarctos (polar bear), etc.; animal family names end in idae; plant family names end in aceae.

fanaloka. See FOSSA.

fang. Any long, sharp tooth by which an animal seizes, poisons, holds, or tears its prey.

Fannia. Important genus of flies in the Family Anthomyiidae; general appearance of small houseflies; breed in dung and decaying material; larvae occasionally produce myiasis.

fan shell. Scallop.

fantail. 1. Commercial breed of goldfish having exceptionally large anal and tail fins. 2. Australian flycatcher with exceptionally broad tail. 3. Variety of domesticated pigeon. 4. Common warbler of the Mediterranean region.

fan-tailed darter. Common sp. of Etheostoma, especially in the central states.

fanworm. Any of many kinds of tube-building marine Polychaeta in which an anterior tuft of gills projects from the tube and has the appearance of a multiple fan; Bispira typical.

Farancia. See MUD SNAKE.

far-sightedness. Hypermetropia; failure of eye lens to bend light rays sufficiently to produce a sharp retinal image; focal plane actually behind the retina.

fascia. Sheet of connective tissue; superficial fascia of mammals is the loose connective tissue beneath the dermis, and deep fascia consists of tough sheets which enclose one or more muscles. Pl. fasciae.

fascia lata. Wide, dense, connective tissue sheath of the thigh muscles in man and other mammals.

fasciation. Faciation.

fascicle. 1. One of several discrete bundles of nerve fibers as seen in the cross section of a nerve. 2. In the spinal cord, a bundle of fibers derived from more than one tract.

Fasciola. Genus of trematodes in the Order Digenea; F. hepatica is the common "liver rot" parasite of sheep, cattle, goats, and other wild and domestic ruminants; also an occasional parasite in the liver of man; the definitive host ingests the metacercariae from aquatic vegetation or the surface film of ponds and swamps; aquatic snails are intermediate hosts.

Fasciolaria. Genus of large carnivorous tropical marine gastropods; band shells, tulip shells; shell thick and spindle-shaped, with an oval aperture terminating in a long, open canal.

Fascioloides. Genus of large distome parasites of the liver of herbivores.

Fasciolopsis. Genus of trematodes in

the Order Digenea; F. buski (to 3 in. long) is a common parasite in the small intestine of man and hogs over a large portion of the Orient; aquatic snails are the first intermediate hosts, and the definitive host ingests metacercariae on water chestnuts, water hyacinth, water bamboo, etc.

fat. 1. Ester of fatty acids and glycerol. 2. Any oily or greasy substance in plants and animals. 3. Adipose tissue.

fat body. Any special large mass of visceral or subcutaneous connective tissue consisting mostly of stored fat; occurs in some invertebrates and vertebrates, but especially in amphibians and lizards; in vertebrates it is in the abdomen and is utilized during hibernation; present also in many insects where it extends along and around the digestive tract.

fathead minnow. See PIMEPHALES.

fatigue. Physiological condition of a tissue or organ which is unable to respond normally to a stimulus without rest; usually accompanied by an accumulation of excretory materials in the area affected.

fat tissue. Adipose tissue.

fatty acid. Any of a series of organic acids having the generalized formula $C_nH_{2n}O_2$; they occur in animals and plants, chiefly bound up with glycerol; stearic and palmitic acids are common examples.

fauces. 1. Short passage between palate and pharynx. 2. That interior portion of a spiral gastropod shell that can be seen by looking into the aperture.

fauna. 1. Collectively, the animal life of any particular area or of any particular past time. 2. A list of animal spp. and descriptions for a particular area or time.

fauna hydropetrica. That fauna living in the thin film of water flowing over stones, especially in seepage areas, along waterfall streamsides, or in splash areas.

faunula. Association of spp. or a community occupying only a small space; e.g. a culture jar of protozoans or the organisms in a rotting tree trunk.

Favia. Genus of corals which includes the star corals; spp. whose polyps make star-shaped patterns in the skeleton.

fawn. Deer during the first year of life.

feather. One of the light, horny, complex, epidermal outgrowths which collectively form the protective covering of birds.

feathered bittersweet shell. See PETUNCULUS.

feather-legged spider. See ULOBORIDAE.

feather mite. See ANALGESIDAE.

feather stars. Group of crinoid echinoderms which are free-swimming as adults; see CRINOIDEA.

feather-tailed phalanger. See DISTOECHURUS.

feather worm. Any of many kinds of tube-building marine Polychaeta in which an anterior tuft of gills projects from the tube and has the appearance of a miniature feather duster.

febrile. Pertaining to fever.

fecal. Pertaining to feces.

feces. Indigestible residue which is expelled from the anus; usually contains a small amount of excretory material and large numbers of bacteria as well as materials resulting from bacterial metabolism.

fecundate. To impregnate or fertilize.

fecundity. Relative number of eggs, sperm, or young produced by an animal.

feeblemindedness. State of retarded mental development; idiots have an IQ of 0 to 25, never advance beyond a mental age of three years, and require constant supervision; imbeciles have an IQ of 26 to 50, range from three to seven years in mental age, and can be taught to perform simple tasks; morons have an IQ of 51 to 70, range from eight to 12 years in mental age, and can be taught to do manual labor of moderate complexity; some types of mental deficiencies are transmitted as Mendelian recessives, but others are caused by organic disturbances which are not genetic.

Fehling's solution. Solution used in testing for the presence of glucose, fructose, and other sugars; contains copper sulfate and sodium potassium tartrate; when heated in the presence of such a sugar a precipitate of copper oxide is formed.

Felidae. Family of carnivorous mammals of medium to large size; tiger, lion, cougar, lynx, wildcat, housecat, leopard, jaguar, etc.; trim, muscular spp. with rounded head, five front toes and four hind toes; claws long and retractile; shearing teeth; pelage short to long.

feline. Pertaining to cats or any member of the cat family.

Felis. Genus of carnivorous mammals in the Family Felidae; includes the typical cats; F. catus is the domestic cat and the European wildcat; see LYNX, BOBCAT, SERVAL, GOLDEN CAT, OCELOT, MARGAY, JAGUARUNDI, COUGAR. Cf. PANTHERA.
femoral artery. Chief artery of the thigh.
femoral vein. Large vein draining the leg in certain vertebrates.
femorotibialis. Dorsal muscle of the hind limb of reptiles and birds.
femur. 1. Large thigh bone of tetrapods. 2. The third segment from the base of an insect leg; also, one of the larger segments in the legs of certain other arthropods. Adj. femoral. Pl. femora.
fence lizard. See SCELOPORUS.
Fenestella. Genus of fossil Bryozoa in the Order Cryptostomata.
fenestra. 1. In insect wings, a transparent glassy spot. 2. Clear area in a vein of an insect wing. 3. Perforation in a membrane. 4. Window-like opening in a structure.
fenestra cochleae. Fenestra rotunda.
fenestra ovalis. Fenestra vestibuli or oval window; a membrane-covered opening in the bony labyrinth of the ear; the opening is further plugged by the stapes bone; mechanical vibrations of the stapes are here transmitted to the perilymph, membranous labyrinth, endolymph, etc.
fenestra rotunda. Fenestra cochleae or round window; a membrane-covered opening of the bony labyrinth lying between the liquid-filled inner ear and the air-filled tympanic cavity; below and a little behind the fenestra ovalis.
fenestra vestibuli. Fenestra ovalis.
Fenestrellina. Genus of fossil Bryozoa in the Order Cryptostomata.
Feniseca tarquinius. Sp. of butterfly in the Family Lycaenidae; larva preys on woolly aphids.
Fennec. Genus of small, fluffy, foxlike, nocturnal mammals (fennecs) found in deserts of northern Africa and Arabia.
feral. 1. Gone wild, having once been tamed. 2. Undomesticated, untamed, wild, savage.
fer-de-lance. See BOTHROPS.
ferment. Enzyme.
fermentation. Decomposition of carbohydrates, fats, and certain other simple organic compounds by bacteria and yeasts. Cf. PUTREFACTION.
ferret. See MUSTELA.
Ferrissia. Genus of small fresh-water limpets.
fertilization. Union of male and female gametes to form a diploid zygote.
fertilization cone. Small protuberance from an ovum at the point of contact with a fertilizing spermatozoan; occurs in many animals.
fertilization membrane. Thin membrane formed rapidly by certain animal ova immediately after the penetration of one sperm cell; has the effect of excluding other sperm cells.
fertilizin. Substance secreted by the ovum of certain animals; has an attracting effect on sperm.
Ferungulata. Taxon of mammals which includes about eight extinct orders and the following living orders: Artiodactyla, Carnivora, Hyracoidea, Perissodactyla, Proboscidea, Sirenia, Tubulidentata.
Festschrift. Memorial volume to a scientist, usually published at his retirement, 70th birthday, 80th birthday, etc.; composed of technical articles contributed by students, colleagues, and friends.
fetlock. The joint, hair tuft, or projection just above the rear margin of the hoof of a horse.
fetus. Late stage of embryological development of vertebrates; the term is used especially with reference to mammals, and in man refers to the embryo between the end of the third month of development and birth. Adj. fetal.
F_1 generation. First generation offspring of a particular mating.
F_2 generation. Offspring derived from a mating of two F_1 individuals.
fibril. Any minute threadlike organelle forming part of a cell.
fibrilla. Fibril, especially a very fine fibril. Pl. fibrillae.
fibrillar system. Neuromotor system.
fibrin. Protein which has coagulated into a mass of fine entangling fibrils during blood clotting. See CLOTTING.
fibrinogen. Blood protein which is concerted into fibrin during blood clotting. See CLOTTING.
fibroblast. Irregular branching cell found in connective tissue; forms and maintains the fibers in such tissue.
fibrocartilage. Type of cartilage containing an abundance of connective tissue fibers; in mammalian vertebral pads and the pubic symphysis.
fibroglia. Fine fibrils in the cytoplasm of connective tissue fibroblasts.
fibroin. Chief protein of silk.

fibrous connective tissue. Type of connective tissue composed of scattered spherical or branched cells in a matrix of delicate secreted fibers. See WHITE FIBROUS CONNECTIVE TISSUE and YELLOW FIBROUS CONNECTIVE TISSUE.

fibula. Posterior of the two bones between the knee and ankle in hind limbs of tetrapods.

fibulotarsalis. Ventral muscle of the foot in urodeles.

Ficina. See HOOK-NOSED SNAKE.

Ficus papyracea. Paper fig shell, a large marine gastropod with a pear-shaped, thin, yellowish shell; to 4 in. long; N. C. to Fla.

fiddler crab. See UCA.

fiddlerfish. See RHINOBATOS.

field bee. Worker bee which gathers pollen and nectar.

fieldbird. Upland plover.

field cricket. Any of several crickets especially common in field habitats.

fieldfare. Medium-sized European thrush.

field mouse. Meadow mouse, meadow vole; see MICROTUS.

field skipper. See ATALOPEDES.

Fierasfer. Carapus.

fig eater. See COTINIS.

Figitidae. Cosmopolitan family of small to minute Hymenoptera; parasites on aphids and coccids.

fig shell. See FICUS.

fig wasp. See AGAONTIDAE.

filament. Delicate fibril or threadlike structure.

filaria. One of a general group of nematodes parasitic in various vertebrates and requiring an intermediate arthropod host.

filariasis. See WUCHERERIA.

filariform larva. Larval form of many free-living and a few parasitic spp. of nematodes; body very long, and esophagus long, cylindrical, and without any conspicuous muscular bulb.

filator. Silk-spinning structure in caterpillars.

filefish. See MONACANTHIDAE.

file shell. See LIMA.

file snake. Any of several large nonpoisonous African snakes; triangular in section.

filet of sole. Large, flat piece of flesh from one side of a flatfish; not necessarily from the sole, and most commonly from a flounder in the U. S.

filial regression law. Offspring of parents exhibiting extremes of a variable character tend to regress toward the average for the whole population; e. g. offspring of tall parents may be shorter than the parents but taller than the average.

Filibranchia. Order of pelecypod mollusks; gills W-shaped in cross section, with long filaments; usually two adductor muscles, but anterior muscle sometimes reduced or absent; Arca, Mytilus, Ostrea, and Pecten typical.

Filicrisia. Genus of marine Bryozoa; low, delicate growth form.

filiform. Filamentous or threadlike.

filiform papillae. Abundant conical or cylindrical papillae on the surface of the tongue; usually 0. 5 to 3. 0 mm. high.

Filinia. Genus of fresh-water plankton rotifers characterized by three long spinelike appendages.

Filistata hibernalis. Common snare-building spider of southern states.

Fillicollis. Genus of Acanthocephala parasitic especially in ducks.

Filograna. Genus of small, marine, tube-building Polychaeta; usually eight branchial filaments.

filoplume. Minute hairlike type of feather sparsely distributed over the body of many birds; shaft bare except for a cluster of barbs at the tip.

filopodium. Filamentous pseudopodium consisting almost entirely of ectoplasm; it may be branched but the branches do not anastomose; common in many Testacea. Pl. filopodia.

filter feeder. Any animal that obtains its food (usually small particles) by filtering it from water; e. g. Daphnia, clams, and tunicates.

filtration pressure. High pressure of the blood in a kidney glomerulus; especially effective in filtering elements of the blood into Bowman's capsule.

filum terminale. Fine posterior end of the spinal cord of many vertebrates.

fimbria. Frilled or fringelike opening of an oviduct, especially in mammals.

fimbriate. Having a frilled edge.

fin. Thin projection or extension on the body of an aquatic animal; used for locomotion and steering.

finback whale. See BALAENOPTERA.

finch. 1. General term for birds in the Family Fringillidae. 2. Specifically, a group of closely related spp. in the Family Fringillidae; common American spp. are the house finch (Carpodacus mexicanus), purple finch (C. purpureus), brown-capped rosy finch

(Leucosticte australis), common goldfinch (Spinus tristis), and the pine siskin (S. pinus).

fin fold. Median fold of tissue in a fish embryo; gives rise to the dorsal, caudal, and anal fins.

fin fold theory. Theory which accounts for the development of pectoral and ventral fins from long lateroventral folds of tissue in primitive vertebrates.

finfoot. See HELIORNITHIDAE.

fingerling. Length designation for immature fish; from 1 in. long (or disappearance of yolk sac) to whatever size is attained at the end of the first year's growth.

fingernail clam. Common name of small fresh-water pelecypods, especially those in the genus Sphaerium.

finger shell. See PTEROCERA.

finger sponge. Any large marine sponge (especially Chalina) which has a growth form resembling clusters of fingers.

finless porpoise. See NEOMERIS.

fin rays. Numerous horny or cartilaginous rodlike supports of fins.

Fiona. Genus of nudibranchs.

fire ant. See SOLENOPSIS.

fire-bellied toad. See BOMBINA.

fire bird. Any of several bright red or scarlet birds; e.g. scarlet tanager.

firebrat. Thermobia domestica.

fire bug. See PYRRHOCORIDAE.

firefly. See LAMPYRIDAE.

fire salamander. Common European spotted salamander.

fireworm. Larva of a small moth which feeds on cranberry leaves.

firmisternal shoulder girdle. Shoulder girdle in which the two coracoid plates meet ventrally; present in some frogs.

first antennae. Anterior pair of antennae on the head of Crustacea.

first filial generation. See F_1.

first form male. Mature male cambarine crayfish in which the first pleopods are corneous, sculptured at the tip, and capable of transferring sperm during copulation; such an instar is followed by a molt and a second form male instar.

first intermediate host. Host in which the earliest stage(s) of a parasite occur; e.g. a snail is the first intermediate host of the human liver fluke.

first maturation division. First meiotic division; see MEIOSIS.

first maxillae. 1. Fourth pair of head appendages in Crustacea; variously modified for feeding. 2. Third pair of head appendages in millipedes and centipedes; used for feeding.

first meiotic division. See MEIOSIS.

first polar body. Small cell formed during oogenesis and containing a haploid nucleus and little cytoplasm; one of the two haploid cells resulting from the first meiotic division of oogenesis, the other being the secondary oocyte.

first ventricle. Cavity of the left half of the telencephalon.

fish. See PISCES.

fish crow. See CROW.

fish duck. Merganser.

fisher. See MARTES.

fishfly. Alderfly.

fish hawk. Osprey; see PANDION.

fishing frog. See LOPHIIDAE.

fish lice. Copepod parasites of fishes; members of the Order Branchiura.

fish tapeworm. Dibothriocephalus latus.

fission. Asexual reproduction in animals by a division of the body into two or more parts. See MULTIPLE DIVISION and BINARY FISSION.

fissiparous. Reproducing by fission.

Fissipedia. Suborder of mammals in the Order Carnivora; includes the dogs, raccoons, cats, bears, badgers, skunks, and hyaenas; chiefly flesh eaters although some may also feed on vegetable material; toes separate from each other; originally found everywhere except the Australian region and oceanic islands, but some spp. have been introduced into these places.

fissirostral. Pertaining to an anatomical feature of swifts and nighthawks; bill broad and deeply cleft toward the sides of the head.

Fissurella. Common genus of keyhole limpets.

fissure of Rolando. Deep furrow running down each side of the cerebral cortex; separates anterior motor area from posterior sensory area.

fistula. Deep, channel-like ulcer, often having an opening to an internal hollow organ or surface of the body. Pl. fistulae.

Fistulana. Genus of burrowing marine bivalve mollusks.

Fistularia. Genus of long, slim marine fishes in the Family Fistulariidae; cornet fishes; head enormously elongated and tubelike; tropical and temperate seas.

Fistulariidae. See FISTULARIA.

fixation. Process of rapid killing and chemical fixing of tissues without distortion so that subsequent study and interpretation may be easy and accurate; common fixatives are various mixtures of alcohol, mercuric chloride, formalin, picric acid, and osmic acid.

fixation disc. Preoral lobe by which crinoid and sea star larvae attach to the substrate just before the onset of metamorphosis.

fixation papillae. Group of three glandular papillae at the anterior end of certain tunicate larvae; used for attachment to the substrate just preceding the onset of metamorphosis.

Flabelligera. Genus of Polychaeta.

Flabellina. Genus of nudibranchs.

flabellum. 1. Fan-shaped structure. 2. Distal lateral lobe of certain branchiopod thoracic appendages. 3. Epipodite of certain crustacean appendages. 4. Terminal lobe of the glossa in certain insects. Pl. flabella.

Flabellum. Genus of solitary deep-sea corals.

Flagellata. Mastigophora.

flagellated chamber. One of the many internal cavities of a sponge; lined with an epithelium of choanocytes.

flagellispore. Free-swimming protozoan spore having one or more flagella.

flagellum. 1. Long whiplike living process present in some types of plant and animal cells, especially certain Protozoa and male reproductive cells; a particular cell usually has only one or a few flagella; each flagellum is capable of beating or lashing movements which produce locomotion or a current past the cell; the action of each flagellum is governed by a basal granule in the peripheral cytoplasm of the cell; a rhizoplast connects the basal granule to the nucleus (or parabasal body); except for relative size and abundance, the extracellular portions of a flagellum and a cilium are fundamentally similar; each flagellum consists of two central fibrils surrounded by a ring of nine fibrils which, in turn, is enveloped by a thin membrane (corresponding to the cell membrane). 2. Whiplike tip of the penis or copulatory cirrus in certain invertebrates, especially gastropods. Pl. flagella.

flagfish. See JORDANELLA.

flame bulb. Special, hollow, club- or bulb-shaped structure formed of one or several small cells; the flame bulb contains a single flagellum (solenocyte) or a tuft of cilia which beats delicately and produces a slight outward current in the fluid filling the bulb and its associated duct; each flame bulb is situated at the end of a minute duct, and these ducts are connected and ultimately open to the outside of the animal by means of one or more pores (nephridiopores); flame bulbs may be scattered throughout the tissues or localized in clusters. See FLAME BULB SYSTEM and SOLENOCYTE.

flame bulb system. Protonephridial system; collectively, all of the flame bulbs, their collecting ducts, "bladder" (if present), and external pores in an animal; present in flatworms, nemertines, priapulids, entoprocts, acanthocephalans, cephalochordates, rotifers, kinorhynchs, gastrotrichs, and some annelids; also in many larval annelids, echiuroids, phoronids, and mollusks; although the flame bulb system may have an incidental excretory function, its chief role seems to be the maintenance of the proper internal salt-water balance by means of secretion of excess body water into the collecting ducts by their surrounding cells.

flame cell. Flame bulb.

flamingo. See PHOENICOPTERIDAE.

flannel moth. See MEGALOPYGIDAE.

flannelmouth sucker. _Catostomus latipinnis_; a large sucker of swift waters of the Colorado River system.

flash color. Bright conspicuous patch of color on an otherwise drab animal; presumed momentarily to distract the attention of a pursuer.

flatfish. See HETEROSOMATA.

flathead catfish. See GOUJON.

flatworm. Any member of the Phylum Platyhelminthes.

flavin. Any of a group of water-soluble pigments widely distributed in both plant and animal kingdoms, e.g. riboflavin.

flavoproteins. Large class of respiratory enzymes; a conjugated protein in which the prosthetic group contains a flavin; e.g. cytochrome _c_ reductase.

flea. See SIPHONAPTERA.

flea beetle. Common name of many small beetles in the Family Chrysomelidae, especially those that have hind legs modified for leaping.

fledge. 1. To acquire flight feathers. 2. To care for an immature bird until it is able to fly.

fledgling. Young bird with newly-acquired flight feathers.

Fleming, Sir Alexander. Scotch bacteriologist (1881-1955); discovered antibiosis, penicillin, and lysozyme; shared 1945 Nobel prize in physiology and medicine.

Flemming, Walther. German cytologist (1843-1915); founder of animal cytology; discovered many details of the process of mitosis; described longitudinal splitting and chromosome constancy.

Flemming's solution. Killing and fixing agent in microtechnique; 15 parts 1% aqueous chromic acid, 4 parts 2% aqueous osmic acid, and 1 part glacial acetic acid.

fleshflies. See SARCOPHAGIDAE.

flexor. Muscle that bends one part of the body on another.

flexor accessorius. Muscle of the bird wing.

flexor antibrachii. Muscle which flexes the elbow joint in certain tetrapods.

flexor carpi radialis. Muscle of the forelimb of certain tetrapods.

flexor carpi ulnaris. Muscle of the forelimb of certain tetrapods.

flexor digitorum communis. One of the muscles in the forearm of certain tetrapods; flexes the digits.

flexor digitorum longus. Muscle of the hind limb of tetrapods; flexes toes and extends foot.

flexor digitorum profundus. Muscle of the forelimb of certain tetrapods; flexes the digits.

flexor digitorum sublimis. Muscle which originates on adjacent portions of the humerus, radius, and ulna, and inserts on the second phalanges of the fingers; flexes phalanges; in mammals and birds.

flexor hallucis longus. Muscle of the hind limb of mammals and birds; flexes great toe and extends foot.

flexor palmaris profundus. Muscle which flexes the palm in Amphibia.

flexor tarsi. Muscle on the upper surface of the hind foot of certain tetrapods; flexes the foot.

flexor tibialis. Group of two ventral hind leg muscles of reptiles.

flicker. See COLAPTES.

flickertail. Richardson ground squirrel, Citellus richardsoni; Mont., N.D., and adjacent Canada.

flier. See CENTRARCHUS.

flight feathers. Those larger feathers of a bird which support it during flight.

flittermouse. Common or colloquial name for any insectivorous bat.

floating ribs. Ribs having free ends and not connected to the sternum or cartilages of other ribs; 11th and 12th pairs in man.

floatoblast. Free-floating statoblast; annulated, without peripheral processes; e.g. in some spp. of Plumatella.

flocculus. 1. Small lobe on the lower side of each cerebellar hemisphere. 2. Tuft of fine setae on the metacoxa of certain Hymenoptera.

Florida thula. Snowy egret of temperate and tropical Americas; in the Family Ardeidae; smaller and extending farther inland than the American egret (Casmerodius); to 24 in. long; plumage white, long crest on crown and back, bill black; along shores of fresh and brackish waters; nest a platform of sticks.

Floscularia. Genus of sessile, case-building, fresh-water rotifers.

Flosculariacea. Order of rotifers in the Class Monogononta; often sessile; no lorica, but often in a secreted tube; corona with two concentric wreaths of cilia having a ciliated furrow between; sometimes with long spines or armlike structures.

flounder. See HETEROSOMATA.

flour beetle. See TRIBOLIUM.

flour mite. Any of several mites which occasionally infest flour.

flour moth. See EPHESTIA.

flour worm. Larva of any of numerous insects infesting flour.

flower bees. See HALICTIDAE.

flower bugs. See ANTHOCORIDAE.

flower-faced bat. See ANTHOPS.

flower flies. See SYRPHIDAE.

flowerpecker. See DICAEIDAE.

fluke. 1. Member of the Class Trematoda in the Phylum Platyhelminthes. 2. Lateral expansion of the tail of a whale.

Fluminicola. Genus of fresh-water snails; several stream spp. west of the Continental Divide.

Flustra. Genus of encrusting marine Bryozoa.

Flustrella. Genus of marine Bryozoa.

Fluta. See SYMBRANCHII.

flute mouth. See AULOSTOMUS.

fly. 1. Strictly speaking, an insect in the Order Diptera. 2. In combination, an insect in several other orders, e.g. caddisfly, damselfly, ichneumon fly.

flycatcher. Any of numerous birds that

feed chiefly on flying insects; Myiarchus and Empidonax common; also, see MUSCICAPIDAE.

flying dragon. See DRACO.

flying fish. See EXOCOETIDAE.

flying fox. See MEGACHIROPTERA.

flying frog. Any of several East Indian frogs (Polypedates) having exceptionally large webbed feet; in leaping, they have a parachute effect; see POLYPEDATIDAE.

flying gurnard. See DACTYLOPTERUS.

flying lemur. See DERMOPTERA.

flying phalanger. See PHALANGER.

flying squirrel. See GLAUCOMYS.

flytrap. Any of several carnivorous plants which may entrap and digest flies and other insects; e.g. Venus' flytrap and pitcher plant.

fly-up-the-creek. Any of several small American herons, especially the green heron.

flyway. Geographically established air route of migratory birds, e.g. Mississippi flyway.

foal. 1. Young animal of the horse family (Equidae). 2. In the horse family, to give birth to young.

Foettingeria. Aberrant genus of ciliates; parasitic in the gastrovascular cavity of sea anemones; cyst forms found on small Crustacea.

foetus. Fetus.

foliaceous appendage. Flat, leaflike appendage, often lobed and subdivided; present in cladocerans, phyllopods, etc.

Folia parallela. Species of Venus' girdle in the Phylum Ctenophora; body ribbon-like, 15 mm. high, and 15 cm. broad; abundant along East Coast.

folic acid. Pteroylglutamic acid, PGA; vitamin of the B complex; essential to blood cell formation and concerned with synthesis of deoxyribonucleic acid; deficiency produces anemia and sprue; abundant in green leaves, liver, yeast, and eggs; $C_{19}H_{19}O_6N_7$.

follicle. A small sac, cavity, gland, or pit, such as a Graafian follicle, hair follicle, feather follicle, etc.

follicle mite. See DEMODICIDAE.

follicle-stimulating hormone. FSH; gonadotropic hormone secreted by the anterior pituitary of female mammals; stimulates growth of Graafian follicles.

follicular fluid. Liquid in a Graafian follicle.

folliculin. See ESTROGENS.

Folliculina. Genus of marine Ciliata; sessile and loricate.

Folsom man. Prehistoric man associated with stone implements and bones of extinct bison, camel, and mastodon; first found near Folsom, N.M. and subsequently in many areas in western and southwestern states; no human bones have thus far been found.

fontanel. Fontanelle.

fontanelle. 1. Gap in the bony covering of the brain; covered only by skin and membranes. 2. Any unossified area in an otherwise bony surface. 3. Shallow depression on the surface of the head in termites; bears the opening of the frontal gland.

Fontaria. Large genus of millipedes.

food chain. Food web.

food cup. Temporary concavity formed on the surface of an amoeboid protozoan; the concavity contains a food organism which sinks into the cell until the "cup" closes at the top, thus creating a food vacuole.

food groove. Median ventral groove in which food particles are concentrated in such crustaceans as fairy shrimps and cladocerans.

foodstuffs. In the biological sense, proteins, carbohydrates, and lipids as ingredients of foods.

food vacuole. A fluid-filled (usually) spherical cavity containing food particle(s) in a cell; the food undergoes digestion in the vacuole, and soluble digestion products are absorbed into the surrounding cytoplasm; the boundary of the food vacuole is a plasma membrane; present in holozoic Protozoa and in the cells lining the digestive cavities of many lower invertebrates.

food web. Food chain; complex spp. interrelationships in any community with special reference to feeding habits; a typical food web includes plants, herbivores, carnivores, omnivores, and detritus feeders.

fool hen. See CANACHITES.

foot-and-mouth disease. Acute, contagious, virus disease of cattle, sheep, and swine; externally marked by ulcers in and around the mouth and hoofs.

foramen. Any opening, orifice, or perforation, especially through bone. Pl. foramina.

foramen magnum. Posterior opening of the skull by which the nerve cord leaves the brain and skull.

foramen of Munro. Small opening between the lateral ventricles and third ventricle of the vertebrate brain.

foramen of Winslow. Epiploic foramen.

foramen ovale. Opening between the right and left auricles in embryonic mammals; normally closes at birth.

foraminal tubule. Small projecting tubule on some sponge gemmules; at germination the mass of amoebocytes is extruded through the aperture of the tubule.

Foraminifera. One of the orders of the Class Sarcodina; main bulk of the cell is enclosed within a simple or chambered and/or coiled shell or test composed of secreted calcium carbonate (usually), silicon dioxide, or bits of foreign material cemented together with an organic secretion; pseudopodia project from openings or pores in the shell; almost exclusively on the sea bottom in deep water, although a few spp. are pelagic.

forcipate mastax. Rotifer mastax in which the trophi are elongated for protrusion from the mouth and seizing prey.

Forcipulata. Order of echinoderms in the Class Asteroidea; marginal plates inconspicuous; pedicellariae with crossed jaws; includes the common genus _Asterias_.

forebrain. Prosencephalon; large anteriormost of the three primary divisions of the vertebrate brain; includes the telencephalon (olfactory lobes and cerebral lobes) and the diencephalon (thalamus and associated structures).

foregut. 1. Anterior portion of the digestive tract of arthropods; it is lined with cuticle, and this lining is shed when the arthropod molts; in some arthropods the foregut is very short, in others it is nearly half as long as the body. 2. Short blind tube in the anterior portion of many vertebrate embryos; it becomes a part of the anterior digestive tract.

Forelia. Widely distributed genus of Hydracarina.

foreskin. Prepuce.

forest-hog. See HYLOCHOERUS.

forewings. Anterior pair of wings on the second thoracic segment of typical insects.

forfex. In certain insects, a transverse paired scissors-like structure near the anus.

Forficula auricularia. European earwig which has been widely distributed in temperate and subtropical regions; household and garden pest; to 15 mm. long.

forktail. Any of numerous fishes and birds having a markedly forked tail.

formaldehyde. Colorless, penetrating gas, usually marketed as formalin, a 40% solution of the gas in water; disinfectant and preservative; CH_2O.

formalin. See FORMALDEHYDE.

formalin-aceto-alcohol. FAA.

formation. Major plant community or community-type.

formative cells. Special cells scattered between muscles and visceral organs of certain lower invertebrates; they produce new structures during regeneration.

Formica. Large and common genus of ants; most spp. build anthills of soil and debris over nests in ground; _F. sanguinea_ raids nests of _F. fusca_ (Negro ant) and makes slaves of them; _F. rufa_ is the common red ant of Europe.

formic acid. Colorless, liquid organic acid; found naturally in some ants, a few other insects, and in certain plants; H_2CO_2.

Formicariidae. Large family of insectivorous birds in the Order Passeriformes; antbirds; dull-colored forest spp.; some feed largely on ants; Central and South America.

formicary. Ants' nest.

Formicidae. Large family of polymorphic Hymenoptera comprising the ants; 2 to 25 mm. long; large pronotum, thin abdominal pedicel with scale or node; social, colonies of few to very many individuals; usually one functional queen per colony; scavengers, predaceous, or herbivorous; familiar genera are _Ponera_, _Eciton_, _Monomorium_, _Solenopsis_, _Pogonomyrmex_, _Atta_, _Formica_, _Iridomyrmex_, and _Camponotus_.

formicology. Myrmecology.

fornix. 1. Band of fibers connecting the cerebral hemispheres. 2. Arched recess or cavity. 3. Strengthening ridge above the base of each antenna in the Cladocera. Pl. fornices.

fossa. 1. Pit, depression, or trench, especially with reference to bone anatomy. 2. See CRYPTOPROCTA.

Fossa. Fanalokas; genus of genet-like mammals found in Madagascar; gray, black, and dirty whitish coloration.

fossa ovalis. Thin, partly membranous, shallow depression in the right atrium of the adult heart; marks the site of the aperture between the two atria in the fetus.

fossil. Remains or relic of an organism living at some former time; a fossil may be: an unaltered hard part (tooth, bone); a mold in rock; petrifaction (wood, bone, etc.); cast; tracks, burrows, or tubes; impression (especially of fossil plants); unaltered or partially altered soft parts (mammoths in frozen Siberian and Alaskan soils, skin and hair of ground sloths in dry caves).

fossorial. Pertaining to burrowing or digging animals, or to any structure adapted for burrowing or digging.

foul brood. Any of three bacterial diseases of honeybee larvae.

fouling organism. Any macroscopic organism that attaches to the submerged surfaces of boats, piling, and other underwater structures.

four-eyed fish. 1. Any of several tropical reef fishes having a dark eyelike spot on the posterior part of the dorsal fin or trunk. 2. See ANABLEPIDAE.

fourth ventricle. Cavity of the vertebrate hindbrain; especially evident in the medulla oblongata.

four-toed salamander. Hemidactylium scutatum; a salamander averaging only 2.5 in. long; upper surface reddish brown; found from Maine to Va., west to Minn. and Ark.

fovea. Small depression.

fovea centralis. Small depression in the retina of some vertebrates; point of especially acute vision; marked by the absence of rods and nerve fibers; lies within the macula lutea in man and some other primates.

Fovia. Genus of small marine triclad Turbellaria.

Fowler's toad. A subspecies of Woodhouse's toad; see BUFO.

fox. See UROCYON and VULPES.

fox snake. See ELAPHE.

fox sparrow. See PASSERELLA.

fox squirrel. See SCIURUS.

fractional mutant. Mosaic individual resulting from a dominant mutation occurring in one of the two daughter chromosomes formed during one of the early mitotic divisions of the embryo.

fragmentation. More or less mechanical breaking of an animal into two or more parts, each of which regenerates into a complete new animal; the term is occasionally used for Protozoa, but more commonly for nemertines, turbellarians, polychaetes, etc.

Frankliniella. Genus of thrips which contains several important pest spp. that feed on field and truck crops, cereals, and ornamental shrubs and trees.

Franklin's gull. Common gull of prairie lakes, Larus pipixcan; winters on South American Pacific Coast.

Fraser Darling effect. The larger the bird colony, the greater the percentage of breeding members and the percentage of young produced, and the shorter the breeding season.

Fratercula. Genus of sea birds in the Family Alcidae; puffins; common along shores of North Atlantic; to 1 ft. long; black above, white below; bill very deep, almost parrot-like, and banded with bright colors; expert swimmers and divers.

fraternal twins. Individuals born at the same time but originating from two different fertilized eggs; they may be of the same or different sexes. Cf. IDENTICAL TWINS.

freckles. Small brownish pigmented spots on the skin; often accentuated by exposure to bright sunlight; tendency for freckles is a Mendelian dominant.

Fredericella. Genus of fresh-water Bryozoa.

free-living. Carrying on an independent, non-parasitic existence.

Freemania. Genus of polyclad Turbellaria.

freemartin. Sterile female member of twins of unlike sexes in cattle; produced by the hormone effects of the male in the placental circulation.

free-swimming. Actively moving about in water or capable of moving about in water; opposite of sessile or attached.

free-tailed bat. See MOLOSSIDAE and TADARIA.

Fregata. See FRIGATE BIRD.

Fregatidae. Family of birds in the Order Pelecaniformes; includes the frigate birds.

Frenatae. One of the three suborders of insects in the Order Lepidoptera; moths and millers; mouth parts modified into a long, coiled sucking proboscis; antennae not enlarged at tip; venation of forewings and hind wings unlike; hind wing with one or more strong bristles attached to forewing during flight; wings folded rooflike over body when at rest; mostly nocturnal; pupa usually covered by a silken cocoon.

Frenesia. Genus of Trichoptera.

frenulum. 1. Any membranous fold;

frenum. 2. Fold of tissue between the lower surface of the tongue and the floor of the mouth. 3. Process on anterior margin of hind wing of Lepidoptera; attaches to posterior margin of forewing. 4. Thickened subumbrellar area of certain jellyfish.

frenum. Frenulum; fold of skin or mucous membrane which limits the movements of an organ, e.g. the membrane on the under surface of the tongue.

frequency distribution. The grouping of all individuals in various classes or categories, according to size or some other criterion or measurement.

Freyella. Genus of sea stars with many long slender arms.

friar bird. Any of several Australian honey eaters having no feathers on the head.

Fridericia. Genus of semiaquatic and fresh-water oligochaetes.

frigate bird. Man o' war bird; five large, tropical, oceanic spp. with a great wing expanse and long hooked beaks; nest on islands; Fregata.

frilled organ. See CESTODARIA.

frilled shark. See CHLAMYDOSELACHUS.

fringed tapeworm. See THYSANOSOMA.

fringed worm. See CIRRATULUS.

Fringillidae. Largest family of birds in the Order Passeriformes (about 700 spp.); found everywhere except in the Australian region; includes cardinals, grosbeaks, finches, sparrows, towhees, juncos, and buntings; bill short, stout, adapted for seed-eating.

fringing reef. Large, thick coral mass at the edge of an island or other land mass in tropical seas; it begins at the water's edge and extends out a few yards to a quarter mile; surface more or less exposed at low tide. Cf. BARRIER REEF and ATOLL.

frit fly. See OSCINOSOMA and CHLOROPIDAE.

Fritillaria. Genus of pelagic prochordates (Larvacea).

fritillaries. Group of several genera of butterflies in the Family Nymphalidae; usually orange-brown above and silver-spotted beneath.

frizzle fowl. Inherited defect in chickens marked by constant molting.

frog. Any of a large number of tail-less Amphibia; usually have smooth skin and are customarily found in water or damp places; more streamlined and better jumpers than toads; most frogs are in the Order Diplasiocoela. Cf. TOAD.

frogfish. See ANTENNARIIDAE.

froghopper. See CERCOPIDAE.

frogmouth. Any of certain Australian and Oriental goatsuckers; mouth opening exceptionally wide; ground insectivores; nest in trees; 12 spp. forming the Family Podargidae.

frog shell. See BURSA.

frons. Front or anterior surface of an insect head.

frontal bone. Large dermal bone covering the anterior surface of the brain in tetrapods and bony fishes; forms the forehead in man and contains frontal sinuses.

frontal glands. Glands having various functions, imbedded in the anterior end of certain Platyhelminthes and Nemertea.

frontalis muscle. Broad, flat muscle which covers the forehead and part of the top of the skull; pulls the scalp forward.

frontal lobe. Anterior portion of a cerebral hemisphere, especially in the higher vertebrates.

frontal lobotomy. Operation in which the white matter of a frontal lobe is cut by means of holes drilled through the skull; used for certain epileptic conditions.

frontal lunule. Small crescent-shaped area just above the base of the antennae of Diptera.

frontal plane. Horizontal longitudinal plane through the body of a bilaterally symmetrical animal, so as to separate dorsal and ventral halves.

frontal sinus. Especially in man, one of two cavities in the frontal bone above the eyes, nearly touching at the midline; they open into the nasal cavity.

Frontipoda. Uncommon genus of Hydracarina.

Frontonia. Large genus of marine and fresh-water holotrich Ciliata; ovoid to ellipsoid; 15 to 600 microns long.

frontoparietal. United frontal and parietal bones covering the general dorsal surface of the brain in the frog.

frostfish. See MICROGADUS.

fructose. Levulose.

fruit bat. See MEGACHIROPTERA.

fruit fly. 1. Any of a group of small flies in the Family Trypetidae; often breed in fruit and other parts of plants; many pest spp. 2. Specifically, Drosophila melanogaster (in the Family Drosophilidae) which is widely used in experimental genetics.

frustules. Nonciliated planula-like buds in some Hydrozoa; develop into polyps.

FSH. Follicle-stimulating hormone.

fucilia. Free-swimming euphausid larval stage just preceding the adult.

Fuertes, Louis Agassiz. American biological artist (1874-1927); famous for his bird paintings and murals and habitat groups in the American Museum of Natural History.

fulcrum. 1 Small scale or spine on the anterior margin of fins of many ganoids and a few teleosts. 2. Median portion of the incus in the trophi of rotifers. 3. Chitinous envelope at the base of the mouth in Diptera and Hymenoptera.

Fulgoridae. Lantern flies, a family of insects in the Order Homoptera; head rounded and more or less prolonged, often grotesque and hollow dorsally; tropical and subtropical herbivores.

Fulgur. Busycon.

Fulica americana. Common North American coot or mudhen in the Family Rallidae; general gray coloration with blackish head and neck, and white bill; often seen swimming and diving but not a true duck; feet not webbed, but each toe with lateral scallops.

fulmar. See FULMARUS.

Fulmarus glacialis. Marine bird in the Family Procellaridae; fulmar; common in Arctic regions but found as far south as N.J. coast; general coloration light to dark gray; settle on water to feed.

fumarase. Cellular enzyme which converts fumaric acid to malic acid.

fumaric acid. Intermediate metabolic compound formed in the citric acid cycle of cellular metabolism.

Funambulus. Genus of palm-squirrels; widely distributed tropical spp.; feed chiefly on palm tree fruits, leaves, and insects.

Funariidae. Large family of insectivorous birds in the Order Passeriformes; ovenbirds; mostly forest spp.; habits varied, some build clay nests; Mexico to Patagonia.

function. Characteristic role or action of any structure or process in the maintenance of normal metabolism and behavior of an animal.

fundatrix. Female aphid (usually parthenogenetic) which gives rise to a new colony by oviposition (or vivipary).

fundic gland. Type of gastric gland in the stomach wall; composed of three types of cells; secretes hydrochloric acid, pepsin, and mucin.

Fundulus. Killifishes, topminnows; genus of small omnivorous fishes in the Family Cyprinodontidae; to 4 in. long; mouth small, body compressed, and with dusky bars or bands on the sides; several widely distributed stream and pond spp.; F. heteroclitus (mummichog) is a sp. of coastal brackish waters from Newfoundland to Tex.

Fungia. Genus of solitary corals.

fungiform papillae. Mushroom-shaped papillae scattered over the surface of the tongue; usually 0.5 to 1.5 mm. high; a few have taste buds.

fungus ants. See ATTA.

fungus gnats. See MYCETOPHILIDAE and SCIARIDAE.

Funiculina. Genus of sea pens; stem short and stocky; polyps crowded in oblique rows; to 2 ft. tall; Atlantic Ocean.

funiculus. 1. Strand of fibrous connective tissue in Bryozoa; connects the lower end of the U-shaped digestive tract to the body wall. 2. Column of white matter in the spinal cord. 3. Umbilical cord. 4. Spermatic cord. 5. Slender, stalklike middle portion of the antenna in certain insects.

funnel-web spider. See AGELENIDAE.

furca. Collectively, the two caudal rami of copepods. 2. Anal leaping appendage in the Thysanura. 3. Forked sternal process in higher insects. 4. Any forked process.

Furcaster. Genus of extinct serpent stars.

furcate. Forked or branched.

furcilia. Larval stage of Euphausiacea.

furcula. 1. Embryological eminence on the floor of the pharynx which gives rise to the epiglottis. 2. V-shaped "wishbone" of birds; formed of the two clavicles and interclavicle. 3. See COLLEMBOLA.

furculum. Furcula.

Furipteridae. Furies; a small family of Tropical American bats.

Furnarius. See OVENBIRD.

furuncle. Boil.

fury. See FURIPTERIDAE.

fuscin. Accessory brown retinal pigment; functions chiefly in absorption and thus cuts down reflection and scattering; has no true visual function.

Fusconaia. Common genus of unionid mollusks; N.Y. to Minn. and south.

fusiform. Spindle-shaped, tapering at each end.

fusula. Small knoblike tip of a spider spinneret. Pl. fusulae.

G

g. 1. Abbr. for gram or grams. 2. Abbr. for genus.

gadfly. Horsefly or other nuisance fly that bites stock animals.

Gadidae. Family of codfishes in the Order Anacanthini; bottom spp. of the Northern Hemisphere; great commercial importance; fins soft-rayed; ventral fins large and located under pectorals or in front of them; to 100 lbs.; Gadus callarias the common North Atlantic codfish; Pollachius virens the common American pollack or bluefish; Microgadus tomcod the tomcod; Melanogrammus aeglefinus the common haddock; Urophycis includes several important spp. of marine hakes and longs; Lota lota (burbot, lawyer, fresh-water ling, eelpout, cusk) is the only fresh-water sp., Holarctic, in cold northern waters, to 30 in.

Gadiformes. Order of fishes which includes the Gadidae and related forms; equivalent to the Anacanthini.

Gadus. See GADIDAE.

gadwall. See ANAS.

gafftopsail. See BAGRE.

gag. Common grouper of the South Atlantic and Gulf coasts; fine food fish, to 40 lbs.

gaggle. Flock of geese, especially when on water.

galactagogic. Increasing the flow of milk.

galactose. Hexose sugar derived from the hydrolysis of milk sugar and some plant polysaccharides.

Galagidae. Family of nocturnal lemur-like mammals in the Suborder Lorisoidea; bush babies, galagos, ojams; tail long and bushy, ears and eyes large, fur soft and woolly; nocturnal; African forests; Galago common.

Galago. See GALAGIDAE.

Galbula. Genus of jacamars, a group of northern South American and Central American birds; bill long and pointed, coloration coppery or golden green.

Galbulidae. See JACAMARS.

galea. 1. Outer lobe or process attached to the stipes of a typical insect maxilla; often greatly modified. 2. Knob on the movable portion of a pseudoscorpion chelicera.

Galeichthys felis. Sea catfish in the Family Ariidae; a small marine sp. of southeastern U. S. coast; male incubates the eggs in its mouth.

Galen. Greek and Roman physician writer (130-200 A.D.); wrote many treatises on anatomy and physiology; showed that arteries carry blood (rather than air) and discovered much about the nervous system and pulse.

Galeocerdo. See TIGER SHARK.

Galeodes. Genus of Old World solpugids; one sp. attains nearly 3 in.

Galeopithecus. See DERMOPTERA.

Galeorhinus. See SOUPFIN SHARK.

Galeropygus. See CASSIDULOIDA.

Galerucella. Genus of chrysomelid beetles; adults and larvae feed on floating and emergent aquatic plants, especially lily pads.

Galidia. Genus of obscure primitive Madagascar mammals related to the civets.

gall. 1. Abnormal growth or swelling found on a wide variety of plants; produced chiefly by the presence and excretions of insect eggs and larvae imbedded in the plant tissues; each gall has a characteristic form and growth peculiar to the plant organ and the insect producing it; gall insects are mostly certain Hymenoptera, Diptera, and Homoptera but some mites and nematodes also produce plant galls. 2. Sore on the skin produced by irritation, especially in horses.

gall bladder. Small sac associated with the liver in many vertebrates; used for storing bile; has a duct which empties the bile into the digestive tract.

Galleria mellonella. Small gray or brown pest moth inhabiting hives of the honeybee; caterpillar feeds on the wax; adults are bee moths, larvae are waxworms.

Galleriidae. Small family of inconspicuous moths whose larvae are scavengers in nests of bees and wasps or feed on dead animal and vegetable matter; waxworms and bee moths; Galleria common.

gallfly. Fly or other insect which produces plant galls.

gall gnats. See CAECIDOMYIIDAE.

Galliformes. Large order of birds which includes the grouse, quail, pheasants, turkeys, etc.; bill short, feet specialized for scratching and running; nest mostly on ground; feed chiefly on plant material.

gallinaceous. Pertaining to birds in the Order Galliformes.

Gallinula. One of several genera of gallinules in the bird Family Rallidae; henlike spp. which both swim and wade in fresh-water shallows; temperate and tropical America; G. chloropus is the common gallinule of eastern U.S.

gallinule. See GALLINULA and PORPHYRULA.

galliwasp. 1. Any of several harmless Central American lizards. 2. Lizard fish (Synodus foetens), a spindle-shaped marine sp. of the Caribbean.

gall midges. See CAECIDOMYIIDAE.

gall mites. See ERIOPHYIDAE.

gallstone. Concretion formed in the gall bladder or bile duct.

Gallus. Genus of birds in the Family Phasianidae; includes the wild Asiatic jungle fowl (G. bankiva) and the modern domestic varieties of chickens (G. domesticus).

gall wasps. See CYNIPIDAE.

Galumna. Genus of mites in the Family Oribatidae; usually in moss or other vegetation.

Galumnidae. Oribatidae.

Galvani, Luigi. Italian surgeon and anatomist (1737-1798); made contributions to comparative anatomy and bioelectricity.

galvanotropism. Tendency of an aquatic organism to turn or move in a particular direction when subjected to an electric current in its surroundings.

Gamasidae. Family of mites parasitic on birds and mammals; spider mites, beetle mites.

Gambel's quail. See LOPHORTYX GAMBELI.

Gambusia. See POECILIIDAE.

gamete. Mature, haploid, functional sex cell (egg or sperm) capable of uniting with the alternate sex cell to form a zygote; the two types of gametes are usually very unequal in size, the egg being the larger. (A few plants and certain animals, such as rotifers, gastrotrichs, and cladocerans, produce "gametes" which are actually diploid and invariably develop into female offspring.)

gametocyte. Cell which undergoes meiosis and forms gametes; the term is used especially with reference to certain stages in the life history of many sporozoan Protozoa.

gametogenesis. Complex of processes by which oogonia become ova and spermatogonia become sperm.

gamma globulin. One of several blood proteins which are especially effective against measles and epidemic hepatitis.

Gammarus. Genus of marine and fresh-water Amphipoda.

gammexane. Powerful insecticide, $C_6H_6Cl_6$.

gamogenesis. Sexual reproduction by means of male and female gametes.

gamont. See SPORONT.

ganglion. 1. In invertebrates, a discrete mass of nervous tissue containing an abundance of nerve cell bodies and located either within or outside the central nervous system. 2. In vertebrates, a mass of nerve cell bodies (often indistinct) outside the central nervous system; a center of nervous influence. (Some of the "nuclei" of the vertebrate brain are sometimes inappropriately called ganglia.) Pl. ganglia.

gangrene. Death and necrosis of a large mass of tissue; often produced by failure of the blood supply.

gannet. See MORUS and SULA.

ganoid. Pertaining to primitive palaeopterygian fishes, such as Polypterus and Acipenser.

ganoid scales. Thick, bony, rhombic scales, not overlapping, which cover the body of some primitive bony fishes, e.g. the gars; surface covered with enamel-like material, ganoin; similar to cosmoid scales but all three layers grow by increasing in thickness. Cf. COSMOID SCALES.

ganoin. See GANOID SCALES.

Ganonema. Uncommon genus of Trichoptera; eastern U.S.

gape. Distance between the tips of the open jaws of vertebrates.

gaper clam. West Coast marine bivalve which lies deeply buried with the long siphons extending to the surface of the mud; Schizothaerus common.

gapeworm. Parasitic nematode (Syngamus trachealis); in trachea and bronchi of birds.

gar. Garfish, gar pike; see LEPISOSTEUS and BELONIDAE.

garden centipede. See SYMPHYLA.

garden flea. See PAPIRIUS.

garden fleahopper. See HALTICUS CITRI.

garden spider. See ARGIOPE.

garfish. Gar pike, gar; see LEPISOSTEUS and BELONIDAE.

Gari. Genus of marine bivalves; sandy bottoms near low tide line.

Garibaldi. See HYPSYPOPS.

gar pike. See LEPISOSTEUS and BELONIDAE.

Garrulus. Typical genus of Old World jays; includes the common European jay, G. glandarius.

Garrupa nigrita. Black jewfish or Warsaw, a heavy-bodied sport and food fish found south of the Va. coast; to 6 ft. long.

garter snake. See THAMNOPHIS.

Garveia. Genus of marine hydrozoan coelenterates.

Garypus. Genus of pseudoscorpions found along sea coasts.

gas gangrene. Infection of a wound with one or more spp. of bacteria in the genus Clostridium; often a fatal infection in which muscles and subcutaneous tissues become filled with gas.

gas gland. Any gland which secretes a gas (or air); e.g. in the wall of a fish swim bladder and in the float of some large pelagic coelenterates.

Gasserian ganglion. Semilunar ganglion; a ganglion on the larger root of the fifth cranial nerve of vertebrates; gives off the ophthalmic and maxillary nerves.

gaster. Rounded abdominal region behind the basal node in ants.

Gasterophilidae. Family of insects in the Order Diptera; horse botflies; larvae occasionally in dog, rabbit, and man; Gasterophilus common.

Gasterophilus intestinalis. Stomach bot of the horse; maggots parasitic on the stomach wall of the horse; in the fly Family Gasterophilidae.

Gasteropoda. Gastropoda.

Gasterosteidae. Family of bony fishes which includes the sticklebacks; small spp. with 2 to 11 stout spines along the midline in front of dorsal fin; skin scaleless, sometimes bony; fresh waters and along shores of cold seas in the Northern Hemisphere; construct elaborate oriole-like nests in which eggs are deposited and guarded; Gasterosteus and Eucalia common.

Gasterosteiformes. Order of fishes which includes the Gasterosteidae and related families; Scleroparei.

Gasterosteus. See GASTEROSTEIDAE.

Gasterostomata. Group of families of digenetic trematodes in which the proctostome is in the middle of the ventral surface.

Gasterostomum. Bucephalus.

gastraea. Hypothetical ancestral organism from which primitive flatworms are assumed to have arisen; it had the structure of a gastrula larva but was an abundant adult organism.

Gastranella. Genus of very small marine bivalve mollusks; East Coast.

gastric. Pertaining to the stomach.

gastric caecum. One of a series of elongated, hollow projections of the digestive tract, chiefly in the midgut of certain insects; variously for absorption, digestion, and storage.

gastric filaments. Short gastrodermal tentacles in the gastrovascular cavity of scyphozoans, especially near the proctostome and in the gastric pouches; they bear large numbers of nematocysts.

gastric glands. Digestive glands imbedded in the wall of the stomach, or special glandular organs associated with the stomach. See CARDIAC, FUNDIC, and PYLORIC GLANDS.

gastric juice. Fluid secreted into the stomach by gastric gastric glands; in man gastric juice contains a mucin, hydrochloric acid, pepsin, rennin, and perhaps gastric lipase.

gastric lipase. Fat-digesting enzyme which is thought to be present in small amounts in human gastric juice.

gastric mill. Complex arrangement of grinding teeth, ossicles, and stout setae in the pyloric chamber of the stomach of decapod Crustacea and many other Malacostraca; grinds, macerates, and sieves food.

gastric pouch. One of four main enlargements of the gastrovascular cavity arranged around the center of a jellyfish body.

gastric shield. Tough, sclerotized plate lining a part of the stomach in many bivalve mollusks; serves as a surface against which the crystalline style rubs and is worn away, thus releasing carbohydrate-digesting enzymes.

gastric tentacles. Gastric filaments.

gastrin. Hormone secreted by the mucosa of the stomach; stimulates production of gastric juice.

Gastrioceras. Extinct genus of Cephalopoda.

Gastrochaena. Genus of burrowing marine bivalve mollusks; able to bore into sandstone and limestone.

gastrocnemius. Largest muscle of the tetrapod shank; flexor of the leg and extensor of the foot.

gastrocoel. Archenteron.

gastrocolic reflex. Peristaltic colonic wave resulting from the entrance of food into an empty stomach.

Gastrocopta. Large genus of land snails.

gastrodermis. Single layer of cells lining the gastrovascular cavity of the Coelenterata, Ctenophora, and Platyhelminthes; many of the individual cells carry on intracellular digestion of food particles; in older usage this adult tissue is usually called endoderm or entoderm.

Gastrodes. Genus of aberrant ctenophores; a minute, bowl-shaped parasite in certain tunicates for a portion of the life history; larva a planula.

gastroenteritis. Inflammation of both stomach and intestinal tract.

gastrohepatic ligament. Lesser omentum.

gastrolith. 1. One of two calcareous bodies in the wall of the cardiac chamber of the stomach of the crayfish and certain other Malacostraca; thought to store lime salts in readiness for hardening the new exoskeleton after a molt. 2. Any of a group of small pebbles or gravel particles in the stomach of certain fishes and reptiles; presumably function in macerating food.

Gastrophilus. Gasterophilus.

Gastrophryne. Genus of southern U.S. and Central American toads in the Family Microhylidae; toe webs reduced or absent.

Gastropoda. One of the six classes in the Phylum Mollusca; snails, slugs, limpets, whelks, nudibranchs, etc.; usually with an asymmetrical spiral one-piece shell lined with mantle and containing much of the viscera; foot well developed and flattened; head with one or two pairs of tentacles; radula present; monoecious or dioecious; development variable, marine spp. usually with trochophore and veliger larvae; marine, fresh-water, and terrestrial; about 80,000 spp.

gastropores. Openings on the surface of a millepore coelenterate skeleton through which the gastrozooids protrude.

Gastropteron. Genus of sea slugs in the Suborder Tectibranchia.

Gastropus. Genus of plankton rotifers; lorica rigid, foot annulated.

gastrosplenic vein. Vein which drains the stomach and spleen.

gastrostege. One of the large ventral scales anterior to the anus in most snakes.

Gastrostyla. Genus of marine and freshwater hypotrich Ciliata.

Gastrotaenia. See APORIDEA.

gastrotheca. That part of an insect pupal case which covers the abdomen.

Gastrotheca. Genus of South American "marsupial frogs" in the Family Hylidae; eggs carried in a skin sac on the back.

Gastrotricha. Phylum which includes the gastrotrichs; microscopic wormlike animals found on substrates in fresh and salt waters; ventral surface flattened and with (usually) two longitudinal bands of cilia which produce gliding locomotion; body covered with a translucent cuticle which is often scaly or spinous; there are two posterior toes, a pseudocoel, several longitudinal muscle fibrils, flame bulbs in fresh-water spp., and a complete digestive tract; marine spp. hermaphroditic, fresh-water spp. parthenogenetic; about 400 spp.; Chaetonotus most familiar.

gastrovascular cavity. An extensive body cavity in the Coelenterata, Trematoda, and Turbellaria; functions in both digestion and circulation and has a single opening (proctostome) which serves as both mouth and anus (with the possible exception of certain Ctenophora).

gastrozooid. 1. Ordinary hydrozoan feeding polyp. 2. Food-gathering individual in certain pelagic colonial tunicates.

gastrula. Early embryonic stage of many animals; usually it is a double-layered cup, produced by the inpushing of the vegetal pole and/or overgrowth of the animal pole; the blastocoel is nearly obliterated, and the new cavity of the cup is the archenteron, the opening to the archenteron being the blastopore.

gastrulation. Process of gastrula formation in most embryos.

gata. Nurse shark.

Gattyana. Genus of Polychaeta; often in tubes of other polychaetes.

gaur. Large wild ox of India (Bibos); a forest sp. attaining 6 ft. at the shoulder.

Gause's rule. Theory which maintains that two spp. having essentially the same niche cannot coexist in the same habitat; the population of one of the two spp. soon dies out.

Gavia. See GAVIIFORMES.

gavial. See GAVIALIS and TOMISTOMA.

Gavialidae. Family of reptiles in the Order Loricata; includes the gavials.

Gavialis gangeticus. Sp. of gavial in the Family Gavialidae found in rivers of northern India; to 30 ft. long; general habits and structure similar to those of alligators and crocodiles, but the snout is extremely long and slender, like the handle of a frying pan; a timid sp.

Gaviidae. Family of birds which includes the loons; see GAVIIFORMES.

Gaviiformes. Small order of birds which includes the loons; four large, fish-eating spp. found in lake districts of northern half of Northern Hemisphere; occasionally on ocean bays; legs short, toes fully webbed; strong swimmers, divers, and fliers; Gavia immer the common American loon, to 3 ft. long.

gayal. Large wild domesticated ox (Bibos) of the Assam valley.

Gazella. Genus of ruminants in the Family Bovidae; includes a wide variety of gazelles having many local names; small to medium-sized graceful antelopes found in northern and eastern Africa, Arabia, western and central Asia, India, and Mongolia; open plains and desert regions, occasionally in brushy country.

GDH. Growth and differentiation hormone secreted by prothoracic glands of certain insects.

Gecarcinus. Genus of tropical land crabs.

gecko. See GEKKONIDAE.

Gegenbaur, Karl. German zoologist (1826-1903); influential university teacher who emphasized homology and evolution; demonstrated that vertebrate eggs are single cells; author of the classical Vergleichende Anatomie der Wirbeltiere.

Gekkonidae. Family of lizards which includes the geckos; small, climbing spp. with adhesive pads on toes; tongue protrusible; mostly nocturnal; worldwide in warm areas; Phyllodactylus, Coleonyx, Hemidactylus, and Sphaerodactylus typical.

gel. See COLLOIDAL SYSTEM.

gelada. See THEROPITHECUS.

Gelastocoridae. Family of small hemipterans found along the shores of lakes and streams; toad bugs; robust, warty bodies.

gelatin. An amorphous protein material prepared commercially from collagen in tendons, bones, ligaments, and skin; soluble in hot water, forms a jelly on cooling, and hard and flexible when dry; many uses in biology.

gelatinous connective tissue. Type of connective tissue found in the mammalian umbilical cord; the cells have branching processes, and the spaces between are filled with a gelatinous matrix.

gelation. Conversion of the more liquid phase of protoplasm into the more solid, jelly-like phase.

Gelechiidae. Large family of small moths which includes many pest spp.; Sitotroga and Pectinophora are important genera.

Gellius. Genus of marine sponges in the Subclass Monaxonida; irregular, 1 to 2 in. across; common on Atlantic shores.

Gelochelidon nilotica. Cosmopolitan gull-billed tern; in U.S. breeds along coast from Va. to Tex. and in the Salton Sea.

gem clam. See GEMMA.

Gemellaria. Genus of marine Bryozoa; colonies to 10 in. high.

Gemma gemma. Small, smooth, marine bivalve; gem clam; yellow, white, or pink; common in sandy bottoms of East Coast; introduced on West Coast.

gemellus. One of two tetrapod muscles which rotate the extended thigh and abduct the flexed thigh.

gemmule. 1. Special wintering-over cystlike or budlike device of freshwater sponges; such spherical bodies are formed at the base of the sponge in late summer or autumn; a central mass of amoebocytes is surrounded by one or two secreted organic membranes which, in turn, are covered with an external layer of spicules; diameter about 1 mm.; able to withstand freezing and drying, and when environmental conditions become favorable they germinate to produce new sponges. 2. Fine dendritic process at a synapse. 3. Any bud from which a young organism develops.

gemsbok. See ORYX.

Gemuendina. Genus of extinct placoderm fishes.

gena. 1. Side or lateral surface of an insect head. 2. Feathered portion of a bird's mandible. 3. Lateral area of a trilobite cephalic shield. Pl. genae.

gene. Determiner of one or more hereditary characters; a discrete unit of inheritance carried on a chromosome or a definitive region of a chromosome; transmitted from one generation to the next by a gamete; consists chiefly of nucleo-protein and is submicroscopic in its largest dimension; each chromosome complement often consists of thousands of genes; chemically, a gene appears to be composed of deoxyribonucleic acid (DNA), histone (a simple protein), residual protein, and a small amount of ribonucleic acid (RNA).

gene flow. Spreading of particular genes within a population or between two adjacent populations.

gene frequency. Relative occurrence of a given allele in any population.

gene pool. Collectively, all of the alleles of all the genes in a population; a measure of evolution of the population in terms of changing gene ratios in the population as a whole.

genera. Pl. of genus.

general color resemblance. Color pattern and coloration of an animal which produce a blending with its general background coloration and design, as in the ptarmigan, cottontail rabbit, desert reptiles, etc.

generic name. Particular name used for all members of a genus; invariably capitalized; see GENUS.

Gene's organ. Glandular structure in female ticks; functions at oviposition.

genet. See GENETTA.

genetic. Pertaining to genes, heredity, or development.

genetic drift. Sewall Wright effect; chance of random fixation or loss of an allele in a small population during the course of a few generations, or large fluctuations in the allele between any two consecutive generations.

genetic isolation. Reproductive incompatibility of two or more populations owing to differences in chromosome and gene makeup; such differences may be major or only very minor.

genetic map. Linkage map.

genetics. Study of heredity and variations in organisms; from a practical standpoint, it deals with the development of improved strains and breeds.

Genetta. Genus of small catlike mammals in the Family Viverridae; includes the genets; body and tail long; the common sp. is black and gray; related to the civets; Africa and southern Europe.

Genetyllis. Genus of Polychaeta.

genic balance. Phenotypic balance of genes in two to many different chromosomes, especially with reference to sex phenotype; the abnormal occurrence of an extra male or female chromosome causes an unbalance and the formation of a supermale or superfemale.

geniculate. Bent sharply, as a knee.

geniculate ganglion. Ganglion of the facial nerve.

genital. Pertaining to reproduction or the organs of reproduction.

genital atrium. Small cavity which opens to the outside by means of a genital pore in certain invertebrates, especially the flatworms and snails; either or both male and female reproductive ducts open into it.

genital field. Group of structures closely associated with the genital pore in Hydracarina; plates, cupules, folds, knobs, setae, etc.

genital fold. Transverse flap on the anterior ventral portion of the abdomen of certain spiders; covers the median genital pore.

genitalia. Reproductive organs, especially the external reproductive structures in higher vertebrates.

genital plate. One of four specialized skeletal plates surrounding the periproct in sea urchins; each plate bears a genital pore; (the fifth plate in the series is the madreporite).

genital pore. Any pore which permits the release of eggs, sperm, (both), or zygotes from the body; the term is usually restricted to invertebrates.

genital ridge. Early ridgelike swelling of embryonic peritoneum marking the location of the future gonad.

genitourinary. Pertaining to the genital and excretory organs, especially in vertebrates.

genome. Genom; one complete haploid set of chromosomes in any sp.; a normal somatic cell contains two genomes.

genotype. 1. In taxonomy, the type sp. of a genus. 2. In genetics, the category to which an individual belongs on the basis of its genetic make-up, without regard to its external appearance.

gentle lemur. See HAPALEMUR.

genus. Taxonomic category including one or more spp. which have certain fundamental characteristics in common; e.g. the genus Felis includes all of the cats: domestic cat, lynx, ocelot, mountain lion, etc.; each genus name is invariably capitalized. Adj. generic, pl. genera.

geobiont. Organism spending all of its life in soil.

Geocapromys. Genus of short-tailed hutias; nocturnal spp. with rabbit-like habits; West Indies.

Geocentrophora. Genus of small Turbellaria.

Geococcyx californianus. Common road-runner or chaparral cock in the Order Cuculiformes; a long, slender bird with long tail and strong legs; usually runs rather than flys; southwestern U. S. and Mexico.

Geodia. Genus of sponges in the Order Tetractinellida.

geoduck. Gweduc; extremely large burrowing clam (Panope generosa) of intertidal beaches and mud flats of Alaska to Calif.; an excellent food sp.; siphons united and very long.

Geogale. Genus of water-tenrecs; rat-sized mammals with webbed feet and long compressed tail; Madagascar streams.

geographic barrier. Landscape feature which prevents the further spread of a taxon and/or prevents interbreeding of two closely related taxa separated by that feature; e.g. a desert, large river, mountain range, etc.

geographic distribution. Geographic range.

geographic isolation. Prevention of interbreeding between two or more populations because of their separation by climatic or geographic barriers.

geographic range. Collectively, the entire area of land and/or water over which a particular sp. or other taxon occurs naturally.

geological range. Occurrence of a particular sp. or other taxonomic group during past geological times.

geological time table. Various major divisions of past time, designated in accordance with their general occurrence, geology, sedimentary deposits, and fossils; these are outlined below, with the approximate elapsed time since each one began; each term is defined in the appropriate place in this volume.

Cenozoic era
- Quaternary period
  - Recent epoch (20,000)
  - Pleistocene epoch (1,000,000)
- Tertiary period
  - Pliocene epoch (12,000,000)
  - Miocene epoch (25,000,000)
  - Oligocene epoch (40,000,000)
  - Eocene epoch (60,000,000)
  - Paleocene epoch (70,000,000)

Mesozoic era
- Cretaceous period (120,000,000)
  - Upper epoch
  - Lower epoch
- Jurassic period (155,000,000)
  - Upper epoch
  - Middle epoch
  - Lower epoch
- Triassic period (190,000,000)
  - Upper epoch
  - Middle epoch
  - Lower epoch

Paleozoic era
- Permian period (215,000,000)
- Carboniferous period (250,000,000)
  - Pennsylvanian epoch
  - Mississippian epoch
- Devonian period (350,000,000)
  - Upper epoch
  - Middle epoch
  - Lower epoch
- Silurian period (390,000,000)
- Ordovician period (480,000,000)
  - Upper epoch
  - Middle epoch
  - Lower epoch
- Cambrian period (550,000,000)
  - Saratogan epoch
  - Acadian epoch
  - Waucoban epoch

Proterozoic era (1,000,000,000)

Archeozoic era (1,800,000,000)

Geometridae. Large family of fragile moths; naked larvae often called inchworms, cankerworms, or measuring worms because of their looping locomotion; feed on foliage; many pest spp.; Palecrita an important genus.

Geomyidae. Family of herbivorous rodents which includes the pocket gophers; robust spp. with short neck and legs; forelegs modified for digging; fossorial, seldom seen on the surface of the ground but in evidence through the piles of earth thrown out by its excavations; Thomomys and Geomys typical.

Geomys. Genus of pocket gophers in the Family Geomyidae; about four spp. in the plains and prairies from the Mississippi Valley to the foothills of the Rockies and the coastal plain; upper incisors grooved.

Geonemertes. Genus of nemertines in the Order Hoplonemertea; some spp. parasitic in the branchial cavity of tunicates; others in moist tropical soil.

Geophilomorpha. Order of centipedes; 31 to 181 pairs of legs; Geophilus common.

Geophilus. Common genus of centipedes.

Geoplana. Very large genus of tropical South American terrestrial Turbellaria; introduced into U.S. greenhouses.

Geospizidae. Family of land birds found on the Galapagos Islands; mostly finchlike; about 40 spp.

Geospizinae. Subfamily of finches endemic to the Galapagos Islands; "Darwin's finches."

geotaxis. Behavioral response of an animal to gravity; spp. that move or climb upward, away from the earth, are negatively geotactic; those that move downward or remain on the ground are positively geotactic.

Geothlypis. See YELLOWTHROAT.

Geotrupes. Genus of lamellicorn beetles in the Family Scarabaeidae.

Gephyrea. Sipunculoidea.

Gephyroneura. Tribe of Diptera in the Suborder Orthorrhapha.

gerbil. Small burrowing and jumping rodent, found especially in Egypt and South Africa; about as large as a rat, with large eyes and ears, and long hind legs and tail; Gerbillus iateronia.

Gerbillus. See GERBIL.

gerenuk. See LITOCRANIUS.

gerfalcon. Gyrfalcon; see FALCO.

geriatrics. Medical and clinical study of old age, aging, senility, senescence, and associated diseases.

germ. Common name of any minute disease-producing organism, including bacteria, viruses, protozoans, molds, rickettsias, and spirochaetes.

germarium. 1. Among certain insects, the end of an ovarian or testicular tube containing oogonia or spermatogonia. 2. Egg- or sperm-producing part of a gonad, especially in certain invertebrates. 3. Rarely, an ovary.

germ band. In a developing insect egg, the ventral portion of the blastoderm which thickens and grows into the embryo.

germ cells. Gametes, or the direct antecedent cells of gametes.

germinal. Pertaining to a germ cell or embryonic structure.

germinal disc. 1. Small area on a large fertilized egg in which the first trace of an embryo can be seen, as in a hen's egg. 2. Small, disclike, cytoplasmic area of a meroblastic egg; undergoes cleavage.

germinal epithelium. Outermost layer of cells in a testis, testicular tubule, or ovary; gives rise to the primordial germ cells by proliferation inward.

germinal layer. 1. Lowermost layer of cells in a stratified epithelium; gives rise to the cells above it by proliferation. 2. Thin lining of a hydatid cyst; proliferates scolices and daughter cysts.

germinal vesicle. Greatly enlarged nucleus of an oocyte during prophase of first meiotic division.

germinative layer. Germinal layer.

germ layer. Ectoderm, endoderm, and mesoderm as distinguished in an animal embryo during and immediately following gastrulation; all phyla above Protozoa are now recognized by most zoologists as having all three germ layers.

germ nucleus. Nucleus of egg or sperm.

germ plasm. 1. Collectively, the germ cells of an organism as distinct from the body cells, or somatoplasm. 2. As propounded by Weismann, a particular kind of protoplasm transmitted essentially unchanged from one generation to the next through the germ cells, and giving rise to the surrounding body cells of each individual.

germ ring. General area of surface cells surrounding the blastopore in the gastrula stage of development of some animals.

gerontology. Study of old age, its phenomena, diseases, etc.

gerontomorphosis. Theory suggested by de Beer to explain the appearance of certain major taxonomic groups during evolution by means of paedomorphosis and neoteny; e.g. insects may have arisen from millipede larvae which had their embryological development arrested at an early six-legged stage and which then became neotenic.

Gerrhonotus. Genus of alligator lizards or plated lizards in the Family Anguidae; pugnacious spp. with tail about twice the body length; western U.S.,

Mexico, and Central America.

Gerridae. Family of insects in the Order Hemiptera; water striders; slender, long-legged, predaceous spp. of medium size which stride about rapidly on the surface film of ponds and streams; a very few salt-water spp.; Gerris and Halobates common.

Gerris. Common genus of insects in the Family Gerridae.

Gersemia. Genus of sea anemones in the Order Alcyonacea; to 8 in. long.

Geryonia. Genus of marine hydrozoans in the Order Trachylina; medusa with six radial canals; polyp reduced or absent.

Gesner, Konrad. Swiss physician and zoologist (1516-1565); wrote an immense four-volume Historia Animalium, a descriptive natural history of vertebrates and insects.

gestation. Period of carrying the young in the uterus, from conception to delivery, especially with reference to the length of this period, such as the gestation period of nine months in man.

ghost crab. See OCYPODE.

ghostfish. Any of a wide variety of whitish or translucent marine fishes.

ghost shrimp. Callianassa affinis, a small, white, burrowing, crayfish-like crustacean of muddy tide pools; to 2.5 in. long; Calif. to Mexico.

giant armadillo. Tatuasu; largest living armadillo, Priodontes giganteus; 3 ft. long, not including tail; South America.

giant axon. Giant fiber.

giant band shell. Horse conch, Fasciolaria gigantea; a massive conch to 2 ft. long; N.C. to Brazil.

giant chromosomes. Unusually large chromosomes occurring especially in the salivary glands of certain Diptera; useful for detailed cytological study.

giant clam. See TRIDACNA.

giant coccid. See MARGARODIDAE.

giant crab. See MACROCHEIRA.

giant fiber. Nerve fiber of unusually large diameter having the property of transmitting impulses much more rapidly than normal fibers; found in some annelids, crustaceans, squids, and a few vertebrates; in the earthworm there are three giant fibers on the dorsal surface of the nerve cord.

giant panda. See AILUROPODA.

giant salamander. See MEGALOBATRACHUS.

giant silkworm moth. See SATURNIIDAE.

giant tortoise. See TESTUDO.

giant water bug. Any member of the hemipteran Family Belostomatidae.

giant water-shrew. See POTAMOGALIDAE.

Giardia. Genus of pyriform flagellate Protozoa; ventral surface with an anterior sucker disc; eight flagella, two nuclei; in vertebrate digestive tract; G. lamblia is a common commensal in the human intestine although some parasitologists are convinced that it causes "flagellate diarrhea."

gibbon. See HYLOBATES.

gibbous. Markedly convex, rounded.

Gibb's mole. See NEUROTRICHUS.

gid. Disease of sheep produced by a larval tapeworm (coenurus) in the brain; see MULTICEPS.

Giemsa stain. Stain for blood smears; composed of methylene azure and methylene blue.

gigantism. See PITUITARY GIANT.

Gigantopithecus blacki. Fossil anthropoid known only from three giant molars, possibly originating from caves in southern China.

Gigantorhynchus gigas. Macracanthorhynchus hirundinaceus.

Gigantostraca. Merostomata.

Gigantura. See GIGANTUROIDEA.

Giganturida. Giganturoidea.

Giganturoidea. Small order of marine bony fishes; lower lobe of tail elongated; pectoral fins with many long soft rays; pelvic fins lacking; Indian and Atlantic oceans; Gigantura typical.

Gila. Difficult genus of chubs in the Family Cyprinidae; Great Basin, Pacific, and Rio Grande drainages.

gila monster. See HELODERMATIDAE.

gill. Aquatic respiratory organ; usually a thin-walled projection from some part of the external body surface or from some part of the digestive tract.

gill arch. 1. Tissue separating adjacent gill slits in chordates. 2. Cartilaginous supporting material in a typical fish gill.

gill bailer. Scaphognathite.

gill bar. 1. One of many dorsoventral rodlike thickenings of pelecypod gill lamellae. 2. Gill arch.

gill book. Book gill.

gill chamber. 1. Space on either side of the cephalothorax in many malacostracan Crustacea; contains gills; formed by the expansion and downward projection of the branchiostegite. 2. Occasionally, pharyngeal gill area of tunicates, Amphioxus, etc.

gill cleft. 1. In early chordate embryology, the visceral furrow indicative of a gill pouch and its subsequent gill slit. 2. Gill slit.

gill cover. Operculum of fish.

gill filament. One of many finger-like subdivisions of a gill, especially in fishes, cephalopods, gastropods, etc.

Gillia altilis. Fresh-water snail found from N.J. to S.C.

gill lamella. One of two thin plates making up a pelecypod gill.

gill plate. 1. Thickened area on each side of the anterior region of an early amphibian embryo; destined to become a gill and gill slit. 2. Gill lamella.

gill pouch. 1. In early chordate embryology, an evagination of the pharyngeal wall indicative of a gill cleft and subsequent gill slit. 2. Pouchlike gill slit of cyclostomes.

gill raker. One of many wartlike, filamentous, finger-like, or branched projections on the inner border of each gill arch in a typical fish; prevents hard particles and food from passing out through the gill slits.

gill ray. One of several to many short cartilaginous rods which project laterally in gill arches and stiffen the arches.

gill separator. Epipodite.

gill slits. Vertical, elongated, slitlike openings between the pharynx and the external surface in chordates; present in all chordate embryos but persistent only in adult prochordates, fishes, and a few amphibians.

Gilson's fluid. General killing and fixing agent; consists of 5 g. mercuric chloride, 5 ml. 80% nitric acid, 1 ml. glacial acetic acid, 25 ml. 70% alcohol, and 220 ml. water.

gingiva. The gum in the oral cavity.

Ginglymodi. Order of voracious bony fishes; includes the gar pikes and alligator gars; body very long, with long narrow jaws bearing stout conical teeth; diamond-shaped ganoid scales in oblique rows; Lepisosteus and Atractosteus typical.

Ginglymostoma. See NURSE SHARK.

Giraffa. Genus of ruminants in the Family Giraffidae; includes the two familiar giraffes of Africa; the northern sp. has a red chestnut coat with a fine network of white lines, and there are three short horns; the southern sp. has a creamy or yellowish-white ground color marked by irregular blotches, and only two horns; to 19 ft. tall; feed mostly by browsing on trees.

Giraffidae. Family of African ruminants which includes the giraffes (Giraffa) and okapi (Okapia); neck and legs very long.

girdle. 1. Peripheral portion of the mantle which is not covered by the calcareous plates in typical chitons. 2. Equatorial furrow of dinoflagellates. 3. See PECTORAL GIRDLE and PELVIC GIRDLE.

Girella nigricans. Opaleye, green perch, or Catalina perch; favorite sports fish for surf casters from Monterey Bay south; to 6 lb.

gizzard. Anterior part of the digestive tract in many animals; the wall is thick and muscular, and the organ functions in softening and macerating the food previous to digestion further on in the digestive tract; e.g. well developed in birds, reptiles, and the earthworm.

gizzard shad. Compressed, silvery, herring-like fish found in ocean shallows and fresh waters from Cape Cod to Colo. and Tex.; mouth inferior; chiefly a plankton feeder; important forage fish; to 12 in. long; Dorosoma cepedianum.

glabella. 1. Thickened median portion of the head of a trilobite; continues posteriorly as the rachis. 2. In man, the area between the inner ends of the eyebrows.

glabrous. With a smooth surface, hairs and projections being absent.

glacial periods. Various long periods during earth's history when extensive ice sheets covered much of the surface of the land, especially during Pleistocene, Carboniferous, Permian, and Proterozoic times.

gladiolus. Mesosternum.

gladius. Internal shell or "pen" of a cuttlefish or squid.

glair. 1. White of an egg. 2. Viscous material secreted by a female crayfish at the time her eggs are extruded.

gland. Structure elaborating one or more excretions or secretions which are discharged to the outside of the gland; ranges from a single cell to a very large aggregate; e.g. the liver.

glands of Bartholin. Vestibular glands.

glands of Meibom. Branched acinous glands in the eyelids; ducts open onto the eyelid margins.

glands of Moll. Large modified sweat glands along the margin of each eyelid.

glandular epithelium. Any epithelium

forming the secretory surface of a gland.

glandulomuscular cells. Cells of the basal disc of fresh-water. hydroids; secrete mucus.

Glans. Genus of small marine bivalves.

glans clitoridis. Tip of the clitoris in female mammals.

glans penis. Tip of the penis in male mammals.

Glareola. See PRATINCOLE.

Glareolidae. Family of birds in the Order Charadriiformes; coursers and pratincoles; aberrant insect-eating shorebirds of sandy deserts in Africa, southern Asia, and Australia.

Glaridacris. Genus of tapeworms found in fresh-water fishes.

glass shrimp. Any of several small marine shrimps which are quite translucent or transparent.

glass snake. See OPHISAURUS.

glass sponge. Common name for any sponge in the Class Hexactinellida.

glassworm. Any member of the Phylum Chaetognatha.

Glaucidium. Genus which includes the pygmy owls; small grayish or rufous spp. of western forested country.

Glaucionetta. Bucephala.

glaucoma. Pathological condition of the eye, marked by increased pressure within the eyeball and increase in size; often results in blindness; some cases may be relieved by surgery; cause of the disease is obscure but at least one variety is a Mendelian dominant.

Glaucoma. Genus of ovoid to ellipsoid fresh-water holotrich Ciliata.

Glaucomys. Genus of rodents in the Family Sciuridae; flying squirrels; small arboreal squirrels with large eyes and broad, lateral folds of skin extending from wrists to ankles; tail flat and broad; mostly herbivorous but sometimes taking insect food; two spp. and many subspecies over most of forested N. A.

glaucous gull. Large circumpolar sea gull, Larus hyperboreus; often winters on inland waters in north temperate latitudes; white and pale gray.

Glaucus. Genus of tropical pelagic sea slugs.

Glebula rotundata. Unionid mollusk of slow streams in southern and southeastern U. S.

Glenodinium. Genus of biflagellate fresh-water Protozoa covered with a cellulose membrane having an annular groove; spherical and somewhat flattened; several yellow to brown chromatophores.

glenoid cavity. Cup-shaped cavity on each side of the pectoral girdle into which the head of the humerus fits in tetrapods.

glenoid fossa. Glenoid cavity.

glia. Neuroglia.

gliding feather-tail. See ACROBATES.

gliding mice. See IDIURUS.

Glires. Taxon of mammals which includes only two orders: Lagomorpha and Rodentia.

Gliridae. Old World family of rodents which includes the dormice; insectivorous, nocturnal spp. resembling miniature squirrels; hibernation period long; see GLIS.

Glis. Genus which includes the common European dormouse; body somewhat mouselike but 6 in. long and with a 5 in. bushy tail; nests in holes on ground or in trees; omnivorous; hibernates in winter.

Glisson's capsule. Liver capsule.

globefish. See DIODONTIDAE.

Globicephala. See BLACKFISH.

Globigerina. Common genus in the Order Foraminifera; an important constituent of globigerina ooze.

globigerina ooze. Deep-sea deposit characterized by a preponderance of foraminiferan tests, especially those of the genus Globigerina; covers about one-third of the ocean floor.

globin. Protein portion of a hemoglobin molecule.

globin-zinc-insulin. Substance used by diabetics in place of insulin because of its slower and longer-lasting effects.

globulin. Any of a group of common proteins which are insoluble in water and soluble in salt solutions.

globuloproteins. Blood proteins of a globulin nature which can become converted into specific antibodies.

glochidium. Bivalve larval stage characteristic of fresh-water mussels; upon release from the parent it must come in contact with a fish, then quickly encysts in gills, fins, and other superficial tissues; after several weeks to a year, the cyst ruptures and the small juvenile mussel is released and falls to the substrate. Pl. glochidia.

Gloger's rule. Races of birds or mammals in cool, dry regions are lighter in color than those in warm, humid regions.

glomerulus. 1. See MALPIGHIAN BODY. 2. Small mass of spongy tissue in the proboscis of hemichordates; presumed to have an excretory function. Pl. glomeruli.

Glomeridesmus. Genus of tropical millipedes in the Order Limacomorpha.

Glomeris. Genus of Old World millipedes in the Order Oniscomorpha.

glomus. Small mass of cavernous blood vessels.

glossa. 1. One of the two median lobes of the insect labium. 2. Coiled mouth parts of Lepidoptera. 3. A loose synonym for tongue, both in vertebrates and invertebrates. Pl. glossae.

Glossina. Genus to which tsetse flies belong.

Glossinidae. Small family of medium-sized flies; adults suck blood and transmit human trypanosome infections; Glossina typical.

Glossiphonia. Common genus of fresh-water leeches; snail leeches; free-living and carnivorous on snails, oligochaetes, etc.; to 30 mm. long; body broad and very depressed.

Glossobalanus. Balanoglossus.

Glossodoris. Genus of nudibranchs.

Glossograptus. See GRAPTOLOIDEA.

glossohyal. Tongue bone, especially in fishes.

Glossophaga. Genus of Tropical American leaf-nosed bats.

glossopharyngeal nerve. One of the pairs of cranial nerves of vertebrates; originates on the medulla oblongata and innervates the first gill and palate of fishes but the pharyngeal and tongue region of tetrapods; a mixed nerve.

Glossosoma. Genus of Trichoptera.

glossy snake. Faded snake, Arizona elegans, comprising a group of blotched, spotted, gray-brown constrictor subspecies of the Southwest; average 33 in. long.

Glottidia. Genus of brachiopods in the Class Inarticulata; common on Atlantic and Pacific coasts.

glottis. Opening on the upper surface of the trachea or larynx; between vocal cords in mammals.

glove sponge. Sp. of Euspongia, an American bath sponge of little commercial value.

glowworm. See LAMPYRIDAE.

glucagon. Hormone elaborated by certain cells of the islets of Langerhans in the pancreas; increases blood sugar concentration.

glucocorticoids. Group of adrenal cortex hormones which facilitate the conversion of proteins to carbohydrates in cell metabolism.

glucosamine. Alpha-amino derivative of dextrose; obtained by hydrolyzing mucin and chitin.

glucose. Dextrose; common monosaccharide of fundamental importance as an energy source in both plant and animal metabolism; formed as an intermediate product of photosynthesis, and as the end product of alimentary carbohydrate digestion in most animals; the blood stream always contains glucose as a readily available source of energy for the tissues; for storage, glucose is converted to starch in plants and to glycogen in most animals; it occurs in two optically different forms, d-glucose and l-glucose, but only the former occurs naturally in living organisms.

glucoside. Naturally occurring compound which upon treatment with dilute acid yields a monosaccharide (usually glucose) and one or more of a wide variety of other substances such as benzaldehyde, mustard oil, and aromatic compounds; widely distributed in plants as anthocyanins.

glue cells. See COLLOBLASTS.

glutamic acid. Amino acid especially abundant in cereals; $C_5H_9O_4N$.

glutathione. Compound of cysteine, glycine, and glutamic acid; present in many living systems; a hydrogen acceptor in its oxidized form.

glutelins. Simple proteins, soluble in very dilute acids and alkalis.

gluteus. Extensor, abductor, and rotator muscle or group of muscles of the thigh in tetrapods.

glutinant. See NEMATOCYST.

glutton. See GULO.

glutton bird. Giant fulmar.

Glycera. Genus of marine Polychaeta found on sand and mud flats; important as fish bait.

glycerin. Glycerol.

glycerin jelly. Jelly-like substance used especially for making permanent microscope slide mounts of small invertebrates or parts; dissolve 65 g. pure gelatin in 400 ml. water with gentle heat and add 200 ml. glycerin and 2 g. carbolic acid; this mixture is solid when cool, and should be heated gently so that it can be dropped on objects as a fluid.

glycerol. Glycerin; syrupy, colorless, alcohol obtained by the saponification of natural fats and oils; $C_3H_5(OH)_3$.

Glycinde. Genus of Polychaeta.

glycine. Simple and common amino acid; gelatin contains more than 25% of this substance; $C_2H_5O_2N$.

glycocol. Glycine.

glycogen. Animal starch; soluble polysaccharide composed of many united glucose molecules and having the general formula $(C_6H_{10}O_5)_{12}$ to $(C_6H_{10}O_5)_{18}$; carbohydrates are stored in this form in both animals and fungi; in vertebrates, glycogen is especially abundant in muscles and in the liver.

glycogenesis. Physiological formation of glycogen.

glycogenolysis. Breakdown of glycogen.

glycolysis. Complex series of biochemical steps beginning with glucose and involving a variety of enzymes, coenzymes, and phosphate compounds; the end product is lactate, alcohol, pyruvic acid, or carbon dioxide and water.

glycoprotein. Organic compound consisting of united protein and carbohydrate groups, e. g. mucin.

glycosuria. Abnormally large amount of glucose in the urine, e. g. more than one gram in 24 hours.

glycyl glycine. Simplest polypeptide, composed of only two amino acids.

Glycymeris. Genus of tropical marine bivalves; bittersweet shells; sandy shores.

Glyphotaelius hostilis. Caddisfly of northeastern U. S.

glyptodont. Fossil mammal having the general appearance of a large armadillo, but with a single boxlike shell extending dorsally and laterally over the body; tail often armored; Glyptodon common.

Glyptotendipes. Genus of tendipedid midges.

g. n. Abbreviation for new genus.

Gnaphosa. Common genus of spiders in the Family Gnaphosidae.

Gnaphosidae. Ground spiders, a family of active spp. which build a silken tubular retreat under stones or in rolled leaves.

gnat. Any of a wide variety of small insects, especially in the Order Diptera; some bite man, others do not.

gnatcatcher. See POLIOPTILA.

gnathites. Arthropod oral appendages.

gnathobase. Median basal process found on certain appendages in trilobites and some crustaceans and arachnoideans; usually used for biting, crushing, and handling food.

Gnathobdellida. Arhynchobdellida.

gnathochilarium. Appendage composed of united second maxillae or millipedes; lower lip.

gnathopod. 1. First thoracic appendage, especially in the Isopoda and Amphipoda; the terminal segment is slender and can be bent back against the broad, enlarged penultimate segment; used for seizing and holding; sometimes described as subchelate. 2. In the broad sense, a crustacean appendage in the mouth region; modified for food handling.

Gnathostoma. Genus of nematodes parasitic in the stomach and intestine of fishes, reptiles, and mammals.

Gnathostomata. Taxon often used to include those groups of vertebrates having true jaws and usually paired appendages, as contrasted with the vertebrate Class Agnatha which have no true jaws or paired appendages.

gnotobiotics. Sterile culture of animals.

gnu. See CONNOCHAETES and GORGON.

goa. See PROCAPRA.

goat. Any of many hollow-horned ruminants in the Family Bovidae; related to sheep but more lightly built, with straight hair, and backward arching horns; see OREAMNOS and CAPRA.

goatfish. See MULLIDAE.

goat-gazelle. See PROCAPRA.

goat moth. See COSSIDAE.

goatsucker. Any member of two similar families of birds represented by the whip-poor-will and chuck-will's-widow in the U. S.

Gobiesocida. Xenopterygii.

Gobiesox. See XENOPTERYGII.

Gobiidae. Large family of small carnivorous percomorph fishes; gobies; mostly tropical and subtropical; marine and brackish, few in fresh waters; usually on bottom in shallows; ventral fins sometimes forming a sucking disc.

Gobioidei. Suborder taxon sometimes used to include the percomorph fish families Eleotridae and Gobiidae.

goblet cell. Swollen or pear-shaped cell present in many types of epithelial tissues and containing a droplet of mucin which is released to the exterior at intervals; a unicellular gland.

goblin fish. Any of several grotesque and ugly marine bottom fishes; mostly tropical.

goblin shark. See MITSUKURINA.

goby. See GOBIIDAE.

godwit. See LIMOSA.

Goera. Genus of Trichoptera.

goiter. Enlargement of the thyroid gland; simple goiter occurs when there is insufficient iodine in the diet and the thyroid gland compensates by enlarging and thus usually produces insufficient quantities of the hormone for normal or near-normal metabolism. See EXOPHTHALMIC GOITER.

gold beetle. Any of a large number of beetles having a more or less gold coloration; popularly, a goldbug.

golden cat. Large gold-colored wildcat of Sumatra.

golden eagle. See AQUILA.

goldeneye. See BUCEPHALA.

golden mole. See CHRYSOCHLORIDAE.

golden nematode. Minute Old World nematode which has become important as a potato pest in the eastern U.S.; Heterodera rostochiensis.

golden plover. See PLUVIALIS.

golden trout. See SALMO.

goldeye. See HIODON.

goldfinch. Small, brightly colored European finch, often caged as a pet. Also see FINCH.

goldfish. See CARASSIUS.

Goldschmidt, Richard B. German and American zoologist (1878-1958); head of Kaiser Wilhelm Institute until 1936 and professor at the University of California thereafter; famous for his many important contributions in sex determination, evolution, embryology, and physiological genetics.

goldsmith beetle. See CETONIA.

Golfingia. Genus in the Phylum Sipunculoidea; small, slender spp.; includes some spp. formerly in Phascolosoma.

Golgi, Camillo. Italian neurologist and histologist (1843-1926); discovered Golgi apparatus, developed a silver nitrate stain for Golgi cells, and recognized the individuality of the three common types of human malaria; shared 1906 Nobel prize in physiology and medicine for his work on the structure of the nervous system.

Golgi apparatus. Localized irregular network of fibrils, membranous structures, and granules in animal cells and perhaps certain plant cells; it is stained only by silver techniques and osmic acid and is probably a lipid material; has a suspected role in secretion.

Golgi bodies. Golgi apparatus.

Golgi cells. Astrocytes; nerve cell bodies having very short processes; in the posterior horns of a spinal cord cross section.

Golgi method. Chrome-silver microtechnique method for staining neurons and their processes black.

goliath beetle. See GOLIATHUS.

goliath frog. See RANA GOLIATH.

Goliathus. Small genus of giant African beetles in the Family Cetoniidae; males to 5 in. long.

Gomphaeschna. Common genus of dragonflies; eastern U.S.

Gomphoides williamsoni. Dragonfly of southeastern U.S.

Gomphotherium. Fossil genus of ancestral elephants; may have given rise to both elephants and mastodons; Miocene and Pliocene.

Gomphus. Common genus of dragonflies in the Order Odonata.

Gonactinia. Genus of sea anemones; commonly reproduce by transverse fission.

gonad. Gamete-producing organ in an animal; the ovary or testis.

gonadectomy. Surgical removal of the testes and/or ovaries.

gonadotropic hormones. Hormones secreted by the anterior lobe of the pituitary; induce normal growth of the Graafian follicles in the ovary and stimulate growth of both the seminiferous tubules and interstitial tissue.

gonangium. 1. Asexual reproductive polyp in certain hydrozoan coelenterates; consists of an axial blastostyle and often a surrounding gonotheca. 2. Occasionally the term is restricted to the gonotheca surrounding the medusa-producing polyp. Pl. gonangia.

gonapod. Male decapod crustacean appendage modified for copulation; occasionally applies to similar appendages in certain other Malacostraca.

gonapophyses. 1. Posterior structures or valves facilitating copulation or egg-laying in insects. 2. Parts of an insect sting. Sing. gonapophysis.

Gonatodes. Genus of Central American and West Indian geckos; one sp. in southern Fla.

Gonaxis. Genus of predaceous pulmonate snails.

Gongylonema. Genus of spirurid nematodes parasitic in the esophagus and stomach of birds and mammals.

Gonia. See TACHINIDAE.

Goniada. Genus of small Polychaeta.

Goniaster. Widely distributed genus of broadly pentagonal sea stars in the Order Phanerozonia.

Gonidea angulata. Common unionid

mollusk found in West Coast states.

Goniobasis. Genus of small, elongated, fresh-water snails; most abundant in southeastern states but scattered over much of U. S. except for the Rocky Mountain area.

Goniodes. Genus of bird lice in the Order Mallophaga.

Goniodoma. Genus of marine dinoflagellates.

Goniodoris. Genus of sea slugs; usually on rocky shores.

Gonionemus. Common genus of marine hydrozoans; medusa to 30 mm. in diameter; in shallow seas on or near the bottom; polyp stage reduced.

Gonium. Genus of ovoid green biflagellates; 4 or 16 cells arranged in a flat plate; fresh water.

gonocalyx. Bell of a medusoid gonophore.

gonocoel. 1. Cavity containing a gonad. 2. Body cavity resulting from the expansion of the cavity of a gonad.

Gonodactylus. Large genus of mantis shrimps.

gonoduct. Duct of a gonad leading to the exterior.

gonophore. 1. Bud formed asexually on the hydroid generation of many coelenterates; such buds produce gametes or develop into medusae. 2. Occasionally, the blastostyle of a hydrozoan. 3. Collectively, the reproductive parts in colonial hydroids.

gonopod. Gonopodium.

gonopodium. 1. Copulatory anal fin of of male fishes in the Family Poeciliidae. 2. Clasper of a male insect or millipede, etc.; a modified leg in many cases. Pl. gonopodia.

gonopore. 1. Distal opening in the gonotheca of hydrozoan reproductive polyps through which the medusae are released to the outside. 2. General term for an opening through which eggs and/or sperm are released, especially in invertebrates.

gonorrhea. Contagious venereal disease of the genital mucous membranes; caused by a bacterium, Neisseria gonorrhoeae, transmitted during coitus; disease may run its course or become chronic with many secondary effects.

gonosome. In hydroid coelenterates, that part which produces medusae or gametes.

Gonostomum. Genus of fresh-water hypotrich Ciliata.

gonostyle. 1. Blastostyle. 2. Sexual individual in the siphonophore coelenterates. 3. Male insect clasping organ at the tip of the abdomen, especially in certain Diptera.

gonotheca. Thin, translucent, vaselike exoskeleton surrounding the blastostyle of certain hydrozoan coelenterates.

Gonothyraea. Genus of marine coelenterates in the Suborder Leptomedusae.

gonotype. Progeny of a type specimen.

gonozooid. 1. Polyp of certain hydrozoan coelenterates which is specialized for reproduction. 2. Type of individual produced during the life history of certain tunicates; eventually liberated and reproduces sexually.

Gonyaulax. Genus of fresh-water and marine dinoflagellates; test consists of numerous flat plates arranged edge to edge; sometimes extremely abundant in the sea; G. catenella produces an alkaloid toxic to man, and when man eats mollusks that have fed on a preponderance of this sp., severe illness and sometimes death follow.

Gonyostomum. Genus of green ovoid biflagellate Protozoa with numerous small chromatophores and peripheral trichocyst-like structures; fresh water.

goosander. Merganser.

goose. Any of a large number of birds in the Subfamily Anserinae (Family Anatidae); somewhat like swans, but neck shorter than body and lores feathered; walk more readily than ducks; chiefly vegetarians; strong fliers; migrate long distances; American representatives in Chen, Branta, and Anser.

goose barnacle. See LEPAS and MITELLA.

goosebeak whale. See ZIPHIUS.

goosefish. See LOPHIIDAE.

gooseflesh. Involuntary erection of the skin papillae induced by cold, shock, or emotions.

gooseneck barnacle. See LEPAS and MITELLA.

gopher. See CITELLUS and GEOMYIDAE.

gopher frog. See RANA AESOPUS.

gopher snake. See DRYMARCHON.

gopher turtle. See GOPHERUS.

Gopherus. Genus of land tortoises; gopher turtles; carapace high; dig deep burrows; south Atlantic and Gulf states, and southwestern desert areas into Mexico; chiefly vegetarians.

goral. See NAEMORHEDUS.

Gordiacea. Nematomorpha.
gordian worm. Any member of the Phylum Nematomorpha.
Gordioidea. Class of Nematomorpha which includes the great majority of spp. in the phylum; without swimming bristles; pseudocoel of adult female filled with parenchyma and two ovaries; Paragordius and Gordius common.
Gordionus. Genus of Nematomorpha.
Gordius. Common name of Nematomorpha; adult female to 1 mm. in diameter and 30 in. long.
Gorgodera. Genus of trematodes commonly found in frog urinary bladder.
Gorgon. Genus of African ruminants in the Family Bovidae; includes the gnu or wildebeeste; large spp. with massive head, maned neck, tail of a horse, and legs and feet of an antelope. Cf. CONNOCHAETES.
Gorgonacea. Order of alcyonarian anthozoans; the horny corals, including the sea fans and red coral; skeleton axial and composed of calcareous spicules, hornlike gorgonin, or both; polyps small and short; Corallium and Gorgonia familiar.
Gorgonia. Genus of horny corals in the Order Gorgonacea; the sea fans; brightly colored, open network, fan-shaped colonies to 10 ft. high; polyps small; widely distributed.
gorgonin. Organic, horny material forming the exoskeleton of some corals of the Order Gorgonacea.
Gorgonocephalus. See EURYALAE.
Gorilla gorilla. Common gorilla of forested lowlands in West and Central Africa; the largest anthropoid ape; males to 500 lbs. and 65 in. tall; coloration blackish; a secretive, herbivorous sp. living on the ground in family groups; G. berengei is the closely related African mountain gorilla.
goshawk. See ASTUR.
gossamer. See LYCAENIDAE.
Götte's larva. Free-swimming ciliated larva of some spp. of Stylochus, a genus of marine Turbellaria; with four posteriorly-directed lappets.
goujon. Flathead catfish, Pylodictis olivaris; a large, yellow catfish of the Mississippi Valley; to 100 lb.
gourami. Any of several southeastern Asiatic fishes; brightly colored spp., many adapted to fresh-water aquaria; at least one sp. is a valuable food fish.
Graafian follicles. Fluid-filled cavities in the ovary of higher mammals; a single maturing ovum occurs in each cavity, and the cavity is lined with several layers of hormone-secreting cells; after release of the ovum by follicle rupture, the space occupied by the fluid is replaced by a corpus luteum.
gracilis major. Large muscle on the inner posterior surface of the thigh in certain tetrapods; inserted on the tibia and flexes and adducts the leg.
gracilis minor. Small muscle on the posterior surface of the thigh of certain tetrapods.
grackle. 1. Quiscalus quiscula, which includes several N.A. subspecies variously known as the bronze grackle, purple grackle, etc. 2. Any of several Old World birds in the Family Sturnidae.
gradual metamorphosis. See PAUROMETABOLA.
graft. Small part of an animal which is removed from its normal position on the donor and placed on another area (host or recipient) where it may fuse and grow. See AUTOPLASTIC GRAFT, HOMOPLASTIC GRAFT, and HETEROPLASTIC GRAFT.
grain alcohol. Ethyl alcohol; C_2H_5OH.
grain beetle. Any of several beetles whose larvae feed on stored grain.
grain itch. See TARSONEMIDAE.
Grallinidae. Small family of passerine birds; mudnest builders; open woodland spp. that build deep bowl-like mud nests; New Guinea and Australia.
gram. Metric unit of weight; 28.35 grams in one ounce avoirdupois.
Grammatophora. Genus of tropical lizards notable for their bipedal locomotion.
gram-negative. Staining reaction common to certain groups of bacteria; after being stained with gentian violet and iodine solution, such bacteria lose their coloration when treated with 95% alcohol.
Grampidelphis griseus. Risso's dolphin; a non-beaked sp. of tropical and temperate seas.
gram-positive. Staining reaction common to certain groups of bacteria; after being stained with gentian violet and iodine solution, such bacteria do not lose their coloration when treated with 95% alcohol.
grampus. Local name for the whip scorpion (Mastigoproctus giganteus).
Grampus. Genus of widely distributed killer whales; aggressive spp. which attack fishes, porpoises, seals, and other small whales; slate-gray, to 14

ft. long.

gram stain. Common and diagnostic biological stain used especially for bacteria; see GRAM-NEGATIVE and GRAM-POSITIVE.

granary weevil. See SITOPHILUS.

granivorous. Grain-eating or seed-eating.

Grantia. Genus of small, vase-shaped, marine sponges in the Class Calcarea; European coasts (not in U. S.); (textbook references to Grantia usually refer in fact to Scypha.)

granulation tissue. Vascular tissue which covers an open wound and becomes scar tissue.

granulocytes. 1. Collectively, the neutrophils, eosinophils, and basophils of blood; marked by granules in the cytoplasm. 2. Mature leucocytes of vertebrate blood; contain prominent cytoplasmic granules.

grape leafhopper. Erythroneura comes, an important pest of grapes; a yellow, red, and brown hemipteran in the Family Cicadellidae.

grape phylloxera. See PHYLLOXERIDAE.

Graphiurus. Genus of African dormice; fur thick and woolly; total length to 2 ft. ; forested areas.

Graphoderus. Common genus of dytiscid beetles.

Grapsidae. Common family of crabs found especially along marine shores; carapace square; eyes near the two anterior corners.

Grapta. Same as Polygonia; see ANGLEWING.

Graptemys. Genus of fresh-water turtles found in ponds, swamps, and streams of the eastern half of U. S. ; map turtles; with a faint yellow pattern on the carapace.

Graptocorixa. Genus of water boatmen in the Family Corixidae; mostly southwestern states.

Graptoleberis testudinaria. Common and widely distributed fresh-water cladoceran; on bottom in shallows.

Graptolita. Graptozoa.

Graptoloidea. Group of stalked or compound colonial graptozoans; Tetragraptus and Glossograptus typical.

Graptozoa. The graptolites; a fossil invertebrate phylum usually placed in the Class Hydrozoa but sometimes in the bryozoans, echinoderms, or chordates; tentatively considered a prochordate phylum in this volume; occur mostly as carbonized outlines in shales; colonial polyps commonly encased in a branching tubular skeleton composed of series of rings.

grass frog. See RANA PIPIENS.

grasshopper. Essentially any solitary and resident orthopteran in the Family Acrididae; often abundant but only occasionally and incidentally migratory. Cf. LOCUST.

grasshopper mouse. See ONYCHOMYS.

grasshopper sparrow. See SPARROW.

grass pickerel. Esox vermiculatus, chain pickerel.

grass snake. Any of several common snakes found among grasses, especially Thamnophis and Natrix.

grass spider. See AGELENA.

grass sponge. Large, dark brown sponge found off Fla. and the West Indies; no commercial importance.

Grave's disease. Exophthalmic goiter.

gravid. 1. Same as pregnant, but used more commonly to apply to invertebrates. 2. Containing fertile eggs, e. g. a posterior tapeworm proglottid.

grayback. 1. Vernacular name for human body louse, Pediculus humanus corporis. 2. Common name for several ducks, shorebirds, fishes, and whales.

gray crescent. Narrow crescent-shaped area on the surface of a fertilized egg; located on a meridian 180 degrees from the point of sperm entrance.

grayfish. See SQUALUS.

gray goose. Graylag; see ANSER.

graylag. See ANSER.

grayling. See THYMALLIDAE.

gray matter. Nervous tissue consisting chiefly of nerve cell bodies in vertebrates; most concentrated in the outermost layer of the cerebral hemispheres, cerebellum, and brain "nuclei."

gray shark. See CARCHARIAS.

gray squirrel. Any of several spp. of Sciurus; pelage gray; eastern and western states.

gray whale. See RHACHIANECTES and ESCHRICHTIDAE.

greaser blackfish. See ORTHODON.

great ape. Any anthropoid ape, e. g. gorilla and chimpanzee.

great auk. See PLAUTUS.

Great Barrier Reef. Very large and complex coral reef formation extending for hundreds of miles along the northeastern coast of Australia; famous for its wide variety of exotic tropical marine invertebrates.

great blue heron. See ARDEA.

great glider. See SCHOINOBATES.

great horned owl. See BUBO.
great northern pike. See ESOCIDAE.
Great Plains. Collectively, the flat to rolling grasslands extending from about 95° west longitude to the foothills of the Rockies.
Great Plains toad. See BUFO.
grebe. See COLYMBIFORMES.
green abalone. Haliotis fulgens, a small West Coast abalone.
green anemone. See CRIBRINA.
green-back trout. Subspecies of Salmo clarki originally native to the headwaters of the Arkansas and Platte rivers.
green bottle flies. See LUCILIA.
green crab. See CARCINIDES.
green frog. See RANA CLAMITANS.
green gland. Typical excretory apparatus of most Malacostraca; one in the anteroventral portion of the cephalothorax on each side; consists of a minute terminal end sac, a spongy cortical mass, a convoluted tubule, a bladder, and a short duct which opens on the basal segment of the first antenna.
greenhouse white fly. See TRIALEURODES.
greenling. See HEXAGRAMMUS.
green monkey. See CERCOPITHECUS.
green perch. See GIRELLA.
green salamander. Black salamander with yellowish-green blotches; Aneides aeneus; 4 in. long; damp places in the middle Appalachians.
green snake. See OPHEODRYS.
green sunfish. See LEPOMIS.
green turtle. See CHELONIA.
green-winged teal. See ANAS.
Gregarina. A large genus of Protozoa in the Class Sporozoa; digestive tract parasites of arthropods.
Gregarinida. Order of sporozoan Protozoa; parasitic in the digestive tract of many invertebrates, especially arthropods and annelids; obtain nourishment solely by absorption from the surrounding tissues and material in the cavity of the digestive organs; reproduction usually by spore formation, without any schizogony.
grenadier. See MACROURIDAE.
Grevy's zebra. See DOLICHOHIPPUS.
gribble. See LIMNORIA.
gridiron-tailed lizard. See CALLISAURUS.
griffon vulture. Any of several large vultures (Gyps) in the Family Buteonidae; mountainous open country of southern Eurasia and northern Africa.
grilse. Young salmon when it returns to fresh water from the sea.
Grimaldina brazzai. Fresh-water cladoceran sometimes found among littoral vegetation.
Grison. Genus of Central and South American terrestrial weasel-like mammals.
gristle. Cartilage or material of cartilaginous consistency.
grivet. Monkey of east central Africa.
grizzly bear. See URSUS.
groin. Lowest part of the abdominal wall, near the base of the thigh.
Gromia. Genus of fresh-water and marine foraminiferans; test aperture terminal; filopodia branching and anastomosing.
grosbeak. 1. Any of several finchlike birds having a heavy conical bill; common American spp. are the evening grosbeak (Hesperiphona vespertina) and pine grosbeak (Pinicola enucleator). 2. See PHEUCTICUS.
ground beetles. See CARABIDAE.
ground dove. See COLUMBIGALLINA.
groundhog. See MARMOTA.
ground pearls. See MARGARODES.
ground rattlesnake. See SISTRURUS.
ground shark. Any of numerous voracious sharks (Carcharius) found along warm coastal areas.
ground skink. See LYGOSOMA.
ground snake. Any of several southwestern snakes in the genus Sonora; variously colored or banded; 8 to 19 in. long; see CARPHOPHIS.
ground spider. See GNAPHOSIDAE.
ground squirrel. See CITELLUS.
ground thrush. See PITTA.
grouper. Any of a wide variety of large fishes of tropical and subtropical seas, especially Epinephelus and Mycteroperca; some are important food fishes.
grouse. See BONASA and DENDRAGAPUS.
grouse locust. See TETRIGIDAE.
growth. In the biological sense, the enlargement of organisms and the development and differentiation of new parts, especially by the process of incorporation of food materials to form additional protoplasm.
growth-stimulating hormone. Growth-controlling hormone secreted by the anterior lobe of the pituitary; see ACROMEGALY, PITUITARY GIANT, and PITUITARY DWARF.
grub. 1. Larva of a terrestrial beetle. 2. General term for any thick, terrestrial insect larva.
grubby. As applied to stock animals in

the western states, infested with warble or botfly larvae.

Grubia. Genus of marine Amphipoda.

Gruidae. Family of birds (14 spp.) in the Order Gruiformes; cranes of Northern Hemisphere and Africa; fly with necks extended; always with some bare skin around the head; usually tall wading birds superficially similar to the herons; Grus typical.

Gruiformes. Large order of birds which includes the cranes, rails, coots, limpkins, gallinules, bustards, etc.; large to small spp. of various habits; feathers with aftershaft.

grunion. See LEURESTHES.

grunt. Any of numerous marine fishes in the Family Haemulidae related to the snappers; make a grunting sound when taken from the water; often brightly colored; mostly tropical and subtropical; some are valuable food fishes; Anisotremus virginicus is the brightly colored porkfish or grunt of Fla. waters; Haemulon includes the white and bluestripe grunts; Bathystoma aurolineatus is the small Caesar; Orthopristis chrysopterus is the common pigfish of Long Island Sound to Tex.

Grus. Large and widely distributed genus of cranes in the Family Gruidae; long feathers on the back curl over the ends of the wings; bare red skin about face; G. americana, whooping crane, a rare large white sp. which summers in Canada and winters on Tex. coast; G. canadensis pratensis, the sandhill crane, is a rare gray sp. which summers mostly in Canada and winters in southern states to Mexico.

Gryllacrididae. Stenopelmatidae.

Gryllidae. Insect family in the Order Orthoptera; crickets and tree crickets; short, mostly dark-colored insects, with long antennae; hind legs specialized for jumping; males chirp by rubbing the bases of the forewings together; common genera are Gryllus, Acheta, and Oecanthus.

Grylloblatta. See GRYLLOBLATTODEA.

Grylloblattodea. Order of pale, wingless insects to 30 mm. long; biting mouth parts; simple metamorphosis; antennae long, compound eyes small, cerci long; in vegetation, debris, and under stones and logs in mountainous areas, especially near melting snow, streams, and lakes; rare, only several spp. of Grylloblatta known.

Gryllotalpa. Common genus of mole crickets in the Family Gryllotalpidae.

Gryllotalpidae. Family of robust insects in the Order Orthoptera; mole crickets; forelegs specialized for burrowing in the ground; body brown or blackish and covered with hairs; common in tropical and subtropical climates everywhere but also found in temperate areas; Gryllotalpa common.

Gryllus. Genus of crickets; includes most of the common tree and house crickets.

Gryphus. Genus of Brachiopoda.

grysbok. See RAPHICERUS.

guacharo. Oilbird, Steatornis steatornis; the only sp. in the Family Steatornidae; a large barred and penciled brownish bird, spotted with white; nocturnal, lives in caverns in northern South America; young birds sometimes used as a source of fat by natives.

guan. Any of several brownish to greenish Central and South American birds; somewhat chicken-like, but feed on tree-top fruits; Penelope common.

guanaco. Wild South American herbivorous mammal (Lama guanicoe) related to the camel, llama, vicuna, and alpaca; large, graceful, and having the appearance of a long-legged sheep; about 40 in. at the shoulder; brown back and sides; plains and highlands of southern South America.

guanase. Enzyme which catalyzes the conversion of guanine to xanthine and ammonia.

guanine. White crystalline substance found in RNA, and especially in guano, fish scales, some plant seedlings, and various animal tissues; decomposition product of nuclein; used in making artificial pearls; $C_5H_5N_5O$.

guano. 1. Specifically, a thick, terrestrial deposit of seafowl or bat excrement; valuable as fertilizer; extensive deposits of seafowl guano occur along the Peruvian coast. 2. Generally, any bird excrement.

guanophores. Chromatophores containing an abundance of whitish pigment granules or crystals.

guara. Aguara.

guard hairs. Outer coat of coarse protective hairs in many mammals.

gubernaculum. 1. Posterior rudder-like flagellum in certain Protozoa. 2. Fetal cord extending between epididymis and scrotum wall. 3. Strand of tissue extending between gonophore and gonotheca in certain hydrozoan coelenterates. 4. In tooth development, a strand of connective tissue between the fibrous gum tissue and the sac of a permanent tooth. Pl. gubernacula.

gudgeon. 1. Small European fresh-water fish related to carp. 2. Local name for certain fishes, especially the gobies and killifishes.

guemal. See HIPPOCAMELUS.

guenon. Any of several long-tailed African monkeys, especially those in the genus Cercopithecus.

guereza. See COLOBUS.

guillemot. See CEPPHUS.

guinea fowl. See NUMIDIDAE.

guinea pig. Robust, short-eared, short-tailed rodent (Cavia) in the Family Caviidae; to 8 in. long; domesticated and much used as an experimental animal; many colorations.

Guinea worm. See DRACUNCULUS MEDINEN

Guiraca caerulea. Blue grosbeak; male deep blue with tail and wings mostly black and chestnut; southern half of U. S. to Central America.

guitar fish. Any of a small group of rays in the Suborder Batoidea; body guitar-shaped; usually about 2 ft. long; Rhinobatos common.

gular. Pertaining to the throat region.

gular pouch. Pouch in the throat region, as in birds of the Order Pelicaniformes.

Gulf Stream. Massive current of warm tropical sea water originating in the Gulf of Mexico, rounding the tip of Fla., and proceeding northeasterly until its expanded and weakened subdivisions disappear as far north as Iceland, Spitsbergen, and North Cape.

Gulf Strip of Lower Austral zone. One of Merriam's North American life zones; a strip 100 to 300 miles wide along the Gulf of Mexico and including all of Fla. except the lowermost one-third of the peninsula.

gull. See LARUS.

gull-billed tern. See GELOCHELIDON.

gullet. General term referring to the pharyngeal or postoral region.

Gulo. Genus of powerful, sturdy, dark brown to black carnivores in the Family Mustelidae; to 3 ft. long; commonly called the glutton in Europe and the wolverine or carcajou in N. A. where it occurs from the northern states to the Arctic Ocean.

gulper. See LYOMERI.

gumma. Tumor-like structure formed in the tissues as the result of a spirochaete infection.

gundi. See CTENODACTYLIDAE.

Gundlachia. Genus of small fresh-water limpets.

gunnel. See PHOLIS.

guppy. See LEBISTES.

gurnard. See DACTYLOPTERUS.

gustatory. Pertaining to the sense of taste.

gut. 1. Digestive tract or alimentary canal, or part of it. 2. Intestine. 3. Occasionally, the whole visceral digestive system, including the digestive glands.

gweduc. Geoduck.

Gyalopion. See HOOK-NOSED SNAKE.

Gymnamoebae. General taxonomic term used to include naked amoeboid Protozoa. Cf. THECAMOEBAE.

Gymnarchus niloticus. African river fish having a long pointed tapering tail and a soft dorsal fin extending the whole length of the trunk; emits continuous electrical impulses into the water which are presumed to operate as a sensory system for detecting objects in the vicinity.

Gymnoblastea. Anthomedusae.

gymnoblastic. In the Anthomedusae, having naked medusa buds.

Gymnocerata. One of the two suborders in the Order Hemiptera; antennae long; terrestrial and subaquatic insects.

Gymnodinium. Genus of subspherical biflagellate Protozoa; numerous variously colored chromatophores; cell with one annular and one posterior longitudinal groove; fresh-water and marine; sometimes responsible for "red tides."

Gymnogyps californianus. California condor in the Order Falconiformes; rare in isolated mountain areas of southern Calif.; wingspread to 11 ft.; head bare, yellow to orange; plumage black with white wing patches; a carrion feeder; a related sp. occurs in the high Andes.

Gymnolaemata. One of the three classes in the Phylum Bryozoa; lophophore circular; zooecium of various construction; chiefly marine; statoblasts absent; examples are Paludicella, Bugula, and Membranipora.

Gymnomera. Small suborder of Cladocera in which a shell is absent and the legs are prehensile and not foliaceous; Polyphemus and Leptodora.

gymnopedic. Having a young bird hatched naked (without feathers).

Gymnophiona. Apoda; order of stegocephalian Amphibia; caecilians; wormlike spp. without limbs or girdles; ribs long, skin with glands which secrete an irritating fluid; small tentacle between eye and nostril; numerous modern spp; see CAECILIIDAE.

Gymnopis. Genus of caecilian Amphibia; burrow in moist soil of tropics; eggs deposited on ground near water.

Gymnorhinus cyanocephalus. Pinyon jay; a large grayish-blue bird with a long slender bill, flat at the tip; pinyon and juniper forests of western states.

Gymnosomata. Group of genera of Pteropoda in which a shell is absent.

gymnospores. Dense clusters of minute cells formed in the digestive tract of crustaceans in the life history of certain sporozoans; voided from the body and then taken into a molluscan host.

Gymnostomata. Suborder of holotrich Ciliata; cytostome on body surface or in peristome, without strong cilia; many common genera, e.g. Didinium, Coleps, Lionotus, Chilodonella.

Gymnotidae. Family of fresh-water fishes of Central and South America; slender body, long tapering tail; dorsal and pelvic fins absent; emit continuous electrical pulses into the water which are presumed to operate as a sensory system for objects in the vicinity.

Gymnura. Echinosorex.

Gynacantha nervosa. Dragonfly found in Fla. and Calif.

gynaecophoral canal. Groove formed by the inrolling of the lateral margins of certain male flatworms, e.g. Schistosoma; used for carrying the female.

gynaecotelic type. Among certain social insects, those prototype infertile worker females which, if sufficiently cared for and properly fed, are capable of laying eggs and replacing queens.

gynandromorph. Individual having one part of the body with male features and the other part with female features; gynandromorphs are usually bilateral in which the left and right halves are of different sexes; especially common in arthropods.

Gynembia. Genus of insects in the Order Embioptera; southwestern U.S.

Gyps. See GRIFFON VULTURE.

gypsy moth. See LYMANTRIA.

Gyratrix hermaphroditus. Common fresh-water and brackish rhabdocoel turbellarian; whitish, 2 mm. long.

Gyraulus. Large genus of fresh-water snails.

Gyretes. Genus of whirligig beetles in the Family Gyrinidae; western states.

gyrfalcon. See FALCO.

Gyrineum. Genus of tropical and subtropical marine snails.

Gyrinidae. Small family of gregarious blackish beetles found gliding and darting about on the surface of ponds and sluggish rivers; whirligig beetles; eyes divided into two horizontally; adults and larvae predaceous; Gyrinus common.

Gyrinophilus porphyriticus. "Purple salamander" of hilly parts of eastern states; actually brownish in color.

Gyrinus. See GYRINIDAE.

Gyrocotyle. Genus of tapeworms in the Subclass Cestodaria; elongated, unsegmented, and without a scolex; body margin ruffled; with a minute anterior sucker; parasitic in the intestine of chimaeroid fishes.

Gyrodactylus. Genus of viviparous monogenetic trematodes parasitic on fresh-water fishes ;posterior attachment organ with one pair of large anchors and 16 hooklets; sometimes epidemic in hatcheries.

Gyrodinium. Genus of marine and freshwater dinoflagellates; holozoic, chromatophores rarely present.

Gyrotoma. Genus of snails found in the Coosa River, Ala.

gyrus. One of the many ridges on the surface of the cerebral cortex. Pl. gyri.

gyttja. Very finely divided organic ooze forming a large portion of the bottom of many lakes; results from the feeding action of the bottom fauna on detritus.

H

habenular body. Group of small nuclei through which olfactory stimuli pass in going from hemispheres to brain stem; present in all vertebrates and involved in the correlation of olfactory impulses.

habitat. The specific place where a particular plant or animal lives; usually used in a much more restricted sense than environment, and refers to a smaller area; e.g. spring brook, treetop, weedy pond, and sandy beach.

Habronema megastoma. Nematode parasite of the horse stomach.

Habrobracon. Genus of long-legged parasitic wasps; important in experimental genetics research.

Habrophlebia. Genus of Ephemeroptera; eastern and southeastern U.S.

Habrophlebiodes. Genus of Ephemeroptera.

Habrotrocha. Genus of creeping, leechlike rotifers.

hackle-banded weavers. See DICTYNIDAE and ULOBORIDAE.

hadal. Pertaining to that part of the ocean and the bottom at depths exceeding 6000 m.

haddock. See GADIDAE.

Hadromerina. Order of sponges in the Subclass Monaxonida; spicules pin-shaped, spongin absent.

Hadropterus. Genus of Percidae (fish) now included in Percina.

Hadrosaurus. Genus of N.A. duck-billed herbivorous dinosaurs; Cretaceous.

Hadrurus hirsutus. Large, hairy scorpion; often stings human beings.

Haeckel, Ernst Heinrich. German biologist (1834-1919); exponent of Darwinism; developed a theory of embryonic recapitulation; noted for his studies on marine invertebrates, especially sponges, radiolarians, and medusae.

haem-. Prefix equivalent to hem-.

Haemadipsa. Genus of tropical land leeches.

Haemagogus. Small genus of tropical American mosquitoes; transmit a jungle form of yellow fever.

haemal spine. Single median spine which projects ventrally from most caudal vertebrae in lower vertebrates.

Haemaphysalis. Common genus of ixodid ticks found on mammals and occasionallv on birds.

Haematobia irritans. Horn fly in the Family Stomoxyidae; superficially similar to the housefly; adults are pests on cattle and suck blood, especially at the base of the horns and on the neck and rump.

Haematococcus. Genus of oval or ellipsoidal fresh-water flagellates; two anterior flagella, reddish pigmentation, and a gelatinous envelope surrounding the cell.

Haematopinus suis. Hog louse in the Order Anoplura; a large cosmopolitan parasite of domestic hogs; to 6 mm. long.

Haematopodidae. Cosmopolitan family of birds in the Order Charadriiformes; oyster-catchers; stout-bodied shore birds found along marine coasts where they feed chiefly on small bivalves by forcing the two valves apart with their strong bill; the common American sp., Haematopus palliatus, is mostly black above and white below.

Haematopus. See HAEMATOPODIDAE.

Haematosiphon inodora. Adobe bug; a bedbug which infests poultry in southwestern U.S. and Mexico.

haematoxylin. Hematoxylin.

haemocoel. Hemocoel.

haemocyanin. Hemocyanin.

Haemogamasus liponyssoides. Mole mite; a small mite in the Family Parasitidae; parasitic on moles.

haemoglobin. Hemoglobin.

Haemogregarina. Genus of Sporozoa in erythrocytes of turtles, frogs, and fishes.

haemolysis. Hemolysis.

haemolyze. Hemolyze.

Haemonchus contortus. Rhabditid nematode parasitic in the stomach and small intestine of sheep, cattle, and other ruminants; the twisted stomach worm or wireworm.

Haemonia nigricornis. Chrysomelid beetle; adults feed on floating and emergent rooted aquatics; larvae feed on the roots and rhizomes.

Haemopis grandis. American freshwater leech; green, gray, or slaty colored, with dark blotches; to 12 in. long.

Haemopis marmoratis. Common "horse leech" of the U. S.; occasionally takes blood from man; to 160 mm. long.

Haemoproteus. Genus of Sporozoa found in erythrocytes and viscera of birds and reptiles; transmitted by bloodsucking flies.

Haemosporidia. Order of sporozoan Protozoa; minute intracorpuscular parasites of vertebrates; schizogony occurs in the definitive host, and sporozoite formation occurs in the digestive tract of some intermediate blood-sucking arthropod host; the human malaria organism, Plasmodium, is a familiar example.

Haemulidae. See GRUNT.

Haemulon. See GRUNT.

Hagenius. Genus of dragonflies; abdomen of nymph circular; eastern and southern U. S.

hagfish. See MYXINOIDIA.

Hahnia. Common genus of small ground spiders.

Haideotriton wallacei. Small blind salamander known only from a deep artesian well at Albany, Ga.

hair. 1. True hairs occur only in mammals where they are cylindrical, filamentous outgrowths of the epidermis; each hair consists of numerous cornified epidermal cells. 2. Fine cylindrical exoskeletal outgrowth of an arthropod; cf. SETA.

hair cells. Special ciliated cells located in the scala media chamber in the inner ear of higher vertebrates; thought to be receptors for various vibration frequencies received by the cochlea.

hair follicle. Small cylindrical cavity in mammalian skin filled by the base of a hair.

hair papilla. Microscopic protuberance at the bottom of a hair follicle, below the root of the hair.

hairstreaks. Any of numerous small butterflies having one or two thin filaments projecting from the posterior margin of the hind wings; see LYCAENIDAE.

hair worm. Common name of any member of the Phylum Nematomorpha.

hairy flower bees. See ANTHOPHORIDAE.

hairy flower wasps. See SCOLIIDAE.

hairy frog. Any of several large West African frogs having fine hair-like outgrowths of the skin on the lateral surfaces of the trunk and thighs.

hairy-tailed mole. See PARASCALOPS.

hairy woodpecker. See DRYOBATES.

hake. See MERLUCCIIDAE and GADIDAE.

Halacaridae. Small family of mites; most spp. marine, few in fresh waters; in algae and debris of bottom; do not swim; Halacarus a common marine genus.

Halacarus. See HALACARIDAE.

Halcampa. Genus of elongated creeping and burrowing sea anemones.

Haldea. Genus of earth snakes in the Family Colubridae; secretive, uniformly colored, to 12 in. long; southeastern quarter of U. S.

Halecium. Genus of marine coelenterates in the Suborder Leptomedusae.

Halecostomi. Order of Mesozoic bony fishes.

halfbeaks. See HEMIRAMPHIDAE.

Haliaeetus leucocephalus. Bald eagle; a sp. of oceanside, rivers, and lakes; snow-white head and tail; wingspread to 7. 5 ft.; nests in large trees; feeds chiefly on fish; H. albicilla is the European white-tailed sea eagle.

halibut. See HETEROSOMATA.

Halichoerus grypus. Gray seal in the Family Phocidae; large, uncommon, grayish seal found from Nova Scotia to Greenland.

Halichondria. Genus of marine sponges in the Subclass Monaxonida; crumb of bread sponge.

Halichondrina. Order of sponges in the Subclass Monaxonida; spicules double-pointed and arranged in a plumose or confused fashion, e.g. Halichondria.

Haliclona. Genus of sponges in the Order Monaxonida; deadman's fingers.

Haliclystus. Genus of sessile scyphozoan coelenterates in the Order Stauromedusae.

Halicore. Dugong.

Halicryptus. See PRIAPULIDA.

Halictidae. Sweat bees, flower bees; a family of small to medium-sized short-tongued bees; construct burrows in ground; valuable pollinizers; Halictus and Sphecodes common.

Halictus. See HALICTIDAE.

Haliotis. Marine gastropod genus which includes the common abalones or ear shells; shell ear-shaped, the spire being very flat and the aperture enormously enlarged; shell to 1 ft. long and with a series of holes near the left margin; foot very large; shells used for button and buckle manufacture, and for inlaying; soft parts often used as human food; more than 100 spp. in the Pacific and Indian oceans, and along European coast; several important West Coast spp.

Haliplidae. Small family of crawling water beetles; crawl about on submerged vegetation and debris in fresh water shallows; yellowish or brown spp., 2 to 5 mm. long; Haliplus common.

Haliplus. See HALIPLIDAE.

Halisarca. Genus of sponges in the Order Tetractinellida; skeleton absent.

Halistaura. Genus of marine hydrozoan coelenterates.

Haller's organ. Small sense organ at the tip of each first leg in ticks.

Hallezia. Genus of fresh-water suctorian Protozoa; tentacles in bundles; stalk present or absent.

Hallopora. Genus of fossil Bryozoa in the Order Treptostomata.

hallux. Median "big toe" of pentadactyl hind limb.

Halobates. Genus of insects in the hemipteran Family Gerridae; unique because of their occurrence on the surface of warm oceans, often far from land.

Halobisium. Genus of pseudoscorpions.

Halocynthia. Genus of ovate or barrel-shaped sessile ascidians; often peach-colored.

Halocypris. Common genus of marine Ostracoda.

halolimnic. Pertaining to marine organisms which are also able to survive in fresh water.

Halopsyche. Genus of shell-less pteropod mollusks.

Halosoma. Genus of Pycnogonida.

Halosydna. Genus of Polychaeta.

halteres. Movable filaments with knobbed tips representing non-functional hind wings in the insect Order Diptera; organs of equilibration. Sing. halter.

Halteria. Genus of small fresh-water ciliate Protozoa; cell fusiform or spherical, with conspicuous anterior adoral zone and small equatorial grooves containing cirri.

Halticus citri. Garden flea hopper; a small black bug in the Family Miridae; a general farm pest on cultivated plants.

Hamacantha. Genus of encrusting marine sponges in the Subclass Monaxonida.

Haminea. Genus of small opisthobranch snails in which the thin shell is largely covered by flaps of the foot.

Hamite. Ethiopian.

Hamitermes. Genus of Australian termites which build large slablike mounds whose long axis is almost always precisely north and south; magnetic termites.

hamlet. 1. Large West Indian grouper. 2. Yellow and black West Indian moray.

Hamm, Ludwig. Dutch biologist who first observed sperm with a microscope in about 1677.

hammer. Malleus.

hammer-head. See UMBRETTE and SCOPIDAE.

hammer-headed bat. See HYPSIGNATHUS.

hammerhead shark. See SPHYRNA.

hamster. See CRICETUS.

hamster-rat. See CRICETOMYS.

hamula. See COLLEMBOLA.

hamulus. Any hook-shaped process.

Hancockia. Genus of nudibranchs.

hand. Collectively, the propodus and dactylus, or pincer, of the first leg of a crayfish or other decapod crustacean.

hanging drop. Drop of water, culture solution, or other fluid on the under side of a coverslip and suspended in the concavity of a hollow-ground microscope slide.

hanging fly. One of several insects in the Order Mecoptera; suspends itself from vegetation by the forelegs and with the other four legs seizes small insects flying past.

Hannemania hylae. Tree-toad chigger; a mite in the Family Trombidiidae; parasitic on toads in southwestern U.S.

hanuman. Especially sacred monkey of the Hindus; genus Presbytis.

Hapale. Callithrix; see CALLITHRICIDAE.

Hapalemur. Genus of large gray lemurs; gentle lemurs; face short and catlike; total length to 4 ft.; in Madagascar bamboo thickets.

Hapalidae. Callithricidae.

Hapalogaster. Genus of hairy marine crabs in the Suborder Anomura.

Haplobothrium. Genus of tapeworms in the Order Trypanorhyncha; adults in fresh-water fishes, larvae in Cyclops.

Haplodoci. Order of bony fishes which includes the toadfishes; head large and flat, often with fleshy flaps; mouth very wide; pectoral fins jugular in position; scales absent, body slimy; voracious and carnivorous; mostly in shoals of tropical and subtropical marine littoral; Opsanus the common toadfish, Porichthys the midshipman.

haplodont. Having molar teeth with simple crowns and no tubercles.

haploid. Having a single set of dissimilar chromosomes in each nucleus; the n number of chromosomes, as found in germ cells; among spp. having alternation of generations, there are certain stages in which the body is composed of haploid cells; some sexually mature animals in certain of the lower phyla likewise have a haploid constitution, e.g. male Rotatoria.

Haplomi. Order of fresh-water bony fishes, including the pikes and mud minnows; fins soft rayed, scales cycloid, pelvic fins abdominal; air bladder with duct to pharynx.

haplont. That portion of the life cycle of an organism in which the body cells have the haploid number of chromosomes.

Haplopoda. Tribe of Cladocera which includes the limnetic carnivore Leptodora.

Haplosclerina. Order of sponges in the Subclass Monaxonida; with double-pointed spicules and reticulate spongin.

Haploscoloplos. Genus of Polychaeta.

Haplosporidia. One of the four subclasses of Sporozoa; spores without polar capsules; parasitic in invertebrates and fishes; an example is Haplosporidium, chiefly in aquatic annelids and mollusks.

Haplosporidium. See HAPLOSPORIDIA.

Haplosyllis. Genus of Polychaeta.

Haplotaxis. Genus of minute semiaquatic and fresh-water annelids.

Haplotrema. Genus of land snails; Pacific Coast states.

haptophore. That portion of a toxin molecule which unites with active chemical groups elaborated by cells or body fluids; not thought to be the toxic element of a toxin. Cf. TOXOPHORE.

harbor seal. See PHOCA.

Harderian gland. Gland which secretes an oily fluid onto the surface of the eyeball in amphibians, reptiles, and birds.

hardhead. 1. See MYLOPHARODON. 2. Any of several marine and fresh-water fishes. 3. Type of commercial marine sponge. 4. Gray whale. 5. Any of several ducks.

hard palate. Anterior portion of the palate supported by bone, especially in man.

hardshell clam. See VENUS.

hard-shelled crab. U.S. local term for an edible marine crab which has not recently shed the exoskeleton.

Hardy-Weinberg law. Mathematical expression of the frequencies of the members of a pair of allelic genes in a biparental population; an expansion of a binomial equation; implies the genetic stability of a population in the absence of selection.

hare. 1. In the broad sense, any of numerous members of the mammal Order Lagomorpha. 2. In the strict sense, any member of the Order Lagomorpha which does not burrow and which produces young with a full coat of hair.

harelip. Congenital cleft or similar defect in the upper lip owing to failure of the premaxillary and maxillary bones to unite properly; may or may not be associated with a cleft palate.

hare-lipped bat. See NOCTILIONIDAE.

Harenactis. Genus of long, slender, burrowing sea anemones; Calif. coast.

harlequin bug. See MURGANTIA.

harlequin duck. See HISTRIONICUS.

harlequin snake. See MICRURUS and MICRUROIDES.

Harmolaga. See SPRUCE BUDWORM.

Harmoleta grandis. Sp. of wasp in the Family Eurytomidae; larva a pest in wheat stems, often known as a joint-worm.

Harmothoë. Genus of depressed Polychaeta.

Harnischia. Genus of tendipedid midges.

Harpa. Genus of large tropical gastropods; harp shells; surface of shell ribbed and variously patterned.

Harpacticoida. Suborder of marine and fresh-water copepods; body more or less cylindrical, with a movable articulation between fifth and sixth thoracic segments; first antennae short, both prehensile in male; small spp., often less than 0.5 mm. long; usually associated with a substrate; almost always free living.

Harpacticus. Genus of marine, brackish,

and fresh-water harpacticoid copepods.

Harpalus. Very common genus of black ground beetles.

harpes. Lepidoptera claspers.

Harpiocephalus. Genus of tube-nosed bats; nostrils project from the snout as two small tubes; eastern Asia.

harp seal. See PHOCA.

harp shell. See HARPA.

harpy. Harpy eagle; powerful, crested bird of prey of southern Mexico to Brazil; to 40 in. long.

harrier. See CIRCUS.

Harrimania. Genus of Enteropneusta of Calif. coast; to 4 in. long.

Harrison, Ross Granville. American zoologist (1870-1959); well known for his pioneer work in tissue culture, nerve growth and regeneration, and embryonic nerve-muscle interrelationships.

Harris's hawk. See PARABUTEO.

hart. See STAG.

hartebeest. See ALCELAPHUS.

Hartmannella. Genus of small amoeboid Protozoa; nucleus vesicular; in foul water and wet soil.

harvester ants. See POGONOMYRMEX.

harvest flies. See CICADIDAE.

harvestman. See PHALANGIDA.

harvest mite. See TROMBIDIIDAE.

harvest mouse. 1. Small European field mouse. 2. See REITHRODONTOMYS.

Harvey, William. English physiologist (1578-1657); gave the first detailed account of blood circulation; wrote a book on the embryology of the chick and other vertebrates.

Hastaperla. Genus of small Plecoptera.

Hastatella. Genus of free-swimming fresh-water peritrich Ciliata; cell with two to four rings of long conical processes.

hatchet-back. Lasmigona complanata, a thick-shelled unionid clam of central U.S.

hatchetfish. See STERNOPTERYX and ARGYROPELECUS.

Hathrometra. Antedon.

Hatschek's pit. In cephalochordates, a small, thin-walled coelomic cavity which extends to the extreme anterior end of the embryo beneath the notochord.

hat shell. See ANATINA.

haustellum. Insect mouthparts specialized for sucking up liquids.

Haustorius. Genus of marine Amphipoda; appendages heavily setose.

Havers, Clopton. English physician and anatomist (1650-1702); contributed to the structure and method of bone growth.

Haversian system. Arrangement of a Haversian canal and its associated surrounding lamellae in bone; most of the Haversian canals in bone run longitudinally although there are numerous connecting canals, and many communicate with the marrow cavity and the surface of the bone; each Haversian canal has a diameter of about 50 microns and contains a nerve and blood vessels; each canal is surrounded with about 2 to 10 concentric lamellae of bone; the lamellae are more or less separated from each other by series of small cavities (lacunae) containing the individual living bone cells; minute channels (canaliculi) radiate out irregularly from each lacuna and interconnect with canaliculi from other lacunae.

Hawaiia. Very common genus of land snails found over much of U.S.

hawfinch. Common European grosbeak.

hawk. Any of a large number of predatory birds in the Order Falconiformes; smaller than eagles and vultures; see ACCIPITER, ASTUR, BUTEO, and FALCO.

hawkmoth. See SPHINGIDAE.

hawk owl. See SURNIA.

hawksbill turtle. See ERETMOCHELYS.

hazel mouse. See MUSCARDINUS.

HCl. See HYDROCHLORIC ACID.

head fold. Projecting fold at the anterior end of the primitive streak in early bird embryos; first evidence of the future brain and foregut.

heart. Specialized muscular organ or blood vessel used for pumping blood.

heart leakage. Failure of heart valves to close tightly, thus allowing some blood to leak back into the chamber from which it has just passed.

heart murmur. Abnormal heart sounds, especially those produced by faulty valve action.

heart muscle. Cardiac muscle.

heart shell. Any of several bivalve mollusks having a heart shape when viewed from the anterior or posterior end.

heart sounds. Characteristic sounds produced by the contractions of the heart chambers and the closing action of the valves; heard with the ear placed on the chest and much more precisely with a stethoscope.

heart urchin. See SPATANGOIDA.

heartworm. See DIROFILARIA.

heat. Estrus.

heath hen. See TYMPANUCHUS.

heat rigor. Permanent contraction and death of a muscle produced by excess heat.

Hebella. Genus of marine coelenterates in the Suborder Leptomedusae.

Hebridae. Velvet water bugs; a family of minute hemipterans found along the margins of ponds and slow streams; body covered with fine pile; usually run about on the surface of the water; Neogaeus common.

Hebrus. Neogaeus.

hectocotylized arm. Heterocotylized arm; see HETEROCOTYLUS.

hectocotylus. Heterocotylus.

hedgehog. 1. Local name for the American porcupine. 2 See ERINACEIDAE.

hedgehog tenrecs. Small hedgehog-like tenrecs of Madagascar; snout pointed, tail stumpy; mostly able to roll into a ball; Setifer and Echinops typical.

hedonic gland. Any of several types of glands in certain salamanders and reptiles; believed to function in sexual stimulation.

Hedymeles. Pheucticus.

Heidelberg man. See HOMO HEIDELBERGENSIS.

Helarctos malayanus. Sun bear, a small black bear of southeastern Asia.

Helea. Genus of minute biting midges in the Family Ceratopogonodae.

Heleidae. Ceratopogonidae.

Heleodytes brunneicapillus. Campylorhynchus brunneicapillus.

heleoplankton. Pond plankton.

helical. Spiral-shaped.

Helichus. Common genus of dryopid beetles.

Helicina. Very large genus of terrestrial snails; shell depressed or conical; tropical America and Pacific islands but one sp. as far north as Iowa.

Helicodiscus. Genus of land snails; eastern U. S.

Helicopsyche. Genus of Trichoptera; larval case spiral, composed of sand grains.

Helicosporidia. Small and poorly known order of sporozoan Protozoa parasitic in insects.

helicotrema. Tip of cochlea spire where the scala vestibuli and scala tympani meet.

Helictis. Tree badgers; genus of omnivorous, tree-dwelling, forest mammals; distant relatives of badgers; Oriental zoogeographic region.

Heliochona. Genus of marine chonotrichs.

Heliopora. See COENOTHECALIA.

Heliornithidae. Family of three spp. in the bird Order Gruiformes; finfoots; shy marsh birds of tropical America, Africa, and southeastern Asia.

Heliosphaera. Genus in the Order Radiolaria.

Heliosporidium. Genus of sporozoan parasites in fly and mite larvae; in the Order Helicosporidia.

Heliothis obsoleta. Important moth in the lepidopteran Family Noctuidae; caterpillar known as corn earworm but feeds also on other field and garden crops.

Heliozoa. Order of the Class Sarcodina; spherical amoeboid protozoans having numerous filamentous radiating pseudopodia; mostly fresh-water; sun animalcules; examples are Actinophrys and Clathrulina.

Helisoma. Widely distributed genus of discoidal fresh-water snails; common in quiet waters.

Helius. Genus of crane flies in the Family Tipulidae.

Helix. Genus of European land snails, including several edible spp.; introduced into the U. S. and now found in many areas.

hellbender. See CRYPTOBRANCHUS.

helldiver. See PODILYMBUS.

hellgrammite. 1. Larva of the common dobsonfly, Corydalus cornutus; under stones and other objects in lakes, ponds, and streams; a large predaceous sp. with segmented lateral abdominal tracheal gills; much in demand by fishermen for bait. 2. Sometimes stonefly nymphs are incorrectly and locally called hellgrammites.

Helly's fluid. General killing and fixing agent; consists of 2.5 g. potassium dichromate, 5 g. mercuric chloride, 1 g. sodium sulfate, 100 ml. water, and 5 ml. formalin.

helmet shell. Any of numerous very large, triangular, thick-lipped, tropical marine gastropods used in making cameos; Cassis to 10 in. long, common from N. C. to Caribbean area.

helminth. 1. In the narrow sense, the parasitic nematodes, all trematodes, and all cestodes. 2. in the broad sense, any worm or wormlike animal.

Helminthes. Old taxon which included all groups of worms and wormlike animals.

Helminthoglypta. Genus of terrestrial snails; Pacific Coast states.

helminthology. Study of parasitic worms,

especially nematodes, cestodes, and trematodes.

Helmis. Elmis.

Helmitheros vermivora. Worm-eating warbler; bill stout and wedge-shaped; eastern half of U.S. as far as Conn. and Nebr.

Helobdella. Common genus of fresh-water leeches; 5 to 20 mm. long; scavengers, or feed on small invertebrates; occasionally parasitic on fish and frogs, rarely on man.

Helochares. Genus of hydrophilid beetles; eastern and southwestern U.S.

Helocombus bifidus. Hydrophilid beetle of eastern U.S.

Helocordulia. Genus of dragonflies; eastern U.S.

Heloderma. See HELODERMATIDAE.

Helodermatidae. Family of lizards which includes only three spp., two of which are poisonous; trunk and tail heavy and cylindrical; dorsal scales beadlike; total length to 2 ft.; teeth bluntly fanglike, with poison glands on either side of the lower jaw; bite fatal to small animals when held tightly in the jaws so that the poison can reach the prey; rarely fatal to man; pink and black Heloderma suspectum is the Gila monster of Ariz., N.M., southern Nev., Utah, and Mexico; yellow and black H. horridum is the beaded lizard of Mexico and Central America.

Helodes. See HELODIDAE.

Helodidae. Small family of pubescent terrestrial beetles; less than 6 mm. long; larvae are all thought to be aquatic; Helodes, Scirtes, and Cyphon common.

Helodrilus. Common genus of earthworms.

Helogale. Small genus of dwarf mongooses; arid parts of eastern Africa.

Helophorus. Common genus of hydrophilid beetles.

hemacytometer. Apparatus used in making counts of the various kinds of red and white cells in a blood sample.

hemal. Pertaining to blood or blood vessels.

hemal arch. Arch on the ventral surface of posterior vertebrae in fishes; carries an artery and vein.

hemal system. Poorly defined system of small vessels in echinoderms; presumed to function in circulation; typically consists of one or two circular vessels and five radiating canals; usually contained within a special tubular portion of the coelomic cavity which is often incorrectly called the perihemal system.

hematein. See HEMATOXYLIN.

hematin. Heme.

hematochrome. Red pigment found in many flagellate Protozoa.

hematocrit. 1. Centrifuge for determining the relative volumes of corpuscles and fluid in blood. 2. The determinations obtained from such a procedure.

hematopoiesis. Hemopoiesis.

hematoxylin. Crystalline material obtained from logwood and used extensively as a biological stain when oxidized to hematein.

heme. Nonprotein, organic, iron-containing fraction of the hemoglobin molecule and certain other respiratory pigments.

hemelytron. Anterior wing of insects in the Order Hemiptera; basal half leathery, distal half membranous. Pl. hemelytra.

Hemerobius. Common genus of small brownish neuropterans; see APHIS WOLF.

Hemerocampa leucostigma. White-marked tussock moth; larva feeds on foliage of deciduous trees.

hemerythrin. Any of several invertebrate iron-containing respiratory pigments; in a few polychaetes and some sipunculids, Priapulida, and Brachiopoda.

Hemiarthrus. Genus of marine isopods parasitic on Crustacea.

hemiazygous vein. Mammalian vein which drains the left thoracic wall.

Hemibelidius. Genus of large phalangers; Australian Region.

hemibranch. Gill with filaments on only one side.

hemicellulose. Any of a group of polysaccharides less complex than cellulose and which can be hydrolyzed to monosaccharides.

Hemicentetes. Genus of striped tenrecs; size of small rats, but tail absent; with yellow spines arranged in several longitudinal rows.

Hemichordata. Same as Enteropneusta; sometimes the Enteropneusta and Pterobranchia are grouped together as the Subphylum Hemichordata of the Phylum Chordata.

Hemicidaris. Genus of extinct echinoids in the Order Stirodonta.

Hemicyclaspis. See OSTEOSTRACI.

Hemicycliophora. Genus of microscopic, free-living nematodes.

Hemidactylium. See FOUR-TOED SA-

LAMANDER.

Hemidactylus. Genus of Mediterranean geckos; introduced into Fla.

Hemidinium. Genus of naked fresh-water dinoflagellates.

Hemifusus. Genus of giant Australian gastropods; to 18 in. long.

Hemigale. Genus of mammals somewhat similar to genets; hemigales; Malaya and Borneo; see CHROTOGALE also.

Hemigaleus. Small genus of sharks.

Hemigrapsus. Genus of marine mud crabs in the Suborder Brachyura.

Hemimetabola. Group of orders of insects having a series of immature aquatic forms, or naiads, which have gills but which are somewhat similar to adults except for size and structural proportions; adults are aerial and winged; stages in the life history are: egg, a series of naiads, and adult; examples are Ephemeroptera, Odonata, and Plecoptera. Adj. hemimetabolous. See also AMETABOLA, METABOLA, PAUROMETABOLA, and HOLOMETABOLA.

Hemimysis. Genus of marine crustaceans in the Order Mysidacea.

Hemioniscus. Genus of marine Isopoda parasitic on barnacles.

hemipenis. Half of a double copulatory structure in most male snakes and lizards. Pl. hemipenes.

Hemipholis. Genus of brittle stars in the Order Ophiurae.

hemipneustic respiration. Larval respiration in which one or more pairs of spiracles are closed.

hemipode. See TURNICIDAE.

Hemipodus. Genus of Polychaeta.

Hemiprocnidae. Small family of birds in the Order Micropodiformes; crested swifts; long forked tail and crested head; nest in trees; southeastern Asia.

Hemiptera. Very important order of insects which probably contains more than 50,000 described spp.; the true bugs; food habits highly varied but mouth parts always modified into a beak for piercing and sucking fluid foods; forewings thick and leathery at bases, membranous distally; hind wings entirely membranous, folded under forewings; metamorphosis incomplete; this term is sometimes used to include the Order Homoptera as well as the true hemipterans; same as Heteroptera and Hemiptera-Heteroptera.

Hemiramphidae. Family of marine bony fishes which includes the half-beaks; lower jaw greatly elongated, upper jaw short; mostly in warm seas; herbivorous; <u>Hemiramphus</u> typical.

Hemiramphus. See HEMIRAMPHIDAE.

Hemiseptella. Genus of marine Bryozoa.

Hemithyris. Circumpolar genus of Brachiopoda.

Hemitragus. Genus of goatlike ruminants in the Family Bovidae; includes the tahr of Himalayan forests.

Hemitripterus americanus. Sea raven; a sculpin of the Atlantic Coast; head with numerous fleshy tabs, dorsal fin with ragged structure, and skin prickly; to 20 in. long; voracious bottom feeder.

hemixis. Process occasionally thought to have been observed in certain ciliate Protozoa; involves fragmentation and reorganization of the macronucleus without any accompanying changes in the micronucleus.

Hemizonida. Order of extinct sea stars; Ordovician to Carboniferous; <u>Taeniactis</u> typical.

hemocoel. Cavity present in Onychophora, Mollusca, Arthropoda, Linguatulida, and Tardigrada(?); consists of a complex of spaces between tissues and organs through which blood circulates freely without restriction to blood vessels; such animals have a reduced coelom; a hemocoel never contains germ cells and never has an opening to the outside.

hemocyanin. Any of a variety of blue copper-containing respiratory pigments; found in most Mollusca, <u>Limulus</u>, all Malacostraca, and some Arachnoidea.

hemocyte. Any formed element in the blood.

hemocytoblasts. Generalized connective tissue precursors of blood cells; large cells with an open nucleus.

hemocytozoon. Protozoan parasite of blood corpuscles.

hemoflagellate. Any flagellate protozoan which is parasitic in the blood stream for some part of its life cycle.

hemoglobin. Any of a large group of respiratory pigments occurring in the erythrocytes of vertebrates and dissolved in the blood plasma of a wide variety of invertebrates (where they are often called erythrocruorins); each sp. has its own specific hemoglobin; the molecule is a complicated protein containing a small amount of iron which is the element forming a loose temporary combination with oxygen during transport of the latter.

hemoglobin carbamate. Compound formed in erythrocytes or plasma when hemoglobin takes up carbon dioxide from the tissues.

hemolymph. Fluid in invertebrate tissues and body cavities; may act as the equivalent of blood (e.g. arthropods) or it may be a fluid in addition to blood (e.g. earthworm).

hemolysin. Substance which disintegrates erythrocytes.

hemolysis. 1. Disintegration of erythrocytes or their cell membranes with the liberation of hemoglobin. 2. Alteration of the membrane of erythrocytes so that hemoglobin leaves the cells; may be induced osmotically, by the use of various solvents, or by the use of substances that attack the membranes.

hemolyze. To disintegrate, as erythrocytes, with the liberation of the contained hemoglobin.

hemophilia. Failure of blood to clot properly in human males; determined by sex-linked recessive transmitted from carrier mother to affected son and to carrier daughter.

hemopoiesis. Formation or production of blood cells.

hemorrhage. Copious bleeding.

hemorrhagin. Cytolysin found especially in certain snake venoms; produces disintegration of capillary endothelium.

hemotoxin. Any toxin specific for blood corpuscles.

hemotrophic nutrition. Passage of nutritive materials from the circulating blood of the mother to that of the fetus via the placenta.

hen clam. Common name for a large sp. of Spisula.

hen louse. Any sp. of louse infesting poultry, especially Menopon.

Henricia. Genus which includes the familiar blood sea star, or red sea star, of the Atlantic Coast.

Hensen's line. H line; a fine light line running along the center of an isotropic disc in striated muscle tissue.

Hensen's node. Small elevation at the anterior end of the primitive streak, as in an early chick embryo.

heparin. Antiprothrombin.

hepatic. Of or pertaining to the liver.

hepatic artery. Blood vessel in vertebrates and many higher invertebrates; supplies blood to the liver or other large digestive gland.

hepatic caecum. 1. One of ten large digestive and storage glands in a typical sea star; there are two such caeca in each arm and they open into the pyloric stomach. 2. General term for a glandular caecum of the digestive tract in many invertebrates and prochordates.

hemorrhoid. Painful tumor or venous dilatation, especially in the region of the rectum and anus.

hepatic portal system. System of blood vessels carrying blood from the stomach and intestines to the liver in vertebrates.

hepatic vein. Vein draining the liver in vertebrates and a few invertebrates.

hepatitis. Liver inflammation.

hepatopancreas. Large digestive and storage gland in many decapod Crustacea and in a few other invertebrates.

Hepatozoon. Genus of Sporozoa occurring in mammalian marrow, liver, and other tissues, as well as the circulating blood cells; transmitted by ticks and mites.

Hepatus. Genus of tropical and subtropical brachyurans (marine crabs).

Hepialidae. Family of moths; caterpillars often bore into plants and are pests on economic plants; Hepialus common.

Hepialus. See HEPIALIDAE.

Heptagenia. Common genus of Ephemeroptera.

Heptagyia. Genus of tendipedid midges.

Heptanchus. Genus of sharks.

herbivore. Any animal which relies chiefly or solely on vegetation for its food.

herbivorous. Pertaining to an herbivore; phytophagous.

Hercothrips. Common insect genus in the Order Thysanoptera; H. fasciatus (bean thrips) is an important pest of many cultivated plants; H. hemorroidalis is a greenhouse thrips which feeds on roses and tomatoes.

Hercules beetle. Very large tropical American beetle; pronotum produced into a very long downcurved horn.

heredity. 1. Physiological, physical, and psychological resemblances among individuals related by descent. 2. Transmission from one generation to the next of factors causing offspring to resemble parents.

Hermaea. Genus of nudibranchs; feed mostly on green algae.

hermaphrodite. Individual having both male and female functional reproductive organs; it is unusual for an individual to produce both eggs and sperm at the same time; more commonly the male reproductive system produces mature male gametes at one time and the female reproductive system mature female gametes at another time; hermaphrodites may occur as aberrations among unisexual animals.

hermaphroditic. Of or pertaining to a single individual which possesses both male and female reproductive organs, e.g. earthworms, trematodes, bryozoans, and at least one marine fish.

hermaphroditic duct. Convoluted duct which carries sperm and ova from the ovotestis in many gastropods.

Hermetia illucens. Sp. of hardy soldier fly which breeds in decaying plant and animal material; occasionally causes human intestinal myiasis.

Hermissenda. Genus of brightly-colored nudibranch mollusks.

hermit crab. See PAGURUS.

hermit thrush. See THRUSH.

hermit warbler. Common warbler in the genus Dendroica; Pacific coniferous forests.

heron. See ARDEIDAE.

Heronimus. Genus of monostome trematodes found in turtles.

Herophilus. Greek anatomist (2nd and 3rd centuries B.C.); taught human anatomy at Alexandria medical schools; made careful dissections and descriptions of much of the human body, but only a few of his writings persist.

Herpestes. Genus of Old World carnivores; includes common mongooses.

Herpetocypris. Genus of fresh-water ostracods.

herpetology. Study of reptiles and amphibians.

Herpetomonas. Genus of protozoans in the Family Trypanosomidae; parasitic in various invertebrates.

Herpobdella. See ERPOBDELLA.

herpon. Community of microscopic plants and animals living on the surface of soft substrates in aquatic habitats.

Herpyllus vasifer. Common black spider in the Family Gnaphosidae.

herring. See CLUPEA, SARDINOPS, and SARDINIA.

herring gull. Larus argentatus, a Holarctic marine and fresh-water sp.; coloration gray, black, and white.

Hertwig, Oscar. German embryologist (1849-1922); first to determine that fertilization involves fusion of male and female nuclei; emphasized the importance of the nucleus as a basis of heredity; investigated germ layer theory and embryo formation.

Hertwig, Richard. German zoologist (1850-1937); emphasized the fact that a nucleus influences only a limited amount of cytoplasm.

Hertwigia. Genus of rotifers living as parasites in colonies of Volvox.

Hesione. Genus of short, cylindrical Polychaeta.

Hesperagrion heterodoxum. Damselfly of extreme southwestern U.S.

Hesperalia. Genus of Polychaeta.

Hesperiidae. Family of Lepidoptera which includes the skippers; abdomen heavy, antennae with hooked tips; common and abundant butterflies with a rapid, darting flight; Atalopedes common.

Hesperiphona. See GROSBEAK.

Hesperocorixa. Common and widely distributed genus of water boatmen in the Family Corixidae.

Hesperoleon. Genus of insects in the Order Neuroptera having a superficial resemblance to damselflies; larvae are predaceous ant lions.

Hesperoleucas. Genus of small freshwater fishes in the minnow family; Calif., Ore., Nev., and Utah.

Hesperonoë. Genus of polychaetes; common commensals in burrows of decapod Crustacea.

Hesperophylax. Genus of Trichoptera.

Hesperornis. Genus of fossil birds from Niobrara chalk deposits of Kan.; in the Order Hesperornithiformes; specialized for swimming and diving; flightless, with webbed hind feet.

Hesperornithiformes. Order of fossil birds in the Superorder Odontognathae.

Hessian fly. See PHYTOPHAGA.

Hetaerina. Common genus of damselflies in the Order Odonata.

Heterakis. Genus of rhabditid nematodes parasitic in the digestive tract of birds and a few mammals; H. gallinae common in the chicken and turkey caecum.

Heterandria formosa. Small killifish in the Family Poeciliidae; coastal swamps from S.C. to Fla.

heterauxesis. Allometry.

Heterelmis. Genus of elmid beetles.

Heterocephalus. Sand-puppies, a genus of burrowing rodents in the Family Bathyergidae; body almost naked and mouse-sized; burrow in Somaliland deserts.

Heterocera. Suborder of insects in the Order Lepidoptera; includes the moths; usually divided into the Suborder Jugatae and the Suborder Frenatae.

heterocercal. Pertaining to the tail fin of certain fishes, especially sharks, in which this fin is not dorsoventrally symmetrical; posterior end of spinal column turned upward and dorsal portion of tail fin longer than ventral portion.

heterochromatin. That portion of a chromosome containing few or no genes and staining more deeply than the euchromatin; includes DNA and RNA.

heterochromosome. Sex chromosome, such as the X or Y chromosome in man or Z and W in birds.

Heterocloeon. Uncommon genus of Ephemeroptera; eastern U.S.

Heterocoela. One of the orders of sponges in the Class Calcarea; tissues and walls thickened and folded internally; choanocyte epithelium not continuous; examples are Grantia and Sycon.

heterocoelous. Pertaining to a type of vertebra in the neck of birds; saddle-shaped articular faces between the centra.

Heterocotylea. Monogenea.

heterocotylus. Heterocotylized arm; specialized tip of one of the arms in males of certain cephalopod mollusks; used to transfer spermatophores to the mantle cavity of the female and in some spp. breaks off during the process.

Heterocrypta. Genus of marine crabs; body triangular.

heterodactyl. In birds, having the first and second toes projecting backward and the third and fourth forward.

Heterodera. Important genus of tylenchid nematodes parasitic on the roots of many plants of economic importance, especially the sugar beet and potato; H. schachti is the common sugar beet nematode; H. radicicola (H. marioni), the root knot nematode, is common in a great many spp. of plants.

Heterodon. Puff adders or hog-nosed snakes; a genus of non-poisonous snakes of central and eastern U.S.; when alarmed they flatten the neck and head while hissing loudly; often thresh about or feign death; feed on toads and insects.

heterodont. Having a variety of tooth types, as in most mammals.

Heterodontida. Superfamily of sharks in the Suborder Squali; bullhead sharks.

Heterodontus. Genus of primitive sharks found especially off Australia; Port Jackson shark; head and eyes large; mouth small, with crushing teeth.

heteroecious. Having different stages of the life history in different host spp.; e.g. most tapeworms and many flukes.

heterogametic. Sex whose body cells have a pair of unlike sex chromosomes, or one unpaired sex chromosome; former condition in human males (X and Y chromosomes), the latter (X and O chromosomes) in certain male insects; males are usually heterogametic, but females are heterogametic in Trichoptera, Lepidoptera, birds, some amphibians, reptiles, and fishes, and a few plants.

heterogamy. 1. Anisogamy; condition in which two uniting gametes are unlike in size or structure. 2. Occasionally, pertaining to animals having different types of sexual reproduction in two successive generations, such as parthenogenesis and syngamy in certain aphids.

heterogenesis. See ALTERNATION OF GENERATIONS.

heterogony. 1. Allometry. 2. Heterogenesis; see ALTERNATION OF GENERATIONS.

heterogynous. Having two different kinds of females; e.g. queens and workers among bees.

Heterlimnius. Genus of elmid beetles.

Heteromastus. Genus of Polychaeta.

Heterometabola. Same as Exopterygota, or Hemimetabola plus Paurometabola.

Heteromeyenia. Very common and widely distributed genus of fresh-water sponges; largest fresh-water genus.

Heteromi. Order of bony fishes which includes the spiny eels; long, slim body and tail but caudal fin absent; deep oceanic waters; Notacanthus common (Family Notacanthidae).

Heteromida. Heteromi.

heteromorphosis. Regeneration of a body part different from that lost or removed; in the crayfish, for example, removal of an eye and its stalk results in its replacement by an abnormal antenna-like growth during subsequent molts.

Heteromyidae. Family of rodents which includes the pocket rats, kangaroo rats, and pocket mice; small spp. with external fur-lined cheek pockets;

hind legs more or less elongated for jumping; tail often longer than head and trunk; Perognathus and Dipodomys typical.

Heteromys. Genus of Central and South American ratlike rodents; most spp. with spines and bristles instead of hair; habits variable.

Heteromysis. Marine genus of Mysidacea; to 10 mm. long.

Heteronema. Genus of elongated plastic fresh-water Protozoa; one flagellum directed forward, the other trailing and shorter.

Heteronemertea. Order in the Phylum Nemertea; dermis fibrous; body wall with three muscle layers.

heteronereis. Name used for certain spp. of marine Polychaeta in which the posterior part of the body containing the gonads differs markedly from the anterior part.

Heteroneura. Frenatae.

heteronomous segmentation. Relative dissimilarity and specialization of certain body segments, as in most insects. Cf. HOMONOMOUS SEGMENTATION.

heterophagous. Pertaining to an animal which eats two or more kinds of foods.

Heterophrys. Genus of marine and fresh-water spherical heliozoan Protozoa; pseudopods and chitinoid spicules radially arranged.

Heterophyes heterophyes. Minute fluke occurring in the small intestine of man, cat, dog, and other domestic animals; intermediate hosts are a snail and fresh-water mullet; Near East and Far East.

Heterophyllidea. Diphyllidea.

Heteropia. Genus of very small cylindrical calcareous marine sponges.

heteroplastic grafting. Transplantation of a graft from one donor sp. to a different host sp.

Heteropoda. Division of taenioglossid snails which includes the pelagic sea snails; foot developed into a finlike swimming device; body translucent, shell reduced, visceral sac small.

Heteropoda venatoria. Banana spider; a large rufous and yellow spider in the Family Sparassidae; common in the tropics and often imported with fruit.

Heteroptera. Hemiptera.

heteropycnotic body. Minute dark-staining mass on the inner surface of the nuclear membrane of cells in the females of most spp. of mammals.

heteropycnosis. Exceptionally dark-stained condition obtaining in certain chromosomes and parts of chromosomes.

Heteroscelus incanus. Wandering tattler, a brownish gray sandpiper-like bird; breeds in Alaska, winters from Mexico to the Galapagos, and in the southwest Pacific.

heterosis. Hybrid vigor.

Heterosomata. Order of marine bony fishes which includes several families of flatfishes and soles; young bilaterally symmetrical but adults asymmetrical and swim on one (left or right) side, and have eyes on the upper (right or left) side; body greatly compressed; dorsal and anal fins elongated and fringing both margins; on bottom in coastal water; some important food spp.; Hippoglossus hippoglossus (American halibut), Paralichthys (flounder), Symphurus (sole), Achirus (round sole and hogchoker), and Hippoglossoides (plaice and sand dab).

heterosome. Heterochromosome.

Heterostelea. Carpoidea.

Heterostraci. Order of ostracoderms; nostrils paired and ventral; gills opening through a common slit on each side behind the main armor plates; Anglaspis typical.

Heterotardigrada. Class in the Phylum Tardigrada; head with anterior cirri and lateral filaments; e.g. Echiniscus.

Heteroteuthis. Genus of deep-sea squids.

heterotopic bone. Any of a wide variety of extra or accessory bones formed in many vertebrates in addition to the normal skeletal complement; e.g. baculum, os palpebrae of crocodilians, and the os cordis of deer.

Heterotricha. Suborder of the ciliate Order Spirotricha; ciliation complete and more or less uniform; e.g. Balantidium and Stentor.

heterotrophic bacteria. Consumer bacteria which oxidize or ferment organic materials into simple organic and inorganic compounds; decomposers and mineralizers.

heterotrophic nutrition. Type of nutrition occurring in animals and some bacteria and true fungi; characterized by a dependence on complex organic food materials which have originated in other plants and animals.

heterozygous. Pertaining to an individual having two different alleles at the same locus on a pair of homologous chromosomes.

hexacanth. Oncosphere.

Hexacorallia. Zoantharia.

Hexactinellida. One of the three classes of Porifera; skeleton composed of six-rayed silicon dioxide spicules, either separate or arranged in a complicated network; usually cylindrical to funnel-shaped; exclusively marine; Regadrella and Venus' flower basket typical.

Hexacylloepus. Genus of elmid beetles.

Hexagenia. Very common genus of Ephemeroptera; nymphs fossorial.

Hexagrammus. Genus of large, carnivorous, commercial and sport fishes of rocky northern Pacific coasts; greenlings.

Hexamita. Genus of pyriform, colorless flagellates; with two nuclei, eight flagella, and two axostyles; stagnant water or parasitic.

Hexanchida. Superfamily of sharks in the Suborder Squali; frilled and cow sharks.

Hexapanopeus. Genus of marine mud crabs in the Suborder Brachyura.

hexaploid. Organism having 6n chromosomes instead of 2n in the somatic cells.

Hexapoda. Insecta.

Hexasterophora. Order of sponges in the Class Hexactinellida; amphidisks absent; star- shaped spicules present; e. g. Euplectella.

Hexatoma. Genus of crane flies in the Family Tipulidae.

hexokinase. Enzyme which facilitates the transfer of phosphate from adenosine triphosphate to hexose molecules, the end products being hexose-6-phosphate and adenosine diphosphate.

hexose. Any sugar containing six carbon atoms in the molecule, e. g. glucose and galactose.

hexylresorcinol. White or yellow crystalline phenol; a powerful germicide; especially useful as a drug for infections of Ascaris and hookworms, and less useful for other helminths.

hibernaculum. 1. Case or covering secreted or otherwise made by terrestrial animals for protection during an unfavorable environmental period, especially winter. 2. In a few freshwater Bryozoa, a sclerotized case which encloses a small mass of undifferentiated cells; a wintering-over device. 3. Epiphragm of a snail. Pl. hibernacula.

hibernation. 1. In the broad sense, a type of short to protracted dormancy and inactivity which is usually induced by dry or cold weather in warm, temperate, and cold climates; occurs in at least one bird and in many mammals, reptiles, amphibians, and invertebrates; metabolic processes are greatly slowed, and in mammals the body temperature drops to approximately that of the immediate surroundings; the physiological state varies from a "deep sleep" to severe unconsciousness. 2. In the narrow sense, a condition occurring in a relatively small number of mammals in which, at some time during the year, the body temperature drops and there is a decrease in metabolism during a relatively long, torpid, inactive period (usually in winter).

hickory borer. Any of several beetles whose larvae live in the cambium or xylem of hickory trees.

hickory shad. Gizzard shad.

hilum. 1. Small notch or depression where blood vessels, nerves, etc. enter and/or leave an organ. 2. Thin spot in the protective layers of a sponge gemmule.

Himalayan bear. See SELENARCTOS.

Himantopus mexicanus. Common black-necked stilt in the Family Recurvirostridae; large, slim, wading bird with very long legs; mostly around inland waters of temperate N. A. and northern South America.

hind. 1. In European usage, an adult female red deer. 2. Any of various tropical and subtropical marine fishes, especially spotted and speckled spp. related to the groupers.

hindbrain. Posterior of the three primary divisions of the vertebrate brain; includes the metencephalon (cerebellum and pons Varolii) and myelencephalon (medulla oblongata).

hindgut. 1. Posterior part of the digestive tract of arthropods; lined with cuticle, this lining being shed when the arthropod molts; in some arthropods the hindgut is short, in others it is more than half the body length; excretory tubules commonly enter the digestive tract at the junction of the midgut and the hindgut. 2. Short, blind tube in the posterior portion of many vertebrate embryos; becomes a portion of the posterior part of the digestive tract.

hind wings. Posterior pair of wings situated on the third thoracic segment of typical insects.

hinge ligament. Strong elastic band by which the two valves of a bivalve mollusk are attached to each other along the dorsal margin.

hinge tooth. Toothlike projection or ridge on the inner surface of a typical bivalve mollusk shell near the hinge.

Hinnites. West Coast genus of scallops; right valve fixed to substrate.

hinny. Hybrid offspring of a stallion and a female ass.

Hiodon. Genus of silvery, herring-like fishes in the Family Hiodontidae; toothed herrings, mooneyes, and goldeye; two spp. in N. A. from Hudson Bay to Mississippi Valley and southeast; unimportant as food fish; unique because of the absence of cones in the retina and because the rods are arranged in bundles.

Hiodontidae. Family of fishes in the Order Isospondyli; the mooneyes; small, silvery, herring-like spp. in U. S., tropical Africa, and the East Indies; Hiodon common in U. S.

hip girdle. Pelvic girdle.

Hippa. Emerita.

Hippasteria. Broadly pentagonal genus of sea stars in the Order Phanerozonia; widely distributed.

Hippidion. Genus of fossil ancestral horses; South American Pleistocene.

Hippidium. Hippidion.

Hippoboscidae. Small family of insects having a flat, louselike body; bird ticks, louse flies; adults ectoparasitic, suck blood from birds, mammals, and occasionally man; Melophagus, Lynchia, and Ornithoica common.

Hippocamelus. Genus of small deer of the high Andes; guemals; body heavy and hair coarse.

Hippocampidae. Family of bony fishes in the Order Solenichthyes; includes the sea horses; body grotesque, rigid, compressed, ridged, and covered with rings of bony plates; head bent ventrally but anterior-posterior axis is normally kept in a vertical position by undulations of the dorsal fin; tail long, coiled, and prehensile; most spp. to 6 in. long; often cling to eelgrass and algae with tail; Hippocampus common.

hippocampus. In the brain of higher vertebrates, a large curved olfactory mass on the floor of each lateral ventricle.

Hippocampus. See HIPPOCAMPIDAE.

Hippodamia. Common genus of coccinellid beetles; feed chiefly on aphids.

Hippodiplosia. Common genus of marine Bryozoa.

Hippoglossus. See HETEROSOMATA.

Hippolysmata. Genus of intertidal shrimps.

Hippolyte. Large genus of small marine shrimps found among vegetation in the shallows; decapod Suborder Macrura.

Hippomorpha. Suborder of the Order Perissodactyla; includes the modern horses and zebra in addition to three extinct families.

Hippopotamidae. Small family of mammals in the Order Artiodactyla; includes the spp. of hippopotamus; enormous piglike herbivores of African rivers and lakes; usually in the water during the daytime; feed on coarse grass and reeds; mouth large and muzzle bulbous; able to remain submerged for 1 to 2. 5 minutes; Hippopotamus and Choeropsis.

Hippopotamus. Genus of mammals which includes the common African hippopotamus; to four tons; live in family groups or herds.

Hippopus. Genus of large tropical marine clams; sedentary, with the valves gaping upward; foot degenerate, shell to 12 in. long.

Hipposideridae. Family of Old World bats; horseshoe-nosed bats.

Hippospongia. Genus of sponges in the Order Keratosa; the "horse" sponge and grass sponge are of no commercial importance, but the wool sponge or sheep's wool sponge is especially valuable.

Hippothoa. Genus of marine Bryozoa.

Hippotigris. Generic name sometimes used for the zebras which are usually placed in Equus.

Hippotragus. Genus of African ruminants in the Family Bovidae; includes the sable antelope and roan antelope; large, handsome spp. of plains, glades, and open forests.

hippuric acid. Detoxication product formed by kidneys and liver from benzoic acid and glycine.

Hircina. Genus of marine sponges in the Subclass Keratosa; shape of a plate standing on edge; N. C. to West Indian waters.

hirudin. Secretion of salivary glands of many leeches; prevents coagulation while the leech is taking a blood meal.

Hirudinea. One of the five classes of Annelida; leeches; body pigmented, depressed, and most of the 34 segments subdivided by means of annuli; with a posterior sucker and usually an anterior sucker surrounding the mouth; clitellum poorly defined; coelom largely obliterated; monoecious, eggs usually deposited in cocoons; no special larval stage; fresh-water, marine, and terrestrial; predaceous, free-living, or parasitic; about 300 spp.

Hirudo medicinalis. European medicinal leech, introduced into northeastern states; usually green, with four or six brown stripes; to 4 in. long; a blood sucker.

Hirundinidae. Cosmopolitan family of passerine birds which includes the swallows and martins; sparrow-sized birds with long, slim wings and graceful flight; Riparia riparia is the bank swallow; Hirundo erythrogaster the barn swallow; Petrochelidon albifrons cliff swallow; Progne subis the common purple martin.

Hirundo. See HIRUNDINIDAE.

His, Wilhelm. German zoologist (1831-1904); invented the microtome and studied early embryological development.

hispid. Rough, owing to a covering of bristles, stiff hairs, or small spines.

histaminase. Tissue enzyme which destroys histamine; flavoprotein found in rat lung and intestine, and in the kidneys of some other mammals.

histamine. Organic base or amine occurring in animal and plant tissues; dilates and increases permeability of capillaries.

hister beetle. See HISTERIDAE.

Histeridae. Family of small, flattened, hard beetles often found in decaying animal matter; hister beetles.

histidine. Common amino acid; $C_6H_9O_2N_3$.

histiocyte. Macrophage.

histogenesis. Origin, development, and differentiation of tissues.

histology. Study of tissues, especially their structure and arrangement.

histolysis. Dissolution or breakdown of tissue.

Histomonas meleagris. Flagellate parasite of the intestinal mucosa and liver of the chicken, turkey, quail, etc.

histone. Simple protein; part of gene and chromatin structure.

histotrophic nutrition. See EMBRYOTROPH.

Histricopsylla gigas. Wood rat flea in the Order Siphonaptera; to 5 mm. long; in wood rat nests, especially in Calif.

Histrio. See ANTENNARIIDAE.

Histriobdella. Genus of Polychaeta; usually in branchial chamber of lobsters.

Histrionicus histrionicus. Harlequin duck; male with dark body, reddish sides, and white spots and streaks on head and neck; Holarctic and mostly Arctic and Subarctic; eastern subspecies as far south as New England, western subspecies as far south as Calif. and Colo.; flesh not palatable.

hitch. See LAVINIA.

H line. Hensen's line.

hoary bat. See LASIURUS.

hoary marmot. Marmota caligata, a sp. of high mountains of Mont., Ida., and Wash., north to middle Alaska; head and foreparts mixed black and white.

hoatzin. See OPISTHOCOMIDAE.

hock. Tarsal joint or tarsal region in the hind leg of horses and cattle.

hog. See SUS.

hogchoker. See HETEROSOMATA.

hogfish. Any of numerous fishes whose head has a fancied resemblance to that of a hog; see LABRIDAE.

hog louse. See HAEMATOPINUS.

hog-nosed skunk. See CONEPATUS.

hog-nosed snake. See HETERODON.

hogsucker. See HYPENTELIUM.

Holarctic realm. Major zoogeographic realm consisting of the combined Palaerctic and Nearctic realms; this terminology is sometimes used because of the similarity of Palearctic and Nearctic faunas.

Holbrookia. Genus of earless lizards in the Family Iguanidae; body small, slender, flattened; Great Plains and southwestern states.

Holectypoida. Order of echinoids with Aristotle's lantern and keeled teeth; most genera extinct; Echinoneus modern and circumtropical.

Holectypus. Genus of extinct echinoids.

holobenthic. Pertaining to a marine animal which completes its entire life cycle in deep water. Noun holobenthos.

holoblastic cleavage. Complete and approximately equal division of cells in the early embryology of animals having a minimum or moderate amount of yolk in the fertilized egg; found in many aquatic invertebrates, Amphioxus, and mammals.

holoblastic egg. Fertilized egg that undergoes complete division during early embryonic cleavage.

holobranch. Fish gill with two rows of filaments on the branchial arch.

Holocene. Recent epoch of geological time; past 20,000 years.

Holocentridae. Soldier fishes, squirrel fishes; a large family of carnivorous percomorph fishes found along rocky tropical and subtropical shores; body covered with rough prickles; coloration brilliant; Holocentrus common.

Holocentrus. See HOLOCENTRIDAE.

Holocephali. See CHIMAERA.

holocrine gland. Any gland in which the secretory cells disintegrate and form part of the secretion.

holoenzyme. Complete enzyme, involving a coenzyme and a protein.

hologamete. Protozoan which becomes modified to form a gamete, without any special cell division(s).

hologamy. Union of two protozoan hologametes to form a zygote.

Hololepis. Genus of darters now usually included in Etheostoma.

Holomastigina. Small order of flagellates in which the whole cell is amoeboid and with many flagella; usually included in the Order Hypermastigina.

holomastigote. Pertaining to a protozoan which has many flagella scattered evenly over the surface.

Holometabola. Group of orders of insects having complex, indirect, or complete metamorphosis in which the stages are egg, larva, pupa, and adult; the larval stage has a series of instars which are very different from the pupa and adult; pupa usually non-motile and non-feeding, enclosed within a cell, cocoon, or last larval skin; includes the higher orders of insects such as Trichoptera, Coleoptera, Diptera, Siphonaptera, Lepidoptera, and Hymenoptera. Adj. holometabolous. See also AMETABOLA, METABOLA, PAUROMETABOLA, and HEMIMETABOLA.

Holomyaria. Taxon sometimes used to include those nematodes having no discrete muscle cells when the body is seen in cross section.

holonephros. Idealized primitive vertebrate kidney; archinephros.

Holopagurus. Genus of hermit crabs in the Suborder Anomura.

Holopedium. Genus of fresh-water plankton Crustacea; animal nearly enclosed in a large gelatinous case.

Holophrya. Genus of small globose to ellipsoidal holotrich Ciliata; fresh and salt water.

holophytic nutrition. Type of nutrition occurring in green plants and certain protozoans, and involving the synthesis of carbohydrates from carbon dioxide and water in the presence of light, chlorophyll, and certain enzymes.

holopneustic. Pertaining to an insect larva having all spiracles functional.

Holosteica. Subclass of fishes which includes the gars and bowfins; spp. in this taxon are usually included in the Protospongyli and Ginglymodi.

Holosticha. Genus of marine and fresh-water hypotrich Ciliata.

Holostomata. Group of families of digenetic trematodes; anterior oral sucker, ventral acetabulum, and a large ventral adhesion disc posterior to the acetabulum.

Holothuria. Genus of sea cucumbers found in the Fla. and West Indies area.

Holothuroidea. Class in the Phylum Echinodermata; sea cucumbers; body elongated on the oral-aboral axis so there are "anterior" and "posterior" ends; sausage-shaped or wormlike; body wall thin to leathery, flexible; skeleton of microscopic plates or spicules imbedded in the body wall; mouth surrounded by tentacles; madreporite internal; tube feet present; arms, spines, and pedicellariae absent; to 20 in. long; on all types of substrates at all depths; often imbedded except for the anterior end.

Holotricha. Order of ciliates characterized by uniform ciliation over the whole surface; holozoic or saprozoic; no adoral membranelles; Paramecium the most familiar example.

holotrichous isorhiza. Isorhiza nematocyst which is large and has the tubule covered with spinules.

holotype. Single specimen designated as the "type" by the author in his published description of a new sp. or genus.

holozoic nutrition. Ingestion and metabolic utilization of liquids or solid food particles, including debris as well as living plant and animal material; the term is usually used with reference to certain Protozoa but applies equally well to most other animals.

Homalopoma. Genus of small, dark-colored, marine snails.

Homalozoon. Genus of long, thin, fresh-water holotrich Ciliata.

Homaridae. Family which includes the American and European lobsters.

Homarus. Genus of decapod crustaceans which includes the American lobster of the Atlantic Coast, *H. americanus*; rostrum with teeth along each side; first pair of pereiopods developed into large pinching and crushing claws; body elongate, abdomen large and not reflexed under cephalothorax; rare specimens exceed 24 in. but the usual market size is 8 to 12 in.

homeosis. 1. Transformation of one organ into another by mutation; in *Drosophila* examples are the transformation of an antenna into a leg and balancers into a second pair of wings. 2. Occurrence of a homologous appendage on a segment where it does not normally occur.

homeostasis. 1. Tendency to stability in normal internal fluids and metabolic conditions. 2. Steady internal state. 3. Ecological stability.

home range. General area traversed during the normal daily or seasonal activities of an animal.

homing instinct. Poorly-understood instinct found especially in birds which are able to fly to the nest quickly, even when released a long distance away from it; also true of some fishes, e.g. return to home stream by migrating salmon.

Hominidae. One of the two families in the primate Superfamily Hominoidea; includes modern man and several fossil spp.; the other (Pongidae) includes all anthropoid apes.

Hominoidea. Superfamily of the mammal Order Primates; includes the anthropoid apes and man; tail and cheek pouches absent; divided into the Pongidae and Hominidae.

homocercal tail. Type of tail characteristic of most bony fishes; upper and lower portion of tail fin similar in shape, with the vertebral column ending at the middle of the base.

Homocoela. One of the orders of sponges in the Class Calcarea; organization simple, with tissues and walls having no internal folding; spongocoel lined continuously with choanocytes; an example is *Leucosolenia*.

homodont. Having all teeth of the same general kind, as in certain primitive mammals.

homogametic. Pertaining to individuals of a sex characterized by cells having a pair of similar sex chromosomes, e.g. the two X chromosomes in human females.

homogenistic acid. Organic acid normally oxidized to carbon dioxide and water in the body; in certain individuals lacking the necessary oxidizing enzymes this acid accumulates in the blood and leaves the body in the urine; upon standing, such urine turns black (alkaptonuria).

Homo heidelbergensis. Heidelberg man; fossil mid-Pleistocene sp. found in Germany; one mandible and teeth suggest a powerful type intermediate between ape and early man; no associated implements.

homoiotherm. Animal having a relatively constant body temperature regardless of the temperature of its environment; warm-blooded animals (birds and mammals only). Adj. homothermous.

homolecithal egg. Ovum which has yolk evenly distributed throughout its mass, as in mammals.

homologous. 1. Basic similarity of organs or other structures as the result of similar embryonic origin and development; e.g. the forelimbs of bat, bird, horse, mole, and man are all homologous. 2. Term occasionally used with reference to similar function, such as clawing of a cat and clawing of a bear. Noun homology.

homologous chromosomes. Two chromosomes which normally pair at meiosis and which are found in all somatic cells; have identical gene loci and may have identical or homologous alleles.

Homo neanderthalensis. Neanderthal man; fossil Pleistocene sp. found in many areas of Europe and Asia; represented by abundant skeletal material; brain capacity 1300 to 1600 cc.; jaw protruding, posture slightly stopped; simple chipped stone tools and weapons; probably dates to 100,000 years ago.

Homoneura. Jugatae.

homonomous segmentation. Relative similarity in body segments, as in the earthworm. Cf. HETERONOMOUS SEGMENTATION.

homonym. In biological nomenclature, one of two or more identical but independently proposed names for the same or different taxa.

homoplastic grafting. Transplantation of tissue or organ from one donor to a different nost individual of the same sp.

homoplasy. Analogy; correspondence of structure acquired independently and not the result of common ancestry.

Homoptera. Large order of herbivorous insects which includes aphids, scale insects, cicadas, etc.; mostly small spp. with piercing and sucking mouth parts; wings entirely membranous; metamorphosis usually gradual but sometimes complete in the male; many pest spp. Cf. HEMIPTERA.

Homo rhodesiensis. Rhodesian man; fossil Pleistocene sp. found in Rhodesia, represented by one skull; brain capacity about 1280 cc.; jaws heavy but human rather than apelike.

Homo sapiens. Scientific name of modern man; a highly variable sp. divided into numerous races but all capable of interbreeding.

Homo soloensis. Fossil anthropoid (Ngandong man) known from 11 crania found in Java in 1933; somewhat similar to H. neanderthalensis.

homozygous. Pertaining to an individual having identical alleles, one of each pair under consideration being located on each member of a chromosome pair.

homunculus. Miniature but completely formed human being presumed by some early embryologists to occur in the headpiece of a human sperm or in the ovum. Pl. homunculi.

honey. Sirupy carbohydrate solution made by bees and representing chemically altered nectar from flowers; some of the chemical changes occur in the digestive tract of the bee before the material is regurgitated at the hive; other changes occur in the hive where the material is mouthed by other bees.

honey ant. Honey pot.

honey badger. See RATEL.

honeybee. Apis mellifera (in the hymenteran Family Apidae).

honeycomb. Mass of hexagonal wax compartments constructed by bees.

honey creeper. See DREPANIDIDAE.

honeydew. Sweet excretion from the anus of most insects in the families Cicadellidae and Aphididae; often eaten by ants and bees.

honey eaters. See MELIPHAGIDAE.

honey guide. Any of a small group of small, drab, insectivorous birds in the Family Indicatoridae (Order Piciformes); leads man and other mammals to trees containing bee nests and supplies of honey and wax; lay eggs in nests of other spp.; Africa, East Indies, and the Himalayas.

honey pot. One of the castes (replete) in certain spp. of ants; such individuals become swollen storehouses of a sweet carbohydrate food reserve used by the whole colony.

honey stomach. Crop of the honeybee and its relatives; used for transporting nectar.

honey stopper. Special valve at the posterior end of the crop of the honeybee and its relatives; prevents the loss of nectar into the stomach.

honey-sucker. See NOOLBENGER.

hooded crow. European crow (Corvus cornix); back and underparts gray.

hooded merganser. See LOPHODYTES.

hooded rat. Laboratory variety of the black rat; body white, head black.

hooded seal. See CYSTOPHORA.

hoof. 1. Hard horny casing of the foot or ends of the digit(s) in certain mammals, e.g. horse, cow, deer, etc. 2. Whole foot of a hoofed mammal.

Hooke, Robert. English physicist, mathematician, and inventor (1635-1703); first described and figured plant cells in his book Micrographia in 1665.

hooked mussel. See BENT MUSSEL.

hook-nosed snake. Any of several snakes in Ficima and Gyalopion; tip of nose slightly up-turned; southern Tex., N.M., and Ariz. and into Mexico; 5 to 19 in. long.

hookworm. See ANCYLOSTOMA and NECATOR.

hoolock. Slender, black gibbon (Hylobates) of the central Himalayas.

hoopoe. Any of a group of odorous tropical and temperate Old World birds in the Family Upupidae having a slender, curved bill; head with a pronounced erectile crest; fawn-colored; feed on insects secured by probing in the ground.

hoop snake. See COLUBER and MUD SNAKE.

Hopkins' Bioclimatic Law. Other conditions being equal, the variations in the time of a given periodic biological event in eastern temperate N.A. are later by four days for each: one degree latitude northward, five degrees longitude eastward, and 400 ft. altitude upward in spring and early summer; true only in a very general way since there are many local departures from this "law."

Hopkins-Cole reaction. Test for the presence of protein; when acetic acid containing a small amount of glyoxylic acid is carefully added to a protein suspension, a violet ring forms at the interface, indicating tryptophane.

Hopkinsia. Genus of nudibranchs.

Hoplias. See AIMARA.

Hoplitophrya. Genus of ovoid to ellipsoid holotrich Ciliata; with an anterior chitinoid attachment organelle; parasitic in earthworm gut.

Hoplocampa. Genus of sawflies in the Family Tenthredinidae; larvae pests on apples, cherries, pears, etc.

Hoplocarida. One of the four subdivisions of the crustacean Division Eumalacostraca; mantis shrimps; first three thoracic segments fused with head; first five thoracic appendages modified as gnathopods (subchelate); second thoracic appendages enlarged and raptorial; abdomen large, broad; burrow in sea bottom; to 12 in. long; some spp. edible; includes a single order, the Stomatopoda, with 200 spp.; Squilla and Chloridella common.

Hoplolaimus. Genus of microscopic, free-living nematodes.

Hoplonemertea. Order in the Phylum Nemertea; proboscis with one or more stylets; intestine straight, without diverticula.

Horatia micra. Small fresh-water snail of southern Tex.

Hormiphora. Genus of ctenophores; elongated spheroidal, with two retractile tentacles; in warm waters.

hormones. Organic secretions of endocrine glands; such secretions are usually released in minute amounts into the blood stream and exercise a wide variety of regulatory roles over many physiological activities; hormones are elaborated by the pituitary, thyroid, parathyroids, adrenals, islets of Langerhans, gonads, etc. in vertebrates, and by various small glandular areas in certain invertebrates.

horn. Structure projecting from the head of a mammal and usually used for offense and defense; cattle, sheep, Old World antelopes, etc. have permanent hollow keratin horns growing over bony cores; branched antlers of the deer family are composed of bone and are shed annually; rhinoceros horns are permanently fused hairs of the snout.

hornbill. Any of a group of large omnivorous forest birds having an enormous and grotesque down-curved bill which is usually surmounted by a casque; body 2 to 5 ft. long; 45 spp. in the Family Buceratodidae; Old World tropics.

horned dace. See SEMOTILUS.

horned grebe. See COLYMBUS.

horned lark. See OTOCORIS.

horned lizard. Horned toad; see PHRYNOSOMA.

horned owl. Any of several owls having two earlike tufts of feathers.

horned pout. See AMEIURUS.

horned rattlesnake. See CROTALUS.

horned toad. See PHRYNOSOMA.

horned viper. See CERASTES.

hornet. See VESPIDAE.

horn fly. See HAEMATOBIA.

horn shell. Any of numerous small dark-colored marine snails with high, sharply-spired shells; genus Cerithidea.

horntail. See SIRICIDAE.

hornworm. Larvae of certain moths having a thornlike process near the posterior end.

horny corals. Group of anthozoans in the Order Gorgonacea; skeleton extensive and composed of a hornlike material, gorgonin.

hornyhead chub. Hybopsis biguttata, a small minnow of clear gravelly streams; N.Y. to Wyo. and south to Tenn. and Okla.

horny mite. See ORIBATIDAE.

horse. See EQUIDAE.

horse bot. See GASTEROPHILUS.

horse conch. See PLEUROPLOCA.

horse flies. See TABANUS and TABANIDAE.

horsehair worm. Member of the Phylum Nematomorpha; so-called because they were thought to have originated from a horsehair by spontaneous generation.

horse leech. See HAEMOPIS.

horse mackerel. See THUNNUS.

horse mussel. See MODIOLUS.

horse oyster. Ostrea equestris, a small inedible oyster of southeastern U.S. mangrove areas.

horseshoe crab. See MEROSTOMATA.

host. 1. Organism being parasitized by an organism of another sp.; usually the host is subject to some degree of damage by the parasite. 2. An embryo into which a graft is experimentally transplanted.

Hottentot. African negroid offspring of a Bushman and Bantu mating.

houndfish. See BELONIDAE.

house centipede. See CERMATIA.
house finch. See FINCH.
housefly. See MUSCIDAE and MUSCA.
house sparrow. See PASSER.
house spider. See THERIDION.
house wren. See TROGLODYTES.
Houssay, Bernardo Alberto. Argentine physiologist and endocrinologist (1887-); shared 1947 Nobel prize in physiology and medicine for his work on the interrelations of the pituitary and insulin.
hover flies. See SYRPHIDAE.
Howell-Joly bodies. Nuclear fragments appearing in blood smears.
howler. See ALOUATTA.
Howship's lacunae. Pockets in bone occupied by osteoclasts.
Hudsonian zone. One of Merriam's North American life zones, located between the Arctic-Alpine and Canadian zones; in the U.S. it is restricted to the forested higher elevations of the Rocky Mountain system, and to the tops of a few high peaks in the eastern U.S.
Hudson seal. Muskrat fur which has been trimmed, plucked, and dyed to resemble sealskin.
humblebee. Bumblebee.
humerus. Single bone in the proximal part of the tetrapod forelimb; the upper arm bone in man.
hummingbird. See TROCHILIDAE.
hummingbird moth. See SPHINGIDAE.
humor. In the broad sense, any body fluid or semifluid; e.g. aqueous humor.
humpbacked flies. See PHORIDAE.
humpbacked sucker. See XYRAUCHEN.
humpback salmon. Mature male salmon, *Oncorhynchus gorbuscha*; when it ascends rivers to spawn the jaws are distorted and the back has a pronounced hump.
humpback whale. See MEGAPTERA.
Hungarian partridge. See PERDIX.
Hunter, John. Scotch surgeon and anatomist (1728-1793); noted for his extensive investigations in comparative anatomy.
Hunter's butterfly. See PYRAMEIS.
hunting cat. Cheetah, hunting leopard.
hunting spider. See LYCOSIDAE.
Huntington's chorea. Dominant hereditary defect marked by irregular movements, speech disturbances, and dementia.
Hutchinsonella. See CEPHALOCARIDA.
hutia. Any of several West Indian rodents; large, fat, ratlike spp.; *Capromys* and *Geocapromys* typical.
hutiacarabali. Arboreal hutia in the genus *Capromys*; fur long, body large and ratlike, tail prehensile; West Indies.
hutiacouga. Arboreal hutia in the genus *Capromys*; fur long, body rat-shaped, 22 in. long; West Indies.
Huxley, Thomas Henry. English biologist (1825-1895); influential exponent of Darwinism; made many important contributions to evolution, vertebrate and invertebrate anatomy, and physiology.
Hyaena. Genus of mammals which includes the Indian and African striped hyena.
Hyaenarctos. Genus of Pliocene bears.
Hyaenidae. Family of mammals which includes the hyenas and aard-wolf; doglike spp. with gray or brown coloration and dark spots or stripes; teeth large, jaws powerful and used in cracking large bones; front legs considerably longer than hind legs; notable as scavengers; Africa and Asia; typical genera *Hyaena*, *Crocuta*, and *Proteles*.
Hyale. Genus of marine Amphipoda.
Hyalella. Common genus of fresh-water Amphipoda; in streams, lakes, and ponds.
Hyalina. Genus of small, smooth-shelled tropical and subtropical marine gastropods.
hyaline cartilage. Homogeneous, bluish-white, translucent cartilage covering joint surfaces and forming the ends of ribs, a part of the nose, and rings in the trachea; it is the skeletal cartilage of all vertebrate embryos and adult sharks and rays.
Hyalinella. Genus of fresh-water Bryozoa.
Hyalinoecia. Widely distributed genus of Polychaeta; in translucent horny tubes which are dragged about on the sea bottom.
Hyalobryon. Genus of solitary or colonial fresh-water biflagellate Protozoa; cell surrounded by a translucent lorica.
Hyalocyclis. Genus of pelagic pteropods; shell stout and conical.
Hyalodiscus. Genus of fresh-water amoeboid Protozoa.
Hyalogonium. Genus of spindle-shaped biflagellate Protozoa; stagnant fresh waters.
Hyalomma. Genus of ticks found on domestic and wild mammals in Europe, Asia, and Africa.
Hyalonema. Genus of hexactinellid

sponges.

hyaloplasm. Clear ground substance of protoplasm as distinguished from its contained granules, fibrils, and other formed bodies.

Hyalosphenia. Genus of fresh-water amoeboid Protozoa; most of cell enclosed in a vase-shaped test.

Hyalospongiae. Hexactinellida.

hyaluronic acid. Organic acid found in many connective tissues; aids in holding together the protective layer of corona radiata cells surrounding a human ovum.

hyaluronidase. Enzyme contained in human sperm cells; has the ability to inactivate hyaluronic acid; also found in leeches, in snake and spider venom, and in many normal human tissues in minute quantities.

Hyas coarctatus. Toad crab, a common brachyuran crab of the Atlantic Coast; body slightly notched and violin-shaped.

Hybocodon. Genus of marine coelenterates in the Suborder Anthomedusae; usually in tide pools.

Hybognathus. Small genus of minnows widely distributed east of Rockies.

Hybopsis. Large genus of chubs in the Family Cyprinidae; widely distributed over U.S.; in large muddy rivers to clear mountain brooks; 2 to 12 in. long.

hybrid. 1. Offspring of two parents that differ in one or more heritable characters. 2. An individual heterozygous for one or more pairs of genes. 3. The offspring of two different varieties or (rarely) two different spp.

hybridization. 1. Production of individuals from different populations.

hybrid offspring. See MENDELISM.

hybrid sterility. General inability of spp. hybrids to mate and/or produce young.

hybrid swarms. Segregating progenies of spp. hybrids occurring at geographic boundaries separating the ranges of related spp.

hybrid vigor. Relatively better physiological and physical condition evident in the offspring of individuals not closely related; result of the heterozygous condition in which many defective features are masked by dominant normal characters.

Hyctia pikei. Long slim spider in the Family Salticidae; light gray with brown markings; common east of Rockies.

Hydaticus. Common genus of dytiscid beetles.

hydatid. Hollow, fluid-filled cyst which forms in the tissues of the intermediate host of certain tapeworms; develops from a single oncosphere and may be small or as large as a grapefruit, and sometimes very irregular in shape; lined with a germinal layer which produces numerous scolices and daughter cysts which are more or less suspended in the hydatid fluid; the hydatid cyst stage of Echinococcus granulosus occurs in many animals, including man; infections are obtained by the accidental ingestion of feces of a dog, wolf, jackal, etc. that has the mature tapeworms in the intestine.

Hydatina. 1. Epiphanes. 2. See BUBBLE SHELL.

Hydnoceras. Genus of fossil siliceous sponges.

Hydra. Common genus of small fresh-water hydrozoan coelenterates; organization simple, gastrodermis without zoochlorellae, medusoid stage absent, reproduction syngamic or by budding; common spp. are H. americana ("white" hydra) and H. oligactis ("brown" hydra).

Hydracarina. Hydrachnoidea.

Hydrachna. See HYDRACHNIDAE.

Hydrachnellae. Hydrachnoidea.

Hydrachnidae. One of many families of fresh-water mites in the Superfamily Hydrachnoidea; Hydrachna common in ponds.

Hydrachnoidea. Large superfamily of mites; water mites; mostly fresh-water, few marine; mostly free living, few parasitic.

Hydractinia. Small genus of marine hydrozoans usually found on rocks and on mollusk shells inhabited by hermit crabs; form fields of polymorphic polyps connected by a network of stolons.

Hydraena. Genus of small aquatic beetles in the Family Hydraenidae.

Hydraenidae. Small family of beetles; usually 1 to 2 mm. long; some spp. aquatic; Hydraena common.

Hydrallmania. Genus of marine coelenterates in the Suborder Leptomedusae.

Hydranassa tricolor. Louisiana heron, a slaty blue heron of the southern states and tropical America.

hydranth. Large vase-shaped, cylindrical, or bottle-shaped terminal portion of a typical hydroid polyp not including the base and stalk (if present); bears proctostome and tentacles.

hydrarch succession. Complicated series of animal and plant community changes occurring during the slow conversion of a newly-formed pond or lake to dry land covered with climax vegetation typical of the region; involves the development of aquatic vegetation, slow filling in of the basin, and the encroachment of shore vegetation. See ECOLOGICAL SUCCESSION, CLIMAX COMMUNITY.

hydratuba. Simple hydra-like stage in the life history of many jellyfishes.

Hydrellia. Genus of shore flies in the Family Ephydridae; larvae mine in aquatic plants.

Hydrichthys. Genus of marine coelenterates in the Suborder Anthomedusae; hydroid parasitic on small fishes.

Hydridella. Genus of Australian freshwater mussels.

Hydrobates pelagicus. Stormy petrel, a small sooty-black and brownish oceanic bird marked with white on the upper tail coverts; North Atlantic and Mediterranean.

Hydrobatidae. Family of small oceanic birds in the Order Procellariiformes; petrels; 22 spp.

Hydrobia. Genus of very small marine snails often found in salt marshes.

hydrobiology. Study of all phases of aquatic organisms and their habitats.

Hydrobius. Common genus of hydrophilid beetles.

Hydrocanthus. Common genus of dytiscid beetles.

hydrocarbon. Any chemical compound containing only carbon and hydrogen.

hydrocaulus. Branched upright portion of the colonies of most colonial hydrozoan coelenterates (not including hydranths); often consists of both coenosarc and perisarc.

hydrocephalus. Abnormal increase in the amount of cerebral ventricular or subarachnoid fluid; marked by enlargement of the head, brain atrophy, mental weakness, convulsions, and death.

Hydrochara. Genus of hydrophilid beetles.

hydrochloric acid. Inorganic acid having the formula HCl; important in general metabolism; constituent of gastric juice.

Hydrochoeridae. Small family of rodents which includes the capybara.

Hydrochoerus capibara. South American rodent; the capybara; an aquatic sp. resembling a giant guinea pig; to 4 ft. long; feeds on aquatic vegetation in larger rivers; largest living rodent.

Hydrochoreutes ungulatus. Widely distributed sp. of Hydracarina.

Hydrochus. Common genus of hydrophilid beetles.

hydrocoel. That portion of the embryonic echinoderm coelom which is destined to develop into the water vascular system.

Hydrocorallina. Order of hydrozoan coelenterates; usually subdivided into the Order Milliporina and the Order Stylasterina.

Hydrodamalis stelleri. Steller sea cow in the Order Sirenia; shores of Bering Island and adjacent region; related to the dugong and manatee; extinct since 1854.

Hydrodroma. Diplodontus.

hydrofuge pubescence. Pile; a fine dense coat of hairs on many aquatic beetles; holds a thin film of air (plastron) to the body when submerged.

hydrogen acceptor. In physiology, any substance which will readily combine with the electrically charged hydrogen atoms of an organic fuel, thereby separating these hydrogen atoms from the molecule of that fuel and affording energy; such processes are usually enzymatic; oxygen is the most familiar hydrogen acceptor (to form water), but there are many examples in which oxygen is not involved, and the process is anaerobic; pyruvic acid, for example, may be a hydrogen acceptor and is thereby converted to lactic acid.

hydrogen-ion concentration. Concentration of hydrogen ions in any solution; usually expressed as the reciprocal of the logarithm of grams of hydrogen ions per liter; scale ranges from pH 0 to 14, pH 7.0 being neutral, at which point the number of hydrogen ions and hydroxyl ions are of equal concentration; values below pH 7.0 are acid, above pH 7.0 are alkaline.

hydroid. 1. The polyp form of coelenterates as distinguished from the medusa or jellyfish form; solitary or colonial, usually attached; some individuals of colonial spp. are specialized for protection, etc. 2. Any coelenterate belonging to the Class Hydrozoa.

Hydroida. Order of hydrozoan coelenterates; polyp well developed, solitary or colonial, and usually forming free-swimming medusae by budding; marine and fresh-water; examples are Hydra, Obelia, and Plumularia.

Hydroides. Genus of polychaete annelids living in contorted tubes; head with a large cluster of tentacular gills.

Hydrolagus colliei. Ratfish; chimaera fish of the North Pacific.

hydrolase. Any enzyme which functions in a hydrolytic reaction.

hydrolysis. Chemical decomposition into simpler substances by the taking up of H^+ and OH^- ions of water; important enzymatic process in digestion.

hydrolytic enzyme. Any enzyme which catalyzes a hydrolysis.

Hydromantes. Small genus of salamanders; head and body depressed; toes partly webbed; tongue free on all margins; Calif. and southern Europe.

hydromedusa. Free-swimming, jellyfish-like stage of hydrozoan coelenterates.

Hydromedusae. Hydrozoa.

Hydrometra. See HYDROMETRIDAE.

Hydrometridae. Small family of carnivorous Hemiptera; marsh treaders, water measurers; exceedingly slender brownish or greenish spp. with long thin legs; walk about on the surface of ponds and puddles; Hydrometra common.

Hydromys. Genus of Australian water rats; large muskrat-like spp. with webbed hind feet.

Hydromyza confluens. Dung fly in the Family Scatophagidae; larvae imbedded in submerged petioles of yellow water lilies.

Hydrophiidae. Family of venomous sea snakes; brightly-colored spp., found especially in the Indian Ocean and tropical western Pacific; feed on fish; one sp. (Hydrus platurus) off the west coast of Mexico, Central America, and South America; Hydrophis typical, some spp. to 8 ft. long.

Hydrophilidae. Family of minute to very large black beetles; water scavenger beetles; amphibious, aquatic, and subaquatic; adults and larvae mostly vegetable scavengers; Hydrophilus common.

Hydrophilus. Common genus of aquatic beetles in the Family Hydrophilidae; some spp. are more than 30 mm. long.

Hydrophis. See HYDROPHIIDAE.

hydrophobia. Rabies.

hydrophyllium. Shield-shaped structure covering a cormidium in certain pelagic siphonophores.

Hydroporus. Common genus of dytiscid beetles.

Hydropotes. Genus of river deer; small hornless spp. in which the male has tusklike upper canines; marsh dwellers in Korea and China.

Hydropsyche. Common insect genus in the Order Trichoptera; larva builds a clumsy case in swift water and obtains its food by means of a fine net made from a salivary secretion.

Hydroptila. Common genus of caddisflies.

hydrorhiza. Collectively, the branched rootlike basal portions (stolons) of certain colonial hydrozoan coelenterates; affixed to the substrate.

Hydroscapha. See HYDROSCAPHIDAE.

Hydroscaphidae. Very small Holarctic family of ovate aquatic beetles; only one minute U.S. sp., Hydroscapha natans, found in running waters of the southwestern states.

hydrosere. Sere which begins in water or moist places and ends in a climax terrestrial community typical of that region; involves the development of aquatic vegetation, slow filling in of the basin, and the encroachment of surrounding terrestrial vegetation; see ECOLOGICAL SUCCESSION and CLIMAX COMMUNITY.

hydrosphere. Collectively, all of the bodies of water on the surface of the globe; the aqueous envelope.

hydrostatic organ. Swim bladder of bony fishes.

hydrotheca. Thin translucent vaselike exoskeleton surrounding the feeding polyps of certain hydrozoan coelenterates.

Hydrous. Hydrophilus.

Hydrovatus. Genus of dytiscid beetles.

hydroxycorticosterone. Hormone of the adrenal cortex; see CORTICOSTERONE.

hydroxydeoxycorticosterone. Deoxycortisol.

hydroxyglutamic acid. Amino acid; not definitely known to occur in natural proteins; $C_5H_9O_5N$.

hydroxyproline. Uncommon amino acid; $C_5H_9O_3N$.

Hydrozoa. One of the three classes in the Phylum Coelenterata; stomodaeum absent; mesoglea with few or no cellular elements; usually metagenetic; medusa small, with a velum; Obelia, Hydra, and Gonionemus familiar.

Hydrurga. See SEA LEOPARD.

Hydrurus foetidus. Species of flagellate Protozoa; flagellate stage transient; cells imbedded in large, macroscopic, gelatinous masses with an offensive odor; cold streams.

Hydrus. See HYDROPHIIDAE.

Hydryphantes. Widely distributed genus of Hydracarina.

Hyemoschus. Genus of water-chevrotains which are small, shy, spotted and striped mouse-deer of African rain forests; semi-aquatic; muzzle pointed; a peculiar group having certain affinities with pigs and camels.

hyena. See HYAENIDAE, HYAENA, and CROCUTA.

Hygrobates. Genus of Hydracarina.

hygropetric. Pertaining to an animal inhabiting steep wet or splash zones of rocky surfaces.

hygroscopic water. Small amount of water tightly bound to soil particles and removable as vapor only by high temperatures.

Hygrotus. Common genus of dytiscid beetles.

Hyla. Large genus of amphibians which includes many small frogs and toads in the Family Hylidae; H. versicolor, the common tree toad, from Great Plains eastward; H. regilla, Pacific tree frog, from British Columbia to Lower California; H. crucifer, spring peeper, appears early each spring in the eastern half of the U.S.

Hylemya. Important genus in the fly family Anthomyiidae; the larva of H. brassicae feeds on cabbage, and the larva of H. antiqua is the onion maggot.

Hylidae. Large family of frogs in the Order Procoela; cricket frogs, tree frogs, tree toads, chorus frogs; body and head usually less than 2 in. long; teeth in both jaws; terminal bone of each digit clawlike; each toe with expanded adhesive disc used in climbing; chiefly New World; Pseudacris, Hyla, Acris, and Gastrotheca typical.

Hylobates. Genus of small anthropoid apes which includes some of the gibbons; arms especially long; arboreal spp. of Asiatic tropical forests.

Hylocharis. Genus of hummingbirds in the Family Trochilidae; Ariz. south into South America.

Hylochoerus. Genus of forest-hogs; large, tusked, wild hogs of African tropical forests.

Hylocichla. See THRUSH.

Hylodes martinicensis. West Indian frog which lays its eggs in moss or under stones; tadpole stage absent, but the young frog has a large vascularized tail.

Hylomys. Genus of odorous mammals in the Family Erinaceidae; small, rat-shaped, with long, scaled, bristly tails; coat of thick woolly fur and an overcoat of coarse bristles; Malaya, Sumatra, Borneo, and Java.

Hyman, Libbie H. American invertebrate zoologist (1890-); associated with the American Museum of Natural History and noted for many important papers on lower invertebrates, invertebrate zoology texts, and comparative vertebrate anatomy laboratory manual.

hymen. Fold of mucous membrane which partially closes the vaginal opening in immature or virginal human females.

Hymenamphiastra. Genus of red or orange marine sponges.

Hymeniacidon. Genus of slimy, papillated marine sponges.

Hymenolepis. Common genus of small tapeworms in the Order Cyclophyllidea; H. nana is the dwarf tapeworm of the human intestine (to 2 in. long); it has no separate intermediate host; H. diminuta is a common parasite of the rat, with larval stages in several insects; H. fraterna is a closely related sp. in rats and mice.

Hymenomonas. Genus of fresh-water biflagellate Protozoa; membrane brownish, often sculptured; chromatophores present.

Hymenoptera. Order of insects which includes ants, bees, wasps, etc.; usually a pronounced constriction at junction of thorax and abdomen; mouth parts biting, chewing, lapping, or sucking; usually four small membranous wings with few veins, anterior pair larger; female with conspicuous ovipositor; complete metamorphosis; larva apodous or with true thoracic legs and abdominal prolegs; pupa herbivorous or parasitic, may or may not be enclosed in a cocoon; many social spp.; more than 120,000 described spp.

Hymenostomata. Suborder of Ciliata in the Order Holotricha; cytostome in peristome; peristome with a membrane and with or without free cilia.

Hynobiidae. Family of small Asiatic land salamanders; eyelids present, vomerine teeth forming a V behind internal nares; Chungking to Ural Mountains; Hynobius and Ranodon typical.

Hynobius. See HYNOBIIDAE.

hyoglossal. Pertaining to the tongue and

and hyoid arch.

hyoid. Pertaining to cartilages or bones at the base of the tongue; part of the hyoid arch.

hyoid arch. Second visceral arch, just behind the jaws in vertebrates; the dorsal portions form the hyomandibular, and the ventral portion forms the hyoid bones, usually supporting the base of the tongue. See VISCERAL ARCH.

hyomandibular arch. See VISCERAL ARCH.

hyomandibular cleft. Slitlike external invagination formed on either side of the embryonic vertebrate pharyngeal region; located between the first and second visceral arches; does not become perforate to the inside.

hyomandibular pouch. Slitlike evagination on either side of the embryonic pharynx of vertebrates; located between the first and second arches.

Hyostrongylus rubidus. Nematode parasite of the hog stomach.

hyostylic jaw suspension. Jaw suspension involving the hyomandibular; present in most modern fishes.

hypaxial musculature. Ventro-lateral trunk musculature of most vertebrates.

hypemia. Anemia.

Hypentelium nigricans. Hog sucker or stoneroller; a sucker-like fish of swift or rocky streams in the eastern half of the U.S.

hyperbranchial groove. Middorsal furrow in pharynx of cephalochordates.

hypercholesterolemia. Pathological condition marked by an excess of cholesterol in the blood.

hyperdactyly. Polydactyly.

hyperemia. Excess of blood in any part of the body.

hyperglycemia. Abnormally high concentration of blood sugar.

Hyperia. Genus of amphipods usually found on large medusae; eyes enormous.

hyperinsulinism. 1. Condition produced by the injection of an overdose of insulin into a diabetic; remedied by drinking or injecting a sugar solution. 2. Production of excess insulin by the pancreas.

Hypermastigina. Order of flagellates having many flagella; inhabit the digestive tract of termites and other insects; bits of cellulose and other particulate materials ingested by pseudopodia; no cytostome; some spp. are true symbionts which digest cellulose for their own and their host's use.

hypermetamorphosis. Complex type of metamorphosis occurring especially in certain families of some of the insect orders, e.g. Neuroptera, Coleoptera, Strepsiptera, Diptera, and Hymenoptera; the first larval instar is minute and active, the second is robust and sluggish, and the third (if present) is apodous; some spp. also have an additional prepupa.

hypermetropia. Farsightedness; hyperopia.

hypermorph. Mutant allele which has a greater effect on the development of a character than the normal ancestral allele.

Hyperoodon ampullatus. Bottlenose whale; a beaked whale (Ziphiidae) of the North Atlantic and Arctic oceans; forehead bulging; only two concealed teeth at tip of lower jaw; total length to 30 ft.

hyperopia. Farsightedness; hypermetropia.

hyperparasite. Organism which parasitizes a parasite; e.g. a protozoan parasite living in the digestive tract of a flea.

hyperpharyngeal band. Thickened ridge along the inner dorsal surface of a tunicate branchial chamber.

hyperpituitary. Overactive pituitary gland, often causing acromegaly.

hyperplasia. Great increase in number of normal cells arranged normally in a tissue.

hyperpnea. Abnormally rapid breathing.

hypertensin. Angiotensin.

hyperthyroidism. Secretion of abnormally large amounts of thyroxin by the thyroid gland, resulting in high metabolic rate, nervous excitability, and high tension; it may or may not be characterized by an enlargement of the thyroid. See EXOPHTHALMIC GOITER.

hypertonic. Situation obtaining in a solution having a concentration of dissolved materials such that it gains solvent through a differentially permeable membrane separating it from another solution which has a higher solvent concentration, and therefore a lower solute concentration.

hypertrichosis. Retention and growth of embryonic hair throughout life; a Mendelian dominant in man.

hypertrophy. Enlargement of a tissue or organ due at least in part to an increase in the size of its cells.

hypervitaminosis. Disruption of the

physiological steady state by means of massive doses of vitamin(s).

Hyphantria cunea. Common moth whose larvae are known as fall webworms; these gregarious caterpillars spin loose silken webs which envelop whole limbs of trees.

hypnotoxin. Toxic substance released when certain types of nematocysts are extruded.

hypoblast. Endoderm, especially in very early embryos.

hypobranchial. Below the gill(s).

hypobranchial gland. Gland on the inner surface of the mantle near the ctenidium in many marine snails.

hypobranchial groove. Ventral ciliated groove in the pharynx of Amphioxus and tunicates.

hypochord. Subnotochordal rod; transient, slender, longitudinal rod appearing just below the notochord in amphibian embryos.

hypocleidium. Small median process on the wishbone of many birds.

Hypoconcha. Genus of short, broad, marine crabs in the Suborder Brachyura.

hypocone. 1. Inner posterior cusp of an upper mammalian molar tooth. 2. That part of a dinoflagellate which is posterior to the equatorial groove. Cf. EPICONE.

hypoconid. Outer posterior cusp of a lower mammalian molar tooth.

Hypoderma bovis. Common cattle warble fly in the Family Hypodermatidae; ox bot, ox warble; larvae parasitic in tissues just beneath hide and maintain contact with the outside through a hole in the hide.

Hypodermatidae. Family of large, hairy flies; warble flies; larvae parasitic under skin of cattle, deer, etc.

hypodermis. Any epidermis which secretes an overlying cuticle or other dead material.

Hypoeulalia. Genus of Polychaeta.

hypogastric. Pertaining to the lower median abdominal region.

Hypogastrura. Common genus of Collembola; several semiaquatic spp.

hypogean. Pertaining to an animal which burrows or otherwise lives below the surface of the earth. Cf. EPIGEAN.

hypoglossal nerve. One of a pair of cranial nerves in higher vertebrates; a motor nerve supplying the tongue muscles.

hypoglottis. 1. Lower or under surface of a tongue. 2. In many Coleoptera, a sclerite between mentum and labium.

hypoglycemia. Abnormally low concentration of blood sugar.

hypoglycemic shock. Coma and sometimes fatal condition produced by excessive insulin secretion and the associated absorption of blood sugar from the tissues.

Hypohippus. Genus of Miocene and Pliocene horses.

Hypohomus. Genus of Percidae (fish) now included in Percina.

hypohyals. Four small bones serving for the attachment of the hyoid apparatus.

hypolimnion. Lowermost cold layer of a thermally stratified lake during the summer months; thickness ranges from 1 m. or less in shallow lakes to a very great thickness in deep lakes; not usually evident in lakes less than 5 m. deep. Cf. EPILIMNION, METALIMNION.

hypomere. 1. Basal portion of certain sponges containing no flagellated chambers. 2. In embryology, a lower mesodermal plate zone from which the walls of the pleuro-peritoneal cavity develop.

Hypomesus. Small genus of smelts; enter fresh water to spawn; an excellent food fish; Japan to Alaska and south to Calif.

hypomorph. Mutant allele which has a weaker developmental effect than the normal allele.

hyponeural sinus. Special tubular portion of the coelom of many echinoderms; contains the hemal vessels.

hyponeural system. Part of the echinoderm nervous system; consists of a ring of motor nerve tissue around the mouth and five pairs of radiating nerves; lies just below the ectoneural system.

Hyponeura lugens. Damselfly of extreme southwestern U.S.

hyponome. Funnel of a cephalopod.

hyponym. Generic name not based on a type sp.

Hypopachus. Genus of Central and South American toads in the Family Microhylidae; one fossorial sp. reaches southern Tex.

hypopharynx. Elongated tongue-like structure attached to the inner basal surface of the insect labium; variously modified according to food habits.

hypophysectomy. Surgical removal of the pituitary.

hypophysis. Median ectodermal fold

just in front of the mouth of the vertebrate embryo; it pushes upward and fuses with the infundibulum to form the pituitary body on the ventral surface of the brain; pituitary body.

hypopituitary. Underactive pituitary gland, often resulting in obesity and the retention of early adolescent characteristics.

hypoplastron. One of four large median plates of a tortoise plastron.

hypoploidy. Condition involving less than the normal euploid chromosome number.

hypopneustic. In certain insects, having a reduced number of spiracles.

hypopus. Inactive cystlike stage in a few minute non-parasitic mites.

hypopygium. 1. Anus of insects, especially the lower part of the anal opening. 2. Male insect copulatory organs. 3. In beetles, the last segment behind the elytra.

hypostasis. Condition shown by certain gene loci whose expression is suppressed by one or more genes which are non-allelic to it; e.g. the black-brown locus in rabbits is hypostatic to the albino allele at another locus. Cf. EPISTASIS.

hypostome. 1. Conical raised area surrounding the proctostome in many coelenterate polyps. 2. Median anterior mouth part in Acarina; elongated, cylindrical, often clublike, and sometimes armed with spines or thorns; thrust into tissues of prey or host during feeding. 3. In Diptera, the general anterior part of the head included between antennae, eyes, and mouth. 4. In Hemiptera, the lower part of the anterior surface (face) of the head.

Hypostomides. Order of very small bony fishes of tropical seas; body enclosed in a bony boxlike exoskeleton; dragonfishes; Pegasus typical.

hypothalamus. That ventral portion of the vertebrate brain just below the cerebral hemispheres; in mammals it is a coordinating center and has much to do with body temperature control.

hypotheca. Theca covering the hypocone of certain dinoflagellates.

hypothesis. 1. Temporary working explanation or supposition based on accumulated facts and suggesting some general principle or relation of cause and effect; sometimes called a working hypothesis. 2. A mere assumption or guess. See THEORY and LAW.

hypothyroidism. Secretion of abnormally low amounts of thyroxin. See CRETINISM and MYXEDEMA.

hypotonic. Situation obtaining in a solution having a concentration of dissolved materials such that it loses solvent through a differentially permeable membrane separating it from another solution which contains a greater concentration of dissolved materials and therefore a lower concentration of solvent.

Hypotremata. Batoidea.

Hypotrematica. Batoidea.

Hypotricha. Suborder in the ciliate Order Spirotricha; cirri only, on ventral surface; dorsal surface usually with rows of short bristles; e.g. Stylonychia, Euplotes.

hypotype. Specimen used to amplify or correct the original description of a sp.

hypoxia. State of having too little oxygen in the tissues for normal metabolism.

Hypsagonus. See AGONIDAE.

Hypsibius. Large and common genus of Tardigrada.

Hypsiglena. See NIGHT SNAKE.

Hypsignathus. Genus of hammer-headed bats; West African fruit bats with large "horselike" heads.

hypsodont. Type of dentition found in grazing mammals; teeth prism-shaped and high-crowned.

Hypsypops rubicundus. Garibaldi; a small golden pomacentrid tide pool fish of the southern Calif. coast.

Hyptiotes. Genus of gray triangle spiders; build flat triangular webs in trees and shrubs; common in U.S.

hypurals. Enlarged hemal spines which support the base of the homocercal tail fin of teleost fishes.

Hyracoidea. Small order of ungulate mammals which includes the coneys or hyraxes; small herbivores, superficially similar to guinea pigs; four front toes, three hind toes, all with small hooflike nails; in trees and rocky areas in Africa and Middle East.

hyrax. See HYRACOIDEA.

hysterectomy. Surgical removal of the uterus.

Hysterocarpus traski. Valley perch; a small percomorph fish in the Family Embiotocidae; rivers of central Calif.

Hystrichis. Genus of enoplid nematodes parasitic in the intestine of birds.

Hystricidae. Family of terrestrial rodents which includes the Old World porcupines (Hystrix).

Hystricomorpha. Suborder of the Rodentia; includes the New World porcupines and the nutria.

Hystrix. See HYSTRICIDAE.

hythergraph. Instrument which records both temperature and precipitation.

I

Iais. Genus of marine Isopoda.
Ianiropsis. Genus of marine Isopoda.
Ianthina. Violet snails; a genus of delicate, pelagic, carnivorous, marine gastropods which secrete supporting bubbles into a floatlike operculum and thereby hang upside down at the surface film.
Iapygidae. Family of insects in the Order Aptera; fragile, slender, wingless, whitish spp. with a pair of unsegmented terminal forceps; in debris on the floor of damp forests and meadows; Iapyx common.
Iapyx. See IAPYGIDAE.
Ibero-insular. Race of short caucasoids living in northern Spain; skin tawny, hair dark, head long; same as Pyrenean.
ibex. Any of several gregarious wild goats (Capra) living near the snow line in the mountains of Asia and Europe; brown to gray, to 42 in. high at the shoulder, and with long ridged horns curved backward; rare in the Alps.
ibid. Large monitor lizard of the Philippines.
ibis. Any of a cosmopolitan group of large wading birds related to the herons; in the Family Threskiornithidae; bill slender and evenly curved; found around fresh waters; feed on fish and aquatic invertebrates; Threskiornis aethiopica is the sacred ibis of Egypt; several spp. in other genera occur in the U.S., especially Mycteria americana, the wood ibis (a true stork), a large white heron-like southern and western sp. with decurved bill; Plegadis guarana (or P. chichi) is the white-faced ibis, a bronze-chestnut, medium-sized, heron-like western bird with a decurved bill.
Icerya. See COTTONY CUSHION SCALE.
Ichneumia. Genus of African carnivores; includes the white-tailed mongoose.
ichneumon. Egyptian sp. of mongoose.
ichneumon flies. See ICHNEUMONIDAE.
Ichneumonidae. Family of minute to large slender wasps; ichneumon flies; ovipositor usually very long; female deposits eggs in or on the larvae and pupae of other insects; larvae are ectoparasites or endoparasites of insect hosts; adult ichneumon emerges from host pupal stage; Thalessa, Therion, and Ephialtes are examples.
Ichthydium. Common genus of fresh-water gastrotrichs; cuticle smooth.
Ichthyobdellidae. Family of leeches in the Order Rhynchobdellida; body segments subdivided into three or more subsegments; parasitic on marine and fresh-water fishes and turtles.
ichthyology. The study of fish.
Ichthyomyzon. Genus of small, nonparasitic brook lampreys; central and eastern U.S. and Canada.
Ichthyophis. Genus of caecilian amphibians; burrow in moist ground of tropics; eggs deposited on ground near water.
Ichthyophthirius. Genus of oval holotrich ciliates parasitic on the integument of fresh-water fishes, especially under crowded conditions.
Ichthyopterygia. One of the six subclasses of reptiles; all spp. fishlike, marine, and extinct; a single Order Ichthyosauria; examples are Merriamia, Ichthyosaurus, and Ophthalmosaurus.
Ichthyornis. Genus of fossil birds from Mont. and Kan.; in the Superorder Odontognathae and Order Hesperornithiformes; gull-like spp. with teeth and well developed wings.
Ichthyornithiformes. Order of fossil birds in the Superorder Odontognathae.
Ichthyosauria. See ICHTHYOPTERYGIA.
Ichthyosaurus. Genus of Jurassic fish-like reptiles in the Subclass Ichthyopterygia.
ichthyosis. Peculiar and rare abnormality in which the human skin becomes thick, scaly,. and blackish; the abnormal allele is carried on the Y chromosome.

Ichthyosis. Genus of limbless tropical amphibians in the Order Gymnophiona.

Ichthyostega. Genus of extinct stegocephalian Amphibia.

iconotype. Drawing or photograph of a type specimen.

Icosteida. Malacichthyes.

Icosteus. See MALACICHTHYES.

ICSH. Interstitial cell-stimulating hormone.

Ictaluridae. Ameiuridae.

Ictalurus. Genus which includes the N. A. catfishes; head conical, mouth small; I. lacustris (channel or spotted catfish) to 36 in. and 25 lb., common in large streams of Great Lake, Gulf, and Mississippi drainages; I. furcatus (blue cat) to 4.5 ft. and 150 lb., lower Mississippi Valley and Gulf states; I. catus (white cat) olive blue above, silvery below, 20 24 in., drainages of N. Y. to Fla., introduced in Nev. and Pacific Coast drainages.

Icteria virens. Yellow-breasted chat; olive green above, belly white; in most of U.S. and Mexico.

icteric index. Test of liver activity as measured by the amount of bilirubin (yellow color) in the blood.

Icteridae. Family of New World passerine birds; includes the bobolinks, blackbirds, orioles, grackles, cowbirds, and meadowlarks; bill conical, sharp-pointed.

icterus. Jaundice.

Icterus. Genus of birds in the Family Icteridae; orioles and chats; males usually black and yellow or orange; several common U.S. spp.

Ictidosauria. Order of Triassic reptiles in the Subclass Synapsida; presumably ancestral to primitive mammals.

Ictinia. See KITE.

Ictiobius. Small genus of medium to large fishes found especially in the larger rivers of the U.S. and Mexico; buffalo fish; body stout, head large; used as human food in some areas.

Idarcturus. Genus of marine Isopoda.

identical twins. Two individuals resulting from the division of a single fertilized egg or from the division of an early embryo; another theory postulates their origin from one blastocyst with two embryonic axes; they are always of the same sex and are remarkably similar in all physical, physiological, and mental features.

idiochromatin. Dormant but partially active chromatin material.

idiot. See FEEBLE MINDEDNESS.

I disc. Isotropic disc.

Idiurus. Genus of gliding mice in the Family Anomaluridae; glide in the manner of flying squirrels; African forests.

Idmonea. Genus of bryozoans in the Order Cyclostomata.

Idotea. Idothea.

Idothea. Large and widely distributed genus of marine Isopoda.

iguana. 1. In the broad sense, any lizard in the large Family Iguanidae. 2. In the narrow sense, any of numerous large tropical American lizards; usually greenish, 2 to 6 ft. long, and with a crest of spiny scales from neck to long tail.

Iguana. Genus of inoffensive lizards of Mexico and southward; to 6 ft. long.

Iguanidae. Family of New World lizards; iguanas, American chameleons, swifts, etc.; tongue non-protrusible; scales fine; eyelids well developed, pupil round; tropical and subtropical in the Americas, but a few East Indian spp.; most of the 50 U.S. spp. are found west of the Mississippi River; examples are Anolis, Sceloporus, Phrynosoma, Uta, Crotaphytus, Sauromalus, and Iguana.

Iguanodon. Genus of large robust biped vegetarian dinosaurs with horny beak and powerful tail; to 34 ft. long.

ilangurra. See WYULDA.

ileocaecal valve. Muscle sphincter at the junction of ileum and large intestine; ileocolic valve.

ileofibularis. Biceps muscle in the frog.

ileum. Last portion of the small intestine in certain of the higher vertebrates; in man it is about 15 ft. long.

iliac artery. Artery supplying the leg and posterior part of the vertebrate body.

iliacus internus. Muscle which originates on the ventral border of the ilium and inserts on the proximal half of the femur; draws the thigh forward; present in the frog.

iliac vein. Vertebrate vein which brings blood from a leg into the trunk.

iliocostalis. In reptiles, a thin sheet of dorsal musculature which extends laterally and attaches to the ribs.

iliofemoralis externus. Large dorsal muscle of the bird leg.

iliofemoralis internus. Large dorsal muscle of the bird leg.

iliofibularis. Dorsal muscle of the hind leg of tetrapods, not including mammals; corresponds to the mammalian gluteus maximus.

iliopsoas. Compound muscle on the anterior surface of the body where the leg joins the trunk.

iliotibialis. Dorsal muscle of the reptilian hind leg.

iliotrochantericus. Dorsal muscle of the bird leg.

ilium. Dorsal part of the pelvic girdle which is attached to the vertebral column in tetrapods.

Illex illecebrosus. Short-finned squid; very common, Greenland to Gulf of Mexico; 12 to 18 in. total length.

Illinobdella. Genus of fresh-water fish leeches.

illoricate. 1. Pertaining to a rotifer that does not have an especially rigid cuticle (lorica). 2. Pertaining to a protozoan that is not surrounded by a lorica.

Ilybiosoma. Genus of dytiscid beetles.

Ilybius. Common genus of dytiscid beetles.

Ilyocryptus. Genus of Cladocera; usually found creeping on substrate.

Ilyocypris. Common genus of fresh-water Ostracoda.

Ilysia. Anilius.

Ilysiidae. Anilidae.

imaginal disc. Thickened area enclosed in a sac of the body wall of the pupa of holometabolous insects; gives rise to an adult organ at metamorphosis.

imago. Sexually mature adult stage in the life history, especially in insects.

imbecile. See FEEBLE MINDEDNESS.

imbibition. Mechanical absorption of water by capillarity and other mechanisms, especially these processes in cellulose and other dead organic materials.

imbricate. Overlapping, as the shingles of a roof.

imbrication. Especially orderly and linear overlapping, such as the scales of a fish.

immersion lens. See OIL IMMERSION OBJECTIVE.

immunity. General ability of any organism to resist infections caused by viruses, bacteria, fungi, and multicellular parasites. See NATURAL IMMUNITY and ACQUIRED IMMUNITY.

immunology. Science of the phenomena of immunity; has many medical, bacteriological, and biochemical interrelationships.

Imogene. Genus of polyclad turbellarians; body extremely thin.

Imostoma. Genus of Percidae (fish) now included in Percina.

impala. See AEPYCEROS.

imperforate. 1. Lacking a normal opening. 2. In certain snails, having the umbilicus obliterated by most recent shell growth.

imperial moth. See BASILONA.

implant. Tissue or some other object, either living or dead, artificially placed in an embryo or adult animal.

implantation. 1. Attachment and embedding of the early mammalian embryo on the uterine wall. 2. The artificial placing of living tissue or some other object in an embryo or adult animal.

import. A poorly-understood method of ingestion by amoeboid Protozoa; e. g. a long algal filament appears to "glide" into the cell until it is coiled up in a food vacuole within the cytoplasm.

impregnation. 1. Introduction of sperm into the female reproductive tract, usually with the implication that fertilization follows; same as insemination (Def. 1). Verb impregnate.

impulse. See NERVE IMPULSE.

Inachoides. West Coast genus of marine crabs.

inanition. Type of inactivity and dormancy induced by insufficient food.

Inarticulata. One of the two classes in the Phylum Brachiopoda; valves chitinoid and similar in appearance; anus present.

inbreeding. Mating of closely related individuals, especially brothers and sisters; opposite of outbreeding.

Inca dove. See SCARDAFELLA.

inchworm. Measuring worm; see GEOMETRIDAE.

incisures of Schmidt-Lantermann. Diagonal clefts in myelin sheaths of nerve fibers as seen in longitudinal section.

incisors. Teeth at the front of the jaws in mammals; chiefly adapted for chisel cutting but may be considerably modified in various orders (e. g. the elephant), or sometimes absent; eight present in man.

incomplete dominance. Occurrence of F_1 individuals intermediate between the homozygous dominant character of one parent and the homozygous recessive of the other parent; caused by incomplete masking of the recessive character by the dominant character in the heterozygous F_1; if a black and a splashed white Andalusian fowl are mated, the offspring (F_1) are all an intermediate "blue" shade; the F_2 generation consists of 25% black, 25% splashed white, and 50% "blue."

incomplete metamorphosis. See HEMIMETABOLA.

inconnu. See STENODUS.

incubation. 1. Artificial rearing and hatching of eggs, often with the aid of heat. 2. Induction of development by means of heat.

incubation period. Interval between introduction of an infective agent and the first appearance of symptoms.

incudate mastax. Rotifer mastax in which the trophi are prehensile.

incurrent canal. One of the many canals in a sponge which carry a current of water from the external surface to the flagellated chambers.

incurrent siphon. See SIPHON.

incus. 1. Middle of three minute auditory ossicles in the middle ear of mammals; it has the vague shape of a miniature anvil. 2. Median Y-shaped portion of rotifer trophi.

indeciduate placenta. Nondeciduate placenta; type of placenta which, upon removal at birth, is not accompanied by any loss of maternal uterine endometrial tissue.

independent assortment. See MENDELISM.

indeterminate cleavage. Type of early embryological cleavage in which individual blastomeres do not have a predetermined fate in the way of tissues or organs, e.g. vertebrates.

index species. Indicator sp.; any sp. which is typical or peculiar to a particular kind of habitat.

Indian meal moth. See PLODIA.

Indian python. See PYTHON.

Indicatoridae. See HONEY GUIDE.

indicator species. Index species.

indigenous. Native to a particular area, and not introduced by man.

indigo bunting. See BUNTING.

indigo snake. See DRYMARCHON.

indirect metamorphosis. See HOLOMETABOLA.

Indo-Afghan. Race of tall caucasoids living in northwestern India; skin light brown, hair black and wavy, head long.

indole. Putrefaction product of tryptophane; found in feces.

Indonesian. Race of mongoloids found from southern China to the East Indies; skin yellow to light brown, head long, hair black, cheekbones prominent.

Indri. Genus of black and white silky lemurs found in Madagascar; the indris; body slender, tail lacking, arms long, hands very large and woolly.

Indridae. Family of woolly or silky lemurs in the Suborder Lemuroidea; muzzle broad and short, face naked; vegetarians.

inductorium. Simple apparatus used in applying electrical stimuli to responsive tissue such as nerve or muscle.

indumentum. Bird plumage.

industrial melanism. Recent increasing frequency of melanism, especially in insects in industrial areas; most striking in parts of England.

inelastic fibrous connective tissue. See WHITE FIBROUS CONNECTIVE TISSUE.

infantile paralysis. Poliomyelitis.

infauna. Fauna consisting of burrowers in the bottom deposits of the sea. Cf. EPIFAUNA.

infection. 1. Invasion of tissues by pathogenic organisms. 2. Presence of pathogenic organisms in or on the body.

inferior. 1. Below or toward the lower side. 2. In human anatomy, toward or pertaining to the foot end of the body.

inferior mesenteric artery. Artery supplying the lower colon and rectum of certain vertebrates.

inferior mesenteric vein. Vein which drains the large intestine and mesentery of certain vertebrates.

inferior vena cava. Posterior vena cava.

inflammation. Condition of tissues as the result of injury or infection; characterized by dilated blood vessels, swelling, pain, invasion of leucocytes, and exudation.

influents. Common animals of a community; not so abundant or ecologically important as the dominant spp.

infraciliature. Collectively, the subsurface basal granules and fibrils of ciliate Protozoa.

infraclass. In taxonomy, a subdivision of a subclass.

infraneuston. Collectively, those animals living on the underside of the surface film in an aquatic habitat; e.g. dipteran pupae, hydras, ostracods, planarians, etc.

infraorbital gland. Small salivary gland below the eye in certain mammals, as in the cat and rabbit.

infrared. Radiation beyond the visible red end of the spectrum; 7700 to 500,000 angstrom units; heat rays.

infraspecific. Of or pertaining to categories within a sp., e.g. subspecies and varieties.
infraspinatus. Muscle located on the back of the shoulder; fills infraspinous fossa; rotates head of humerus outward, and aids in carrying arm backwards.
infraspinous fossa. That portion of a mammalian scapula which lies posterior to the longitudinal ridge of the scapula.
infundibulum. 1. Any funnel-shaped organ or passage. 2. Swelling on the floor of the diencephalon in the vertebrate brain; in higher vertebrates it fuses with the hypophysis to form the pituitary body. 3. One of the divisions (calyx) of the pelvis of a kidney. 4. Flattened stomach-like part of the gastrovascular cavity in ctenophores. 5. One of the small cavities at which a bronchial tube terminates in the lung. 6. Small pouchlike portion of the right ventricle where the pulmonary artery originates. 7. Opening of an oviduct in the coelom. 8. Subgenital pit in certain scyphozoan medusae.
infusion. A heterogeneous laboratory culture solution usually consisting of water and decaying vegetation; rich in bacteria, Protozoa, and micrometazoa.
Infusoria. Term formerly used to include all microscopic animals (and sometimes plants) found in stagnant ponds, puddles, and laboratory cultures of water and organic substances, such as hay infusions; also used in a more restricted sense to include only ciliate Protozoa.
infusorigen. See MESOZOA.
ingestion. Taking in of food.
ingluvies. 1. Crop of birds. 2. First stomach of ruminants.
inguinal. Pertaining to the region of the groin.
inguinal canal. Small opening in the trunk musculature of male mammals on either side at the base of the scrotum.
inguinal hernia. Projection of a loop of the intestine into the scrotum via the inguinal canal.
inhalation. Inspiration.
inhibition. Any arrest or restraint of a process or activity; e.g. nervous inhibition is the prevention of the activation of an effector by means of inhibiting impulses through a different set of fibers.
inhibitor center. Cardioinhibitor center.
inhibitor genes. Genes which suppress or mask the effects of genes usually on another chromosome and always at another locus.
Inia geoffroyensis. Bouto; a dolphin-like mammal of the Amazon drainage; total length to 7 ft.; snout long.
Iniomi. Order of bony fishes which includes the lantern fishes and lizard fishes; usually small spp., in deep oceanic waters; Myctophum and Synodus typical.
ink sac. Sac situated near the anus in most cephalopod mollusks; contains a blackish fluid which can be squirted into the water in clouds and thus conceal the mollusk from its enemies; the secretion is also reputed to have a deadening effect on the olfactory senses of predators.
inner ear. Membranous labyrinth.
innervation. Nerve supply to a particular organ or tissue.
innominate artery. Short artery which subdivides into a subclavian artery (to the arm) and a carotid artery (to the head) in many birds and mammals.
innominate bone. Each half of the pelvic girdle when the pubis, ilium, and ischium are fused into a single mass; in birds and mammals.
innominate vein. Vein draining much of the shoulder region in tetrapods.
inoculation. 1. Introduction of microorganisms, infective material, serum, etc. into living tissues of the body, or into a culture medium. 2. Introduction of a disease agent into a healthy individual, usually in order to produce antibodies and immunity.
inoculum. Substance used in making an inoculation. Pl. inocula.
inositol. Bios I; presumed vitamin of the B complex found in muscle, urine, viscera, and some plant tissues; said to prevent loss and graying of hair; essential for some small mammals but not for man; $C_6H_{12}O_6$.
Inostemma. Genus of Hymenoptera; larvae parasitic in dipterous gall midges.
inquiline. 1. Commensal organism which lives on or within its host but does not derive its food from the host. 2. Any animal living habitually in the nest or burrow of another animal. 3. Insect developing in a gall produced by another sp. of insect.
Insecta. One of the eight classes of the Phylum Arthropoda; probably contains in excess of 1,000,000 described spp.; includes such familiar examples as

grasshoppers, termites, dragonflies, lice, bugs, moths, butterflies, flies, fleas, beetles, weevils, ants, bees, and wasps; all insects have head, thorax, and abdomen, three pairs of legs on the thorax, one pair of antennae, tracheal system, and Malpighian tubules; the mouth parts are variously modified for chewing, sucking, lapping, etc.; two pairs of wings are typically present on the second and third thoracic segments, but in several orders wings are absent and in the Diptera there is only a single pair; insects abound in all habitats except the sea.

insectarium. Place where insects are raised for commercial or experimental purposes. Pl. insectaria.

insecticide. Any chemical used for destroying insects.

Insectivora. Order of mammals which includes the moles, shrews, and hedgehogs; small spp. with long tapered snout and rooted, tuberculate teeth; usually insectivorous; feet usually with five clawed toes; Northern Hemisphere and Africa.

insectivorous. Insect-eating.

insectivorous plants. Any of a variety of terrestrial and fresh-water plants which are specialized for trapping and digesting insects and other small arthropods; e.g. Venus' flytrap, sundew, and pitcher plant.

insemination. 1. Impregnation; introduction of sperm into the female reproductive tract. 2. Release of sperm onto an extruded egg or egg mass. Verb inseminate.

insertion. More movable end of a voluntary muscle.

insolation. 1. Exposure to sunlight. 2. Rate of delivery of solar energy per unit of earth's surface.

inspiration. Inhalation; mechanical process of drawing in air during respiration; in mammals, produced by raising the ribs forward and outward and by flattening the diaphragm, the result being a potential enlargement of the thoracic cavity and hence inflation of the lungs by air pressure.

instar. 1. Period or stage between molts of an insect during larval or nymph portion of the life history; instars are usually numbered, the first larval instar being that stage hatching from the egg and extending to the first ecdysis. 2. In the broad sense, the period between any two successive molts in an arthropod, nematode, tardigrade, etc.

instinct. A simple to complex and invariable behavior pattern, usually assumed to be an innate pattern of successive reflexes which is not based upon previous experience; reproductive activities, migration, nest building, care of the young, etc. are all examples.

insula. Triangular area of cerebral cortex forming the floor of the lateral cerebral fossa; hidden from view by the parietal and frontal lobes.

insulin. Vertebrate hormone secreted into the blood stream by the islets of Langerhans; concerned with synthesis of glucose-6-phosphate from glucose as a necessary preliminary to glycogen synthesis and oxidation; the absence or insufficiency of insulin in man is responsible for diabetes mellitus which is characterized by an accumulation of glucose in the blood and its partial release in the urine; insulin excess produces so much liver and muscle glycogen at the expense of blood glucose that the brain has insufficient glucose for its normal metabolism, and "insulin shock" and death may result.

integument. Coating or external skin; investment.

integumentary system. Collectively, the skin or investment of an animal together with all of its appendages and outgrowths.

intelligence. Mental capacity characteristic, in varying degrees, of the higher vertebrates; marked by an ability to associate facts, truths, and meaning, and a capacity for understanding and adaptive behavior.

interaction of supplementary factors. Distinctly variable form of one body structure owing to the effects of two or more pairs of alleles affecting that single structure; a good example is the type of comb in fowl: rose comb is produced by RRpp and Rrpp genotypes; pea comb is produced by rrPP and rrPp genotypes; single comb is produced by an rrpp genotype; walnut comb is produced by RRPP, RrPP, RrPp, and RRPp genotypes.

interambulacral areas. Five broad radially arranged areas or series of plates, especially in the Echinoidea, which do not bear rows of tube feet; alternate with ambulacral areas.

interarcual. Small muscle of the gill arch; connects the epibranchials and pharyngobranchials; e.g. in the shark.

interatrial septum. Interauricular septum.

interauricular septum. Muscular partition separating left and right atria of the heart.

interbranchial. Broad muscle of a gill arch, as in a shark.

interbreed. To cross two different varieties.

intercalary segment. Transitory embryonic segment between the antennae and mandibles of early embryos in insects, centipedes, and millipedes.

intercalated discs. Dark transverse bands in voluntary muscle fibrils.

intercellular. Between adjacent cells.

intercellular bridge. One of several to many cytoplasmic strands between adjacent cells in certain tissues; often called plasmodesmata in plants.

intercellular spaces. Cavities between neighboring cells in certain tissues; such spaces are filled with body fluid(s).

intercentra. Series of small crescentic elements wedged between successive centra in reptiles and the tail of mammals.

interclavicle. Median bone lying between the tips of the clavicles and on the ventral surface of the sternum, as in lizards.

intercostal. Located between the ribs.

intercostal muscle. Muscle extending from one rib to another.

interglacial. Between glacial periods, especially in Pleistocene times.

interkinesis. Period between mitotic divisions of a cell.

interlamellar concrescences. Transverse cords of tissue which unite the two lamellae of a pelecypod gill.

interlamellar junctions. Cross partitions between the two lamellae of a pelecypod gill.

interlobular vein. 1. Vein forming a network around a liver lobule. 2. Vein which drains blood from the capillary network of the kidney cortex.

intermandibular. Thin sheet of muscle between the two lower jaw rami in fishes.

intermaxillary gland. Large mucous gland situated in the anterior median palate area of amphibians.

intermediate host. That host which harbors an immature stage in the life cycle of a parasite; e.g. the hog is the intermediate host of the human pork tapeworm; some parasites have two or three intermediate hosts; e.g. the human liver fluke has a snail as its first intermediate host and a fish as its second intermediate host. Cf. DEFINITIVE HOST.

intermediate pituitary. Small intermediate section of the pituitary gland; embryologically derived from a part of the anterior lobe.

intermedin. Hormone secreted by the intermediate lobe of the pituitary gland in certain vertebrates; among other functions, it is known to control, in part, the size and activity of certain chromatophores.

intermuscular. Between muscles.

internal carotid artery. Artery supplying a portion of the head in many vertebrates.

internal environment. Environment afforded internal cells of the body by the blood, lymph, and other fluids bathing cells.

internal fertilization. Union of sperm and egg inside the body of the female parent.

internal nares. Two openings on the upper surface of the oral cavity of certain vertebrates; lead into respiratory passages opening to the outside via the external nares. Sing. internal naris.

internal oblique muscle. Flat abdominal muscle of certain vertebrates.

internal respiration. 1. Cell respiration. 2. The sum total of all osmotic, enzymatic, and chemical processes involved in the oxidation of materials within individual cells.

internal reticular apparatus. Golgi apparatus.

International Code of Nomenclature. International Rules of Zoological Nomenclature.

International Congress of Zoology. Periodic meeting of zoologists from all over the world; research papers and symposia are presented and major zoological problems are discussed; the Fourth Congress created a permanent Commission on Nomenclature in 1898 to prepare an International Code of Nomenclature; meetings last up to two weeks or more and, except during times of war and international uncertainties, have been held every three to five years; the First Congress met in Paris (1889), the Fifth in Berlin (1901), the Tenth in Budapest (1927), the Fifteenth in London (1958), and the Sixteenth in Washington, D.C. (1963).

International Rules of Zoological Nomen-

clature. Involved set of regulations and practices governing zoological nomenclature and taxonomic procedures drawn up by the Fifth International Zoological Congress in Berlin in 1901; the original Rules consisted of 41 Articles and 20 Recommendations, but subsequently many of the statements have been revised and clarified by special Opinions; a complete revision of the rules was authorized at the Paris Zoological Congress in 1948 and the Copenhagen Zoological Congress in 1953, and the definitive International Code of Zoological Nomenclature appeared in 1961.

internuncial neuron. Adjustor neuron.

interoceptor. Receptor which receives stimuli from within the body, such as the hunger and thirst stimuli. Cf. PROPRIOCEPTOR.

interopercle. In fishes, a flat bone below and behind the preopercle in the operculum.

interosseus. 1. One of numerous small tetrapod muscles of the feet and hands; variously flex, abduct, adduct, and extend. 2. Ligament between radius and ulna, and between tibia and fibula.

interpetaloid. In Echinoidea, an area between two successive ambulacral areas.

interphase. Interkinesis; so-called "resting stage" of a cell when the nucleus is not undergoing any mitotic changes.

interplant. In experimental embryology, a part which has been removed from one embryo and placed in an indifferent environment in another embryo, e. g. abdominal cavity for chorio-allantoic membrane.

interproglottidal glands. Transverse row of cell clusters or small glands along the posterior margin of the proglottids of certain tapeworms, e. g. Moniezia.

interpterygoid vacuity. Midline space between the two pterygoid bones in certain tetrapods.

interradial canal. 1. One of the several types of gastrovascular canals extending outward to meet the ring canal in jellyfish. 2. One of the gastrovascular canals extending outward to meet a meridional canal in ctenophores.

interradius. One of five general areas on the surface of echinoderms between the five areas (radii) which bear the tube feet; most evident in the Holothuroidea and Echinoidea. Pl. interradii.

interrenal body. Adrenal cortex or tissue of comparable embryonic origin.

interscapular. Pertaining to the dorsal region between the shoulders.

intersex. Individual which is more or less intermediate between a true male and a true female in some characteristics, but not necessarily a gynandromorph or hermaphrodite; may originate through some abnormality of the sex chromosome mechanism, or through some hormone abnormality during development.

interspecific. Between different spp.

interspersion of types. See EDGE EFFECT.

interspinous bone. One of several to many small bones projecting into the musculature from the basal region of the spines of the dorsal fine of certain fishes.

interstitial cells. 1. Small, narrow, pyramidal, or spherical undifferentiated epidermal cells of coelenterates; give rise to cnidoblasts, nematocysts, and perhaps buds and gametes. 2. Small undifferentiated connective tissue cells lying between ovarian follicles and testicular tubules of vertebrate gonads; the source of testosterone in the testes.

intertemporal. Small dorsolateral paired bone in the primitive tetrapod skull.

intertentacular organ. Intertentacular tube.

intertentacular tube. Small ciliated opening between the coelom and the base of the lophophore in certain Bryozoa; presumably functions in excretion and reproduction.

intertidal zone. That portion of the sea bottom between high and low tide lines; depending on tidal amplitude and slope of the bottom, the intertidal zone may be narrow or very wide.

intervertebral disc. Flat disclike piece of fibrous cartilage between two adjacent vertebrae.

intestinal gland. Tubular gland in the wall of the small intestine of many vertebrates; secretes intestinal juice.

intestinal juice. Succus entericus; enzyme-containing fluid secreted by the wall of the small intestine of vertebrates; in man it contains enterokinase (an activator of trypsinogen), aminopeptidase, dipeptidase, disaccharases, amylase, lipase, and nucleases.

intestinal villi. Minute finger-like projections which make a velvety lining in the small intestine of many vertebrates; serve to increase the intestinal absorptive surface greatly; each contains a lacteal and a capillary network. Sing. villus.

intestine. Principle organ of digestion, transport, and absorption for food; elongated, tubular, and forms the more posterior portion of the digestive tract in many invertebrates and vertebrates.

intima. Innermost coat or investment membrane of an organ or other part; commonly composed of epithelium and underlying connective tissue.

intracellular. Occurring within individual cells.

intracellular digestion. Digestion of particulate or dissolved food materials within individual cells.

intracorpuscular. Occurring within corpuscles.

intradermal. Between layers of the skin.

intramuscular. Within or into muscle tissue, especially with respect to a hypodermic injection.

intramuscular bone. Any bone, especially in fishes, which is completely imbedded in muscle and often extends lengthwise from one rib to another.

intraspecific. Within members of the same sp.

intrathoracic pressure. Pressure within the pleural cavity, slightly less than that of the atmosphere.

intravenous. Into a vein, especially with respect to a hypodermic injection or withdrawal of blood.

intra vitam stain. Non-toxic stain used for certain living tissues and microorganisms.

introgressive hybridization. 1. With two freely interbreeding spp. in a population, hybrids tend to back cross with the more abundant sp.; this process, with repetition, results in a population of individuals most of which resemble the more abundant parental stock. 2. Infiltration of genes of one sp. into the gene pool of another sp.

intromittent organ. Any male copulatory organ used in transferring sperm into the female genital tract.

introvert. 1. Anterior narrow portion of a typical sipunculid; by means of long muscles which are inserted internally at the anterior end, the whole introvert may be withdrawn into the trunk. 2. Distal end of a bryozoan which bears the lophophore and is capable of being withdrawn into the zooecium. 3. Invagination of one tubular part into another.

intussusception. 1. Growth of an organism by metabolic incorporation of food materials to form additional protoplasm. 2. Slipping of one part of the intestine into an adjacent part by folding of the wall; a special kind of invagination.

invagination. Pushing-in process of a sheet or layer of cells, especially in animal embryos, so as to form a depression or pocket opening to the outside.

inversion. Reversal of part of a chromosome so that the linear arrangement of certain genes is in reverse order; e.g. if a series of genes are normally arranged in the order ABCDEF, an inversion might produce the order ADCBEF, ABCFED, etc.

invertase. Saccharase, sucrase; enzyme produced by the intestinal mucosa; converts sucrose to glucose and fructose.

Invertebrata. Collectively, all animals without a vertebral column.

invert sugar. Mixture of dextrose and levulose, as in certain fruits.

in vitro. Term applied to biological processes, tissues, and organs maintained experimentally and isolated from the whole organism; e.g. tissue culture.

in vivo. Occurring within the living organism; opposite of in vitro.

involuntary muscle. Nonstriated muscle.

involution. 1. In-turning process of the germ ring area during the formation of the gastrula stage of development. 2. Decrease in size of an organ, e.g. reduction in thymus tissue after puberty. 3. Degeneration of a tissue or organ. 4. Appearance of bacteria, yeasts, etc., with abnormal physiology and morphology in old cultures.

Io. Genus of snails found in rapid streams of Tenn. and Va.

Ioa. Genus of darters now usually included in Etheostoma.

Iodamoeba butschli. Small intestinal commensal amoeboid protozoan of man.

iodogorgoic acid. Amino acid in corals and sponges.

iodopsin. Pigment in the retinal cones of those vertebrates having color vision.

io moth. See AUTOMERIS.

Ione. Genus of isopods parasitic on Callianassa.

Iotichthys phlegethontis. Least chub, a very small chunky cyprinid of the Utah Bonneville Basin.

Iotonchus. Genus of microscopic, free-living nematodes.

iridocyte. Cell with a silvery or chalky-white appearance produced by crystals of guanine; especially in the skin of fishes.

Iridomyrmex humilis. Argentine ant; house and garden pest in tropical and subtropical America; occasionally in houses in temperate localities; especially common in sea ports.

iridophore. Iridescent chromatophore.

Iridoprocne bicolor. Tree swallow in the Family Hirundinidae; blue-black or green-black above, clear white below; nests in Canada and U.S.

iris. Pigmented, muscular anterior part of the choroid coat in the vertebrate eye; contains a central opening, or pupil, and the relative size of the pupil is governed by the contraction of the iris.

iris diaphragm. Device composed of thin overlapping metal leaves located below a microscope stage; by means of a lever, the position of these peripheral leaves may be changed so that the size of a circular opening to admit substage light may be made small to large.

Irish elk. Large extinct deer (Megaceros) found in European Pleistocene deposits.

Iron. Genus of Ephemeroptera; generally distributed.

Ironodes. Western genus of Ephemeroptera.

Ironopsis. Uncommon western genus of Ephemeroptera.

iron porphyrin. Iron-containing prosthetic group forming part of the molecule of, e.g. cytochrome proteins; essential for the electron transport process.

Ironus. Genus of microscopic free-living nematodes.

irradiation. Exposure of organisms to some artificial source of electromagnetic or other rays, e.g. X-rays and atomic radiation.

Irrawaddy dolphin. See ORCAELLA.

Irregularia. One of the two subclasses of echinoid echinoderms; anus marginal, on either the oral or aboral surface; slight bilateral symmetry.

irritability. Fundamental ability of protoplasm to respond to stimuli, or changes in the environment.

Irus. Genus of short, globose, marine bivalves; among rocks.

Irwin loop. Small, flattened, metal loop used for picking up and handling individual micrometazoans and large protozoans.

isabelita. Brightly colored marine fish in the genus Angelichthys; West Indies.

isabella moth. Common U.S. moth in the genus Isia; see ARCTIIDAE.

ischiadic artery. Artery supplying the hind leg of primitive tetrapods.

ischial callosities. Two bare hardened pads on the buttocks of Old World monkeys; often brightly colored; present in the mandrill, baboon, langur, etc.

ischiofemoralis. Ventral muscle of the hind leg of urodeles and birds.

ischioflexorius. Ventral muscle of the hind leg of urodeles and birds.

ischiopodite. Third segment of pereiopod in most malacostracan Crustacea.

ischiotrochantericus. Ventral muscle of the reptile hind leg.

ischium. 1. Posterior ventral bone on each side of the pelvic girdle. 2. Ischiopodite.

Ischnochiton. Genus of Pacific Coast chitons.

Ischnura. See ZYGOPTERA.

Ischyrocerus. Genus of marine Amphipoda.

Isia. See ARCTIIDAE.

islets of Langerhans. Groups of special cells scattered throughout the pancreas of most vertebrates; they constitute an endocrine gland which secretes insulin.

isoagglutinin. Agglutinin in blood plasma which will agglutinate the red corpuscles of certain other individuals of the same sp.; may appear spontaneously or be induced by repeated injections of blood.

isoalleles. 1. Two or more different dominant mutants which occupy the same locus on a chromosome pair and are dominant over the same recessive; e.g. A'a and A"a; difficult to distinguish. 2. Two or more alleles with similar phenotypic effects in the homozygous state but which show different dominant-recessive relations with a third allele.

Isobactrus. Genus of mites in the Family Halacaridae; usually in brackish water.

Isocapnia. Common genus of Plecoptera.

isocercal. Pertaining to a type of fish tail in which the terminal vertebrae become progressively smaller and end in the midline of the caudal fin.

isochela. Type of chela sponge spicule in which the two ends are alike.

Isociona. Genus of encrusting marine sponges; usually vermilion to orange.

Isocirrus. Genus of tube-building Polychaeta.

isodactylous. Having similar digits, especially with reference to length.

isodont. Having all teeth alike.

isoelectric point. Definite pH value at which a dissolved or colloidal substance will migrate neither to the cathode nor the anode in an electrolytic system; in biology, the term is usually used with reference to amino acids and proteins.

isogametes. Two uniting gametes similar in size and structure; see ISOGAMY.

isogamy. Uncommon condition in which two uniting gametes are similar in size and structure (but presumably different in their genetic make-up); in certain algae, fungi, and protozoans.

Isogenus. Genus of Plecoptera; western and northern U. S. and Canada.

Isognomon. Genus of small Caribbean oysters; tree oysters; in clusters on rocks or mangroves.

isolation. Effective separation and the prevention of interbreeding of a group of individuals of any sp. from another group of the same sp. or different spp.; such separation may be geographic, ecological, seasonal, physiological, behavioral, or genetic; it may also involve a combination of any two or more of these factors.

isolecithal. Pertaining to an ovum or zygote which has yolk evenly distributed throughout the whole cell, e. g. sea urchin and Amphioxus ova.

isoleucine. Amino acid; $C_6H_{13}O_2N$.

isomerase. Enzyme which catalyzes internal changes in the configuration of atoms in the molecules of certain intermediate metabolic compounds.

isometric contraction. Muscle contraction against such great resistance that the muscle cannot shorten, all energy being given off as heat.

Isonychia. Common genus of Ephemeroptera; forelegs of nymphs with fringes of long hairs.

Isoperla. Common genus of Plecoptera.

Isophyllia. Genus of reef-building corals of the tropical Pacific and Indian oceans.

Isopoda. Order of malacostracan Crustacea; includes pill bugs, sow bugs, and wood lice; body depressed; first thoracic segment fused with head; abdomen short, some or all segments fused; 3100 spp.; in bottom debris of salt and fresh waters; some terrestrial spp.; some marine spp. parasitic on fishes and crustaceans.

Isoptera. Order of insects which includes the termites or white ants; mostly tropical and subtropical; small to medium-sized, soft-bodied, pale social insects having a highly developed caste system; metamorphosis simple, with a series of nymphs; mouth parts biting, sometimes vestigial; wings large, small, or absent in adult; colonies of a few to thousands of individuals living in closed tunnels in earth, wood, or nests; each colony consists of three or more castes: (1) sexuals (kings and queens), which swarm, settle in pairs, and then mate and produce a new colony, (2) wingless soldiers, with large mandibles, used in defending the colony, and (3) wingless workers, which construct tunnels, build nests, collect food, and care for young; supplementary reproductives, derived from certain nymphs, may become sexually mature and replace the king and queen; food consists of cellulose which is digested by symbiotic intestinal flagellates in some spp.; termites do great damage to wooden structures in the tropics; common genera are Reticulotermes, Zootermopsis, Kalotermes, Amitermes, Macrotermes, and Nasutitermes.

isorhiza nematocyst. Type of nematocyst whose tubule secretes a paralyzing substance when discharged but which has no enlarged base; tubule sticky.

isosmotic. Having the same osmotic pressure.

Isospondyli. Order of bony fishes which includes the tarpons, herring, salmon, and trout; head not scaly, fins not spinous; scales cycloid, ctenoid, or lacking; ventral fins abdominal; air bladder, if present, with a duct to pharynx; tail homocercal.

Isospondylida. Isospondyli.

Isospora hominis. Intestinal sporozoan parasite of man; produces a mild diarrhea.

Isotoma. Common genus of Collembola; several spp. semiaquatic.

Isotomurus palustris. Common springtail; often found on puddles and ponds.

isotonic. Static relationship between two solutions when they have osmotically equal concentrations of solutes on either side of a selectively permeable membrane. Cf. HYPERTONIC and HYPOTONIC.

isotonic contraction. Muscle contraction in which the muscle shortens and work is done; cf. ISOMETRIC CONTRACTION.

isotope. Any one of two or more kinds of atoms of a chemical element which are nearly identical in their properties and occupy the same place in the periodic table; they differ, however, in the number of neutrons in the atomic nucleus.

Isotricha. Genus of commensal holotrich ciliates occurring in the stomach of cattle and sheep.

isotropic discs. Microscopic light-colored transverse stripes or discs as seen in surface view of striated muscle fibers; alternate with dark-colored anisotropic discs; same as I discs or J discs; see KRAUSE'S MEMBRANE.

Istiophoridae. Family of marine fishes which includes the spearfishes and sailfishes; with a swordlike beak which is rounded at the tip and not flattened; dorsal fin long and capable of being depressed into a groove; Istiophorus and Makaira.

Istiophorus. Genus of sailfishes; large oceanic spp. related to the swordfish but with a much higher sail-like dorsal fin; in the Family Istiophoridae.

Isurus. Genus of voracious mackerel sharks.

itch mites. Group of mites in the Family Sarcoptidae; attack skin of birds and mammals and produce intense itching.

iter. Any tubular passage, but especially the passage from the third ventricle to the infundibulum.

Ithytrichia. Genus of Trichoptera.

ivory. 1. In the broad sense, the material of exceptionally large teeth of certain mammals, such as the walrus, narwhal, wild boar, and elephant. 2. In the strict sense, a type of dentine found only in elephant tusks.

ivory barnacle. Balanus eburneus, a large, white, smooth barnacle found along the low tide line; Atlantic Coast.

ivory-billed woodpecker. See CAMPEPHILUS.

Ixodes ricinus. Castorbean tick; 3 to 10 mm. long; common parasite on mammals.

Ixodidae. Family of Acarina which includes the hard ticks; mouth parts anterior; body with dorsal shieldlike plate; usually parasitic on mammals; familiar genera are Ixodes, Dermacentor, and Boophilus.

Ixodoidea. Superfamily of the arachnid Order Acarina; ticks; hypostome conspicuous, armed with teeth; usually more than 2 mm. long; mostly parasitic on mammals.

Ixoreus naevius. Varied thrush; a robin-like bird of western coniferous forests.

J

jabiru. 1. Wood ibis, Mycteria americana. 2. See JABIRU and EPHIPPIORHYNCHUS.

Jabiru mycteria. Jabiru stork, a large sp. of southern Mexico to Argentina; white with black featherless neck and head, and a deep red patch at base of neck.

jacamar. Any of a small group of forest birds in the Family Galbulidae (Order Piciformes); tropical South and Central America; bill long, stout, and slightly curved; body slim, with bright metallic coloration; catch flying insects, nest in burrows.

jacana. Any of several tropical and subtropical birds which have extremely long toes and are thereby able to run about on floating lily pads and similar aquatic plants; Jacana spinosa occurs as far north as southern Tex.

Jacanidae. Family of birds in the Order Charadriiformes; includes the jacanas; toes very long, legs long, bill plover-like; insectivorous swamp spp.

jack. See CARANX.

jackal. Any of several small Old World nocturnal wild dogs (Canis); eat carrion and plants but often hunt prey in packs; seldom exceed 25 lbs.

jackdaw. Common glossy-black grackle-like bird of Europe (Corvus).

jackfish. Pike, especially the great northern pike.

jackknife clam. See TAGELUS.

jack rabbit. See LEPUS.

jacksnipe. 1. Old World snipe (Limnocryptes). 2. In the U.S., any of several sandpipers and snipes.

Jacobson's organ. Olfactory diverticulum found in many vertebrates; often develops into a sac which opens into the mouth.

Jaculus. Genus of African desert jerboas.

jaeger. See STEROCORARIUS.

Jaera. Genus of broad, flat, marine Isopoda.

Jaeropsis. Genus of marine Isopoda.

jaguar. Largest American mammal in the Family Felidae (Panthera hernandesii); coloration tawny, marked with black spots or rosettes; secretive, preferring jungle and thicket areas; to 250 lbs.; southern Tex., N.M., and Ariz. to Paraguay; sometimes included in Felis.

jaguarundi. Small unspotted cat (especially Felis cacomitli) with slender body and long tail; the cacomitl cat; Tex. to Paraguay.

Janira. Genus of marine Isopoda.

Janiralata. Genus of marine Isopoda.

Janolus. Genus of nudibranchs.

Janssen, Zacharias, and C. Janssen. Dutch spectacle-makers (father and son) credited with constructing the first compound microscope in about 1590.

Janthina. Ianthina.

Japanese beetle. See POPILLIA.

Japanese oyster. Ostrea gigas, a commercial sp. imported to the Pacific Coast from Japan.

Japygidae. Iapygidae.

Japyx. Same as Iapyx; see IAPYGIDAE.

Jassa. Genus of marine Amphipoda.

Jassidae. Leafhoppers; a large and important family of small Hemiptera which feed on plants; Empoasca and Erythroneura typical.

jaundice. Icterus; yellowing of skin, eyeballs, and urine owing to the presence of bile pigments in the blood and urine.

Java ape man. See PITHECANTHROPUS.

Java stork. See LEPTOPTILUS.

javelina. Peccary; see TAYASSUIDAE.

jay. Any of a group of birds in the Family Corvidae; smaller than crows, more arboreal, and more highly colored; crest often present; Perisoreus and Cyanocitta typical.

J disc. Isotropic disc.

Jefferson's salamander. Ambystoma jeffersonianum; an eastern U.S. sp. which is aquatic only during the breeding season.

jejunum. Portion of the small intestine immediately following the duodenum in certain of the higher vertebrates; in man it is about 10 ft. long; the main absorptive area of the small intestine.

jellyfish. 1. Free-swimming umbrella-shaped medusa stage of scyphozoan coelenterates ("true"jellyfish) having a series of marginal tentacles. 2. The term is also sometimes used to apply to the free-swimming craspedote medusa stage of certain hydrozoan coelenterates, and incorrectly to certain massive, floating hydrozoans such as the Portuguese man o' war.

jelly of Wharton. Type of gelatinous connective tissue in the umbilical cord.

Jenner, Edward. English physician (1749-1823); first demonstrated that inoculation with cowpox virus produces an immunity to smallpox in man.

jennet. Female ass.

Jennings, Herbert S. American zoologist (1868-1947) professor at Johns Hopkins University; made important contributions and wrote textbooks on genetics, cat anatomy, invertebrate behavior, and evolution.

jerboa. See DIPODIDAE.

jewel box shell. See CHAMA.

jewel wasp. See CHRYSIDIDAE and PTEROMALIDAE.

jewfish. Any of numerous large, sluggish, heavy-bodied, marine fishes of the tropics and subtropics; some to 12 ft. long; see STEREOLEPIS, GARRUPA, and PROMICROPS.

jigger. See TUNGA.

jingle. See ANOMIA.

Joenia. Genus of flagellates symbiotic in termite guts; cell ellipsoid, covered with numerous short processes, and with many anterior flagella.

joey. Young kangaroo, especially while it is still young enough to be carried in the mother's pouch.

John Dory. See ZEOMORPHI.

Johnny darter. Etheostoma nigrum, a small colorful darter (fish) found in central U.S. streams.

Johnston's organ. Auditory organ on the second segment of each antenna in many insects.

joint cavity. Narrow cavity surrounding the ends of two long bones in the region of a joint; filled with a clear fluid and enclosed by a capsular ligament.

jointworm. See HARMOLITA.

Jolietoceras. Genus of extinct cephalopods having a coiled shell.

Jonah crab. Large, red, deep-water crab of the U.S. East Coast; Cancer borealis.

Jordan, David Starr. American educator and zoologist (1851-1931); famous for his vigorous organizational policies as first president of Stanford University; an outstanding ichthyologist and (with Evermann) the author of The Fishes of North and Middle America.

Jordanella floridae. Flagfish; a small cyprinodontid from Fla. swamps and streams.

Jordan's rule. Fishes in cold waters have more vertebrae than the same or closely related spp. in warm waters.

Juday, Chancey. American limnologist (1871-1944); made many fundamental contributions to limnology, especially with reference to Wisconsin lakes; associated with the Wisconsin Geological and Natural History Survey, professor of zoology at the University of Wisconsin.

jugal. 1. Pertaining to the cheek or bones in the cheek region of the face. 2. Zygomatic. 3. Dermal bone between maxilla and squamosal.

Jugatae. One of the three suborders of insects in the Order Lepidoptera; venation of forewings and hind wings similar; process of forewings hooked to hind wings during flight; without a coiled proboscis.

jugular. Pertaining to the throat.

jugular vein. One of two large veins draining the head of higher vertebrates.

jugum. Small process on the basal posterior margin of the forewing of certain Lepidoptera and Trichoptera; serves to hold forewing and hind wing together during flight.

Jukes family. Fictitious name of a New York State family tree studied by Dugdale and Estabrook; through several generations the family was shown to have a very high percentage of familial feeble-mindedness, pauperism, prostitution, and criminal tendencies.

Juliformia. Order of millipedes; 40 or more body segments; male with one or two pairs of gonopods on seventh segment; spinning glands absent; Julus and Narceus typical.

Julus. Common genus of small millipedes in the Order Juliformes.

jumping beans. See LASPEYRESIA SALTITANS.

jumping mouse. See ZAPODIDAE.

jumping plant lice. See PSYLLIDAE.

jumping spider. See SALTICIDAE.

jumprock. Common name for several southeastern spp. of fishes in the genus Moxostoma.

Junco. Genus of North and Central American birds in the Family Fringillidae; juncos; sparrow-like birds with white outer tail feathers and gray or black heads.

June beetle. Large robust beetle in the Family Melolonthidae; see PHYLLOPHAGA.

June bug. June beetle; see PHYLLOPHAGA.

jungle fever. Common name for several diseases typical of tropical jungle areas, especially a kind of East Indian malaria.

jungle fowl. Any of several wild chicken-like Asiatic birds; one sp. is believed to be the ancestor of the common domestic fowl; see GALLUS.

Junonia coenia. The buckeye; a common brownish butterfly with conspicuous blue spots on the wings; larva feeds on plantain and snapdragon.

Jurassic period. One of three geological subdivisions of the Mesozoic era, between the Cretaceous and Triassic periods; lasted from about 155 million to 120 million years ago; an age of level continents, deserts, and cycads; characteristic animals were bivalves, modern bony fishes and sharks, first lizards and crocodiles, many giant reptiles, and first toothed birds.

juvenal plumage. First true plumage of a bird; appears after the down.

juvenile hormone. See CORPORA ALLATA.

Jynx. See WRYNECK.

K

Kaburakia. West Coast genus of polyclad Turbellaria.

Kadiak bear. Kodiak bear; see URSUS.

kagu. See RHYNOCHETIDAE.

kaka. Local name for a New Zealand parrot; talks and mimics well in captivity.

kakapo. Burrowing parrot of New Zealand; owl parrot.

Kakatoe. See COCKATOO.

kala azar. Protozoan disease of man caused by Leishmania donovani which parasitizes the endothelium of blood vessels and lymphatics; found mostly in southern Europe, northern Africa, and southern Asia; transmitted by Phlebotomus, a minute sandfly.

Kallikak family. Family tree studied by Goddard; descendents of Martin Kallikak and a feeble-minded girl showed a high incidence of feeble-mindedness and criminal tendencies.

kallikrein. Proteolytic substance found in urine, pancreas, and kidneys; a shock-inducing compound.

Kallima. Genus of butterflies found especially in India; under surface of wing has a striking resemblance to a dry leaf.

kalong. A kind of flying fox in the Family Megachiroptera; a fruit-eating bat of the Old World tropics.

Kalotermes. Genus of termites living in wood; southern half of U. S.

Kammerer. Austrian zoologist who attempted to "prove" the doctrine of acquired characteristics by fraud; injected the limbs of several generations of individual salamanders with a dark dye; the resulting black welts were stated to be transmitted from the two original injected parents.

Kamptozoa. Entoprocta.

kangaroo. See MACROPUS and MACROPODIDAE.

kangaroo mouse. See ZAPODIDAE and MICRODIPODOPS.

kangaroo rat. See DIPODOMYS.

Kannemeyeria. Genus of large herbivorous Triassic reptiles.

kappa particles. Cytoplasmic particles found especially in Paramecium; in the presence of a dominant K gene, able to form a substance, paramecin, which slowly diffuses into the surroundings and is toxic to other strains of Paramecium; kappa particles are self-perpetuating in the presence of K and are in the nature of plasmagenes.

karakul. Broadtail sheep of the Asiatic province of Bokhara.

karyogamy. Union of male and female nuclei in the process of syngamy.

karyokinesis. Mitosis.

karyolymph. Nuclear sap.

karyoplasm. Nuclear cytoplasm.

karyosome. Spherical mass of chromatin in a cell nucleus; distinct from the nucleolus.

karyotype. General appearance of somatic chromosomes with reference to size, shape, and number.

katabolism. Destructive metabolism; the breakdown of protoplasm and food materials with the release of energy.

Katharina. West Coast genus of dark-colored chitons; plates two-thirds covered by mantle.

katharobe. Any inhabitant of water in which the dissolved concentration remains moderate to high.

katharobic. 1. Pertaining to aquatic habitats where the oxygen concentration does not go below the minimal requirements of most aerobic organisms. 2. Pertaining to a katharobe.

katydid. Nocturnal insect in the orthopteran Class Tettigoniidae; color often green; common in eastern and southern U. S.; male produces stridulating chirping sounds by rubbing a "scraper" on the base of one forewing against a "file" on the other; Microcentrum common.

kea. Large, green, uncommon parrot of southern New Zealand; normally insectivorous, it occasionally attacks live sheep, tears at the back, and feeds on kidney fat. See PSITTACIFORMES.

Keber's organ. Pelecypod pericardial gland.

ked. See HIPPOBOSCIDAE.

keel. See CARINA.

Kellia. Genus of minute marine bivalves.

Kellicottia. Common genus of limnetic rotifers; lorica with one long posterior spine and four or six anterior spines.

kelp bass. See PARALABRAX.

kelp crab. See PUGETTIA.

kelpfish. Any of many spp. of small colorful fishes found along the tropical and temperate Pacific Coast in kelp beds and on rocky shores; body elongated, mouth protruding; includes Family Clinidae and representatives of other families.

kentrogon. Larval stage of barnacles in the Order Rhizocephala.

Keratella. Genus of loricate plankton rotifers; extremely common in all types of fresh waters.

Keraterpeton. Genus of Carboniferous amphibians.

keratin. One of a group of tough, fibrous proteins which are especially abundant in skin, claws, hair, feathers, hooves, etc.

keratinase. Keratin-digesting enzyme.

Keratoisis. Genus of corals in the Order Alcyonacea; colony irregularly branched, to 4 in. high.

Keratosa. One of the three subclasses of sponges in the Class Demospongiae; skeleton composed of spongin only, no spicules; often quite large; commercial sponges; Spongia and Hippospongia common.

keratosis. Abnormal, tough, horny, thickened condition of human skin; a Mendelian dominant.

Kerivoula. Genus of painted bats; coloration and patterns variable; Africa and southern Asia.

kermes. Hemipterous insect found on the leaves of certain oak trees in the Mediterranean area; they are dried, and a purplish-red commercial dye is extracted from them; genus Kermes.

Kermidae. Family of highly modified scale insects in the Order Homoptera; mature females are hemispherical to spherical and usually covered with cottony wax; Eriococcus spurius, the elm scale, is a common pest of elm trees in Europe and N.A.

Kerona pediculus. Flat, kidney-shaped ciliate protozoan found as a commensal on the surface of fresh-water hydras.

kestrel. See FALCO.

keto acid. See DEAMINATION.

ketogenic. Pertaining to certain foods, e.g. fats, which are capable of being converted to ketone bodies in metabolism.

ketone body. Organic compound resulting from the incomplete oxidation of fatty acids.

ketonuria. Abnormal accumulation of ketones in the urine; occurs in some cases of diabetes; indicates incomplete fat metabolism.

key. Tabular, dichotomous arrangement of diagnostic features used for identifying organisms in various taxonomic categories.

keyhole limpet. One of a group of marine limpets which have a small hole at the top of the low conical shell.

Khawkinea. Genus of saprozoic fresh-water Protozoa; similar to Euglena but colorless.

kiang. See EQUIDAE.

kidney. Organ of excretion and/or water balance in vertebrates as well as in certain invertebrates; in the former it consists of many nephrons and their associated blood vessels and is the source of urine formation.

kidney stones. Concretions formed in the urinary passages; contain relatively insoluble uric acid and calcium phosphate.

kidney threshold. Concentration of a particular substance in the blood above which that substance will pass through the kidney and be found in the urine.

kidney worm. Any of several large nematodes parasitic in the kidneys of mammals.

killdeer. Type of plover (Oxyechus vociferus) widely distributed in North and South America; see CHARADRIIDAE.

killer paramecium. Strain of Paramecium which can release a substance (paramecin) into water which is sufficiently toxic to kill other strains of Paramecium.

killer whale. See ORCINUS and GRAMPUS.

killifish. See CYPRINODON and FUNDULUS.

kilocalorie. Same as Calorie.

kilogram. Metric unit of weight; 1000 grams; 2.2 lbs.

kinesis. Undirected locomotor response proportional to the intensity of the stimulus; the stimulus does not control direction of movement.

kinesthesis. Internal sense by which motion, balance, weight, and position are perceived; sensory endings for this sense are in vertebrate muscles, tendons, and joints.

kinetic energy. In the broad sense, energy of molecular motion such as that possessed by any moving object, heat, chemical reactions, electricity, and light.

kinetochore. Centromere.

kinetodesma. One of many fine strands running parallel to the surface of a ciliate protozoan and lying just below the surface; closely associated with the kinetosomes (basal granules).

kinetoplast. Accessory cytoplasmic apparatus in many flagellate Protozoa; composed of the blepharoplast, the parabasal body, and a uniting fibril.

kinetosome. Basal granule of a cilium or flagellum; a self-reproducing organelle. See KINETODESMA and KINETY.

kinety. In ciliate Protozoa, a kinetodesma and its closely associated kinetosomes; all of the kinetia collectively form the neuromotor system.

kingbird. Any of several American flycatchers noted especially for their pugnacity; eastern sp. often called the tyrant flycatcher or bee martin; Tyrannus common.

king crab. See MEROSTOMATA.

king crows. See DICRURIDAE.

kingfish. Any of a wide variety of fishes; in the U.S. a common Atlantic Coast sp. is the king mackerel (Scomberomorus regalis); another is the whiting (Menticirrhus saxatalis), an Atlantic Coast gamefish weighing to 3 lbs.

kingfisher. Any of a worldwide group of non-passerine birds, mostly crested and brightly colored, with a long, stout bill and short body and tail; usually fly or perch along the edges of rivers, lakes, and ponds; 250 spp. and subspecies; see MEGACERYLE.

kinglet. See REGULUS and CORTHYLIO.

king mackerel. See SCOMBEROMORUS.

king of the herrings. See REGALECUS.

king rail. See RALLUS.

king salmon. See ONCORHYNCHUS.

kingsnake. See LAMPROPELTIS.

kinkajou. See POTOS.

kinomere. Centromere.

Kinorhyncha. Phylum composed of microscopic wormlike forms living on the sea bottom; the external cuticle is arranged in 13 (14) folded consecutive ringlike structures (zonites) giving the appearance of segmentation; head and trunk bear spines and teeth; cilia are absent but there is a pseudocoel, complete digestive tract, two flame bulbs, brain, and ventral nerve cord; sexes are separate but difficult to distinguish; growth by molting; about 50 spp.

Kinosternidae. Family of turtles which includes the musk and mud turtles; both ends of plastron more or less hinged; with a musklike odor; North and Central America; Sternotherus and Kinosternon common.

Kinosternon. Common American genus of mud turtles; in sluggish waters and ponds of U.S., Mexico, and Central America; in U.S. from N.Y. south to Fla. and west to southeastern Calif.; carapace yellow or brownish, usually with bands; commonly less than 4 in. long.

Kiricephalus. Genus of tongue worms parasitic in water snakes.

Kirtland's warbler. Dendroica kirtlandi; a bluish gray, black, and white warbler; rare, Mich.

kissing bug. Common name applied to insects in the hemipteran Family Reduviidae.

kite. Any of several graceful falcon-like birds of prey with long pointed wings; especially common in southern states, Central, and northern South America; Elanoides forficatus is the swallow-tailed kite of the eastern U.S. and southward; Ictinia misisippiensis is the Mississippi kite occurring in the same general area; Elanus includes the white-tailed kite, an open country sp. found from the southern U.S. to Guatemala; Rostrhanus includes the Everglades kite of fresh-water marshes.

kit fox. See VULPES and DESERT FOX.

kitol. Provitamin A, obtained from whale liver oil.

kitten's paw. See PLICATULA.

kittiwake. See RISSA.

kiwi. See APTERYGIFORMES.

kiyi. See LEUCICHTHYS.

Klattia. Genus of rhabdocoel Turbellaria.

Klebsiella. Genus of bacteria normally occurring in the human large intestine.

klipspringer. See OREOTRAGUS.

kneecap. Patella.

knee jerk. Simple reflex straightening of the relaxed knee joint when the tendon of the knee cap is tapped.

knobbed Littorina. Littorina dilatata, a conical southern periwinkle.

knobbed scallop. Pecten nodosus, a scallop ranging from N. C. to the Gulf of Mexico.

knobbed whelk. Busycon caricum, a giant whelk of the Atlantic Coast.

knobby top shell. Chestnut top shell; Turbo castaneus, a variegated orange and brown tuberculous marine snail; N. C. to Gulf of Mexico.

knot. See CALIDRIS.

koala. See PHASCOLARCTUS.

kob. See ADENOTA.

Kobus. Genus of ruminants in the Family Bovidae; includes several spp. of waterbuck; large brownish antelopes which frequent rivers and swim well; eastern and southern Africa.

Koch, Robert. German physician and bacteriologist (1843-1910); developed bacterial stains and techniques; discovered the bacteria causing many diseases, including anthrax, wound infections, tuberculosis, and cholera; also studied sleeping sickness, malaria, and bubonic plague; received Nobel prize in physiology and medicine in 1905 for developing tuberculin.

Kodiak bear. Giant brown bear of Kodiak and adjacent islands; see URSUS.

Koenenia. See PALPIGRADI.

Koenikea. Genus of Hydracarina.

Kofoid, Charles A. American zoologist (1865-1947); professor at universities of Illinois and California; made many contributions to fresh-water and marine plankton, protozoology, and marine invertebrates.

Kofoidia. Genus of symbiotic flagellates found in termite guts; cell spherical, with a single nucleus and anterior tufts of many flagella.

Kogia breviceps. Pygmy sperm whale in the Family Physeteridae; a cosmopolitan sp. to 13 ft. long.

kokanee salmon. Oncorhynchus nerka kennerlyi, a dwarf form of the sockeye salmon; spawns and matures in fresh water; a plankton feeder, found in western N. A. and planted in the Rocky Mountain region; to 18 in. long.

Kölliker, Rudolf. Swiss anatomist and physiologist (1817-1905); the "father of modern histology" and made notable contributions to embryology.

Kölreuter, Joseph Gottlieb. German botanist (1733-1806); founder of experimental hybridization; performed the first cross-pollination experiments and emphasized the importance of insects and wind in natural pollination of flowers.

Komodo dragon. Giant carnivorous lizard of southeastern Asiatic jungles; to 250 lbs. and 9 ft. long.

kongoni. A sp. of hartebeest.

kookaburra. Laughing jackass; a large kingfisher-like bird of Australia; its cry is abrupt and discordant and has the elements of an inane laugh.

Kophobelemnon. Genus of slender, club-shaped sea pens in the Order Pennatulacea.

korrigum. See DAMALISCUS.

Kowalevskia. Genus of pelagic prochordates in the Class Larvacea.

krait. See BUNGARUS.

Krause's end bulb. Microscopic granular spherical capsule functioning as a special nerve ending; found in mucous membranes of nose, eyes, mouth, and external genitals.

Krause's membrane. Z line; telophragma; a fine dark line running along the center of an isotropic disc in striated muscle tissue.

Krebs cycle. Citric acid cycle.

krill. Common name of Euphausia, a genus in the Order Euphausiacea.

Krogh, August. Danish comparative physiologist (1874-1949); professor at the University of Copenhagen; noted for his work on excretion, respiration, osmoregulation, and circulation; in 1920 he won the Nobel prize in physiology and medicine for work on blood supply to muscle tissue.

Krumbachia. Genus of rhabdocoel Turbellaria.

kryoplankton. "Plankton" imbedded in snow masses and glaciers of high altitudes and latitudes.

kudu. See STREPSICEROS.

Kuhne, Willy. German physiologist (1837-1900); proposed the name enzyme.

Küpffer cells. Special stellate phagocytic cells found in sinusoids of the liver.

Kurzia latissima. Cladoceran common in aquatic vegetation.

kymogram. Record produced by a kymograph.

kymograph. Instrument which makes a linear record of various physiological activities, such as heart beat, pulse, and muscle twitches; such records are usually made on a sooted paper covered drum revolving at an appropriate speed.

kyphosis. Abnormal "humpbacked" dorsal curvature of the spine.

L

labellum. 1. Prolongation of the labrum in Coleoptera and Hemiptera. 2. One of two small lobes at the tip of the labium in certain Diptera. 3. Small lobe at the tip of the glossa in the honeybee. Pl. labella.

labia. Lips; pl. of labium.

labial. Of or pertaining to a lip.

labial palp. 1. One of two thin fleshy palps on either side of the mouth region in many bivalve mollusks; function in separating inedible from edible material. 2. A one- to four-segmented sensory appendage on each side of a typical insect labium.

labia majora. Two large lateral folds of skin which border and cover the general vulva area, especially in female mammals.

Labia minor. Small earwig having a world-wide distribution.

labia minora. Two lateral folds of skin which cover the vaginal opening and are largely covered by the labia majora, especially in female mammals.

Labidesthes. See ATHERINIDAE.

Labidognatha. Araneomorphae; group of spider families having vertical chelicerae.

Labidosaurus. Genus of primitive reptiles.

Labidura. Genus of brown and black earwigs; tropical, subtropical, and temperate, especially in the Northern Hemisphere.

labiella. 1. Median mouth part of myriapods. 2. Hypopharynx of insects.

labium. 1. Lower lip of insects; a complex structure forming the floor of the mouth in mandibulate insects; opposed to the labrum; in insects with sucking mouth parts it is greatly modified and often elongated. 2. Any of several liplike structures in certain other arthropods. Pl. labia.

labor. Childbirth.

labor pains. Pains caused by rhythmic contractions of the uterus during childbirth.

Labrador duck. See CAMPTORHYNCHUS.

La Brea tar pits. Asphalt pits in Los Angeles in which a large number of Pleistocene animals were trapped and from which their bones are now being removed.

Labridae. Family of brightly colored percomorph fishes which includes the wrasses; mouth small, teeth and scales large; usually edible; in tropical and temperate seas; *Lachnolaimus maximus* (hogfish) is a bright red food fish of the Carolinas to Fla.; *Tautoga onitis* is the common tautog, an im- Atlantic Coast wrasse; *Pimelometopon pulcher* is the Pacific sheepshead, a red wrasse of the Calif. coast.

labrum. 1. Upper lip of insects; variously modified, depending on food habitats, but in typical mandibulate insects covers the base of the mandibles and forms the roof of the mouth; articulates with the frons. 2. Liplike structure in certain other arthropods.

Labyrinthodonti. Order of extinct stegocephalian amphibians; Late Devonian to Triassic; small to large spp. resembling salamanders or crocodiles; to 15 ft. long; jaws with large teeth having greatly convoluted dentine; body usually covered with dermal armor plates; *Eugyrinus*, *Eryops*, and *Capitosaurus* typical.

Labyrinthula. Genus of minute amoeboid fresh-water protozoans which feed on algae.

Laccifer. See LAC INSECTS.

Lacciferidae. See LAC INSECTS.

Laccobius. Common genus of hydrophilid beetles.

Laccophilus. Common genus of dytiscid beetles.

Laccornis. Genus of dytiscid beetles.

lace bug. See TINGIDAE.

Lacerta. See LACERTIDAE.

Lacertidae. Family of lizards found in Europe, Asia, and Africa; *Lacerta* especially common in Europe.

Lacertilia. Sauria.

lacewings. Common name for most insects in the Order Neuroptera.

Lachesis. Genus of venomous snakes in the Family Crotalidae occurring in tropical America; L. muta, to 12 ft. long, is the deadly bushmaster.

Lachnolaimus. See LABRIDAE

lachrymal. Of or pertaining to tears or to various structures associated with the eye.

lachrymal bone. Small thin bone at the anterior (median) and upper (lower) part of the orbit.

lachrymal caruncle. Reddish eminence at the inner angle of the eye.

lachrymal duct. One of two small ducts extending from inner corner of each eye to the nasal cavity; serves as drain for tears.

lachrymal glands. 1. Small glands associated with the eyes of tetrapods; secrete a fluid which keeps the surface of the eyeball moist. 2. Tear glands.

lachrymal sac. Small sac located medially from the inner corner of each eye; receives tears and moisture from each eye via the two lachrymal ducts and discharges into the nose through a single nasolachrymal duct.

lacinia. Inner lobe or process attached to the stipes of a typical insect maxilla; often tipped with hairs or spines.

lacinia mobilis. Small movable process near the mandibular teeth of certain higher crustaceans and a few insects.

lac insects. Highly specialized group of hemipterous insects in the Family Lacciferidae; bodies encased in a perforated case of secreted resin or lac; great masses of females (especially Laccifer laca of India) occur on limbs of certain trees in the tropics and subtropics; these masses are gathered, and commercial shellac is processed from the material.

Lacinularia. Genus of sessile rotifers; in colonies with adhering gelatinous tubes.

lacrimal. Lachrymal.

Lacrymaria. Genus of marine and freshwater holotrich ciliates; polymorphic, cylindrical to flask-shaped.

lactalbumin. Abundant protein of milk; contains all essential amino acids.

lactase. Milk sugar-digesting enzyme contained in the intestinal juice of certain mammals; hydrolyzes lactose to glucose and galactose.

lactation. Production of milk by mammary glands.

lacteal. Minute lymph duct contained in each intestinal villus of vertebrates; functions in the absorption of fatty acids and glycerol.

lactic acid. Simple organic acid formed during the preliminary enzymatic anaerobic oxidation of glucose via pyruvic acid in many animal cells; regularly produced by skeletal muscle during vigorous exercise; also found in sour milk and formed by bacterial fermentation of glucose; $C_3H_6O_3$.

Lactobacillus. Genus of gram-positive, non-motile, non-sporeforming, rod-shaped bacteria which ferment carbohydrates and produce acids (usually lactic acid); found in milk, pickle vats, soil, dung, intestinal tract, mouth, etc.

lactogen. Prolactin, a lactogenic hormone.

lactogenic hormone. Prolactin, lactogen; hormone secreted by the anterior lobe of the pituitary; stimulates milk production in the mammary glands after birth of the young.

lactophenol. Special mounting medium having a favorable index of refraction for studying cuticular structures under the microscope; consists of 20 g. phenol crystals, 40 g. glycerin, 20 g. lactic acid, and 20 ml. water.

Lactophrys. See OSTRACIIDAE.

lactose. A 12-carbon atom sugar present in the milk of mammals; $C_{12}H_{22}O_{11}$.

lacuna. 1. Small space in the matrix of bone or cartilage containing the living bone or cartilage cell. 2. One of a complex series of spaces and channels in the hypodermis of Acanthocephala. 3. Follicle in the urethral mucosa. 4. In many invertebrates, a space in the tissues serving in place of vessels for the circulatory fluid. Pl. lacunae.

Lacuna. See CHINK SHELL.

lacunar system. Open circulatory system.

lacustrine. Pertaining to lakes; living in lakes.

ladder shell. See WENTLETRAP.

Ladona. Genus of dragonflies; eastern U.S.

ladybird beetle. See COCCINELLIDAE.

ladybug. See COCCINELLIDAE.

lady crab. See OVALIPES.

ladyfish. See ALBULIDAE.

Laelaptidae. Dermanyssidae.

Laeospira. Genus of Polychaeta.

laeotropic cleavage. Type of spiral cleavage in which the micromeres are displaced in a counterclockwise direction when viewed from the animal pole. Cf. DEXIOTROPIC CLEAVAGE.

Laetmonice. Genus of Polychaeta; dorsal surface with a covering of gray feltlike setae.

Laevicardium. Common genus of cockles; small marine bivalve mollusks; U.S. coasts.

Lafoea. Genus of marine coelenterates in the Suborder Leptomedusae.

Lafystius. Genus of Amphipoda parasitic on marine fishes.

Lagadon rhomboides. Pinfish; a small, fine-flavored porgy found along the coast from Cape Cod to the Gulf of Mexico.

lagena. Small blind tubular projection from the lower surface of the sacculus in the inner ear of lower vertebrates; in higher vertebrates it is elaborated into a cochlea.

Lagenophrys. Genus of peritrich ciliates; cell within an unstalked lorica which has a flat lateral attachment surface; ectocommensals on marine and fresh-water invertebrates.

Lagenorhynchus. Genus of striped or white-sided dolphins; also called piebald dolphins; coloration black and white; in all except extreme polar seas, sometimes in large schools.

Lagidium. Genus of mountain-chinchillas; large spp., to 20 in. long, found at highest altitudes of Peru, Bolivia, and Chile.

Lagisca. Genus of Polychaeta.

Lagodon. See SAILORS' CHOICE.

Lagomorpha. Large order of herbivorous mammals which includes the pikas, hares, and rabbits; body small to medium size; toes with claws, tail short, and incisors chisel-like, growing continually; canines absent; distinguished from rodents by two pairs of upper incisors.

Lagopus. Genus which includes the several spp. of ptarmigan in the Family Tetraonidae; feet completely feathered; mottled brown plumage becomes white in winter; northern latitudes and areas above timberline at middle latitudes.

Lagorchestes. Genus of Australian marsupials in the Family Macropodidae; hare wallabies; similar to kangaroos but much smaller.

Lagostomus. Genus of viscachas in the Family Chinchillidae; rodents having the appearance of giant chipmunks; in large communal warrens in pampas of South America.

Lagothrix. Genus of South American monkeys which includes the woolly monkeys.

lag phase. See LOGISTIC CURVE.

Lagurus curtatus. Sagebrush vole; a short-tailed, thick-bodied mouse of western U.S. sagebrush areas.

Lagynophrya. Genus of ovoid to cylindrical holotrich ciliates; stagnant marine and fresh waters.

Laila. Genus of nudibranchs.

lake herring. See LEUCICHTHYS.

lake sturgeon. See ACIPENSER.

lake trout. See CRISTIVOMER.

laking. 1. Separating hemoglobin from erythrocytes. 2. Causing erythrocytes to swell, burst, and lose their contents by placing in a hypotonic solution.

Lama. Genus of mammals in the Suborder Tylopoda and Family Camelidae; includes the llama, alpaca, vicuna, and guanaco.

Lamarck, Jean Baptiste. French naturalist (1744-1829); best known for his theory that characteristics developing in an organism as a result of a need created by its environment are passed on to the offspring, the result being the gradual evolution of new spp.; the concept became known as "the inheritance of acquired characteristics"; also widely known for his seven-volume "Histoire naturelle des animaux sans vertèbres."

Lamarckism. Theory advocated by Lamarck to explain evolution on the basis of the inheritance of acquired characteristics and habits.

Lambis. Genus of spider crabs; large tropical marine gastropods with prominent spines at edge of aperture.

lamella. 1. Any thin, platelike structure. 2. One of the thin layers of bone laid down concentrically around a Haversian canal. 3. One of the two plates forming a gill in a bivalve mollusk. Pl. lamellae. See HAVERSIAN SYSTEM.

Lamellaridae. Family of small, carnivorous, marine snails; shell thin, operculum absent.

Lamellibranchiata. Pelecypoda.

lamellicorn beetle. Any beetle having the last few antennal segments expanded laterally in the form of flat plates and therefore included in the Superfamily Lamellicornia.

Lamellidoris. Genus of small sea slugs;

dorsal surface tuberculous; anus surrounded by branchial plumes.

lamina terminalis. Septum which bounds the third ventricle anteriorly in higher vertebrates.

Lamippe. Genus of copepods; often parasitic on the coelenterate Renilla.

lammergeier. Large bird of prey of European mountains; has certain features of both eagles and vultures; Gypaëtus barbatus grandis.

Lamna nasus. Common pelagic shark; mackerel shark, porbeagle, blue dog.

Lamnida. Superfamily of sharks in the Suborder Squali; e.g. thresher and hammerhead sharks.

Lamnidae. Family of large voracious sharks; mackerel sharks.

lampbrush chromosome. Contracted diplotene chromosome showing fuzzy irregular outlines; observed in certain vertebrate oocytes.

Lampetra. Genus of small river and brook lampreys; parasitic or free-living; only a few small posterior teeth; Eurasia, eastern and western N.A.

Lampornis. Genus of humming birds in the Family Trochilidae; mountains of southwestern U.S. to Panama.

lamprey. See PETROMYZONTI.

Lampris luna. Short, deep-bodied fish in the Order Allotriognathi; opah, moonfish; common and widely distributed.

Lamprochernes. Genus of Pseudoscorpionida.

Lampropeltis. Genus of non-poisonous constrictor snakes in the Family Colubridae; king snakes, scarlet snake, milk snake, chain snakes; feed on rodents and reptiles; to 7 ft. long, often brightly colored; southeastern Canada, much of U.S., and south to Ecuador.

lamp shells. Any of certain members of the Phylum Brachiopoda; so-called because of the resemblance of a valve to an ancient Roman oil lamp.

Lampsilis. Common genus of N.A. fresh-water mussels.

Lampyridae. Family of cylindrical nocturnal beetles; glow-worms, fireflies; most adults and some immature stages are bioluminescent and exhibit flashing light at night; light-generating organs on abdomen; predaceous; usually in damp habitats; Lampyris and Photinus common.

Lampyris. European genus of fireflies in the Family Lampyridae.

lancelet. Common name of cephalochordates.

lanceolate. Elongated and lance-shaped.

lancetfish. See ALEPISAURIDAE.

land bridge. Former occurrence of land connections between two land masses now separated by water; the Bering Strait area was formerly a broad land connection between Asia and N.A. at various times; postulated to account for the present distribution of many spp. of plants and animals.

land crab. Any of several very large and powerful crabs found along tropical seashores.

landlocked salmon. See SALMO.

Landsteiner, Karl. American medical researcher (1868-1943); made many contributions to immunology, antigens, and serological chemistry; awarded the Nobel prize in 1930 for his discovery of human blood groups.

languets. Small finger-like processes projecting into the branchial chamber from the dorsal edge of the pharynx in certain tunicates.

langur. See PRESBYTIS.

Languria. See EROTYLIDAE.

Laniidae. Family of predatory passerine birds which includes the shrikes or butcherbirds; widely distributed but not in South America, Central American, or Australia; coloration usually plain gray, brown, or rufous, sometimes varied with black or white; feed on large insects, frogs, reptiles, and small birds and mammals; prey often held by being impaled on a thorn; Lanius excubitor borealis is the northern shrike and L. ludovicianus the loggerhead shrike.

Lanius. See LANIIDAE.

Lankester, Sir Edwin Ray. English zoologist (1847-1929); director of the British Museum of Natural History; especially noted for his contributions to the comparative biology of Protozoa, Mollusca, and Arthropoda.

Lankesterella. Genus of sporozoan parasites of frog erythrocytes; transmitted especially by leeches.

lanternfish. See MYCTOPHUM and INIOMI.

lantern fly. See FULGORIDAE.

lantern shell. Periploma leanum, a small spoon shell of the Atlantic Coast.

Lanthanotus. See HELODERMATIDAE.

Lanthus. Genus of dragonflies.

lanugo. 1. Covering of fine, soft hair on the human fetus at about the 20th week of development; lost shortly before or at birth. 2. Fine hair found on nearly all parts of the body.

Lanx. Genus of small fresh-water limpets; West Coast states.

Laodicea. Genus of marine hydrozoan coelenterates with well developed medusae; hydroids sometimes parasitic on other hydrozoans.

Laonice. Genus of Polychaeta.

Laophonte. Large and widely distributed genus of marine Copepoda.

Laphygma frugiperda. Important sp. of moth; fall armyworm; caterpillars feed on wheat, oats, and alfalfa.

Lapp. Nomad mongoloid race of Arctic Scandinavia and nearby Russia; see UGRIAN.

lappet. 1. Small flaplike projection on the periphery of a jellyfish; one such lappet occurs on each side of a tentaculocyst. 2. Small earlike process on each side of the head or a planarian.

lappet moth. See LASIOCAMPIDAE.

lapwing. Common crested plover of the Old World.

Laqueus. Common genus of Brachiopoda; deep waters of Pacific Coast.

Lara. Genus of elmid beetles; found along western streams.

larch sawfly. See PRISTIPHORA.

large intestine. The colon; a large posterior part of the digestive tract of mammals; in man it is about 5 ft. long and consists of the caecum, appendix, ascending colon, transverse colon, descending colon, sigmoid colon, and rectum; in lower vertebrates the corresponding part of the digestive tract is relatively much shorter.

largemouth black bass. See MICROPTERUS.

Lari. Suborder of the bird Order Charadriiformes; includes gulls and their relatives.

Laridae. Cosmopolitan family of long-winged, web-footed birds in the Order Charadriiformes; terns and gulls; on seashores and inland waters; Sterna and Larus especially common.

Lariidae. Large family of small robust beetles; seed weevils, bean weevils; about 50 spp. are economic pests on legumes.

lark. Any of numerous songbirds in both the Old and New World; often nest on the ground; see STURNELLA and ALAUDIDAE.

lark bunting. See CALAMOSPIZA.

lark sparrow. See CHONDESTES.

Larridae. Family of small to medium-sized burrowing wasps; nests usually in sand and provisioned with other insects; Tachysphex is a large cosmopolitan genus.

Larus. Common and widely distributed genus of birds in the Family Laridae; includes some of the gulls; robust, long-winged swimming spp. with great powers of flight; bill slightly hooked; tail not forked; especially abundant along seacoasts, but often far inland in large numbers; alight on water to feed; important as scavengers.

larva. General term for any independent, active, immature stage of an animal which is morphologically quite unlike the adult; grows into an adult by a complicated metamorphosis in most cases.

Larvacea. Class of transparent, free-swimming tunicates; appendicularians; minute to 5 mm. long, body divided into trunk and long solid tail which serves in locomotion; tunic voluminous, so that the animal can sometimes move about within it; Appendicularia common.

larvarium. 1. Nest or other type of shelter made by certain insect larvae. 2. Cage or other type of container used for rearing insect larvae. Pl. larvaria.

larviparous. Bringing forth larvae (rather than eggs); e.g. certain insects and mollusks.

laryngeal. Pertaining to the larynx.

laryngitis. Infection in the larynx which usually results in inflammation and swollen vocal cords.

larynx. Dilated uppermost part of the trachea in tetrapods (except birds) at the point of junction with the pharynx; the walls contain small cartilages and muscles, and the two thin vocal cords project inward from the lateral walls.

Lasaea. Genus of tiny reddish marine bivalves; among kelp holdfasts and barnacles.

Lasiocampidae. Large family of moths, especially abundant in tropics; adults with atrophied proboscis; many pest spp.; larvae feed on leaves and spin massive common webs among branches of trees; Malacosoma is the most important American genus.

lasion. Periphyton assemblage in which the spp. are interdependent and physically crowded on the substrate.

Lasionycteris noctivagans. Silver-haired bat; many dark hairs tipped with silver; mostly in Canada and Alaska, rare in U.S.

Lasiurus. Genus which includes red

and hoary bats; three spp. in North and South America.

Lasius. Common genus of ants found in Europe and N. A.

Lasmigona. Common genus of fresh-water bivalve mollusks in the Family Unionidae.

Laspeyresia saltitans. Mexican jumping bean moth; larvae inhabit the thin-walled seed of Sebastiana, and when the kernel is consumed, the larvae throw themselves forcibly about in the cavity, causing the light shell to "jump" or jerk about.

lasso cells. Colloblasts.

latebra. Small central whitish mass of the yolk of a bird's egg.

latent period. 1. Time interval between the application of a stimulus and the first detectable response in any irritable tissue, such as muscle or nerve; frequently less than a millisecond. 2. Time interval between the application of a stimulus and the first visible reaction in a reflex action. 3. Incubation period for any infection.

lateral. Located away from the midline; near the side(s).

lateral canal. One of many minute canals in the water vascular system of echinoderms; each such canal connects a tube foot with a radial canal.

lateral cricoarytenoid. One of two small lateral muscles of the larynx.

lateral field. One of the two sides of a fish scale (dorsal and ventral).

lateral line. Faint line appearing along each side of a nematode; produced by longitudinal thickened hypodermis area.

lateral line system. Series of small sense organs along the head and sides of cyclostomes, fishes, and some amphibians; such organs are located in pores or canals arranged in a longitudinal line on each side of the body as well as in a complicated pattern of lines on the head; system probably detects currents, vibrations, and pressure.

lateral plate. Unsegmented mesodermal mass on lateral and ventral areas of a vertebrate embryo.

lateral teeth. Set of longitudinal ridges below the hinge area in many bivalve mollusks; they serve to keep the two valves in juxtaposition.

lateral ventricle. One of the two cavities in the telencephalon of the vertebrate brain.

laterigrade. Walking or running sideways; e. g. certain spiders and marine crabs.

lateroneural blood vessel. Blood vessel along the lateral surface of the ventral nerve cord in many annelids.

laterosphenoid. Paired bone forming part of the lower anterior portion of the braincase in some reptiles and birds.

lathe shell. Any of several small to minute marine snails in which the shape is nearly cylindrical and the body whorl is so large as nearly to obscure the slightly raised spire; e. g. Tornatina.

Lathonura rectirostris. Littoral cladoceran.

Latigastra. Taxon sometimes used to include those orders of arachnids in which the mesosoma and prosoma are broadly joined.

Latimeria chalumnae. Deep-sea coelacanth (lobe-finned fish); member of the Order Coelacanthini which was thought to be extinct until the capture of a live specimen off the east coast of South Africa in 1938; more recently other specimens have been taken, especially between Madagascar and Africa; to 180 lbs. and 5. 5 ft. long; six of the seven fins paddle-like; scales large and circular.

latissimus dorsi. Muscle extending from the vertebral column to the humerus in certain vertebrates; it functions in drawing the arm downward and backward and rotating it.

Latona. Genus of Cladocera; in littoral vegetation.

Latonopsis. Genus of littoral Cladocera; common in southern states.

Latreutes. Genus of very small marine shrimps often found in floating brown algae.

Latrodectus mactans. Black widow spider; female shiny black with small red markings on ventral surface of abdomen; web irregular, with a funnel-shaped retreat; the venom is virulent, but this sp. is timid and has no instinct to bite man; unless a person is already ill and weakened, the bite is seldom fatal; in all continental states.

laughing jackass. See KOOKABURRA.

launce. See AMMODYTES.

Laura. Genus of barnacles in the Order Ascothoracica; parasitic on black coral.

Laurer's canal. Narrow canal leading from the oviduct of certain flukes to the ventral surface of the body where it may or may not open to the outside; in the former case it is thought to function as an accessory vagina or an outlet for excess sperm.

Lauterborniella. Genus of tendipedid midges.

Lavinia exilicauda. Hitch; a large minnow in Clear Lake and rivers entering San Francisco Bay; to 12 in. long.

law. In the biological sense, a statement of cause and effect, order, or relation that has been found to be invariably true under repeated similar conditions; a further substantiation of a theory by the accumulation of a large body of incontrovertible evidence.

law of independent assortment. See MENDELISM.

Law of Parsimony. "Entities should not be multiplied beyond necessity," i.e. do not suppose more forces or causes than are necessary in order to account for a phenomenon (in biology); sometimes called "Occam's razor"; first stated by William of Occam (?-1349), and English Franciscan philosopher.

law of priority. Provision that the name under which a genus or sp. is first designated under international rules of nomenclature is the name that must apply.

law of segregation. See MENDELISM.

law of superposition. In paleontology, the simple idea that the deepest layers of rock were deposited first and the other layers successively during later periods of time.

law of the minimum. See LIMITING FACTOR.

law of toleration. See LIMITING FACTOR.

lawyer. 1. See GADIDAE. 2. Name occasionally applied to: a stilt, the bowfin, or one of the marine snappers.

laxative. Mild cathartic; substance used to induce a bowel movement.

lazuli bunting. See BUNTING.

leaf beetle. See CHRYSOMELIDAE.

leafbird. See CHLOROPSEIDAE.

leaf bugs. In the narrow sense, the common name of hemipteran insects in the Family Miridae.

leaf butterfly. Any of numerous tropical butterflies whose wing coloration and venation bear a remarkable resemblance to a leaf.

leaf coral. See AGARICA.

leaf-cutting bees. See MEGACHILE and MEGACHILIDAE.

leaf fish. See MONOCIRRHUS.

leaf-footed bug. Any of numerous hemipterans in the Family Coreidae; hind legs with curious leaflike dilations; most abundant in tropics.

leafhopper. See CICADELLIDAE and JASSIDAE.

leaf insect. See PHASMIDA.

leaf-lipped bat. See CHILONYCTERIS.

leaf-nosed bat. See MACROTUS.

leaf-nosed snake. Any of several small chunky snakes in the genus Phyllorhynchus; with a prominent triangular shield over the tip of the nose; 6 to 20 in. long; deserts of southern Calif., Ariz., and adjacent Mexico.

leaf rollers. Group of moths in the families Tortricidae and Olethreutidae; many larvae form nests of rolled leaves.

Leander. Palaemon.

Leanira. Genus of Polychaeta.

least chub. See IOTICHTHYS.

least flycatcher. See EMPIDONAX.

least sandpiper. See EROLIA.

least shrew. See CRYPTOTIS.

least tern. See STERNA.

leatherback. See DERMOCHELIDAE.

leather carp. Scale-less variety of common carp.

leatherside chub. See SNYDERICHTHYS.

leather star. See DERMASTERIAS.

Lebertia. Common genus of Hydracarina; mostly in cold waters.

Lebistes reticulatus. Small fresh-water fish of Venezuela, Barbados, and Trinidad; feeds largely on mosquito larvae and pupae; widely cultivated and easily kept as an aquarium fish; many varieties; to 2 in. long.

Lecane. Large and common genus of loricate rotifers; with two exceptionally long toes and a short foot.

Lecanicephaloidea. Order of tapeworms in the Subclass Cestoda; lower part of scolex with four suckers, upper part disclike, globular, or subdivided into tentacle- or petal-like structures; adults in the intestine of elasmobranch fishes; Polypocephalus typical.

lechwe. See ONOTRAGUS.

lecithin. One of a group of phospholipid compounds present in all cells, and especially abundant in nervous tissues; upon hydrolysis, the lecithin molecule yields fatty acids, glycerol, phosphoric acid, and choline.

Leconte's sparrow. Ammospiza lecontei, a sparrow of swales of the northern Great Plains.

lectotype. One of a series of syntypes which, subsequent to the publication of an original description of a taxon, is selected (by publication) to serve as the type specimens for that taxon.

Lecythorhynchus. Genus of Pycnogonida.

leech. 1. See HIRUDINEA. 2. To apply leeches to the body for purposes of blood-letting.

Leeuwenhoek. See VAN LEEUWENHOEK.

left atrium. Receiving chamber of the heart on the left side of the body; it transmits blood to a ventricle; fishes have a single atrium but tetrapods have two.

left ventricle. Left posterior chamber of the heart in birds, mammals, and a few higher reptiles; it pumps blood to all of the body except the lungs.

leg bud. Simple embryonic protuberance in higher vertebrate embryos; develops into a leg.

legionary ants. See ECITON.

Leidy, Joseph. American biologist and paleontologist (1823-1891); associated with Univ. of Pennsylvania, Swarthmore, and the Academy of Natural Sciences of Philadelphia; outstanding anatomist, vertebrate paleontologist, parasitologist, and protozoologist.

Leidyi. Genus of isopods parasitic on marine decapods.

Leiochone. Boreal genus of tube-building Polychaeta.

Leiochrides. Genus of Polychaeta.

Leioptilus. Genus of thick fleshy sea pens.

Leiosella. Genus of marine sponges.

leishman body. See LEISHMANIA.

Leishmania. Genus of minute Mastigophora parasitic in vertebrates and invertebrates; in the vertebrate host the organism is intracellular, has no flagellum, and is ovoid in shape, with a diameter of 2 to 5 microns, and is called a leishman body; in the invertebrate vector the parasite usually is elongated and has a flagellum and undulating membrane; L. donovani causes kala azar, L. tropica causes Oriental sore, and L. braziliensis causes American leishmaniasis (espundia); a genus of blood-sucking flies, Phlebotomus, is the transmitter of these diseases.

leishmaniasis. Infection with a sp. of Leishmania.

lelwel. Type of tropical African hartebeest; horns very large.

Lembadion. Genus of oval fresh-water holotrich ciliates.

Lembos. Genus of marine Amphipoda.

lemming. See LEMMUS and DICROSTONYX.

lemming mouse. See SYNAPTOMYS.

Lemmus. Genus of small, herbivorous, circumpolar rodents in the Family Cricetidae; common brown lemmings; 4 to 6 in. long; thickset spp. with small ears, very short tail, and soles of feet hairy; burrow and make runways in grassy areas; populations cyclic and often exhibiting mass migrations.

Lemnaphila scotlandae. Shore fly in the Family Ephydridae; larvae mine in aquatic plants.

Lemniscomys. Genus of West African white-spotted mice.

lemniscus. 1. One of two long spongy bodies attached internally in the neck region of acanthocephalans and hanging free in the pseudocoel; they are presumed to act as reservoirs for lacunar fluid when the proboscis is retracted into the body. 2. Any band of fibers, especially nerve fibers. Pl. lemnisci.

lemon shark. See NEGAPRION.

Lemur. See LEMUROIDEA.

Lemuridae. Family of true lemurs in the Suborder Lemuroidea.

Lemuroidea. Suborder of the mammal Order Primates which includes the lemurs; characters superficially between those of true monkeys and squirrels; tail usually long but not prehensile; eyes large; nocturnal or crepuscular, sluggish during daylight; omnivorous; Africa, Madagascar, and the southeastern Asiatic area; Lemur typical.

lenitic. Lentic.

lens. Biconvex transparent structure used for focusing light in eyes or eyelike structures; it is best developed and most versatile in mammals in which a sharp image is focused upon the retina; in most invertebrates it merely serves to concentrate light on a receptor area without the formation of an image.

lens placode. Ectodermal plate from which the embryonic vertebrate eye lens develops.

lens vesicle. Thickened area of ectoderm on either side of the embryonic head of vertebrates; produces lens of the eye.

lentic. Collectively, pertaining to the standing-water series; includes lakes, ponds, and swamps.

lenticular. Having the shape of a double-convex lens.

lenticular aperture. Minute pore connecting the early invaginated lens vesicle to the outside, as in a chick embryo.

Leodice fucata. West Indian palolo worm; a polychaete which has a mating swarm in the marine shallows during the third quarter of the June-July moon.

Leontocebus. Genus of maned marmosets; pelage long, silky, and golden or dark gray; tusks large; southeastern Brazil.

leopard. Large carnivore in the cat family; common sp. is Panthera pardus; smaller than the tiger; usually tawny with black spots, but many other varieties are known, including all black; chiefly tree-living and nocturnal; Africa and southern Asia but one variety is found as far north as Siberia; sometimes placed in the genus Felis. Cf. PANTHER.

leopard cat. Ocelot.

leopard frog. See RANA PIPIENS.

leopard moth. See ZEUZERA.

leopard seal. Sea leopard.

leopard shark. Any of several mottled or spotted sharks.

Lepadella. Genus of small depressed loricate rotifers.

Lepas. Genus of barnacles characterized by a long fleshy basal stalk; goose or gooseneck barnacles.

Lepeophtheirus. Genus of marine copepods in the Suborder Caligoida; body depressed and highly modified; parasitic on fishes.

Lephthyphantes. Common genus of web-building spiders in the Family Linyphiidae.

Lepidametria. Genus of polychaetes; often commensal in tubes of Amphitrite.

Lepidasthenia. Genus of Polychaeta.

Lepidoblepharis. Genus of lizards; one Panama sp. is only 2 in. long.

Lepidocaris. Genus of minute, blind, fresh-water phyllopod crustaceans of Middle Devonian times.

Lepidocentroida. Order of echinoids; test flexible, with overlapping plates; teeth grooved; Paleodiscus (fossil) and Phormosoma typical.

Lepidochiton. Genus of chitons.

Lepidodermella. Genus of fresh-water gastrotrichs.

Lepidomeda. Spinedace; small genus of minnows found in Ariz., Nev., and Utah; scales present but small.

Lepidomuricea. Genus of marine coelenterates in the Order Alcyonacea; colony highly branched and fanlike; temperate to cold seas.

Lepidonotus. Genus of elongated marine polychaete worms in which the dorsal surface is covered with large overlapping scales.

Lepidopa myops. Mole crab, a burrowing decapod of West Coast sandy beaches; appendages short, flat.

Lepidoptera. Very large and important order of insects which includes the moths and butterflies; appendages covered with microscopic overlapping scales; eyes large, antennae long, four membranous wings with few cross veins; mouth parts of adult modified into a long coiled sucking proboscis; complete metamorphosis; larva a terrestrial caterpillar, with biting mouth parts, three pairs of thoracic legs, and two to four pairs of abdominal prolegs; pupa exposed or covered by a silken cocoon; more than 130,000 spp.

Lepidopus. See SCABBARD FISH.

Lepidosaphus. See OYSTERSHELL SCALE.

Lepidosauria. One of the six subclasses of reptiles; lizards, snakes, and related forms.

Lepidoscelis. Genus of small wasplike hymenopterans; females ride about on the body of female grasshoppers, and when the latter lays her eggs, Lepidoscelis oviposits in the grasshopper eggs.

Lepidosiren. See LEPIDOSIRENIDAE.

Lepidosirenidae. Family of lungfishes; scales small, fins filamentous; when swamps and stagnant bodies of water dry up during the dry season they burrow into the mud and use the lung in getting air through an aperture at the surface of the mud; Lepidosiren in South American tropics; three spp. of Protopterus in tropical Africa.

Lepidosteus. Lepisosteus.

Lepidostoma. Genus of Trichoptera.

lepidote. Covered with very small scales.

lepidotrichia. Elongated supporting rays of the fins in higher bony fishes. Sing. lepidotrichium.

Lepidurus. Common genus of tadpole shrimps.

Lepilemur. Genus of weasel-lemurs; nocturnal forest spp. which build large nests in trees; eastern Madagascar.

Lepisma saccharina. Common insect in the Order Thysanura; "silverfish"; sometimes a minor pest when it feeds on starch in books, paper, and clothing.

Lepisosteidae. The only family of fishes

in the Order Ginglymodi; includes the gars.

Lepisosteiformes. Taxon equivalent to the fish Order Ginglymodi.

Lepisosteus. Small genus of N.A. bony fishes in the Order Ginglymodi; gar pikes; jaws and face extended forward into a long beak supplied with sharp conical teeth; to 6 ft. long; warm lakes and streams, especially in eastern half of U.S.; voracious; L. osseus is the northern longnose gar.

Lepomis. Genus of N.A. fresh-water centrarchid fishes generally called sunfishes; often brightly colored; many important game spp.; includes the bluegill (L. macrochirus), pumpkinseed (L. gibbosus), green sunfish (L. cyanellus), longear sunfish (L. megalotis), redear sunfish or shellcracker (L. microlophus), and the spotted bream or stumpknocker (L. punctatus).

Lepophidium. See OPHIDIIDAE.

Leporidae. Family of herbivorous mammals in the Order Lagomorpha; hares and rabbits; ears long, hind legs adapted for jumping, tail short; Lepus, Sylvilagus, and Oryctolagus typical. See HARE and RABBIT.

Lepospondyli. Amphibian taxon (subclass) sometimes used to include the salamanders and the Gymniophiona; with direct deposition of bone around the notochord.

Lepraea. Genus of tube-building Polychaeta.

Lepralia. Genus of encrusting marine Bryozoa.

leprosy. Chronic disease presumably caused by the bacterium Mycobacterium leprae; marked by granular skin, mucous membrane lesions, and degeneration of nervous tissue; method of transmission obscure; tropical and subtropical; about 500 lepers are known in the U.S.

Leptasterias. Common genus of northern sea stars.

Leptestheria. Genus of clam shrimps found in ponds in the southwestern states.

Lepthemis. Genus of dragonflies.

Leptinotarsa decemlineata. Colorado potato beetle; common yellow and black striped beetle; now of wide distribution; grubs feed on leaves of potato and other plants.

Leptocardii. Cephalochordata.

Leptocella. Genus of Trichoptera.

leptocephalus. Small, translucent, leaf-shaped larva of fresh-water eels (Anguilla); migrates from the Sargasso breeding areas to continental fresh waters of N.A. and Europe; after reaching a length of about 60 mm., it becomes more slender and elongated and is known as an elver. Pl. leptocephali.

Leptocheirus. Genus of marine Amphipoda.

Leptochelia. Common genus of marine Isopoda.

Leptochiton. Genus of chitons.

Leptoconops. Genus of minute biting flies in the Family Ceratopogonidae; no-see-ums or punkies; especially common in southwestern states; larvae and pupae aquatic.

Leptocottus armatus. Small sculpin in the Family Cottidae; found along the coast from Alaska to Baja California; occasionally enters coastal streams.

Leptocuma. Genus of crustaceans in the Order Cumacea.

Leptodactylidae. Family of amphibians which includes a group of frogs without parotid glands and in which the toes have no intercalary bones or cartilages; eggs often with terrestrial development; N.A., South America, Africa, Australia, and New Guinea; Leptodactylus, Eleutherodactylus, and Syrrophus in the U.S.

Leptodactylus. Genus of Tropical American toads; see WHITE-LIPPED FROG.

Leptodea. Genus of unionid mollusks; eastern half of U.S.

Leptocerus americanus. Genus of Trichoptera; larval case a silken cone.

Leptodeira annulata. Cat-eyed snake, a nocturnal colubrid found from southern Tex. to South America.

Leptodesmus. Genus of millipedes.

Leptodora. Genus of carnivorous limnetic water fleas in the Order Cladocera; to 18 mm. long.

Leptolucania ommata. Ocellated killifish of Ga. and Fla. swamps.

Leptomedusae. Suborder of the Coelenterata Order Hydroida; hydrotheca present, gonophores in gonothecae; medusa with statocysts; Obelia, Sertularia, Plumularia, and Polyorchis typical. Cf. CALYPTOBLASTEA.

leptomeninx. Pia-arachnoid membrane.

Leptomonas. Genus of small monoflagellate protozoans parasitic in various invertebrates.

Lepton. Coin shells; genus of very small and symmetrical bivalves.

Leptonychotes. Weddell's seal; common shore seal of extreme southern South America and islands of the seas surrounding Antarctica; pale gray spotted with yellow or white above, pure yellow below.

Leptopecten. See SCALLOP.

leptopel. Finely particulate suspended and dispersed organic and inorganic particles in any natural water; includes only dead materials although bacteria are inseparable from it.

Leptophis. Genus of slender harmless tree snakes; Central and South America.

Leptophlebia. Common genus of insects in the Order Ephemeroptera.

Leptoplana. Notoplana.

Leptoptilus. Genus of storks which includes the gray and white adjutant stork of India and southeastern Asia, the smaller Java stork of southeastern Asia, and the marabou stork of tropical Africa.

Leptorhynchoides. Genus of Acanthocephala; adult parasitic in fishes, larval stages in amphipods.

Leptospira. Genus of motile spirochaetes (bacteria) in which the shape is that of a tightly-wound spiral; ends of the cell especially flexible; found in sewage and natural waters; several spp. cause minor diseases of man and domestic animals.

Leptospondyli. Order of extinct stegocephalian amphibians; small spp. resembling salamanders or eels; Carboniferous and Permian; Lysorophus and Diplocaulus typical.

Leptostraca. One of the two divisions in the crustacean Subclass Malacostraca; small shrimplike forms living in marine algae in the littoral zone; seven abdominal segments; most of body covered by a bivalve carapace; common spp. 10 to 15 mm. long; contains a single order, the Nebaliacea; Nebalia the most common genus.

Leptosynapta. Synapta.

leptotene stage. Preliminary stage of meiosis during which the chromosomes first become apparent as individual beaded filaments in the nucleus just before their union in synapsis.

Leptothrips. Common genus of predaceous insects in the Order Thysanoptera.

Leptothrix. Genus of sheathed bacteria found especially in iron-rich waters; oxidize ferrous to ferric iron.

Leptotyphlopidae. Small family of harmless burrowing snakes found in south-central and southwestern U. S. to Argentina, and Africa and southwestern Asia; worm snakes, blind snakes; teeth only in lower jaw; vestiges of pelvic girdle and femora present internally; Leptotyphlops representative.

Leptotyphlops. See LEPTOTYPHLOPIDAE.

Leptychaster. Widely distributed circumboreal genus of small sea stars in the Order Phanerozonia.

Lepus. Common and widely distributed N. A. genus of hares and rabbits; L. americanus is the snowshoe hare which changes coloration with the seasons; L. californicus is the large western jackrabbit with rapid gait and very long ears.

Lernaea. Genus of fresh-water copepods parasitic on fishes; elongated and wormlike.

Lernaeenicus. Genus of marine copepods in the Suborder Caligoida; adults greatly elongated and thin; anterior portion buried in tissues of fish host.

Lernaeopodoida. Suborder of copepods; adults all parasitic on marine and fresh-water fishes; female sessile and enormously modified, often with no trace of segmentation and few or no paired appendages; body commonly variously lobed, sometimes having the appearance of a tumorous growth; male a pygmy which crawls about on the body of the female.

Lernanthropus. Genus of copepods parasitic on the gills of marine fishes.

lesion. Any injury or any structural abnormality resulting from disease.

lesser omentum. Omentum between liver and stomach.

lesser panda. See AILURUS.

lesser rorqual. Small whalebone whale, Balaenoptera acutorostrata, found in North Atlantic; to 30 ft. long.

lesser yellowlegs. See TOTANUS.

Lestes. Common genus of damselflies in the Order Odonata.

lethal factor. A gene which stops development at some immature point and causes death when present in the homozygous dominant condition (usually); e. g. if two heterozygous yellow (Yy) mice are bred, the resulting living offspring will occur in the ratio of 2Yy (yellow) to 1yy (non-yellow), the YY (yellow) individuals having died in the fetus stage.

lethal gene. See LETHAL FACTOR.

Lethocerus. Common genus of giant water bugs in the hemipteran Family

Belostomatidae.

lettered olive. Oliva litterata, a slender olive gastropod to 3 in. long; covered with fine markings; warm American marine waters.

Leucania unipuncta. Important moth in the Family Noctuidae; green, brown, and yellow caterpillar known as armyworm; feeds on alfalfa, small grains, and corn; sometimes in enormous numbers.

Leucauge. Genus of spiders.

Leucichthys. Large and difficult genus of fresh-water fishes in the Family Coregonidae; chubs, ciscos, lake herring, kiyi, bloater, tullibee, etc.; in N. A. found mostly in cold waters of northeastern states and Canada.

leucine. Common amino acid; $C_6H_{13}O_2N$.

Leuckart, Karl, G. F. Rudolph. German zoologist (1823-1898); founder of animal parasitology; made numerous fundamental contributions to comparative invertebrate zoology.

leucocyte. Any colorless nucleated corpuscle found in blood or other body fluid; human blood usually contains 6,000 to 10,000 leucocytes per cu. mm.

leucocytosis. Abnormally high white blood cell count, usually the result of infection or tissue damage.

Leucocytozoon. Genus of Sporozoa occurring in endothelium and visceral cells of vertebrates, especially birds; transmitted by blood-sucking insects.

leucon. Most complex type of canal system found in sponges; canals extensively branched, the choanocytes being restricted to many small internal chambers; in commercial and fresh-water sponges.

Leuconia. Genus of very small tapered calcareous marine sponges.

leucopenia. Pronounced reduction in leucocytes in the blood.

Leucophrys. Genus of pyriform fresh-water holotrich ciliates.

Leucorrhinia. Genus of dragonflies; northern U. S.

Leucosolenia. Genus of sponges in the Class Calcarea; consists of finger-shaped masses rising from a basal portion.

Leucosomus corporalis. Fallfish; small fish in the minnow family; streams and lakes of northeastern U. S.

Leucosticte. See FINCH.

Leucothea. Genus of tropical ctenophores with large oral lobes.

Leucothoë. Genus of marine Amphipoda.

Leucotrichia. Genus of Trichoptera; larval case translucent, ovoid, and flat.

Leucozonia. Genus of small robust gastropods; on rocks in Caribbean.

Leuctra. Common genus of small Plecoptera.

leukemia. Fatal blood disease characterized by a great increase in leucocytes and growth of the spleen, lymph glands, and bone marrow; accompanied by anemia and internal hemorrhaging.

leukocyte. Leucocyte.

Leuresthes tenuis. Grunion; a small silvery fish of the Calif. coast; spawns in wet sand at the water's edge at high tide on second, third, and fourth nights after the spring tide.

Leurognathus marmorata. Shovel-nosed salamander; a completely aquatic sp. in the Family Plethodontidae; to 90 mm. long; southern Appalachian area.

levator. Muscle that raises a part of the body.

levator palatoquadrati. Muscle which elevates the upper jaw in sharks.

levator palpebrae superioris. Small muscle which raises the upper eyelid; in most amniotes.

levator scapulae. Muscle which originates on the transverse processes of cervical vertebrae and inserts on the posterior edge of the scapula; raises upper angle of scapula and rotates head.

levulose. Simple sugar occurring especially in honey; fruit sugar or fructose; $C_6H_{12}O_6$.

Lewis's woodpecker. See ASYNDESMUS.

Lexingtonia. Genus of U. S. fresh-water bivalve mollusks in the Family Unionidae.

Leydigia. Uncommon genus of littoral Cladocera.

l-glucose. See GLUCOSE.

LH. luteinizing hormone secreted by the anterior pituitary; promotes ovulation and the formation of the corpus luteum in the ovary of mammals.

Liacarus. Common genus of mites in the Family Oribatidae.

Libellula. Common genus of dragonflies.

Libellulidae. Large and widely distributed family of brightly colored dragonflies; wings frequently partly clouded; abdomen triangular in cross section.

libido. 1. Instinctive or psychic excitation. 2. Sexual energy. 3. General energy, drive, and motivation.

Libinia. Genus of decapods; includes the common spider crabs; legs long, body compact and hairy, more or less covered with algae, hydroids, etc., and usually 1 to 3 in. long; especially abundant on mud flats, in kelp, and in tide pools.

Lichanotus. Genus of nocturnal silky lemurs; avahis; Madagascar.

Lichanura. See ROSY BOA.

Lichenopora. Common genus of marine bryozoans; colony calcareous, thin, discoid, and laminate.

Lichnophora. Genus of heterotrich ciliate protozoans; free-swimming or commensal in marine or fresh-water animals; cell discoid, with an attachment disc, flattened neck, and several concentric ciliary coronas.

Lieberkuhnia. Genus of marine and fresh-water amoeboid protozoans in which the animal is largely enclosed by an ovoid or spherical test.

life belt. Ecological altitudinal zone, as in mountainous areas.

life cycle. 1. Series of morphological changes and activities of an organism from time of zygote formation until death. 2. Series of morphological stages in an organism exhibiting alternation of generations and associated sexual and asexual reproduction; the complete cycle begins with, e.g., a zygote in one stage and proceeds through all stages until a return to the next zygote stage; same as life history.

life form. 1. Characteristic vegetative appearance of a mature sp., e.g. tree, shrub, herb, etc. 2. Among animals, a characteristic way of living plus a group of modifications for that way of living; e.g. the following life forms may be recognized in steppe habitats: subterranean zoophagous burrowers, subterranean phytophagous burrowers, herbivorous burrowers which feed above ground, phytophagous runners, zoophagous runners, and phytophagous jumpers.

life history. Life cycle.

life table. Statistical table showing, for a particular sp., such data as age groups, mortality rate, survival rate, and life expectation.

life zones. Series of latitudinal and altitudinal biological zones (especially in N.A.) based chiefly on temperature characteristics; C. H. Merriam (1898) divided the U.S. into the following life zones: Arctic-Alpine, Hudsonian, Canadian, Transition, Upper Austral, Lower Austral, Gulf Strip of Lower Austral, and Tropical; in a very general way, these zones have characteristic plants and animals. See BIOTIC PROVINCE.

ligament. 1. Band of inelastic whitish fibrous tissue attaching one bone to another. 2. Longitudinal strand of connective tissue in the pseudocoel of some Acanthocephala; holds certain visceral organs in place.

ligamentum nuchae. Median dorsal neck ligament in many quadrupeds; aids in supporting the head; rudimentary in man.

ligamentum patellae. Flat ligament extending from the lower margin of the patella to the tibia; about 3 in. long.

ligate. To tie with a ligature.

ligature. Thread or wire for tying a vessel or strangulating a part.

lightning beetle. Firefly.

lightning bug. Firefly.

lightning shell. Busycon perversum, a left-handed whelk found from N.C. to Tex.

Ligia. Genus of marine Isopoda.

Ligidium. Genus of terrestrial Isopoda.

lignin. Cyclic unsaturated organic compound found especially in woody plant tissues where it often amounts to 20 35% of the cell walls.

ligula. 1. Collectively, the terminal lobes of an insect labium. 2. Band of white matter in the dorsal wall of the fourth ventricle. 3. Lobe of a parapodium in certain Polychaeta.

Ligula. Genus of tapeworms in the Order Pseudophyllidea; larval stages parasitic in fresh-water copepods and fishes; adults in the intestine of fish-eating birds.

ligule. Small lobe of a polychaete parapodium.

Ligumia. Genus of large unionid mollusks; eastern half of U.S.

Liguus. Genus of brightly-colored tropical tree snails.

Ligyda. Common genus of large active isopod crustaceans; along seashores, especially among rocks and on piling.

Lillie, Frank R. American zoologist and educator (1870-1947); best known for his embryology text and as teacher and administrator at the University of Chicago, Woods Hole Oceanographic Institution, and the Marine Biological Laboratory at Woods Hole.

Lima. File shells, a genus of marine bivalve mollusks having the valves markedly ridged.

Limacina. Genus of pelagic sea butterflies having a thin coiled shell; abundant in Arctic seas near the surface.

Limacomorpha. Order of tropical millipedes; 22 body segments; last one or two pairs of legs are male gonopods; Glomeridesmus typical.

Limapontia. Genus of nudibranchs.

Limax. Genus of land slugs found in gardens, moist woodlands, etc.; shell reduced to a thin calcareous plate which is completely imbedded in the small mantle; several American spp.

limb bud. One of four simple budlike outgrowths of an embryo vertebrate; first appearance of a limb.

lime cells. Microscopic hyaline spherical or oval bodies in the tissues of some tapeworms; composed of calcium carbonate and secreted by mesenchyme cells; not true cells.

limicoline. Shore-inhabiting, especially with reference to shore birds such as plovers and sandpipers.

limicolous. Living in mud, especially the mud of the seashore.

limiting factor. 1. Law of toleration, law of the minimum; when one environmental factor is near the limits of toleration, that one factor will be the controlling one and will determine whether or not a sp. will be able to maintain itself. 2. Any factor which limits the rate of a physiological process.

limivorous. Mud-eating, as earthworms and certain aquatic animals.

Limnadia. Genus of clam shrimps; rare in U.S.

Limnaoedus ocularis. Small tree frog in the Family Hylidae; to 18 mm. long; southeastern states.

Limnatis nilotica. Small African freshwater blood-sucking leech; other spp. in this genus are larger.

Limnephilus. Common genus of insects in the Order Trichoptera; the aquatic larva builds a conical case of sand and other materials.

Limnesia. Common genus of Hydracarina.

limnetic. Pertaining to the open waters of lakes, especially areas too deep to support rooted aquatic plants.

Limnetis. Lynceus.

Limnias. Genus of sessile rotifers; most of body hidden within a secreted chitinoid tube.

Limnebius. Genus of small aquatic beetles in the Family Hydraenidae.

Limnius. Genus of elmid beetles.

limnobiology. 1. In the broad sense, fresh-water biology. 2. In the narrow sense, lake biology.

Limnocalanus macrurus. Large limnetic copepod found in a few large northern lakes; thought to be a marine relict.

Limnochares. Common genus of creeping Hydracarina.

Limnocnida. Genus of fresh-water hydrozoan "jellyfish" found chiefly in India and central African lakes.

limnocrene. Spring in which water emerges from the ground and forms a pool.

Limnocythere. Common genus of freshwater Ostracoda.

Limnodrilus. Genus of fresh-water Oligochaeta.

limnodromous. Pertaining to fishes that move about freely in lakes; e.g. onshore and offshore for feeding or breeding.

Limnodromus griseus. Dowitcher; a long-legged shore bird common on beaches, marshy shores, and open meadows near water.

Limnogale. Genus of water-tenrecs; rat-sized mammals with webbed feet and long compressed tail; Madagascar streams.

Limnogonus. Small genus of water striders in the hemipteran Family Gerridae.

limnology. 1. In the broad sense, the study of all types of fresh waters and their inhabitants. 2. In the narrow sense, the study of all types of standing fresh waters and their inhabitants.

Limnomedusae. Suborder of the coelenterate Order Hydroida with alternation of generations; polyp sessile, with or without tentacles; medusae with velum and hollow tentacles; Gonionemus and Craspedacusta typical.

Limnophila. Genus of crane flies.

Limnophora. Genus of flies in the Family Anthomyiidae; larvae aquatic.

limnoplankton. Lake plankton.

Limnoria lignorum. Marine isopod which burrows into submerged timbers and piling; causes great damage to docks; the gribble.

Limnoscelis. Genus of primitive, unspecialized Permian reptiles.

limnosere. Sere which begins in a large (usually cold) lake.

Limnothlypis swainsoni. Swainson's warbler; brown above and dirty white tinged with yellow below; southeastern quarter of U.S.

Limonia. Genus of crane flies.

Limonius. See ELATERIDAE.

Limosa. Common genus of shore birds in the Family Scolopacidae; godwits; very long, slightly upcurved bill; coloration mostly reddish or brownish; shores of fresh and salt waters in North and South America.

limpet. Type of gastropod mollusk in which most of the body is covered with a low conical tentlike shell that has little or no evidence of coiling; mostly marine; Acmaea especially common.

limpkin. See ARAMIDAE.

Limulus. See XIPHOSURA.

limy. Consisting chiefly of calcium carbonate.

Linckia. Genus of red Pacific Coast sea stars.

Lincoln index. Estimate of total population of animals in a habitat based on capture, marking, releasing, and recapture of a certain percentage of the marked individuals.

Lincoln's sparrow. Melospiza lincolni, a sp. breeding in deciduous thickets of northern coniferous forests.

Lindia. Uncommon genus of marine and fresh-water rotifers; body cylindrical; closely associated with substrates.

linea alba. Whitish median line of tendon in the abdominal and lower chest region of man; serves as an attachment for oblique and transverse abdominal muscles.

linea aspera. Rugose line representing a vestigial adductor crest on the femur of mammals.

linea nigra. Faint dark-colored line running down the center of the abdomen after pregnancy in human beings; replaces the linea alba.

line breeding. Domestic animal breeding involving an intensification and continuity of the contributions of especially desirable individuals to a stock while keeping the contributions of contemporary individuals at a minimum.

lined snake. See TROPIDOCLONION.

Lineidae. Family of nemertine worms having a pilidium stage in the life history.

Lineus. Genus of nemertines in the Order Heteronemertea; usually in bottom materials along marine shores; noted for their great powers of regeneration when fragmented.

ling. See GADIDAE.

lingua. 1. Tongue or tongue-like organ. 2. Glossa. 3. Coiled mouth parts of a butterfly or moth.

lingual. Pertaining to the tongue.

lingual artery. External carotid artery.

lingual cartilage. Cartilaginous skeletal structure imbedded in the base of the tongue in certain lower vertebrates.

lingual ribbon. Radula.

Linguatula. See LINGUATULIDA.

Linguatulida. Small phylum which includes the tongue worms; soft, wormlike segmented parasites of the respiratory ducts of vertebrates; to 4 in. long; anterior end ("head") with two pairs of ventral hooks near the mouth; trunk with numerous annuli; no circulatory, respiratory, or excretory systems; larva vaguely resembles an immature mite, and the group is sometimes therefore included with the arachnids; about 80 spp.; Linguatula is common in the respiratory system of many wild and domestic animals.

Lingula. Genus of modern brachiopods which has persisted unchanged since Cambrian times; common in the Pacific and Indian oceans; peduncle long and contractile; mostly in vertical burrows in the sea bottom.

Lingulella. Genus of primitive fossil brachiopods.

linin. Fibrillar network of a cell nucleus which does not take a chromatin stain.

linkage. Tendency of certain alleles to be inherited together, or collectively, owing to the fact that they are located on the same chromosome.

linkage groups. Occurrence of alleles in groups which tend to be inherited together, or collectively, owing to the fact that they are located on the same chromosome; thus, an organism has as many linkage groups as it has pairs of chromosomes in the body cells. See CROSSING OVER and MEIOSIS.

linkage map. Genetic map; diagram showing the relative linear positions of genes on the chromosomes; most thoroughly worked out for Drosophila.

Linnaeus, Carolus. Swedish botanist (1707-1778); founder of modern systematics and the first to use binomial nomenclature consistently for both plants and animals; the tenth edition (1758) of his greatest work, Systema Naturae, is taken as the starting point for the binomial nomenclature used today; he recognized classes, orders, genera, and spp., and each of the many plants and animals considered was given a genus and sp. name along with a definitive diagnosis.

linnet. Small Old World finch.

Linognathus. Common genus of sucking lice (Anoplura); parasitic on sheep, goats, dogs, and rabbits.

linoleic acid. Linolic acid; unsaturated fatty acid required in the diet of many insects and mammals; abundant in linseed oil and some animal tissues; $C_{18}H_{32}O_2$.

linolenic acid. Unsaturated fatty acid especially abundant in linseed oil; $C_{18}H_{30}O_2$.

linolic acid. Linoleic acid.

Linotaenia. Common genus of small centipedes.

linsang. Long-tailed catlike mammal in the Family Viverridae; Asia, Africa, and the East Indies; more or less arboreal; related to the civets; also called the delundungs; Prionodon gracilis is the most common form; Pardictis is a large Burmese linsang; Poiana is a West African genus.

Linuche. Genus of scyphozoan jellyfish in the Order Coronatae; tropical and subtropical oceans.

Linyphia. See LINYPHIIDAE.

Linyphiidae. Sheet-web weavers; a family of small spiders which build a horizontal dome-shaped or curved sheetlike web in vegetation and litter; Linyphia marginata common.

Liobunum. Common genus of daddy longlegs.

Liocyma fluctuosa. Wavy carpet shell; a small, thin, white marine bivalve; Greenland to Nova Scotia.

Liodytes alleni. Striped swamp snake in the Family Colubridae; stout aquatic sp. to 25 in. long; southern Ga. and Fla.

lion. 1. Large familiar carnivore (Panthera leo) in the cat family; tail with a tuft; male with a shaggy mane; found in Africa and occasionally in extreme southwestern Asia; sometimes included in the genus Felis. 2. Locally, the cougar.

lionfish. See SCORPAENIDAE.

Lionotus. Genus of elongated, flask-shaped, ciliate protozoans with a necklike anterior region; fresh and salt water.

lion's paw. See LYROPECTEN.

Liopelma. Genus of grayish primitive frogs found in New Zealand; two spp.; in the Family Ascaphidae (Order Amphicoela).

Liopelmidae. See ASCAPHIDAE.

Lioplax. Genus of fresh-water snails; east of Mississippi River.

Liparidae. Family of marine fishes which includes the sea snails or snailfishes; small, tadpole-shaped, soft-bodied spp. with a smooth skin and a ventral sucking disc with which they attach to kelp or stones; cold and cold-temperate seas; Neoliparis and Liparis common.

Liparis. See LIPARIDAE.

Liparocephalus. Genus of staphylinid beetles.

lipase. Digestive enzyme which hydrolyzes fats to fatty acids and glycerol.

Lipeurus. Genus of bird lice in the Order Mallophaga; especially common on domestic fowl.

lipid. One of a large variety of fats and fatlike compounds occurring in living organisms; all are insoluble in water and soluble in ether and hot alcohol; in addition to true fats, they include sterols, phospholipids, steroids, carotenes, and terpenes.

lipide. Lipid.

lipin. Lipid.

lipoblast. Mesenchymal precursor of a fat cell.

lipocaic. Hormone elaborated by the pancreas; prevents fat deposition in the liver; not critically established.

lipoid. Lipid.

Liponyssus. Genus of mites parasitic on rats and fowl; occasionally some spp. infect man.

lipophore. Chromatophore containing red or yellow pigment granules.

lipoprotein. Conjugated protein containing a lipid group in the molecule.

Liposcelis divinatorius. Death watch, book louse, or cereal psocid; a wingless insect in the Order Corrodentia; 1 mm. long; feeds on cereals, books, organic debris, old wood, molds, etc. and sometimes is an important pest; produces a faint ticking sound by striking the ventral surface against paper or dry wood.

Lipostraca. Small order of Crustacea in the Division Eubranchiopoda; minute, blind, fresh-water spp.; antennae biramous; males with maxillae adapted for clasping; trunk appendages without branchiae, first three pairs modified for feeding; Middle Devonian; Lepidocaris typical.

Lipotes. Small fresh-water dolphin; a mammal restricted to Tung Ting Lake on the Yangtse River; bluish gray above, white below; beak upcurved and about a foot long; to 7 ft. long and 300 lbs.

Lirceus. Genus of fresh-water isopods; common in U.S.

Liriope. Genus of marine hydrozoan coelenterates; common and widely distributed; medusa large and free-swimming; polyp stage absent.

Lissodelphis borealis. Right whale dolphin; a black dolphin of the North Pacific Ocean; to 8 ft. long.

Lissodendoryx. Genus of marine sponges in the Subclass Monaxonida; growth form variable; temperate to cold seas.

Lissorhoptrus simplex. Rice water weevil; larvae feed on roots and adults on leaves of the rice plant and many spp. of wild aquatic plants.

Lister, Joseph. English surgeon (1827-1912); developed aseptic surgery and antisepsis in general; first used absorbable ligatures and drainage tubes; perfected many new types of operations.

Listriolobus. Rare genus of small to medium-sized echiurid worms; proboscis long, soft, sometimes deciduous.

Listronotus. Genus of aquatic weevils; feed on aquatic vegetation.

Lithacolla. Genus of spherical, amoeboid, fresh-water Protozoa; radiating pseudopodia; gelatinous envelope covered with sand grains and diatoms.

Lithadia. Genus of small, stony-looking brachyuran marine crabs.

Lithasia. Genus of fresh-water snails; Ken., Tenn., Ala., and Ind.

lithite. Same as statolith; see STATOCYST.

Lithobiomorpha. Order of centipedes; 15 pairs of short legs, spiracles lateral; 3 to 30 mm. long; Lithobius common.

Lithobius. Common genus of centipedes.

Lithocircus. Genus of radiolarians.

lithocyst. Essentially same as statocyst; enclosed granules usually calcium sulfate crystals.

lithocytes. See STATOCYST.

Lithodes. Stone crabs; a genus of spiny marine decapod crustaceans; typically crablike but with a small, asymmetrical, bent-under abdomen.

Lithodomus. Lithophagus.

Lithophaga. Widely distributed genus of marine mussels; shell cylindrical, inflated, wedge-shaped posteriorly.

lithophagous. 1. Stone-eating or gravel-eating, as certain birds. 2. Rock-burrowing, as certain mollusks and echinoids.

Lithophagus. Genus of marine mussels which are able to bore slowly into a rock so that an old individual rests in a cylindrical depression.

lithosere. Succession developing on exposed rock surfaces.

lithosphere. First portion of the evolution of the earth and life; includes only that period before the appearance of living protoplasm; followed by the biosphere period.

Litocranius. Gerenuks; a genus of antelopes of east African scrublands; legs long, neck very long; male with small horns.

Litopterna. Order of extinct early South American mammals; horselike and camel-like; Paleocene and Pleistocene.

Litorina. Littorina.

Litosomoides. Genus of filariid nematodes; most common sp. (L. carinii) is a parasite of the cotton rat.

little auger shell. Terebra dislocata, a common gastropod with a very long, slender spire; shallow sand bars from Va. to Tex.

little brown bat. See MYOTIS.

little chief hare. See OCHOTONIDAE.

little chink. See PUNCTURELLA.

little moon shell. Natica clausa, a pale brown snail found over a wide depth range in the North Atlantic.

little neck clam. See VENUS and PROTOTHECA.

little olive. Oliva mutica, a small olive shell to 0.5 in. long; marine, N.C. to Tex.

little sun shell. Margarites obscura, a very small marine snail of the Atlantic Coast.

little surf clam. See MULINIA.

littoral. In the sea, that shallow portion of the bottom extending from the shoreline to a depth of 200 m.; in lakes, that shallow portion of the bottom from the shoreline to the lakeward limit of rooted aquatic growths; the term is also used to include both the bottom and the water above the bottom at the depths indicated. (Same spelling for both noun and adj.)

littoral cells. Large phagocytic cells lining the sinusoids of bone marrow.

Littoral-Mediterranean. Race of caucasoids native to the lowlands from western Ireland to Spain and around much of the periphery of the Mediterranean; skin tawny whitish, hair black, face long.

Littorina. Marine gastropod genus which includes the periwinkles; small, heavy-shelled snails of the intertidal zone; many spp., some edible; widely distributed.

live bearers. See POECILIIDAE.

liver. Large digestive gland occurring in many invertebrates and vertebrates; in some vertebrates the liver has a secondary excretory function, and also removes foreign particles from the blood and lymph; usually the largest gland in the body.

liver fluke. See CLONORCHIS and FASCIOLA.

liver rot. Disease of sheep, elk, etc.; fluke infestation of the liver.

living fossil. Any sp. or genus which has persisted, unchanged, for a very long time in geological history; e.g. Limulus, common oyster, Amia, Latimeria.

Livoneca. Genus of isopods parasitic on marine fishes.

lizard. Any reptile in the Suborder Sauria.

lizard bird. See ARCHAEORNITHES.

lizard fishes. See SYNODONTIDAE and INIOMI.

llama. 1. Domesticated South American herbivorous mammal in the genus Lama; related to the camel and presumably a descendent of the guanaco; smaller than a camel, lacking a hump, and somewhat sheeplike in appearance, with a long shaggy coat; brown, black, white, or a mixture; in Andean areas used as a beast of burden and for its flesh, wool, and milk. 2. General term which includes the llama, alpaca, guanaco, and vicuña.

loach. Any of several small slender Asiatic and European fresh-water fishes having several barbels around the mouth.

loach minnow. See TIAROGA.

Loa loa. Spirurid nematode occurring in subcutaneous tissue and the human eye in equatorial Africa and the West Indies; transmitted by tabanid flies.

lobate. Provided with lobes or arranged in lobes.

Lobates. One of the orders of the ctenophore Class Tentaculata; body compressed, with two oral lobes, and four large auricles at the oral end; Mnemiopsis common.

lobe-finned fishes. Fishes in the Superorder Crossopterygii; see RHIPIDISTIA and COELACANTHINI.

Lobipes lobates. Northern phalarope; circumarctic and subarctic; winters at sea in southern oceans.

Lobodon. Genus of small seals of the Antarctic coast; feed chiefly on crabs.

Lobophyllium. Genus of corals; tentacles capable of enormous extension.

lobopodium. Finger-like or tonguelike pseudopodium; the distal end is usually rounded and not branched; common in Amoeba. Pl. lobopodia.

Lobosa. Amoebina.

Lobotes surinamensis. Tripletail, a widely distributed marine fish with compressed body and prominent dorsal and anal fins which give it the appearance of having three tails.

lobotomy. Frontal lobotomy.

lobster. Large marine crustacean used as food and included in the Suborder Macrura; with well developed pincerlike claws on the first walking legs; feed on living and dead animal food on the sea bottom; see HOMARUS and PANULIRUS.

Loch Leven trout. Brown trout; see SALMO.

lociation. Local variations in the sp. composition of a faciation.

lock-and-key theory. Concept proposed to account for the specificities of interaction between compounds, as for example between antigen and antibody, or between enzyme and substrate; it suggests that the molecular configurations of the reacting substances must have such surface contours that they will fit architecturally into each other in some particular portions of their surfaces, thus limiting the number of compounds which will interact to a specific pair or to relatively few compounds.

lockjaw. See TETANUS.

locular. Having a series of small spaces or cavities.

locus. Precise position of a gene in a chromosome. Pl. loci.

locust. A gregarious and migratory Old World member of the orthopteran Family Acrididae. Cf. GRASSHOPPER.

Locusta migratoria. Winged migratory locust of the Old World.

locust borer. See CYLLENE.

Locustidae. Tettigoniidae.

Loeb, Jacques. American physiologist (1859-1924); best known for his experimental work on artificial parthenogenesis and regeneration.

Loewi, Otto. German and American physiologist (1873-); made important contributions to the physiology and pharmacology of the heart, kidneys, and nervous system; shared the 1936 Nobel prize in physiology and medicine for his work on the chemical transmission of nerve impulses.

logarithmic phase. See LOGISTIC CURVE.

loggerhead shrike. See LANIIDAE.

loggerhead sponge. Massive Caribbean sponge; of little commercial importance.

loggerhead turtle. See CARETTA.

logistic curve. Typical sigmoid or S-shaped population growth curve; shows an initial slow growth (lag phase), rapid intermediate growth (logarithmic phase), and later slow growth (upper asymptote or stationary phase).

logperch. See PERCINA.

Lohmannella. Genus of marine mites in the Family Halacaridae.

Loimia. Genus of Polychaeta.

Loligo. Common genus of squids; body elongated and torpedo-shaped, usually less than 11 in. long; fins rhombic and united at the posterior tip of the body.

Loligopsis. Genus of squids; rare along U.S. coasts.

Lonchopteridae. Family of Palaearctic flies with wings pointed at the tips.

lone star tick. Amblyomma americanum, a common tick (Ixodidae) on man and cattle.

long-beaked dolphin. See STENO.

long clam. See MYA.

long-eared owl. See ASIO.

long-ear sunfish. Lepomis megalotis, a small sunfish of eastern U.S.; unimportant as a game sp.

longfin dace. See AGOSIA.

long-horned beetles. See CERAMBYCIDAE.

long-horned grasshopper. Any grasshopper in the Family Tettigoniidae.

longicorn beetle. See CERAMBYCIDAE.

longissimus dorsi. One of the long flat muscles in the dorsal body wall of certain vertebrates; supports the vertebral column and bends the trunk backward.

longitudinal muscle. Any discrete muscle or layer of muscle tissue whose fibers extend in a longitudinal direction; such tissue may be voluntary or involuntary.

longjaw. 1. Any of several Great Lakes ciscoes (Leucichthys). 2. Any of several marine needlefish or billfish.

long-neck clam. See MYA.

long-nosed snake. One of several burrowing snakes in the genus Rhinocheilus; back with dark patches broken by red, yellow, or white bands; 18 to 41 in.; southwestern states.

longnose gar. See LEPISOSTEUS.

longspur. See CALCARIUS.

long-tailed salamander. Eurycea longicauda longicauda, a thin salamander, 5 in. long; yellow to orange; widely distributed in the Ohio River drainage, and as far as southern N.Y. and northern Ala. and Ga.

long-tailed weasel. Mustela frenata; upper parts brown, underparts pale; body to 11 in.; generally distributed from southern Canada to Bolivia.

Longurio. Genus of crane flies in the Family Tipulidae.

lookdown. See SELENE.

Loomelania melania. Black petrel, a sooty black oceanic bird ranging from Calif. to Peru.

loon. See GAVIIFORMES.

loop of Henle. Long horseshoe-shaped bend or loop of a kidney tubule in higher vertebrates; situated between the proximal convoluted tubule and the distal convoluted tubule.

Lophaster. Widely distributed genus of sun stars in the Order Spinulosa; five arms.

Lophiidae. Family of marine bony fishes in the Order Pediculati; angler fishes, monkfishes, fishing frogs, goosefish; head extremely large and depressed; mouth enormous; body scaleless and flaccid; pectoral fins with fleshy base; first dorsal spine modified as a movable lure; to 50 lbs.; on the bottom from shallows to deep water; voracious and carnivorous; Lophius americanus common.

Lophiomys. Genus of crested hamsters in the Family Cricetidae; body twice that of a large rat; fur gray, with a prominent median crest of long hair extending from the head to the tip of the tail; arboreal leaf eaters; East Africa.

Lophius. See LOPHIIDAE.

Lophocharis. Genus of small rotifers with a boxlike sculptured lorica.

lophodont. Type of dentition in which the molar cusps fuse to form ridges, as in the horse.

Lophodytes cucullatus. Hooded merganser, an American fish-eating duck.

Lophohelia. Genus of colonial branching corals found as far north as Norwegian fjords.

Lopholatilus. See BRANCHIOSTEGIDAE.

Lophomonas. Genus of complex flagellates living as symbionts in the intestine of cockroaches; ovoid or elongated; with a dense anterior tuft of flagella.

Lophopanopeus. West Coast genus of brachyuran marine crabs; rocky shores.

lophophore. Horseshoe-shaped, three-part, coiled or variously shaped distal part of the body bearing numerous ciliated tentacles; in Bryozoa, Brachiopoda, and Phoronidea.

Lophopodella carteri. Uncommon freshwater bryozoan.

Lophopsetta. See WINDOWPANE.

Lophopus crystallinus. Common freshwater bryozoan; zooecium gelatinous; attached to vegetation.

Lophortyx californica. Valley quail or California quail; a small, plump, grayish sp. with a short plume that curves forward; open brushy and cultivated areas; native to southern Ore. and Calif. but introduced elsewhere in the West.

Lophortyx gambeli. Gambel's quail, a small, plump, chick-like quail with a short plume that curves forward; similar to the California quail but restricted to southwestern desert areas (introduced in Hawaii).

Lora. Genus of small marine snails.

lordosis. Spinal abnormality marked by an exaggerated forward convexity of the lumbar region and protruding buttocks.

lore. Space between the eye and the base of the bill in birds; sometimes also used to designate the corresponding facial area of reptiles and fishes; customarily used in the plural form (lores).

lorica. Enveloping rigid case or secreted exoskeleton in invertebrates, as in certain protozoans and many rotifers.

Loricata. 1. Order of reptiles which includes the alligators, caimans, gavials, and crocodiles; large, ponderous, amphibious, carnivorous lizard-like reptiles with powerful jaws bearing numerous conical teeth; short legs bearing clawed, webbed toes; tail long, heavy, and compressed; skin thick, reinforced with dermal plates dorsally; ear opening covered with small flap; heart four-chambered, bladder absent; eggs deposited on shore in masses of decaying vegetation; most spp. hibernate for two or three months or more of each year; about 23 spp. in Australia, tropics, and Gulf states of U.S.; genera are Alligator, Crocodylus, Gavialis, Tomistoma, Osteolaemus, and Caiman. 2. Polyplacophora.

loricate. 1. Of or pertaining to a rotifer having a lorica. 2. Having a protective case or shell.

Loripinus chrysostoma. Apricot shell, buttercup shell; a small, circular bivalve having yellow inner and white outer surface; tidal flats of Atlantic Coast.

loris. See LORISOIDEA.

Lorisidae. Family of the Suborder Lorisoidea; includes the lorises and pottos.

Lorisoidea. Suborder of mammals in the Order Primates; loris, a sluggish type of nocturnal lemur, found in southern Asia to the Philippines; also includes the galagos, pottos, and several spp. of African lemurs.

lory. Any of numerous small brightly colored birds of the Australian region; tongue with a brushlike tip; feed on soft fruits.

Lota. See GADIDAE.

lotic. Pertaining to running-water habitats.

Lottia. Genus of West Coast limpets; one sp., L. gigantea (owl limpet), reaches 3 in. and is prized as food by Mexicans.

louse. 1. Common name of members of the insect orders Mallophaga (biting lice) and Anoplura (sucking lice). 2. Any of wide variety of arthropods which suck blood or juice of various animals or plants. 3. Name applied to numerous small insects which are not parasitic, such as the book louse, bark louse, and wood louse.

louse fly. See HIPPOBOSCIDAE.

lousy. Infested with lice.

lovebird. Any of numerous small greenish African and South American parrots that show great affection for their mates; Agapornis common; see PSITTACIFORMES.

Lovenella. Genus of marine coelenterates in the Suborder Leptomedusae.

Lovenia. Genus of heart urchins.

Lower Austral zone. One of Merriam's life zones, located between the Upper Austral and Tropical zones; a broad strip from N.C. through Ga. and west through Tex. but not including a narrow strip along the Gulf of Mexico; occurs also in large areas of Calif., Nev., and Ariz.; east of 100° west longitude it is often called the Austroriparian zone, and west of 100° it is often called the Lower Sonoran zone.

Lower Sonoran zone. See LOWER AUSTRAL ZONE.

Loxia. Genus of finches having the two parts of the bill curved and crossing each other; crossbills; feed mostly on coniferous tree cones.

Loxocalyx. See LOXOSOMATIDA.

Loxocemus. Genus of dwarf pythons of southern Mexico and Central America.

Loxocephalus. Large genus of marine and fresh-water holotrich Ciliata; ovoid to cylindrical.

Loxoconcha. Genus of marine Ostracoda.

Loxodes. Genus of fresh-water holotrich Ciliata; lancet-shaped and compressed.

loxodont. Having shallow depressions between the ridges of the molar teeth.

Loxodonta. See ELEPHANT.

Loxophyllum. Genus of large marine and fresh-water holotrich ciliate Protozoa; ventral side flat.

Loxorhynchus. Genus of spider crabs and masking crabs in the Suborder Brachyura; sluggish spp., often covered with "planted" algae, hydroids, sponges, etc.; presumably this habit has a survival and camouflage value.

Loxosceles. Genus of spiders; found especially in Mo. and Ar,.

Loxosoma. See LOXOSOMATIDA.

Loxosomatida. Class of the Phylum Entoprocta; solitary and minute spp., with the stalk attached to marine invertebrates by means of an adhesive disc; Loxosoma and Loxocalyx common.

Loxosomella. Genus of marine Entoprocta.

LTH. Luteotropic hormone.

lubber grasshopper. Large, clumsy, short-winged, flightless grasshopper, especially Romalea microptera of the southeastern states and Brachystola magna of the southwest.

Lucania parva. Rainwater fish; a small cyprinodontid fish of coastal swamps and rivers from Mass. to Fla. and N.M.

Lucanidae. Family of large, robust beetles; stag beetles; mandibles of males often enormously enlarged and antler-like; first tibiae enlarged for digging; larvae in damp, decaying wood; Lucanus common.

Lucanus. Genus of stag beetles in the Family Lucanidae; to 2 in. long.

luce. British term for a full-grown pike.

Lucernaria. Genus of sessile scyphozoan coelenterates in the Order Stauromedusae.

Lucernariida. Stauromedusae.

Lucifer. Genus of small pelagic shrimps.

luciferase. See BIOLUMINESCENCE.

Lucifer humming bird. See CALOTHORAX.

luciferin. See BIOLUMINESCENCE.

Lucilia. Genus of flies in the Family Calliphoridae; greenbottle flies; larval food habits variable, but some spp. attack the flesh of domestic animals.

Lucina. See LUCINA SHELLS.

lucina shells. Whitish tropical and subtropical marine bivalves of shallows to moderate depths; surface rough, often ridged, and to 3.5 in. across; Lucina, Phacoides, and Codakia typical.

Ludwig, Karl. German anatomist and physiologist (1816-1895); wrote Lehrbuch der Physiologie des Menschen.

lugworm. See ARENICOLA.

Luidia. Genus of sea stars in the Order Phanerozonia; disc small and arms long, narrow, and flexible; diameter to 14 in.; able to swim.

lumbar. Of or pertaining to the waist region of tetrapods.

lumbar plexus. Nerve plexus in the psoas muscle.

lumbar vertebrae. Vertebrae of the waist region, located between thoracic and sacral vertebrae; they have no associated ribs; five in man.

lumbricalis. Any of four small muscles in the palm and sole of the foot.

Lumbricidae. Common family of oligochaetes; includes many terrestrial and a few aquatic earthworms.

Lumbricillus. Genus of semiaquatic and fresh-water oligochaetes.

Lumbriculus. Genus of small oligochaetes found chiefly in fresh waters.

Lumbricus. Genus of earthworms; L. terrestris especially common.

Lumbrinereis. Genus of Polychaeta.

lumen. Cavity of any organ, sac, duct, or passageway.

lumirhodopsin. Intermediate compound in the biochemical reactions occurring when light strikes the rhodopsin of rods in the retina of the eye; converted into metarhodopsin.

Lumpenidae. Fish family now usually included in Stichaeidae.

Lumpenus lumpretaeformis. Serpent blenny, a very slender fish (Lumpenidae or Stichaeidae) of the littoral regions of the Arctic and North Atlantic oceans, to 19 in. long; other spp. in this genus in the Pacific and Bering seas.

lumper. Taxonomist who tends to unite two or more related taxa into a single taxon, such as combining two or more

genera into a single genus, etc.; opposite of a splitter.
lumpfish. See CYCLOPTERUS.
lump-nosed bat. See CORYNORHINUS.
lumpsucker. Lumpfish; see CYCLOPTERUS.
luna moth. See TROPAEA.
lunate. Shaped like a new moon.
lunate bone. Small proximal wrist bone in mammals.
Lunatia. Genus of globular marine gastropods; moon shells; sand flats and bottoms; East and West coasts.
lung. Spongy or thin-walled elastic respiratory organ capable of being filled with air; a true lung is restricted to amphibians, reptiles, birds, mammals, and a few fishes; lunglike structures are also found in a few snails, arthropods, etc.
lung book. Book lung.
lungfish. See DIPNOI, CERATODONTIDAE, and LEPIDOSIRENIDAE.
lung fluke. See PARAGONIMUS WESTERMANI.
lungworm. Any of numerous nematodes parasitic in the lungs and tracheae of hogs, cattle, sheep, deer, etc.
lunule. 1. Small crescent-shaped area just above the base of the antennae in certain Diptera. 2. Heart-shaped depression just in front of the umbo in many bivalve mollusks. 3. Any crescent-shaped part or mark.
Luscinia. Nightingales; genus of Old World thrushes noted for their sweet song.
luteal tissue. See CORPUS LUTEUM.
lutein. Yellow fatty material in cells of a corpus luteum.
luteinizing hormone. Hormone secreted by the anterior pituitary of mammals; stimulates ovulation and the formation of the corpus luteum.
luteotropic hormone. LTH; hormone secreted by the anterior pituitary; activates and maintains corpus luteum in a functional state.
Lutjanidae. Family of marine percomorph fishes; snappers; mostly in warm seas; some edible; Lutjanus includes the red snapper, gray snapper, and schoolmaster; Ocyurus includes the yellowtail snapper of the Atlantic Coast.
Lutjanus. See LUTJANIDAE.
Lutra canadensis. Common river otter found over most of U.S. and Canada; long, slim, glossy brown carnivore in the Family Mustelidae; pelage dense, legs short, feet webbed; superb swimmers. feed mostly on fish and crayfish; fur rich and durable.
Lutrochus. Common genus of dryopid beetles.
Lutrogale. Genus which includes a large otter of southeastern Asia; the simung; aquatic but often travels overland for considerable distances.
luxation. Complete dislocation of bones of a joint.
Luxilus cornutus. Redfin; a small minnow which is especially abundant in sluggish streams east of the U.S. Continental Divide.
Lycaena hypophloeas. Copper colored butterfly in the Family Lycaenidae; common and widely distributed in U.S.
Lycaenidae. Family of small delicate butterflies; blues, coppers, hairstreaks, and gossamers; upper surface of wings usually metallic blue, green, copper, or bronze; common genera are Lycaena, Feniseca, and Strymon.
Lycaenops. Genus of Permian mammal-like reptiles.
Lycaon. Genus of southeastern African wild dogs; coat mottled black, white, and orange; hunt in packs.
Lycidae. Small family of tropical and subtropical net-winged beetles.
Lycosa. See LYCOSIDAE.
Lycosidae. Family of wolf spiders or hunting spiders; large, active spp. which make tunnels in ground, use natural crevices, or are wanderers and have no retreat; Lycosa a large common genus.
Lyctidae. Family of small reddish-brown beetles which bore in wood, especially furniture, beams, and flooring; powder post beetles; some are important pests.
Lyell, Sir Charles. Scottish geologist (1797-1875); dispelled the theory of catastrophism by showing that sedimentation, uplift, and erosion are continuing and regular processes.
Lygaeidae. Large family of hemipteran insects which includes the chinch bugs; mostly herbivorous; Blissus leucopterus is the common chinch bug which is a serious pest on corn, wheat, and other cereals; black, 5 mm. long, and with two generations per year.
Lygosoma laterale. Ground skink in the Family Scincidae; a small skink, to 2 in. long; N.J. to Fla. and west to Tex.
Lygus. Important genus in the Hemipteran Family Miridae; pest on many cultivated plants.
Lymantria dispar. Palaearctic moth

which has spread to the northeastern states; gypsy moth; male olive brown, female whitish; caterpillar strips leaves from ornamental and forest trees.

Lymantriidae. Family of moths containing several important pest spp.; caterpillars (tussock moth caterpillars) large and densely covered with long and conspicuous tufts of hairs; Lymantria and Euproctis familiar genera.

Lymnaea. World-wide genus of fresh-water snails; the pond snails; usually in quiet waters.

lymph. Colorless fluid occurring in intercellular spaces, cavities, and special lymph ducts; derived from blood by filtration through the blood capillary walls; lymph contains dissolved salts and small amounts of organic compounds, chiefly proteins; it also contains varying numbers of suspended lymphocyte cells; lymph circulates through the tissues very slowly and eventually is brought back into the blood stream by way of special, large lymph ducts; lymph serves as an intermediary for the transport of materials between cells and the blood stream; although lymph is most characteristic of vertebrates, a similar substance occurs in many invertebrates, especially those having a closed circulatory system. Adj. lymphatic.

lymphatic cisterns. Thin-walled expansions of large lymph ducts.

lymphatic system. Complex system of thin-walled ducts, lymph nodes, and intercellular spaces plus the contained fluid (lymph); found especially in vertebrates but present in many invertebrates also; vertebrate lymph is derived from blood by filtration through the blood capillary walls; it wets all cells and enters the extensive lymph capillary system and thence into larger lymph vessels (sldom more than 3 mm. in diameter) which eventually empty the lymph back into the venous blood stream; lymph nodes (birds and mammals only) are contained on the larger lymph vessels; these are spongy masses from 1 to 15 mm. in diameter which strain foreign bodies from the lymph and contribute lymphocytes to it; lymph moves along its channels by the squeezing action of adjacent voluntary muscles and by an arrangement of one-way valves; fishes, amphibians, and reptiles have flimsy lymph hearts. See LYMPH.

lymphatic tissue. Lymphoid tissue.

lymph duct. See LYMPHATIC SYSTEM.

lymph gland. Lymph node; see LYMPHATIC SYSTEM.

lymph heart. One of several flimsy pulsating lymph sacs occurring in fishes, amphibians, reptiles, and birds; in the frog there are four such hearts on the dorsal surface of the trunk.

lymph node. See LYMPHATIC SYSTEM.

lymphoblasts. Lymph node cells which are the precursors of lymphocytes.

lymphocyte. Type of non-granular white blood cell originating in vertebrate lymphoid tissues; non-phagocytic, comprising 20 to 25% of human leucocytes.

lymphocytosis. Excess of lymphocytes in the blood.

lymphoid tissue. Type of vertebrate tissue which produces lymphocytes and macrophages; especially prominent in spleen, lymph nodes, tonsils, etc.

lymph vessel. Lymph duct; see LYMPHATIC SYSTEM.

Lynceus. Common and widely distributed genus of clam shrimps.

Lynchia. Blood-sucking dipteran parasite of pigeons; in the Family Hippoboscidae.

Lyncodon. Genus of Patagonian weasels.

lynx. Any of a group of cats in the genus Felis (Family Felidae) having long ear tufts and legs, and a stubby tail; N.A., Europe, Asia, and Africa; the common N.A. lynx, F. canadensis, occurs in boreal latitudes and attains a maximum length of 39 in.; a shy, nocturnal prowler with large feet; sometimes placed in the genus Lynx. Cf. BOBCAT.

Lyogyrus. Genus of minute fresh-water snails; Atlantic Coast states.

Lyomeri. Order of deep-sea bony fishes which includes the gulpers; body eel-like and naked, mouth very large, tail long and thin; Saccopharynx typical.

Lyomerida. Lyomeri.

Lyonet, Pierre. French and Dutch naturalist (1707-1789); noted for his careful anatomical studies of the sheep tick and the goat moth.

Lyonetiidae. Small family of widely distributed minute moths; adults brilliantly colored; larvae usually leaf miners, sometimes pests.

Lyonsia. Genus of small marine clams; shell thin, valves unequal; usually on sandy bottoms.

lyophile colloid. Any colloid in which

the dispersed phase shows considerable affinity for the dispersion medium.

lyophilization. Formation of a stable biological material (plasma, serum, etc.) by quick-freezing and dehydration under a high vacuum.

lyophobe colloid. Any colloid in which the dispersed phase shows little affinity for the dispersion medium.

Lyperosia. Genus of blood-sucking flies in the Family Stomoxyidae; commonly attack stock animals in the Old World.

lyre bird. One of two Australian passerine birds forming the Family Menuridae; males spread the long tail feathers into the shape of an expanded lyre; genus Menura.

lyre crab. Hyas lyratus; a marine crab found from Puget Sound to the Bering Sea; carapace lyre-shaped.

lyre snake. Any of several black-fanged poisonous snakes (harmless to man) in the genus Trimorphodon; head large, neck thin; 24 to 43 in. long; rocky areas of the southwestern states.

lyriform organs. Minute lyre-shaped structures on the body and appendages of many spiders; presumably centers for odor reception.

Lyropecten nodosus. Large heavy-shelled scallop; lion's paw; Carolinas to West Indies.

Lyrula. Genus of marine Bryozoa.

lyse. To cause or undergo lysis.

Lysenkoism. Doctrine of the Russian Lysenko; heredity is not based on germ plasm, and acquired (somatic) characters are inherited.

Lysianopsis. Genus of marine Amphipoda.

lysin. Antibody which may cause lysis or disintegration of cells; usually named according to its specific action, such as hemolysin, bacteriolysin, etc.

lysine. Amino acid; $C_6H_{14}O_2N_2$.

Lysiosquilla. Large genus of mantis shrimps; common along Gulf coast.

Lysiphlebus testaceipes. Braconid wasp; important parasite of aphids.

lysis. 1. Destruction of cells by a specific lysin. 2. Gradual abatement of disease symptoms.

Lysorophus. Genus of extinct stegocephalian amphibians.

lysosomes. Small cytoplasmic particles consisting of several enzymes involving digestion and synthesis of certain large cellular molecules; present in many types of cells.

lysozyme. Crystalline enzyme-like protein present in tears, saliva, and many other animal fluids; able to destroy bacteria by disintegraion.

Lytechinus. Genus of small sea urchins.

Lytta. Genus of beetles in the Family Meloidae; larvae feed on grasshopper eggs; see SPANISH FLY.

M

m. Abbr. of meter or meters.

Macaca. Genus of small southern Asiatic monkeys (plus one Gibraltar sp.); macaques or rhesus monkeys; the common monkeys of zoos and biological research.

macaco. Any of several lemurs, especially in the genus Lemur.

Macandrevia. Genus of Brachiopoda.

macaque. See MACACA.

macaw. Any of a group of large, brightly colored parrots of Central and South America; tail long, beak large and hooked; Ara common; see PSITTACIFORMES.

maceration. Softening or dissociating of a solid or tissue by soaking in various solutions of potassium hydroxide, potassium bichromate, acetic acid, nitric acid, etc.

Machilidae. Widely distributed family of Thysanura; insects with cylindrical scaly bodies, long antennae, caudal cerci, and caudal filaments; live in duff of grassy and wooded areas.

machopolyp. Machozooid; a type of polyp in some marine hydrozoans; specialized for protection and food capture; gastrovascular cavity reduced or absent.

machozooid. Machopolyp.

mackerel. See SCOMBER, PNEUMATOPHORUS, and SCOMBEROMORUS.

mackerel shark. See LAMNA and ISURUS.

Mackinaw trout. See CRISTIVOMER.

Macleod, John James R. Scottish and Canadian physiologist (1876-1935); along with Banting, first isolated insulin in 1922 for which he shared the 1923 Nobel prize in physiology and medicine.

Macoma. Large, cosmopolitan genus of small marine clams; thin, glossy, whitish shells; generally on mud bottoms in protected waters.

Macracanthorhynchus. Genus of acanthocephalan parasites of the small intestine of mammals; the large M. hirudinaceus is common in the hog intestine and is occasionally found in other mammals (rare in man); beetle larvae are intermediate hosts.

Macrobdella decora. American medicinal leech, common in northern states; feeds almost exclusively on vertebrate blood; to 9 in. long.

Macrobiotus. Common fresh-water genus in the Phylum Tardigrada.

Macrobrachium. Common genus of river shrimps; N.C. to Tex., and lower Mississippi and Ohio river systems; one sp. with a body to 9 in. long.

Macrocallista. Callista.

Macrocentrus. Genus of hymenopterans in the Family Braconidae; larvae parasitic on Lepidoptera.

Macrocheira kämpferi. Giant Japanese crab; to 13 ft. across, measured to the tips of the legs; body only about 1 ft. across.

Macrochelidae. Family of mites; usually parasitic on invertebrates; Macrochelus on insects.

Macrochelus. See MACROCHELIDAE.

Macrochelys. Macroclemys.

Macroclemys temmincki. Alligator snapping turtle of sluggish waters of the Gulf Coast and up the Mississippi Valley to Mo.; to 30 in. long and 150 lbs.; differs from common snapper in having three high ridges on carapace.

Macrocyclops. Common genus of littoral fresh-water copepods.

Macrodactylus. Genus of beetles in the Family Scarabaeidae; rose chafers; adults feed on leaves of cultivated trees and shrubs; larvae feed on roots.

Macrodasyoidea. Class in the Phylum Gastrotricha; with anterior, lateral, and posterior adhesive tubules; flame bulbs absent; hermaphroditic; marine littoral, especially in sandy beaches; Cephalodasys and Macrodasys common.

Macrodasys. See MACRODASYOIDEA.

Macrodiplax balteata. Uncommon dragonfly of Gulf Coast states.

macrodont. Having large teeth.

Macrodontia cervicornis. Giant Brazilian cerambycid beetle; black, tan, and brown coloration; one of the largest living insects.

macroevolution. Evolution based on major genetic changes (mutations) within a sp., often resulting in individuals which are presumably in the category of a new sp. (rather than only a subspecies); this concept is not generally accepted. Cf. MICROEVOLUTION.

Macrogalidia. Palm-civet genus in the Family Viverridae; a small gray-brown civet of the Celebes.

macrogamete. Female gamete, as distinguished from the smaller, more abundant microgametes (male gametes); the term is usually used with reference to certain Protozoa which exhibit sexual reproduction.

macrogametocyte. Cell which undergoes meiosis and forms a macrogamete (egg), especially in sporozoan Protozoa.

Macroglossus. Genus of long-tongued fruit-bats; small spp. which feed on pollen; Oriental, Indonesian, and Australian regions.

Macrolaimus. Genus of microscopic, free-living nematodes.

macrolepidoptera. Collectively, the larger butterflies and moths.

macromere. One of the distinctly larger cells resulting from unequal cleavages during early embryology of polychaetes, archiannelids, polyclad turbellarians, nemertines, mollusks, and some chordates, such as the frog and Amphioxus.

Macromia. Common genus of large U.S. dragonflies.

macromutation. As defined by Goldschmidt, a major mutation which may account for the appearance of a new sp. in one step (generation). See MACROEVOLUTION.

Macronemum. Genus of Trichoptera.

macronucleus. Large nucleus, especially as distinguished from the much smaller micronucleus, in ciliate and suctorian Protozoa.

Macroperipatus. Common genus in the Phylum Onychophora; to 6 in. long.

Macropetalichthyida. Order of placoderms; bony armor reduced; transitional between arthrodires and shark-like fishes.

macrophage. One of many kinds of phagocytic cells in connective tissue, the walls of the lymphatic system, bone marrow, spleen, liver, etc. of vertebrates.

macrophagous. Pertaining to animals which feed on relatively large pieces of food compared to their own body size, e.g. carnivorous land mammals. Cf. MICROPHAGOUS.

macroplacoids. Small sclerotized pieces imbedded in the pharyngeal musculature of a tardigrade; cf. MICROPLACOID.

macroplankton. Plankton organisms the taxa of which can be recognized with the naked eye.

Macropodidae. Family of marsupials which includes the kangaroos and wallabies; mostly large herbivorous spp. with strong hind legs and feet adapted for leaping; tail large, muscular, and used as a support in standing and walking; Macropus, Petrogale, Lagorchestes, and Dendrolagus typical.

Macropoma. Genus of extinct coelacanth fishes.

macropterous. 1. In insects, having large, well-formed, normal wings. 2. Occasionally, having large fins in fishes or large wings in birds.

Macropus. Common genus of kangaroos in the Family Macropodidae; about 24 spp. in the Australian region; large, leaping forms with small forelegs and powerful hind legs; head small, ears large.

Macroscelides. See MACROSCELIDIDAE.

Macroscelididae. Small family of mammals in the Order Insectivora; includes the African elephant shrews; small spp. with long hind legs, large eyes, and pointed muzzle; general appearance of miniature delicate kangaroos; Macroscelides common in northern Africa.

Macrosetella. Common genus of marine copepods in the Suborder Harpacticoida.

Macrostomum. Large genus of turbellarians in the Order Rhabdocoela.

Macrotermes. Genus of tropical termites which build massive nests (termitaria) above ground.

Macrothrix. Common genus of fresh-water Cladocera; usually in aquatic vegetation.

Macrotis. Genus of Australian marsupials; the rabbit-bandicoots or bilbies; long-haired, bushy-tailed, nocturnal burrowers.

Macrotrachela. Genus of creeping bdelloid rotifers.

macrotrichia. Large hairs on the surface of insect wings, especially in Diptera. Sing. macrotrichium.

Macrotus californicus. Common leaf-nosed bat of southwestern states and Mexico; muzzle with nose leaf; caves and mine tunnels.

Macrouridae. Grenadiers; a family of deep-sea fishes related to the cods; head large, snout projecting, body slender and tapering to a whiplike tail; Macrourus common.

Macrourus. See MACROURIDAE.

Macrovelia. See MACROVELIIDAE.

Macroveliidae. Small family of hemipterans; Macrovelia horni occurs in vegetation and debris at water's edge in western U. S.

Macrozoarces americanus. Ocean pout or eelpout; an eel-like fish of the ocean bottom; with long dorsal and anal fins.

Macrura. One of the three suborders in the crustean Section Reptantia; lobsters, crayfishes, shrimps, and prawns; large abdomen with five pairs of pleopods and a tail fan; marine and freshwater.

macrurous. Pertaining to the Macrura.

Mactra. Genus of marine pelecypods; includes most of the surf clams; large spp., especially abundant in tropics.

maculae. 1. Sensory areas of the sacculus, utriculus, and cochlea (lagena) of the membranous labyrinth. 2. Any anatomical spots or spotlike features. Sing. macula.

macula lutea. Small yellowish area on the retina of man and some primates; an area of especially acute vision, surrounds the fovea.

macula neglecta. Small sensory spot in the utriculus of some fishes and lower land vertebrates.

Madoqua. Genus of ruminants in the Family Bovidae; includes the African dikdiks; small antelopes about the size of a large rabbit.

Madreporaria. Order of zoantharian anthozoans; the true, or stony, corals; exoskeleton a compact calcareous mass with polyps situated in small cup-shaped cavities on the surface; mostly colonial, forming large coral reefs; Fungia, Astrangia, Orbicella, Acropora common.

madreporite. Small sievelike porous plate on the surface of most echinoderms (internal in sea cucumbers); a device through which fluid is able to pass slowly back and forth between the surrounding sea water and the internal water vascular system. See WATER VASCULAR SYSTEM.

madtom. See SCHILBEODES.

Maeandra. Genus of reef-building corals of the tropical Pacific and Indian oceans; includes brain corals and rose corals.

Maera. Genus of marine Amphipoda.

Magellania. Genus of common brachiopods in the Order Telotremata.

Magelona papillicornis. Peculiar long polychaete of sandy bottoms; prostomium spoon-shaped; two very long anterior tentacles; first eight segments everted as a respiratory device.

maggot. Larva of insects in the Order Diptera; without true legs although one or more pairs of prolegs may be present.

Magicicada septendecem. See CICADIDAE.

magnolia warbler. Dendroica magnolia, a warbler which breeds in northeastern coniferous forests.

Magnus, Olaus. Sixteenth century Swede who wrote a fanciful natural history.

magpie. See PICA.

maguari stork. See EUXENURA.

major coils. Series of large spirals or coils of chromatin granules making up a chromonema; such coils are loose or absent in an interphase cell and tight in a definite chromosome at metaphase. Cf. MINOR COILS.

major genes. A typical pair of dominant and recessive genes which have pronounced alternate genetic effects; they contrast with the small graded effects of multiple factors upon quantitative characters.

Makaira. Genus of large oceanic fishes in the Family Istiophoridae; marlins; upper jaw prolonged into a beak which is slender and cylindrical; not flat as in the swordfish; first dorsal fin very long and falcate; blue marlin (M. ampla) to 1000 lbs.; white marlin (M. albida) to 160 lbs.

Malacichthyes. Order of bony fishes; ragfishes; skeleton poorly ossified, body limp, median fins long; Pacific Ocean; Icosteus typical.

Malaclemys. Genus which includes the terrapins, diamondback terrapins, or salt marsh turtles; with angular rings on the carapace; coastal tidewaters and brackish areas from Mass. to Mexico; highly prized as food; carapace to 8 in. long.

Malacobdella. Genus of nemertines in the Order Bdellonemertea; three spp. living in the mantle cavity of snails.

Malacoclemmys. Malaclemys.

malacology. Study of mollusks.

Malacosoma americana. Common American moth whose tent caterpillar defoliates trees and spins extensive communal webs among branches and twigs; in the Family Lasiocampidae.

Malacostraca. One of the subclasses in the Class Crustacea; lobster, crab, crayfish, shrimps, pill bugs, sideswimmers, etc.; body typically composed of five fused head segments, eight thoracic segments, and usually six abdominal segments; one or more thoracic segments fused with the head to form a cephalothorax; marine, fresh-water, and a few terrestrial spp.; abdominal segments with appendages; 17,000 spp.

malaise. General or vague feeling of physical discomfort or illness.

Malania. Genus of lobe-finned fishes recently discovered off the African coast; now included in Latimeria.

malaria. Infection of erythrocytes by sporozoan parasites, especially in birds and man; Plasmodium is the prevailing genus, and the numerous spp. all have complicated life cycles involving transmission from one definitive host to another by Anopheles or Culex mosquitoes; in man the infection is tropical and subtropical and is characterized by a series of periodic high fevers and numerous other pathological symptoms; malaria is often considered the most important of all human diseases; benign tertian malaria in man has a chill and fever cycle every 48 hours and is caused by Plasmodium vivax; quartan malaria has a 72-hour cycle and is caused by P. malariae; malignant tertian malaria has an irregular 24 to 72 hour cycle and is caused by P. falciparum; a fourth type of mild malaria is caused by P. ovale and usually runs its course in 15 days; human malaria is transmitted only by Anopheles.

malaxation. Method employed by certain wasps; consists of biting and squeezing insect prey in the neck region and elsewhere, thus making it inactive and suitable for larval food.

Malaysian. Indonesian.

Maldane. Genus of long, slender, tube-building Polychaeta.

malignant. Spreading, increasing, and tending to go from bad to worse; e.g. a tumor.

malignant neoplasm. Any neoplasm which invades adjacent tissues and may be carried to other parts of the body by blood and lymph.

mallard. See ANAS.

malleate mastax. Rotifer mastax specialized for horizontal grinding of ingested food; feebly prehensile.

malleolus. 1. Tapered projection on the median portion of the lower end of the human tibia (medial malleolus). 2. Swollen lower end of the human fibula (lateral malleolus).

malleoramate mastax. Rotifer mastax specialized for grinding; teeth larger than in the ramate type.

malleus. Outermost of three minute auditory ossicles in the middle ear of mammals; has the rough shape of a hammer or mallet.

Mallomonas. Genus of uniflagellate Protozoa; cell elongated, with minute scales and spines; two chromatophores.

Mallophaga. Order of insects which includes the bird lice or biting lice; ectoparasitic or commensal on birds and a few mammals; to 6 mm. long; biting mouth parts; without wings; simple metamorphosis; body broad, depressed, and strongly sclerotized; head broad, thorax small; important genera are Menopon, Goniodes, Lipeurus, and Trichodectes.

Mallotus villosus. Capelin; small slender olive and silvery marine fish in the Family Osmeridae; spawns inshore on gravel bottoms of Arctic and Subarctic coasts; to 8 in. long; important bait fish.

malnutrition. Pathological condition resulting from insufficient food, vitamins, or minerals, or faulty assimilation.

Malpighi, Marcello. Italian anatomist (1628-1694); pioneer micro-anatomist who discovered many details of plant and animal structure; completed Harvey's account of blood circulation by observing the role of capillaries; wrote a classical account of the anatomy of the silkworm.

Malpighian body. Minute double-walled globular capsule (Bowman's capsule) surrounding a knot of capillaries (glomerulus) in the kidney of reptiles, birds, and mammals, as well as certain fishes and amphibians; the protein-free fluid fraction of the blood passes through the glomerular endothelium into Bowman's capsule and down a long urinary tubule leading away from the capsule; as this fluid traverses the urinary tubule, an investing capillary network reabsorbs most of the material which is useful to the body; the residual fluid in the urinary tubule is urine.

Malpighian corpuscle. Malpighian body.

Malpighian layer. Basal layer of stratum germinativum.

Malpighian tubules. Blind tubular glands, sometimes branched, opening into the anterior part of the hindgut of insects and most other terrestrial arthropods; chief function is excretion.

maltase. Digestive enzyme which converts maltose into glucose.

Malthus, Thomas Robert. English economist, sociologist, and pioneer in modern studies of human populations (1766-1834); in _An Essay on the Principle of Population_ (1798), he emphasized the fact that the world's population was growing more rapidly than the available food supply.

maltose. A 12-carbon atom sugar formed by the digestion of starches; occurs in both plants and animals; a molecule is readily hydrolyzed to two molecules of glucose.

mamba. See DENDROASPIS.

mamma. Mammary gland. Pl. mammae.

Mammalia. Class of the Subphylum Vertebrata; mammals; body covered with true hair (sometimes scanty); skull with two occipital condyles; teeth specialized for particular food habits; eyes with movable lids, and each ear with an external pinna; each of the four feet with five or fewer digits, modified for various types of locomotion and other functions; heart of two atria and two ventricles; only the left aortic arch persistent; erythrocytes platelike, without nuclei; lungs and larynx present; pleural and abdominal cavities separated by muscular diaphragm; cerebrum and cerebellum large and highly developed; homothermous; fertilization internal; developing embryo usually retained in uterus, surrounded by embryonic membranes, and obtaining nourishment from maternal tissues by means of a placenta (except in monotremes); after birth, the young are fed by milk secreted from female mammary glands; 4500 living spp.

mammalogy. In its broadest aspects, the study of all phases of the biology of mammals.

mammary gland. Ventral milk-producing gland peculiar to female mammals; growth and activity governed by hormones of the ovary, uterus, and pituitary.

mammilla. Nipple.

mammillary body. One of a pair of small swellings on the posterior ventral wall of the infundibulum in the mammalian brain; has fiber connections with the olfactory organ.

mammillate. With nipples or nipple-like projections.

mammoth. See MAMMUTHUS.

mammotropic effects. In many mammals, the stimulating effects of lactogen and estrogens on the mammary glands.

Mammut. Common genus of mastodons; extinct elephant-like mammals having conical projections on the molar teeth; forest-dwelling browsers; found in Asia, Europe, and N.A. to the close of the Pleistocene. Cf. MAMMUTHUS.

Mammuthus. Common genus of mammoths; extinct elephant-like mammals of circumpolar distribution; found to the close of the Pleistocene; thick coat of brown woolly hair covered by an outer coat of long black shaggy hair; many-ridged molar teeth. See MAMMUT.

Mammuthus columbi. Pleistocene elephant inhabiting much of N.A.; with large, incurved tusks and a height of about 11 ft.

Mammuthus imperator. Plains-inhabiting elephant of Pliocene and Pleistocene times; 13.5 ft. tall.

Mammuthus primigenius. Common hairy mammoth of Pleistocene times; differed from the mastodon in its heavier coat of hair, less elongated head, and more curved tusks; circumpolar and contemporary with prehistoric man; smaller than modern elephants.

manakin. See PIPRIDAE.

manatee. See TRICHECHUS.

Manayunkia. Genus of small fresh-water polychaete annelids; _M. speciosa_ was found many years ago in rivers in N.J. and Penn.; _M. eriensis_ has been found more recently in the Great Lakes area.

Manculus quadridigitatus. Dwarf four-toed salamander in the Family Plethodontidae; to 35 mm. long; coastal plain of N.C. to Tex.

mandarin duck. See AIX GALERICULATA.

mandibles. 1. Third pair of appendages on the head of Crustacea; more or less modified for crushing or grinding. 2. Pair of mouth parts of insects; situated behind the labrum and in front of the maxillae; variously modified for different food habits. 3. Pair of feeding appendages on the head of millipedes and centipedes. 4. Jaws of beaked animals, such as birds. 5. Lower jaws of vertebrates. 6. General term for jaws.

mandibular arch. First visceral arch of vertebrate embryos.

Mandibulata. Arthropod taxon (subphylum) often used to include Crustacea, Insecta, and Myriapoda; usually have a pair of true mandibular mouth parts.

mandrill. See MANDRILLUS.

Mandrillus. Genus of Old World monkeys; includes the West African drills and mandrills; tail-less doglike spp. with brightly colored cheeks.

maned marmosets. See LEONTOCEBUS.

mangabey. See CERCOCEBUS.

mange. Skin infection of domestic and and other mammals produced by mites in the Family Sarcoptidae; characterized by epidermal scaling, itching, redness, and loss of hair.

Mangelia. Genus of small marine snails.

Mangora. Common genus of spiders in the Family Araneidae.

manic-depressive psychosis. Mental disorder marked by alternation of elation and depression, and sometimes accompanied by delusions.

Manidae. Single family of mammals in the Order Pholidota (q. v.); pangolins.

Manis. See PHOLIDOTA.

mannitol. A hexa-hydroxy alcohol used in measuring the rate of glomerular filtration.

manometer. Physiological instrument used in measuring the pressure or tension of liquids or gases; a column of mercury with a calibrated scale is often used.

man o' war bird. See FRIGATE BIRD.

Mansonia. Genus of mosquitoes; respiratory tube of larva modified for piercing plant tissues.

Manta. See DEVIL RAY.

mantid. See MANTODEA.

Mantidae. Family of insects in the Order Mantodea; includes the typical mantids.

Mantispidae. Small family of insects in the Order Neuroptera; forelegs raptorial; adults rare.

mantis shrimp. See HOPLOCARIDA.

mantle. 1. Soft extension of the body wall or outermost layer of the body wall in certain invertebrate groups, such as brachiopods and mollusks; usually secretes a shell. 2. Thin body wall of tunicates.

mantle cavity. Special cavity found in certain invertebrate groups, such as brachiopods and mollusks, which is lined with epidermis; such a cavity is usually freely exposed to sea water, fresh water, or air, depending on the habitat; a mantle cavity may or may not contain part of the viscera.

mantle fibers. Continuous spindle fibers; cytoplasmic fibrils extending from one aster to the other in a cell undergoing mitosis.

Mantodea. Order of insects which includes the preying (praying) mantids; medium to large insects with elongated prothorax; forelegs greatly modified and raptorial; chewing mouth parts; metamorphosis simple; wings folded flat and extending over sides of body; mostly tropical spp.; *Stagmomantis carolina* (Carolina mantid) is the most common sp. in southern half of U. S.

manubrium. 1. Large process which hangs down from the center of a medusoid coelenterate; often variously lobed; bears the proctostome. 2. Large swollen process in certain hydrozoan polyps which bears the proctostome. 3. Uppermost piece of the sternum in man and many other mammals. 4. Anterior sternal process in many birds.

manus. 1. Hand. 2. Collectively, the carpus, metacarpus, and digits of a tetrapod foreleg. 3. Claw or other prehensile appendage of an arthropod.

Manx cat. Variety of cat having a stubby tail.

manyplies. Omasum.

maple borer. Any of several insects whose adults or larvae bore into maple trees; especially includes certain moths and beetles.

map turtle. See GRAPTEMYS.

mara. See DOLICHOTIS.

marabou. Large African stork (*Leptoptilus*) to 5 ft. tall, with a huge bill; head and neck almost devoid of feathers; upper parts dark green; scavenger but also eats small birds and mammals.

maral. Large sp. of red deer; Asia Minor and southwestern Asia.

marbled salamander. See AMBYSTOMA.

marbled spider. See ARANEUS.

march fly. See BIBIONIDAE.

Mareca. Genus of fresh-water ducks; *M. penelope* is the European widgeon which occurs occasionally in N. A.;

M. americana is the common baldpate, widely distributed in N.A., and marked by a shiny white crown in the male.

Margaritana. Margaritifera.

Margarites groenlandica. Wavy top shell, a very small marine snail of the Atlantic Coast.

Margaritifera. Widely distributed genus of bivalve mollusks; three spp. occur in American fresh waters; other spp. are the important marine pearl oysters of eastern and southern Asia.

Margarodes. Unique genus of Hemiptera in the Family Margarodidae; females fossorial and enclosed in overlapping lustrous wax plates often worn by natives as ornaments.

Margarodidae. Small but widely distributed family of Hemiptera which includes the giant coccids, marsupial coccids, and ground pearls; females large and hidden beneath or in waxy secretions; see especially MARGARODES.

Margaronia. See MELON MOTH.

Margaropus. Boophilus.

margay. Spotted cat (Felis tigrina) similar to the ocelot; Central America to Brazil.

Margelopsis. Genus of marine coelenterates in the Suborder Anthomedusae; hydroid free-swimming.

marginal lappet. Small flat peripheral projection on either side of each tentaculocyst in many jellyfishes.

marginal plates. Set of peripheral plates of a tortoise carapace.

Marginella. Genus of small tropical marine snails; shell highly polished, pear-shaped, and with a short or hidden spire; aperture long and narrow.

Maricola. Taxon which includes the marine spp. of triclad Turbellaria.

Marikina. Genus of small marmosets of northern South America.

marine zoogeographic realms. Major geographic divisions of the oceans according to their distinctive faunas.

Marionina. Genus of semiaquatic and fresh-water Oligochaeta.

marita. Adult fluke.

markhor. Very large Himalayan wild goat (Capra).

marlin. Any of several large oceanic game fishes related to the sailfishes; bones of lower jaw somewhat elongated, those of upper jaw much more elongated into a long beak; see MAKAIRA.

Marmosa. Genus of mouse-opossums; small mouselike marsupials; omnivorous but mostly insectivorous; usually in trees; Tropical America.

marmoset. See CALLITHRICIDAE.

marmot. See MARMOTA.

Marmota. Genus of rodents in the Family Sciuridae; woodchucks, marmots, or groundhogs; large, herbivorous, terrestrial rodents with heavy body and short tail; upper parts brownish or yellowish; live in burrows, piles of rock, and trash; several spp. and many subspecies over much of U.S. and Canada.

Marphysa. Genus of brightly colored Polychaeta.

marrow. See BONE MARROW, RED MARROW, and YELLOW MARROW.

marrow cavity. Central cavity of bones, especially long bones, containing marrow.

Marsenina. Genus of small marine snails; shell with only two whorls.

marsh deer. See BLASTOCERUS.

marsh fly. See TETANOCERIDAE.

marsh hare. Marsh rabbit, Sylvilagus palustris; a small cottontail of lowland thickets and marshes; Va. to Fla.

marsh hawk. See CIRCUS.

marsh hen. Any of various birds associated with marshy areas, e.g. rail and coot.

Marshia. Genus of fresh-water harpacticoid copepods.

marsh rabbit. See MARSH HARE.

marsh treader. See HYDROMETRIDAE.

marsh turtle. See MALACLEMYS.

marsh wren. See CISTOTHORUS.

Marsipobranchii. Cyclostomata.

marsupial. See METATHERIA.

marsupial bones. Prepubes.

marsupial coccid. See MARGARODIDAE.

marsupial frog. See GASTROTHECA.

Marsupialia. The single order of mammals in the Infraclass Metatheria.

marsupium. 1. Gill of bivalve mollusk when it is used to contain the embryos and early larval stages. 2. External pouch of marsupials and Echidna; fold of skin supported by projections of pelvic girdle; contains mammary glands and teats and receives the newborn young (eggs in Tachyglossus). 3. Ventral area next to body in isopods and amphipods; used to carry eggs and young. 4. External pouch for eggs and/or young in fish, e.g. sea horse. Pl. marsupia.

marten. See MARTES.

Martes. Genus of long, slender, carnivorous mammals in the Family Mustelidae; martens and fishers; legs short, feet digitigrade; terrestrial, semiarboreal, or semiaquatic; the several spp. of martens are weasel-like but rich yellowish brown, slightly smaller than a house-cat, and more or less arboreal; typical of northern evergreen forests; pelt of the marten often called sable; the fisher (M. pennanti) or pekan of northern evergreen forests is much larger and darker than the martens; good swimmer and tree-climber; mostly nocturnal.

Martesia cuneiformis. Wood piddock; small wedge-shaped bivalve which bores into wood; widely distributed in the North Atlantic.

martin. See PROGNE.

Maruina. Western genus of moth flies in the Family Psychodidae; larvae in swift streams, depressed, with ventral sucker-like discs.

Maryland yellowthroat. See YELLOW-THROAT.

masculinization. Abnormal condition produced by adrenal cortical hyperfunction and other hormonal conditions in human females; characterized by deep voice, growth of a beard, and regression of the genitalia.

masking crab. See LOXORHYNCHUS.

mason bees. See ANTHIDIUM.

mason wasps. See EUMENIDAE.

massasauga. See SISTRURUS.

masseter. Adductor muscle of the mandible in certain vertebrates.

Mastacembelus. See OPISTHOMI.

mastax. Muscular thickened portion of a rotifer pharynx; contains trophi.

mast cells. Large cells found in connective tissue and containing basophil granules; in a wide variety of invertebrates and vertebrates.

mastication. Chewing or shredding food with teeth or oral appendages.

Masticophis. Genus of slender non-poisonous snakes in the Family Colubridae; whipsnakes; extremely active spp., some being good climbers; to 6 ft. long; Wash. to Mo. south to northern South America.

mastiff bat. See EUMOPS.

Mastigamoeba. Genus of colorless amoeboid flagellates with a single flagellum; fresh water and soil.

mastigate. Mastigote.

Mastigella. Genus of fresh-water monoflagellate Protozoa with numerous pseudopodia.

Mastigina. Genus of colorless amoeboid flagellates with a single flagellum; in intestine of frogs and tadpoles.

mastigium. One or more telescopic anal organs in certain caterpillars; presumably used to repel parasite attacks. Pl. mastigia.

mastigobranchia. Slender process on an epipodite in certain decapod crustaceans.

Mastigophora. One of the classes of Protozoa; characterized by having one or more long, whiplike flagella used chiefly in locomotion; the surface of the cell is covered by a more or less firm, dead, secreted pellicle; certain spp. are photosynthetic and are classed as algae by botanists; free living, parasitic, marine, and fresh-water. Adj. and common noun mastigophoran.

Mastigoproctus. See WHIP SCORPIONS.

mastigote. Provided with one or more flagella; mastigate.

mastodon. One of a group of extinct elephant-like mammals; widely distributed in the Northern Hemisphere in Oligocene and Pleistocene times; see MAMMUT and MASTODON.

Mastodon americanus. American mastodon; a hairy, elephant-like mammal about 9.5 ft. tall inhabiting forests of N.A. during Pleistocene times; less hairy, with straighter tusks, and longer head than the mammoth.

mastodont. Mastodon.

mastoid. Process of the temporal bone just behind the ear of certain mammals; contains air chambers which communicate with the middle ear.

mastoid cells. Numerous small cavities in the mastoid process of the temporal bone, especially in higher vertebrates.

matamata. See CHELYS.

maternal impressions. Erroneous idea that experiences of the pregnant mother can influence the growth or appearance of the fetus.

mating dance. Any of many types of repeated rhythmical displays and movements associated with mating; mostly stimulatory, chiefly to prepare the female for mating; examples are: peacock displays, sage grouse display, male spider dancing, etc.

mating types. Especially among ciliate Protozoa, two or more physiological (or genetic) forms that will readily unite in conjugating pairs with each other.

matrix. Any dead intercellular substance in which living cells are imbedded, as in bone, cartilage, etc.

matroclinous. Pertaining to a type of inheritance in which the offspring particularly resemble the female parent in certain traits.

maturation. Gametogenesis; see MEIOSIS.

Matus. Genus of dytiscid beetles.

maw worm. Ascaris megalocephala.

maxillae. 1. Pair of appendages on the head of an insect; situated between the mandibles and labium; variously modified for different food habits. 2. One or two pairs of head appendages of Crustacea; used for handling, shredding, or straining food. 3. Pair of head appendages in millipedes. 4. Upper jaw (sing.) of higher vertebrates, representing the fusion of several smaller bones. 5. (Sing.) One of the dermal bones in the upper jaw of vertebrates. Sing. maxilla. See FIRST MAXILLAE and SECOND MAXILLAE.

maxillary. Of or pertaining to a jaw, jawbone, or maxilla.

maxillary glands. Two small glands imbedded in the tissues just posterior to the mouth in ostracods and a few other crustaceans; presumably excretory.

maxillary organ. Capitulum; anterior ventral headlike structure, especially in Hydracarina; bears the first two pairs of appendages.

maxillary palp. One- to seven-segmented process on the typical insect maxilla; variously modified.

maxillary shield. Ventral surface of the capitulum, or maxillary organ, especially in Hydracarina.

maxillary sinus. Especially in man, a large flat dorsoventral cavity in the maxillary bone; opens into the nasal cavity.

maxillary teeth. Row of teeth around the margin of the upper jaw in certain lower vertebrates.

maxilliped. Paired thoracic appendage modified for feeding, food handling, and locomotion; found on the first one to three thoracic segments in many Crustacea.

maxillulae. 1. First maxillae in a crustacean having two pairs of maxillae. 2. Non-functional appendages between mandibles and first maxillae of primitive insects.

Mayatrichia. Genus of Trichoptera.

May beetle. Large robust beetle in the Family Melolonthidae; see PHYLLOPHAGA,

mayfish. Killifish.

mayfly. Any member of the insect Order Ephemeroptera.

Mazama. Genus of ruminant mammals in the Suborder Ruminantia; see BROCKET.

McClung, Clarence E. American zoologist and cytologist (1870-1946); made extensive studies of germ plasm and chromosomes; discovered that sex of the offspring is determined by the chromosome constitution of the gametes.

meadowlark. See STURNELLA.

meadow mouse. See MICROTUS.

mealworm. See TENEBRIO.

mealybug. See PSEUDOCOCCIDAE.

Meandrina. Genus which includes the brain corals; spp. whose massive calcium carbonate skeleton has a fancied resemblance to the convolutions of the cerebral hemispheres.

Meantes. One of the three orders in the amphibian Subclass Caudata; sirens; body slender, eel-shaped, without hind legs; three pairs of gills persistent throughout life; entirely aquatic; a single family, Sirenidae.

Mearns's quail. See CYRTONYX.

measuring worm. See GEOMETRIDAE.

meatus. Any small passage or canal.

mechanoreceptor. Any sense organ which receives mechanical stimuli such as those involved in touch, pressure, hearing, and balance.

meconium. 1. Dark green material in the digestive tract of a full-term fetus; mixture of intestinal secretions and swallowed amniotic fluid. 2. Waste products of an immature quiescent animal, e.g. insect pupa.

Mecoptera. Order of insects which includes the scorpion flies; small to large predaceous spp. with long legs and antennae; biting mouth parts; with four similar, slender, many-veined wings which are held rooflike over the body when at rest; complete metamorphosis; larva burrowing and caterpillar-like, with three pairs of true legs and usually paired abdominal prolegs; Panorpa, Bittacus, and Boreus familiar genera.

Meda fulgida. Meda spinedace; a small minnow of the Gila River system; scales absent; with two dorsal spines, the anterior one being grooved at the back.

media. Fourth longitudinal vein in an insect wing.

medial. Median.

median. Medial; lying in or near the plane dividing a bilaterally symmetrical animal into two mirror-image halves in the midline.

Mediaster. Common genus of Pacific

Coast sea stars.

mediastinum. Median space between the two pleural cavities in the chest region of mammals; contains the heart, aorta, trachea, esophagus, etc.

medical zoology. Study of animals that affect the health of man.

medicinal leech. See HIRUDO.

Medina worm. Dracunculus medinensis, the guinea worm.

mediocubital. Cross-vein in an insect wing; between cubitus and posterior media.

Mediomastus. Genus of Polychaeta.

Medionidus. Genus of unionid mollusks; southeastern U.S.

Mediorhynchus. Genus of Acanthocephala parasitic in birds.

Mediterranean fruit fly. See CERATITIS.

Medlicottia. Genus of extinct cephalopods.

medulla. 1. Inner part of the adrenal gland of vertebrates; secretes adrenalin. 2. Inner mass of the kidney of higher vertebrates. 3. Inner portion of an organ. 4. See MYELENCEPHALON.

medulla oblongata. Myelencephalon.

medullary groove. Neural groove.

medullary plate. Neural plate.

medullary sheath. Whitish fatty material which forms a covering around medullated axons of the nervous system.

medullary tube. Neural tube.

medullated nerve fiber. Axon surrounded by a medullary (myelinated) sheath which, in turn, is covered by a membranous neurilemma; whitish color.

medusa. 1. Free-swimming, bell- or umbrella-shaped stage in the life history of many hydrozoan coelenterates; it has a central projecting manubrium bearing the proctostome, and few to many tentacles on the periphery; reproduces by syngamy, sometimes by budding. 2. Common name given to any jellyfish or jellyfish-like animal. Pl. medusae, adj. medusoid.

medusa bud. Protuberance formed on a hydrozoan coelenterate polyp or medusa; develops into a mature, free-swimming medusa.

meerkat. See SURICATA.

Megaceros. Genus of giant European Pleistocene elks.

Megaceryle alcyon. Common belted kingfisher of N.A. and northern South America; large head with a ragged crest and a long, stout bill; blue-gray above; under parts white, with one or two broad transverse abdominal bands; noisy, rattling flight; feeds chiefly on small fishes caught by diving from a perch at the edges of rivers, ponds, and lakes.

Megachile. Genus of bees in the Family Megachilidae; burrows are in ground or wood, lined with bits of leaves cut from plants.

Megachilidae. Large family of robust medium to large bees resembling honeybees; leaf-cutting bees, mason bees; nest variable, in earth or plant tissues; Megachile and Anthidium common.

Megachiroptera. Suborder of the mammal Order Chiroptera; includes the fruit bats or flying foxes; large, gregarious spp. which feed on fruits at night, dawn, and dusk; during daylight they sleep in trees by hanging head downward with wings folded around body; Africa, southern Asia, and Australian region; Pteropus edulis of Java is the largest sp., with a body length of 1 ft. and a wingspread of 5 ft.

Megacyllene. Genus of long-horned beetles.

Megadermatidae. Family of false vampire bats; ears about 20 times the size of the head; nose with a large leaflike appendage; carnivorous but not bloodsuckers; tropics.

Megadrili. Order of oligochaetes; includes the larger spp. with many segments, never reproducing asexually.

megagamete. Macrogamete.

megagametocyte. Macrogametocyte.

megakaryocytes. Giant cells of bone marrow; blood platelets are fragments of megakaryocytes.

megalecithal ovum. Egg which is strongly telolecithal, e.g. hen's egg.

Megaleuctra. Genus of Plecoptera.

Megalichthys. Genus of primitive crossopterygians.

Megalobatrachus japonicus. Giant salamander of eastern China and Japan; in cold streams; some specimens nearly 6 ft. long.

Megalocephalus. Genus of Paleozoic labyrinthodonts.

Megalomma. Genus of Polychaeta.

Megalonaias gigantea. Large, heavy-shelled unionid mollusk; Mississippi system and Ala. to Tex.

Megalopidae. Family of bony fishes which includes the tarpons; medium to very large; scales very large; Tarpon atlanticus an important game fish of the southeastern U. S. coast south to Brazil; most abundant in salt water passes, channels, cuts, and estuaries; occasionally exceeds 200 lbs.

megalops. Larval stage of marine crabs; just precedes the adult and has carapace, thoracic appendages, and abdomen which are relatively crablike.

Megaloptera. Small order of insects which includes the dobsonflies, alderflies, fishflies, and sialids; medium to large spp. with complete metamorphosis; biting mouth parts and long antennae; wings large, similar; larvae aquatic, carnivorous, and with segmented lateral abdominal gills.

Megalopygidae. Small family of moths; flannel moths; wings covered with a loose coat of soft scales.

Megalosaurus. Genus of giant carnivorous dinosaurs; poorly known.

meganephridium. Typical annelid nephridium with nephrostome and nephridiopore; one pair on most segments. Cf. MICRONEPHRIDIUM.

Meganeura. Genus of giant fossil insects resembling dragonflies; wingspread to 28 in.

Meganthropus paleojavanicus. Fossil anthropoid known only from part of a large jaw found in Java in 1941.

meganucleus. Macronucleus

Megapodiidae. Small family of birds (10 spp.) in the Order Galliformes; mound builders; eggs incubated in mounds of earth or rotting vegetation; Australian and East Indian region.

Megaptera novaeangliae. Humpback whale; a large stout sp. to 50 ft. long; blackish above, gray below; in all oceans.

Megapus. Genus of Hydracarina.

Megarhinus. Genus of tropical flower-feeding mosquitoes.

Megarhyssa. Genus of ichneumon flies; larvae parasitic on the larvae of other hymenopterans.

megascleres. Larger supporting skeletal spicules of certain sponges.

Megascolecidae. Large family of terrestrial earthworms; Northern and Southern hemispheres.

Megascolides australis. Giant earthworm of Australia; attains a length of more than 6 ft.

Megatebennus. Genus of keyhole limpets.

Megatherium. Genus of extinct giant sloths of North and South America; to 18 ft. tall; Pleistocene ground sloths.

Megathura. Genus of Pacific Coast keyhole limpets.

Megninia. Large genus of feather mites in the Family Analgesidae; on pigeons and other birds.

Mehli's glands. Shell glands; a cluster of unicellular glands surrounding the ootype in many trematodes and tapeworms; presumably aid in the formation of the shell which is laid down around the zygote.

Meibomian glands. See GLANDS OF MEIBOM.

meiofauna. Collectively, the microscopic and small macroscopic metazoan fauna inhabiting the surface of the sea bottom; includes nematodes, kinorhynchs, ostracods, copepods, halacarids, turbellarians, gastrotrichs, oligochaetes, etc.

meiosis. Set of two special nuclear divisions; begins with one diploid cell and ends with four haploid cells; in the animal kingdom it occurs regularly during gametogenesis, and in the plant kingdom it usually results in the formation of four spores from one spore mother cell; in a few algae it occurs during gamete formation, and in some other algae, some fungi, and some Protozoa it occurs shortly after formation of the zygote in the life history. In prophase and in preparation for the first meiotic division the two homologous chromosomes of each chromosome pair become closely associated side by side in a condition known as synapsis, or pairing; then each partner appears duplicated and each associated pair of duplicated chromosomes constitutes a tetrad, or four chromatids; at this point transverse breaks may occur whereby parts of one of the two chromatids derived from one chromosome may be exchanged for equivalent parts of one of the two chromatids derived from the corresponding homologous chromosome; this process is known as crossing over, and the points of crossing over are known as chiasmata; there are sometimes one or more chiasmata per tetrad in each meiosis. At anaphase two chromatids from each tetrad go (by chance assortment) to opposite poles of the spindle; except where crossing over has occurred, the two chromatids passing to each pole are derived from a single chromosome. After anaphase, and possibly a short interphase (according to some authors), each of the two daughter cells undergoes the second meiotic division; at metaphase the equatorial plate has the haploid number of chromosomes, but each chromosome consists of two closely associated chromatids derived from the first meiotic division; at anaphase one chromatid from each pair moves (again by chance assortment) to one of the poles of the spindle.

Meissner's corpuscles. Special tactile structures, 50 to 100 microns long, especially abundant on the tactile surfaces of the fingers and toes; each corpuscle is ovoid and consists of concentric layers of connective tissue cells lying transverse to the long axis of the corpuscle; dendrites ramify among these connective tissue layers.

Meissner's plexus. Flat layer of nerve fibrils in the submucosa of the stomach and small intestine.

Melamphaidae. Family of deep-water oceanic fishes; head deep, snout blunt, skull soft and spongy, coloration blackish; Melamphaes typical.

Melampus. Genus of small pulmonate snails found along the high tide line.

Melanella oleacea. Small, long-spired snail which parasitizes the sea cucumber, Thyone.

Melanerpes erythrocephalus. Common red-headed woodpecker of U.S.; entire head red.

Melanesian. Race of negroids having chocolate-colored skin and long head; southwestern Pacific islands, especially New Guinea, Fijis, and Marshalls.

Melania. Common genus of fresh-water snails; widely distributed in tropics and subtropics.

melanin. One of a series of dark brown or blackish organic pigments responsible for the yellowish to blackish coloration and color patterns of many animals; in man, melanins are especially abundant in the skin and hair.

melanism. Unusual darkening of the normal coloration due to the deposition of abnormally large amounts of melanins in superficial tissues.

Melanitta. Genus of New and Old World dark-colored sea ducks in the Family Anatidae; scoters; occasionally found far inland; M. deglandi is the American white-winged scoter; M. perspicillata is the American surf scoter; sometimes called sea coots.

melanoblast. In early embryology, a cell originating in the neural crest and giving rise to a melanophore.

Melanocetidae. Family of bizarre bony fishes in the Order Pediculati; deep sea anglers; mouth enormous, otherwise generally similar to Ceratiidae in structure and habits; Melanocetus typical.

Melanogrammus. See GADIDAE.

melanoma. Tumor composed of cells containing much melanin.

melanophore. Chromatophore containing an abundance of black or brown pigment granules.

Melanoplus. Common genus of grasshoppers; M. femur-rubrum is the red-legged grasshopper; M. spretus is the Rocky Mountain grasshopper.

Melanostomiatidae. Dragonfishes; a small family of slender abyssal marine spp.; mouth large, color dark, photophores present; Bathophilus typical.

Melanotus. Large genus of click beetles in the Family Elateridae.

Melapterurus electricus. Electric catfish; fresh waters of central and northern Africa.

Meleagrididae. Very small family of birds in the Order Galliformes; contains the American turkey, Meleagris gallopavo, and its several subspecies; originally found from eastern N.A. to the Mexican plateau; now domesticated and widely distributed.

Meleagrina. Genus of marine bivalves used extensively in the culture pearl industry of Japan.

Meleagris. See MELEAGRIDIDAE.

Meles. Genus which includes the Old World badgers.

Melibe. Common genus of nudibranchs commonly found in kelp.

Melicerta. Discarded generic name of certain rotifers; most such spp. are now in the genus Floscularia.

Melinna. Genus of Polychaeta; in tubes of hardened mucus coated with mud.

Meliphagidae. Large family of small passerine birds; honey eaters; down-curved pointed bills used in feeding on nectar; coloration and habits variable; Australian region and East Indies.

Meliponidae. Tropical family of minute to small stingless bees; in colonies in honey combs.

Melita. Genus of marine Amphipoda.

Melitaea phaeton. The Baltimore, a common butterfly in the northeastern U.S.; blackish wings marked with red and yellow.

Mellita. Common genus of sand dollars in the Order Clypeasteroida.

Mellivora. See RATEL.

Meloë. Large genus of beetles in the Family Meloidae.

Meloidae. Family of medium-sized beetles with narrow prothorax; oil beetles, blister beetles; adults feed on foliage and flowers; some spp. exhibit hypermetamorphosis with extra, specialized larval stages and a prepupa; cantharadin, a crystalline solid, is extracted from the dried bodies of certain spp.; important genera are Epicauta, Lytta, Meloë, and Sitaris.

Melolonthidae. Widely distributed family of beetles which includes the cockchafers, May beetles, and June beetles; small to large spp., variously colored; larvae largely root feeders, some of great economic importance; Melolontha typical; this family sometimes considered a subfamily of the Scarabaeidae.

Melonanchora. Widely distributed genus of marine sponges in the Subclass Monaxonida; encrusting or massive.

melon aphid. See APHIS.

Melonechinoida. Order of Carboniferous echinoids with a high, rigid test; Melonechinus typical.

Melonechinus. See MELONECHINOIDA.

Melongena. Genus of marine gastropods; conchs; variable spinous spp. to 4 in. long; Fla. to Tex.

melon moth. Melon worm, Margaronia hyalinata; white and black winged moth whose larvae feed on foliage of melon and cucumber plants; in the Family Pyraustidae.

melon worm. See MELON MOTH.

Melophagus ovinus. Blood-sucking dipteran parasite in the Family Hippoboscidae; sheep tick, ked; irritates the host, soils the wool, and transmits at least two protozoan parasites.

Melopsittacus undulatus. Common parakeet, or budgerigar, originally derived from Australian stock; small, parrot-like bird which can be taught to mimic the human voice.

Melospiza. See SPARROW.

Melursus ursinus. Sloth bear, a blackish nocturnal sp. of the Far East; muzzle long and mobile, commonly used for feeding on ants; claws stout and strongly curved.

Membracidae. Family of small bizarre insects in the Order Homoptera; treehoppers; protectively colored, pronotum greatly enlarged and projecting over scutellum and head; feed on plant sap; Platycotis and Ceresa bubalus (buffalo treehopper) typical.

membrane bone. Dermal bone.

membranelles. Structures used in locomotion and feeding in many Ciliata; each consists of two to four rows of cilia completely fused into a flat plate.

membrane potential. Electrical potential across the membrane of living cells, resulting from differences in concentrations and mobilities of ions, usually potassium and sodium; the outside is positive compared with the interior, and there is higher concentration of sodium on the outside and potassium on the inside; the condition is maintained by a normally lower permeability of the membrane for sodium than for potassium, plus the existence of some unknown "sodium pump" mechanism which uses metabolically released energy to transport to the outside such sodium ions as may diffuse into the cell.

membrane theory. Theory of impulse transmission by nerve fibers and other excitable cells; at the point of stimulation, the permeability for sodium ions is increased, allowing sodium ions to diffuse inward, and this permits outward movement of potassium ions; the difference in mobility between the two types of ions produces a depolarization, or in some cases a reversal of polarity at the point of stimulus; with the change of potential at the point of stimulus, a local circuit produces a flow of current between the point of stimulus and adjacent regions, thereby stimulating the latter; in this manner, the electrical changes spread as a wave from the initial point of stimulus; recovery of the original resting state of the membrane involves the active removal of sodium ions which previously penetrated plus an inward diffusion of potassium ions.

Membranipora. Genus of bryozoans in the Order Cheilostomata.

membranous labyrinth. Delicate, complex organ of equilibrium forming the inner ear of vertebrates; in all vertebrates beginning with amphibians it has the additional function of an organ of hearing; typically consists of a hollow utriculus, sacculus, three semicircular canals, and cochlea (in higher vertebrates only); the whole labyrinth is imbedded in a bony or cartilaginous labyrinth having mostly the same shape as the membranous labyrinth, the small space between the membranous and bony labyrinths being filled with perilymph.

menadione. Vitamin K analogue, prepared synthetically.

menarche. In girls, the beginning of the functional menstrual cycle, usually at an age of 12 to 15.

Mendel, Gregor Johann. Austrian scientist and Roman Catholic priest (1822-1884); especially noted for pioneer work in heredity at Brno; carried out hybridization experiments on garden peas and kept accurate statistical records of each inherited character studied; results were published in 1866 an obscure journal, but the fundamental value of the research was not recognized until 1900. See MENDELISM.

Mendelism. Fundamentals of heredity as first revealed by the experimental work of Mendel on garden peas; separate characters (e.g. seed color and seed shape) are inherited as units (factors); some factors are dominant over others (recessives); the factors for a pair of alternate characters are separate and only one may be carried on a particular gamete; a recessive character may be masked in the F_1 generation but appears in some of the subsequent generation (principle of segregation); the inheritance of one pair of characters is independent of the simultaneous inheritance of other traits, such characters "assorting independently" as though there were no other characters present (later modified by the discovery of linkage). When an individual purebred with respect to one character showing dominance is bred to an individual purebred for the alternate (recessive) character, the offspring (hybrid) all exhibit the dominant character, but their cells all carry the masked recessive factor; a mating between two such hybrids produces offspring averaging 25% bearing two dominant factors, 50% bearing both factors, and 25% bearing two recessive factors; the hybrid offspring therefore show the dominant character in a ratio of 3 : 1.

Mendel's laws. See MENDELISM.

Menetus. Genus of fresh-water snails.

menhaden. See BREVOORTIA.

Menidia. See ATHERINIDAE.

meninges. Series of membranes surrounding the vertebrate brain and spinal cord; serve for protection and nourishment. Sing. meninx.

meningitis. Inflammation of the meninges.

meninx primitiva. Single membrane covering the nerve cord of fishes.

Menipea. Genus of bryozoans in the Order Cheilostomata.

Menippe mercenaria. Stone crab, a common edible marine sp. of shallows along the coast from N.C. to Mexico; dark purplish blue.

meniscus. 1. Fibrous cartilage found in certain joints. 2. Intervertebral disc. 3. Disc in a tactile corpuscle.

Menoidium. Genus of fresh-water monoflagellate Protozoa; cell curved, rigid.

menopause. Time of cessation of menstruation in women, usually at 45 to 50 years.

Menopon. Genus of bird lice in the Order Mallophaga; _M. gallinae_ is the common hen louse, and _M. stramineum_ the large hen louse.

menses. Menstruation.

menstrual cycle. See MENSTRUATION.

menstrual period. See MENSTRUATION.

menstruation. Menses; periodic discharge of blood and disintegrated lining of the uterus through the vagina in certain primates; in mature human females the cyclic process occurs about every 28 days and the period of discharge lasts an average of four or five days; menstruation ceases temporarily during pregnancy and permanently at menopause.

mental. Pertaining to the chin.

mental foramen. Small opening on each side of the lower jaw near the chin; carries a nerve and blood vessel.

menthol. Anesthetic used occasionally for small aquatic invertebrates; a few crystals are sprinkled on the surface of the water in a small container.

Menticirrhus. Genus of croakers, kingfishes, or whiting in the Family Sciaenidae; on both coasts of U.S.; the corbina is a Pacific Coast sp. often taken by surf-casting; most forms are good food fish.

mento-Meckelian cartilage. Anterior end of Meckel's cartilage at the tip of the lower jaw; ossified in the Anura.

mentum. 1. Distal sclerite of a typical insect labium; attached to the submentum and bears the labial palps. 2. Projection below the mouth in a few mollusks.

Menura. See LYREBIRD.

Menuridae. See LYREBIRD.

Mephitis. Genus of carnivores in the Family Mustelidae; includes the common or striped skunks; about as large as a housecat; coloration black, with two longitudinal dorsal white stripes; widely distributed over most of U.S. and much of Canada; insectivorous and carnivorous; most active at night; the offensive odorous secretion is sprayed from two perianal glands.

mercaptan. One of a group of organic compounds similar to alcohols but with the oxygen atom replaced by sulfur; many such compounds are foul-smelling.

Mercenaria. Genus of heavy-shelled edible marine gastropods; quahogs; to 6 in. long; Canada to Tex.

Mercierella. Genus of Polychaeta.

merganser. Goosander; any of three American and European fish-eating ducks; mandibles with toothed edges; head crested; see LOPHODYTES and MERGUS.

Mergus. Holarctic genus of fish-eating ducks; bill serrated; M. merganser is the American merganser (goosander), and M. serrator is the red-breasted merganser.

meridional canal. One of the eight canals running in an oral-aboral direction just under the external surface of a ctenophore; a part of the gastrovascular system.

Meringodixa. Genus of dixa midges in the Family Dixidae; Pacific Coast states.

meristic. Segmented or serially divided.

Merizocotyle. Genus of external monogenetic trematode parasites of marine fishes.

Merkel's disc. Flat disclike tactile body in the submucosa of the tongue and oral cavity of man; also found in some other mammals, birds, and reptiles.

Merlangus merlangus. European whiting, an important marine food fish.

merlin. Small European falcon, Falco aesalon.

Merlucciidae. Family of marine fishes closely related to the cods; hakes, true whitings; chin barbels absent; to 5 lbs; Merluccius bilinearis is the common New England hake or whiting.

Merluccius. See MERLUCCIIDAE.

mermaid. Mythical animal having the head, arms, and trunk of a woman and the tail of a fish.

mermaid's purse. Horny capsule in which one or more eggs are deposited by certain cartilaginous fishes; often supplied with tendrils that attach it to seaweeds, etc.

merman. Mythical animal having the head, arms, and trunk of a man and the tail of a fish.

Mermis. Genus of long, thin, enoplid nematodes; adults free-living in soil or fresh water; immature stages parasitic in terrestrial and fresh-water invertebrates, especially insects.

meroandry. Having less than the normal number of testes. Adj. meroandric.

meroblastic cleavage. Type of cell division and embryo formation occurring in fertilized eggs having a very large amount of yolk concentrated at the vegetal pole; cleavage planes are restricted to a small disclike cytoplasmic area on the surface of the egg; occurs in bird eggs and many invertebrates.

merocrine gland. Type of gland which releases its secretion without any obvious damage to the cells.

merocyte. 1. Nucleus in the unsegmented portion of an early meroblastic embryo. 2. One of many nuclei formed by divisions of extra sperm nuclei often present in a fertilized egg of reptiles, birds, and selachians. 3. Schizont.

Merodon. Genus of flies in the Family Syrphidae; larvae feed on bulbs of narcissus, onion, etc.

merogametes. Especially in certain Protozoa, gametes which are smaller than vegetative cells and result from multiple division of vegetative cells; usually male gametes. Cf. HOLOGAMETE.

merogamy. 1. Union of gametes which differ from the parent (vegetative) cells. 2. Having such gametes.

merogony. Development of an ovum which has been (experimentally) deprived of its nucleus and fertilized by a sperm; performed especially in sea urchins and amphibians.

Meromyaria. Taxon sometimes used to include those nematodes having only two to five longitudinal rows of muscle cells in the dorsal and ventral muscle masses.

Meropidae. See BEE EATER.

meroplankton. Collectively, certain of those plankton organisms which exhibit daily vertical movements and are found at or near the surface only part of the time.

meropodite. Fourth segment of the pereiopod in most malacostracan Crustacea.

Merostomata. Subclass of the arthropod Class Arachnoidea; king crabs and eurypterids; cephalothorax with lateral compound eyes; cephalothorax and abdomen broadly joined; abdomen with six pairs of appendages bearing exposed gills; terminal spinous telson; see XIPHOSURA and EURYPTERIDA.

merozoite. One of many cells resulting from multiple fission of a schizont in the haemosporidian Sporozoa; each merozoite enters an erythrocyte and develops into a trophozoite.

Merragata. Common genus of velvet water bugs in the Family Hebridae.

Merriam, C. Hart. American naturalist (1855-1942); founder and chief of the U. S. Biological Survey from 1885 to 1910; best known for his division of N. A. into "life zones" based chiefly on temperature criteria.

Merriamia. Genus of Triassic fishlike reptiles in the Subclass Ichthyopterygia.

Mertensia ovum. Common American ctenophore; ovoid, 2 in. long, compressed.

merus. Meropodite.

Merychippus. Genus of Miocene ancestral horses.

mesectoderm. 1. Mass of mesenchyme cells in certain fish and amphibian embryos; gives rise to the anterior part of the braincase and the gill bar system. 2. In the chick embryo, the primitive streak before mesoderm and ectoderm are distinct.

mesencephalon. Midbrain.

mesenchyme. Embryonic or unspecialized connective tissue; usually the living cells are irregular in shape, sometimes amoebocytic, and often imbedded in a gelatinous matrix; in invertebrates and in vertebrate embryos.

Mesenchytraeus. Genus of semiaquatic and fresh-water Oligochaeta.

mesenteric filament. Septal filament.

mesenteron. 1. Midgut. 2. Central part of the anthozoan gastrovascular cavity, as distinct from the cavities between mesenteries.

mesentery. 1. In the broad sense, a thin supporting membrane or partition in the body cavity of many animals. 2. In the strict sense, as in vertebrates, it is a double thickness of the coelomic peritoneum through which blood vessels and nerves connect with the visceral organs.

mesethmoid. Anterior braincase bone in primates, rodents, and carnivores.

mesic. Characterized by moderate moisture or precipitation conditions.

Mesitornithidae. Family of three spp. of brownish birds in the Order Gruiformes; seldom fly; to 11 in. long; forests and brush of Madagascar.

mesoblast. Embryonic mesoderm.

mesocardium. Embryonic mesentery which connects the embryonic heart to the ventral body wall and encloses the aorta and pulmonary artery.

Mesochaetopterus. Genus of large polychaetes inhabiting curved tubes.

mesochondrium. Matrix of hyaline cartilage.

mesocoel. Original embryonic cavity of the mesencephalon.

mesocolon. Mesentery which supports the colon.

mesocoracoid. Arch-shaped bone in front of the coracoids in certain soft-rayed fishes.

Mesocyclops. Common plankton genus of fresh-water copepods.

mesoderm. Primitive germ layer or mass of cells formed between the primitive gut (endoderm) and the outermost layer (ectoderm) of an embryo; it gives rise to the skeleton, circulatory system, musculature, excretory system, most of the reproductive system, etc. of vertebrates; the term is properly applied to this mass of cells while it is a distinct region (after gastrulation) but before differentiation of its derived tissues.

mesodermal pouch. Coelomic pouch.

mesodermal somites. Blocklike masses of mesoderm arranged as two linear strips, one on each side of the notochord of a vertebrate embryo; each somite becomes a myotome, sclerotome, and dermatome.

Mesodesma. Genus of small northern marine Pelecypoda.

Mesodinium. Genus of marine and fresh-water holotrich ciliate Protozoa; ovoid, with an equatorial ciliated furrow; anterior cytostome with tentacle-like processes.

mesoduodenum. Embryonic mesentery which supports the duodenum.

mesofurca. Furca of the mesothorax in insects.

mesogaster. Dorsal fold of the peritoneum which supports the stomach in many vertebrates.

mesoglea. Layer of (mesodermal?) jellylike or cement material between epidermis and gastrodermis in coelenterates and ctenophores; it may be very thin or thick and bulky; it may contain no cell elements, very few, or abundant cells and fibers; the term is also properly used to include the jelly-like matrix between epithelial layers in the Porifera.

mesogloea. Mesoglea.

Mesogonistius chaetodon. Blackbanded sunfish in the Family Centrarchidae; N.J. to Fla. coastal streams.

Mesohippus. Genus of Oligocene ancestral horses; about the size of a sheep.

mesolecithal. Pertaining to an ovum with a moderate amount of yolk.

mesolimnion. Thermocline, discontinuity layer, metalimnion (q. v.).

mesomere. 1. Mesoblastic somite. 2. Blastomere of average size. 3. Middle portion of coelomic pouches in an embryo.

mesonephric duct. Functional (Wolffian) duct for the mesonephric kidney of adult fishes and amphibians, and also carries sperm in male fishes and amphibians; the functional vas deferens in male adult reptiles, birds, and mammals.

mesonephridium. Nephridium originating from mesodermal tissue.

mesonephros. Functional kidney of fishes and amphibians; the component tubules have terminal nephrostomes but also have connections with glomeruli somewhere along their length; the tubules empty their contents into a collecting (Wolffian) duct. Adj. mesonephric.

mesonotum. Dorsal surface of an insect mesothorax.

mesopelagic. Pertaining to mid-depths of the open ocean, especially between depths of 200 and 1000 m.

Mesoperipatus. Congo genus of Onychophora.

mesophragma. 1. Old term for M line in striated muscle fibers; 2. Mesothoracic phragma in certain insects; an internal chitinous projection from the post-scutellum. 3. In a few crustaceans, an endosternal process arching over the sternal canal.

mesoplankton. 1. Collectively, those plankton organisms below the photosynthetic stratum. 2. Collectively, those plankton organisms retained by a plankton net.

mesopleuron. Lateral portion of an insect mesothorax.

Mesoplodon. Genus of beaked whales and cowfish in the Family Ziphiidae; to 16 ft. long; Atlantic, Pacific, and Indian oceans; five or six spp. along N.A. shores.

mesopterygium. Middle heavy cartilaginous rod supporting the base of each pectoral fin in elasmobranch fishes.

mesopterygoid. Bone of the suspensorium.

mesorchium. Special part of the peritoneum which supports the vertebrate testes.

mesosalpinx. Fold of the peritoneum which suspends the vertebrate oviduct.

mesosaprobic. Pertaining to aquatic habitats with a reduced quantity of dissolved oxygen and considerable organic decomposition; intermediate between polysaprobic and katharobic.

mesoscutellum. Scutellum of the insect mesothorax.

mesoscutum. Scutum of the insect mesothorax.

mesosoma. 1. Middle part of the body. 2. In arachnids, the anterior portion of the abdomen, often clearly set off from the more posterior metasoma.

mesosternum. 1. Gladiolus; central element of the tetrapod sternum. 2. Chief ventral sclerite of an insect mesothorax.

Mesostigmata. Suborder taxon sometimes used to include certain families of Acarina; one pair of stigmata; free-living and predaceous, or commensal.

Mesostoma. Large and widely distributed genus of small fresh-water rhabdocoel turbellarians.

mesothelium. Lining of a coelomic cavity, especially the thin membrane in vertebrates.

mesothorax. Second segment of an insect thorax. Adj. mesothoracic.

Mesothuria. Genus of sea cucumbers in the Order Aspidochirota; in shallow to deep waters.

mesotroch. Ciliated band around the middle of a marine annelid larva.

mesotrophic. Category occasionally used to include lakes of intermediate productivity, i.e. between clearly defined eutrophic and oligotrophic conditions.

mesovarium. Special part of the peritoneum used in supporting the vertebrate ovary.

Mesovelia. See MESOVELIIDAE.

Mesoveliidae. Water treaders; a small family of carnivorous Hemiptera found around ponds and on floating vegetation; Mesovelia mulsanti common over most of U.S.

Mesozoa. Phylum of parasites of the nephridia of squids and octopuses and rare parasites in various tissues and body cavities of other marine invertebrates; considered by some authorities to be a degenerate class of Platyhelminthes; body typically wormlike, 0.5 to 10 mm. long when mature, and consists of an outer layer of 16 to 42 ciliated somatic cells enclosing one or more very large axial reproductive cells; the eight anterior cells form a compact polar cap; such an individual is called a nematogen, and the axial cell(s) by repeated asexual fission give rise to vacuolated agamete cells which escape from the parent and grow into nematogens; under other circumstances a nematogen may become modified and release minute ciliated infusorigens which swim about and then enter a new host; Dicyema and Rhopalura are best-known genera; about 50 spp. in the phylum.

Mesozoic era. One of the five major geological subdivisions of time; the "Age of Reptiles," lasting from about 190 million to 70 million years ago; divided into Cretaceous, Jurassic, and Triassic periods.

Metabola. Taxon which includes most orders of insects having distinct external changes during the various stages in the life history; subdivided into Paurometabola, Hemimetabola, and Holometabola (q.v.). Adj. metabolous.

metabolic gradient. Axial gradient.

metabolic water. Water formed in the tissues as the result of oxidation of food substances and other metabolic activities.

metabolism. Collectively, all physiological processes in any organism, including such processes as feeding, digestion, respiration, excretion, reproduction, defecation, etc.

metabolite. 1. Any substance produced as the result of metabolic processes. 2. Any substance involved in metabolism, whether taken in or produced within the body.

Metacaprella. Genus of caprellid Amphipoda.

metacarpal. Metacarpus.

metacarpus. 1. Region of the forefoot of tetrapods. 2. One of the bones in the forefoot. 3. Region of the palm in man.

metacercaria. Trematode cercaria which has lost its tail and become encysted in or on an intermediate aquatic host or some inanimate object; the metacercaria is usually rounded and enveloped by a secreted covering.

Metacheirus. Genus of small tree-dwelling Central and South American opossums.

metachronal rhythm. Pattern of coordinated wavelike beating of fields of cilia or flagella in any tissue.

Metacineta. Genus of fresh-water suctorians; cell in a vase-shaped lorica.

metacneme. Secondary mesentery of many anthozoan coelenterates; extends from the body wall inward a considerable distance but does not attach to the stomodaeum.

metacoel. Cavity of the embryonic metencephalon.

metacone. Outer posterior cusp of a mammalian upper molar tooth.

metaconid. Inner anterior cusp of a mammalian lower molar tooth.

metaconule. Minor cusp near the posterior margin of a mammalian upper molar.

Metacrinus. Genus of sessile crinoid echinoderms.

metagenesis. Alternation of generations among animals.

metagnathous. 1. Having the mandible tips crossed, as in the crossbills. 2. In certain insects, having larvae with biting mouth parts and adults with sucking mouth parts, e.g. Lepidoptera.

Metagonimus yokogawai. Minute intestinal fluke found in southeastern Asia, East Indies, and the eastern Mediterranean area; in the small intestine of man and other mammals; intermediate hosts are a fresh-water snail and various fishes.

metalimnion. Thermocline (c.f.), mesolimnion, discontinuity layer; middle layer of a thermally stratified lake; temperature drops at least 1° C. per meter throughout this stratum; usually 1 to 8 m. thick, depending on lake morphology.

metallic wood borers. See BUPRESTIDAE.

metamere. One of the repeated units of structure in the body of typical Annelida, Arthropoda, and Chordata (faintly developed); a body segment. Adj. metameric.

metamerism. Linear repetition of body plan and certain parts (metameres), either external or internal or both; the condition in typical Annelida, Arthropoda, and Chordata (obscure).

metamorphosis. Period of abrupt transformation from one distinctive stage in the life history to another, such as the transition of a frog tadpole to an adult, the transition of a larval insect to a pupa, the transition of one insect stage to another, the transition of an annelid trochophore into the typical worm stage, etc. See AMETABOLA, METABOLA, PAUROMETABOLA, HEMIMETABOLA, and HOLOMETABOLA.

metamyelocyte. Late stage in the differentiation of a myelocyte.

metanauplius. One to several larval stages subsequent to the first (nauplius) larva in copepods, ostracods, barnacles, etc.; characterized by increasing size and the appearance of additional appendages.

Metandrocarpa. Genus of ascidians.

metanephridium. Type of nephridium; in the broad sense, one of few to numerous paired osmoregulatory and excretory tubules in certain phyla; structure simple and tubular to large and glandular; the ciliated internal opening (nephrostome) is located in the coelom or pericardial cavity, and the external opening (nephridiopore) lies on the surface of the body; present in most annelids, sipunculids, echiurids, mollusks, and onychophorans. Pl. metanephridia. See also COELOMODUCT.

metanephromixium. Type of nephromixium in which the coelomostome is grafted on to an open nephrostome; serves as genital and/or excretory duct; in certain Polychaeta.

metanephros. Functional kidney of reptiles, birds, and mammals; its collecting duct is a ureter. Adj. metanephric.

metaphase. That stage in mitosis during which the duplicated chromosomes are arranged in a flat plate in the center of the mitotic spindle midway between the two asters (if present).

Metaphelenchus. Genus of microscopic, free-living nematodes.

metaplasia. 1. Change in form of an adult tissue to an abnormal condition. 2. Transformation of one tissue into another, as cartilage into bone.

metaplasm. Collectively, all those particles or inclusions in a cell which are not essential parts of the living protoplasm.

metapleural fold. Lateroventral finlike fold on each side of a cephalochordates; extends from anterior end to region of the atriopore.

metapleuron. Lateral portion of an insect metathorax.

metapneustic. Pertaining to certain Diptera larvae in which only the last abdominal spiracles are open.

metapodial. Proximal bony element of a toe, e.g. metacarpal or metatarsal.

metapodium. 1. Posterior portion of a mollusk foot. 2. Metacarpus or metatarsus of a tetrapod. 3. That portion of the foot between tarsus and digits.

metapterygium. Heavy cartilaginous rod supporting the base of each of the pectoral and pelvic girdles in elasmobranch fishes.

metapterygoid. Bone of the suspensorium; supports the lower jaw.

metascutellum. Scutellum of insect third thoracic segment.

metascutum. Scutum of insect third thoracic segment.

metasoma. 1. Posterior part of body in Arachnoidea. 2. Postabdomen. 3. Abdomen in certain crustaceans, e.g. isopods and copepods. 4. That part of the trunk of a copepod anterior to the especially flexible articulation; includes cephalothorax and usually all of the free thoracic segments. 5. Posterior part of the body of a mollusk. Also spelled metasome.

metastasis. Transfer of disease from one area to another one not adjacent; results from circulatory movement of microorganisms or groups of malignant cells. Pl. metastases.

metasternum. 1. Ventral metathoracic plate of insects. 2. Posterior part of sternum in tetrapods.

metastoma. Lower lip of certain Crustacea and eurypterids; often cleft into two paragnaths.

Metastrongylus. Genus of rhabditid nematodes; several spp. are important parasites in the hog lung, especially M. apri and M. elongatus.

metatarsus. 1. Sole; general region of the hind foot in tetrapods, containing a series of metatarsal bones. 2. One or more of the terminal segments in the legs of certain arthropods. 3. Second last segment in the legs of most arachnids. Adj. metatarsal, pl. metatarsi.

Metatheria. One of the three infraclasses of the mammal Subclass Theria; marsupials; female with a ventral pouch (marsupium) or folds surrounding the nipples; uterus and vagina double, placenta absent; early development in uterus, but premature young leave the uterus and crawl into marsupium where each one becomes attached to a nipple with the mouth until fully formed; includes the opossum, koala, wombat, wallaby, kangaroo, etc.; great majority of spp. restricted to the Australian region.

metathorax. Last or third segment of an insect thorax. Adj. metathoracic.

metatroch. Postoral ciliary band in a trochophore larva.

Metazoa. Subkingdom or division of the animal kingdom which includes all phyla except the Protozoa.

metazoan. Any multicellular animal.

metazoea. Marine crustacean larval stage occurring between the zooea and the megalops stages, especially in decapods.

Metchnikoff, Élie. Russian biologist (1845-1916); made pioneer observations on the behavior of leucocytes and developed the theory of phagocytosis; also worked on lactic acid bacteria and syphilis; shared the Nobel prize in physiology and medicine in 1908.

metecdysis. Especially in decapod crustaceans, the molting recovery period, when the animal is still somewhat soft.

metencephalon. Front part of the hindbrain of vertebrates; includes the cerebellum and pons Varolii.

metepipodite. Epipodite borne on the coxopodite, as in many crustaceans.

methemoglobin. Modified form of oxidized hemoglobin, resulting from the action of various drugs and oxidizing agents, such as ozone, nitrites, or chlorates; chocolate brown and incapable of functioning in respiration.

methionine. Amino acid; $C_5H_{11}O_2NS$.

methyl alcohol. Methanol; alcohol derived from fermentation of wood and other vegetable substances; toxic to man; CH_3OH.

methyl cellulose. Synthetic organic compound readily available commercially and excellent for studying living protozoans; a 10% syrupy solution slows protozoan movements greatly with no obvious physiological effects.

methylene blue. Important biological stain and mild antiseptic; used as a hydrogen acceptor in physiological work.

Metis. Common genus of marine harpacticoid Copepoda.

Metopa. Genus of marine Amphipoda.

Metoporhaphis. Genus of spider crabs.

Metopus. Common genus of elongated, plastic, heterotrich ciliates; peristome conspicuous, slightly spiral; marine, fresh-water, and parasitic.

Metridia. Genus of marine copepods in the Suborder Calanoida.

Metridium. Common genus of sea anemones.

Metriocnemus. Genus of tendipedid midges.

Metrobates. Uncommon genus of water striders in the Family Gerridae.

Mexican bean beetle. See EPILACHNA.

Mexican jumping bean moth. See LASPEYRESIA SALTITANS.

Meyenia. Common and widely distributed genus of fresh-water sponges.

Miana bug. Argas persicus.

Miastor. Genus of minute flies in the Family Cecidomyiidae; commonly reproduce by paedogenesis by which an immature larva may give birth to 7 to 30 daughter larvae.

Miathyria. Genus of tropical and subtropical dragonflies.

Micaria. See CLUBIONIDAE.

micelle. Micell; physical submicroscopic unit of structure in living protoplasm; molecular aggregate; term now seldom used.

Michigan grayling. See THYMALLIDAE.

Michtheimysis. Genus of marine crustaceans in the Order Mysidacea.

Michurinism. Heredity principles of the Russian horticulturist Michurin; dominance is modified and controlled by the environment; transmission of characters depends on comparative parental vigor; genetic make-up of a stock may be modified by a graft.

Mico. See TAMARIN.

Micrarionta. Genus of land snails; Pacific Slope.

Micrasema. Genus of Trichoptera.

Micrathena. Genus of brightly colored spiders in the Family Araneidae.

Micrathene whitneyi. Elf owl, a small brownish-gray sp. of southwestern U. U. to central Mexico; about as large as a sparrow; commonly nests in saguaro.

micresthete. See MICROPORE.

microbe. Microscopic vegetable microorganism, especially pathogenic bacteria.

microbiology. Study of microorganisms, including Protozoa, Algae, Fungi, Bacteria, and viruses.

microbiota. Collectively, the microscopic plants and animals of a habitat or region.

Microbuthus. Genus of small scorpions; one sp. only 13 mm. long.

Microcebus. Genus of dwarf lemurs of

Madagascar.

Microcentrum. Common genus of U.S. katydids.

microcephalic idiot. Type of human idiot in which the head is relatively small.

microcephalus. Abnormally undersized cranium.

microcercous. Having a very short or vestigial tail.

Microchiroptera. Suborder of the mammal Order Chiroptera; includes the smaller insectivorous bats; fly about at night and feed on insects in flight; hang head downward during day in crevices, hollow trees, caves, buildings, etc.; typical genera are Myotis, Eptesicus, Tadarida, and Desmodus.

Microciona. Genus of sponges in the the Class Demospongiae; redbeard sponges; clusters of bright orange-red finger-like lobes.

microclimate. Climatic conditions as related to small organisms living on or in the ground or in a microhabitat; such conditions are often greatly different from those obtaining in the circulating air several feet above the ground.

Micrococcus. Genus of gram-positive bacteria; spherical cells in irregular masses; widely distributed wherever decay is in progress; one sp. produces boils, carbuncles, and abscesses in man and domestic animals.

Microcorycia. Small genus of fresh-water amoeboid protozoans; cell mostly covered by a discoidal or hemispherical test.

Microcotyle. Genus of monogenetic trematodes parasitic on gills of marine fishes.

Microcylloepus. Genus of elmid beetles.

Microdalyellia. Large genus of small rhabdocoel turbellarians.

Microdeutopus. Genus of marine Amphipoda.

Microdipodops. Small genus of kangaroo mice or dwarf pocket rats; nocturnal spp. with long hind legs and feet; body to 3 in. long; deserts of west and southwest. Cf. ZAPODIDAE.

microdissection. Technique involving dissection, manipulation, or operation on a small bit of tissue or a single cell with mechanical instruments while being viewed through a microscope.

Microdrili. Order of oligochaetes; includes small spp. with few segments, often reproducing asexually.

microevolution. Evolution based on the gradual accumulation of small genetic changes within a sp., leading to the formation of a slightly differing new subspecies, but not leading directly to the formation of a new sp. Cf. MACROEVOLUTION.

microfilaria. Minute larval stage of filaria nematode parasites. Pl. microfilariae.

Microgadus tomcod. Frostfish or tomcod; inshore codlike fish of the northern Atlantic Coast; commercially important; see GADIDAE.

Microgale. Genus of long-tailed tenrecs; mouselike insectivores of Madagascar.

microgamete. Male gamete, as distinguished from the larger, less abundant macrogamete (female gamete); term usually used with reference to certain Protozoa which exhibit sexual reproduction.

microgametocyte. Cell which undergoes meiosis and forms microgametes (sperm), especially in sporozoan Protozoa.

Microgaster. Genus of Hymenoptera in the Family Braconidae; small flies whose larvae parasitize Lepidoptera.

microgyne. Dwarf female ant.

microhabitat. Small or restricted habitat, e.g. a dead animal, fallen acorn, protozoan culture, etc.

Microhydra. Generic name formerly used for a minute fresh-water hydroid; now known to be the hydroid generation of Craspedacusta sowerbyi.

Microhyla carolinensis. Common narrow-mouthed toad, found from Va. to Ind. and south to Fla. and Tex.; head and body to 1 in. long; without teeth; nocturnal, in ground during day.

Microhylidae. Cosmopolitan family of amphibians in the Order Diplasiocoela; narrow-mouthed toads or frogs; head small, narrow, wedge-shaped; burrowing to arboreal; Microhyla a familiar U.S. genus of the southeastern states.

micro-incineration. Burning of bits of tissue or microorganisms on a microscope slide so that the nature and distribution of the residual ash may be studied.

Microkalyptorhynchus. Genus of rhabdocoel Turbellaria.

Microlaimus. Genus of microscopic, free-living nematodes.

microlepidoptera. Collectively, the smaller moths and butterflies.

Microlinypheus. Genus of minute spiders, less than 1 mm. long.

micromanipulator. Finely built device

for handling microtools used in operating on single cells, etc. under the microscope.

micromelia. Abnormally small size of the limbs.

micromere. One of the distinctly smaller cells resulting from unequal cleavages during early embryology of polychaetes, archiannelids, polyclad turbellarians, nemertines, mollusks, and some chordates.

micromutation. One of many slight mutations the sum of which may result in the formation of a new subspecies (or rarely a new sp.).

Micromya. Genus of unionid mollusks; eastern half of U.S.

micron. Measure of distance; 0.001 mm. or 0.00003937 inch.

micronephridium. Minute mephridium, often without a nephrostome; large numbers forming clusters often found in a single segment in certain oligochaetes.

micronucleus. Small mass of nuclear material, especially as distinguished from the adjacent large macronucleus in ciliate and suctorian protozoans; one or more micronuclei may be present in such a protozoan; micronuclei typically divide by mitosis during ordinary fission.

microorganism. Any microscopic organism, e.g. protozoan, bacterium, rotifer, etc.

Micropalama himantopus. Stilt sandpiper; breeds in Arctic America, winters in Argentina.

Micropallas. Micrathene.

Microperca. Genus of darters now usually included in Etheostoma.

microphages. Neutrophils.

microphagous. Pertaining to animals which feed on relatively small particles of food compared to their own body size, e.g. clams. Cf. MACROPHAGOUS.

microphonic impulses. Those nerve impulses generated by vibrations in the cochlea and carried by the auditory nerve.

microphotograph. Strictly speaking, a photograph which is of microscopic size; often misused in place of photomicrograph.

microphthalmos. Abnormally small eyes; a Mendelian recessive in man.

microplacoid. Single, very small sclerotized piece imbedded in the posterior part of the pharyngeal musculature of a tardigrade. Cf. MACROPLACOID.

microplankton. Nannoplankton.

Micropodidae. Same as Apodidae; large family of birds in the Order Micropodiformes or Apodiformes; swifts; bill small and weak, used for taking insects on the wing; do not perch but cling to vertical surfaces; wings long, slender, and stiff; mucilaginous salivary secretion used in gluing twigs together for the nest; Chaetura pelagica is the common chimney swift of eastern U.S.

Micropodiformes. Order of birds which includes the swifts and hummingbirds; small spp.; legs and feet small, bill short and weak (swifts) or long and slender (hummingbirds); same as Apodiformes; see MICROPODIDAE and TROCHILIDAE.

Micropogon. See SCIAENIDAE.

micropore. 1. One of many small pores in the dorsal plates of certain chitons; contains a sense organ(micresthete). 2. In certain coccids, a minute tubule which serves as a duct for wax glands.

Microporella. Genus of marine Bryozoa; colonies form small silvery discs.

Microprotopus. Genus of marine Amphipoda.

micropterous. Having small or vestigial wings.

Micropterus. Genus of centrarchid fishes which includes the smallmouth black bass (M. dolomieu), largemouth black bass (M. salmoides), spotted bass, and redeye bass of American lakes; all are important game fishes; (the largemouth black bass is known by at least 43 local names).

Micropterygidae. Family of small moths in the insect Order Lepidoptera; mandibles functional; live chiefly on liverworts; Micropteryx and Epimartyria typical.

Micropteryx. Old World genus of small moths in the lepidopteran Family Micropterygidae.

micropyle. 1. Minute pore in the chorion of an insect egg through which sperm enter. 2. Small area of a sponge gemmule which is not covered by the gemmule membrane; germination of the gemmule occurs through this area.

Microsauria. Group of late Paleozoic amphibians sometimes included in the Subclass Stegocephalia.

microscleres. Small spicules scattered about in the mesenchyme of certain sponges.

microscope. In the broad sense, any optical instrument containing one or more lenses for magnifying the image of small objects. See BINOCULAR MICROSCOPE, COMPOUND MICROSCOPE.

microsere. Successional series of plant and animal communities in a microhabitat.

microsome. Ribosome; a fine granular element of cytoplasm, usually invisible with the ordinary laboratory microscope but easily seen with the electron microscope; composed of protein and RNA and thought to function in protein synthesis.

Microsorex. Genus which includes the pygmy shrews; the smallest N. A. mammals, body less than 3 in. long; dark brown above, tail short; eastern U. S.

microspecies. Reproductively isolated chromosomal population; e. g. a population of Artemia salina which reproduces only by parthenogenesis.

microspore. Smaller of two types of spores formed by certain amoeboid protozoans.

Microsporidia. One of the three orders in the Subclass Cnidosporidia; each spore with one polar capsule; intracellular parasites of fishes, arthropods, and a few other invertebrates; example is Nosema.

Microstomum. Common genus of marine and fresh-water turbellarians (Order Rhabdocoela); often found in asexual chains of four individuals.

Microtendipes. Genus of tendipedid midges.

Microthelyphonida. Palpigrada.

microthorax. 1. In Odonata, a minute area at the anterior margin of the thorax. 2. Neck of an insect.

Microthorax. Large genus of small, flattened, fresh-water holotrich Ciliata; with delicate, ridgelike, keeled armor.

Microtinae. Subfamily of the mammal Family Cricetidae; voles and lemmings.

microtome. High precision machine used for cutting frozen or embedded tissues into sections for mounting on slides to be studied under the microscope; for routine study, sections are cut about 10 microns thick, but some special instruments can cut sections only 1 micron or less thick.

microtrichia. Small fixed hairs of an insect wing; no basal articulation. Sing. microtrichium.

Microtus. Very large genus of essentially herbivorous rodents in the Family Cricetidae; includes the meadow mice, field mice, or voles; medium-sized, short-tailed mice, common everywhere in N. A. in a wide variety of habitats; commonly make ramifying runways on ground which is covered by grasses and debris.

Microvelia. Common and widely distributed genus of broad-shouldered water striders in the Family Veliidae.

microvilli. Minute and abundant fingerlike irregularities in cell membranes of some tissues; visible only with the electron microscope.

Micrura. Genus of small, flat nemertines of the intertidal zone; caudal cirrus present.

micrurgy. Surgery on a microscopic basis; done while viewing the object through a microscope, and often with the aid of microdissection apparatus.

Micruroides euryxanthus. Sonoran coral snake of southern N. M. and Ariz.; coloration consists of red, black, and yellow bands; burrower which feeds on lizards and snakes; venom presumably dangerous to man, although there appear to be no published records of such bites.

Micrurus fulvius. American coral snake or harlequin snake; coloration consists of red, black, and yellow bands; to 3 ft. long; burrower which feeds mostly on lizards and snakes; highly venomous to man although the fangs are short and used with "chewing" movements; Gulf States; also several South and Central American spp.

mictic. 1. Of or pertaining to a female rotifer whose haploid eggs are capable of being fertilized by a male. 2. Of or pertaining to fertilized eggs released by a female mictic rotifer.

micturation. Urination.

midbrain. Mesencephalon; middle of the three fundamental divisions of the vertebrate brain; includes the corpora quadrigemina and peduncles.

middle ear. Cavity between the tympanic membrane and inner ear (membranous labyrinth); present in all tetrapods except the Gymnophiona, Caudata, and snakes; air-filled and communicates with the pharynx by means of a eustachian tube; one or three minute auditory ossicles traverse the cavity of the middle ear and transmit vibrations from the tympanic membrane to the endolymph of the inner ear.

middle lacerate foramen. Opening at the back edge of the alisphenoid in the mammalian skull; carries the internal carotid artery.

middle piece. Midpiece; that portion of a spermatozoan just behind the head; structure variable but often contains the centrioles and mitochondrial sheaths.

Mideopsis. Genus of Hydracarina.

midge. Any of numerous, small, delicate dipterans; some spp. bite.

midgut. 1. Middle portion of the digestive tract of arthropods; it is not lined with cuticle and is preceded by the foregut and followed by the hindgut; may be very short, or more than half as long as the body. 2. In vertebrate embryos, that part of the embryonic gut which gives rise to the intestine (between stomach and rectum).

midshipman. See HAPLODOCI.

midwife toad. See ALYTES.

Migratory Bird Act. Agreement reached by Great Britain and the U.S. on 3 July 1918 for the protection of migratory game birds.

Miliola. Genus of foraminiferans; skeletons form an important part of certain limestones.

military monkey. See ERYTHROCEBUS.

milk snake. See LAMPROPELTIS.

milk teeth. Those teeth in juvenile mammals which are lost and replaced by permanent teeth; deciduous teeth.

milkweed bug. One of many black and red or orange bugs which feed chiefly on milkweed plants.

milkweed butterfly. See DANAUS.

millepede. Millipede.

Millepora. See MILLEPORINA.

Milleporina. Suborder of colonial hydrozoan coelenterates; the polyps are embedded in a massive upright, leaflike or branching calcareous exoskeleton; Millepora the only genus ("stinging coral") a regular component of coral reefs.

miller. Common name applied to many spp. in several families of moths, especially Tineidae and Noctuidae; medium-sized moths which come into houses and flutter about lights; the wings are covered with fine light-colored scales which come off easily and have the appearance of flour. See FRENATAE.

miller's thumb. Any of certain small fresh-water fishes in the genus Cottus.

milliliter. One-thousandth of a liter and essentially equivalent to a cubic centimeter. Abbr. ml.

millimeter. Unit of distance; 0.001 meter or 0.039 inch.

millimicron. Unit of distance; one-thousandth of a micron.

millions. See ACANTHOPHACETUS.

millipede. See DIPLOPODA.

millisecond. A thousandth of a second.

Millon's test. Addition of Millon's fluid to proteins and other compounds containing a hydroxy-phenyl group gives a diagnostic red color; Millon's fluid contains mercuric nitrite and nitrate in nitrous and nitric acids.

Milne-Edwards, Henri. French zoologist (1800-1885); comparative anatomist and physiologist; made important studies on crustaceans, mollusks, and corals.

Milnesium tardigradum. Cosmopolitan sp. of tardigrade.

milt. 1. Mass of sperm suspended in fluid released by a male fish. 2. Testis of a fish. 3. The spleen.

milu. See ELAPHURUS.

mimetic. Mimicking another sp.; see MIMICRY.

mimic. See MIMICRY.

mimicry. Imitation in form, color, or behavior by a comparatively defenseless and edible sp. (mimic) of another sp. (model) which has qualities that cause it to be avoided by predatory animals; especially evident among insects. See BATESIAN MIMICRY and MÜLLERIAN MIMICRY.

Mimidae. Family of passerine songbirds which includes the thrashers, mockingbirds, and catbirds; most spp. found in N.A.; robin-sized, tail long, bill slender; Mimus, Dumetella, and Toxostoma typical.

Mimulus. Genus of marine crabs in the Suborder Brachyura.

Mimus polyglottos. Common American mockingbird in the Family Mimidae; tail long; coloration brownish gray above, white and gray below; exceptional singer and mimic.

mineral cycle. World-wide circulation and reutilization of atoms of various "minerals," chiefly due to metabolic processes of plants and animals and their disintegration by bacterial action after death; calcium, phosphorus, sulfur, etc. are all circulated in this manner.

mineralocorticoids. Group of adrenal cortex hormones which regulate sodium and potassium metabolism.

minim. 1. Small liquid measure, about

a drop. 2. Ant of the smallest worker class.

minimal air. That small amount of air remaining in the lungs even after complete collapse.

mining bee. See ANDRENIDAE.

Miniopterus. Genus of long-winged bats; southern Eurasia and northern Africa

minivet. See CAMPEPHAGIDAE.

mink. See MUSTELA.

minnie. Slang for minnow.

minnow. 1. Strictly speaking, any fish in the Family Cyprinidae. 2. In common usage, any small or young fish.

minor coils. Large number of minute coils of chromatin granules making up a chromonema; in addition, each chromonema is twisted into a relatively small number of major coils; coiling is loose and drawn out (if present at all) in an interphase cell and tight in a metaphase chromosome.

minuten nadeln. Extremely fine, short, headless pins used in mounting minute insects or parts of insects.

Minytrema melanops. Sp. of sucker occurring in much of the eastern half of U.S.; usually in small streams and impoundments.

Miocene epoch. One of the five geological subdivisions of the Tertiary period, between the Pliocene and Oligocene epochs; lasted about 13 million years; especially characterized by an abundance and variety of mammals, including carnivores.

Miohippus. Genus of fossil ancestral horses; about 24 in. tall at the shoulder; Oligocene to Miocene.

Miomastodon. Genus of Miocene mastodons.

miracidium. Minute, ciliated free-swimming larva which hatches from a trematode egg; sometimes hatches internally from an egg ingested by a snail, sometimes hatches in the water and must find and penetrate the tissues of a snail; usually develops into a sporocyst within the snail. Pl. miracidia.

Miranda. Argiope.

Miridae. Large family of hemipteran insects which includes the plant bugs, leaf bugs, or capsid bugs; mostly herbivorous, a few carnivorous; many spp. are pests of cultivated plants; Lygus and Halticus are important genera.

Mirounga angustirostris. Elephant seal or sea elephant in the Family Phocidae; largest of true seals; males to 18 ft. long, with a long inflatable proboscis; almost extinct, found only on Guadalupe Island and certain isolated areas of Lower California; a related sp. occurs in the Southern Hemisphere.

mirror carp. Variety of common carp with exceptionally large scales.

miscarriage. Abortion.

Misgurnis fossilis. Thunderfish; European loach which burrows and feeds in muddy ponds and stream bottoms; when it swims actively in the water, a thunderstorm is said to be imminent.

misreference. Sensation of pain in a part of the body other than that in which the nerve stimuli originate; e.g. the sensation of pain in the right shoulder in persons having gall bladder trouble.

missing link. Popular expression to denote hypothetical intermediate between man and a presumed apelike ancestor; a more credible idea holds that manlike and apelike creatures had common ancestors.

Mississippian epoch. Earlier of the two epochs of the Carboniferous period.

Misumena. Common genus of brightly-colored crab spiders, usually associated with flowers.

Misumenoides. Common genus of spiders in the Family Thomisidae; especially on flowers.

mite. General term for the great majority of spp. in the arachnid Order Acarina; all Acarina except the Ixodoidea (ticks); hypostome small, unarmed; usually less than 2 mm. long.

Mitella. Genus of barnacles having a long fleshy basal stalk; goose or gooseneck barnacles; carcaleous plates more numerous than in Lepas.

mithan. Gayal.

mitochondria. Minute granular or rodlike bodies in the cytoplasm of all cells except those of bacteria and blue-green algae; usually 1 to 7 microns in greatest dimension and appear to consist of protein and lipid materials; contain enzymes and coenzymes and are closely associated with fat, glycogen, and protein metabolism; the chief source of ATP. Sing. mitochondrion.

mitogenetic rays. Short wave-length rays presumed to emanate from living tissues and stimulate mitosis in other tissues exposed to them in the immediate vicinity; modern biologists doubt the existence of such a mechanism.

mitosis. Usual process of nuclear division in which there is an equal qualitative and quantitative division of the chromosome material between the two resulting nuclei; each chromosome has a duplicate, double structure in the form of two closely associated chromatids some time before the onset of mitosis; during mitosis each chromatid of a pair assumes the size of a chromosome and goes into one of the daughter nuclei; the process of mitosis is customarily divided into prophase, metaphase, anaphase, and telophase. Adj. mitotic, pl. mitoses. See CHROMATID.

mitotic spindle. Mitotic figure; barrel-shaped or spindle-shaped arrangement of fibrils and gel appearing in cells during mitosis and meiosis with the chromosomes arranged at its equatorial region; in animal cells a centriole may be located at each pole of the spindle; it is likely that the separation of chromatids and subsequent movement of daughter chromosomes toward opposite poles is produced by pulling action of some of the spindle fibers which are attached to the centromeres of the chromosomes.

Mitra. Genus of mitre shells (Gastropoda); long, slender, with large aperture; Carolinas to Caribbean.

mitral valves. Tricuspid and bicuspid valves.

mitraria. Peculiar lobate type of trochophore found early in the life history of some Polychaeta.

Mitrella. Genus of small gastropods of all U. S. coasts; to 0. 25 in. long.

mitre shell. Any of a large number of tropical gastropods; shell elongate, spindle-shaped, and often brightly colored and large.

Mitsukurina owstoni. Goblin shark; a Japanese sp. having the snout greatly prolonged and bladelike.

mixed nerve. Nerve containing both sensory and motor neurons.

mixonephridium. Type of nephromixium having the coelomostome completely fused to the inner end of the nephridium so as to form a large funnel-shaped organ having both excretory and genital functions; in many Polychaeta.

mixotrophic nutrition. Term usually applied to certain Protozoa and unicellar algae that exhibit two or three of the following types of nutrition: holozoic, holophytic, and saprozoic.

M-N antigens. Two proteins (M and N) present in the red blood cells of man; not antigenic for humans but can be demonstrated by causing the production of their specific antibodies in rabbits; M and N are determined by a single pair of alleles acting independently of each other without dominance.

Mnemiopsis leidyi. Common bioluminescent ctenophore of the North Atlantic; with two oral lobes and four large auricles; to 5 in. in height.

Mniobia. Genus of bdelloid rotifers.

Mniotilta. See MNIOTILTIDAE.

Mniotiltidae. Large family of passerine birds which includes the warblers; restricted to North and South America; small, active, brightly-colored, insectivorous spp. with thin bills; common genera are Mniotilta, Vermivora, and Dendroica.

moa. See DINORNITHIFORMES.

Moapa coriacea. Small, relict minnow known only from Warm Springs, Clark Co., Nev.

Mobilia. Suborder of Ciliata in the Order Peritricha; free-swimming but with the aboral end highly modified for temporary attachment; e. g. Trichodina.

moccasin. See AGKISTRODON.

Mochlonyx. Genus of phantom midges in the Family Culicidae.

mockingbird. See MIMUS.

model. See MIMICRY.

modifying genes. Complex of genes which influences the relative intensity of expression of another particular pair of alleles; e. g. if Jersey cattle are homozygous recessive for "spotting," the amount of spotting is dependent upon the relative numbers and kinds of homozygous and heterozygous modifying genes.

Modiolaria. Common and widely distributed genus of marine mussels.

modiolus. Central bony pillar of the cochlea of the ear.

Modiolus. Common and widely distributed genus of marine mussels; includes the horse mussel (M. modiolus) and the ribbed mussel (M. demissus).

Moeritherium. Original and most ancient fossil genus of ancestral elephants; about 24 in. high at the shoulder, with a long, narrow head and 36 teeth; true trunk absent; late Eocene.

mohr. North African gazelle; each horn with 11 or 12 prominent rings.

Moina. Common genus of heavy, thick-bodied Cladocera usually found in pools and ponds.

Moira. Genus of heart urchins in the Order Spatangoida.

mojarra. Any of numerous small tropical marine shore fishes; silvery spp. with protrusible jaws and lips; to 6 in. long.

Mola. See MOLIDAE.

Molanna. Common genus of caddis flies; larva builds a case of sand grains.

molar. Pertaining to the most posterior teeth in mammals; usually more or less modified for crushing and grinding; 12 molars in man.

mole. See TALPIDAE.

mole crab. See EMERITA and LEPIDOPA.

mole cricket. See GRYLLOTALPIDAE.

molecular biology. That phase of biology which treats the ultimate physical, chemical, and physiological nature of living matter, especially on an ultramicroscopic basis.

mole mite. See HAEMOGAMASUS.

mole-rat. Any of a wide variety of molelike rodents found in Europe, Asia, and Africa; see SPALACIDAE.

mole-shrew. See ANOUROSOREX.

Molgula. Large cosmopolitan genus of tunicates; simple and more or less spheroidal; often mostly buried in sand or mud; common spp. usually called sea grapes.

Molidae. Family of massive oceanic fishes; ocean sunfish; body shortened; soft dorsal and anal fins short and very high; caudal peduncle absent; caudal fin reduced to a flap of skin around rear periphery of trunk; Mola mola common, to 1800 lbs.

Mölleria. Genus of minute snails of the North Atlantic.

Mollienisia. Genus of small American tropical and subtropical fresh-water fishes in the Family Poeciliidae; the mollies; males unknown in some spp.; S.C. to Tex. in the U.S.

mollusc. Mollusk; any member of the Phylum Mollusca.

Mollusca. Phylum including soft-bodied animals usually partly or wholly enclosed within a calcium carbonate shell and having a muscular "foot" for locomotion; body covered by a mantle which secretes the shell; usually a head; heart and some blood vessels present, but with an open circulatory system; gills usually present; one, two, or four nephridia (coelomoducts); nervous system composed of few paired ganglia and connectives; marine, fresh-water, and terrestrial; includes snails, clams, chitons, squids, octopus, etc.; about 96,000 living spp.

Molluscoidea. Phylum category used by some zoologists; includes the Bryozoa, Brachiopoda, and Phoronidea, all of which have a lophophore.

mollusk. Mollusc; any member of the Phylum Mollusca.

molly. See MOLLIENISIA.

Moloch. Genus of Australian lizards having a particularly spiny head and body.

Molossidae. Family in the Order Chiroptera; includes the free-tailed bats; tropical and subtropical; Tadarida typical.

Molothrus ater. Small blackbird in the Family Icteridae; bill short, conical; female gray; eggs mostly laid in the nests of other birds which then hatch and care for the young; temperate N. A.

Molpadia. See MOLPADONIA.

Molpadonia. Echinoderm order in the Class Holothuroidea; tentacles finger-shaped, tube feet represented only by anal papillae, respiratory tree present; Molpadia and Caudina typical.

molt. 1. Periodic shedding of the exoskeleton to permit an increase in size; especially characteristic of nematodes, arthropods, and tardigrades. 2. Process of shedding and feather replacement in a bird, or the shedding of hair or horns.

molting hormone. Hormone secreted by gland cells associated with insect ganglia; induces molting and maturation.

Monacanthidae. Family of tropical to temperate marine fishes; includes the filefishes; body deep, compressed, and covered with tiny hard scales set in a tough skin; small terminal mouth and one stout dorsal spine; Monacanthus, Alutera, and Stephanolepis.

Monocanthus. See MONACANTHIDAE.

Monachus. Genus of tropical and subtropical seals in the Family Phocidae; large spp. with coarse pelage; one sp. (M. tropicalis) reaches the coast of Fla. and Tex.

monad. 1. Small, simple flagellate protozoan, such as Monas. 2. Any minute simple organism or organic unit. 3. In meiosis, one of the four units of a tetrad.

Monadenia. Genus of wide, depressed land snails; northwestern U.S.

monarch butterfly. Common N.A. nymphalid butterfly (Danaus) with rich brown wings patterned with black and white markings; distasteful to birds and thought to be mimicked by the viceroy butterfly.

Monas. Genus of minute colorless flagellates; two flagella of unequal length; plastic and actively motile; in stagnant fresh water.

Monaxonida. One of the three subclasses in the Class Demospongiae; spicules monaxon, spongin present or absent; marine and fresh-water; examples are Suberites and Spongilla.

monaxon spicules. Simple, needle-shaped spicules occurring in some sponges.

Monera. Taxon set up by Haeckel to include those spp. of amoeboid protozoans thought to have no nucleus; a nonexistent group.

money shell. See SAXIDOMUS.

Mongol. Race of mongoloids living in northcentral Asia, China, and southwestern Asia; skin pale yellow to brown; cheekbones prominent.

Mongolian fold. Fold of skin extending from the upper eyelid down over the inner corner of the eye; a Mendelian dominant in man.

Mongoloid. Group of races of modern man characterized by straight, stiff hair, sparse beards, and skin color ranging from yellow to red and brown; includes the following races: Eskimo, Amerind, Patagonian, Ugrian, Mongol, Indonesian, Polynesian, and Turko-Tatar.

mongoose. Any of numerous spp. of small slender carnivorous mammals in the Family Viverridae; notable for their ability to kill rats and snakes; Africa and southern Asia, and introduced into West Indies, Hawaiian Islands, and elsewhere; Herpestes the common Indian genus. Pl. mongooses.

Monhystera. Common genus of chromadorid nematodes; marine, fresh-water, and terrestrial.

Monhysterina. Suborder of chromadorid nematodes; esophago-intestinal valve flat or circular; mouth cavity with one or three small teeth or six outwardly acting teeth; aquatic and terrestrial; Monhystera, Plectus, and Cylindrolaimus common.

Moniezia expansa. Common tapeworm of sheep and goats; larval stage in free-living mites.

moniliform. Constricted at regular intervals so as to have the appearance of a string of beads.

Moniliformis. Genus of Acanthocephala found in rodents; beetles and cockroaches are intermediate hosts.

monitor. See VARANIDAE.

monkey. 1. Any member of the Order Primates excepting man and the lemurs; New World monkeys are arboreal, with prehensile tails, flat nose, and nostrils directed laterally, e.g. capuchin and howler; Old World monkeys occur in Africa, Asia, and adjacent islands; nostrils directed downward, tail not prehensile, or absent; e.g. macaques and rhesus. 2. In a narrower sense, any long-tailed monkey.

monkfish. See LOPHIIDAE and SQUATINA.

Monochilum. Genus of ovoid to ellipsoid fresh-water Ciliata.

Monocirrhus. Genus of tropical fresh-water fishes having a superficial resemblance to a dried leaf.

monoclimax hypothesis. Idea that all climax plant and animal communities are basically determined by climate. Cf. POLYCLIMAX.

Monocotyle. Genus of small monogenetic external trematode parasites of skates.

Monocystis. Sporozoan parasite living in the seminal vesicles of earthworms.

monocyte. 1. Large type of phagocytic blood cell in certebrates; from 2 to 7% of human white blood cells are monocytes. 2. Large type of cell in blood of certain invertebrates.

Monodella. See THERMOSBAENACEA.

Monodon monoceros. Narwhal; a large cetacean of Arctic seas; gray above, whitish below; body to 13 ft. long; with only two teeth; one tooth in male to 9 ft. long, tusklike, slender, and twisted (one of the "unicorns" of mythology).

monodont. Having a single tooth, e.g. a male narwhal.

Monodontidae. Family of whales; includes the white whale (Delphinapterus) and narwhal (Monodon).

monoecious. Having both male and female reproductive organs in the same individual.

Monogenea. One of the three orders of the Class Trematoda; oral sucker weak or absent; anterior end sometimes with a pair of weak adhesive devices; posterior end expanded into a ventral adhesive organ bearing one or more suckers and usually provided with sclerotized hooks; endo- or ectoparasites, usually of cold-blooded vertebrates; no alternation of hosts; Gyrodactylus the most familiar genus.

monogenetic. 1. Of or pertaining to the trematode Order Monogenea. 2. Of or pertaining to an animal that does not have alternation of sexual and asexual reproduction in its life history.

Monogononta. One of the two classes of rotifers; with a single ovary.

monograph. Exhaustive treatment of a taxon, with special reference to all available information on taxonomic interpretation and intrarelationships; commonly this systematic treatment is correlated with anatomical, ecological, and distributional data.

monohybrid. With reference to a single pair of alternate characters, the offspring of one parent possessing two dominant characters and another parent possessing two recessive characters.

monohybrid cross. Mating between parents with reference to a single pair of alternate hereditary characters; one parent possesses a pair of dominant characters and the other a pair of recessive characters.

Monommata. Genus of small littoral rotifers; toes unequal and very long.

monomolecular layer. Layer of an organic compound at the interface between air and water or oil and water; formed of materials having markedly hydrophilic and markedly hydrophobic portions of the molecule, so that these parts tend to enter the appropriate adjacent media; fatty acids, proteins, and certain alcohols commonly form such layers; often used in experimental biochemical and physiological studies; thought to occur naturally on surfaces of lakes and ponds.

Monomorium. Genus of small black or red ants which live in soil or wood.

Monomyaria. Out-dated taxon sometimes used to include those bivalve mollusks with a single adductor muscle. Cf. DIMYARIA.

Mononchulus. Genus of microscopic, free-living nematodes.

Mononchus. Common and widely distributed genus of microscopic, carnivorous nematodes.

mononuclear leucocyte. See LYMPHOCYTE and MONOCYTE.

mononucleosis. Infectious disease marked by fever and abnormally large numbers of mononuclear leucocytes in the blood.

Monoöphorum. Marine genus of broad fusiform rhabdocoel flatworms; proctostome posterior.

monophagous. Having or utilizing only a single kind of food, such as many insects which feed on a single sp. of plant.

monophyletic. Pertaining to a taxon whose units are all part of a single immediate line of descent. Cf. POLYPHYLETIC.

monophyodont. Having a single set of teeth, none of which are replaced.

Monoplacophora. Class of mollusks established in 1940, and, until 1952 believed to be represented only by early Paleozoic shells; in that year some "living fossil" specimens were dredged up along the west coast of Mexico, and a description was first published in 1957; limpet-like mollusks but with six pairs of symmetrically-arranged muscle scars, a posterior anus, and segmentally-arranged excretory organs and ctenidia; on the bottom of the deep sea; <u>Neopilina</u> <u>galatheae</u> and <u>N. ewingi</u> the only known living spp.; off coast of Mexico and Peru.

Monopylaea. Suborder of radiolarians; central capsule not uniformly perforate; capsule monaxonic.

monorchid. 1. Having a single testicle. 2. Having a single testicle in the scrotum.

monosaccharide. Six-carbon sugar or carbohydrate having the formula $C_6H_{12}O_6$; fructose, galactose, and glucose are monosaccharides all having this generalized formula, but their peculiar properties are dependent on the configuration and arrangement of the individual atoms in the molecule; occur naturally or result from the hydrolysis of disaccharides or polysaccharides.

Monosiga. Genus of colorless, stalked, fresh-water flagellates; a thin projecting conical collar surrounds the base of the single flagellum.

monosomic. Organism lacking one chromosome of a diploid complement, the somatic number being 2n - 1.

monospermy. Condition in which only one spermatozoan penetrates the outer membrane of an ovum during fertilization.

Monostaechas. Genus of marine coelenterates in the Suborder Leptomedusae.

Monostomata. Group of families of digenetic trematodes; only an oral sucker present.

Monstrilla. Common genus of marine monstrilloid Copepoda.

Monstrilloida. Suborder of marine copepods; adults cylindrical and free-swimming, but second antennae and mouth parts lacking and digestive tract vestigial; immature stages parasitic on marine invertebrates.

Monostyla. Very common genus of loricate rotifers having a short foot and a single long toe.

Monotocardia. Pectinibranchia.

Monotremata. The single order of mammals in the Subclass Prototheria; monotremes, duck-billed platypus, spiny anteater, and echidna; external ear absent; adults with horny beak; teeth in young only; cloaca present; testes abdominal, scrotum absent; penis carrying sperm only; uterus and vagina absent; mammary glands without nipples; females lay fertilized eggs; Australian region; representative genera are Ornithorhynchus, Tachyglossus, and Zaglossus.

monotrophic. Monophagous; using only one kind of food.

monotypic. Pertaining to a taxonomic category which contains only one immediately subordinate botanical or zoological unit, such as an order containing only one family, a genus containing only one sp., etc. Noun monotype.

monovoltine. Univoltine.

monoxenic culture. Culture consisting of one or more individuals of a single sp. with only one associated (food) sp.

Monozoa. Cestodaria.

monozygotic twins. Identical twins.

mons pubis. Mons veneris.

Monstrilla. Genus of marine Copepoda.

mons veneris. Rounded prominence at the lowest point (pubic symphysis) of the human female abdomen.

Montana grayling. See THYMALLIDAE.

montane. 1. Pertaining to the flora and fauna of mountainous regions. 2. An altitudinal zone of middle mountain elevations, usually beginning at timberline or 1000 to 1500 ft. below timberline and extending downward for a vertical distance of 1500 to 2000 ft.

Montipora. Genus of reef-building corals of the tropical Pacific and Indian oceans.

mooneye. See HIODON.

moonfish. Any of a wide variety of deep-bodied, compressed marine or freshwater fishes; coloration usually silvery or yellow; see LAMPRIS.

moon jelly. A common Atlantic Ocean jellyfish, Aurelia aurita.

moon shell. Any of numerous widely distributed marine snails; live in sand and feed on bivalves by engulfing them with their large foot; deposit eggs inside a flat, collar-like ring composed of sand grains cemented together; Natica, Polinices, and Lunatia common.

Moorish idol. See ZANCLIDAE.

moose. See ALCE.

Mopalia. Genus of chitons of the Pacific area.

moray. See COLOCEPHALI.

Mordellidae. Tumbling flower beetles; large family of small spp.; body silky, compressed, and arched; cosmopolitan.

Morgan, Thomas Hunt. American zoologist (1866-1945); especially noted for his demonstration of the physical basis of heredity in the fruit fly, Drosophila; received the 1933 Nobel prize in physiology and medicine for his theory that hereditary characters are dependent upon genes in the chromosomes, the behavior of which he studied and mapped.

Morgan's canon. Principle of animal behavior; activities of an animal should be interpreted in terms of the simplest possible mental processes.

moribund. In a dying state; term used especially with reference to invertebrates.

Mormon cricket. See ANABRUS SIMPLEX.

Mormyrida. Mormyrii.

Mormyrii. Order of bony fishes which includes African electric fishes.

Mormyrus. Genus of sacred fish of the Nile; tail short, snout elongated; worshipped by ancient Egyptians.

moron. See FEEBLE-MINDEDNESS.

Morone americana. White perch, a basslike fish in the Family Serranidae; in shallow water near shore, estuaries, and in ponds and lakes near the sea; a commercial and sport sp. to 15 in. long; Atlantic Coast.

Morone chrysops. White bass, a game fish of the larger rivers and lakes of the Mississippi Valley and the Great Lakes; in the Family Serranidae; to 18 in. long.

Morone interrupta. Yellow bass; a game and food sp. in the sea bass family (Serranidae); northcentral lakes and rivers; to 15 in. long.

Moronidae. Serranidae.

Morpho. Genus of tropical and subtropical butterflies in the Family Morphoidae; large, handsome spp. with metallic iridescence and bright bluish coloration on the upper surface of the wings.

Morphocorixa. Uncommon genus of water boatmen in the Family Corixidae; southwestern U.S.

morphogenesis. Evolution and development of form in ontogeny or in regeneration.

Morphoidae. Family of large, handsome, metallic butterflies of tropical America and northern India.

morphology. Study of the form of an organism considered as a whole or in its gross aspects.

morphotype. Type specimen for only one of the forms of a polymorphic sp.

mortality. Death rate; ratio of total deaths to total population per unit time.

mortality rate. Ratio of number of deaths resulting from a given disease to the total number of cases of that disease.

morula. Loose spherical group of cells during the early stage of segmentation in many embryos, especially before the appearance of a distinct blastocoel.

Morus bassanus. Large, white, gregarious, cliff-breeding, fish-eating oceanic bird in the Family Sulidae; the gannet, a gooselike sp. and the largest bird of the north Atlantic Coast; excellent underwater swimmer.

mosaic. 1. Different genetic make-up in adjacent tissues owing to mutations during embryology, or other causes. 2. Embryological theory which states that the make-up of the cytoplasm of the early blastomeres determines which tissues and organs will develop from each cell.

mosaic development. Type of embryology chracterized by independent differentiation of each part of the early embryo, as opposed to the principle of embryonic regulation.

mosaic image. See COMPOUND EYE.

Moschus moschiferus. Musk deer in the Family Cervidae; male has a musk bag under the abdominal skin, and its secretion is used as a perfume base; antlers absent but there are well developed canine tusks; high altitudes in central Asia.

mososaur. Any of numerous Cretaceous marine lizards.

mosquito. Any of a large group of small dipterous insects in the Family Culicidae; female usually able to take a human blood meal and thus transmit malaria, yellow fever, and other diseases.

mosquito fish. Any of numerous small fresh-water and brackish fishes which feed extensively on mosquito larvae and pupae; e.g. Gambusia.

mosquito hawk. See ODONATA and ANISOPTERA.

moss animalcules. Common name of Bryozoa owing to the superficial resemblance of the colonies of some spp. to a low growth of mosses.

mossbunker. See BREVOORTIA.

Motacillidae. Widely distributed family of passerine songbirds; pipits and wagtails; superficially similar to larks and sparrows; insectivorous spp. which hop and run about on the ground in open country; American pipits (Anthus) breed chiefly in Canada and high altitudes of western states; wagtails are Old World spp.

moth. See FRENATAE.

Mother Carey's chickens. See OCEANODROMA.

mother-of-pearl. Nacre; see PEARL.

moth fly. See PSYCHODIDAE.

motmot. See MOTMOTIDAE.

Motmotidae. Small tropical American family of birds in the Order Coraciiformes; motmots; greenish, long-tailed, jaylike spp. which tunnel in river banks.

motor area. That portion of the cerebral hemispheres which initiates impulses to the skeletal muscles.

motor end plate. End plate.

motorium. Special concentration of fibrils and basal granules in the cytopharyngeal region of many ciliates; apparently acts as a governing and coordinating center of the neuromotor system and ciliary action.

motor nerve. Nerve consisting of fibers which carry impulses only outward or away from the central nervous system; such impulses are transmitted to effector organs or tissues.

motor neuron. Neuron carrying stimuli away from the central nervous system and whose fibers connect with an effector tissue or organ.

motor root. Short ventral root of a spinal nerve as it leaves the spinal cord before it unites with its corresponding dorsal (sensory) root; carries only motor fibers.

moufflon. Wild sheep of mountains of Corsica and Sardinia.

moult. Molt.

mountain beaver. See APLODONTIDAE.

mountain bluebird. Sialia currucoides,

a turquoise blue bird which breeds in mountains of the western states.

mountain boomer. 1. Common name for a small collared lizard. 2. Common name for Aplodontia rufa.

mountain coati. See NASUELLA.

mountain crab. Black or deep blue crab of the Bahamas and southern Fla.

mountain goat. See OREAMNOS.

mountain lion. See COUGAR.

mountain midge. See DEUTEROPHLEBIIDAE.

mountain plover. See EUPODA.

mountain porcupine. See ECHINOPROCTA.

mountain quail. See OREORTYX.

mountain sheep. See BIGHORN.

mountain sickness. Illness occurring in persons upon reaching high mountain altitudes; characterized by fatigue, labored breathing, nausea, and vomiting.

mountain sucker. See PANTOSTEUS.

mountain trout. In the U.S., any trout caught or hatchery-reared in a mountainous area.

mourning cloak. See NYMPHALIS.

mourning dove. See ZENAIDURA.

mouse. General term to include a wide variety of small rodents, such as members of the genera Mus, Perognathus, and Microtus; there is no general distinction between a rat and a mouse except for size. See CRICETIDAE and MURIDAE.

mouse-bird. See COLIIFORMES.

mouse deer. See TRAGULIDAE.

mousefish. See ANTENNARIIDAE.

mouse-opossum. See MARMOSA.

mouse-tailed bat. See RHINOPOMATIDAE.

mouth cavity. Buccal cavity, oral cavity; cavity just inside mouth orifice in vertebrates and certain invertebrates.

mouth parts. Paired and unpaired appendages more or less surrounding the mouth in arthropods; variously modified for obtaining, handling, and processing food.

Moxostoma. Genus of catostomid fishes which includes various spp. of redhorse and jumprocks; body and habits sucker-like; common in lakes and rivers of eastern and central U.S.; usually considered rough fish but used as human food in some areas.

MS. 1. Multiple sclerosis. 2. Manuscript.

MSS. Manuscripts.

mucin. Protein lubricating material forming mucus in solution; e.g. secretion of cardiac glands in the stomach wall.

mucket. Any of several American unionid clams (especially Actinonaias) with an unusually thick shell, useful for button manufacture.

mucoprotein. Any one of numerous compounds composed of conjugated carbohydrates and proteins, such as mucin.

mucosa. Mucous membrane.

mucous. Of or pertaining to mucus.

mucous gland. Any gland which secretes mucus and is associated with a mucous membrane; may be unicellular or multicellular and large.

mucous membrane. Any moist surface in animals characterized by the secretion of a watery, slimy fluid; in vertebrates the digestive tract, respiratory tract, and much of the urinogenital system are lined with mucous membrane.

mucro. Any abrupt pointed tip, process, or toothlike structure.

mucronate. Ending in a spinelike point or tip.

Mucronella. Genus of crustose marine Bryozoa.

mucus. Any viscous, slimy, or lubricating secretion of a mucous gland.

mud crab. Any of numerous marine crabs inhabiting muddy littoral areas, especially spp. in the Suborder Brachyura.

mud-dauber wasp. See SPHECIDAE.

muddler. See COTTIDAE.

mud eel. See SIREN.

mudfish. Common or local name for any of several fishes found in muddy water or actually on the surface of a muddy bottom, e.g. bowfin and killifish.

mudhen. See FULICA.

mud minnow. See UMBRIDAE.

mudnest builder. See GRALLINIDAE.

mud puppy. See NECTURUS.

mud siren. See PSEUDOBRANCHUS.

mudskipper. See PERIOPHTHALMUS.

mud snail. See NASSA.

mud snake. Long, bluish-black, colubrid snake (Farancia abacura) with red spots along each side; wet woods of southeastern states; with a spike-like "stinger" at the tip of the tail; often called the hoop snake.

mud star. See CTENODISCUS.

mud turtle. See KINOSTERNON.

Muggiaea. Genus of pelagic siphonophore hydrozoans.

Mugil. Common genus of mullets; chiefly marine but some spp. enter coastal fresh waters.

Mugilidae. Marine family of percomorph fishes which includes the gray mullets; herbivorous and scavengers; mostly tropical; some important commercial spp.

Mugiliformes. Order sometimes used to include the fish families Atherinidae and Mugilidae (here included in the Percomorphi).

Muhlenberg turtle. Small semi-aquatic turtle in the genus Clemmys; with a large orange spot on each side of the head; eastern states.

mulatto. 1. Person having one Negro and one white parent. 2. The intermediate grade of skin color of such a person or of some members of later generations.

mule. See EQUIDAE.

mule deer. See ODOCOILEUS.

mule killer. In the southern states, any of several arthropods reputed to kill livestock by stinging, biting, or being swallowed; e.g. whip scorpions, mantids, walking sticks.

Mulinia lateralis. Little surf clam; a small triangular sp. with smooth valves; Atlantic Coast and Caribbean.

Müller, Fritz. German biologist (1821-1897); early proponent of the principle of "ontogeny repeats phylogeny."

Muller, Hermann J. American geneticist (1890-); best known for his discoveries regarding mutations induced by X-rays; awarded the 1946 Nobel prize in physiology and medicine.

Müller, Johannes Peter. German physiologist and anatomist (1801-1858); made major contributions to experimental and human physiology, including the law of specific energies.

Müller, Otto Friedrich. Danish invertebrate zoologist (1730-1784).

Müllerian duct. Oviduct of a female gnathostome vertebrate; except in birds, two such ducts are typically present, but the lower portions are often modified into uteri which may empty into a cloaca, or there may be fusion into a single uterus.

Müllerian mimicry. Form of animal mimicry; a number of unrelated spp., more or less distasteful to their predators, resemble one another and thus presumably come to be easily recognized and avoided; all are protected by an "education of predator" mechanism. See BATESIAN MIMICRY and MIMICRY.

Müller's larva. Free-swimming ciliated larva of certain marine polyclad turbellarians; eight posteriorly-directed lappets.

mullet. See MULLIDAE and MUGILIDAE.

Mullidae. Marine family of percomorph fishes which includes the red mullets or goatfishes; one or two barbels on chin; Fla. and West Indies.

multicellular. Composed of many cells.

Multiceps multiceps. Cyclophyllidean tapeworm; adult in dog; cysticercus in brain or spinal cord of sheep and cattle where it causes gid or staggers.

Multicilia. Genus of marine and freshwater Protozoa; plastic, with 40 to 50 radiating flagella.

multihybrid. The F_1 resulting from parents differing in three or more pairs of characters, e.g. a mating between a homozygous black, rough-haired, short-haired guinea pig and a white, smooth-haired, long-haired guinea pig results in heterozygous offspring which are all black, short-haired, and rough-haired.

multinucleate. Having two or more nuclei.

multiple alleles. More than two alternative factors which have the same chromosomal locus and affect the same genetic character; e.g. in mice there are four allelomorphs in the albino series: C (full color), c^r (chinchilla), c^d (extreme dilution), and c^a (albino); various paired combinations of these four allelomorphs may exist.

multiple factors. Two or more pairs of genes having a cumulative and similar phenotypic effect, and producing quantitative inheritance; e.g. skin color in man.

multiple fission. Asexual reproduction in animals by a division of the body into three or more parts, each of which then reconstitutes into a complete new animal; common in some Protozoa, sea anemones, annelids, etc.

multiple sclerosis. Nervous system disease produced by disintegration of myelin sheaths and their replacement by scar tissue; results in abnormal sensations and muscle control.

multipolar neuron. Neuron having two or more (branched) dendrites and one axon.

Multituberculata. Order of small Jurassic mammals in the Subclass Allotheria; single pair of rodent-like incisors above and below; molars with two or three longitudinal rows of tubercles; premolars with a single cusp or in the form of sharp-edged cutting teeth.

multivoltine. Pertaining to spp. having two or more generations during a growing season or year's time.

mummichog. See FUNDULUS.

Mungos. Genus of small banded mongooses; drier parts of East Africa.

Munidopsis. Genus of deep-water marine decapod crustaceans in the Suborder Anomura.

Munna. Genus of marine Isopoda.

Muntiacus. Genus of small deer in the Family Cervidae; muntjak or barking deer; jungle spp. in which the male has tusklike canines; southeastern Asia and some East Indian islands.

muntjak. See MUNTIACUS.

Muraenidae. See COLOCEPHALI.

Murex. Large genus of marine snails; includes many murex or rock shells; shell heavy, with three or more rows of protuberances or spines; aperture round and ending in a long canal; carnivorous, especially on bivalves; the royal purple dye of the ancients was made from a fluid secreted from the anal glands of several spp.; tropical and subtropical; about 1000 spp., many of which are handsomely formed and colored.

Murgantia histrionica. Insect in the hemipteran Family Pentatomidae; the harlequin cabbage bug; a red and black bug which is a serious pest on cabbage and other cruciferous plants.

Muricea. Genus of gorgonian coelenterates of the Pacific Coast; treelike growth form.

Muridae. Large family of rodents which includes the Old World rats and mice; mostly nocturnal spp. with no premolars, and tail generally nearly naked or scaly; Mus and Rattus introduced into U.S. and common everywhere; this family name is sometimes used to include the Cricetidae (native U.S. rats and mice).

Murrayite. Synthetic cement used for sealing coverslips and museum jars.

murre. See URIA.

murrelet. See ENDOMYCHURA.

Mursia. Genus of marine crabs in the Suborder Brachyura.

Mus. Genus of mice in the Family Muridae; Mus musculus musculus is the common and widely distributed house mouse; omnivorous, but prefers grain and other vegetable products.

musang. Medium-sized carnivore of southeastern Asia; grayish coloration, with dark face, legs, and tail; to 2 ft. long; Paradoxurus common.

Musca. Genus of houseflies containing about 30 spp.; M. domestica is the common cosmopolitan housefly; by no means as abundant now as it was when horses were more common; aside from its importance as a nuisance insect, it transmits typhoid, diarrhea, dysentery, cholera, trachoma, and tuberculosis; maggots chiefly in manure and garbage.

muscae volitantes. Small translucent specks, rods, or fibrils seen in the field of vision; produced by transient imperfections in the lens and vitreous humor.

Muscardinus. Genus which includes the hazel mouse, the common dormouse of western Europe; a mouselike animal having a body to 3 in. long; fur brownish; nests in bushes.

Muscicapidae. Passerine family of Old World flycatchers; insectivorous tree-dwelling spp. with broad, flat bill; some with exceptionally long tail feathers.

Muscidae. Small family of small to medium-sized flies in the Order Diptera; houseflies; larvae in decomposing organic matter; many obnoxious and harmful spp.; important genera are Musca and Muscina.

Muscivora forficata. Scissor-tailed flycatcher; a bird of the southern states and Mexico; tail extremely long and forked; pale gray, white, and pink coloration.

muscle. 1. Organ whose special functions are support and the production of motion by contraction and relaxation; consists of an abundance of fibers which contract when stimulated. 2. A type of tissue consisting of highly contractile cells. See NONSTRIATED MUSCLE, VOLUNTARY MUSCLE, and CARDIAC MUSCLE.

muscle glycogen. Small reserve of glycogen in muscle cells; ready for instant phosphorylation and oxidation by the muscle tissues; derived from blood glucose; major source of energy in muscle.

muscle plate. Myotome.

muscle scar. Slightly depressed area on the inner surface of a bivalve shell; denotes the attachment of a muscle.

muscle tone. Relative amount of contraction or tenseness of a muscle as contrasted with a completely flaccid condition; even when at rest and during sleep a muscle is contracted to a certain extent by means of groups of fibers working in relays.

muscovy duck. See CAIRINA.

muscular dystrophy. Progressive dystrophy of muscles without any discoverable lesions of the spinal cord.

muscularis. 1. Collective muscle layers of the vertebrate digestive tract; commonly it consists of a layer of circular muscle fibers and a layer of longitudinal muscle fibers; the term is also occasionally applied to a similar layer in the digestive tract of certain invertebrates. 2. Muscular coat of an organ.

muscularis mucosae. One or two thin layers of smooth muscle at the base of the mucosa of the vertebrate digestive tract or other mucosa.

muscular system. Collectively, all of the muscles of an animal.

musculature. Collectively, the muscles of the body, or any special group of those muscles.

Musculium. Large and generally distributed genus of small fresh-water bivalve mollusks; to 12 mm. long.

musculo-epithelium. Epithelial cells whose base is elongated and contains contractile fibril(s); such cells have roughly the shape of an inverted T; especially characteristic of coelenterates and nematodes.

museum beetle. Any of several beetles which may be pests by feeding on animal skins, dried insects, etc.; some spp., however, are used to clean dried flesh from bones; e.g. Dermestes.

musk. Secretion from a variety of special glands in many kinds of animals, but especially that obtained from a sac (musk bag) under the abdominal skin of the male musk deer; used as a perfume base; another example is the group of secretions from the musk glands in the mouth and cloaca of alligators.

musk deer. See MOSCHUS.

muskeg. In northern N.A., an extensive boggy area marked by growths of sphagnum mosses and often subsurface permafrost.

muskellunge. See ESOCIDAE.

musk gland. Any gland secreting a musky material; see MUSK.

muskox. See OVIBOS.

muskrat. Common N.A. semiaquatic rodent; see ONDATRA ZIBETHICA.

musk turtle. See STERNOTHERUS.

Musophagidae. See TOURACO.

mussel crab. See PINNOTHERES.

mussel poisoning. Type of poisoning caused by eating sea mussels and certain other bivalves, especially during warm months when the water contains an abundance of dinoflagellates which produce the toxic materials.

mussels. 1. Marine bivalves belonging especially to the genera Mytilus and Modiolaria; attached to rocks and piling with byssus threads. 2. Collectively, all large fresh-water pelecypods.

mussel scale. Any of numerous scale insects which secrete a scale having the shape and configurations of a miniature mussel shell.

mussel worm. Any of several spp. of Nereis.

mustard gas. Highly toxic gas sometimes used to induce mutations in experimental genetics.

Mustela. Genus of small, slender mammals in the Family Mustelidae; mink, weasels, stoats, ermines, ferrets; highly carnivorous and predatory; the many spp. of weasels are found in the Eastern and Western hemispheres, but they are essentially northern; coloration yellowish to brown, but in regions of cold winters the coat turns white during the winter months and any such white weasel is properly an ermine; stoats are common European weasels; ferrets are a semidomesticated European sp. used for hunting rabbits and rats, although some western U.S. spp. are also called ferrets; the American mink is a large weasel, M. vision, of semiaquatic habits; found around streams and lakes and feeds on terrestrial mammals and birds as well as fish; feet partly webbed.

Mustelidae. Large family of carnivores which includes the weasels, martens, minks, otters, skunks, badgers, and wolverines; slender, small to medium size; feet digitigrade to subplantigrade; anal scent glands more or less developed.

Mustelus. Common genus of sharks, especially along Atlantic Coast; dogfish, smooth dogfish; to 5 ft. long.

Mutabilia. One of the three orders of caudate amphibians; true salamanders; adults usually without gills; lungs usually present; paired vomerine teeth.

mutagen. Any mechanism or substance markedly increasing the normal mutation rate, such as X-rays, mustard gas, etc. Adj. mutagenic.

mutant. 1. Gene which has undergone a mutation. 2. Organism carrying a gene which has undergone a mutation. 3. Character produced as the result of a mutated gene; most mutants are recessives and may pass unnoticed through many generations; over 1000 mutants have been observed in Drosophila.

mutation. New, abrupt genetic feature in any individual produced by changes in the individual genes or chromosomes; the most important mutations occur in the gametes, since they can produce the visible and heritable changes in the characteristics of the whole organisms developing from them; a somatic mutation, on the other hand, occurs in a body cell and is transmitted only to those cells derived from it by mitosis.

mutation rate. Relative frequency of mutations in any sp.; "persistent types" appear to have essentially a zero mutation rate.

Mutica. Taxon of mammals which includes only the Order Cetacea.

Mutilla. Cosmopolitan genus of velvet ants in the Family Mutillidae.

Mutillidae. Family of wasps in the Order Hymenoptera; velvet ants; dark-colored body covered with long, light-colored hairs; in dry, sunny, sandy places; larvae parasitic on larvae of ground bees and wasps; Mutilla common.

mutualism. Term sometimes used as a synonym of symbiosis; a close association between two different spp. whereby each sp. derives some benefit; an example is termites and the cellulose-digesting protozoans in their digestive tracts; another example is the Yucca plant and the Tegeticula moth, the former being absolutely dependent upon the moth for pollination and the latter being absolutely dependent on the fruit of the Yucca as a food source for the larvae developing from eggs deposited in the ovary of the Yucca flower.

mutually supplementary factors. Modified two-factor Mendelian ratio where two dominant alleles are necessary for the occurrence of the dominant character; e.g. in color of Duroc-Jersey swine: red is produced by RRSS, RRSs, and RrSs genotypes; sandy color is produced by RRss, Rrss, rrSS, and rrSs genotypes; white is produced by the rrss genotype.

Mya arenaria. Common soft-shell clam, long clam, long-neck clam, sand clam, steamer clam, or mud clam; siphons forming a very long tubular projection; shell a dull chalky white partially obscured by the brown periostracum; to 6 in. long; edible; common on East Coast, introduced on West Coast.

Myadestes townsendi. Townsend's solitaire; slim, grayish bird in the Family Turdidae; mountains of western states.

Mycale. Genus of small marine sponges in the Subclass Monaxonida; commonly form growths on Bryozoa.

Mycetophila. See MYCETOPHILIDAE.

Mycetophilidae. Family of small, mosquito-like gnats in the Order Diptera; fungus gnats; larvae feed on decaying vegetation and fungi; Mycetophila contains many mushroom-infesting spp., some of which are economic pests; it is often difficult to find a mushroom or toadstool that is not infested.

Mycetozoa. Order of the Class Sarcodina in which the vegetative form is a large, multinucleate mass of protoplasm to several inches in diameter; slime molds; usually feed on decaying vegetation; Ceratiomyxia and Badhamia typical; same as Myxomycetes in the plant kingdom.

Mycobacterium. Genus of gram-positive, slender, non-motile, rod-shaped bacteria; includes certain important pathogenic spp. causing, e.g. leprosy (M. leprae) and tuberculosis (M. tuberculosis).

mycosis. Any disease produced by a fungus.

Mycteria. See IBIS.

Mycterothrix. Genus of fresh-water holotrich ciliates; in a gelatinous case; anterior end conical.

Myctoperca. See GROUPER.

Myctophidae. See MYCTOPHUM.

Myctophum. Genus of lanternfishes; small, silvery, oceanic spp. with large eyes, large mouth, and luminous organs; in the Family Myctophidae.

Mydaus. Genus of nocturnal mammals which includes the teledu, a relative of the badgers; dark brown with a median white stripe; snout piglike; large anal stink glands from which the fluid can be squirted; feeds on small invertebrates; mountains of Borneo, Sumatra, and Java.

myelencephalon. Medulla oblongata; posteriormost portion of the vertebrate brain; has a thick floor and walls but a thin roof; in lower vertebrates contains coordination centers for the sensory receptors, but in higher vertebrates it is more important for coordination and contains control centers of respiration, blood vessel contractions, heart contraction, etc.

myelin sheath. Medullary sheath.

myelitis. Inflammation of bone marrow.

myelocoel. 1. Cavity of the embryonic myelencephalon. 2. Cavity of the spinal cord.

myelocyte. 1. Amoeboid cell of red bone marrow; source of granular leucocytes of the blood. 2. Cell of the gray matter of the nervous system.

Mygale. Genus of very large tropical spiders; reputed to attack and kill small birds.

Mygalomorphae. Orthognatha.

Myiarchus. Common genus of flycatchers; four U.S. spp.

myiasis. Infection, disease, or injury caused by Diptera larvae, especially with reference to man and other mammals; depending on the infected organ, the myiasis is characterized as gastric, intestinal, urethral, etc.

Mylocheilus caurinum. Peamouth, a minnow of the Columbia River system and north to the Fraser River; to 12 in. long.

mylohyoideus. Flat transverse muscle on the lower surface of the lower jaw of certain vertebrates.

Mylopharodon conocephalus. Hardhead; a large, slender minnow of central and northern Calif. rivers; to 25 in. long.

Mymaridae. Small family of minute black and yellow hymenopterans; wings narrow and fringed with setae; larvae parasitic in the eggs of other insects; cosmopolitan.

mynah bird. Any of numerous starling-like birds of southeastern Asia; some can be taught to talk.

myoblast. Cell which develops into muscle.

myocardium. Collectively, the heart musculature.

Myocastor. See CAPROMYIDAE.

myochordotonal organ. Minute proprioceptor in the third segment of each leg of decapod Crustacea.

myocoel. Cavity of a myotome.

myocomma. Membranous partition between each two myotomes, especially in primitive fishes. Pl. myocommata.

myocyte. 1. One of a group of contractile cells which form poorly-organized sphincters around the external pores of sponges. 2. Contractile ectoplasmic layer in certain gregarine protozoans. 3. Muscle cell or contractile cell.

Myodocopa. One of the four orders of ostracods; with a permanent anterior opening between the shell valves through which the antennae are extended; only second antennae used in locomotion; heart present; three pairs of trunk limbs; caudal rami compressed; all spp. marine; <u>Conchoecia</u> common.

myofibrils. Long, minute, contractile fibrils of cardiac and skeletal muscle tissue; many such fibrils make up a single fiber and are all enclosed within the sarcolemma of the fiber.

myogenic rhythm. Especially in insects, an extremely rapid rhythm of wing muscle contraction which originates and is determined within the muscle itself; although nerve impulses reaching such muscle have a high frequency, this frequency is much lower than that of the wing beats. Cf. NEUROGENIC RHYTHM.

myoglobin. Type of protein found in muscle tissue.

myology. 1. The study of muscles. 2. collectively, the muscles of an animal or a part.

myomeres. Successive segmental trunk musculature; especially obvious in fish.

myometrium. Especially in human females, the thick muscular layers of the uterus, as contrasted with the endometrial lining.

Myomorpha. Suborder of the Order Rodentia; includes mice, voles, rats, and jumping mice.

myonemes. Long contractile fibrils in certain ciliate Protozoa.

myoneural junctions. Points at which terminal nerve fibers are in contact with muscle fibers.

myophrisk. In certain radiolarians, a cluster of contractile fibrils surrounding a spine at its junction with the peripheral cytoplasm.

myopia. Nearsightedness; light rays from distant objects are brought to a sharp focus just in front of the retina. Adj. myopic.

Myosciurus. Genus of dwarf palm squirrels of Africa.

myoseptum. Connective tissue septum between two myotomes, especially in fishes.

myosin. Most abundant protein in vertebrate muscle; largely responsible for contraction and relaxation.

Myosoma. Genus of marine Bryozoa.

Myospalax. Genus of sokhors; subterranean lemming-like rodents; India and central Asia.

Myotis. Common genus of N.A. bats, including especially the little brown bats; less than 4 in. long; upper parts dull brown, under parts paler.

myotome. 1. Discrete voluntary muscle segment in cephalochordates and vertebrates. 2. Group of muscles innervated by a single spinal nerve. 3. One of the paired masses of mesodermal tissue in a vertebrate embryo which is destined to form musculature.

Myoxocephalus. Genus of sculpins occurring in Arctic and Subarctic areas of the Pacific.

myriapod. Any member of the arthropod classes Chilopoda, Diplopoda, Symphyla, or Pauropoda.

Myriapoda. Taxonomic term (class) sometimes used to include the classes Chilopoda and Diplopoda.

Myriochele. Genus of simple, tube-building Polychaeta.

Myriophrys. Genus of spherical fresh-water Protozoa; cell with numerous flagella, axopodia, and a gelatinous envelope containing scales.

Myrmeciplana. Genus of elongated marine and brackish rhabdocoel flatworms.

Myrmecocystus. Genus of honey ants of southwestern U.S. and Mexico; abdomen of workers becomes enormously enlarged with stored honey.

myrmecology. Study of ants.

Myrmecophagidae. Family of mammals in the Order Edentata; includes the anteaters; tropical American spp. with elongated head and snout and shaggy hair; teeth absent; tongue sticky, slender, and protrusible; used for lapping up ants and termites after the anteater has torn open the nest with its powerful claws; Myrmecophagus typical.

myrmecophagous. Feeding on ants.

Myrmecophagus. See MYRMECOPHAGIDAE.

Myrmedonia. Genus of rove beetles living in ant nests and superficially resembling ants.

Myrmeleon. Genus of insects in the Order Neuroptera having a superficial resemblance to damselflies; larvae are predaceous ant lions.

myrtle warbler. Dendroica coronata, a warbler which breeds in American northern coniferous forests.

Mysidacea. Order of malacostracan Crustacea that includes the opossum shrimps; all thoracic appendages biramous; carapace over much of thorax; 300 spp.; mostly marine; Mysis common.

Mysis. Common marine and fresh-water genus in the malacostracan Order Mysidacea.

mysis larva. Schizopod larva.

Mystacides. Genus of Trichoptera.

Mystacinidae. Family of New Zealand short-tailed bats.

Mystacocarida. Primitive subclass in the Class Crustacea; wormlike; mouth parts primitive; four pairs of thoracic legs; genital pore on first thoracic segment; to 500 microns long; in interstitial waters of sandy marine beaches; Derocheilocaris the only genus; several spp.

mystax. In certain Diptera, a patch of setae or bristles above the mouth.

Mysticeti. Suborder of the mammalian Order Cetacea which includes the toothless whales or whalebone whales; feed by means of numerous slatlike parallel horny fringed plates (whalebone or baleen) on sides of upper jaw; used to strain schizopod crustaceans (krill) from water. common examples are Balaena, Balaenoptera, and Rhachianectes.

Mystriosuchus. Genus of Triassic reptiles in the Order Thecodontia.

Mytilimeria. Genus of marine bivalves; often imbedded in compound tunicates.

Mytilina. Genus of fresh-water loricate rotifers.

Mytilopsis. Genus of small wedge-shaped bivalve mollusks; brackish waters; Md. to Fla.

Mytilus. Genus of which includes most marine mussels; wedge-shaped bivalves which occur abundantly on all substrates near the low-tide line; M. edulis is the edible European mussel, also found on American coasts; it reaches a length of 3.5 in.; certain other spp. attain 8 in.

myxedema. Pathological condition produced by inadequate quantities of thyroxin in adults; characterized by low metabolic rate, reduction of physical and mental vigor, loss of sex drive, loss of hair, and puffiness of the skin; may be more or less alleviated by the administration of thyroxin.

Myxicola. Genus of thick-bodied, tube-building Polychaeta.

Myxidium. Genus of Myxosporidia occurring in body cavities and tissues of fishes, amphibians, and reptiles.

Myxilla. Genus of marine sponges in the Subclass Monaxonida; growth form variable; cold seas.

Myxine. Common genus of blind hagfishes and slime eels in the cyclostome Order Myxinoidia; M. limosa is the slime eel which is uncommon in the Atlantic Ocean north and east of New England; M. glutinosa is the common hagfish of both coasts of the North Atlantic; feeds chiefly on dead, dying, and netted fish by boring into the body and consuming viscera and musculature; secretes copious mucus.

Myxini. Myxinoidia.

Myxinoidia. Order of marine cyclostomes which includes slime eels and hagfishes; mouth terminal and not funnel-like; few teeth; nasal sac terminal, connected to pharynx; 10 to 14 pairs of gill chambers; no special larvae; about 25 spp.; Myxine, Bdellostoma, and Polistotrema are examples.

Myxobolus. Genus of Myxosporidia occurring in tissues of fresh-water fishes and amphibians.

Myxocephalus. Genus of marine sculpins found especially in the North Pacific.

myxopodia. Anastomosing rhizopods in certain amoeboid Protozoa.

myxopterygium. Clasper, in fishes.

Myxosporidia. One of the three orders in the Subclass Cnidosporidia; each spore with two valves; one, two, or four polar capsules; parasitic in cavities or tissues of fresh-water fishes, amphibians, and reptiles; examples are Myxobolus and Myxidium.

Myzopodidae. Family of Madagascar bats; thumbs and ankles with suction cups.

myzorhynchus. Central stalked sucker arising from the midst of the bothria in some tapeworms.

Myzostoma. One of the five classes in the Phylum Annelida; highly modified, small, oval, disc-shaped parasites of sea stars and other echinoderms; five pairs of parapodia; about 120 spp.; common genus is Myzostomum; some authorities include this group with the Polychaeta.

Myzus. See PEACH BORER.

N

N.A. Numerical aperture.

Nabidae. Damsel bugs; a family of medium-sized Hemiptera; predaceous on plant bugs and therefore beneficial; Nabis cosmopolitan.

Nabis. See NABIDAE.

nacre. Nacreous layer, mother of pearl; innermost lustrous layer of the shell of mollusks, especially in gastropods and pelecypods; secreted by mantle epithelium; see PEARL.

nacreous layer. Nacre.

Naegleria. Dimastigamoeba.

Naemorhedus. Genus of ruminants in the Family Bovidae; includes gorals, several spp. having certain features of both goats and antelopes; China to southern Himalayas.

Naeogeus. Common genus of fresh-water hemipterans in the Family Hebridae.

nagana. Ngana; disease of horses and cattle in Central Africa; produced by Trypanosoma brucei and transmitted by the tsetse fly.

naiad. 1. Fresh-water mussel. 2. See HEMIMETABOLA. Pl. naiads or naiades.

naid. Any of many small fresh-water oligochaetes, especially in Nais and related genera.

Naididae. Family of small oligochaetes found chiefly on fresh-water substrates; microscopic to 1 in. long; reproduction chiefly by transverse fission.

Naidium. Genus of fresh-water Oligochaeta.

nail. Flat horny epidermal translucent growth protecting the upper portion of the tips of the digits; a claw is a modified nail.

Nainereis. Genus of Polychaeta.

Nais. Genus of translucent microscopic oligochaetes found everywhere on substrates in ponds, lakes, and streams; few segments; usually reproduce by fission.

Naja. Genus of snakes which includes cobras; the habit of spreading the neck into a flat spectacular "hood" before striking is variously developed; when the prey is grasped in the jaws it is chewed so that a series of fang wounds are made; large spp. exceed 6 ft.; Africa, India, southeastern Asia, and the Philippines.

Najadicola ingens. Sp. of Hydracarina parasitic or commensal in gills or mantle cavities of fresh-water mussels.

namaycush. Lake trout, Cristivomer namaycush.

Nandinia. Common genus of West African false palm-civets; small mammals about as large as a cat; fur long and brown; nest in hollow trees.

Nannolene. Genus of small millipedes.

nannoplankton. Plankton organisms having their maximum dimension less than 0.040 mm.

Nannosciurus. Genus of minute Asiatic tree squirrels about as large as a small shrew.

Nannothemis bella. Dragonfly found especially in eastern U.S.

Nansen bottle. Oceanographic device used for taking a water sample at any predetermined depth.

Napaeozapus. See ZAPODIDAE.

napu. Any of several southeastern Asiatic and East Indian chevrotains, especially in the genus Tragulus.

Narceus. Common genus of large millipedes, especially in southern states; same as Spirobolus.

Narcomedusae. Suborder of the coelenterate Order Trachylina (sometimes a separate order); medusa margin scalloped; gonads in floor of gastric pouches; Cunina, Aegina, and Solmaris typical.

narcosis. State of unconsciousness or inactivity produced by a drug.

narcotize. To place under the influence of a narcotic substance.

nares. Openings into the nasal cavity. Sing. naris. See EXTERNAL NARES and INTERNAL NARES.

Narpus. Genus of elmid beetles.

narrow-mouthed frog. See MICROHYLIDAE.

narrow-mouthed toad. See MICROHYLIDAE.

narwhal. See MONODON.

nasal. 1. One of two small bones forming the arch of the nose. 2. Of or pertaining to the nose.

nasal capsule. Bone, cartilage, and other tissues enclosing vertebrate olfactory epithelium.

nasal cavity. Cavity in the tetrapod head which contains the olfactory nerve endings and communicates with the oral cavity or pharynx and external surface of the head by means of internal and external nares.

Nasalis. Genus of arboreal proboscis monkeys in the Superfamily Cercopithecoidea; nose elongated into a grotesque proboscis; restricted to Borneo.

Nashville warbler. Vermivora ruficapilla, a common N. A. warbler which winters in Mexico and Guatemala.

Nasiaeschna pentacantha. Dragonfly found especially in the eastern U.S.

nasolachrymal duct. Short duct which carries tears from the surface of the eye into the nasal cavity.

nasopharynx. That part of the pharynx which lies above the level of the soft palate.

Nassa. Widely distributed genus of predaceous and scavenger marine snails; includes certain dog whelks and mud snails; shell small, with a rotund body whorl and tapered spire; 0.5 to 1 in. long.

Nassarius. Genus of marine snails; shell ribbed and ridged, to 1 in. long.

Nassula. Genus of marine and freshwater holotrich ciliate Protozoa; oval to elongate and brightly colored; ventral surface flat.

Nasua. Genus of mammals in the Family Procyonidae; includes the coati or coati mundi; arboreal, with some characters of both the raccoon and squirrel, but with longer body and tail and a piglike snout. Mexico, Central and South America.

Nasuella. Genus of mountain coatis; a small uncommon coati of the northwestern Pacific forests of South America.

nasute. 1. Having prominent nostrils. 2. Having a large nose.

nasute soldier. Type of termite caste having small jaws and a noselike snout on the head from which a sticky fluid can be squirted onto an enemy.

Nasutitermes. Genus of tropical termites in the Order Isoptera; build massive nests of wood debris above ground.

natal. 1. Pertaining to birth. 2. Pertaining to the buttocks.

Natalidae. Family of Tropical American long-legged bats; Nyctiellus typical.

Natantia. Section in the Order Decapoda; free-swimming shrimps and prawns; compressed cephalothorax and long antennae; pleopods for swimming; 2000 spp.

Natica. Common genus of globose marine snails; live in sand and feed on bivalves by engulfing them with the large foot.

native protein. Any naturally-occurring protein which has not been altered by chemical or physical agents.

native trout. Name usually applied to the cutthroat trout, Salmo clarki, of the western states.

Natrix. Widely distributed genus of thick-bodied non-poisonous snakes in the Family Colubridae; water snakes; to 6 ft. long; feed mostly on fish and amphibians; viviparous and semiaquatic; about ten spp. in U.S.

natural classification. Classification based on characters indicating phylogenetic relationships.

natural history. Activities and behavior of organisms in their natural surroundings; occasionally the meaning of the term is extended to include field and identification aspects of geography, geology, mineralogy, paleontology, etc.

natural immunity. Any immunity mechanism possessed by a plant or animal which is present naturally and is not contingent upon any previous infection; examples are the skin, acid stomach contents, phagocytosis, and the presence of certain antibodies dating from birth.

naturalistic theory. Theory to account for the beginning of simple primordial life on earth by the occurrence and fortuitous association of proper chemical compounds under proper environmental conditions early in the history of the earth.

natural selection. Complex of processes by which the collective factors of the environment eliminate those individuals least fitted to that environment; survival of the fittest. See DARWINISM.

nature study. Study of all phenomena of nature, whether animate or inanimate, especially on an elementary level.

Naucoridae. Creeping water bugs; a family of broad, flattened hemipterans found in ponds, streams, and lakes of the Old World and in the New World tropics and warm temperate areas; feed on aquatic metazoans; four genera in the U. S., mostly in the southwest; Ambrysus typical.

Naucrates ductor. Pilotfish; tropical and subtropical marine fish which sometimes follows ships and is commonly associated with sharks; feeds on scraps from the shark's meal or picks parasites from the surface of the shark.

nauplius. Free-swimming microscopic larval stage characteristic of copepods, ostracods, decapods, barnacles, etc.; typically with only three pairs of appendages.

nausea. Dizziness, tendency to vomit, and stomach sickness.

Naushonia. Genus of burrowing marine decapod crustaceans in the Suborder Anomura.

Nausithoë. Genus of scyphozoan jellyfish in the Order Coronatae; tropical and subtropical oceans.

Nautiloidea. One of the two suborders in the cephalopod Order Tetrabranchia; includes only a single living genus, Nautilus, with several spp. (but many fossil genera and spp.); the pearly nautilus or chambered nautilus; shell with a few whorls and many septa, to 10 in. in diameter; about 90 tentacles; in deep waters of Pacific and Indian oceans.

nautilus. See NAUTILOIDEA and PAPER NAUTILUS.

Navanax. West Coast genus of opisthobranch gastropods; shell hidden; mantle produced into recurved caudal wings or lobes.

navel. Area in center of the abdomen representing the point of attachment of the umbilical cord during fetal life.

navicular bone. Small ankle bone of mammals.

Neanderthal man. See HOMO NEANDERTHALENSIS.

Neanthes. Nereis.

neap tide. Exceptionally low high tide which occurs twice per lunar month, when there is a first or third quarter moon and the earth, sun, and moon are at right angles to each other.

Nearctic realm. One of the six major zoogeographic realms of the world; consists of all of N. A. as far south as central Mexico and includes Greenland; some characteristic animals are: mountain goat, prong-horned antelope, caribou, muskrat, bison, prairie dog, moose, pocket gopher, and turkey.

Nebalia. See LEPTOSTRACA.

Nebaliacea. See LEPTOSTRACA.

Nebela. Large genus of fresh-water amoeboid protozoans; most of cell enclosed in a thin test composed of platelets.

Necator americanus. "New World" hookworm; widely distributed in southern U. S., Central and South America, Africa, India, and the Orient. (See ANCYLOSTOMA DUODENALE for life history pattern.)

necrobiosis. Series of tissue changes occurring at death. Adj. necrobiotic.

necrophagous. Feeding on carrion.

Necrophorus. Genus of beetles in the Family Silphidae; most spp. oviposit on small animals and then undermine and bury them.

necrosis. Death of tissue which is in direct contact with healthy tissue. Adj. necrotic.

Nectariniidae. See SUNBIRD.

nectocalyx. Nectophore.

nectochaeta. Type of free-swimming larva found in some Polychaeta; three pairs of parapodia as well as rings of cilia.

Nectogale. Genus of web-footed shrews; aquatic spp. of northern Himalayan mountain streams; eyes hidden, external ears absent.

Nectonema. See NECTONEMATOIDEA.

Nectonematoidea. Class of Nematomorpha which includes only a few pelagic marine spp.; natatory bristles present; adult female with an open pseudocoel and one ovary; Nectonema to 8 in. long.

Nectonemertes. Genus of pelagic marine nemertine worms; body flattened, with finlike margins and caudal end.

nectophore. One of the simple swimming bell-shaped medusae of a siphonophore colony.

Nectridia. Group of late Paleozoic amphibians sometimes included in the Subclass Stegocephalia.

Necturus. Genus of American salamanders; includes two spp. of mud puppies or water dogs; N. maculosus occurs generally in rivers and lakes over the eastern half of the U.S.; entire life cycle aquatic; gills red, body usually dark brown with black spots; to 15 in. long; carnivorous, sometimes taken on baited hooks; N. punctatus in N.C. and S.C. coastal rivers; to 6 in. long.

Needham, John. English Catholic priest and biologist of the 18th century who supported the theory of spontaneous generation.

Needham's sac. A terminal reservoir in the male squid reproductive system; holds mature spermatophores.

needlefish. See BELONIDAE and SCOMBERESOCIDAE.

needle shell. Turritella acicula, a very slender white or brownish marine snail to 0.25 in. long; Labrador to Mass.

Negaprion brevirostris. Lemon shark; a yellowish brown and greenish sp. of the warm Atlantic.

negative eugenics. Prevention of the transmission of undesirable hereditary traits by discouraging matings between individuals carrying such traits, e.g. feeblemindedness.

Negrillo. Race of negroids which includes the Congo pygmys; skin yellow, hair dark brown.

Negrito. Race of short negroids having medium to dark skin; Philippines, Malaya, and East Indies.

Negro. 1. Race of negroids having dark brown to black skin and long head; originally tropical Africa. 2. General term used to include any of the dark negroid races of Africa.

negro ant. See FORMICA.

negro bug. See THYREOCORIDAE.

Negroid. Any of a general group of races of modern man characterized by woolly hair and black to yellow skin; includes the following: Bantu, Bushman, Melanesian, Negro, Negrillo, Nigritian, and Negrito.

Nehallenia. Genus of damselflies; eastern U.S.

Neisseria. Genus of biscuit-shaped, gram-negative, non-motile, parasitic bacteria; e.g. cause gonorrhea (N. gonorrhoeae) and epidemic meningitis.

nekton. Collectively, the macroscopic animals suspended in the waters of ponds, lakes, rivers, and seas; they move about independently of currents and include such forms as fishes and whales.

nema. Any member of the Phylum Nematoda.

Nemasoma. Common genus of small millipedes.

Nemata. Nematoda.

Nemathelminthes. 1. Nematoda. 2. Term sometimes used to designate a phylum including the Nematoda, Nematomorpha, and Acanthocephala.

Nematocera. One of the two tribes in the dipteran Suborder Orthorrhapha; antennae long, usually with 6 to 16 segments, rarely up to 39; larvae with complete head.

nematocysts. Minute spherical or elongated capsules imbedded chiefly in special epidermal cells (cnidoblasts) of coelenterates; each capsule contains a minute coiled thread tube which, upon proper stimulation, is rapidly everted to aid in protection or the capture of prey; sometimes the external surface of a cnidoblast bears a small, pointed projection (cnidocil) which acts as a trigger for the sudden discharge of the contained nematocyst; various kinds of nematocysts are recognized, depending on their biochemical and physical action and their minute anatomy; most contain a toxin; four common types are: penetrant, volvent, streptoline glutinant, and stereoline glutinant.

Nematoda. Phylum which includes all the true roundworms; body slender, cylindrical, often tapered near ends, and covered with a cuticle; no segmentation, cilia, or circular muscle fibers; with a true digestive tract and pseudocoel; sexes separate, male with a cloaca; 100 microns to 1 m. long; marine, fresh-water, terrestrial, and parasites of plants and animals; about 13,000 spp.

nematogen. See MESOZOA.

Nematognathi. Siluroidei.

nematology. Study of nematodes.

Nematomorpha. Phylum closely related to the Nematoda; the "horsehair" worms or "hairworms"; adult extremely slender and long (to 3 ft.); body wall consists of a heavy cuticle, hypodermis, and longitudinal muscle fibers; pseudocoel more or less filled with loose connective tissue and reproductive organs and cells; there is a ventral nerve cord and a degenerate but complete digestive tract; adults terrestrial and in fresh and salt water, but the spiny larvae are parasitic in terrestrial and aquatic insects and in other aquatic invertebrates; about 100 spp.; Gordius a common genus.

Nematomorphoidea. Order of millipedes; 26 to 60 body segments; male with one or two pairs of gonopods on seventh segment; posterior spinning glands; Striaria typical.

Nematonereis. Genus of Polychaeta.

nematozooid. Hydrozoan zooid specialized for defense; bears numerous nematocysts.

Nemertea. Phylum consisting of elongated, muscular, soft, unsegmented worms that are somewhat flattened or oval in cross section; mostly marine (and especially intertidal), but a few occur in fresh waters and in tropical terrestrial habitats; body covered with a thick glandular epithelium; dorsal to the digestive tract is a long cavity containing a highly muscular and protrusible proboscis which is often armed with a sclerotized stylet and is used in food getting and defense; with a flame bulb system, circulatory system, brain, and two latero-ventral nerve cords; 5 mm. to 20 m. long; often brightly colored; about 750 spp.

nemertean theory. Phylogenetic theory of Hubrecht (1883) which sought to derive chordates from modified nemerteans.

nemertine. Any member of the Phylum Nemertea.

Nemertinea. Nemertea.

Nemertopsis. Genus of marine nemertine worms.

Nemesis. Genus of marine copepods in the Suborder Caligoida; parasitic on fishes.

Nemichthyidae. Family of snipe eels; oceanic fishes with extremely slender body and tail tapering to a thread; beak long and sometimes recurved; Nemichthys common.

Nemichthys. See NEMICHTHYIDAE.

Nemopsis. Genus of marine coelenterates in the Suborder Anthomedusae; hydroids in clusters on floating objects.

Nemotelus. Genus of soldier flies in the Family Stratiomyiidae.

Nemoura. Genus of stoneflies lacking cerci in the adult stage; several common spp. in eastern U.S.

nene. Local name for a Hawaiian goose.

neoarsphenamine. Neosalvarsan; organic arsenic compound similar to arsphenamine but less toxic; an anthelminthic.

Neobalaena. Genus of rare pygmy right whales of New Zealand and Australian waters; 20 ft. long.

Neobisium. Genus of Pseudoscorpionida.

Neoceratodida. Sirenoidei.

Neoceratodus. See CERATODONTIDAE.

Neochordodes. Genus of Nematomorpha.

Neocomatella. Genus of feather stars; more than ten arms; in warm seas.

Neoconocephalus. Genus of cone-headed grasshoppers.

neo-Darwinism. Concept of evolution which reconciles the original Darwinian ideas with modern genetics and speciation.

Neodasys. Common genus of marine Gastrotricha.

Neoechinorhynchus. Genus of acanthocephalan worms; parasitic in the intestine of marine and fresh-water fishes and N.A. fresh-water turtles; larval stages in insects and crustaceans.

Neoelmis. Genus of elmid beetles.

Neofiber. Genus of Fla. rodents in the Family Cricetidae; includes the Florida water rat and Everglades water rat; similar to the muskrat but smaller, not so distinctly aquatic, and with a cylindrical tail.

Neogaea. One of the three primary zoogeographic areas of the world; includes South America, Central America, and southern Mexico. Cf. ARCTOGAEA and NOTOGAEA.

Neognathae. One of the three superorders of birds; includes most modern birds; teeth absent; sternum with a prominent keel, wings well developed, pygostyle present; except for penguins and a few other spp., all are able to fly.

Neohydrophilus castus. Common hydrophilid beetle.

Neolampas. Genus of small echinoids in the Order Cassiduloida; test convex above, concave below; Aristotle's lantern absent; common in Caribbean.

Neolenus. Genus of Cambrian trilobites.

Neoliodes. Genus of mites in the Family Oribatidae.

Neoliparis. See LIPARIDAE.

Neolithic. Post-glacial phase of human history which began about 10,000 years ago; featured by domestication of animals and cultivation of plants.

Neo-Mendelism. Mendelism as modified and extended by many more recent discoveries in genetics such as linkage, multiple factors, sex chromosomes, crossing over, etc.

Neomenia. Genus of amphineurans found in North Atlantic; body covered with spicules; plates absent.

Neomeris. Genus of finless porpoises; small gray spp. found from the Cape of Good Hope to Japan; dorsal fin absent; to 5 ft. long; sometimes ascend rivers.

neomorph. Mutant allele having a new type of developmental effect not found in the original ancestral allele.

Neomys. Genus of water-shrews of mountain streams of Europe and western Asia.

Neomysis. Genus of crustaceans in the Order Mysidacea; often abundant in estuaries, tide pools, and adjacent fresh waters.

neonatae. Newly-born or newly-hatched individuals. Sing. neonata.

Neon nellii. Common brownish gray spider in the Family Salticidae; to 2.5 mm. long.

Neonura. Genus of damselflies.

Neonympha. Common genus of dark brown butterflies; wood satyrs.

neopallium. Non-olfactory portion of the cerebral cortex.

Neopanopeus. Genus of marine brachyuran mud crabs.

Neoperla clymene. Common and widely distributed stonefly.

Neophasganophora capitata. Common stonefly; eastern U.S.

Neophylax. Genus of Trichoptera.

Neopilina. See MONOPLACOPHORA.

Neoplanorbis. Genus of fresh-water snails.

neoplasm. Any new and abnormal localized cell growth. See MALIGNANT NEOPLASM and BENIGN NEOPLASM.

Neopterygii. Large subclass of bony fishes which includes most modern spp.; vertebrae usually amphicoelous; each ray of dorsal and anal fins supported by a skeletal rod; caudal fin homocercal, scales cycloid or ctenoid, nostrils not connected to mouth cavity.

Neornithes. One of the two subclasses of birds; includes all fossil and modern spp. except the fossil Archaeopteryx.

neosalvarsan. Neoarsphenamine.

Neoscutopterus. Genus of dytiscid beetles; northern states and Canada.

Neoseps reynoldsi. Sand skink in the Family Scincidae; a tan or brown sp. of Fla.

Neosphaeroma. Genus of marine Isopoda.

Neosporidia. Out-dated taxon which included many of the Sporozoa.

neosynephrin hydrochloride. Vasoconstrictor; also an anesthetic for small aquatic metazoans (1% solution).

neoteny. 1. Attainment of functional sexual maturity in an animal otherwise immature, e.g. in the tiger salamander. 2. Retarded development of individual structures.

Neotermes. Common and widely distributed genus of termites.

Neothunnus macropterus. Yellowfin tuna of the Pacific Coast; excellent game fish, to 500 lbs.

Neotoma. Genus of herbivorous rats in the Family Cricetidae; variously known as wood, pack, trade, mountain, or brush rats; often build large nests of vegetable debris in which they may incorporate metal and other bright objects picked up from their surroundings; North and Middle America.

Neotragus. Genus of royal antelopes; minute spp. of western and northeastern Africa.

Neotremata. Order of brachiopods in the Class Inarticulata; common example is Crania.

Neotrichia. Genus of Trichoptera.

Neotropical realm. One of the six major zoogeographic realms of the world; includes South America, Galapagos Islands, West Indies, and Central America as far north as central Mexico; some characteristic animals are: sloth, armadillo, American monkeys, rhea, anaconda, hoatzin, agouti, llama, and toucan.

neotype. Single specimen selected as the type for a particular taxon after the original type has been destroyed or definitely lost.

Nepa. Common genus of water scorpions in the insect Family Nepidae.

Nephelopsis obscura. Common and widely distributed fresh-water leech; feeds on invertebrates, also a scavenger; to 4 in. long.

Nephila. Genus of large spiders which build large nests in shaded woods of the southern states and tropics; legs I, II, and IV with conspicuous hair tufts.

nephridiopore. One of the paired external openings of the nephridia of flatworms, annelids, Onychophora, etc.

nephridium. In the broad sense, any tubule specialized for excretion and/or osmoregulation; with an external opening and with or without an internal opening. See PROTONEPHRIDIUM, METANEPHRIDIUM, NEPHROMIXIUM, PROTONEPHROMIXIUM, METANEPHROMIXIUM, MIXONEPHRIDIUM, and

COELOMODUCT.

nephritis. Kidney inflammation.

nephrocytes. 1. Migratory cells found near the fat body or pericardium of insects; take up and concentrate nitrogenous wastes from the hemocoel. 2. Term occasionally used for similar cells in other invertebrate groups.

nephrogenic ridge. In the early embryology of mammals, a narrow band of mesoderm on each side between somites and lateral plate; it differentiates into kidney and is associated with gonad formation.

nephromixium. Combination of nephridium and coelomoduct into one compound organ having mixed genital and excretory functions, e.g. as in Brachiopoda and certain Polychaeta. See PROTONEPHROMIXIUM, METANEPHROMIXIUM, and MIXONEPHRIDIUM.

nephron. Unit of kidney structure in reptiles, birds, and mammals; each usually consists of a Malpighian body and its contained glomerulus plus an associated urinary tubule; a human kidney is composed of about a million nephrons.

nephropore. Nephridiopore.

Nephrops. Genus which includes the small Norway lobster.

Nephroselmis. Genus of fresh-water biflagellate Protozoa; all reniform; chromatophores present.

nephrostome. 1. Ciliated internal opening of a nephridium. 2. Ciliated internal opening of certain kidney ducts of lower vertebrates; functional in immature fishes and in both immature and mature amphibians.

nephrotome. In an early vertebrate embryo, a strip of tissue laid down between the body somites and the coelom; gives rise to the segmental pronephric tubules.

Nephthys. Large and common genus of marine Polychaeta.

Nepidae. Family of insects in the Order Hemiptera; water scorpions; slender, sticklike insects to 2 in. long; crawl about in the shallows of ponds and obtain air at the surface film through a long anal respiratory tube; Ranatra and Nepa common.

nepionic. Pertaining to infantile postembryonic structures or processes.

nepionic valve. Small portion of a bivalve shell immediately surrounding the umbo and more or less distinct from the older portion of the shell; represents the extent of the juvenile shell.

Neptunea. Widely-distributed genus of large carnivorous marine snails; shell usually ridged.

Neptune's goblet. Common name of either of two very large spp. of sponges in the genus Poterion; to 4 ft. high.

neptune shell. Any of numerous small to large carnivorous marine snails; Neptunea, Colus, and Busycon typical.

Nereis. Same as Neanthes; common genus of marine polychaetes; clam worms; body elongated, usually 1 to 20 in., and cylindrical or more or less flattened; active and carnivorous; protrusible pharynx with two large jaws; actively swimming or in burrows or tubes in sand and other types of sea bottom; extensively used as fish bait; several fresh-water spp. have occasionally been collected in drainages near the sea; N. brandi of Calif. attains almost 6 ft.

Nerilla. Genus of Archiannelida.

Nerine. Genus of small Polychaeta.

Nerinides. Genus of Polychaeta.

Nerita. Large genus of snails; shell thick, globose, with a small spire and large body whorl; mostly marine, but a few spp. terrestrial or in fresh water.

neritic. Pertaining to that part of the sea where the water is 20 to 200 m. deep; the term includes both the bottom and the water between surface and bottom.

Neritina. Genus of heavy globose snails; fresh and brackish waters of Fla. and Gulf Coast.

neritopelagic. Pertaining to the communities suspended in the sea between the shoreline and a depth of 200 m.

Nerocila. Genus of marine Isopoda.

Nerthridae. Gelastocoridae.

nerve. Bundle of nerve fibers with its accompanying connective tissue.

nerve cell. Neuron.

nerve center. Any specific area or mass in the central nervous system having chief control over the regulation of a particular physiological process, e.g. respiratory center.

nerve cord. Prominent longitudinal bundle of nerve fibers forming much of the central nervous system in animals; it may or may not contain ganglia; depending on the phylum, one, two, or more may be present.

nerve eminence. Epithelial sense organ

in many cyclostomes, fishes, and amphibians; consists of flat epithelium, raised area, sunken pit, or portion of an enclosed canal.

nerve endings. Cluster of fine branches at the end of a nerve fiber, or the distinct end-organ at the end of a nerve fiber.

nerve fiber. General term for any process of a neuron, either dendrite or axon.

nerve gas. Highly toxic gas intended chiefly as a military weapon; inactivates cholinesterase.

nerve impulse. Impulse which travels from one part of an animal to another by means of nerve fiber(s); generally thought to be achieved electrically, according to the membrane theory; recovery after the transmission of the impulse involves production of a small amount of carbon dioxide and consumption of a small amount of oxygen; more heat is released during recovery than during transmisssion; velocities of transmission range from below 1 m. per second in some mollusk fibers to over 100 m. per second in some mammalian fibers; velocity varies directly with diameter of the fiber, and is typically faster in myelinated fibers than in non-myelinated.

nerve net. Diffuse network of simple branching nerve cells as found in the ctenophores, Coelenterata, and to a lesser extent in the Platyhelminthes; there is no differentiation into dendrites and axons, and impulses are transmitted by all processes through synapses to adjacent nerve cells; located in the lower epidermis and upper mesoglea (mesodermal) region.

nerve ring. Ring of nervous tissue surrounding the oral area in many echinoderms and the pharyngeal area in arthropods and annelids.

nerve root. See DORSAL ROOT and VENTRAL ROOT.

nerve-winged insects. Common name of most insects in the Order Neuroptera.

nervous system. Collectively, all of the nerve cells of an animal; the receptor-conductor-effector system; in man, the nervous system consists of brain, spinal cord, and all nerves.

nervous tissue. Tissue composed of neurons; functions in the reception and transmission of nervous impulses.

nervus terminalis. Tiny cranial nerve extending along the median surface of the olfactory tract between the olfactory sac and olfactory lobe in all vertebrates except cyclostomes and birds; probably both a somatic and sensory nerve.

Nesochen sandvicensis. Hawaiian goose.

Nesolagus. Genus of short-eared rabbits of Sumatra.

Nesomys. Genus of Madagascar voles in the Family Cricetidae.

Nesotragus. Genus of sunis; tiny east African antelopes about 12 in. high at the shoulder.

nestling. Any young bird that has not left the nest.

net plankton. Plankton organisms of such size as will be caught by a number 25 bolting silk net (mesh openings 0.060 to 0.070 mm.).

nettle cell. Nematocyst.

net-winged. Among insects, having wings with a fine network of veins, e. g. Neuroptera.

net-winged midges. See BLEPHAROCERIDAE.

Neumania. Common genus of Hydracarina; active swimmers.

neural. Of or pertaining to a nerve or the nervous system.

neural arch. Dorsal arch of a vertebra, the cavity below being traversed by the nerve cord.

neural canal. 1. Cavity of the brain and spinal cord, especially in vertebrate embryos. 2. Space enclosed by neural arches of the vertebral column.

neural crest. Longitudinal thin line of cells formed in the angle between the neural tube and ectoderm on either side of a vertebrate embryo; produces the dorsal roots of the spinal nerves, sympathetic ganglia, and a few other structures.

neural fold. Longitudinal ridge which forms on either side of the neural plate (neural groove) in a vertebrate embryo; the folds come together and meet in the midline to form the neural tube.

neural gland. Small gland near the ganglion (brain) of tunicates; presumably endocrine.

neural groove. Dorsal longitudinal groove which forms in a vertebrate embryo; bordered by two neural folds; preceded by neural plate stage and followed by neural tube stage.

neural plate. 1. Long median thickened area of ectoderm in the early vertebrate embryo; eventually gives rise to the central nervous system. 2. One of about seven dorsal plates in the carapace of a turtle.

neural spine. Single median spine which projects dorsally from a vertebra.

neural tube. Primitive hollow dorsal nervous system of the early vertebrate embryo.

neurasthenia. Nervous prostration due to prolonged expenditure of energy; marked by fatigue, back pains, loss of memory, insomnia, loss of appetite, etc.

Neureclipsis. Genus of Trichoptera.

neurectoderm. Dorsal area of ectoderm in certain vertebrate blastulas; eventually develops into the embryonic nervous system.

neurenteric canal. 1. Temporary canal between the amniotic cavity and the yolk sac cavity in certain vertebrate embryos. 2. Temporary connection between the posterior end of the canal of the spinal cord and the enteric cavity in vertebrate embryos.

neurilemma. Thin membrane surrounding non-medullated nerve fibers and the medullary sheath of medullated fibers; Schwann's sheath.

neuritis. Inflammation of nervous tissue, especially one or more nerves.

neurobiotaxis. In embryology, the tendency of nerve cell bodies to migrate toward the source of the stimuli whose impulses they carry.

neuroblast. Embryonic cell which will develop into a neuron.

neurocoel. Small cavity in a chordate nerve cord.

Neurocordulia. Genus of dragonflies.

neurocranium. That part of the skull which surrounds the brain and sense organs. Cf. SPLANCHNOCRANIUM.

neurocytolysin. Venom constituent in certain snakes; causes lysis of nerve cells.

neurogenic rhythm. Especially in insects, the rhythm of nerve impulses which corresponds precisely with the rhythm of wing beats. Cf. MYOGENIC RHYTHM.

neuroglia. Special type of delicate packing or connective tissue interspersed among nerve cells and fibers.

neurohormone. Organic compound produced by neurons and released at their endings to act as a hormone; examples are oxytocin and vasopressin.

neurohumor. Organic substance secreted in minute amounts by the terminal endings of an axon when an impulse passes through them; facilitates the transmission of the impulse across the synapse to the processes of an associated neuron or to an effector; e.g. acetylcholine and sympathin.

neurohypophysis. That portion of the pituitary which is derived from downgrowth of nervous tissue from the diencephalon of the brain.

neurokeratin. In histological sections, the precipitated protein component from myelin of myelin sheaths.

neuromast. 1. General term for a sensory papilla as found in many invertebrates and vertebrates. 2. One of many small budlike sensory areas, or pit organs, of the lateral line system of fishes and amphibians; each neuromast cell is elongated, with a projecting hairlike structure; a cupula (q.v.) encloses the tips of these hairs.

neuromeres. Rhombomeres; paired ventral embryonic bulges in the hindbrain.

neuromotor apparatus. Neuromotor system.

neuromotor system. Collectively, the basal granules of the cilia together with their complex arrangement of interconnecting fibrils in many ciliate protozoans; the system governs the coordination of ciliary beating. See MOTORIUM and KINETY.

neuromuscular spindle. Proprioceptor in amphibians and amniotes; consists of nerve endings wound around several muscle fibers, the whole being enclosed in a connective tissue capsule.

neuron. 1. Type of cell forming nerve tissue; a nerve cell body including all its processes, specialized for the transmission of nervous impulses. 2. Occasionally, the nerve cell body without its processes.

neurone. Neuron.

neuropil. 1. In certain invertebrate ganglia, a network of association, motor, and sensory fibrils. 2. In the vertebrate central nervous system, a network of delicate unmyelinated fibrils having many synapses; thought to function in the diffusion of impulses.

neuropodium. 1. Distinctive lower half of a polychaete parapodium. 2. Terminal fibril of a non-medullated neuron.

neuropore. Open anterior or posterior end of the neural canal in an early vertebrate embryo.

Neuroptera. Order of terrestrial carnivorous insects which includes the lacewings, nerve-winged insects, and ant lions; legs long and slender; four similar wings with a fine network of veins, sometimes hairy and mothlike; larvae mostly terrestrial and carnivorous, with piercing and sucking mouth parts; common genera are Chrysopa, Myrmeleon, and Hesperoleon.

neurosecretory cell. Any cell (neuron) of the nervous system which produces one or more hormones.

neurosis. Psychoneurosis; a mental disorder less severe than a psychosis, less incapacitating, and with the personality remaining more or less unchanged; e.g. hysteria, excessive anxiety, shyness, and fear. Pl. neuroses.

neurotendinal spindle. Proprioceptor in mammals; consists of nerve endings wound around several tendon fibers, the whole being enclosed in a connective tissue capsule.

Neuroterus. Genus of gall wasps which form oak galls.

neurotoxin. Toxin which affects the nervous system.

Neurotrichus gibbsi. Gibb's mole; a dark gray mole of swampy and moist areas of western Calif., Ore., Wash., and southern British Columbia.

neurula. Stage of vertebrate embryo development marked by the first appearance of tissues destined to become the central nervous system; stage ends when the neural tube is well formed.

neurulation. First formation of the nervous system in an early embryo.

neuston. Community of microorganisms associated with the surface film of bodies of water; e.g. some Protozoa, insects, planarians, ostracods, etc.

neuter. 1. Sexless; without functional reproductive organs, but other organ systems essentially normal. 2. A worker or undeveloped female in some Hymenoptera.

neutral red. Intra-vitam stain for Protozoa; commonly used in 1% solution.

neutrophil. Type of phagocytic leucocyte occurring in vertebrate blood and formed in bone marrow; large cells with nucleus broken into several large fragments; about 65 to 75% of all human leucocytes are neutrophils; they can make their way through the endothelium of capillaries and into surrounding tissues.

neutrophilic leucocyte. Neutrophil.

Newman, Horatio H. American zoologist (1875-1957); professor at Univ. of Chicago; pioneer in the study of human genetics; wrote many articles and six books.

newt. 1. Type of salamander with vomerine teeth in two longitudinal rows; the American spp. are able to float without swimming movements; in the Family Salamandridae. 2. Eft.

n. g. Abbreviation for new genus.

ngana. Nagana.

Ngandong man. See HOMO SOLOENSIS.

n. gg. Abbreviation for new genera.

niacin. Nicotinic acid.

niche. Ecological niche.

Nichomache. Genus of tube-building Polychaeta.

Nicolea. Genus of tube-building Polychaeta.

nicotine sulfate. Insecticide; anesthetic for small aquatic incertebrates (2% solution).

nicotinic acid. Niacin; vitamin of the B complex; stimulates gastric secretion and essential to nucleotide formation in cell metabolism; deficiency produces pellagra (roughened skin, sore mouth, diarrhea, and nervous disturbances); abundant in green leaves, wheat germ, eggs, meat, and yeast; $C_6H_5O_2N$.

nictitating membrane. Thin membrane at the inner angle of the eye or beneath the lower lid of many vertebrates; in some spp. it can be drawn over the surface of the eyeball; the "third eyelid."

nidamental gland. 1. One of four cylindrical glands in the mantle of a female squid; two are very large, the other two (accessory nidamentals) are much smaller; secretes gelatinous material which envelops the fertilized eggs as they are extruded. 2. Occasionally, any invertebrate gland secreting material that envelops eggs.

nidation. 1. Implantation. 2. Development of uterine mucosa between two successive menstrual periods.

nidicolous. 1. Pertaining to birds which are relatively undeveloped upon hatching and must remain in the nest for some time. 2. Living in a nest, especially the nest of another sp.

nidifugous. Pertaining to birds which are well developed upon hatching and are able to leave the nest almost immediately.

niggerfish. Caribbean grouper, a fish with prominent blue or black spots.

niggerhead. 1. Any of several dark-colored, smooth, fresh-water mussels, especially Quadrula. 2. Shallow water coral head.

night blindness. Inability to see normally in very dim light owing to an insufficient quantity of vitamin A in the diet.

nightcrawler. Any large earthworm which comes out of its burrow at night.

nighthawk. See CHORDEILES.

night heron. Any heron especially active at dawn and dusk, e.g. Nycticorax and Nyctanassa.

nightingale. See LUSCINIA.

nightjar. European goatsucker.

night lizard. See XANTUSIIDAE.

night snake. Any of several spp. and subspecies of western and southwestern snakes in the genus Hypsiglena; light or dark buff color dorsally with about 50 brown dorsal spots; 6 to 26 in. long.

Nigritian. Race of negroids which includes the Nile Negro; skin very dark, head long; northwestern Africa.

nilgai. See BOSELAPHUS.

ninhydrin reaction. When protein suspensions are treated with ninhydrin (triketo-hydrindene-hydrate), a blue, red, or violet color is produced.

nipple. Protuberance of a mammary gland which serves as an outlet point for the milk.

Nippotaeniidea. Order of small tapeworms in the Subclass Cestoda; scolex with an apical sucker only; a few spp. in the intestine of Russian and Japanese fresh-water fishes; Nippotaenia typical.

Nissl granules. Angular nucleoprotein granules in the cytoplasm of unfatigued nerve cell bodies.

nit. 1. Egg of a louse or other parasitic insect; often glued to hair of the host. 2. Occasionally, a very small insect.

Nitidella. Genus of small marine gastropods; dove shells; to 0.5 in. long; tropical and subtropical.

Nitocra. Common genus of marine and brackish copepods.

Nitocris. Genus of fresh-water snails; eastern U.S.

nitrification. Conversion of organic nitrogen compounds into ammonium compounds, nitrites, and nitrates by bacteria in soil and aquatic environments. See NITROGEN CYCLE.

Nitrobacter. Genus of soil and aquatic nitrifying bacteria which convert nitrites into nitrates.

nitrogen cycle. World-wide circulation and reutilization of nitrogen atoms, chiefly due to metabolic processes of plants and animals; plants take up inorganic nitrogen, mostly as ammonium and nitrate compounds, and convert much of it into organic compounds (chiefly protein); such nitrogen atoms then may be assimilated into the bodies of one or more successive animals; excretion, burning, and bacterial action on dead organisms return the nitrogen atoms to an inorganic state. See DENITRIFYING BACTERIA, ROOT NODULE BACTERIA, and NITROGEN-FIXING BACTERIA.

nitrogen-fixing bacteria. Certain aquatic and soil bacteria which are able to utilize atmospheric nitrogen and convert it into nitrate compounds; the nitrates are then available to green plants. See NITROGEN CYCLE and ROOT-NODULE BACTERIA.

nitrogenous waste. Any of a variety of simple nitrogen-containing compounds resulting from the breakdown of proteins, e.g. urea and uric acid.

Nitrosomonas. Genus of soil and water bacteria which convert ammonia to nitrites, and nitrites to nitrates.

Nitzschia. Genus of monogenetic external trematode parasites of marine and fresh-water fishes.

Noah's ark. Arca noae, a common and widely distributed bivalve mollusk of marine shallows.

Noctilionidae. Tropical American family of hare-lipped bats; feed on terrestrial and aquatic vertebrates and invertebrates.

Noctiluca. Genus of holozoic marine flagellates; spherical, bilaterally symmetrical, and with cytostome and flagellum at the bottom of a deep peristome; cell with a large tentacle.

Noctuidae. Large family of medium to large moths; owlet moths; herbivorous caterpillars known as armyworms and cutworms; many pest spp.; important genera are Laphygma, Heliothis, Leucania, and Erebus.

nodal tissue. Special cardiac muscle tissue in the vertebrate heart; determines, initiates, and transmits the intrinsic heart beat stimuli.

noddy. See ANOUS.

node. Nodus.

node of Ranvier. One of the series of regularly repeated constrictions in the medullary substance covering a medullated nerve fiber; occur at junctions of neurilemma cells.

Nodosaria. Genus of Foraminifera; chambers of shell arranged in a gentel curve or a straight line.

nodus. 1. Small protuberance or boss. 2. Prominent cross-vein connecting the first two longitudinal veins in the wings of Odonata. 3. Small segment(s) between thorax and abdomen in ants and other Hymenoptera.

Nomada. Cosmopolitan genus of wasplike bees; lack pollen-collecting apparatus and live in the nests of other bees.

Nomadidae. Family of wasplike bees; bright yellow and red coloration; parasitic or inquilines in nests of other bees.

nomenclature. Application of distinctive name(s) to each taxon in any classification.

nomen conservandum. Scientific name whose usage has been preserved by decision even though it is in conflict with the rules of nomenclature.

nomen dubium. Name of a sp. for which the evidence is insufficient to recognize the organism to which it was first applied.

nomen novum. New name, as applied to a sp. whose taxonomic status is being changed.

nomen nudum. Published scientific name which is already used for another sp. or is otherwise unavailable.

nomen rejectum. Published scientific name which, for one reason or another, is rejected and should not be used for that particular animal.

Nomeus. Genus of small marine fish; at least one sp. lives among the tentacles of the Portuguese man-o'-war with impunity.

non-adaptive characters. Characters or mutants appearing in an animal which have no apparent advantage or disadvantage to the animal.

nondeciduate placenta. Indeciduate placenta.

nondisjunction. Failure of homologous chromosomes to separate during meiosis; results in a nullosomic and/or a disomic gamete.

nongranulocyte. General class of leucocytes without obvious granules in the cytoplasm, especially monocytes and lymphocytes.

nonmedullated nerve fiber. Axon having no surrounding sheath; common among invertebrates, but in vertebrates restricted to the autonomic system, the internal portion of the spinal cord, and the external portion of the brain (gray matter).

nonstriated muscle. In vertebrates, a type of muscle tissue consisting of flat sheets or layers of long delicate spindle-shaped cells held together with fibrous connective tissue; such tissue occurs in the walls of the digestive, respiratory, circulatory, excretory, and reproductive ducts and is not under voluntary control; also, the chief type of visceral and locomotor muscle tissue in most invertebrate phyla.

noolbenger. Honey-sucker; marsupial about the size of a small mouse, found in southwestern Australia; snout proboscis-like, tongue very long; feeds on nectar, pollen, and small insects.

noösphere. Present period of ecological history when vegetational and animal climax communities are undergoing profound transformations as the result of agriculture, grazing, lumbering, and industrial activities; preceded by the biosphere period.

Nordic. Race of tall caucasoids native to northern Europe; skin whitish or pink, hair light, head long.

norepinephrine. Hormone secreted by the adrenal medulla; similar to epinephrine but has weaker effects on blood sugar and heart rate and more pronounced vasoconstrictor effects.

norleucine. Uncommon amino acid; $C_6H_{13}O_2N$.

normal curve. Normal dispersion; symmetrical bell-shaped curve of normal distribution found for many biological events and characteristics; e.g. a height-frequency curve for human males, or a curve showing the number of kernels on numerous ears of corn.

normal dispersion. Normal curve.

normoblasts. Nucleated cells which immediately precede mature erythrocytes.

Norrisia. Genus of intertidal gastropods; turban shells.

northern crab. Stone crab, rock crab; Cancer borealis, a common brick-red crab of rocky New England coastal areas.

northern creek chub. See SEMOTILUS.

northern hairy-keel. See TRICHOTROPIS.

northern phalarope. See LOBIPES.

Norway rat. See RATTUS.

no-see-um. Any of a wide variety of small biting and non-biting midges.

Nosema. Genus of microsporidian intracellular parasites of arthropods and a few other invertebrates; see PEBRINE.

nosepiece. Lower end of revolving turret of a microscope; usually carries two or more objectives.

Nosopsyllus fasciatus. European rat flea; an important transmitter of bubonic plague.

Notacanthidae. Family of bottom-feeding deep-sea fishes; spiny eels; body long, tail tapering, anal fin long, jaws set below protruding snout; Notacanthus common.

Notacanthus. See HETEROMI and NOTACANTHIDAE.

notched sand dollars. Certain spp. of sand dollars in the genus Encope.

Notechis. Genus of Australian snakes which includes the poisonous tiger snake.

Notemigonus crysoleucas. Roach or golden shiner; common fresh-water fish in the Family Cyprinidae; to 12 in. long; shallow lakes and quiet rivers east of the Continental Divide in U. S.; not a desirable food fish.

Notharctus. Small Eocene fossil lemur found in Europe and U. S.; orbits large; 40 teeth.

Notholca. Marine and fresh-water genus of plankton rotifers; lorica spinous anteriorly, composed of two plates immovably fused laterally.

Nothria. Genus of Polychaeta.

Notiosorex crawfordi. Desert shrew; a small sp. of the arid southwest.

notochord. Long, slender skeletal rod composed of large vacuolated cells and lying between the central nervous system and digestive tract; present in adult cephalochordates and larval tunicates; in the vertebrates it is present only in the embryonic stages and later is surrounded and supplanted by the vertebral column.

notochordal groove. Temporary groove in the ventral portion of the embryonic notochord in certain vertebrate embryos.

notochordal rod. Notochord, especially during embryonic development.

Notocotylus. Genus of monostome trematodes; in water birds and mammals.

Notodelphyoida. Suborder of marine copepods; urosome cylindrical and narrower than metasome; eggs carried in an incubatory pouch, not ovisacs; first antennae short, prehensile in male; commensals in or on ascidians or other invertebrates, or external parasites.

Notodelphys. Genus of copepods commensal in the pharynx of certain tunicates.

Notodontidae. Family which includes the puss moths; somber, nocturnal spp.; larvae of many are pests on leaves of cultivated plants.

Notodromas monacha. Actively swimming fresh-water ostracod; usually in aquatic vegetation.

Notogaea. One of three primary zoogeographical areas of the world; includes Australia, New Zealand, New Guinea, Celebes, and nearby islands.

Notomastus. Cosmopolitan genus of large Polychaeta; parapodia rudimentary.

Notommata. Genus of fresh-water illoricate rotifers.

Notommatidae. Large family of ploimate rotifers; trunk flexible; toes well developed; most spp. associated with substrates.

Notonecta. Common genus in the hemipteran Family Notonectidae.

notonectal. Swimming with the dorsal surface downward.

Notonectidae. Family of insects in the Order Hemiptera; backswimmers; swim about in ponds and lakes by means of long hind legs and with the ventral side upward; feed on body fluids of many small aquatic animals; Notonecta common.

Notophthalmus. Small genus of newts in the Family Salamandridae; to 55 mm. long; eastern half of U. S. and northeastern Mexico.

Notoplana. Common genus of polyclad Turbellaria.

notopodium. Distinctive upper half of a polychaete parapodium. Pl. notopodia.

Notoptera. Grylloblattodea.

Notopteroidei. Suborder taxon sometimes used to include the fish Family Hiodontidae.

Notornis. Genus of flightless, gallinule-like birds of New Zealand; recently extinct.

Notoryctes. See NOTORYCTIDAE.

Notoryctidae. Family of Australian marsupials which includes the marsupial moles, Notoryctes; forelimbs greatly modified for burrowing in sandy soils.

Notosolenus. Genus of oval fresh-water biflagellate Protozoa; one flagellum directed forward, the other shorter and trailing.

Notostraca. Order of crustaceans which includes the tadpole shrimps; 10 to 58 mm. long; with a low, arched carapace covering head and much of trunk; 35 to 71 pairs of trunk appendages; two long cercopods; especially common on the mud bottom of vernal prairie ponds; not known east of the Mississippi River in the U.S.; Lepidurus and Apus common.

Nototropis. Genus of marine Amphipoda.

Notoungulata. Order of early mammals; mostly hoofed but a few forms with claws; small and active to large and clumsy; Paleocene to Pleistocene.

Notropis. Large and difficult genus of small lake and stream fishes in the Family Cyprinidae; shiners; usually silvery in color; common and widely distributed in U.S.; extensively used as bait.

notum. 1. Dorsal portion or sclerite of an insect segment. 2. Tergum.

Noturus. Genus of small catfishes related to the bullheads but with the adipose fin continuous with the tail; stonecats; yellowish-brown spp. of clear waters.

Novius. See COCCINELLIDAE.

Novumbra hubbsi. Western mud minnow; a small, brightly colored uncommon sp. of western Wash.

NSH. Neurosecretory hormone of certain insects; among other functions, it is known to trigger the action of GDH.

n. sp. Abbreviation for new species (sing.).

n. spp. Abbreviation for new species (pl.).

nuchal. Pertaining to the nape of the neck.

nuchal flexure. Cervical flexure.

nuchal organs. Two glandular pits on the dorsal surface of the first segment of certain Polychaeta.

nuchal plate. Small median plate at the anterior margin of the carapace of certain turtles.

Nucifraga. Genus of birds in the Family Corvidae; N. columbiana is Clark's nutcracker of the Rocky Mountains; much like a small crow, but with gray body and white patches on black wings and tail; N. caryocatactes is a European sp., dark brown with white spots.

Nuclearia. Genus of fresh-water amoeboid Protozoa.

nuclear membrane. Thin membrane surrounding a cell nucleus; the electron microscope shows it to be double-layered and with minute pores through which the nucleoplasm is continuous with the cytoplasm.

nuclear sap. More liquid homogeneous portion of protoplasm contained within the membrane of a cell nucleus.

nuclease. Enzyme which accelerates the hydrolysis of nucleic acids to mononucleotides and other substances; intracellular and extracellular.

nucleic acid. One of several complex organic acids which combine with proteins to form the nucleoproteins of chromatin; also found in the cytoplasm; examples are deoxyribonucleic acid (DNA) and ribonucleic acid (RNA).

nuclein. Decomposition product of nucleoprotein.

nucleolus. One or more small, dense, more or less spherical bodies in the nucleus of non-dividing cells; apparently produced by chromosomes and contain ribose nucleoprotein; quite distinct from the karyosomes. Pl. nucleoli.

nucleoplasm. Protoplasm of a cell nucleus.

nucleoprotein. One of a group of chemical constituents of chromatin, composed of protein(s) united with a nucleic acid (especially deoxyribonucleic acid and ribonucleic acid).

nucleoside. Any of several compounds obtained from nucleic acids; composed of a carbohydrate joined to a purine or pyrimidine base.

nucleotide. See DIPHOSPHYPYRIDINE NUCLEOTIDE and TRIPHOSPHOPYRIDINE NUCLEOTIDE.

nucleus. 1. Body present in nearly all cells; bounded by a double nuclear membrane and contains the chromatin material and certain other discrete bodies; essential to normal metabolism and reproduction of the cell; variously shaped but often ovoid or spherical. 2. A small aggregation of nerve cell bodies within the central nervous system in chordates. Pl. nuclei.

nucleus-cytoplasm ratio. Ratio of nucleus volume to cytoplasm volume in a cell.

nucleus of Pander. Small whitish area just below the nucleus of an unfertilized bird egg.

Nucula. See NUT CLAM.

Nuculana. See NUT CLAM.

Nuda. One of the two classes of Ctenophora; tentacles absent; body laterally compressed and thimble- or helmet-shaped, with an immense pharynx; Beroë is the principle genus, to 8 in. high, pinkish, and found in all cold seas.

Nudibranchia. Suborder of opisthobranch

gastropods; includes the sea slugs or nudibranchs; shell absent; respiration through general integument, posterior adaptive gills, or symmetrically arranged finger-like projections on the dorsal and lateral surfaces; Dendronotus and Aeolis typical.

nullisomic. Organism in which a chromosome pair is absent from the somatic cells.

numbat. Small, rare marsupial of southwestern Australia; feeds on ants and termites.

Numenius. Cosmopolitan genus of birds in the Family Scolopacidae; curlews and whimbrels; breed in the Holarctic area.

numerical aperture. N.A.; indication of the resolving power and efficiency with which a microscope objective (or condenser, etc.) transmits light rays; expressed as N.A.= n x sine μ, where n is the refractive index of the medium between cover slip and objective, and μ is half the angle of aperture of the objective; the numerical apertures of the low power, high power, and oil immersion objectives of an ordinary laboratory microscope are about 0.25, 0.66, and 1.25, respectively.

Numididae. Family of birds in the Order Galliformes; guinea fowl; native to Africa south of the Sahara but widely domesticated; Numida meleagris most common.

Nummulites. Genus of fossil Foraminifera.

nunatak. 1. Hill or mountain now or formerly surrounded by an ice sheet. 2. Refugium.

nuptial chamber. Small burrow made by a pair of termites as the beginning of a new colony.

nuptial flight. Flight of a virgin queen bee during which she is followed by many drones and copulates with one of them; similar behavior in some ants.

nuptial pads. Roughened swellings on the fingers of male Anura; appear during breeding season.

nuptial plumage. An especially bright-colored plumage assumed by birds during the breeding season.

nurse cell. 1. In female sponges, a choanocyte or amoebocyte which engulfs an ingested sperm cell and transfers it to the vicinity of the ripe ovum where the sperm is released and fertilizes the ovum. 2. One of a few to many cells closely associated with an ovum; presumed to have a nutritive function.

nursery-web weavers. See PISAURIDAE.

nurse shark. Gata; a large, sluggish shark of warm waters; Ginglymostoma cirratum.

nut clam. Any of many spp. of small marine bivalves having many sawlike teeth along the hinge; often brownish; widely distributed and commonly very abundant; Acila, Nucula, and Nuculana typical.

nutcracker. See NUCIFRAGA.

nuthatch. See SITTIDAE.

nutria. See CAPROMYIDAE.

nutrition. Sum of all processes involved in taking in, converting, and using food substances by living organisms.

nut shell. Common name of marine bivalves in the genus Nucula.

Nuttallina. Common West Coast genus of small chitons.

Nuttalornis borealis. Olive-sided flycatcher in the Family Tyrannidae; nests in coniferous forests of N.A.; winters in South America.

nyala. See STREPSICEROS.

Nyctanassa violacea. Yellow-crowned night heron; body bluish gray, bill black; southeastern U.S., West Indies, Central America, and northern South America.

Nyctea nyctea. Common snowy owl of northern part of Northern Hemisphere; large whitish sp., to 26 in. long; uncommonly winters as far south as Calif., Colo., and Va.

Nyctereutes. Genus of raccoon-dogs; omnivorous burrowers with the appearance of raccoons; Japan.

Nycteribiidae. Bat ticks; a small family of highly modified wingless Diptera ectoparasitic on bats.

Nycteridae. Family of hollow-faced bats; mostly African; Nycteris common.

Nycteris. See NYCTERIDAE.

Nyctibiidae. Family of five birds in the Order Caprimulgiformes; potoos; Central American goatsuckers; hunt insects in flycatcher fashion.

Nycticebus. Genus of slow lorises in the Family Lorisidae; small stout lemur-like nocturnal mammals with short limbs and thick fur; southeastern Asia.

Nycticeius fuscus. Big brown bat; a large sp. common from Canada to Central America.

Nycticorax nycticorax. Black-crowned night heron of North and South America in the Family Ardeidae; feeds in shallows of fresh water, especially in early morning and evening.

Nyctidromus albicollis. Pauraque; a sp. of goatsucker in the Family Caprimulgidae; Neotropical mainland and as far north as Tex.

Nyctiellus. See NATALIDAE.

Nyctiphanes. Common genus of North Atlantic pelagic Euphausiacea; about 15 mm. long.

Nyctoporis. Genus of darkling ground beetles.

Nyctosaurus. Genus of pterosaurs; Mesozoic flying reptiles.

Nyctotherus. Genus of flat oval to kidney-shaped ciliate Protozoa parasitic in the intestine of amphibians and various invertebrates.

Nygmia. *Euproctis.*

Nygolaimus. Genus of microscopic free-living nematodes.

nymph. 1. One of a series of immature stages in certain insects (see PAUROMETABOLA). 2. Immature stage in certain arachnids, especially the instars preceding the sexually mature adult. 3. See SATYRUS.

Nymphalidae. Large family of medium to large brightly colored butterflies; forelegs reduced; familiar genera are *Nymphalis*, *Vanessa*, and *Basilarchia*.

Nymphalis. Common genus of butterflies in the Family Nymphalidae; the mourning cloak, *N. antiopa*, is a Holarctic sp. with brownish-black coloration and yellow wing borders; *N. californica* is the common California Tortoise-shell.

nymphochrysalis. Pupa-like stage between larva and nymph forms in certain mites.

Nymphon. See PYCNOGONIDA.

nymphophan. Quiescent pupa-like stage between the parasitic larva and active nymph stages in typical Hydracarina.

Nymphopsis. Genus of Pycnogonida.

Nymphula. Genus of moths in the Family Pyralididae; larvae semiaquatic.

Nyroca. *Aythya.*

O

oarfish. See REGALECUS.

Obelia. Common genus of sessile, colonial, marine hydrozoan coelenterates; medusae free-swimming.

Obeliscus. Genus of small long-spired marine gastropods.

Oberholseria. See TOWHEE.

obesity. Excessively fat or fleshy condition.

obligate parasite. Organism which can exist only as a parasite.

Obliquaria reflexa. Common unionid mollusk of eastern U. S.

obliquus externus. Large muscle covering most of the sides of the abdomen in certain vertebrates.

obliquus internus. Broad lateral abdominal muscle in man; originates on the crest of the ilium and from lumbar fascia and inserts on the lowest ribs, linea alba, and crest of the pubic bone; compresses the abdomen and flexes the thorax.

obliterative shading. Darker coloration on that part of an animal which is exposed to the most intense illumination; examples are the light-colored ventral surfaces of many fishes and hares as contrasted with the darker patterns of their lateral and dorsal surfaces; result is an effective blending with the background.

Obovaria. Genus of unionid mollusks; eastern half of U. S.

obstetrical toads. Spp. of toads (Alytes) in which the male carries egg strings wound around the his hind legs for several weeks; hatch out when the male enters water.

obtect. Pertaining to an insect pupa having the wings and legs held closely against the body. Cf. EXARATE.

obturator externus. Mammalian muscle which originates on the edge of the obturator foramen and inserts on the base of the great trochanter; rotates thigh outward.

obturator foramen. Large opening on each side of the pelvic girdle in many vertebrates; bounded by the articulated pubis and ischium above and below.

obturator internus. Mammalian muscle which originates on the trochanter foramen, pubis, and ischium, and inserts on the great trochanter; rotates and abducts thigh.

obturator nerve. Branch of the lumbar plexus; innervates muscles and skin of thigh, hip, and knee.

occipital condyle. Protuberance on the posterior portion of the skull which fits into an appropriate groove on the first vertebra; two such condyles in amphibians and mammals and one in reptiles and birds.

occipital foramen. 1. Posterior opening of an insect head in the neck region; carries the digestive tract, nerve cord, and other organs. 2. Foramen magnum.

occipitalis muscle. Broad flat muscle on the back of the skull.

occipital lobe. Posterior ventral portion of a cerebral hemisphere, especially in higher vertebrates.

occiput. 1. Back part of the head of vertebrates. 2. Hind part of the dorsal surface of the head of an insect; sometimes a distinct sclerite.

occlusion. Proper fitting together of upper and lower teeth.

Oceania. Genus of marine hydrozoan coelenterates; hydroid poorly known.

oceanic. Pertaining to the sea, especially where the water is more than 200 m. deep.

oceanic island. Island separated from a continental land mass by a relatively great distance; often of volcanic or coral origin; examples are the Hawaiian Islands, Philippines, and Galapagos Islands; their faunas are usually highly endemic, and amphibians, fresh-water insects, fresh-water fishes, mammals, land birds, etc. are poorly represented or absent.

Oceanites oceanicus. Wilson's petrel, a sooty black oceanic bird which is widely distributed and breeds in southern South America and Antarctica.

Oceanodroma. Genus of oceanic birds in the Family Procellariidae; one of the genera to which the petrels or Mother Carey's chickens belong; strong fliers with long wings.

oceanography. 1. The study of the oceans and all their chemical, physical, and biological features. 2. In the narrower sense, the study of physical features of the oceans.

ocean perch. See SCORPAENIDAE.

ocean pout. See MACROZOARCES.

ocean sunfish. See MOLA.

ocellated turkey. See AGRIOCHARIS.

ocellus. 1. Eye of many types of invertebrates; structure simple to complex; a beadlike lens and various accessory cells may be present but they merely serve to concentrate light, and no image is formed. 2. Eyelike spot of color. Pl. ocelli.

ocelot. Tiger-cat, leopard cat; medium-sized spotted or marbled yellow nocturnal cat (Felis pardalis) found from Tex. to southern South America.

Ocenebra. Genus of marine snails; tritons.

Ochotona. See OCHOTONIDAE.

Ochotonidae. Small family of mammals in the Order Lagomorpha; pikas, conies, or little chief hares; Ochotona found in mountainous western U.S. and Canada, another genus in Asia; to 7.5 in. long; ears rounded, small; tail not visible, legs short; mostly in rocky areas at high altitudes; gather vegetation and store it as small piles of "hay."

Ochromonas. Genus of small green biflagellates; flagella unequally long; fresh waters.

Ochthebius. Genus of small aquatic beetles in the Family Hydraenidae.

ocre star. See PISASTER.

octamerous. Having parts radially arranged in eight or multiples of eight.

Octocorallia. Alcyonaria.

Octodon. Genus of mammals in the Family Octodontidae; degus; small, tufted-tail, ratlike spp. of the high southern Andes.

Octodontidae. Family of ratlike South American rodents; Octodon, Spalocopus, and Aconaemys typical.

Octogomphus specularis. Dragonfly of far western U.S.

Octomitus. Hexamitus.

Octopoda. One of the two suborders in the molluscan Order Dibranchia; includes the paper nautilus and octopus, or devilfish; body spherical or saclike, without fins; head large, with eight arms bearing sessile suckers; shell usually absent.

Octopus. Genus of cephalopod mollusks with a globular body and large head bearing eight similar arms which are more or less webbed; octopuses; nocturnal or crepuscular, and timid; diameter, with outspread arms, from 1 in. to 22 ft.; more than 50 spp.; shallow to deep water; the arms are sometimes used as human food.

ocular. Pertaining to an eye.

ocular plate. One of five small skeletal plates found alternating between the genital plates on the aboral surface of sea urchins; bears an eyespot.

Oculina. Genus of reef-building corals; solid, branched growth form.

oculomotor nerve. One of the pairs of cranial nerves of vertebrates; supplies the eye muscles and acts as a part of the parasympathetic system in the ciliary process; a motor nerve.

Ocypode. Genus of crabs which includes the white crab, ghost crab, or sand crab of the U.S. Atlantic Coast; to 2 in. across; in burrows above high water mark.

Ocypus. Genus of staphylinid beetles; elytra short, abdomen exposed.

Ocyroë. Ocyropsis.

Ocyropsis. Pelagic genus of tropical Atlantic ctenophores; swim by means of two flapping muscular oral lobes.

Ocyurus. See LUTIANIDAE.

Odobenidae. Family of mammals in the Suborder Pinnipedia; includes the walruses (Odobenus) of Arctic seas; very large, ponderous, seal-like spp. with large tusklike upper canines and few hairs; muzzle blunt, set with coarse bristles; mature males to 3000 lbs.; feed on bottom invertebrates.

Odobenus. See ODOBENIDAE.

Odocoileus. Genus of ruminants in the Family Cervidae; includes the Virginia or white-tailed deer of eastern U.S. (O. virginianus) and the common western mule- or black-tailed deer (O. hemionus); the former is smaller.

Odonata. Order of insects which includes the damselflies, dragonflies, darning needles, snake doctors, and mosquito hawks; medium to large spp. with simple metamorphosis and a series of aquatic naiads in the life history; head large, with biting mouth parts and very large compound eyes;

food of adult consists mostly of small insects taken in flight; wings long, narrow, transparent, and net-veined; more than 5000 spp. See ANISOPTERA and ZYGOPTERA.

Odontaster. Common genus of sea stars in the Order Phanerozonia.

odontoblasts. Connective tissue cells forming the outer surface of the tooth pulp adjacent to the dentine.

Odontoceti. Suborder of mammals in the Order Cetacea; includes the sperm whale, dolphins, porpoises, narwhal, killer whale, etc.; toothed whales; with 2 to 40 teeth.

odontoclast. Type of osteoclast associated with cementum and dentine during resorption of deciduous teeth.

Odontognathae. One of the three superorders of birds in the Subclass Neornithes; New World toothed birds of Cretaceous deposits in Kan. and Mont.; teeth in furrows on both jaws; Hesperornis and Ichthyornis.

odontoid. Pertaining to a tooth or toothlike process.

odontoid process. Peg-shaped process on the anterior end of the axis vertebra; projects forward into the cavity of the atlas.

Odontomyia. Genus of soldier flies in Family Stratiomyiidae.

odontophore. 1. Heavy muscular cartilage-like mass on the bottom of the buccal mass of most mollusks, except bivalves; acts as a support for the radula. 2. Rarely, a radula.

Odontosyllis. Common genus of bioluminescent marine Polychaeta.

Odostomia. Common marine genus of very small pyramid shells (Gastropoda).

Odynerus tempiferus var. macio. Bushnell's wasp, a sp. which builds a mud nest provisioned with paralyzed caterpillars for the larvae.

Oecanthus. Common genus of tree crickets.

Oecetis. Genus of Trichoptera.

oedema. Edema.

Oedemeridae. Widely distributed family of slender soft-bodied beetles.

Oedignathus. Genus of marine crabs in the Suborder Anomura; rocky coasts.

Oedipomidas. Genus of pinchés, a group of marmosets found on the Pacific slope of Central America; large and slender, with long furry tails.

Oeneis. See WHITE MOUNTAIN BUTTERFLY.

oenocytes. Large cells concentrated in the hemocoel near the abdominal spiracles of insects; store concentrated nitrogenous wastes.

Oerstedia. Genus of small marine nemertines; common on stones and wharf piling.

Oesophagostomum. Genus of rhabditid nematodes parasitic in the mammalian intestine; O. columbianum is the nodular worm of the sheep large intestine, and O. dentatum occurs in the hog large intestine.

Oestridae. Family of large beelike flies; botflies; larvae parasitic in nasal cavity of domestic stock, deer, camels, etc.; Oestrus important.

oestrin. Estrone; see ESTROGEN.

oestrus. Estrus.

Oestrus. See OESTRIDAE.

Ogcocephalus. See BATFISH.

Oicomonas. Oikomonas.

Oidemia nigra. Black scoter; common blackish duck of Northern Hemisphere coastal areas.

Oikomonas. Genus of small uniflagellate Protozoa of stagnant water and damp soil; cell 5 to 10 microns long.

oikoplast. Glandular epidermal cell secreting the gelatinous tunic in certain tunicates.

Oikopleura. Genus of pelagic tunicates in the Class Larvacea; with a tadpolelike tail.

oil beetle. Any beetle in the genus Meloë and its close relatives; when disturbed they emit an oily liquid from glands in some of the leg joints; see MELOIDAE.

oilbird. See GUACHARO.

oil immersion objective. High power compound microscope objective (usually about 100X), used by placing a drop of special oil between the outer surface of the lens and the top of the coverslip; results in finer resolution and brighter illumination.

oil of chenopodium. Extract of one of the pigweeds or goosefoot (Chenopodium) used in the treatment of Ascaris, hookworm, and Enterobius infections.

Oithona. Common marine genus of copepods in the Suborder Cyclopoida.

ojam. See GALAGO.

Okapia johnstoni. Uncommon African mammal related to the giraffe; okapi; similar to a mule in general appearance but with longer neck, more tapered head, longer legs, and striped hind quarters and striped forelegs; tropical forests.

old-squaw. See CLANGULA.

oldwife. 1. Any of a wide variety of marine fishes. 2. Old-squaw.

Old World lizards. See AGAMIDAE.

Old World monkeys. See CERCOPITHECOIDEA.

olecranon process. Process on the mammalian ulna which forms the elbow and is used for the attachment of muscles which straighten the forearm.

Olenus. Common genus of trilobites.

Olethreutidae. Large family of small cryptically colored moths; many pest spp.; larvae feed on leaves, flowers, and fruits of many plants; larvae of some spp. form nests of rolled leaves and are called leaf rollers; Carpocapsa pomonella (codling moth) is the most destructive.

olfactory. Of or pertaining to the sense of smell.

olfactory bulb. 1. Large olfactory organ of some of the lower vertebrates, e.g. shark. 2. See OLFACTORY LOBE.

olfactory capsule. Nasal capsule.

olfactory hairs. Fine setae sensitive to substances in the air or dissolved in water; in many arthropods.

olfactory lobe. Anterior extension of the vertebrate forebrain supplying nerves to the olfactory areas of the head.

olfactory nerves. Pair of cranial nerves of vertebrates; supply the organs of smell; sensory nerves.

olfactory pit. 1. Anterior middorsal pit of cephalochordates which functions as an organ of smell. 2. Embryonic invagination in chordates which later becomes the functional olfactory organ. 3. Any olfactory sensory area having the form of a small depression.

olfactory placodes. Embryonic pits in the vertebrate embryo; develop into the two olfactory organs.

olfactory sac. 1. One of two double nostrils of most fishes. 2. Embryonic invagination in chordates which later becomes the functional olfactory organ.

olfactory tract. Olfactory nerve.

Oligobranchiopoda. One of the two divisions in the crustacean Subclass Branchiopoda; with a folded carapace which encloses the body but not the head; five or six pairs of trunk appendages; includes the single Order Cladocera (water fleas).

Oligocene epoch. One of the five geological subdivisions of the Tertiary period, between the Miocene and Eocene epochs; lasted about 15 million years; especially characterized by the rise of higher mammals.

Oligochaeta. One of the classes in the Phylum Annelida; earthworms; definite head and parapodia absent; few setae; gills usually lacking; often with a clitellum which secretes cocoons to enclose the eggs; no special larval stage; moist soil, fresh water, and marine; about 2700 spp.

oligodendrocytes. Neuroglia cells having scanty cytoplasm and only a few, thin, sparsely-branched processes; mainly in the central nervous system.

oligopod stage. Embryological stage of some insects; well developed functional thoracic limbs are present, but abdominal limb buds are absent.

oligosaprobic. Pertaining to an aquatic habitat which is high in dissolved oxygen, and with a minimum of organic decomposition.

Oligotoma. Genus of insects in the Order Embioptera; introduced into U.S.

Oligotricha. Suborder of the ciliate Order Spirotricha; ciliation reduced or absent; cell round in cross section; e.g. Halteria and Strobilidium.

oligotrophic. Pertaining to lakes without distinct oxygen stratification, poor in dissolved nutrients, and poor in plankton.

oliguria. Voiding only a small amount of urine.

Olindias. Genus of marine hydrozoans; the medusae are shallow-water, bottom forms with long tentacles.

Oliva. See OLIVE SHELLS.

Olivella. See OLIVE SHELLS.

olive shells. Group of widely distributed marine gastropods; shell smooth and polished; body whorl long and large, hiding most of the spire; aperture long and narrow; infinite variety of color patterns; Oliva and Olivella common on tropical and subtropical shores.

olive-sided flycatcher. See NUTTALORNIS.

olive warbler. See PEUCEDRAMUS.

olm. See PROTEUS.

Olor. Genus name sometimes used instead of Cygnus (q.v.).

Olympic salamander. See RHYACOTRITON.

olynthus. Early stage in the development of calcareous sponges; body mass vase-shaped and with an ascon organization.

omasum. Third chamber of stomach of ruminants; receives food from the reticulum (from second swallowing after being first regurgitated from reticulum); passes it on to the fourth chamber, or abomasum; sometimes called the "true" stomach; same as manyplies and psalterium.

omental bursa. Coelomic pouch situated behind the stomach, lesser omentum, and part of liver, but in front of the pancreas and duodenum; variously developed in vertebrates.

omentum. Free fold of the peritoneum; connects and supports visceral organs.

Ommastrephes. Genus of sea arrows or flying squids; swift spp. which sometimes shoot clear of the water.

ommatidium. One of the numerous tapering units of the compound eye of typical insects and some crustaceans; each consists of (1) an outermost corneal facet, or lens; (2) two corneagen cells which secrete the lens; (3) a long crystalline cone of four cells; (4) four or more distal pigment cells surrounding the cone; (5) a long, tapering retinula of eight cells which constitute a central rhabdome where they meet; (6) basal pigment cells surrounding the rhabdome region; and (7) a tapetum, which is a reflecting surface composed of pigment or densely massed tracheoles; the inner tip of each retinula penetrates the basement membrane and connects with the central nervous system by means of a nerve fiber. Pl. ommatidia.

Ommatophoca. Genus which includes Ross's seal, an Antarctic form which feeds on bottom animals.

omnivore. Any animal which uses a variety of living and dead plants and animals in its diet. Adj. omnivorous.

Omochelys cruentifer. Snipe eel; a small but greatly elongated marine eel in the Family Ophichthyidae; thought to be a parasitic borer in other fishes; to 16 in. long; north and middle Atlantic Ocean.

omohyoid. Ventral muscle in the region of the sternum, larynx, and jaw symphysis of tetrapods.

omosternum. One of the elements of the tetrapod sternum.

Omphalina. Genus of N.A. land snails.

omphalomesenteric arteries. Vertebrate embryonic arteries which take blood from the dorsal aorta to the vitelline arteries.

omphalomesenteric veins. Vertebrate embryonic veins which take blood to the heart from the vitelline veins; heart formed by fusion of these veins.

Omus. Genus of black, wingless, nocturnal tiger beetles.

onager. See EQUIDAE.

Onchidoris. Genus of small sea slugs; dorsal surface tuberculous; anus surrounded by branchial plumes.

Onchocerca. Genus of spirurid nematodes parasitic in the connective tissue and blood vessels of man and other mammals; O. volvulus occurs in fibrous subcutaneous tumors on the exposed parts of the human body; transmitted by blackflies; Central America and Central Africa.

onchocerciasis. Infection with Onchocerca, especially O. volvulus.

onchosphere. Oncosphere.

Oncidiella. Genus of small opisthobranchs resembling shell-less limpets.

Oncomelania. Genus of fresh-water snails which serve as intermediate hosts for Schistosoma japonicum.

Oncopeltus. Hemipteran which transmits a parasitic trypanosome (Phytomonas) from one milk-weed plant to another.

Oncorhynchus. Genus of bony fishes which includes the five spp. of N.A. salmon of the Pacific Coast; adults spawn once in fresh water and then die; young migrate to sea, mature for two to eight years, then return to headwater rivers and streams to spawn; extremely important food fishes; O. tschawytscha, the Chinook, king, quinnat, or tyee salmon averages 30 lbs., but other spp. are smaller; O. nerka, the sockeye or red salmon, is the most valuable sp.; other spp. are the chum or dog salmon (O. keta), pink or humpback salmon (O. gorbuscha), and the small coho or silver salmon (O. kisutch); see KOKANEE SALMON.

Oncoscolex. Genus of short-bodied Polychaeta.

oncosphere. Minute six-hooked larva which hatches from the eggs of many tapeworms; upon becoming lodged in the tissues of the intermediate host, it commonly develops into a cysticercus, cysticercoid, or procercoid.

Oncousoecia. Genus of marine Bryozoa.

Ondatra zebethica. Common N.A. muskrat in the Family Cricetidae; a robust rat with broad webbed hind feet, short legs, and compressed tail; perineal glands secrete a musky substance; live in or near marshes and shallow waters where they live in bank burrows or build domelike houses of rushes and other aquatic plants; feed chiefly on aquatic vegetation, but may eat mussels and a little other animal food.

onion fly. Hylemya antiqua, a fly in the Family Anthomyiidae; larvae are pests on onions.

onion maggot. See ONION FLY and HYLEMYA.

Oniscoidea. Suborder of Isopoda; includes the common terrestrial spp. of isopods.

Oniscomorpha. Old World order of millipedes; 14 to 16 body segments; last one or two pairs of legs are male gonopods; Glomeris typical.

Oniscus asellus. Common sp. of terrestrial sowbugs; 16 mm. long; in damp places, under bark of fallen timber, stones, etc.

Onoba. Genus of minute marine snails usually found around the low-water mark.

Onotragus. Genus of reddish-brown African antelopes with long, thin, undulating horns; lechwes.

ontogeny. Developmental history of an organism from zygote to maturity. Adj. ontogenetic.

Onuphis. Genus of tube-building Polychaeta.

onychium. Pad between the tarsal claws in certain insects.

Onychodromopsis. Genus of fresh-water hypotrich Ciliata.

Onychodromus. Genus of fresh-water hypotrich Ciliata.

Onychogale. Genus of small Australian wallabies.

Onychomys. Genus of grasshopper mice in the Family Cricetidae; thickset, short-tailed; under parts whitish; feed on insects and vegetable material; common in open semiarid areas of western states.

Onychophora. Small phylum of caterpillar-like terrestrial animals which combine certain features of both annelids and arthropods; body soft and annulated but not segmented externally; to 8 in. long; indistinct head bears one pair antennae, one pair hooked jaws, one pair simple eyes, and one pair oral papillae; trunk with 14 to 43 pairs of stumpy annulated legs; paired nephridia (coelomoducts) and coxal glands; extensive hemocoel and reduced coelom; brain and two lateroventral nerve cords; tropics in moist places; about 80 spp.; Peripatus familiar.

Onychopoda. Tribe of fresh-water cladocerans which include Polyphemus pediculus, a small, short-bodied sp. without a carapace; four pairs of stout prehensile appendages; compound eye very large.

Onychoteuthis. Large and common genus of squids.

oocyst. Spore case formed by a zygote and in which numerous spores and sporozoites are formed; especially in sporozoan Protozoa.

oocyte. Cell derived from a mature oogonium; undergoes meiosis and eventually produces an ovum. See PRIMARY OOCYTE and SECONDARY OOCYTE.

ooecium. Ovicell; special compartment in the polypide of a female bryozoan; contains developing embryo(s). Pl. ooecia.

oogamy. Fertilization of a large nonmotile ovum by a small male gamete; typical of all metazoans and some protozoans.

oogenesis. Series of cell divisions and chromosome and cytoplasmic changes involved in the production of functional ova, beginning with undifferentiated germinal epithelium, passing through oogonial cells, and especially the changes of primary oocytes through secondary oocytes to a mature egg and polar bodies.

oogonium. A cell which is the precursor of one or more mature oocytes; usually forms the bulk of ovarian tissue and multiply by ordinary mitosis. Pl. oogonia.

ookinete. Motile zygote, such as occurs in the mosquito stomach in the life history of malaria parasites.

oology. Collection and study of the size, shape, coloration, etc. of bird eggs.

oosphere. Rarely, an unfertilized egg or female gamete.

oostegites. Flat plates extending horicontally and medially from some of the thoracic appendages in the crustacean Subdivision Peracarida; collectively they form a broad chamber for retaining the eggs and young.

ootheca. Egg case such as that produced by certain mollusks and insects.

ootid. Large haploid cell resulting from the second meiotic division of a secondary oocyte; it quickly develops into a functional ovum. See SECOND POLAR BODY.

ootype. Small thickened portion of the oviduct in the vicinity of the ovaries in many trematodes and tapeworms;

functions in forming the shell which is laid down around the zygote.

oozooid. 1. Tunicate zooid which arises from an ovum. 2. Any individual which has developed from an egg, as opposed to an individual which has developed by budding or fragmentation.

opah. See LAMPRIS.

opaleye. See GIRELLA.

Opalia. Genus of marine snails; long-spired wentletraps.

Opalina. Genus of ciliate protozoans parasitic in the intestine of frogs and toads; body greatly flattened and multinucleate; many protozoologists contend that these protozoans are actually flagellates.

opalinid. 1. Any member of the genus Opalina. 2. Any member of the Subclass Protociliata.

Oparin, Alexander Ivanovich. Russian biochemist (1894-); formulated a biochemical theory of the origin of life on earth.

opal worm. See ARABELLA.

open circulatory system. Any circulatory system in which the circulating fluid is not everywhere confined to definite vessels but may circulate freely among tissues, sinuses, or in body cavities during at least some part of its circulatory path; Onychophora, arthropods, and mollusks have an open circulatory system.

open community. 1. Any community in which one or more ecological niches are presumably unoccupied. 2. Any community in which the plant canopy does not cover or obscure the ground.

opercle. 1. Operculum or gill cover. 2. Large flat bone of the gill cover in fishes.

Opercularella. Genus of marine coelenterates in the Suborder Leptomedusae.

Opercularia. Genus of colonial peritrich ciliate Protozoa; cell elongated and vase-shaped; in fresh-water ponds and attached to other invertebrates and submerged objects.

operculate. Having an operculum or lid.

operculum. 1. Collectively, the movable plates of a barnacle shell. 2. Lid of a trematode or cestode egg. 3. Gill cover; a flaplike outer protective covering for the gills of fish, certain immature amphibians, and a few aquatic invertebrates. 4. Prominent membrane on the ventral surface of tadpoles. 5. Horny plate in some gastropods; used for closing the aperture when the animal has completely withdrawn into its shell. 6. Lid closing the hydrotheca or gonotheca in certain hydrozoans. 7. Lid over a cyst pore in certain Protozoa. 8. Plate or flap covering a spiracle or stigma in spiders. 9. Generally, a covering lid or flap. Pl. opercula.

Ophelia. Widely distributed genus of burrowing Polychaeta.

Opheodrys. Genus of small harmless snakes in the Family Colubridae; green snakes; body slender and uniformly green; generally east of the Rockies.

Ophiacantha. Circumboreal genus of brittle stars in the Order Ophiurae.

Ophiacodon. Genus of early Permian reptiles.

Ophiactis. Genus of brittle stars in the Order Ophiurae.

Ophichthyidae. Snake eels; a family of greatly elongated snakelike marine eels; common and widely distributed; Omochelys typical.

Ophidia. Serpentes.

Ophidiidae. Family of elongated, compressed, eel-like marine fishes; cusk eels; ventral fins barbel-like, located in the throat region; mostly in warm seas; Lepophidium cercinum common in U.S. East Coast areas.

Ophidonais serpentina. Fresh-water oligochaete; to 30 mm. long; muddy substrates.

Ophiocistioidea. Class of extinct echinoderms; body discoidal, completely encased in a test except at the peristome; arms absent; with a few giant tube feet in each ambulacral area; Ordovician to Devonian; Sollasina typical.

Ophiocoma. Common genus of brittle stars; chiefly tropical but found as far north as England.

Ophiocomina. Genus of brittle stars.

Ophiocten. Genus of brittle stars.

Ophioderma. Common genus of brittle stars.

Ophiogomphus. Common genus of dragonflies.

Ophiolepis. Genus of brittle stars.

Ophiomyxa. Genus of tropical brittle stars.

Ophion. Genus of ichneumon flies; larvae parasitic on Lepidoptera and Coleoptera.

Ophionereis. Genus of brittle stars.

Ophiopholis. Genus of cosmopolitan serpent stars.

Ophiophragmus. Genus of brittle stars.

Ophioplocus. Genus of brittle stars which have no free-swimming larval stage.

ophiopluteus. Pluteus larva of the Ophiuroidea.

Ophiopteris. Genus of brittle stars.

Ophioscolex. Genus of brittle stars.

Ophiothrix. Large genus of serpent stars found on both U.S. coasts.

Ophisaurus. Genus of limbless snake-like lizards in the Family Anguidae; O. ventralis is the common American glass snake of southern and central states and as far north as Wis.; to 3 ft. long; tail breaks off easily and partially regenerates; O. apodus, to 4 ft. long, occurs over much of southeastern Europe, southwestern Asia, and northern Africa.

Ophiura. Genus of cosmopolitan serpent stars.

Ophiurae. Order of echinoderms in the Class Ophiuroidea; arms not branched; disc and arms covered by plates; brittle stars, etc.

Ophiuroidea. Class in the Phylum Echinodermata; includes the brittle stars, snake stars, serpent stars, and basket stars; slender arms greatly elongated and sharply set off from disc; ambulacral grooves, anus, special sense organs, and ampullae absent; tube feet tactile, respiratory, and excretory; madreporite oral; larva a pluteus; arms highly flexible, produce rapid horizontal locomotion; usually on the sea bottom or among algae and coral growths.

Ophlitaspongia. Genus of red encrusting marine sponges.

Ophrydium. Genus of long cylindrical peritrich ciliates; in ponds and stagnant waters.

Ophryocystis. Genus of Sporozoa occurring in Malpighian tubules of beetles.

Ophryoglena. Large genus of ellipsoidal to cylindrical holotrich Ciliata; marine, fresh-water, and parasitic.

Ophryoscolex. Genus of complex spirotrich Protozoa; commensal in the stomach of ruminants.

Ophryotrocha. Genus of minute Polychaeta.

Ophryoxus gracilis. Cladoceran common in aquatic vegetation.

ophthalmic. Pertaining to the eye.

ophthalmic artery. 1. Anterior dorsal artery in the crayfish and certain other Malacostraca; supplies blood to the eyes, antennae, and brain. 2. One of two arteries of the head in many vertebrates; supplies eyes, orbits, and face.

ophthalmicus profundus. Profundus nerve.

Ophthalmosaurus. Genus of Cretaceous fishlike reptiles in the Subclass Ichthyopterygia.

Opiliones. Phalangida.

Opisthobranchiata. Sea slugs, sea butterflies, sea hares, etc.; mantle and shell usually vestigial or absent; gills posterior; nervous system secondarily untwisted and concentrated around the esophagus; hermaphroditic; two pairs of tentacles; about 3000 living spp.

Opisthocoela. One of the five orders of amphibians in the Subclass Salientia; vertebrae opisthocoelous; either tadpole or adult with some free ribs; includes the families Discoglossidae, Xenopidae, and Pipidae.

opisthocoelous. Type of vertebra having a centrum with a convex anterior face and a concave posterior face, as in certain toads.

Opisthocomidae. Family of birds in the Order Galliformes; includes only Opisthocomus hoazin, the hoatzin; a peculiar, arboreal, crested sp., famous for having claws on the thumb and index finger of the nestling; South American forests.

Opisthocomus. See OPISTHOCOMIDAE.

Opisthoglypha. Group of genera of snakes in the Family Colubridae; one or more pairs of grooved teeth in posterior part of upper jaw; usually venomous but seldom harmful to man.

Opisthomi. Order of marine bony fishes; eel-shaped spp. with the anterior nostril at the tip of a fleshy tentacle; fresh waters of Africa and southeastern Asia; Mastacembelus typical.

Opisthonecta. Genus of conical fresh-water peritrich ciliates; ends broadly rounded.

opisthonephros. Type of kidney found in certain anamniotes; composed of both mesonephros and metanephros.

Opisthopus. Genus of marine crabs in the Suborder Brachura; commensals in mollusks.

Opisthorchis felineus. Sp. of liver fluke common in cats and dogs from central Europe to Japan and the Philippines; occasional in man; similar to Clonorchis; intermediate hosts are snails and fresh-water fishes.

opisthosoma. Posterior part of the body, especially in arachnids.

opisthotic bone. Bone forming a portion of the otic capsule in lower vertebrates.

Opistomum. Genus of rhabdocoel Turbellaria.

Oplonaeschna. Genus of dragonflies.

Oporornis. Genus of warblers in the Family Parulidae; four U. S. spp.

opossum. See DIDELPHIDAE.

opossum shrimp. Any member of the crustacean Order Mysidacea.

Opsanus. See HAPLODOCI.

opsiblastic egg. Heavy-shelled gastrotrich egg produced when environmental conditions are unfavorable; able to withstand desiccation, freezing, etc. Cf. TACHYBLASTIC EGG.

opsin. Protein portion of rhodopsin; formed in the biochemical sequence occurring when light strikes the rods of the retina.

opsonin. Antibody occurring in blood serum; renders bacteria more susceptible to phagocytosis.

Opsopoeodus emiliae. Pugnose minnow; a small cyprinid of lowland streams from the Great Lakes to the Gulf Coast; mouth extremely oblique.

optic. Of or pertaining to an eye.

optic capsule. Cartilaginous capsule enclosing each eye in elasmobranch fishes and embryos of higher vertebrates.

optic chiasma. Crossing of the fibers of the optic nerves on the ventral side of the brain.

optic foramen. Opening in the skull near the orbit; carries the optic nerve and ophthalmic artery. Pl. optic foramina.

optic ganglion. Ganglion near the eye; receives nerve fibers from the eye in many invertebrates, e.g. the crayfish.

optic lobes. 1. Corpora bigemina; two elevations on the roof of the midbrain of the lower vertebrates; correspond to a part of the corpora quadrigemina in man. 2. The two large ganglia closely associated with the compound eyes of many insects.

optic nerve. One of the second pair of cranial nerves of vertebrates; a sensory nerve which supplies the eye.

optic tectum. Dorsal part of the midbrain; receives optic nerve fibers.

optic thalamus. Mass of gray matter at either side of the base of the vertebrate brain; contains large optic centers in man.

optic tract. Optic nerve.

optic vesicle. Protuberance on either side of the embryonic forebrain of vertebrates; by invagination it forms the retina of the embryonic eye.

optimum conditions. Theoretical set of the most favorable ecological conditions which will permit an animal to most nearly realize its full biotic potential.

optogram. Picture printed on the retina of a freshly excised animal eye when it is exposed to a contrasting view; produced by the pattern of the visual purple.

oquassa. Small trout occurring in the Rangeley Lakes area of Maine; Salvelinus oquassa.

oral. Of or pertaining to the mouth or the region near the mouth.

oral arms. Oral lobes; long dangling folds, lobes, or armlike projections extending out from the central proctostome area of many medusoid coelenterates; they bear numerous nematocysts and are used chiefly for food-getting; four such arms are usually present.

oral disc. 1. Distal, more or less flat surface of anthozoan coelenterates; it bears the proctostome and numerous hollow tentacles. 2. Lophophore of Bryozoa. 3. Central portion of the body of a sea star.

oral groove. Groove on the surface of certain Protozoa (e. g. Paramecium); leads to the cytostome and used in concentrating food particles.

oral hood. In cephalochordates, a funnel-like or hoodlike projection above and anterior to the mouth.

oral lobes. Oral arms.

oral spear. Protrusible sclerotized spearlike structure used by certain nematodes to puncture plant cells or their animal prey.

oral valve. Set of two flaps just inside the mouth opening in a typical fish; prevents outflow of water during respiratory movements.

orang utan. See PONGO.

ora serrata. Zigzag anterior margin of the retina.

Orbicella. Genus of reef-building corals in Fla. and the West Indies.

orbicularis oculi. Sphincter muscle surrounding eyelids; closes eye and wrinkles forehead.

orbicularis oris. Muscle which encircles the mouth; closes mouth and wrinkles lips.

Orbina. Genus of Polychaeta.

orbit. 1. Cavity or depression in the skull of vertebrates containing the eyeball. 2. Part of the head surrounding an eye of vertebrates. 3. In decapod crustaceans, the depression containing the eyestalk. 4. Area around an insect compound eye.

orbitosphenoid. Small bone forming a part of the posterior lower median wall of the orbit in many vertebrates.

orb shell. See SKENEA.

Orbulina. Genus of foraminiferans having a thin globular shell.

orb weavers. See ARANEIDAE.

orb web. Circular spider web built in a single plane.

Orcaella. Genus which includes the Irrawaddy dolphin; a small black dolphin which gets as far as 1000 miles up the Irrawaddy; head bulbous, teeth minute and numerous; feeds on bottom invertebrates.

orchard oriole. Icterus spurius; a common oriole of the eastern U.S.; head, neck, back, wings, and tail black; rump, upper tail, wing bands, and under parts chestnut.

Orchestia. Common genus of amphipods found especially under masses of decaying marine algae in the intertidal zone; beach fleas.

Orchestoidea. Genus of marine Amphipoda; especially common in the intertidal zone of Calif. beaches.

Orcinus. Genus of killer whales; aggressive spp. occurring in all seas; attack fishes, porpoises, seals, and other small whales; black coloration, to 25 ft. long.

Orconectes. Genus of fresh-water crayfishes common east of the Continental Divide; in general morphology, they closely resemble the American lobster; body usually 2 to 4 in. long; ponds, streams, and lakes; O. limosus in larger rivers of the Atlantic coastal plain; O. virilis in the Mississippi drainage; O. pellucidus is a blind cave and subterranean sp.

order. Taxonomic category including one or more families having certain features in common; e.g. the Order Carnivora includes the Canidae, Felidae, Ursidae, Mustelidae, etc.; it is sometimes convenient to distinguish two or more suborders within a single order, and similarly it is sometimes convenient to group certain orders together as superorders within the same class.

Ordovician period. One of the six geological subdivisions of the Paleozoic era; between the Silurian and Cambrian periods; lasted from about 480 million to 390 million years ago; an age of submergence; fauna characterized by an abundance of higher invertebrates and the appearance of ostracoderms.

Oreamnos americanus. Mountain goat or Rocky Mountain goat; an antelope with a goatlike appearance; large, white, and with small slender horns which curve backward; coat shaggy; chin with a short beard; to 300 lbs.; uncommon; Rocky Mountains and Coast ranges from Alaska to Mont. and Idaho, introduced locally farther south.

Oreaster. Genus of large, massive sea stars in the Order Phanerozonia; American spp. to 20 in. in diameter.

Oregonia. Genus of spider crabs.

Oreianthus. Genus of Ephemeroptera; nymphs sprawling.

Oreopithecus. Genus of fossil primates found in Italian lignite deposits; 4 ft. tall, with some human skeletal characters; between ape and man but not in a direct line; 10 million years old.

Oreortyx picta. Mountain quail of Wash., Ore., Calif., Ida., and Nev.; upper parts grayish olive brown; with a crest of two long black cohering plumes.

Oreoscoptes montanus. Sage thrasher; a gray-backed bird with straight slender bill; western sagebrush areas.

Oreotragus oreotragus. Small African ruminant in the Family Bovidae; the klipspringer, an antelope of rocky and mountainous regions; less than 2 ft. high at the shoulder.

Oreotrochilus. Genus of South American hummingbirds of high altitudes.

organ. Discrete structure composed of tissues and having specific function(s); morphologically distinct from adjacent or adjoining structures; e.g. a bone, pancreas, tongue, blood vessel, lung, etc.

organelle. Specialized part of a cell having specific function(s); somewhat analogous to an organ or organ system of higher animals; usually the term is used with reference to protozoans; e.g. flagellum, contractile vacuole, myoneme, etc.

organic. 1. Pertaining to the organs of the body. 2. Pertaining to compounds formed by living organisms. 3 Pertaining to the chemistry of compounds containing carbon.

organic compound. With reference to living organisms, one of many different kinds of substances produced by protoplasm and containing carbon, hydrogen, and oxygen, and often nitrogen and certain other elements.

organism. Any living individual.

organizer. 1. Part of an embryo which influences and directs some other adjacent part in its histological and morphological differentiation. 2. Dorsal lip of blastopore of amphibian gastrula and the corresponding part of other vertebrate embryos; when transplanted to another embryo, the surrounding host tissues may be stimulated to develop into a complete or incomplete embryonic axis.

organ of Bojanus. Nephridium, metanephridium; one of two excretory tubules in many bivalve mollusks; located between the suprabranchial chambers and the pericardial cavity.

organ of Corti. Complex arrangement of sensory receptor cells and associated tissues resting on the basilar membrane of the cochlea in higher vertebrates.

organ of Keber. Paired glandular structure arising from the auricles or pericardium in certain bivalve mollusks.

organ of Krause. Specialized microscopic spherical or ovoid nerve ending presumed to be a vertebrate receptor for cold stimuli.

organ of Ruffini. Specialized microscopic nerve ending presumed to be a vertebrate receptor for heat stimuli.

organogenesis. Development and differentiation of organs.

organoid. 1. Distinct body forming part of the cytoplasm of a metazoan cell, e.g. vacuoles, Golgi bodies, cilia, chondriosomes, etc. 2. Organ-like.

organology. Study of the organs of a plant or animal body.

organ-pipe coral. See TUBIPORA.

organ system. Collectively, a group of organs specialized and integrated for particular function(s), e.g. the organs of the digestive or respiratory system.

orgasm. Culmination or crisis of coitus or sexual excitement.

Oribatella. Genus of true terrestrial mites; one sp. occasionally found creeping about on pond substrates.

Oribatidae. Large family of free-living mites which feed chiefly on vegetation and decaying animal matter; horny mites, beetle mites; Galumna and Scutovortex common.

Oribatoidea. Superfamily of free-living mites.

oribi. See OUREBIA.

Oriental fruit fly. See DACUS.

Oriental realm. One of the six major zoogeographic regions of the world; consists of southeastern Asia south of the Himalayas, as well as Sumatra, Java, Borneo, the Philippines, and many associated islands; some characteristic animals are the Indian elephant, Indian rhinoceros, gibbon, orang utan, and gavial.

Oriental sore. Disease of man caused by Leishmania tropica; marked by dermal and epidermal ulcers; occurs mostly in the eastern Mediterranean, southwestern Asiatic, and central African areas; transmitted by Phlebotomus, a minute sandfly; also known as tropical sore, Aleppo boil, and Delhi boil.

Oriental tree-squirrel. See CALLOSCIURUS.

origin. Fixed end of a voluntary muscle.

oriole. See ICTERUS.

Oriolidae. Family of passerine birds; Old World orioles; pugnacious, brightly colored gregarious spp.

ormer. Local name of a small abalone in certain English Channel areas.

ornate scallop. Pecten ornatus, a southern scallop blotched with red or purple.

ornithine. Amino acid found in excrement of birds; $C_5H_{12}O_2N_2$.

ornithine cycle. Chemical cycle occurring in liver and resulting in production of urea among ureotelic vertebrates; the amino acid ornithine takes up ammonia from the deamination of glutamic acid, which together with metabolically produced carbon dioxide gives rise to another amino acid, citrulline; the latter then combines with aspartic acid to form arginosuccinic acid, which breaks down to release fumaric acid, urea, and ornithine; high-energy phosphate (ATP) is required.

Ornithischia. Order of Late Triassic, Jurassic, and Cretaceous dinosaurs; teeth in rear of jaws only; claws depressed, sometimes hooflike; divided into four suborders: Ornithopoda, Stegosauria, Ankylosauria, and Ceratopsia.

Ornithodelphia. Monotremata.

Ornithodorus. Common genus of soft ticks; O. megnini the ear tick of domestic stock animals and man; O. turicata on hogs, cattle, and man; O. moubata transmits African tick fever; O. hermsi of wild rodents is transmitter of Spirochaeta recurrentis, which causes one kind of human relapsing fever.

Ornithoica. Genus of blood-sucking dipterans in the Family Hippoboscidae; parasites of wild birds.

Ornitholestes. Genus of small, bird-catching dinosaurs.

ornithology. In its broad sense, the study of all phases of the biology of birds.

Ornithopoda. Suborder of Jurassic and Cretaceous dinosaurs in the Order Ornithischia; camptosaurs and duckbills; dominantly bipedal herbovires but forelimbs also sometimes used in locomotion; examples are Camptosaurus, Trachodon, and Corythosaurus.

Ornithorhynchus anatinus. Common duck-billed platypus of eastern Australia and Tasmania in the Order Monotremata; a primitive aquatic mammal to 20 in. long, with a ducklike bill, webbed feet, beaver-like tail, small eyes, and burrowing habits; male with sharp spur on heel, connected to a poison gland; feed on small fresh-water invertebrates and some vegetation; females lay one to three small eggs in burrow; newly-hatched young feed on milk secreted by scattered ventral mammary glands.

Ornithosuchus. Genus of Triassic reptiles in the Order Thecodontia; probably one of the precursors of birds.

Orohippus. Genus of small New World Eocene horses; four front and three hind toes; about 1 ft. high.

Oropallene. Genus of Pycnogonida.

Oroya fever. Acute febrile disease of the slopes of the Andes in northern South America; caused by Bartonella bacilliformis which is transmitted by the bite of Phlebotomus, a sandfly.

Ortalis. Genus (13 spp.) of Central and South American brownish chicken-like birds; chachalacas; noisy spp. of jungle edges, clearings, and grasslands; O. vetula as far north as Tex.

Orthasterias. Genus of fragile sea stars; Pacific Coast.

Orthemis ferruginea. Dragonfly of the Gulf Coast states.

Orthoceras. Genus of extinct cephalopods having a conelike shell; to 20 ft. long.

Orthocyclops modestus. Fresh-water copepod common among vegetation.

Orthodon microlepidotus. Greaser blackfish, a large minnow of Calif. drainages; scales very small, body to 16 in. long.

orthogenesis. 1. Straightline evolution. 2. The tendency for certain organisms to evolve consistently along restricted evolutionary paths over a long period of time toward a "predetermined" goal; e.g. the fossil history of horses. Adj. orthogenetic.

Orthognatha. Mygalomorphae; group of spider families having horizontal fangs.

Orthonectida. One of the two classes in the Phylum Mesozoa; mature sexual form to 1 mm. long; with a mass of inner cells; rare endoparasites of marine invertebrates; Rhopalura.

orthopedic. Pertaining to diseases of bones and joints and the correction of deformities.

Orthopodomyia signifera. Common mosquito; from Mass. to Fla. and west to Calif.

Orthopristis. See GRUNT.

Orthoptera. Order of insects which includes grasshoppers, crickets, katydids, mole crickets, and locusts; medium to large spp. with chewing mouth parts and gradual metamorphosis; forewings narrow and parchment-like; hind wings broad, membranous, and folding fanlike beneath forewings; mostly herbivorous.

Orthorrhapha. One of the two suborders in the Order Diptera; with a crescent-shaped sclerite over the base of antennae; larvae usually with a distinct head; pupa naked and emerging through a T-shaped split on dorsal surface of last larval exoskeleton.

orthoselection. Type of natural selection which effects orthogenesis by promoting and continuing adaptive trends in evolution.

Orthotrichia. Genus of Trichoptera.

Orycteropus. See TUBULIDENTATA.

Oryctolagus cuniculus. European gray rabbit, the source of most domestic breeds.

Orygmatobothrium. Genus of tetraphyllidean tapeworms found especially in the spiral valve of sharks.

Oryssidae. Parasitic wood wasps; small and widely distributed family of small spp.; larvae parasitic on wood-boring insects.

Oryx. Genus of African ruminants in the Family Bovidae; includes the oryx, gemsbok, and beisa; large, handsome antelopes with long horns; chiefly on arid and grassy plains.

Oryzomys. Large genus of rice rats in the Family Cricetidae; herbivorous and carnivorous; grassy areas and open brush lands of tropical America and southeastern U. S.

Oryzorictes. Genus of rice-tenrecs; mole-like tenrecs which are often a nuisance in rice fields; Madagascar.

os. 1. A bone. 2. Mouth. 3. Opening.

Osbornictis. Genus of rare water-civets; body 18 in. long; feet webbed, coat reddish brown; semiaquatic; stream banks of the Congo.

Oscarella. Genus of marine sponges in the Class Demospongiae; without spicules.

oscillations. In the ecological sense, more or less regular temporal variations from the mean density of a population.

oscillograph. Instrument for recording various types of rapidly changing events, and usually for recording fleeting electrical phenomena; cathode ray oscilloscope is best known and most widely used; with it, amplified electrical impulses such as action currents of nerve and muscle, or pulses produced by electrical conversion of other forms of energy to electrical, produce movements of an electron beam in a cathode ray tube, the beam producing traces on a flourescent screen lining the end of the tube; other oscillographs involve movements of metal or optical levers, produced electromagnetically or in some other fashion.

Oscinis. Oscinosoma.

Oscinosoma. Genus which includes the frit fly, a minute Holarctic dipteran whose larvae feed on cereals and grasses.

os cordis. An irregular heterotopic bone in the heart of deer and bovids.

osculum. Main opening or pore on the surface of a sponge out of which water passes; depending on size and complexity, a sponge may have one or many. Pl. oscula.

os falciforme. Sickle-shaped heterotopic bone in the digging appendage of a mole; stiffens the supernumerary digit.

Osmeridae. Family of small salmon-like fishes which includes the smelts and capelins; slender spp. with pointed noses and deeply forked tails; chiefly marine; Mallotus and Osmerus typical.

Osmerus mordax. Common smelt; a small slender olive to silvery fish in the Family Osmeridae; an inshore Atlantic Coast marine sp. which runs into fresh waters to spawn in autumn; introduced into the Great Lakes and other large inland lakes where it is landlocked and runs up streams to spawn in the spring; to 14 in. long.

osmic acid. Osmium tetroxide; a rapid histological fixing agent, especially for cytoplasmic details; fumes highly toxic; usually used in 0.25 to 1% solution.

osmium tetroxide. See OSMIC ACID.

osmoregulation. Maintenance of proper internal salt-water concentrations in a single cell or a whole multicellular organism.

osmosis. When two solutions having different concentrations of dissolved materials are separated by a selectively permeable membrane, the solvent passes through the membrane from the solution of lower concentration to the solution of higher concentration.

osmotic pressure. Pressure that would need to be exerted upon a more concentrated solution on one side of a selectively permeable membrane in order to prevent the passage of water into it from a less concentrated solution on the other side of the membrane.

Osmylidae. Family of flylike insects in the Order Neuroptera; medium to large slender spp. with beautifully marked wings; widely distributed but not in N. A.

os palpebrae. Bony plate imbedded in the eyelid of crocodilians.

os penis. Baculum.

osphradium. Small sensory area in the incurrent siphon of pelecypod and gastropod mollusks; presumably a chemoreceptor sensitive to incoming water. Pl. osphradia.

Osphranter. Genus of wallaroos or rock-kangaroos; stocky kangaroos of Australian rocky areas.

Osphranticum labronectum. Uncommon and erratically distributed fresh-water copepod.

osprey. See PANDION.

ossein. The complex of organic materials found in bone.

osse us. Bony, or resembling bone.

ossicles. 1. Plates, spicules, and rods making up the structure of the echinoderm endoskeleton. 2. Small bony plates in the sclerotic coat of some birds and reptiles. 3. Especially small bones, such as those of the middle ear of vertebrates. 4. Teeth and toothlike processes in the malacostracan gastric mill.

ossification. Process of changing into

bone.

ossify. To become bony or hard and bonelike.

Ostariophysi. Very large order of bony fishes which includes the families of characins, electric eels, suckers, minnows, catfishes, etc.; Weberian organ present; air bladder of two or three parts, usually connected to pharynx by a duct; mostly fresh water.

Ostariophysida. Ostariophysi.

Osteichthyes. Class of vertebrates which includes the true bony fishes; skeleton more or less bony; body usually covered with dermal scales of cycloid or ctenoid pattern, a few spp. with ganoid scales; both paired and median fins present; pelvic girdle absent; tail usually homocercal; mouth usually terminal; two dorsal olfactory sacs; eyes lidless; four pairs of aortic arches; nucleated red blood cells; paired gills in a common chamber on each side of pharynx, covered by an operculum; swim bladder usually present; oviparous (occasionally ovoviviparous or viviparous); anus and urogenital openings separate; salt, brackish, and fresh waters; probably more than 32,000 living spp. described.

osteitis. Bone inflammation.

osteoblast. Special type of cell which secretes the matrix of bone.

osteoclast. Large multinucleate cell involved in the absorption and removal of bony tissue.

osteocomma. 1. Vertebra. 2. Segment of a vertebral column.

osteocranium. That part of a cranium which is bony, in contrast to the cartilaginous portion.

osteocytes. Bone cells; occupy small lacunae in bone matrix.

osteogenesis. Bone formation.

osteogenic bud. Small budlike mass of embryonic connective tissue with blood vessels; provides cells that form osteoblasts and osteogenic fibers on which bone is formed.

osteogenic fibers. Loose network of connective tissue fibers supporting mesenchymal cells in fetal centers of ossification.

Osteoglossidae. Family of large, freshwater, tropical river fishes; fins soft, head encased in bone, and scales large, bony, and arranged in a mosaic pattern; includes the arapaima and related spp.

Osteolaemus tetraspis. Broad-nosed crocodile of West Africa; to 6 ft. long; Family Crocodylidae.

Osteolepis. See RHIPIDISTIA.

osteology. Study of bones and their interrelationships in the skeleton.

osteomalacia. See VITAMIN D.

Osteostraci. Order of ostracoderms; head and gills enclosed in a stout bony shield; nostril dorsal, between the eyes; ventral gill openings; _Hemicyclaspis_ typical.

Ostertagia. Genus of nematodes parasitic in the intestine of ruminants.

ostiole. 1. Inhalent pore of a sponge. 2. Lateral metasternal opening of stink glands in certain Heteroptera.

ostium. 1. Any opening to a passage or chamber in animals, such as those of the arthropod heart or in the mesenteries of a sea anemone. 2. One of many small openings on the surface of a sponge; water currents pass inward through such openings. Pl. ostia.

ostium tubae. Abdominal opening of a vertebrate oviduct.

Ostraciidae. Family of bizarre marine bony fishes in the Order Plectognathi; trunkfishes and cowfishes; body short, cuboid, triquetrous, or pentagonal in cross section, and covered with a carapace of fused polygonal bony patches; mouth small; near bottom in shallows in tropics; _Lactophrys_ and _Ostracion_ common.

Ostracion. See OSTRACIIDAE.

Ostracoda. One of the subclasses in the Class Crustacea; ostracods or seed shrimps; usually 1 or 2 mm. long, but one marine sp. up to 20 mm.; compressed, covered by a bivalve carapace with a hinge and adductor muscle; second maxillae lacking; one or three pairs of trunk limbs; abdomen consisting of two long peculiar rami; fresh and salt waters, mostly on or near the bottom; 2100 named living spp. plus a very large fossil fauna.

Ostracodermi. Subclass of extinct fishlike vertebrates; body covered with bony scales, often fused over the head to form a shield; skull of bone or cartilage; one pineal eye, one nostril, simple mouth, and numerous paired gill slits; fins median; some spp. with a pair of finlike pectoral lobes; fresh waters, Ordovician to Devonian; _Cephalaspis_, _Birkenia_, and _Pteraspis_ typical.

Ostrea. Genus of larviparous oysters; _O. edulis_ is the European flat oyster; _O. lurida_ is the small Olympia oyster of the West Coast; the inedible _O. cris-_

tata (horse oyster) is found on the Atlantic Coast of the U.S. See also CRASSOSTREA.

ostrich. See STRUTHIONIFORMES.

Otaria. Genus of sea lions found along coast of southern South America.

Otariidae. Family of mammals in the Suborder Pinnipedia; includes the northern and southern fur seals (eared seals) and sea lions; hind limbs capable of being rotated forward; neck long; small external ears.

otic. Auditory.

otic placode. Thickened area of the embryonic vertebrate rhombencephalon; sinks below the surface and develops into an auditory vesicle.

otic vesicle. Auditory vesicle.

Otididae. Bustards; a family of Old World birds in the Order Gruiformes; large, bulky spp.; three toes, all directed forward; wings well developed, but, except when pursued, the spp. are chiefly terrestrial; plains and steppes.

otoconia. 1. Microscopic prismatic calcium carbonate particles imbedded in the gelatinous film which covers the maculae of the sacculus and utriculus in the membranous labyrinth; function in equilibration and balance by stimulating the underlying hair cells in higher vertebrates. 2. Rarely, otoliths.

Otocoris alpestris. Horned lark; a common N.A. streaked brownish terrestrial bird with two small erectile feather horns, black collar, and yellow throat.

Otocryptops. Common western genus of centipedes.

otocyst. Statocyst.

Otolithidae. Sciaenidae.

otoliths. 1. Calcareous concretions in the membranous labyrinth of lower vertebrates; range from small grains to masses 15 mm. in diameter, depending on the sp.; by variously making contact with their surrounding sensory epithelia, appropriate equilibration stimuli are sent to the brain. 2. Rarely, otoconia.

Otomesostoma. Genus of alloeocoel Turbellaria.

Otoplana. Genus of marine turbellarians in the Order Alloeocoela; with a pair of anterior ciliated pits and tactile bristles.

otoporpae. Thick epidermal tracts extending upward from the lithostyles in the medusae of certain hydrozoan coelenterates.

otter. See LUTRA and ENHYDRA.

otter-civet. See CYNOGALE.

Otus asio. Common N.A. screech owl; a small gray sp. to 10 in. long; ear tufts conspicuous.

ouananiche. Small land-locked salmon of southeastern Canada; Salmo ouananiche.

Ourebia. Genus of small tawny African antelopes in the Family Bovidae; includes the oribis; horns straight, annulated, and 5 in. long.

outbreeding. Mating of individuals not closely related.

outcrossing. Outbreeding.

outer ear. That part of the ear of certain tetrapods which is external to the tympanic membrane; absent in amphibians and some reptiles, but in other tetrapods it consists of a canal (external auditory meatus) leading from the surface of the head to the tympanic membrane, and in mammals there is an additional outer vibration-collecting appendage, the pinna.

ouzel. See CINCLIDAE.

ova. Pl. of ovum.

ovalbumin. Albumin of egg white.

Ovalipes ocellatus. Lady crab or calico crab; a marine sp. of Atlantic Coast sandy beaches; carapace 3 in. across; sometimes eaten by man.

oval window. Fenestra ovalis.

ovarian follicle. Any saccular arrangement of cells enclosing a developing ovum.

ovariole. One of several tapering egg tubes which form the insect ovary.

ovariotomy. Surgical removal of one or both ovaries.

ovary. The female gonad or reproductive organ. Pl. ovaries.

ovejector. Muscular terminal part of a female genital duct, e.g. some nematodes.

ovenbird. 1. N.A. sparrow-sized ground warbler of woodlands; Seiurus aurocapillus (Family Parulidae). 2. Any of several South American perching birds in the genus Furnarius (Family Funariidae).

overdominance. Heterozygous condition (Aa) which is superior to both the AA and aa conditions; the implication is that A and a have different effects, and their sum or some reaction product is superior to the homozygous dominant and recessive conditions.

overturn. See SPRING OVERTURN and FALL OVERTURN.

Ovibos moschatus. Musk ox; a long-haired, shaggy, wild ox in the Family

Bovidae; short legs and broad downward curving horns; nose white, color otherwise black and brownish-black; Arctic America; feeds on lichens, grasses, and mosses.

ovicell. Ooecium; a greatly modified zooecium which serves as a brood pouch for an early embryo in certain marine Bryozoa.

oviducal bulb. Large bulbous portion of the oviduct in many snails.

oviducal gland. Large reproductive gland in certain female invertebrates, especially the squid, where it is a swollen part of the oviduct below the genital pore.

oviduct. Any tubule or duct used in carrying eggs or egg cells away from the ovary.

ovigerous legs. Accessory pair of legs just behind the pedipalps of some Pycnogonida; used in carrying the developing eggs.

oviparous. Pertaining to females that release eggs from which the young later hatch out; birds, most insects, and many aquatic invertebrates are oviparous. Noun oviparity.

oviposit. To lay eggs, especially in invertebrates. Noun oviposition.

ovipositor. 1. Structure at the posterior end of the abdomen in many female insects used for depositing eggs; morphology and size variable. 2. tubular extension at the female genital pore in certain fishes.

Ovis. Large genus of ruminants in the Family Bovidae; includes the many kinds of wild and domesticated sheep; wild spp. usually have an outer coat of coarse hair.

ovisac. 1. Graafian follicle. 2. In some coccids, an ovarian cavity containing eggs. 3. In some coccids, an envelope in which eggs are deposited. 4. Egg capsule or receptacle.

Ovists. Group of eighteenth century biologists who believed that an egg cell contained a complete miniature individual. Cf. SPERMATISTS.

ovogenesis. Oogenesis.

ovospermiduct. Hermaphroditic duct; a duct which transmits either sperm or ova or both, especially in certain gastropods.

ovotestis. Hermaphroditic reproductive gland, especially in many spp. of gastropods; different histological areas produce either sperm or ova, but usually not at the same time.

ovoviviparous. Pertaining to females that produce large, yolky shelled eggs which are retained and hatch in the oviduct, the young then being released to the outside; occurs in some insects, some snakes, sharks, and lizards. Noun ovoviviparity.

Ovula. Genus of tropical cowries.

ovulation. Release of a mature ovum from the ovary.

ovum. Mature but unfertilized egg cell; in most plants and in all higher animals it contains the haploid number of chromosomes. Pl. ova.

Owen, Sir Richard. English zoologist (1804-1892); head of natural sciences in British Museum for 20 years; made many major contributions to vertebrate and invertebrate comparative anatomy; clearly distinguished between analogy and homology.

Owenia. Circumboreal genus of Polychaeta.

owl. See STRIGIFORMES.

owlet frogmouth. See AEGOTHELIDAE.

owl fly. See ASCALAPHIDAE.

owl limpet. See LOTTIA.

owl parrot. See KAKAPO.

ox. 1. Any of several members of the genus Bos or closely allied genera. 2. Castrated male Bos taurus, used chiefly as a draft animal.

ox beetle. See STRATAEGUS.

ox bot. See HYPODERMA.

oxea. Small rod-shaped sponge spicule; sharp and tapered at both ends.

oxidase. Enzyme which catalyzes the transfer of hydrogen ions to oxygen with the formation of water; usually terminal in the series of enzymes involved in aerobic oxidations; cytochrome oxidase is best known.

oxidation-reduction potential. Redox potential.

Oxnerella. Genus of marine heliozoans.

ox pecker. African starling which feeds on ticks infesting large mammals.

Oxus. Genus of fresh-water Hydracarina.

ox warble. See HYPODERMA.

Oxyaena. Genus of large otter-like primitive carnivorous mammals of Paleocene and Eocene times.

Oxybelis. See VINE SNAKE.

oxybiotic. Using oxygen in metabolism.

Oxycera. Genus of soldier flies (Stratiomyiidae).

Oxyechus. See KILLDEER and CHARADRIIDAE.

Oxyethira. Genus of Trichoptera; case of larva flask-shaped.

oxygen debt. Excess of oxygen consumed by an organism after that organism has been respiring with an inadequate oxygen supply; in man, after muscular exertion, the oxygen consumption remains unusually high and accumulated lactic acid is being removed for a time after the period of exertion is ended.

oxygen dissociation curve. An S-shaped or hyperbolic curve showing the relation between the oxygen tension of the medium and the percentage of blood-carrying pigment (hemoglobin, etc.) which is oxygenated; the specific shape of the curve depends on many variables including: the particular sp. and individual, altitude, temperature, amount of carbon dioxide in the blood, etc.

Oxygyrus. Genus of pelagic Heteropoda; shell white, large enough to receive entire body.

oxyhemoglobin. Hemoglobin which is carrying respiratory oxygen; scarlet color.

oxyluciferin. See BIOLUMINESCENCE.

Oxymonas. Genus of oval or pyriform flagellates occurring in the termite digestive tract; six flagella.

Oxyopidae. Family of hunting spiders; legs spinose; eight eyes arranged in a hexagon; abdomen pointed; runners and jumpers.

oxyphil cells. Special granular cells found in the mammalian parathyroids.

Oxyrhyncha. Tribe of marine crabs; carapace triangular, mouth opening square, first antennae folded longitudinally.

Oxyrrhis. Genus of marine dinoflagellates.

Oxyruncidae. Poorly known family of passerine birds; the sharp bill; a single sp. in Tropical American rain forests.

Oxystomata. Tribe of marine crabs; carapace circular, mouth triangular; burrow in sand or gravel bottoms.

oxytocin. Pitocin; a neurohormone secreted by the posterior lobe of the pituitary gland; stimulates smooth muscle contraction.

Oxytricha. Genus of marine and freshwater spirotrich Protozoa; ellipsoid, with flattened ventral surface; few to many cirri.

Oxyura jamaicensis. Small, plump N. A. diving duck; ruddy duck; adult male reddish-brown with white cheeks, black crown, and large blue bill.

Oxyurella. Genus of littoral Cladocera.

Oxyuris equi. Nematode parasitic in the large intestine of the horse and its relatives.

Oxyurostylis. Genus of crustaceans in the Order Cumacea.

oyster. See OSTREA and CRASSOSTREA.

oyster borer. See UROSALPINX.

oyster-catcher. See HAEMATOPODIDAE.

oyster crab. See PINNOTHERES.

oyster drill. See UROSALPINX.

oyster leech. _Stylochus frontalis_, a polyclad turbellarian which feeds on oysters along the Fla. coast.

oyster plover. Oyster-catcher.

oyster seed. Spat of the oyster.

oystershell scale. A widely distributed scale insect, _Lepidosaphes ulmi_; an important pest of temperate deciduous trees and shrubs; the mature insect is covered with a secreted scale which resembles a miniature oyster.

Ozotoceras. Genus of ruminants in the Family Cervidae; includes the pampas deer, a small sp. of Brazil to Argentina.

P

P_1. Parental generation with respect to a breeding experiment, observation, or genetic cross.

PABA. Para-aminobenzoic acid.

paca. Large tropical South American rodent in the Family Dasyproctidae; spotted or otherwise marked; lives in burrows; Cuniculus and Stictomys typical.

pacemaker. 1. In biochemistry, a substance whose reaction rate determines the speed of a series of linked reactions. 2. Sino-auricular node; a small mass of specialized cardiac muscle (nodal tissue) in a higher vertebrate heart which initiates the heart beat; located at the point where the superior vena cava enters the right atrium; beat spreads progressively to atria and ventricles; when pacemaker is removed, the heart continues beating but at a slower rate; see ATRIOVENTRICULAR NODE and ATRIOVENTRICULAR BUNDLE.

Pachycheles. Genus of porcelain crabs in the Suborder Anomura; rocky shores.

pachyderm. Large, thick-skinned, hoofed mammal, e.g. elephant and rhinoceros.

Pachydiplax longipennis. Common and widely distributed dragonfly.

Pachydiscus. Genus of fossil cephalopod mollusks.

Pachygrapsus. Common West Coast genus of rock crabs in the Suborder Brachyura.

Pachylomerides. See CTENIZIDAE.

Pachylomerus. Ummidia.

pachymeninx. Dura mater.

pachytene stage. Condition occurri during early prophase of meiosis; the paired homologous chromosomes are shortened and thickened; follows zygotene stage.

Pachythyone. Genus of sea cucumbers.

Pacific tree frog. See HYLA.

Pacinian corpuscle. Specialized pressure receptor found in deep layers of the skin, around the viscera, in the peritoneum, and elsewhere; each corpuscle consists of onion-like concenlayers of connective tissue fibers and cells; to 4 mm. long.

pack rat. See NEOTOMA.

paddlefish. See POLYODONTIDAE.

pademelon. Any of several Australian rat-shaped marsupials with the appearance of a stocky wallaby or kangaroo; body to 3 ft. long; tunnel in herbage; Setonyx and Thylogale typical.

paedogamy. Autogamy.

paedogenesis. Pedogenesis; parthenogenetic production of eggs, young, buds, or other stages by immature animals; e.g. in certain gall flies and in most trematodes.

paedomorphosis. Theory bearing on evolution put forth by Garstang; larval and developmental adaptations may influence the future evolution of a taxon; e.g. the free-swimming tadpole larvae of sea squirts may be considered forerunners of primitive fishes by loss of the sedentary "adult" stage and the development of gonads and other adult structures in the free-swimming form.

Paedophylax. Genus of Polychaeta.

Paguma. Genus of large, powerful, omnivorous palm-civets; southeastern Asia.

Pagurapseudes. Peculiar genus of marine isopods which live in snail shells in the manner of hermit crabs.

Paguridae. Family which includes most of the hermit crabs; abdomen asymmetrical and soft; first legs strongly chelate; abdominal appendages rudimentary or lacking, but last pair adapted for holding the crab inside its snail shell.

Pagurus. Genus of marine decapod crustaceans which includes many spp. of hermit crabs; the animal lives in a snail shell with only the anterior end and anterior appendages showing at the orifice; as it grows it seeks and uses progressively larger snail shells; abdomen asymmetrical and soft; abdominal appendages rudimentary.

painted bats. See ANAMYGDON and KERIVOULA.

painted beauty. Vanessa virginiensis, a Holarctic butterfly; upper surface tawny orange and brownish-black with white spots.

painted bunting. See BUNTING.

painted lady. See VANESSA.

painted salamander. One of two medium-sized salamanders in the genus Ensatina; reddish to brownish; West Coast.

painted snipe. See ROSTRATULIDAE.

painted turtle. See CHRYSEMYS.

painter. See COUGAR.

pairing. Close association of the two homologous chromosomes of each chromosome pair, e.g. as at the beginning of the first meiotic division of meiosis.

pala. See AEPYCEROS.

Palaeacanthocephala. Class of Acanthocephala; main lacunae lateral; proboscis hooks in alternating radial rows; protonephridia absent; males with six cement glands; hosts usually aquatic; Acanthocephalus typical, in fishes and amphibians.

Palaearctic. Palearctic.

Palaemon. Common genus of marine prawns or shrimp.

Palaemonetes. Genus of small marine and fresh-water prawns; usually less than 40 mm. long.

Palaemonias ganteri. Small, blind, fresh-water shrimp found in Mammoth Cave, Ken.

Palaeobatrachidae. Family of extinct toads in the Order Procoela; Palaeobatrachus a familiar European genus; Jurassic to Miocene.

Palaeobatrachus. See PALAEOBATRACHIDAE.

Palaeodictyoptera. Order of primitive Carboniferous insects which persisted until Permian times; thorax and abdomen similarly segmented; wing venation simple.

Palaeogaea. Faunal region which includes Asia, Australia, Europe, and Africa.

Palaeognathae. One of the three superorders of birds in the Subclass Neornithes; large, walking birds; wings reduced and useless for slight, tail vertebrae free, coracoid and scapula bones small; sternum unkeeled except in the Order Tinamiformes; eight orders, including the ostrich, rhea, emu, kiwi, etc.

Palaeogyrinus. Genus of labyrinthodonts.

Palaeomanon. Genus of fossil cup sponges.

Palaeomastodon. Fossil genus of ancestral elephants; about 42 in. high at the shoulder; a long snout was probably present, and the tusks were fairly long; 26 teeth; Oligocene.

Palaeonemertea. Order in the Phylum Nemertea; dermis gelatinous; body wall with two or three muscle layers.

Palaeoniscus. Genus of early Paleozoic ray-finned fishes.

Palaeophonus. Genus of primitive Silurian air-breathing arachnids.

Palaeopterygii. Actinopterygii; subclass of primitive bony fishes; (Osteichthys); rays of dorsal and anal fins more numerous than their rodlike supports; clavicles present (except in Belonorhynchii); nostrils not opening into mouth cavity; sturgeons, paddlefishes, etc.

Palaeotremata. Extinct order of brachiopods in the Class Articulata.

palama. Web between toes of aquatic birds.

Palamedea. See PALAMEDIDAE.

Palamedidae. Small family of South American gooselike wading and swimming birds in the Order Anseriformes; screamers; bill slender, feet not webbed; a primitive family, possibly ancestral to the birds of prey on the one hand and the ducks and geese on the other hand; Palamedea typical.

palate. Roof of the vertebrate oral cavity.

palatine. One of a pair of bones of the basicranial region in the skull of certain vertebrates.

palatine teeth. Small teeth on the palatine bones of certain lower vertebrates.

palatoquadrate cartilage. Pterygoquadrate cartilage; upper jaw element in the shark.

Paleacrita vernata. Sp. of moth in the Family Geometridae; larva, the spring cankerworm, feeds on fruit and foliage of many wild and cultivated fruit trees.

Paleonotus. Genus of Polychaeta.

Palearctic realm. Largest of the six zoogeographic regions of the world; includes all of Europe, Africa and Arabia as far south as the Tropic of Cancer, all of Asia as far south as the Himalayas, Japan, Iceland, Canary Islands, and Cape Verde Islands;

some characteristic animals are the hedgehog, wild boar, ibex, and chamois.

pale bat. See ANTROZOUS.

Paleocene epoch. Earliest of the five subdivisions of the Tertiary period; lasted about 10 million years; especially characterized by the rise of archaic placental mammals and modern birds.

Paleodictyoptera. Palaeodictyoptera.

Paleodiscus. Genus of extinct echinoids in the Order Lepidocentroida.

paleoecology. Ecology of extinct organisms in their environment.

paleogenesis. Principle which states that descendent ontogenies tend to recapitulate ancestral ontogenies; successor to Haeckel's biogenetic law.

Paleolithic. Old Stone Age of human development; began about 1 million years ago and ended with Neolithic times; featured by hunting, rudimentary social living, stone implements, lack of permanent habitations, and no cultivation.

Paleonemertea. Palaeonemertea.

paleontology. Study of fossil animals and plants and their distribution in time.

paleopallium. Band of nervous tissue along the lateral surface of a cerebral hemisphere; least pronounced in mammals.

Paleoptera. Subclass of insects which includes the Odonata, Ephemeroptera, and five extinct orders.

Paleosimia. Genus of fossil ancestral orang utangs.

Paleospiza. Oldest known genus of fossil passerine birds; Colorado Miocene.

Paleostracha. Merostomata.

Paleozoic era. One of the five major subdivisions of geological time; lasted from about 550 million to 190 million years ago; an age of invertebrates and lower vertebrates; subdivided into Permian, Carboniferous, Devonian, Silurian, Ordovician, and Cambrian periods.

paleozoology. Study of fossil animals and their distribution in time.

pale periwinkle. Littorina palliata; a common thick-shelled, smooth periwinkle of variable coloration.

palingenesis. All or part of the development of an individual which repeats the phylogenetic development of its taxon. Cf. CENOGENESIS.

Palinura. Taxon of decapod crustaceans which includes the sea crayfishes and spiny lobsters; sometimes extended to include the fresh-water crayfishes.

Palinurus. Genus of edible spiny lobsters or rock lobsters of warm seas.

Palio. Genus of sea slugs; body elongated, with a marginal ridge on each side bearing tubercles or finger-like projections.

palla. Pala; see AEPYCEROS.

Pallas, Peter Simon. German naturalist and explorer (1741-1811); made important natural history collections during extensive travels in Asiatic Russia; contributed to the classification of mollusks and certain worm phyla.

Pallene. Genus of Pycnogonida.

pallial chamber. Mantle cavity.

pallial groove. Ventral groove between the foot and mantle of a typical chiton.

pallial line. Faint line near the periphery of each valve of typical bivalve mollusks; it extends along anterior, ventral, and posterior margins, and indicates the internal line of attachment of the mantle to the shell.

pallial nerve. One of a pair of large nerves in mollusks; innervate mantle.

pallial sinus. Indentation in the pallial line in many bivalve mollusks; often indicative of the location of the incurrent siphon.

pallid bat. See ANTROZOUS.

Pallifera. Genus of land slugs.

pallium. 1. Cortex of the cerebral hemispheres along with its underlying white matter. 2. Mantle of a mollusk or brachiopod.

Palmacorixa. Common genus of water boatmen in the Family Corixidae.

palmaris longus. Muscle of the forelimb of certain tetrapods.

palm chat. See BOMBYCILLIDAE.

palm civet. See MACROGALIDIA.

palm crab. Birgus latro.

palmer worm. Any caterpillar which suddenly hatches out in large numbers and feeds on herbage.

palm squirrel. Any of several squirrel-like tropical mammals; feed chiefly on palm tree fruits; includes the ratufas and bush squirrels; Funambulus, Paraxerus, and Myosciurus typical.

Palola. Genus of Polychaeta.

palolo worm. See LEODICE and EUNICE.

Palometa. See POMPANO.

palp. 1. Fleshy cylindrical or tapered projection, often sensory, especially at the anterior end of polychaete worms. 2. In arthropods, a segmented or unsegmented process attached to a head appendage; usually sensory.

palpal organ. In male spiders, the specialized tip of each pedipalp; used to transfer sperm to the female genital pore.

palpifer. Small sclerite bearing the maxillary palp in insects; articulates with stipes.

palpiger. Sclerite bearing a labial palp, especially a mental sclerite in insects.

Palpigradi. Order of minute arachnids; pedipalps leglike; oval abdomen with terminal tail-like process; southwestern states; 20 spp.; Prokoenenia and Koenenia common.

Palpomyia. Genus of ceratopogonid Diptera.

palpon. Tactile zooid in certain siphonophore coelenterates.

palpus. Palp. Pl. palpi.

Paltothemis. Genus of dragonflies; southwestern U.S.

Paludicella. Genus of fresh-water bryozoans in the Order Ctenostomata.

Paludicola. Taxon which includes the fresh-water spp. of triclad Turbellaria.

Paludina. Viviparus.

paludrine. Antimalarial drug.

palustrine. Living in swamps.

palynology. Study of fossil pollen and other microfossils, especially in lake, bog, and pond deposits.

pampas cat. Small wildcat of southern Argentina and Chile.

pampas deer. See OZOTOCERAS.

Pamphagus. Genus of fresh-water amoeboid Protozoa; most of cell encased in a hyaline vase-shaped test.

pan. Shallow pond or lake having an oval, circular, or kidney shape; origin by erosion; perennially dry.

Panagrolaimus. Genus of microscopic, free-living nematodes.

Pancolus. Genus of marine Isopoda.

pancreas. Small digestive gland in vertebrates; in man it lies between the stomach and duodenum and has a duct which opens into the duodenum; certain invertebrates also have a digestive gland referred to as a pancreas.

pancreatic duct. Duct carrying the secretions of the pancreas into the intestine in vertebrates and certain invertebrates.

pancreatic juice. Fluid secreted into the small intestine by the pancreas in vertebrates; in man it contains five digestive enzymes: trypsin, lipase, amylopsin, chymotrypsin, and carboxypeptidase.

pancreatin. Extract of pancreas containing digestive enzymes; prepared commercially from hogs and cattle.

pancreozymin. Hormone secreted by the wall of the small intestine; stimulates the pancreas to discharge pancreatic enzymes

panda. See AILURUS and AILUROPODA.

Pandaka. Genus of very small gobies; one Philippine sp. only 10 mm. long when mature.

Pandalus. Common genus of shrimps found in both Atlantic and Pacific.

Pandarus. Genus of marine copepods in the Suborder Caligoida; body depressed and highly modified; parasitic on fishes.

pandemic. Widely epidemic, such as a disease.

Pandinus. Genus of giant scorpions; one sp. to 6.5 in. long.

Pandion haliaetus. Osprey or fish hawk; large, eagle-like hawk, blackish above, whitish below; flies with decided kink in wings; world wide distribution along seashores, large lakes, and rivers; feeds chiefly on fish.

Pandionidae. Family of birds in the Order Falconiformes; includes only the osprey.

Pandora. Genus of small bivalve mollusks found on both coasts below low tide line; pandora shells; valves thin, unequally flattened, and mostly whitish; mostly less than 1.5 in. long.

Pandorina. Genus of colonial free-swimming green flagellates; 8 to 32 biflagellate cells closely packed in a gelatinous matrix; fresh water.

Paneth cells. Special cells at the base of the crypts of Lieberkühn.

panfish. Generally, any small edible fish suitable for frying whole in a pan; often used more specifically to apply to fresh-water centrarchids.

pangenesis. Discredited theory of heredity; each cell is said to be represented in the blood by minute particles which circulate about freely and come to lie in the sperm and egg cells; when these are transmitted to the next generation the resulting offspring contain particulate materials from every part of both parents.

pangene theory. Pangenesis.

pangolin. See PHOLIDOTA.

panmictic population. Large population of a sp. having random cross-breeding.

Panomya. Northern genus of thick-shelled marine bivalves.

Panope. See GEODUCK.

Panopeus. Genus of small marine decapod crustaceans which includes the mud crabs; sluggish, inconspicuous crabs with the carapace slightly broader than long; on muddy bottoms along shore.

Panoplea. Genus of marine Amphipoda.

Panorpa. Common genus of insects in the Order Mecoptera; includes the common scorpion or hanging fly.

pansporoblast. A myxosporidian sporont which develops into two spores.

Pantala. Common genus of dragonflies.

Panthalops. Genus of sheeplike animals also related to the goats; the chiru; fur reddish-brown, thick, and woolly; male with long horns; muzzle enlarged on each side; Tibet.

panther. 1. Popular name given to large leopards, especially the black varieties. 2. American cougar. 3. Occasionally, the American jaguar.

Panthera. Genus of large cats in the Family Felidae, including the lion, panther, leopard, tiger, and jaguar; certain spp. are sometimes included in the genus Felis instead of Panthera. Cf. FELIS.

Pantodonta. Order of extinct, medium to large ungulate mammals; Paleocene to Eocene.

Pantopoda. Pycnogonida.

Pantosteus. Genus of mountain suckers in the Family Catostomidae; mouth ventral, head small, air bladder two-chambered; western N. A.

Pantostomatida. Rhizomastigina.

pantothenic acid. Vitamin of the B complex; essential to coenzyme A formation; deficiency produces dermatitis in chicks and rats, nerve degeneration in hogs; abundant in yeast, cane molasses, peanuts, eggs, milk, and liver; $C_9H_{12}O_3N$.

Pantotheria. 1. One of the three infraclasses of mammals in the Subclass Theria; primitive Jurassic spp.; jaws long and slender, with numerous cheek teeth. 2. One of the two orders in the Infraclass Pantotheria; molars with four or more cusps.

Pan troglodytes. Chimpanzee of West African forests; in temperament and structure, the most manlike of the anthropoid apes; to 150 lbs.; herbivorous; the common trained ape of circuses, carnivals, etc.

Panulirus. Genus of decapod crustaceans which includes the edible spiny lobsters, sea crayfish, or rock lobsters; abdomen fully developed; none of legs chelate; exoskeleton spiny; to 20 in. long; Calif. and southern states on the Atlantic Coast.

panzootic. Widely epidemic among animals, especially some infection.

papain. Proteolytic enzyme prepared from latex of the tropical pawpaw or papaya tree; used in treatment of dyspepsia.

papal mitre. Mitra papalis, a thick-shelled gastropods of the Indian Ocean.

paper chromatography. Analysis and separation of dissolved materials from each other with the aid of a long strip of filter paper; a small amount of the solution and solutes is allowed to dry out near the end of the strip; when the edge of the strip is dipped into a special solvent, it moves along the filter paper by capillarity and carries each kind of molecule along at a different velocity; the paper is then dried, and each solute is localized in one level of the filter paper strip.

paper fig shell. See FICUS.

paper nautilus. See ARGONAUTA.

paper shells. Thin-shelled fresh-water mussels in the genus Anodonta.

paper wasps. See VESPIDAE and POLISTIDAE.

Papilio. Genus of common and brightly colored butterflies in the Family Papilionidae; swallowtails; larvae of some spp. are pests on economic plants.

Papilionidae. Large family of large, brightly-colored, common butterflies having a tail-like projection on each hind wing; swallowtails; world wide distribution; many spectacular tropical spp.; Papilio a common American genus.

papilla. 1. One of the conical projections in the medullary portion of the kidney of higher vertebrates; a mass of collecting tubules. 2. Any blunt, rounded, or nipple-shaped projection. 3. Any small process projecting into a cavity or above an external surface. 4. One of the small protuberances of the tongue. 5. Breast nipple. See DERMAL PAPILLA and TONGUE PAPILLA. Pl. papillae.

papilla amphibiorum. Special sensory structure near the upper margin of the sacculus of amphibians; has a role in hearing.

papillary cone. Vascular protrusion into the vitreous humor from the optic nerve area in reptiles.

Papio. Genus of Old World monkeys; includes the baboons of Africa and southern Asia; powerful spp. with short tails and doglike heads.

Papirius fuscus. Common insect in the Order Collembola; the garden "flea" which damages young vegetables.

pappus. Tuft or circlet of setae, hairs, or bristles.

papulae. Dermal branchiae.

Papyrula. Genus of marine sponges.

para-aminobenzoic acid. Poorly known member of the vitamin B complex; presumably necessary for proper growth; reverses the bacteriostatic action of sulfanilamide.

parabasal apparatus. Organelle of various shape closely associated with the nucleus of many flagellates; function not definitely known.

Parabathynella. See SYNCARIDA.

parabiosis. 1. Joining together of two animals by one or more major blood vessels. 2. Temporary (and usually experimental) reversible suspension of one or more vital life processes; e.g. suppression of impulse transmission by a nerve, suitable partial air-drying of a clam, etc. 3. Intermingling and living together of two or more spp.; e.g. mixed herds of ungulates, an ant colony consisting of two different spp.

parabiotic twins. Experimental union of two animals to form artificial Siamese twins with circulatory connections.

parablast. 1. Yolk portion of a meroblastic ovum. 2. That part of the mesoblast which gives rise to blood vessels and lymphatics.

Parabuteo unicinctus. Dusky hawk, Harris's hawk; a sooty brown sp. found from southwestern U.S. to Argentina.

paracana. See DINOMYIDAE.

paracasein. Protein formed by the action of rennin on milk in the stomach.

Paracerceis. Genus of marine Isopoda.

parachordal cartilages. Two cartilages which develop below the posterior part of the brain on either side of the notochord during the embryology of many vertebrates.

Parachordodes. Genus of Nematomorpha; poorly known in the U.S.

Paracineta. Genus of fresh-water suctorians; cell within a stalked lorica.

Paraclunio. Genus of midges in the Family Tendipedidae.

paracone. Outer anterior cusp in an upper mammalian molar tooth.

paraconid. Accessory inner anterior cusp of a lower mammalian molar tooth.

Paractis. Genus of sea anemones in the Order Actiniaria; with up to 400 tentacles.

Paracymus. Genus of whale lice, greatly modified amphipods living as external parasites of whales.

Paracyclops fimbriatus. Lake copepod.

Paracymus. Common genus of hydrophilid beetles.

paraderm. Membrane enclosing the pronymph in the dipteran Family Muscidae.

paradidymis. Small closed convoluted tubule above the epididymis; remnant of the posterior part of the Wolffian body.

Paradileptus. Genus of fresh-water holotrich Ciliata; anterior end a long proboscis.

Paradisaeidae. See BIRD OF PARADISE.

paradise fish. Small, handsome, eastern Asiatic fresh-water fish; often kept in aquaria.

Paradoxurus. See MUSANG.

Paraexoglossum laurae. Tonguetied chub; a cyprinid fish of N.Y. to Ohio and south to N.C.

paraganglia. Small masses of chromaffin cells between the kidneys and along the wall of the dorsal body cavity in lower vertebrates (especially fishes); adjoin sympathetic ganglia and produce adrenalin.

paragaster. Spongocoel.

paragastrula. In many sponges, the gastrula formed by the invagination of the flagellated cells of the amphiblastula larva.

Paraglaucoma. Genus of ovoid or ellipsoid fresh-water holotrich Ciliata; anterior end pointed.

paraglossa. One of a pair of lateral terminal lobes of the labium in certain insects. Pl. paraglossae.

paraglossum. Anterior tongue-supporting element of the hyoid apparatus, especially in some birds.

paraglycogen. Polysaccharide food storage material occurring as granules in many ciliates.

paragnath. 1. One of the two lobes forming the lower lip, or metastoma, in certain Crustacea. 2. Small accessory palp on the mandible of certain decapod crustaceans. 3. Sclerotized jaw-like tooth of certain annelids. 4. One of a pair of lateral lobes of the hypopharynx in certain insects.

Paragnetina. Common genus of Plecoptera.

Paragonimus westermani. Human lung fluke which also occurs in many wild and domesticated mammals; aquatic snails are the first intermediate hosts and fresh-water crayfish and fresh-water crabs are the second intermediate hosts; the definitive host becomes infected by ingesting the metacercariae which are encysted in or on the second intermediate host; common in the Far East, and occasional in Africa, Central and northern South America, and in the northeastern U. S.

Paragordius. Common genus in the Phylum Nematomorpha; to 12 in. long.

Paragorgia. Genus of branching colonial anthozoans in the Order Alcyonacea; colony to 5 ft. high.

Parahippus. Genus of Miocene horses.

parahormone. 1. Any hormone-like substance having a physiological effect in some distant organ. 2. Waste product of metabolism which regulates body functions, such as carbon dioxide which affects respiratory ventilation.

Parajulus. Common genus of millipedes.

parakeet. Any of a great many small, slender, parrot-like birds with a long tail; see PSITTACIFORMES, MELOPSITTACUS, and CONUROPSIS.

Paralabrax clathratus. Kelp bass; important game fish of Calif. Coast; to 2 ft. long.

Paraleptophlebia. Common genus of Ephemeroptera.

Paralichthys. See HETEROSOMATA.

Parallorchestes. Genus of marine Amphipoda.

paralysis agitans. Degeneration of ganglion cells of floor of cerebrum; produces involuntary movements, especially of the hands; inherited as a Mendelian dominant of man.

paramecin. See KAPPA PARTICLES.

Paramecium. Common genus of ciliate Protozoa found in stagnant ponds, puddles, infusions, etc.; usually 200 to 300 microns long; often called the "slipper animalcule" because of its characteristic shape; extensively used in experimental work; P. aurelia and P. caudatum common.

paramere. 1. Right or left half of a somite or bilaterally symmetrical animal. 2. Actinomere or either half of an actinomere in radially symmetrical animals. 3. In some male insects, one of the paired lobes exterior to the penis.

Paramonostomum. Genus of monostome trematodes in the intestine of muskrat.

Paramphistomidae. Family which includes the amphistome flukes; acetabulum large, at the posterior end of the body; in vertebrate digestive tracts.

Paramphistomum cervi. Trematode parasite of the stomach of ruminants.

paramylon. Polysaccharide food product formed in certain holophytic protozoans; usually occurs in the form of granules in the cell.

paramylum. Paramylon.

Paranais. Genus of fresh-water Oligochaeta.

Paranemertes. Genus of nemertines in the Order Hoplonemertea.

paranoia. Psychosis characterized by delusions of grandeur or persecution.

Paranthessius. Genus of marine cyclopoid copepods; usually on gills of clams.

Paranthura. Genus of marine Isopoda.

Paraonyx. Genus of otters found along African rivers south of the Sahara; feed on fish.

Parapaguridae. Family which includes some of the hermit crabs.

Paraphanolaimus. Genus of microscopic, free-living nematodes.

Paraphelenchus. Genus of microscopic, free-living nematodes.

Parapholas. Genus of marine bivalves; shell ovate-oblong; anterior gape closed by a thin globose plate.

Parapholyx. Genus of fresh-water snails; several western U. S. spp.

Paraphoxus. Genus of marine Amphipoda.

paraphysis. 1. Non-nervous median dorsal vascularized outgrowth of the telencephalon of adult fishes, amphibians, and reptiles. 2. In some scale insects, one of several thickenings, ingrowths, or club-shaped projections near the edge of the pygidium.

parapineal organ. Parietal organ.

Parapithecus. Genus of Oligocene apes.

parapleuron. Thoracic pleuron of some Coleoptera; located ventrally on each side of the sterna.

Parapleustes. Genus of marine Amphipoda.

parapod. Parapodium.

parapodium. 1. Flat lateral protuberance on each side of most segments in polychaetes; usually has two main parts, a dorsal lobe (notopodium) and a ventral lobe (neuropodium), and both of these have one or more cirri and few to many setae; parapodia are variously reduced or specialized for feeding, locomotion, and respiration. 2. Lobelike extensions of the sole of the foot of certain mollusks.

Parapolia. Genus of marine Nemertea.

Parapolytoma. Genus of fresh-water biflagellate Protozoa; anterior end obliquely truncate.

parapophysis. Lateral projection of the centrum for the articulation of the lower head of a two-headed rib.

paraproct. Sclerite on each side of the anus in millipedes and some insects.

Parapsyche. Genus of Trichoptera.

parapteron. Small sclerite of the insect mesothorax.

Parasabella. Genus of Polychaeta; in membranous tubes in mud and sand.

Parascalops breweri. Hairy-tailed mole; a blackish sp. with the tail densely covered with hair; northeastern U. S. and southeastern Canada.

Parascaris equorum. Nematode parasite of the horse small intestine.

Parasimulium. Genus of blackflies in the Family Simuliidae.

parasite. Organism that lives in or on an organism of another sp. (host) and derives its nutriment therefrom; usually a parasite causes some degree of damage to the host.

parasitic nutrition. Type of nutrition occurring among parasites; depending on the sp., it may be holozoic, it may involve the absorption or ingestion of partially predigested host tissues, or it may be both.

Parasitidae. Family of mites; free-living or parasitic; Parasitus parasitic or commensal on beetles.

Parasitoidea. Superfamily of mites in the arachnid Order Acarina; includes Parasitidae, Dermanyssidae, and other families.

parasitology. Study of animal and/or plant parasites.

Parasitus. Genus of mites commensal or parasitic on insects; see PARASITIDAE.

Parasmittina. Genus of encrusting calcareous marine Bryozoa.

parasphenoid. Median bone on the ventral surface of the brain case in certain vertebrates, especially fishes and amphibians.

Parastenocaris. Large genus of fresh-water harpacticoid copepods; in sandy beaches, damp moss, etc.

parasympathetic nervous system. See AUTONOMIC NERVOUS SYSTEM.

Paratendipes. Genus of tendipedid midges.

Paratenodera sinensis. Chinese mantid, now introduced into eastern U. S.

Paratetranychus. Genus of small mites in the Family Tetranychidae; often harmful to fruit trees.

parathormone. Hormone secreted by the parathyroid glands; controls calcium and phosphate metabolism of the cells and maintains a constant calcium level in the blood; underproduction results in muscular twitching, spasmodic contractions, and convulsions; overproduction results in weakened bones and abnormally low irritability of muscles and nerves.

parathyroid glands. Small endocrine glands occurring in tetrapods; in man there are usually four such glands, 3 to 13 mm. in diameter, and closely associated with the thyroid gland; secrete parathormone.

paratomy. Fission, following preparatory internal tissue reorganization in certain annelids.

Paratriozoa. Genus of insects in the homopteran Family Psyllidae; transmit a disease of the potato.

paratrophic. Deriving food parasitically from the host organism.

paratype. One of a group of specimens from which the type specimen is selected and described for a new sp. or genus.

paratyphoid. Type of fever similar to typhoid fever and caused by related spp. of Salmonella.

Paraxanthias. Genus of marine crabs in the Suborder Brachyura.

Paraxerus. Genus of African bush squirrels or palm squirrels.

Parazoa. Branch of the Metazoa which contains only the Phvlum Porifera; the only group of multicellular animals having no true digestive cavity. See ENTEROZOA.

parchment worm. See CHAETOPTERUS.

parchment worm crab. Pinnixa chaetopterana, a small crab commensal in the tubes of Chaetopterus and Amphitrite.

Pardictis. See LINSANG.

Pardosa. Large genus of wolf spiders in the Family Lycosidae.

parenchyma. 1. In lower animals, a spongy mass of vacuolated mesenchyme cells filling spaces between muscles, epithelia, and viscera; sometimes syncitial. 2. Tissue peculiar to an organ (such as a gland), as distinct from the connective tissue, blood vessels, etc. of that organ.

parenchymula. Stereogastrula.

Paresperella. Genus of thick, encrusting marine sponges.

Pareurythoë. Genus of polychaetes; to 2 in. long.

Paridae. Family of insectivorous passerine birds; includes the chickadees and titmice; mostly smaller than sparrows, with longer tails; absent from South America and the Australian region; Penthestes and Baeolophus typical.

parietal blood vessel. Segmental blood vessel in the body wall of typical annelids; delivers blood to the dorsal vessel.

parietal bone. One of a pair of bones on the roof of a vertebrate skull, between frontals and occipitals; in man the two parietals form much of the top and sides of the cranium.

parietal lobe. Posterior dorsal portion of a cerebral hemisphere, especially in the higher vertebrates.

parietal organ. Small median dorsal outgrowth of the brain closely associated with the pineal body in cyclostomes, some lizards, and Sphenodon; in the latter two groups it has the structure of a degenerate eye.

parietal peritoneum. That portion of the peritoneum which lines the coelomic cavity in higher animals.

parietal pleura. See PLEURA.

Parnassiidae. Small family of alpine, subarctic, and high altitude butterflies; whitish to brownish gray; Parnassius the most common genus.

Parnassius. See PARNASSIIDAE.

paroccipital process. Extension of the occipital bone; closely associated with the mastoid process.

Paroctopus. West Coast genus of octopuses.

paronychium. Bristle-like appendage of an insect pulvillus.

paroophoron. Vestige of a Wolffian body found in the ovarian region of female mammals.

paroquet. Parakeet; see CONUROPSIS.

parotid gland. Large salivary gland in front of the ear in mammals.

parovarium. Epoophoron.

parrakeet. Parakeet.

parr marks. Vertical bars along the sides of the body, especially in young salmonid fishes.

parrot. See PSITTACIFORMES.

parrot fever. See PSITTACOSIS.

parrot fish. Any of various brightly colored tropical marine fishes, some of which have jaws similar to the beak of a parrot.

pars intercerebralis. Special part of the brain in many larval insects; secretes one or more hormones which induce molting and differentiation; see CORPORA ALLATA.

parthenita. Sporocyst, redia, or cercaria stage in the life history of a fluke.

parthenogenesis. Type of unisexual reproduction involving the production of young by females which are not fertilized by males; a common method of reproduction in rotifers, cladocerans, aphids, bees, ants, and wasps; depending on the sp., a parthenogenetic egg may be haploid or diploid. Adj. parthenogenetic.

parthenogonidium. 1. A cell that gives rise to a miniature colony by asexual multiplication in such protozoans as Volvox. 2. Any cell in a colonial protozoan which reproduces only asexually. Pl. parthenogonidia.

partridge. See PERDIX.

partridge cask shell. See PARTRIDGE TUN SHELL.

partridge tun shell. Partridge cask shell, Tonna perdix; a mottled Caribbean gastropod to 9 in. long.

Partula. Genus of land snails of certain islands of the southwest Pacific.

parturition. 1. Bringing forth young, especially in mammals. 2. Childbirth. 3. Delivery.

Parulidae. Mniotiltidae.

Parus. Penthestes.

Passalidae. Family of large shiny black or brown beetles; front of head with a short, bent horn.

passenger pigeon. See ECTOPISTES.

Passer. Genus of sparrows; includes the common house sparrow or English sparrow (P. domesticus) and the European tree sparrow (P. montanus) which has been introduced into St. Louis and adjacent Ill.

Passerculus. See SPARROW.

Passerella iliaca. Fox sparrow; a large dark brown or gray-brown sparrow with streaked under parts; coniferous forests of Canada and much of U.S.

Passerherbulus. Ammospiza.

Passeriformes. Large order of birds;

includes the great majority of all birds; passerine or perching birds and songbirds; about 56 families; foot with three front toes and one hind toe; habits variable.

Passerina. See BUNTING.

passerine. Pertaining to perching birds in the Order Passeriformes.

passive immunity. Artificial immunity.

Pasteur, Louis. Famous French chemist, bacteriologist, and immunologist (1822-1895); best known for pioneer work in fermentation, pasteurization, silkworm disease, chicken cholera, anthrax, hydrophobia, and vaccination.

Pasteurella. Genus of gram-negative, non-sporeforming, non-motile, rod-shaped, parasitic bacteria; among other diseases, certain spp. cause bubonic plague and tularemia; _P. pestis_ causes bubonic plague and is transmitted by the bite of the rat flea, _Xenopsylla cheopis_. See TULAREMIA.

pasteurization. Method of heat-treating certain foods, especially milk, in order to free them of most bacteria; milk is warmed to 145° F. for 30 minutes, then rapidly cooled and kept cool until consumed.

patagium. 1. Extension of skin on the anterior and posterior surfaces of the forelimb of a bird; prevents the flight feathers from twisting during flight. 2. Parachute-like skin extensions in the Dermoptera and flying squirrels. 3. One of a pair of small lateral prothoracic processes in many insects. 4. Bat wing membrane. Pl. patagia.

Patagonian. Race of mongoloids of extreme southern South America; skin brown, head short, face square; same as Tehuelche.

patch-nosed snake. See SALVADORA.

patella.. 1. Small bone at the knee; present in most mammals and some birds and reptiles. 2. Short fourth segment of a walking leg in a typical arachnid. 3. Modified segments of the anterior tarsi in Dytiscidae.

Patella. Genus of marine limpets.

Patellidae. Discarded taxon which includes the limpets.

patelliform. Shaped like a kneecap.

patellula. A patella having ringlike openings, as in certain Dytiscidae.

paternity tests. Tests of the blood of a child and a supposed male parent with respect to M-N and A-B groups in order to see whether the man can be excluded as the father of the child on the basis of known inheritance mechanisms of these blood groups.

pathogen. Any disease-producing organism. Adj. pathogenic.

pathology. Study of the cause and nature of disease, but especially the study of diseased tissues.

Patiria. Genus of Pacific Coast sea stars; commonly red or yellow; to 4 in. in diameter; sea bats.

paunch. Rumen.

pauraque. See NYCTIDROMUS.

Paurometabola. Group of orders of insects having simple, direct, or gradual metamorphosis in which the series of immature forms (nymphs) are considerably like the adult except in size and structural proportions; nymphs acquire wing pads early, but functional wings do not appear until the adult instar; stages in the life history are: egg, a series of nymphs, and adult; typical examples are Orthoptera, Isoptera, and Hemiptera; in a few Paurometabola the first two active nymph instars are called "larvae," especially if the third and fourth nymph instars are markedly different and/or quiescent. Adj. paurometabolous. See also AMETABOLA, METABOLA, HEMIMETABOLA, and HOLOMETABOLA.

Pauropoda. Small class of minute, cylindrical, terrestrial arthropods; head with one pair of branched antennae, one pair mandibles, and two pairs maxillae; trunk of 6 to 12 segments bearing 9 to 11 pairs of legs; respiratory and circulatory systems reduced; under stones and debris in damp places; about 300 spp.; _Pauropus huxleyi_ common in Europe and much of U.S.

Pauropus. See PAUROPODA.

Paussidae. Family of tropical beetles; antennae remarkably thick, the apical segments fused into a club or broad lamina.

Pavlov, Ivan Petrovitch. Russian physiologist and psychologist (1849-1936); well known for his work on conditioned reflexes and digestive gland physiology for which he received the Nobel prize in physiology and medicine in 1904.

Pavo cristatus. Common pea fowl; domesticated since ancient times; male (peacock) with greatly elongated and brightly colored upper tail coverts which can be raised vertically in a fanlike fashion; female (peahen) more drab and without such a device.

paxillae. Modified spines present in some sea stars; calcareous rods having minute spinules at the tip.

PC. Phosphocreatine.

peach aphid. Myzus persicae, a pink or green aphid; cosmopolitan and omnivorous but common on peach trees.

peach borer. See CONOPIA.

Peachia. Genus of sea anemones.

peach-tree borer. See AEGERIA.

peacock. See PAVO.

peacock worm. Any of many kinds of tube-building marine Polychaeta in which an anterior tuft of long food-collecting cirri project from the tube and has the appearance of miniature spread peacock feathers; Sabella and Eudistylia common.

pea comb. Three-ridged type of comb occurring in Brahma fowl.

pea crab. See PINNOTHERES.

pea fowl. See PAVO.

peahen. See PAVO.

peamouth. See MYLOCHEILUS.

peanut worm. Common name of certain spp. of sipunculids; when the introvert is retracted they have the superficial shape of a peanut.

pear conch. Busycon pyrum; a small whelk of sandy shallows from N.C. to Tex.

pearl. Dense concretion, lustrous and varying in color, formed as an abnormal growth in the tissues of many bivalve mollusks; most pearls are formed as the result of a foreign body (parasite, sand grain, etc.) becoming lodged in the mantle or nearby tissues; such a foreign body must have some mantle epithelium associated with it before a pearl will be formed, since the mantle epithelium secretes concentric layers of nacre around the foreign body; depending on the specific location in the mollusk tissues, the pearl may be spherical, button-shaped, ovoid, pear-shaped, discoidal, or any irregular shape (baroque); American oysters occasionally contain pearls, but they are of no commercial value; some fresh-water mussels form pearls but these are seldom marketable; most of the finest "natural" pearls come from the pearl oysters of the Persian Gulf; the majority of commercial pearls are "cultured" in the coastal waters of Japan; skilled technicians insert a bit of shell and mantle into pearl oyster tissues, and the oysters are then put back into the sea; after several years these oysters are harvested and examined, and a percentage of them will be found to have developed pearls; such "cultured" pearls can be told from "natural" pearls only by an expert with the aid of X-rays; "artificial pearls" are made synthetically in factories from various organic and inorganic compounds, and have nothing to do with mollusks.

Pearl, Raymond. American biologist (1879-1940); known for his work on the statistics of human populations, birth rates, mortality, and longevity.

pearl essence. An aqueous suspension of the silvery material of various fish scales; used in the manufacture of artificial pearls.

pearlfish. 1. Any of various fishes from whose scales essence of pearl is made. 2. Any of several whitish tropical marine fishes.

pearl oyster. In the broad sense, any oyster in which pearls are likely to be formed.

pearly nautilus. See NAUTILOIDEA.

pear midge. See CONTARINIA.

pear slug. See CALIROA.

pebrine. Sporozoan disease of the silkworm; the protozoan, Nosema bombycis, occurs in all tissues of all stages in the life history; during the last century it reached epidemic proportions in France and Italy, and Pasteur first worked out methods of control.

peccary. See TAYASSUIDAE.

peck order. Expression of social hierarchy in a flock of chickens or other fowl; a given hen may peck with impunity all hens below her in social rank but may be pecked in turn by all hens above her in the social scale; any hen newly introduced into the flock soon finds her proper rank by a pecking contest; the term is often used with reference to similar phenomena in fishes and mammals.

pecten. 1. Common name of bivalves in the genus Pecten; the scallops. 2. In insects, any comblike arrangement of heavy setae. 3. Comblike structure in the vitreous body of the eyes of most birds and many reptiles. 4. Part of the stridulating organ in certain spiders. 5. Row of spinules on the respiratory tube of a mosquito larva.

Pecten. Bivalve genus which includes most of the scallops or pectens; shell round, with radiating ridges and two wings at the dorsal edge; able to dart and swim by snapping the valves together; only the large adductor muscle is used for food.

Pectinaria. Genus of small polychaete worms which build conical tubes out of sand grains; trumpet worms.

pectinate. Comb-shaped; with a series of small toothlike projections.

Pectinatella. Genus of fresh-water Bryozoa in the Class Phylactolaemata; zooecium a common gelatinous mass.

Pectinator. See CTENODACTYLIDAE.

pectines. Pair of comblike sensory structures on ventral surface of base of abdomen of scorpions.

pectineus. Muscle of the anterior portion of the thigh; flexes hip, adducts and rotates thigh outward.

Pectinibranchia. Suborder of gastropods in the Order Prosobranchiata; nervous system compact; ctenidium with a single row of filaments; one auricle, one nephridium; cowries, whelks, cone shells, and many fresh-water spp.

Pectinophelia. Genus of Polychaeta.

Pectinophora gossypiella. Pink bollworm; a small moth whose larva is an important worldwide pest of cotton bolls.

pectoral. Of or pertaining to the breast or chest.

pectoral fins. Pair of fins associated with the pectoral girdle in fishes.

pectoral girdle. Series of cartilages or bones for the attachment of the front limbs or fins in vertebrates; in man it consists of two clavicles and two scapulae.

pectoralis. Large ventral fan-shaped muscle in certain vertebrates; originates at the midline and inserts on the proximal end of the humerus.

pectoral sandpiper. Pisobia melanotos, a streaked, brown sp. of grassy mud flats and short-grass marshes; summers in northern N. A., winters in southern South America.

pectus. 1. Rarely, the breast of a bird. 2. Rarely, the ventral surface of an insect thorax or prothorax. 3. Chest region. 4. In arthropods, a fused pleuron and sternum region.

pedal. Of or pertaining to the foot.

pedal disc. Basal disc.

pedal ganglion. In many mollusks, a prominent ganglion imbedded in the foot.

Pedalia. The "jumping rotifer"; a genus of plankton spp. having six, stout, muscular, setose appendages.

Pedetes. See PEDETIDAE.

Pedetidae. Family of South African rodents; nocturnal, jumping, squirrel-like spp. with large hind legs and bushy tail; Pedetes common.

pedicel. 1. One or two basal abdominal segments of ants; greatly reduced and stalklike. 2. Second segment in an antenna which has a knee-like bend. 3. Pedicle; any stalk or stemlike part supporting an organ or other structure. 4. Small footlike organ.

pedicellariae. Minute, stalked or unstalked, pincer-like structures arranged around the base of spines and dermal branchiae in certain echinoderms (especially Asteroidea, Echinoidea, and Ophiuroidea); the two or three jaws snap shut when touched and serve to keep the surface free of debris and small organisms. Sing. pedicellaria.

Pedicellaster. Genus of typical sea stars in the Order Forcipulata.

Pedicellina. See PEDICELLINIDA.

Pedicellinida. Class in the Phylum Entoprocta; colonial marine spp. with basal stolons; Pedicellina and Barentsia common.

Pedicia. Common genus of crane flies in the Family Tipulidae; larvae in springs and brooks.

pedicle. 1. Pedicel; a small stalk or stalklike support. 2. Broad process connecting the lamina of a vertebra to the centrum. 3. Thin stalklike connection between cephalothorax and abdomen in certain spiders. 4. Stalk of a brachiopod.

Pediculati. Order of bony fishes; includes a group of bizarre deep-sea families; dorsal fin reduced to a few separate flexible rays, the first being on the head and commonly having a dilated tip; angler fishes, frogfishes, sea devils, etc.

Pediculatida. Pediculati.

Pediculoides. See TARSONEMIDAE.

pediculosis. Lousiness; having lice.

Pediculus. Genus of lice which feed on primate blood; P. humanus capitis is the human head louse, and P. humanus corporis is the body louse.

pedigerous. Pertaining to a body region, segment, or part which bears appendage(s).

Pedina. Genus of extinct crinoids in the Order Aulodonta.

Pedioecetes phasianellus. Sharp-tailed grouse; a tawny brown, blackish, and whitish grouse of interior western N. A., as far north as Alaska and Northwest Territories.

pedipalp. One of the second appendages

on the cephalothorax of Arachnoidea; variously modified for seizing, crushing, sensory functions, etc.

Pedipalpi. Order in the arachnoid Subclass Arachnida; whip scorpions; usually subdivided into two other orders: Amblypygi and Schizomida.

Pedipes. Genus of small West Coast marine snails.

pedogenesis. Paedogenesis.

pedon. Animal community of lake bottom, especially below the littoral zone. Adj. pedonic.

peduncle. 1. Fleshy attachment stalk of brachiopods and goose barnacles. 2. Any stalklike structure supporting another structure or organ. 3. Slender "waist" region of many arachnids and insects. 4. See CRURA CEREBRI.

peeler. Marine crab that has begun to molt.

peewee. Pewee.

Pegantha. Uncommon genus of coelenterates in the Suborder Narcomedusae.

Pegasus. See HYPOSTOMIDES.

Peisidice. Genus of small polychaetes found especially in mussel beds.

pekan. Fisher; see MARTES.

Peking man. See SINANTHROPUS.

pelage. Collectively, all the hairs or coat of a mammal.

Pelagia. Genus of scyphozoan jellyfish belonging to the Order Discomedusae; open ocean.

pelagial. Pelagic.

pelagic. Of or pertaining to the open waters of the sea and lakes, especially where the water is more than 20 m. deep.

Pelagodiscus. Genus of deep-sea brachiopods; valves circular.

Pelagonemertes. Genus of pelagic marine nemertines in the Order Hoplonemertea.

Pelagothuria. Genus of tropical pelagic sea cucumbers with a large swimming fringe around the mouth; rose and violet coloration.

Pelagothurida. Order of pelagic Holothuroidea; Pelagothuria the only genus.

Pelea. Rhebok; a genus of South African antelopes of rocky and grassy areas.

Pelecanidae. Temperate and tropical family of birds in the Order Pelecaniformes; pelicans; large, gregarious, heavy-bodied spp. with a capacious pouched bill used in scooping fish from the water; wingspread to 9 ft.; Pelecanus erythrorhynchos, the white pelican, occurs on inland waters of western N.A.; P. occidentalis, the brown pelican, occurs in the Caribbean area, Gulf Coast, Calif., and is occasional farther north.

Pelecaniformes. Order of large aquatic birds; pelicans, cormorants, gannets, etc.; short legs, large wings, and throat pouch; web includes all four toes; nostrils small and nonfunctional, or absent; usually with a gular pouch.

Pelecanoides. See PELECANOIDIDAE.

Pelecanoididae. Family of birds in the Order Procellariiformes; diving petrels; four spp. of small coastal diving birds; Pelecanoides.

Pelecanus. See PELECANIDAE.

Pelecosauria. Theromorpha.

Pelecypoda. Lamellibranchiata, a class in the Phylum Mollusca; includes all bivalve mollusks, such as clams, oysters, mussels, scallops, cockles, teredo, etc.; shell composed of two lateral calcium carbonate valves, usually more or less symmetrical, and held together by a dorsal hinge, dorsal ligament, and one or two adductor muscles; mantle lining valves of shell and forming posterior siphons which control flow of water in and out of mantle cavity; head, jaws, and radula absent; gills platelike; foot characteristically hatchet-shaped, often protruded between valves during locomotion; most marine spp. with trochophore and veliger larval stages; typical freshwater spp. with glochidium larvae; about 15,000 spp.

Pelia. Genus of spider crabs; legs not especially long.

pelican. See PELECANIDAE.

Pelidnota. Genus of North and South American beetles.

pellagra. Chronic disease brought about by a deficiency of niacin in the diet; marked by epidermal exfoliation, weakness, digestive disturbances, convulsions, spinal pain, and idiocy.

pellicle. Special thin, dead, translucent, secreted envelope which covers many protozoans; usually it is tightly in contact with the plasma membrane, and often cannot be distinguished unless the protozoan is treated with some chemical to effect a separation of the two.

Pelmatohydra. Genus of small, freshwater hydrozoan coelenterates; usually combined and included in the genus Hydra.

Pelmatozoa. One of the two subphyla in the Phylum Echinodermata; body boxlike, formed of calcareous plates, with slender arms and either an aboral stalk for attachment or an aboral cluster of leglike cirri; mouth and anus on upper surface.

Pelobates. Old World genus of spadefoot toads in the Family Pelobatidae.

Pelobatidae. Family of widely distributed toads in the Order Anomocoela; hind foot in many spp. with a broad, sharp-edged tubercle on the inner side; Pelobates and Scaphiopus common.

Pelocoris. Genus of creeping water bugs in the Family Naucoridae.

peloglea. Collectively, the particulate material found on aquatic objects and organisms; adsorbed from the surrounding water.

Pelomedusidae. Small family of turtles; neck and head completely retractile under edge of carapace; South America, Africa, and Madagascar; Podocnemis (tartaruga) typical.

Pelomyxa. Genus of large, sluggish amoeboid Protozoa; few to many nuclei; few, short, broad pseudopods; fresh water; sometimes called Chaos.

Pelonomus. Genus of dryopid beetles.

Peloscolex. Uncommon genus of freshwater Oligochaeta.

Peltaster. Genus of broadly pentagonal sea stars in the Order Phanerozonia.

Peltodytes. Genus of generally distributed crawling water beetles in the Family Haliplidae.

Peltogaster. Genus of barnacles in the Order Rhizocephala; highly modified tumor-like parasites of hermit crabs.

Peltoperla. Common genus of Plecoptera; nymphs broad, flat, and roachlike

peludo. See CHAETOPHRACTUS.

pelvic. Of or pertaining to the hip region or region of attachment of the hind appendages of vertebrates.

pelvic fins. Pair of fins associated with the pelvic girdle in fishes; in some spp. they may be located somewhat anteriorly.

pelvic girdle. Series of cartilages or bones for the attachment of hind fins or limbs in vertebrates.

pelvic vein. Vein draining the leg in certain vertebrates.

pelvis. 1. Basin-like cavity, or the bones forming the basin-like cavity produced by the arrangement of the bones of the pelvic girdle, especially in man and other mammals. 2. The central cavity of the kidney which receives urine before it is passed into the ureter in amniotes.

Pelycosauria. Order of Carboniferous and Permian reptiles in the Subclass Synapsida; most spp. characterized by enormously elongated dorsal vertebral spines; Dimetrodon and Edaphosaurus typical.

pen. Horny skeleton of a squid, imbedded in the mantle; often feather-shaped.

Penaeidea. Suborder of crustaceans in the Section Natantia; shrimps and prawns with first three pairs of legs chelate; pleura of first abdominal segment overlapping those of second; Penaeus typical.

Penaeus. Genus of swimming shrimps and prawns; first three pairs of legs chelate; abdomen compressed, body wall translucent; P. setiferus is the most important market sp. from Va. to the Gulf of Mexico; P. brasiliensis is an East Coast sp.; most spp., however, are tropical, subtropical, and deep-sea.

Penares. Genus of marine sponges.

Penella. Genus of parasitic marine copepods in the Suborder Caligoida; adults threadlike, anterior portion buried in the tissues of fishes.

Peneroplis. Genus of foraminiferans; most spp. fossil only.

penetrance. Relative percentage of individuals in which an expected phenotype actually materializes; a statistical concept of the regularity with which a gene produces its effect when present in the requisite homozygous or heterozygous condition; percentage of penetrance may be altered by changing environmental conditions during embryological development.

penetrant. See NEMATOCYSTS.

Peneus. Same as Penaeus.

penguin. See SPHENISCIFORMES.

penicillate. Having a tufted tip.

penicillin. Antibiotic secreted by the mold Penicillium notatum, now mass-produced as a white powder; especially useful in treating streptococcus and staphylococcus infections, gangrene, meningitis, gonorrhea, and syphilis.

Penilia. Genus of marine Cladocera.

penis. Male copulatory organ through which sperm are deposited in the female reproductive tract; in lower invertebrates the corresponding structure is usually called the cirrus.

penna. Large contour feather of a bird. Pl. pennae.

Pennaria. Common genus of marine hydrozoan coelenterates.

pennate. 1. Feathered or winged. 2. Pinnate. 3. Wing-shaped.

Pennatula. Genus of alcyonarian anthozoans in the Order Pennatulacea; the sea pens; colony with the shape of an upright fleshy feather, to 16 in. long.

Pennatulacea. Order of alcyonarian anthozoans; sea pens, sea feathers; colony an upright fleshy stalk with many polyps along the sides; stalk sometimes with lateral branches; skeleton composed of imbedded calcareous spicules; examples are Pennatula and Renilla.

penniculus. Special closely-arranged rows of basal granules of cilia in the cytopharynx of certain ciliates; a part of the neuromotor system.

Pennsylvanian epoch. Later of the two epochs of the Carboniferous period.

pen shell. See PINNA.

Pentacrinus. Genus of fossil crinoids.

pentadactyl. Of or pertaining to a limb having five digits.

Pentagenia. Common genus of Ephemeroptera; nymphs fossorial.

Pentalagus. Genus of Asiatic rock-hares.

pentamerous. Arranged in fives or multiples of five.

Pentaneura. Genus of tendipedid midges.

Pentapedilum. Genus of tendipedid midges.

Pentastomatida. Linguatulida.

Pentatomidae. Family of hemipteran insects which includes the stink bugs; body oval or shield-shaped, scutellum large; predaceous and herbaceous; many spp. foul-smelling; Murgantia and Perillus are examples.

Penthestes. Genus of passerine birds in the Family Paridae; chickadees; P. atricapillus is the black-capped chickadee of N.A., a familiar small sp. with gray, white, and black coloration.

pentose. Any five-carbon monosaccharide; present in nucleic acids and many plant polysaccharides; $C_5H_{10}O_5$.

Pentremites. See BLASTOIDEA.

penultimate. Second last, such as the second last segment of an antenna.

pepper conch. Fasciolaria tulipa, a small conch to 6 in. long; shallows from N.C. to Tex.

pepper coral. Common coral (Millepora) of the Fla coast; unusual stinging powers.

pepsin. Enzyme (endopeptidase) which hydrolyzes proteins into proteoses and peptones in acid solution; elaborated by the wall of the vertebrate stomach, but similar enzymes occur in many invertebrates.

pepsinogen. Zymogen secreted by the gastric mucosa and converted to active pepsin in the presence of gastric hydrochloric acid.

Pepsis. Tarantula hawks; a genus of large wasps which provision their nests with tarantulas and other spiders; common in southwestern states.

peptidase. Enzyme which digests peptones or peptides by splitting off smaller fragments or amino acids.

peptide. Intermediate protein digestion compound consisting of two or more linked amino acids.

peptide linkage. The -CO-NH- linkage in a dipeptide (or larger peptide), e.g. NH_2-R-CO-NH-R-COOH.

peptone. Complex intermediate product (polypeptide) of protein digestion; more complicated than a peptide.

Peracarida. One of the four subdivisions in the crustacean Division Eumalacostraca; carapace, if present, leaving four or more posterior thoracic segments exposed; females with thoracic brood pouch.

Peradorcas. Genus of Australian rock-wallabies in the Family Macropodidae.

Perameles. See PERAMELIDAE.

Peramelidae. Family of marsupials; includes the bandicoots (Perameles and related genera); insectivorous and omnivorous burrowers; long ears, long tail, pointed snout; Australian region.

Peranema. Common genus of colorless flagellates occurring especially in infusions and stagnant water; with a single anterior flagellum; saprozoic and holozoic.

Perca. Fresh-water genus of small game fishes in the Family Percidae; P. fluviatilis is European; P. flavescens is the widely distributed American perch, ring perch, or zebra perch.

perch. 1. See PERCA. 2. Any of a wide variety of marine fishes with spiny-rayed fins.

Percidae. Family of carnivorous percomorph fishes which includes the fresh-water perches, pikeperches, saugers, and darters; two distinct dorsal fins; Perca, Stizostedion, Etheostoma, and Hadropterus examples.

Perciformes. Very large order of fishes; roughly equivalent to the Percomorphi as used in this volume.

Percina. Large genus of small, brightly colored stream fishes in the Family

Percidae; logperches and a wide variety of darters; common in N. A.

Percoidei. Suborder which includes most of the families of percomorph fishes.

Percomorphi. Very large order of bony fishes containing about 80 families; dorsal and anal fins with both spiny and soft rays; world wide in fresh and salt waters.

Percomorphida. Percomorphi.

Percopsidae. Small family of fresh-water fishes; to 6 in. long, with both spiny and adipose fins; the troutperch (Percopsis omiscomaycus) is widely distributed over Canada and eastern half of U. S.; the sand roller (Columbia transmontana) is in the Columbia River basin.

Percopsiformes. Salmopercae.

Percopsis. See PERCOPSIDAE.

Perdix perdix. European partridge, or Hungarian partridge, which has been successfully introduced into western U. S. and Canadian prairies; small grayish sp. with chestnut-colored markings; in the Family Phasianidae.

Père David's deer. See ELAPHURUS.

pereiopods. Paired appendages on most of the thoracic segments of Malacostraca; usually modified for seizing and handling food and for locomotion.

perfusion. 1. In the broad biological sense, the pouring of a liquid over or through something. 2. Artificial passage of a fluid through blood vessels of an organ or whole animal.

perianal. Near or surrounding the anus.

periblast. Darker marginal area of an early blastodisc, as in a chick embryo.

peribranchial chamber. 1. Large chamber surrounding the pharynx in tunicates and cephalochordates; not a part of the coelom; sea water passes from the mouth into the pharynx, through pharyngeal gill slits, into the peribranchial chamber, and out of the body by way of the atriopore or cloacal aperture. 2. Atrium or atrial cavity.

pericardial cavity. Pericardial sinus; special cavity containing the heart; in vertebrates it is a part of the coelomic cavity; in arthropods and mollusks it is a portion of the hemocoel.

pericardial sinus. Pericardial cavity.

pericardium. Membranous tissue sac which covers the vertebrate heart; in certain invertebrates it is also present but is separated from the heart tissue by a blood-filled space.

pericentric inversion. Chromosome inversion across the centrosome.

perichondral bone. That type of bone laid down on the surface of cartilage, especially in long bones.

perichondrium. Thin membrane covering cartilage; forms new cartilage on the outer surface of the cartilage mass.

perichordal sheath. Early mesodermal sheath surrounding the notochord in a vertebrate embryo; later becomes cartilage and then bone.

Pericoma. Common genus of moth flies in the Family Psychodidae; larvae aquatic.

periderm. 1. Germinative layer of vertebrate skin. 2. Cuticle or cornified outer layer of skin. 3. Transient outermost layer of embryonic epidermis, especially in vertebrates. 4. Hydroid perisarc.

Peridinium. Fresh-water and marine genus of dinoflagellates; subspherical to ovoid.

peridontal membrane. Alveolar periosteum; fibrous membrane which attaches the tooth cementum to surrounding structures.

Perigonimus. Genus of coelenterates in the Suborder Anthomedusae; hydroids found on spider crabs.

perihemal system. See HEMAL SYSTEM.

perikaryon. Cell body, as distinct from the nucleus.

Perillus bioculatus. Insect in the hemipteran Family Pentatomidae; feeds on caterpillars and beetles.

perilymph. Lymph material which fills the narrow space between the membranous labyrinth of the vertebrate inner ear and its bony labyrinth investment.

perimysium. Sheets of connective tissue between groups of muscle fibers.

Perinereis. Genus of Polychaeta.

perineum. 1. Especially in human beings, the region at the lower end of the trunk between the thighs; in the male it includes the anus and the area up to the base of the scrotum; in the female it includes the anus and the area up to the mons veneris. 2. In placental mammals in general, the body surface between the anus and the urinogenital openings.

perineurium. Layer of dense fibrous connective tissue surrounding a fascicle of neurons in a nerve; sometimes extends septa into a fascicle.

period. 1. One of the major time subdivisions of each of the Cenozoic,

Mesozoic, and Paleozoic eras. 2. Menses.

periodical cicada. See CICADIDAE.

Periodicticus. Genus of pottos (Lorisoidea); deliberate spp. with gingery or yellowish-brown dense fur; eyes large; arboreal in tropical Africa.

Periophthalmus. Genus of mudskippers; a group of fishes which spend much of their time climbing mangroves and skipping about on mudflats in search of insects.

periosteal bone. Bone tissue laid down on the surface of a bone after the shaft has been well ossified.

periosteum. Tough adherent fibrous membrane surrounding bone and serving for the attachment of muscles and tendons; builds bone lamellae at the surface of the bone; absent at the cartilaginous extremities.

periostracum. Outermost chitinoid layer of most mollusk shells; protects the underlying calcareous prismatic layer from chemical and mechanical erosion.

periotic. Compact otic capsule bone of mammals.

Peripatopsis. Common genus in the Phylum Onychophora.

Peripatus. See ONYCHOPHORA.

peripharyngeal band. A pair of ciliated ridges which run around the inner surface of a tunicate pharynx; just inside the mouth.

peripheral nervous system. Collectively, the paired cranial and spinal nerves, especially of vertebrates.

peripheral vision. Relative area and sharpness of the image obtained around the acute vision area of the center of the retina.

Periphylla. Genus of scyphozoan jellyfish in the Order Coronatae; North Atlantic.

periphyton. 1. Aufwuchs; entire assemblage of organisms (mostly microscopic) on submerged objects in aquatic environments; such organisms do not penetrate into the substrate and may or may not be sessile. 2. Often in American usage, the assemblage of organisms (mostly microscopic) living on the submerged surfaces of aquatic plants. 3. Entire assemblage of sessile organisms (mostly microscopic) on submerged substrates in aquatic environments. Cf. LASION and EPIPHYTON.

Periplaneta americana. American cockroach; to 2 in. long and light reddish brown; common in southern states.

Periploma. Spoon shells; common genus of small marine bivalves having a spoon-shaped tooth in the hinge of each valve.

peripneustic. Pertaining to an insect having functional spiracles along the sides of the body.

periproct. Special plate or heavy membrane bearing the anus in Echinoidea; typically surrounded by the madreporite and four genital plates.

Peripsocus. Common genus of Corrodentia.

perisarc. Thin translucent to brownish exoskeleton surrounding the stolons and upright branched portions of certain hydrozoan coelenterate colonies; encloses the living coenosarc.

Perisoreus canadensis. Common Canada jay of northern states and Rocky Mountain region; a large gray bird of forests.

Perispira. Genus of fresh-water holotrich Ciliata; oral ridge spirals to the posterior end.

Perissodactyla. Order of mammals which includes the odd-toed hoofed spp. such as the horse, ass, zebra, tapir, and rhinoceros; stomach simple; great majority of spp. in this order are extinct; native to Asia, Europe, tropical America, and Africa.

peristalsis. Wavelike and ringlike muscular contractions in the digestive tract of vertebrates and higher invertebrates; such contractions move progressively along portions of the digestive tract and serve to mix and move the contents of the digestive tract. Adj. peristaltic.

peristome. 1. Groove or depression of various size and shape which leads to the cytostome in many ciliate Protozoa. 2. Heavy membranous area around the mouth in the Class Echinoidea. 3. Area around the mouth of various invertebrates. 4. First complete segment of most annelids; surrounds mouth; peristomium. Adj. peristomial.

peristomial gills. Clusters of modified tube feet in certain Echinoidea; located around the mouth region and function in respiration.

peristomial tentacles. Sensory tentacles located on the peristomium of typical Polychaeta.

peristomium. First whole segment in most annelids, especially Archiannelida, Polychaeta, and Oligochaeta; bears the mouth.

Perithemis domitia. Dragonfly of eastern and southern U.S.

peritoneal cavity. That portion of the coelomic cavity of mammals posterior to the diaphragm.

peritoneum. Thin membrane lining a coelom and covering all visceral organs located in a coelom.

peritonitis. Bacterial inflammation of the peritoneum; caused by perforation of intestinal tract, typhoid fever, and various primary infections.

Peritricha. Order of ciliate Protozoa having an enlarged disklike ciliated anterior region; the main body ciliation is limited; body often vase-shaped or conical and stalked; Epistylis and Vorticella are familiar examples.

peritroch. 1. Larva having a peripheral ciliary band. 2. Peripheral ciliary band.

peritrophic membrane. Non-cellular lining of the midgut of many insects, crustaceans, and Onychophora; often clearly separated from the underlying epithelium.

perivisceral cavity. Vertebrate coelomic cavity.

perivitelline space. Space between a fertilized egg and its enveloping fertilization membrane.

periwinkle. See LITTORINA.

Perla. European genus of stoneflies.

Perlesta placida. Small common stonefly; widely distributed.

Perlidae. Large family of stoneflies; common in N.A.

Perlinella drymo. Stonefly widely distributed east of Rockies.

Perlomyia. Genus of Plecoptera.

permafrost. Permanently frozen subsurface soil beginning at a depth of several inches to several feet at high altitudes and latitudes.

permanent immunity. Any immunity which is effective for essentially the rest of the life of an organism; either a natural or an acquired immunity mechanism.

permanent teeth. Second of two sets of teeth in most mammals, the first being the milk teeth.

permeability. Relative ability of a membrane, cell, plasma membrane, or blood vessel to transmit fluids or dissolved materials.

permeant. In a terrestrial community above ground, any highly motile animal.

Permian period. Last of the six periods of the Paleozoic era; from about 215 million to 190 million years ago; age of glaciation and mountain formation; fauna characterized by disappearance of early amphibians, placoderms, trilobites, and eurypterids, and the appearance of primitive reptiles and modern insect groups.

pernicious anemia. Severe type of anemia produced by a progressive decrease in the number of erythrocytes; marked by degeneration of the central nervous system and digestive tract.

Perognathus. Genus of rodents in the Family Heteromyidae; pocket mice; small nocturnal mice with long tails and fur-lined external cheek pockets; feed chiefly on seeds; unforested areas of southwestern and midwestern N.A.; numerous spp. and subspecies.

Peromyscus. Large genus of nocturnal rodents in the Family Cricetidae; white-footed mice or deermice; common and widely distributed in N.A.; feed mostly on seeds, fruits, and arthropods.

peroneus. Muscle or group of muscles used chiefly to flex the foot.

peronia. Thick epidermal tracts extending from the base of the tentacles to the bell margin of the exumbrella in certain marine hydrozoan medusae. Sing. peronium.

Perophora. Genus of sessile tunicates; colonies of very small individuals loosely connected by a branching stolon.

perosis. Bone deformity in chicks caused by insufficient choline in the diet.

Perotripus. Genus of caprellid Amphipoda.

peroxidase. Enzyme which catalyzes the oxidation of an organic compound, using hydrogen peroxide or some organic peroxide as the oxidizing agent; an iron-porphyrin-protein.

perradial canal. One of several types of gastrovascular canals extending radially and joining the circumferential canal in certain jellyfish.

Perrinites. Genus of extinct Cephalopoda.

Persephone. Genus of small, globular, marine crabs (Brachyura).

persistent type. Genus or sp. that has persisted essentially unchanged over long geological periods of time and has had a negligible mutation rate; e.g. Limulus and Lingula.

perspiration. See SWEAT GLANDS.

pes. Foot of a tetrapod.

pessimum. Collectively, those theoretically least favorable environmental conditions which an organism can endure.

pesticide. Any chemical preparation used to kill a plant or animal pest.

Petaloconchus. Genus of small brownish marine snails.

petaloid area. Petal-shaped ambulacral area of certain echinoids.

Petalomonas. Large genus of colorless monoflagellate Protozoa; stagnant fresh waters.

Petaurista. Taguans; a genus of large nocturnal flying squirrels of the Indian Region.

Petaurus. Genus of marsupials in the Family Phalangeridae; flying phalangers; arboreal spp. having a general similarity to flying squirrels; Australian region.

Peterson dredge. Heavy clamshell dredge used in taking quantitative samples of bottom organisms from lakes or oceans.

petiole. Any stalk or peduncle, such as the slender abdominal attachment of many spiders, ants, wasps, and Diptera.

Petraster. Genus of extinct sea stars.

petrel. See OCEANODROMA.

Petricola. Genus of marine bivalves with greatly elongated shells; bore into clay or soft rocks.

petrifaction. See PETRIFY.

petrify. To convert to stone; when dead plant and animal materials are petrified, their tissues are replaced, molecule by molecule, by mineral matter, so that the final product is an accurate stony reproduction of the original organism; wood has been petrified most abundantly. Noun petrifaction.

Petrobia. Genus of small mites in the Family Tetranychidae; often harmful to wheat.

Petrochelidon. See HIRUNDINIDAE.

Petrogale. Genus of Australian marsupials in the Family Macropodidae; rock wallabies; generally similar to kangaroos, but smaller, often the size of a large rabbit.

Petrolisthes. Genus of marine porcelain crabs (Suborder Anomura).

Petromyidae. Family of South African rock rats, related to the porcupines; feed on flowers in dry, rocky areas; body 6 in. long, covered with long coarse hair; Petromys typical.

Petromys. See PETROMYIDAE.

Petromyzon marinus. Common sea lamprey of north Atlantic; has spread in Great Lakes where it has greatly decimated the lake trout population; to 27 in. long; attaches to surface of other fish with the suctorial mouth and rasps a hole in the tissues through which the host's blood is sucked, soon resulting in death of the host; breeding and a long larval stage occur in streams flowing into the ocean and Great Lakes; toxic chemicals are being used in an attempt to eradicate the larvae.

Petromyzontes. Taxon which includes the lampreys; same as Petromyzontia.

Petromyzontia. Order of cyclostomes which includes lampreys; oral suctorial funnel with many horny teeth; nasal sac dorsal; blind; seven pairs of gill chambers; long larval period; widely distributed along seacoasts and in lakes and streams; about 25 spp.; Petromyzon, Entosphenus, and Ichthyomyzon common.

Petromyzontidae. Single family in the Order Petromyzontia.

petrophilous. Living among or attached to rocks, in preference to other types of substrates.

petrosal ganglion. Ganglion on the glossopharyngeal nerve.

Petrunkevitch, Alexander. Russian-American zoologist (1875-); professor at Yale; made important contributions to blood histology, invertebrate zoology, and especially in the field of spider biology.

Petunculus pennaceus. Feathered bittersweet shell; an ark shell of N. C. to the West Indies.

Peucedramus taeniatus. Olive warbler; mountains of Ariz. and N. M. south to Nicaragua.

pewee. Common U. S. sparrow-sized flycatcher in the Family Tyrannidae; Contopus typical.

Peyer, J. K. Swiss anatomist (1653-1712).

Peyer's patches. Masses of lymphoid tissue in the mucosa of the small intestine; usually 1 to 4 cm. long.

PGA. Pteroylglutamic acid; folic acid.

pH. See HYDROGEN-ION CONCENTRATION.

phacella. Gastric filament; a filament in the gastrovascular cavity and attached to the gastrodermis in certain jellyfishes; arranged in rows and bears nematocysts.

Phacellophora. Genus of scyphozoan jellyfishes.

Phacochoerus. Genus of mammals in the Family Suidae; includes the African wart hog, a wild sp. with greatly enlarged curved tusks in the upper jaw and three large wartlike prominences on each side of the face.

Phacodinium. Genus of fresh-water heterotrich Ciliata.

Phacoides. See LUCINA SHELLS.

Phacops. Genus of trilobites.

Phacotus. Genus of fresh-water biflagellate Protozoa; cell covered with a bivalve shell; chromatophores present.

Phacus. Genus of flattened, asymmetrical, fresh-water green flagellates; with a single flagellum, stigma, and longitudinal or oblique striae.

Phaenacora. Genus of rhabdocoel turbellarians.

Phaenopsectra. Genus of tendipedid midges.

phaeodium. Metabolic waste material formed by certain radiolarians; occurs in small brownish masses.

Phaëthon. Genus of birds in the Order Pelecaniformes; tropic birds; large sea birds with general whitish coloration and a few black markings; two exceptionally long tail feathers.

Phaëthontidae. Family of birds in the Order Pelecaniformes; includes the tropic birds; see PHAËTHON.

Phagocata. Genus of fresh-water Turbellaria; eyes present or absent; one or numerous pharynges.

phagocyte. Any cell which engulfs foreign particles; found in blood and other body fluids as well as in the tissues of the lymphatics and blood vessels; various invertebrate tissue cells are phagocytes. Adj. phagocytic.

phagocytosis. Process of ingesting a foreign particle by a phagocyte.

Phainopepla. See PTILOGONATIDAE.

Phakiella. Genus of marine sponges (Subclass Monaxonida); cup-shaped, to 6 in. across.

Phalacrocoracidae. Large family of gregarious birds in the Order Pelecaniformes; cormorants; many spp., with dark coloration and hooked beaks; world-wide distribution, especially in coastal areas, although some spp. are found on inland waters; fish-eating, excellent divers; Phalacrocorax contains nearly all spp.; in the Far East some spp. are used for catching fish commercially by keeping them on long tethers and placing a ring around the neck so the fish cannot be swallowed.

Phalacrocorax. See PHALACROCORACIDAE.

Phalaenidae. Noctuidae.

Phalaenoptilus nuttali. Common brownish-gray poorwill of the arid west; active especially before dawn and at dusk; feet and legs small and weak; wide mouth margined with long bristlelike feathers to aid in catching insects on the wing; no nest, eggs laid on the ground.

phalange. 1. Phalanx; one of the bones in a finger or toe. 2. Segment of an insect tarsus. Pl. phalanges.

phalanger. Any of a large group of Australian marsupials; mostly arboreal and herbivorous spp. with foxlike ears and long bushy tail.

Phalanger. Genus of marsupials in the Family Phalangeridae; the cuscuses; sluggish arboreal spp. with long prehensile tails.

Phalangeridae. Family of marsupials which includes the cuscuses, flying phalangers, and koalas; arboreal herbivores; Australian region; Phalanger, Petaurus, and Phascolarctus typical.

Phalangida. Order of arachnids which includes harvestmen or daddy longlegs; body ovoid; unsegmented cephalothorax broadly joined to nine-segmented abdomen; two eyes, often on tubercles; pedipalps similar to legs but shorter; legs very long and slender; respiration by tracheae; in buildings, fields, and woods; Phalangidium and Liobunum common.

Phalangidium. Very common genus of daddy longlegs.

Phalangium. Genus of daddy longlegs (Phalangida).

Phalansterium. Genus of small monoflagellate Protozoa; each cell with a small collar and imbedded in a gelatinous mass.

phalanx. Phalange.

phalarope. See PHALAROPODIDAE.

Phalaropodidae. Small family of sandpiper-like birds in the Order Charadriiformes; phalaropes; swimmers and waders; maritime and fresh-water spp.; several American genera, especially Steganopus, Phalaropus, and Lobipes.

Phalaropus fulicarius. Red phalarope; breeds in circumarctic areas.

phallus. 1. Penis. 2. Embryonic structure which eventually becomes the penis or clitoris. 3. Male copulatory organ in certain invertebrates.

Phaner. Genus of catlike lemurs of Madagascar; dwarf lemurs.

Phanerozonia. One of the five orders of the echinoderm Class Asteroidea; dermal branchiae all aboral; two rows of tube feet in each arm; marginal plates in two rows; Astropecten common.

Phanocerus clavicornis. Beetle in the Family Elmidae; usually on the shores of streams.

phantom crane flies. See PTYCHOPTERIDAE.

phantom midges. Small group of midges in the Family Culicidae; larvae predatory and transparent, found in lakes, ponds, bogs; Chaoborus common.

pharmacognosy. Study of the characteristics of crude drugs.

pharmacopeia. Encyclopedic treatise on drugs, their preparation, dosage, and medicinal action.

Pharnacia serratipes. Giant tropical insect in the Order Orthoptera; related to the walking sticks; to 10 in. long.

Pharomacrus moccino. Quetzal of Central American forests (bird Order Trogoniformes); a graceful sp. with brilliant plumage; national emblem of Guatemala.

pharyngeal. Pertaining to the pharynx.

pharyngeal cleft. Gill slit.

pharyngeal slit. Gill slit.

pharyngeal teeth. Lateral teeth present in the wall of the pharynx in many fishes.

pharyngobranchials. Upper elements of fish branchial arches with teeth.

pharynx. 1. In invertebrates, an anterior muscular portion of the digestive tract. 2. In vertebrates, an anterior cavity common to both the digestive and respiratory tracts. 3. In prochordates, that portion of the digestive tract from which the gill slits open. Pl. pharynges.

phascogale. Any of a wide variety of carnivorous Australian marsupials; size ranges from that of a small mouse to a rat.

Phascogale. Genus of omnivorous brush-tailed marsupials in the Family Dasyuridae; about as large as a rat; Australian region.

Phascolarctus. Genus of arboreal Australian marsupials in the Family Phalangeridae; koalas or Australian "teddy bears"; to 2 ft. long; fur gray, ears large, claws large; feed exclusively on eucalyptus leaves.

Phascolion. Genus of Sipunculoidea having an especially long introvert.

Phascolomidae. Family of Australian marsupials which includes the wombats; herbivorous, tail-less, heavy-bodied, burrowing spp. suggesting a cross between a small bear and a capybara; three spp. of Phascolomys.

Phascolomys. See PHASCOLOMIDAE.

Phascolosoma. Common genus of Sipunculoidea.

phase microscope. Special compound microscope which alters the phase relationships of light passing through and around an object being viewed; allows more critical definition.

phase reversal. Transformation of colloids from sol to gel stage and vice versa.

Phasianella. Genus of small, smooth marine snails.

Phasianidae. Large family of birds in the Order Galliformes; includes the quails, partridges, pheasants, peacock, chicken, etc.; native to Asia but widely introduced elsewhere; typical genera are Phasianus, Pavo, and Gallus.

Phasianus colchicus torquatus. Common ring-necked pheasant in the Family Phasianidae; native to Asia but now widely introduced and present over much of U.S. in farm, grass, meadow, and ranch areas; large chicken-like or gamecock-like bird with long tail feathers; male highly colored, with a white neck ring; female brown and grouselike; a most successful farm game sp.; with several local names.

Phasmatidae. Family in the insect Order Phasmida; body cylindrical and extremely linear; widely distributed but mostly tropical

phasmid. 1. One of a pair of minute precaudal sensory structures or glands in certain nematodes; consists of a unicellular gland, short duct, and a pore which opens on a papilla. 2. Insect belonging to the Order Phasmida.

Phasmida. Order of large, sluggish insects having a striking resemblance to small twigs or dried leaves; stick insects, phasmids, walkingsticks, leaf insects; simple metamorphosis and biting mouth parts; herbivorous, mostly tropical; Diapheromera common in U.S.

Phasmidea. One of the two classes of nematodes; phasmids present; amphids porelike; excretory system composed of two long lateral canals.

pheasant. Any of numerous, long-tailed, brightly colored game birds in the Family Phasianidae; originally native to Asia; see PHASIANUS.

Pheidole. Large genus of ants found mostly in the tropics; some are seed-storing spp.

Phenacobius. Genus of suckermouth minnows; Lake Erie, Mississippi, Tennessee, and Alabama river systems.

Phenacomys. Genus of small, mouse-like, herbivorous rodents in the Family Cricetidae; lemming mice; cold and mountainous areas of western and northern N.A.; uncommon.

phenocopy. Morphological or physiological change which resembles a mutation but which is environmentally-produced and cannot be transmitted to the next generation. Pl. phenocopies.

phenology. Appearance of life-history events of an organism as correlated with season and weather.

phenomenon of edges. See EDGE EFFECT.

phenotype. In genetics, the class to which an individual is assigned according to its external appearance. Adj. phenotypic. Cf. GENOTYPE.

phenylalanine. Amino acid; $C_9H_{11}O_2N$.

phenylketonuria. Metabolic defect resulting in feeble-mindedness; Mendelian recessive correctable in early years by special diet.

phenylpyruvic acid. Intermediate metabolic breakdown product of phenylalanine; excreted in the urine of individuals who have an inadequate enzyme system for complete phenylalanine breakdown; genetically determined by a double recessive.

phenylpyruvic amentia. Form of idiocy caused by a metabolic inability to produce an enzyme necessary for the oxidation of phenylalanine.

phenylthiocarbamide. PTC; an organic compound which tastes bitter to 70% of human beings but is tasteless to most of the remainder; genetically determined.

pheromone. Substance secreted by an animal that influences the behavior of other individuals in the same sp.; e.g. odor trails of ants and sex attractants released by female moths.

Pheronema. Genus of hexactinellid sponges.

Pheucticus. Small genus of birds in the Family Thraupidae; includes two U.S. grosbeaks.

Phialidium. Genus of marine hydrozoan coelenterates with well developed medusae.

Phidippus. Common genus of hairy spiders.

Phidolopora. Genus of marine Bryozoa.

Philacte canagica. Emperor goose; a white, rusty, gray, and black sp. of Siberia and Alaska; occasional along U.S. Pacific Coast.

Philander. Genus which includes several arboreal woolly opossums of Central and South America; philanders; eyes large and bulbous.

Philantomba. See DUIKER.

Philaster. Genus of marine and freshwater holotrich Ciliata; cylindrical, with a caudal seta.

Philepittidae. Small family of passerine birds of Madagascar; asities; plump arboreal herbivorous spp.; false sunbirds are small, with a long curved bill used in sipping nectar and picking up small insects.

Philhydrus. Enochrus.

Philine. Genus of widely distributed sluglike marine gastropods; shell rudimentary and entirely enclosed.

Phillipsia. Common genus of trilobites.

Philobdella. Common genus of southern U.S. leeches.

Philodina. Common genus of creeping, leechlike rotifers.

Philodinavus. Genus of bdelloid rotifers.

Philodromus. Common genus of crab spiders; usually on plants; often protectively colored.

Philohela minor. Common woodcock in the Family Scolopacidae; a chunky game sp., to 11 in. long, found especially in swampy habitats of eastern N.A.; brown color above; eyes large; bill very long.

Philomycus. Genus of land slugs.

Philopterus. Genus of bird lice in the Order Mallophaga.

Philorus. Genus of net-winged midges in the Family Blepharoceridae.

Philosamia. Genus of large moths native to Asia but now widely introduced; P. cynthia (ailanthus silk moth) occurs in a few eastern states but is used for silk production only in Asia; feeds chiefly on leaves of the ailanthus tree.

Philoscia. Genus of marine Isopoda.

Phiomia. Genus of ancestral Oligocene elephants.

phlebitis. Inflammation of a vein.

Phlebotomus. Genus of small blood-sucking flies of widespread distribution; transmitters of Oriental sore, kala azar, American leishmaniasis, three-day fever, and Oroya fever.

Phlegmacera. Genus of daddy longlegs (Phalangida).

Phloeodictyon. Genus of marine sponges (Subclass Monaxonida); massive, with finger-like processes.

Phloeomys. Genus of cloud-rats; large, long-haired spp. of southeastern Asia and East Indies.

phlorhizin. Glycoside which blocks the kidney tubular reabsorption of glucose.

Phoca. Genus of seals in the Family Phocidae; P. vitulina is the small gray to brown harbor seal or leopard seal of the Atlantic Coast, P. richardii is the Pacific Coast harbor seal, and P. groenlandica is the larger circumpolar harp seal or Greenland seal, hunted for its oil.

Phocaena. See PORPOISE.

Phocaenidae. Family which includes the porpoises (mammals), especially Phocaena and Neomeris.

Phocidae. Family of mammals which includes the earless seals, hair seals, or harbor seals, as well as the sea elephant; hind legs incapable of rotation forward, larger than forelimbs; external ear absent; common genera are Phoca, Erignathus, Halichoerus, Cystophora, and Mirounga.

phoebe. Any of several American flycatchers in the Family Tyrannidae; usually in wooded areas; head with a slight crest; upper parts gray to brownish; Sayornis common.

Phoenicopteridae. Family of large, gregarious wading birds in the Order Ciconiiformes; a few tropical spp., of which only one, Phoenicopterus ruber (flamingo), occurs in coastal areas of extreme southeastern U.S.; neck and legs very long, beak bent sharply downward, lower mandible boxlike and sievelike, being closed by the lidlike upper mandible; beak thrust into the mud in an inverted position, so that the aquatic animals are sieved out; nest a conical mass of marl or mud.

Phoenicopterus. See PHOENICOPTERIDAE.

Phoenicurus. See REDSTART.

Pholadidea. Genus of marine bivalves; shell ovate-oblong; anterior gape closed by a thin globose plate.

Pholas. Genus of marine clams which burrow into wood and rock; wing shells, angel shells, angel wings; anterior end of valves armed with abrading teeth or sharp edges; widely distributed.

Pholcidae. Family of long-legged, pale spiders which usually build webs in basements and other dark locations; Pholcus phalangioides especially common.

Pholcus. See PHOLCIDAE.

Pholidae. See PHOLIS.

pholidosis. Arrangement of scales on a body or part.

Pholidota. Small order of mammals which includes the pangolins or scaly anteaters of Africa and southeastern Asia; covered with overlapping, pointed, horny plates with a few hairs between; teeth absent, tongue long and wormlike; feed on ants and termites; seven spp.; giant pangolin, Manis giantea, to 6 ft. long.

Pholis. Genus of long, slender, eel-like or blenny-like fishes with the dorsal fin extending along the whole trunk; gunnels, rock eels, rockfishes; along both coasts, usually near the low tide line in pools and in puddles among rocks; typical of the Family Pholidae.

Pholoë. Genus of small, delicate Polychaeta.

Phonorhyncus. Genus of marine rhabdocoel Turbellaria.

phoresy. Type of commensalism in which the host-guest relationship is limited to the mechanical transport of the guest by its larger host; e.g. certain small flies are transported to breeding sites by dung beetles, and mites are often transported on the bodies of a wide variety of insects.

Phoridae. Humpbacked flies; minute to small spp. with humped thorax; adults visit flowers, foliage, and buildings; larvae feed on dead plant and animal matter; widely distributed.

Phoriospongia. Genus of encrusting marine sponges (Subclass Monaxonida).

Phormia regina. Blowfly in the Family Calliphoridae; larvae feed on necrotic tissue in open ulcers and wounds of stock animals; rarely used in stubborn cases of human osteomyelitis and other open sores where the feeding and excreta of the larvae stimulate healing.

Phormosoma. West Indian genus of deep-sea echinoids (Order Lepidocentroida).

Phoronidea. Phylum consisting of sedentary elongated wormlike animals inhabiting vertical chitinous tubes in shallows of the sea, with only an anterior double spiral lophophore of ciliated tentacles projecting above the surface into the water; digestive tract U-shaped; closed circulatory system and extensive coelom; two nephridia (protonephridia); hermaphroditic; larva an actinotrocha; to 14 in. long; about 15 spp.; Phoronis and Phoronopsis familiar.

Phoronis. Common genus of Phoronidea; abundant along the Atlantic Coast; to 5 in. long.

Phoronopsis. Genus in the Phylum Phoronidea; P. californica is especially common on the West Coast; to 14 in. long.

phorozooids. Highly specialized nurse individuals which bear the gonozooids in certain pelagic tunicates.

phosphagen. Creatine phosphate or arginine phosphate; the phosphate splits off readily and is important as a source of phosphate and energy for resynthesis of ATP in muscle contraction.

phosphatase. Enzyme which catalyzes the synthesis of an organic phosphate from some organic compound and phosphate, or the hydrolysis of the organic phosphate to its constituent parts.

phosphatide. Phospholipid.

phosphoarginine. Intermediate compound of invertebrate muscle metabolism.

phosphocreatine. Intermediate compound of vertebrate muscle metabolism; broken down into creatine and an inorganic phosphate; involved in the rephosphorylation of ADP and adenylic acid after muscle contraction.

phosphokinase. Enzyme which transfers phosphate groups from a donor to an acceptor in cell metabolism.

phospholipid. Lipid containing phosphoric acid and a nitrogenous base as a part of the molecule; found in all cells.

phospholipin. Phospholipid.

phospholipoid. Phospholipid.

phosphoprotein. Compound consisting of a protein molecule and a phosphorus-containing substance other than nucleic acid or lecithin; e.g. casein.

phosphopyruvic acid. Intermediate compound in cell metabolism.

phosphorescence. Property of emitting light in the dark after previous exposure to light; not a biological phenomenon and not to be confused with bioluminescence.

phosphoric acid. Important compound in many aspects of cell metabolism; found in chemical combination with a wide variety of organic substances; H_3PO_4.

phosphorus cycle. World-wide circulation and reutilization of phosphorus atoms, chiefly due to metabolic processes of plants and animals; plants take up inorganic phosphorus from water or soil, mostly as phosphate, and convert much of it into organic compounds; such phosphorus atoms may then be assimilated into the bodies of one or more successive animals; excretion, burning, and bacterial action on dead organisms return the phosphorus atoms to the inorganic state.

phosphorylase. Enzyme which catalyzes the splitting of organic molecules, taking up phosphate and producing organic phosphate; e.g. amylophosphorylase, which catalyzes the production of glucose-1-phosphate from starch and inorganic phosphate.

phosphorylated sugar. Monosaccharide temporarily combined with a phosphate radical derived from ATP which is thereby converted to ADP; such a compound is then normally converted to pyruvic acid with the release of hydrogen ions and the reconversion of ADP to ATP.

phosphorylation. Introduction of the phosphate radical into an organic molecule.

photic zone. Uppermost layer of a lake or ocean; extends to the maximum depth at which photosynthesis can occur; in exceptionally clear ocean water it may be 100 m. thick; in certain lakes it may be only 2 m. thick.

Photinus. Common American genus of fireflies in the beetle Family Lampyridae.

Photis. Genus of marine Amphipoda.

photoautotrophic. See PHOTOSYNTHETIC BACTERIA.

Photocorynus spiniceps. Tropical deep-sea angler fish in which the male is minute and becomes permanently attached to the female as a parasite.

photogenic. Light-producing.

photomicrograph. Photograph of a microscopic object; usually taken through a microscope. Cf. MICROPHOTOGRAPH.

photonegative. Negatively phototactic.

photoperiodism. Behavioral or physiological response of an organism to changing light intensities, such as the daily cycle of activities of certain insects and crustaceans.
photophobia. Abnormal sensitivity to light.
photophore. Small beadlike light-producing area in bioluminescent animals.
photopositive. Positively phototactic.
photoreceptor. Any receptor able to detect light, e.g. specialized epithelial cells in some invertebrates and all types of eyes.
photosynthesis. 1. Complex of processes involved in the formation of carbohydrates from carbon dioxide and water in living plants in the presence of light and chlorophyll. 2. Biological synthesis of organic compounds from simpler materials in the presence of light and a catalyst.
photosynthetic bacteria. Photoautotrophic bacteria; group of bacteria (chiefly aquatic) which are able to convert simple inorganic materials such as carbon dioxide, water, and hydrogen sulfide into organic materials in the presence of light and various red, green, etc. pigments.
phototaxis. Behavioral movement response of an animal to light; a moth reacts positively, while an earthworm reacts negatively. Adj. phototactic.
phototopic vision. Ordinary daylight vision involving the retinal cones.
phototrophic. Pertaining to organisms which obtain energy from light, e.g. through photosynthesis.
phototropism. Behavioral turning or bending response to light; positive is turning toward light, negative away.
Phoxichilidium. Genus of pycnogonids; genital pores present on the fourth segment of all eight legs.
Phoxocephalus. Genus of marine Amphipoda.
phragma. 1. Any dividing septum, membrane, plate, or other structure. 2. Internal projecting ridge or other process, especially in the insect thorax. Pl. phragmata.
Phragmatopoma. Genus of Polychaeta.
phragmocone. Chambered conical internal shell of certain fossil Cephalopoda.
phragmosis. Closing the burrow or nest with some special part of the body; e. g. with the flattened head in the tiger beetle larvae.
phrenic artery. Blood vessel which supplies blood to the diaphragm.
phrenic nerves. Pair of nerves supplying the diaphragm, pleura, and pericardium; originate from the third, fourth, and fifth cervical nerves.
phrenic vein. Blood vessel which drains the diaphragm.
Phronima. Genus of pelagic marine Amphipoda; often live on or in tunicates, jellyfish, etc.; translucent, with large head and immature eyes.
Phrurolithus. Common genus of hunting spiders in the Family Clubionidae.
Phryganea. Common genus of insects in the Order Trichoptera; aquatic larvae construct portable cases.
Phryganeidae. Family of large caddisflies; larvae in quiet or slowly flowing water, build cylindrical cases of leaves and small twigs; Phryganea common.
Phrynichida. Taxon sometimes used in place of Amblypygi.
Phrynixalus biroi. Small New Guinea frog found in mountain streams; eggs develop into lunged young and not tadpoles.
Phrynosoma. Genus of lizards which includes the horned "toads"; body and tail depressed; sharp-pointed horns at back of head and with a variable number of spines on trunk; frequent in dry, sandy places; active in daytime, feed on arthropods; from British Columbia to Guatemala and Ark. to Pacific Coast; most abundant in arid regions.
Phryxus. Genus of degenerate marine Isopoda; usually parasitic on hermit crabs.
Phthirus pubis. Insect in the Order Anoplura; the crab louse or pubic louse which parasitizes man, especially in the pubic region.
Phycitidae. Family of small tropical and subtropical moths; Plodia (Indian meal moth) is one of the few pests.
phyla. Pl. of phylum.
phylactocarp. Protective sheathlike structure partly surrounding gonangia in certain hydrozoan coelenterates.
Phylactolaemata. One of the three classes of the Phylum Bryozoa; lophophore horseshoe-shaped; form statoblasts; includes nearly all fresh-water spp.; common genera are Plumatella, Pectinatella, and Cristatella.
phyletic. Pertaining to a phylum or racial and evolutionary line of descent.
Phyllium. Genus of South American Orthoptera; certain spp. are called "dry leaf" insects because the wing color,

shape, and venation bear a striking resemblance to a dry leaf.

Phyllobates. Genus of very small tree toads; one adult Cuban sp. has a head and body length of only 9 mm.; a native arrow poison, batracin, is prepared from the skin of the South American P. chocoensis.

Phyllobothrioidea. Tetraphyllidea.

Phyllobothrium. Genus of tapeworms in the Order Tetraphyllidea; adults in the intestine of sharks and rays.

phyllobranchia. Gill composed of thin lamellae.

Phyllobranchopsis. Genus of nudibranchs; feed on cell fluids of algae.

Phyllocarida. Leptostraca.

Phylloceras. Genus of fossil ammonites.

Phyllochaetopterus. Genus of small tube-building Polychaeta.

Phyllocoptes pyri. See ERIOPHYIDAE.

Phyllodactylus. Genus of gecko lizards found from Calif. southward through all of South America; also in Asia and Africa.

Phyllodoce. Genus of slender, flattened Polychaeta.

Phyllodurus. Genus of isopods parasitic on marine shrimps and anomurans.

Phyllomitus. Genus of protozoans found in stagnant fresh waters; one flagellum projecting forward, one trailing.

Phyllonycteris. Genus of West Indian leaf-nosed bats.

Phyllophaga. Genus of beetles in the Family Scarabaeidae; includes the May beetles or June beetles; attracted to lights at night; adults feed on foliage; larvae are pests in soil and lawns where they feed on grass roots.

Phylloplana. Genus of polyclad Turbellaria.

phyllopod. Any member of the crustacean Division Eubranchiopoda; e.g. fairy shrimps, tadpole shrimps, and clam shrimps.

Phyllopoda. Same as the crustacean Division Eubranchiopoda.

phyllopodium. Flat, leaflike, lobed, thoracic arthropod appendage, as in the Notostraca.

Phyllopteryx eques. Australian sea horse; body covered with leaflike outgrowths and appendages which effectively conceal it among seaweeds.

Phyllorhynchus. See LEAF-NOSED SNAKE.

phyllosoma. Free-swimming larva of Palinurus, the sea crayfish, and its close relatives; extremely depressed, wide, and transparent.

Phyllospondylii. Order of extinct stegocephalians; branchiosaurs; small spp. resembling snakes and salamanders; three pairs of gills during larval stage but none in adults; Ichthyostega and Branchiosaurus examples.

Phyllospongia. Genus of sponges in the Order Keratosa; leathery and leaflike.

Phyllostomatidae. Family which includes the leaf-nosed bats; about 35 genera, four in U.S.; Phyllostomus common.

Phyllostomus. See PHYLLOSTOMATIDAE.

Phyllotreta. Widely distributed genus of leaf beetles in the Family Chrysomelidae.

Phylloxera. Same as Dactylosphaera; See PHYLLOXERIDAE.

Phylloxeridae. Family of minute aphidlike insects in the Order Homoptera; naked or covered with cotton-like wax; Dactylosphaera vitifolii, the grape phylloxera, is a serious pest in Europe but is of minor importance in the U.S.; feeds on leaves and roots of grape vines.

phyllozooid. Hydrophyllium.

Phylocentropus. Genus of Trichoptera.

phylogenetic tree. Concept, diagram, or figure intended to show evolutionary lines of descent based on morphology, palaeontology, and other evidence.

phylogeny. 1. Evolutionary relationships and lines of descent in any taxon. 2. The origin and evolution of higher taxonomic categories. Adj. phylogenetic.

phylum. One of the large principle divisions of the animal kingdom; depending on the method of classification used, there are from 10 to about 33. Pl. phyla.

Phymatidae. Family of insects in the Order Hemiptera; ambush bugs; bizarre, sculptured or spiny predaceous bugs which live among flowers and foliage and capture other insects.

Phymosoma. Common West Coast genus of Sipunculoidea now included in Phascolosoma.

Physa. Common and widely distributed genus of fresh-water snails; especially common in northern states.

Physalia pelagica. Portuguese man-o'-war; a large pelagic hydrozoan coelenterate which floats by means of an air-filled chamber projecting above the surface of the sea; the extensive colonial system of polymorphic hydroids is suspended below the float and long tentacles dangle below the surface as much as 50 ft.; by means of its nematocysts this sp. can inflict serious poisoning when it comes in contact with human skin.

Physcosoma. Genus of Sipunculoidea now included in Phascolosoma.

Physeter. See SPERM WHALE.

Physeteridae. Family of sperm whales; small to large spp. of the Atlantic, Pacific, and Indian oceans; teeth numerous; with a reservoir of clear oil in a hollow above the cranium; see SPERM WHALE.

physiological gradient. Axial gradient.

physiological isolation. Prevention of interbreeding between two or more populations because of incompatibility of copulatory structures, lack of fertilization, etc.

physiological saline. Physiological salt solution.

physiological salt solution. Water solution containing precisely or roughly the same kinds and/or amounts of salts present in the blood of a particular animal; human physiological salt solution contains 0.9% sodium chloride.

physiology. Science dealing with the living functions and activities of organisms and their parts.

Physoclysti. Large taxon of bony fishes having no tubular connection between the digestive tract and swim bladder.

Physocypria. Common genus of freshwater Ostracoda.

Physophorida. Suborder of the Siphonophora; polyps arranged on a long stemlike growth; upper end of colony a floating pneumatophore.

Physostomi. Small group of bony fishes in which there is a functional connection or tube between the digestive tract and the swim bladder, e.g. ganoid fishes.

Phytia. Genus of small snails found in salt marshes.

Phytobius velatus. Aquatic weevil; all stages in life history found below the surface on water milfoil; Europe and U.S.

Phytodinium. Genus of spherical or ellipsoidal fresh-water dinoflagellates; chromatophores yellowish brown.

Phytomastigina. One of the two subclasses of the Class Mastigophora; includes flagellates which contain chromoplasts and are chiefly holophytic; a few colorless flagellates are also included in this subclass, but these closely resemble the pigmented Phytomastigina in their structure.

Phytomonadina. An order of the Class Mastigophora; small, more or less spherical, green flagellates having similarities to algae; with one, two, of occasionally four flagella; Chlamydomonas and Volvox are examples.

Phytomonas. Genus in the Family Trypanosomidae; parasitic in the latex of a variety of angiosperms.

Phytonomus posticus. Alfalfa weevil; larvae feed on leaves.

Phytophaga destructor. Minute fly in the Family Cecidomyiidae; Hessian fly; larva an extremely important pest on wheat.

phytophagous. Pertaining to animals which feed chiefly or exclusively on plant material; herbivorous.

phytoplankton. Collectively, all those microscopic plants suspended in the water of aquatic habitats; includes algae and fungi. Cf. ZOOPLANKTON and PLANKTON.

phytosaur. See THECODONTIA.

Phytotomidae. Small family in the bird Order Passeriformes; plant cutters; finchlike spp. that cut up vegetation; temperate South America.

phytotoxin. Toxin derived from a plant, e.g. ricin.

pia-arachnoid membranes. Collectively, an inner pia mater membrane and an outer arachnoid membrane surrounding the mammalian central nervous system; these two membranes are surrounded by a protective layer of connective tissue, the dura mater.

pia-arachnoid spaces. Interconnecting vascular spaces contained in the network of the arachnoid membrane surrounding the mammalian brain and spinal cord.

pia mater. Thin membrane immediately covering the central nervous system of mammals.

Pica. Genus of large passerine birds which includes the magpies; related to the jays but black and white and with long, sweeping tails; P. pica is the common magpie of western N.A.; a closely related subspecies occurs in Europe.

piceous. The color of pitch.

pichiciego. See CHLAMYPHORUS.

Picidae. Family of birds in the Order Piciformes; woodpeckers; tree-climbing spp. with stiff, spiny tails used as props; flight usually undulating; legs and toes stout; beak robust and large, used for drilling into trees in a search for insects; tongue cylindrical and extremely long; cosmopolitan except for Australia, Madagascar, and oceanic islands; typical U. S. genera are *Dryobates*, *Melanerpes*, *Sphyrapicus*, *Colaptes*, and *Balanosphyra* (same as *Melanerpes*).

Piciformes. Large order of arboreal birds which includes the woodpeckers, jacamars, honey guides, toucans, etc.; two toes in front, two behind; nest holes usually in ground or in trees; forests and brush of temperate and tropical areas.

pickerel. See ESOCIDAE.

pickerel frog. See RANA PALUSTRIS.

Picoides. Genus of three-toed woodpeckers of Holarctic boreal forests.

piddock. Any of a variety of small boring bivalve mollusks; found especially along the East Coast; anterior half of shells often quite rough and sculptured.

piebald. Skin and hair spotted with white; a Mendelian dominant in man.

piebald dolphin. See CEPHALORHYNCHUS and LAGENORHYNCHUS.

pied-billed grebe. See PODILYMBUS.

Pieridae. Large family of very common, medium-sized, light-colored butterflies; widely distributed; whites, yellows, and sulfurs; abdomen slender; some are pests; *Pieris* and *Colias* common.

Pieris. Genus of butterflies in the Family Pieridae; *P. rapae* is the cosmopolitan cabbage butterfly whose caterpillar feeds on cabbage and other crucifers.

pigeon. See COLUMBIDAE.

pigeon hawk. See FALCO.

pigeon's milk. See CROP MILK.

pigfish. Any of numerous marine fishes which are able to produce a grunting sound or have jaws with a fancied resemblance to the snout of a hog; see GRUNT.

pig-footed bandicoot. See CHOEROPUS.

pigment. See PIGMENT GRANULES.

pigment cell. Any cell containing an abundance of pigment granules.

pigment granules. Minute granules which occur in many types of animal cells, but are especially abundant in the dermal chromatophores; such granules are usually metabolic wastes and may be black, brown, yellow, red, or white.

pika. See OCHOTONIDAE.

pike. See ESOCIDAE.

pikeperch. See STIZOSTEDION.

Pilargis. Genus of long whitish Polychaeta.

Pilaria. Genus of crane flies in the Family Tipulidae.

pilchard. See SARDINOPS.

pile. 1. Extremely thick growth of fine short hairs on certain aquatic hemipterans and beetles; holds a film of air in contact with the body when the insect is submerged. 2. A hemorrhoid.

pileated woodpecker. See DRYOCOPUS.

pileous. Hairy or covered with slender setae.

pileum. Top of the head of a bird from the base of the bill to the nape of the neck.

pileus. 1. Pileum. 2. Umbrella portion of a jellyfish body.

pilidium. Specialized helmet-shaped trochophore larva found in some spp. of marine Nemertea. Pl. pilidia.

pilifer. Small sclerite on each side of the clypeus in Lepidoptera.

pillbug. Terrestrial sp. of isopod that is able to roll up into a ball.

pilocarpine. Drug which stimulates the parasympathetic nerve endings.

pilose. Hairy or downy.

pilotfish. See NAUCRATES and PROSOPIUM.

Pilsbry, Henry A. American conchologist (1862-1957); associated with the Philadelphia Academy of Natural Sciences; made many important contributions to the anatomy, phylogeny, and taxonomy of mollusks.

Piltdown man. See EOANTHROPUS.

Pilumnus. Genus of marine mud crabs (Brachyura).

Pimelometopon. See LABRIDAE.

Pimephales. Common genus of minnows in the Family Cyprinidae; widely distributed east of the Rockies; bluntnose minnow (*P. notatus*) and fathead minnow (*P. promelas*) especially common.

pimpleback. Any of certain fresh-water mussels with a knobbed or tuberculated shell, especially *Quadrula*.

pinacocytes. Large flat polygonal epithelial cells in certain sponges.

pinché. See OEDIPOMIDAS.

Pinctada radiata. Thin-shelled Atlantic pearl oyster; to 3 in. long; Fla. and West Indies.

pineal apparatus. Close association of the pineal body and the parietal organ in cyclostomes, ganoids, and certain reptiles.

pineal body. The epiphysis; a small glandular outgrowth on the dorsal surface of the diencephalon in the vertebrate brain; often assumed to secrete hormone(s); in a very few vertebrates it forms a vestigial median eyelike structure. See PINEAL APPARATUS and PINEAL EYE.

pineal eye. Median degenerate eyelike structure on the dorsal surface of the head of the Anura and lampreys; develops from the pineal body. See PINEAL BODY and PINEAL APPARATUS.

pine beetle. See DENDROCTONUS.

pine grosbeak. See GROSBEAK.

pine lizard. See SCELOPORUS.

pine marten. Marten.

pine siskin. See SPINUS.

pine snake. See PITUOPHIS.

Pineus. See ADELGIDAE.

pine vole. See PITYMYS.

pinfeather. Undeveloped feather, especially a small one just emerging from the epidermis.

pinfish. Any of numerous fishes having sharp dorsal spines; see LAGADON.

Pinguinus impennis. Great auk (Family Alcidae); a sp. now extinct but formerly abundant along Atlantic Coast, especially north of New England; flightless, but a superb swimmer and diver; to 30 in. long; exterminated by man.

Pinicola. See GROSBEAK.

pink bollworm. See PECTINOPHORA.

pinkfish. Blind goby, Typhlogobius californiensis; a small sluggish flesh-colored sp. living in the burrows of shrimps along the southern Calif. coast.

pinna. 1. See OUTER EAR. 2. Feather. 3. Rarely, a wing or fin.

Pinna. Genus of large, fragile, wedge-shaped bivalves; pen shells, sea pens; hinge at tip of wedge; secrete stout byssus threads; some shells almost 12 in. long; tropical and warm temperate seas.

pinnaglobin. Brown blood pigment in the mollusk genus Pinna; similar to hemocyanin but has manganese in place of copper.

pinnate. Feather-like; having parts arranged on each side, as in a feather.

pinnated grouse. Greater prairie chickin.

Pinnipedia. Suborder of gregarious carnivorous mammals; includes the seals, sea lions, and walruses; large spp., modified for marine habitats, but spending at least some time on the sea shore; limbs finlike tail and ears reduced; hair coarse to fine and dense; feed mostly on fish.

Pinnixa. Genus of small marine crabs; in tubes and burrows of polychaetes and in the mantle cavities of bivalve mollusks.

Pinnotheres. Genus of marine decapod crustaceans which live as commensals in the mantle cavity of pelecypods or in the tubes of living annelids; carapace usually nearly circular and membranous; to 20 mm. long; pea crabs, oyster crabs, mussel crabs.

pinnules. Numerous, slender, flexible lateral projections on the arms of crinoid echinoderms.

pinocytosis. 1. Engulfing of fluid droplets by amoeboid or other cells. 2. Adsorption of particles, molecules, or ions on cell surface followed by an inward progression of the material in a droplet until it is pinched off and incorporated in the cytoplasm.

pintail. Any of several ducks and grouse having two or more greatly elongated central tail feathers; see ANAS.

pinworm. See ENTEROBIUS VERMICULARIS.

pinyon jay. See GYMNORHINUS.

Piona. See PIONIDAE.

Pionidae. One of many families of water mites in the Superfamily Hydrachnoidea; Piona a common fresh-water genus.

Pionosyllis. Genus of Polychaeta.

Piophila casei. Cheese skipper; a small cosmopolitan fly which commonly breeds on cheeses, fats, meats, and carrion; when disturbed, the larva "jumps" or throws itself about in an aimless manner.

Pipa. See PIPIDAE.

pipefish. See SYNGNATHIDAE.

Pipidae. Small family of northern South American toads in the Order Opisthocoela; tongue and eyelids absent; with a starlike cluster of dermal papillae at the tip of each finger; developing eggs carried in individual dermal pockets on the back of the female; Pipa is the common Surinam toad.

Pipilo. See TOWHEE.

Pipistrellus. Genus of small bats; pipistrelles; N.A. and most of Eastern Hemisphere; two U.S. spp.

pipit. See MOTACILLIDAE.

Pipridae. Large family of birds in the Order Passeriformes; manakins; small, stout, brightly-colored spp. of Tropical America; to 6 in. long; beak stubby.

Pipunculidae. Big-headed flies; a family of small, slender Diptera with spherical heads and enormous eyes; many beneficial spp.; mostly Palaearctic.

Piranga. See THRAUPIDAE.

piranha. Caribe.

Pirata. Common genus of spiders found along stream and pond margins.

pirate perch. See APHREDODERIDAE.

Pisaster. Genus of large Pacific Coast sea stars; includes the ochre stars.

Pisauridae. Nursery-web weavers; a family of large hunting spiders found near water, in grassy areas, and in forests; eggs deposited in a spherical silken sac fastened to vegetation and surrounded with a protective, irregular web (nursery web) which is guarded by the female until the eggs hatch; Pisaurina and Dolomedes common, the latter including the fishing spiders or diving spiders.

Pisaurina. See PISAURIDAE.

Pisces. One of the two superclasses of the Subphylum Vertebrata; includes all "fishes," namely Agnatha, Placodermi, Chondrichthyes, and Osteichthyes; formerly considered a single class of vertebrates.

Piscicola. Genus of leeches parasitic on marine and fresh-water fishes; body cylindrical, often divided into narrow anterior and wide posterior portions.

Piscicolaria. Uncommon genus of fresh-water fish leeches.

Pisidium. Genus of small fresh-water pelecypods; usually 2 to 6 mm. long; seed clams.

pismo clam. See TIVELA.

Pisobia. One of several genera of sandpipers; several American spp.

Pista. Genus of tube-building Polychaeta.

pistol shrimp. See ALPHEIDAE.

Pitangus sulphuratus. Derby flycatcher in the Family Tyrannidae; a large, short-tailed flycatcher; rufous wings and tail, yellow under parts and crown, black and white face; lower Rio Grande Valley to South America.

Pitar. Genus of small, plump marine bivalves; Carolinas to Tex., and West Indies.

pitcher plant. Any of several insectivorous bog plants, especially Sarracenia in N.A.; each leaf forms a small, upright, pitcher-like device holding rain water at the bottom; an insect enters through an arrangement of downward-pointing hairs and eventually drowns; the body is thought to be partially digested by enzymes released into the water by the plant.

Pithecanthropus erectus. Java ape man; parts of skulls, mandibles, and femora found in Pleistocene deposits in Java; brain volume about 750 to 900 cc.; primitive human creature; no associated implements known.

Pithecanthropus robustus. Fossil anthropoid known only from skull fragments found in Java in 1938.

Pithecia. Genus of small tropical South American fruit-eating monkeys; the sakis and sakiwinkis.

pitocin. Discarded name for oxytocin.

pitressin. Proprietary name for vasopressin.

Pittidae. Family of passerine birds; brightly patterned terrestrial songless spp. all in the genus Pitta; ground thrushes; body plump, wings and tail short, bill stout, legs long; eastern and southeastern Asia, Australia and associated islands.

pituicytes. Characteristic branching cells in the posterior lobe of the pituitary.

pituitary. 1. Small stalked endocrine gland on the lower surface of the vertebrate brain; produces numerous hormones; secretions of the anterior lobe affect growth and the activities of the thyroid, mammary glands, gonads, and adrenal cortex; the intermediate lobe secretes intermedin which governs the color of vertebrates having variable skin pigmentation; the posterior lobe secretes vasopressin which controls blood vessel contraction and water excretion, and also oxytocin which stimulates smooth (uterine) muscle contraction. 2. A giant long-boned physique produced by excessive secretion of the pituitary gland.

pituitary dwarf. Dwarfed condition produced by an insufficiency of the growth-stimulating hormone secreted by the anterior lobe of the pituitary.

pituitary giant. Pathological condition produced by excessive secretion of the growth-stimulating hormone of the anterior lobe of the pituitary during youth; such human giants may be 7 to 9 ft. tall.

pituitrin. 1. Commercial extract of pituitary glands which contains numerous hormones. 2. An extract of the posterior lobe of the pituitary. See PITUITARY.

Pituophis. Genus of large, heavy, non-poisonous, constrictor snakes in the Family Colubridae; bull snakes (central U.S.) and pine snakes (southeastern U.S.); to 7 ft. long; feed mostly on rodents; capable of hissing loudly.

pit viper. See CROTALIDAE.

Pitymys. Genus of pine voles; tail short, ears small, habits semi-fossorial; two spp. in eastern and central U.S.

pivotal joint. Skeletal joint which permits a rotary movement in one plane only, such as that between the axis and atlas which allows turning of the head from side to side.

placebo. Inactive substance given to please or gratify an unknowing patient; also used as a control in experimentally determining the activity of a new medicinal substance.

placenta. 1. Mammalian joint embryonic-maternal structure composed of the chorion of the former and usually a part of the uterine wall of the latter; maternal and embryonic blood vessels are closely associated in the placenta, but there is no direct connection between the two circulatory systems; by means of the placenta the embryo receives food and oxygen and gets rid of carbon dioxide and other excretory wastes; at birth the umbilical cord is broken, and the placenta is discharged from the uterus and vagina after the embryo. 2. The afterbirth. 3. In onychophorans, scorpions, viviparous sharks, and salps, a structure of various form used for nutrient absorption by the developing embryo.

Placentalia. Eutheria.

placentation. Manner of formation, arrangement, and structure of the placenta in mammals.

Placephorella. Genus of chitons; shape near circular.

Placobdella. Genus of fresh-water leeches; free-living or temporary parasites on fish, frogs, and turtles; to 4 in, long.

Placocephalus. Bipalium.

placode. Localized ectodermal thickening, especially in vertebrate embryos; forerunner of some organ, e.g. the neural tube, lens, ear, or nose.

Placodermi. Heterogeneous class of extinct fishes; lower jaw separate, upper jaw fused to skull; one gill slit on each side; both median and paired fins; Upper Silurian to Permian.

placoid scales. Type of scales found in typical cartilaginous fishes; with basal plate of dentine in the dermis and a projecting back-pointing spine covered with vitrodentine.

Placopecten. See SCALLOP.

placula. Flattened blastula of urochordates; segmentation cavity small.

Placuna sella. Large marine bivalve of the Philippines; the shell is so thin and translucent that it is sometimes cut, polished, and used for small window panes by natives.

Placus. Genus of small, ellipsoid holotrich Ciliata; marine and fresh waters.

Plagiocampa. Genus of ovoid to cylindrical holotrich Ciliata; marine and fresh waters.

Plagiodontia. Genus of rare guinea-pig-like animals; zagoutis; feed on tree leaves and flowers; mountains of Haiti.

Plagiola lineolata. Unionid mollusk in large rivers; Pa. to Iowa and south to Ala. and Okla.

Plagiophrys. Genus of fresh-water amoeboid protozoans having an ovoid test protecting most of the cell.

Plagiopyla. Genus of ovoid holotrich Ciliata; marine, fresh-water, or endozoic.

Plagiostomum. Large genus in the turbellarian Order Alloeocoela.

Plagopterus argentissimus. Woundfin; a slender silvery scaleless minnow of the middle Colorado River basin and the Gila River system.

plaice. Any of various marine flatfishes such as the flounders and dabs, and especially a northern European sp.; see HETEROSOMATA.

plains pocket gopher. See CRATOGEOMYS.

plaited shell. See PLICATULA.

Planaria. A genus of free-living fresh-water flatworms occurring in springbrooks, ponds, streams, and lakes; usually 3 to 15 mm. long; members of this genus are not known from the U.S. but are common in Europe; Dugesia is the closely related U.S. genus.

planarian. Variously used to designate: (1) any free-living flatworm, (2) any member of the Family Planariidae and related families of macroscopic turbellarians, and (3) any member of the genus Planaria and related genera of macroscopic turbellarians.

Planes. Common genus of marine crabs often found in floating brown algae (Brachyura).

Planigale. Genus of tiny kangaroo-like marsupials of Australian desert grasslands; skull extremely flat; one sp. weighs less than one ounce.

plankter. Individual plankton organism or sp.

plankton. Collectively, all those organisms suspended in the water of an aquatic habitat which are not independent of currents and other water movements; most such organisms are microscopic and commonly include bacteria, algae, protozoans, rotifers, larvae, and small crustaceans.

plankton net. Any fine-meshed net used in straining plankton from water.

plankton pulse. Unusually large and transient population of one or more plankton spp. in a body of water.

planktont. Plankter.

plankton trap. Metal apparatus used for taking quantitative zooplankton samples; it is lowered to the proper depth and closed by means of a metal messenger sent down the line; upon being raised from the water, the enclosed sample (usually 10 liters) is strained through an attached fine net.

Planocera. Common genus of marine polyclad Turbellaria; very flat and leaflike; coloration usually olive-green to yellowish; at least one sp. lives as a parasite in the mouth cavity of the gastropod Busycon.

Planogyra. Genus of very small land snails.

Planorbidae. Orb snails; a large family of fresh-water snails; shell discoidal and orblike, or with a low spiral.

Planorbis. Helisoma.

Planorbula. Genus of fresh-water snails; eastern U.S.

plantain eater. See TOURACO.

plantar. Pertaining to the sole of the foot.

plantar aponeurosis. Pad of tissue beneath the ankle in reptiles; muscles and tendons are attached to both ends.

plantaris. Mammalian muscle which extends the hind foot and flexes the leg.

plant bugs. 1. In the narrow sense, the common name of hemipterans in the Family Miridae. 2. In the broad sense, any hemipteran associated with plants, especially for feeding and oviposition.

plant cutter. See PHYTOTOMIDAE.

plantigrade. Pertaining to animals that walk on the whole surface of the foot, such as man.

plant lice. See APHIDIDAE.

planula. Ciliated free-swimming larval type occurring in many coelenterates; the body is more or less cylindrical or ovoid with a single layer of outermost epidermal (ectodermal) cells and a core of endodermal cells. Pl. planulae.

planulaea. Hypothetical ancestral organism from which the Coelenterata and perhaps the Ctenophora and gastraea are assumed to have arisen; it had the structure of a planula larva but was an abundant adult organism.

planuloid. A frustule; in some hydrozoans, an asexually produced larva with the general structure of a planula; develops into a polyp.

plasma. Clear colorless fluid fraction of blood in which corpuscles are suspended; contains dissolved salts and proteins; may be prepared by centrifuging blood; the term is also used to include a comparable fraction of blood and other body fluids of invertebrates.

plasmagel. The more solid, jelly-like colloidal phase of protoplasm; especially notable in the peripheral portion of amoeboid protozoans. See PLASMASOL and COLLOIDAL SYSTEM.

plasmagene. One of numerous inheritable particulate self-reproducing units in the cytoplasm of a cell having the properties of genes but not inherited through the gametes in a Mendelian manner; poorly known but thought to function chiefly in the control of cell differentiation.

plasmalemma. Plasma membrane.

plasma membrane. Living membrane bounding the cytoplasm of cells; it is thought to consist of lipid and protein molecules and is in the general magnitude of 0.01 micron thick; in plants it is covered by the nonliving cell wall, in animals it is covered by a living cell membrane.

plasma protein. Any one of a wide variety of proteins characteristic of the blood plasma or coelomic fluids of vertebrates and invertebrates.

plasmasol. The more liquid physical state of protoplasm; especially notable in the inner portion of amoeboid protozoans. See PLASMAGEL and COLLOIDAL SYSTEM.

plasmochin. Complex organic compound used as an antimalarial; taken by mouth, often in combination with quinine; acts chiefly on the gametocytes in the erythrocytes.

plasmodesmata. See INTERCELLULAR BRIDGE.

Plasmodiophora. Genus of multinucleate mycetozoan Protozoa; parasitic in roots of cabbage.

plasmodium. 1. Vegetative stage of Myxomycetes. 2. Multinucleate mass of naked protoplasm bounded by a plasma membrane; no definite size or shape. Pl. plasmodia.

Plasmodium. Genus of sporozoan Protozoa which includes malaria parasites of man, some other mammals, certain birds, and a few reptiles.

Plasmodroma. One of the two subphyla of the Phylum Protozoa; includes all Protozoa having pseudopodia, flagella, or no special means of locomotion; only one kind of nucleus.

plasmogamy. Union of the cytoplasm of the male and female gametes in the process of syngamy.

plasmolysis. Contraction or shrinking of an animal cell owing to loss of water by osmosis.

plasmotomy. Division of a multinucleate protozoan into two or more multinucleate masses without concurrent mitosis.

plasmotrophoblast. Plasmotrophoderm.

plasmotrophoderm. Outermost layer of syncitial absorptive tissue of an early mammalian embryo; imbedded in the uterine mucosa and later replaced by chorionic villi.

plastic. Capable of undergoing physiological, evolutionary, or morphological alterations readily.

plastid. One of several kinds of small self-propagating bodies in plant cytoplasm; shape is variable but usually oval or circular; they may be colorless or pigmented; not present in animals. See CHROMOPLAST.

plastospecies. Interbreeding population conspicuously separated from all other populations, morphologically and physiologically, but may occasionally interbreed with one or more other populations; e. g. yellow-bellied and red-breasted sapsuckers are two different plastospecies but occasionally hybridize. Cf. CENOSPECIES.

plastron. 1. Flat ventral skeletal plate of turtles. 2. Sternum and costal cartilages of higher vertebrates. 3. Film of air covering some portion of the body of a submerged fresh-water insect.

Platacanthomyidae. Family of spiny dormice; small arboreal Indian rodents which bore holes in tree branches for their nests.

Platanaster. See PLATYASTERIDA.

Platanista. See PLATANISTIDAE.

Platanistidae. Family which includes the susu (Platanista), a blind dolphin of large Indian rivers; with a distinct neck and a long slender beak; probes the bottom for invertebrates; to 9 ft. long.

plated lizard. See GERRHONOTUS.

platelets. Blood platelets.

Plathemis lydia. Common and widely distributed dragonfly.

Platichthys stellatus. Sp. of Pacific flounder; sometimes enters coastal streams.

Platyasterida. Order of extinct asteroids; Ordovician to Devonian; Platanaster typical.

Platycentropus. Genus of Trichoptera.

Platycopa. One of the four orders of Ostracoda; with two pairs of trunk appendages and a flat process at the posterior end of the body; heart absent; both pairs of antennae large; all spp. marine; Cytherella typical.

Platycordulia xanthosoma. Uncommon dragonfly; central U. S.

Platycotis. See MEMBRACIDAE.

Platycrinites. Genus of extinct crinoid echinoderms.

Platycrinus. Genus of extinct crinoid echinoderms.

Platyctenea. One of the orders of the ctenophore Class Tentaculata; aberrant creeping forms having great reduction of the oral-aboral axis; Coeloplana common.

Platydesmus. See COLOBOGNATHA.

Platydorina. Fresh-water genus of colonial free-swimming green flagellates; 32 cells imbedded in a gelatinous matrix in the form of a flat or twisted plate.

Platydra. Pectinophora.

Platyhelminthes. Phylum consisting of elongated, dorsoventrally flattened worms, and including free-living turbellarians, parasitic flukes, and parasitic tapeworms; with a gastrovascular cavity (absent in tapeworms), a simple brain, and a flame bulb system; usually hermaphroditic; about 10, 000 spp.

Platyias. Genus of plankton rotifers; lorica depressed, boxlike, and with anterior and posterior spines.

Platymyaria. Taxon sometimes used to include those nematodes in which the fibrillar zone of the musculo-epithelium is restricted to the basal zone of these cells, adjacent to the hypodermis.

Platynereis. Genus of Polychaeta.

Platyodon. Genus of marine bivalves; bore into soft rock.

Platypsyllidae. Family of beetles containing a single sp., Platypsyllus castoris, which is ectoparasitic on beaver; depressed, eyeless, and wingless.

Platypsyllus. See PLATYPSYLLIDAE.

platypus. Duck-billed platypus; see ORNITHORHYNCHUS.

Platyrhinii. Ceboidea.

Platysamia. Samia.

platysma. Large broad muscle of the neck; originates in the clavicle area and inserts on the lower border of the mandible, cheek muscles, and the corner of the mouth; depresses mouth and lower lip.

Platysomus. Genus of extinct bony fishes in the Order Archista.

Platysternidae. Family of turtles found in southern Asia; head unusually large, tail long; Platysternum typical.

Platysternum. See PLATYSTERNIDAE.

Plautus alle. Common dovekie in the Family Alcidae; a small sp. with dark head, neck, and upper surface, and light below; excellent swimmer and diver.

playa. Relatively flat-bottomed lake basin in an arid or semiarid region; contains water only occasionally as the result of rains and snows.

Plea. See PLEIDAE.

pleasing fungus beetles. See EROTYLIDAE.

Plecodus. Genus of cichlid fishes found especially in Lake Tanganyika; certain spp. feed exclusively on the scales of other living fishes.

Plecoptera. Order of soft-bodied insects; includes the stoneflies and salmonflies; moderate to large size; wings large and membranous but flight is weak; antennae long; chewing mouth parts, often absent in adult; usually two long cerci at tip of abdomen; simple metamorphosis; adults aerial but all spp. have a series of fresh-water naiad stages which are abundant in streams; common genera are Pteronarcys and Taeniopteryx.

Plecotus. Genus of European and Asiatic long eared bats.

Plectognathi. Order of bony fishes which includes the triggerfishes, trunkfishes, puffers, etc.; body shape various, often globose; strong teeth or beak; scales spiny or bony; warm seas.

Plectognathida. Plectognathi.

Plectomerus trapezoides. Fresh-water bivalve mollusk (Unionidae); central and southern U.S.

Plectrophenax. See BUNTING.

plectrum. 1. Columella of the Anura. 2. Uvula. 3. Malleus. 4. Tongue. 5. Small process of the temporal bone.

Plectus. Common genus of chromadorid nematodes; fresh-water and terrestrial.

Plegadis. See IBIS.

Plehnia. Genus of small, brown, eyeless polyclad Turbellaria.

Pleidae. Pygmy backswimmers; a small family of very small aquatic Hemiptera usually in dense aquatic vegetation; Plea striola the only N.A. sp.

pleiotropy. Pleiotropism; multiple physiological or morphological effects of a single gene.

Pleistocene epoch. Recent geological epoch lasting about 1 million years and ending 15,000 to 20,000 years ago; subdivision of the Quaternary period; characterized by alternating cold glacial and warm interglacial times, as well as the rise of primitive man and the disappearance of many large mammals.

Pleodorina. Genus of colonial free-swimming green flagellates; 32 to 128 biflagellate cells in a gelatinous matrix; fresh water.

pleon. 1. Crustacean abdomen. 2. King crab telson.

pleopod. Swimmeret; biramous appendage of certain abdominal segments in many Malacostraca, especially Amphipoda, Isopoda, and Decapoda; by beating back and forth they create a current and ensure an adequate supply of oxygenated water in contact with the body.

Pleraplysilla. Common genus of marine sponges (Subclass Keratosa); colorless encrustations, often on oyster shells.

plerocercoid. Elongated, solid, wormlike macroscopic larval stage in the life history of certain tapeworms; there are two bothria at the anterior end but no other external structural features; when a fish, amphibian, reptile, or mammal ingests a copepod containing the appropriate procercoid, the procercoid is released in the digestive tract and makes its way to the musculature or viscera where it develops into a plerocercoid; when the second intermediate host is ingested by the definitive host, the plerocercoid grows into a mature tapeworm; found in the life history of e.g. Dibothriocephalus latus.

plerocercus. Typical cestode bladderworm.

Plesiosauria. Order of Jurassic and and Cretaceous fossil marine reptiles; limbs paddle-like, neck long; Plesiosaurus typical.

Plesiosaurus. See PLESIOSAURIA.

plesiotype. Specimen upon which later supplementary descriptive material is based.

Plethobasus. Genus of fresh-water bivalve mollusks (Unionidae); central U. S.

Plethodon cinereus. Red-backed salamander; common terrestrial sp. of northeastern U.S. and adjacent Canada; to 4 in. long.

Plethodon glutinosus. Common slimy salamander; chiefly terrestrial; eastern U.S. from New England to Fla. and westward to Wis. and Tex.

Plethodontidae. Family of small lungless and gill-less salamanders; aquatic, amphibious, and terrestrial spp.; chiefly restricted to the Western Hemisphere; examples are Gyrinophilus, Pseudotriton, Desmognathus, Plethodon, Batrachoseps, Aneides, Eurycea, and Typhlomolge.

Plethodontinae. Subfamily of salamanders in the Family Plethodontidae; lower jaw movable; larva with three gill slits on each side; Plethodon typical.

pleura. 1. One of two membranes, each lining one half of the thoracic (pleural) cavity and folded back over the surface of the lung on the same side; visceral pleura covers the lungs, parietal pleura covers the thoracic wall. 2. Thin, flat lateral part of a trilobite on each side of the central rachis. 3. Pleuron. 4. Lateral portion of a flat structure or organ.

Pleuracanthodii. Order of extinct cartilaginous fishes; late Devonian to Triassic; fresh waters; with very long dorsal fin; Pleuracanthus best known.

Pleuracanthus. See PLEURACANTHODII.

pleural. 1. Of or pertaining to the pleura, pleuron, or pleurite. 2. Of or pertaining to the lung.

pleural cavity. One of a pair of coelomic cavities in the chest region of mammals; separated from the abdominal cavity by the diaphragm and from each other by the heart and mediastinum; each pleural cavity contains a lung which actually obliterates most of the cavity. See PLEURA.

pleural ganglion. In many mollusks, one of two ganglia innervating the mantle and parts of the body behind the head; closely associated with cerebral ganglia.

pleural process. Bony process which unites ribs of turtles.

pleural rib. Riblike projection on either side of a vertebra in the trunk region of a fish; imbedded in musculature of body wall.

pleural sac. Pleural cavity.

pleurapophysis. Lateral rib or riblike process of a vertebra.

Pleurobema. Large genus of thick-shelled fresh-water bivalve mollusks (Unionidae); common in eastern half of U.S.

Pleurobrachia. Genus of ctenophores; spheroidal, with two retractile tentacles; common in North Atlantic.

pleurobranch. Malacostracan gill which is attached to the body just above the origin of the basal leg segment.

Pleurocera. Common genus of fresh-water snails; many and variable spp. with a greatly elongated spire; central and southcentral U.S.

Pleurocystis. Genus of extinct echinoderms in the Class Cystoidea.

Pleurocystites. See CYSTOIDEA.

Pleurodira. Small superfamily of turtles; when withdrawn, neck folds sideways under front edge of carapace; pelvic girdle fused to shell; not in N.A.

pleurodont. 1. Type of dentition in which the teeth are attached to the inner edge of the jaws, without sockets. 2. Animal having pleurodont dentition.

Pleurogona. Order of tunicates in the Class Ascidiacea; body without a constriction; two or more lateral gonads; Styela, Molgula, and Botryllus common.

Pleuromonas. Genus of small colorless Protozoa; one flagellum extending forward, one trailing; stagnant fresh waters.

pleuron. 1. Small lateral plate on either side of a typical body segment in most arthropods. 2. Lateral downward extension of a crustacean exoskeleton. Pl. pleura.

Pleuronectida. Heterosomata.

Pleuronectidae. Family of Heterosomata which includes the flounders.

Pleuronectiformes. Order designation

similar to Heterosomata (q. v.);

Pleuronema. Common genus of marine and fresh-water Ciliata; ovoid to ellipsoid spp. with a large peristomial membrane and no cytopharynx.

Pleurophyllidia. Genus of nudibranchs.

Pleuroploca gigantea. Giant, heavy-shelled conch, with oval aperture terminating in a long canal; horse conch; to 2 ft. long.

pleuropodium. 1. Lateral abdominal glandular band found in some insect embryos; modified first abdominal legs. 2. Either of a pair of conspicuous mantle lobes in certain sea hares. Pl. pleuropodia.

Pleurotomaria. Genus of primitive pyramidal marine snails; uncommon, in deep waters.

Pleurotremata. Squali.

Pleurotricha. Genus of fresh-water hypotrich ciliates.

Pleurotrocha. Genus of illoricate rotifers; usually littoral.

Pleuroxus. Common genus of fresh-water Cladocera; usually on substrates in ponds and littoral areas.

Pleustes. Genus of marine Amphipoda.

pleuston. 1. Community of macroorganisms floating on the surface of the sea, e.g. siphonophores, cirripedes, isopods, gastropods, etc. 2. Mat of algal vegetation on or near the surface of a body of fresh water. Cf. NEUSTON.

Plexaurella. Genus of colonial anthozoan coelenterates; arborescent, with cylindrical trunk; common in West Indies.

plexus. Network of nerves or blood vessels, e.g. the sciatic plexus and solar plexus. Pl. plexi.

plicae circulares. Permanent transverse folds of the small intestine.

plicate. With many small ridges or folds, as a fan.

Plicatula gibbosa. Miniature oyster-like bivalve of the Atlantic Coast; the plaited shell, cat's paw, or kitten's paw.

Pliny the Elder. Roman naturalist (23-79); wrote a 37-volume encyclopedia of natural science; apparently all of his information on the 20,000 items was gathered from other sources.

Pliocene epoch. Most recent of the five geological subdivisions of the Tertiary period; lasted about 11 million years; especially characterized by the decline of mammals.

Pliohippus. Genus of Pliocene horses.

Pliotrema. Genus of saw sharks.

Plocamia. Genus of red encrusting sponges.

Plocamissa. Genus of marine sponges.

Ploceidae. Large family of seed-eating passerine birds; includes goldfinches, finches, linnet, sparrows, weavers, and the common house sparrow or English sparrow (Passer domesticus).

Plodia interpunctella. Indian meal moth; small cosmopolitan moth whose larva feeds on cereals and cereal products, dried fruits, chocolate, and candy; spins a silken web throughout these materials.

Ploesoma. Genus of plankton rotifers; lorica stout and rigid; foot annulated.

Ploiariidae. See EMESA.

ploidy. Addition of one or more chromosomes to the normal diploid set.

Ploima. Order of rotifers in the Class Monogononta; usually free-swimming; with or without a lorica; usually without a secreted tube. Adj. ploimate.

plover. See CHARADRIIDAE.

Plumatella. Genus of fresh-water Bryozoa in the Class Phylactolaemata; zooecium chitinoid and translucent or opaque.

plumbeous. Dull bluish gray; lead colored.

plum curculio. See CONOTRACHELUS.

plumed worm. See DIOPATRA.

plume moth. See PTEROPHORIDAE.

plumose. 1. Feathery. 2. With feathers or plumes. 3. Having a tuft of terminal subdivisions.

Plumularia. Genus of sessile, colonial, marine hydrozoan coelenterates.

plumule. Soft down feather of nestling birds; persistent in some adult birds.

Plusia. Important pest genus of moths in the Family Plusiidae.

Plusiidae. Large family of moths resembling the Noctuidae; includes the underwings.

Plutella. Genus of small, narrow, cryptic-colored moths; larvae feed on cruciferous plants; widely distributed.

pluteus. Free-swimming bilaterally symmetrical ciliated larva of the Ophiuroidea and Echinoidea; characterized by long armlike appendages. Pl. plutei.

Plutonaster. Deep-water genus of sea stars (Order Phanerozonia).

Pluvialis dominicus. American golden plover; breeds in American and eastern Asiatic tundra; one race winters in China to Hawaii, Australia, and New Zealand; another race winters in southern South America.

pneumatic duct. Narrow duct leading from the digestive tract upward along the mesentery to the swim bladder in primitive teleosts and lower ray-finned fishes.

pneumatic layer. Thick, crustlike, honeycombed layer surrounding a sponge gemmule; an abundance of spicules are usually more or less imbedded in this layer.

pneumatophore. Air- or gas-filled float in certain hydrozoan coelenterates; keeps the colony suspended at the surface of the sea.

Pneumatophorus. Genus of mackerels; includes the chub mackerel and Pacific mackerel.

pneumococcus. Diplococcus pneumoniae.

pneumogastric nerve. Vagus nerve.

pneumonia. Inflammation of one or both lungs; human pneumonia is most commonly caused by Diplococcus pneumoniae.

Pneumonoeces. Genus of flukes found commonly in frog lungs.

Pneumonyssus. Genus of mites parasitic in lungs of monkeys.

pneumostome. Pulmonary aperture of the mantle cavity of certain terrestrial and aquatic gastropods.

pneumothorax. Presence of air or other gas in a pleural cavity; condition brought about by an external wound, lung perforation, abscess, or lung rupture.

poacher. See AGONIDAE.

Pocillomonas. Genus of small fresh-water Protozoa; with chromatophores, six anterior flagella, and a gelatinous sheath.

pocket gopher. See GEOMYIDAE.

pocket mouse. See PEROGNATHUS.

pocket rat. See DIPODOMYS.

pod. Group of deposited insect eggs which are cemented together in a single mass.

Podaliriidae. Anthophoridae.

Podargidae. See FROGMOUTH.

Podarke. Common genus of Polychaeta.

podex. 1. Rump. 2. Anal region. 3. Insect pygidium.

podical plates. 1. Lateroventral plates of the tenth abdominal segment of Orthoptera. 2. Anal valves in certain insects. 3. Paranal lobes in certain insects.

Podiceps. Colymbus.

Podicipedidae. Podicipitidae.

Podicipediformes. Podicipitiformes.

Podicipitidae. Family of birds in the Order Podicipitiformes; includes the grebes; same as Podicipedidae.

Podicipitiformes. Small order of birds which includes the grebes; generally similar to small ducks; toes lobed, tail rudimentary; excellent swimmers and divers; helpless on dry land; feed on fish, aquatic vegetation, and aquatic invertebrates; worldwide distribution on lakes and ponds, occasionally on ocean bays; nest usually a raft of water-soaked vegetation; Colymbus and Podilymbus typical.

Podilymbus podiceps. Pied-billed grebe, helldiver, or dabchick (Order Podicipitiformes).

podite. 1. Clearly defined segment of an arthropod appendage, usually definitely correlated with muscle attachment, and situated between two successive areas of flexibility. 2. Arthropod appendage.

podobranch. Malacostracan gill which is attached to the basal segment of a leg.

Podocerus. Genus of marine Amphipoda.

Podochela. Genus of spider crabs (Brachyura).

Podocnemis. Genus of large, fresh-water turtles in the Family Pelomedusidae; carapace to 30 in. long; South America and Madagascar; the tartaruga.

Podocopa. One of the four orders of Ostracoda; shell without a permanent anterior opening; second antennae leglike; heart absent; three pairs of trunk limbs; caudal rami usually well developed and cylindrical; fresh-water and marine; Eucypris, Darwinula, and Entocythere common.

Podocoryne. Genus of coelenterates in the Suborder Anthomedusae; hydroids on gastropod shells occupied by hermit crabs, and also on the carapace of hermit crabs.

Pododesmus macroschisma. Rock oyster, jingle; a West Coast marine pelecypod; valves thin, one adherent to substrate.

Podogona. Ricinulei.

podomere. Podite (1).

Podon. Marine genus of Cladocera; carapace does not cover the abdomen and legs but serves only as a brood pouch.

Podophrya. Genus of protozoans in the Class Suctoria; many stalked spp. attached to substrates in fresh and salt water.

Podothecus. See ALLIGATOR FISH.

pod shrimp. Any conchostracan.

Podura aquatica. Common and widely

distributed springtail (Collembola); usually found on the surface of small ponds and puddles.

Poecilichthys. Genus of darters now usually included in Etheostoma.

Poecilictus. Genus of north and central African striped weasels.

Poeciliidae. Family of small bony fishes in the Order Cyprinodontae; top minnows, live bearers; hardy spp. found chiefly in streams and lakes of tropics and subtropics; Gambusia affinis is especially useful in mosquito eradication in southeastern and southern states.

Poeciliopsis occidentalis. Western topminnow in the Family Poeciliidae; Pacific slope of Central America and as far north as the Gila basin of Ariz.

Poecillastra. West Coast genus of marine sponges.

Poecilogale. Snake weasels; genus of small striped weasels of central and southern Africa; body extremely elongated, legs very short, head small.

poecilogeny. Type of larval dimorphism found in certain Diptera.

Poecilósclerina. Order of sponges in the Subclass Monaxonida; numerous kinds of spicules present and usually some spongin.

Poeobioidea. Formerly, a separate phylum or a class of Echiuroidea containing a single poorly-known pelagic wormlike sp., Poeobius meseres, found off the Pacific Coast; body transparent, without appendages, segmentation, or setae; nervous system of the annelid type; circulatory system well developed; head, digestive system, and nephridia similar to those of some Polychaeta; usually considered an aberrant polychaete.

Poeobius. See POEOBIOIDEA.

Poephagus. Generic name sometimes used for the yak; usually placed in the genus Bos.

Pogonichthys macrolepidota. Splittail; a minnow found in the Sacramento and San Joaquin rivers, Calif.; caudal lobes unequally developed.

Pogonomyrmex. Genus of harvester ants or agricultural ants some of which build large mounds of loose soil and sand; herbivorous, ground around mound usually cleared of vegetation.

Pogonophora. Small, poorly-known phylum containing about 50 spp.; threadlike creatures which remain in long cylindrical cellulose(?) tubes in the sea bottom up to great depths; body consists of a short anterior section bearing few to numerous long tentacles, and a very long trunk; the coelom is extensive and there are only two transverse septa (at the anterior end); the anteriormost one-quarter of the trunk has a ventral groove and some evidence of metamerism owing to the disposition of small surface papillae; the second quarter has only scattered papillae; the posterior one-half of the trunk has serially arranged ventral transverse rows of adhesive papillae bearing chitinoid platelets; there is no alimentary canal and digestion is presumably external; some authorities consider this group a highly specialized class of the Phylum Annelida; others give it phylum status (Brachiata) near the Hemichordata.

Poiana. Genus of West African linsangs which have the superficial appearance of genets.

poikilothermous. Cold-blooded; pertaining to any animal whose body temperature remains close to that of its environment; includes all animals except birds and mammals. Noun poikilotherm.

pointed-tailed wasp. See SERPHIDAE.

poison fish. 1. Any of numerous fishes able to inflict wounds with venomous or non-venomous spines. 2. Any of numerous fishes having poisonous alkaloids in the tissues.

poison gland. Gland which secretes (or excretes) a material toxic or irritating to another animal.

Poisson series. Statistical expression and method used to determine whether the distribution of individuals in any given series of samples is random.

polar bear. See THALARCTOS.

polar body. Minute cell produced during meiotic divisions in oogenesis; contains the proper chromosome complement but very little cytoplasm; primary oocyte typically divides into one large secondary oocyte and one first polar body; secondary oocyte divides into an ovum and one second polar body; same as polocyte.

polar cap. See MESOZOA.

polar capsule. Special cell in the spores of certain Sporozoa; one to four polar capsules may develop within a single spore; a polar capsule usually contains a coiled thread, the polar filament, which is presumably an anchoring device for the germinating disc.

polar filament. See POLAR CAPSULE.
polarograph. Dropping mercury electrode system used for determining dissolved oxygen in liquids.
polecat. 1. European sp. of weasel (Mustela putorius) closely related to the domesticated ferret. 2. In the U. S., a skunk. 3. In Africa, a zoril.
polian vesicles. Blind finger-like caeca projecting from the ring canal in certain echinoderms in the classes Echinoidea and Holothuroidea.
Polinices. See MOON SHELL.
poliomyelitis. Acute virus infection of the gray matter of the spinal cord and or brain; characterized by sudden onset, fever, motor paralysis, muscle atrophy, and deformity.
Polioptila. Genus of birds peculiar to America; includes the gnatcatchers; small, blue-gray spp., much like miniature mockingbirds.
Polistes. Common cosmopolitan genus of large wasps in the Family Polistidae.
Polistidae. Family of large slender wasps resembling yellow-jackets and hornets; adults build small to large paper nests attached to buildings, trees, etc.; Polistes a familiar genus.
Polistotrema. Genus of hagfishes on the West Coast.
Pollachius. See GADIDAE.
pollack. See GADIDAE.
pollard. Any male mammal which is normally horned but is hornless for a variety of reasons.
polled. Shaved or hornless.
pollen basket. Corbiculum or pollen plate; in honeybees and their relatives, the broad hind tibia and its peripheral curved hairs; used for carrying masses of pollen.
pollen brush. Pollen comb; arrangement of stout setae on metatarsi in the honeybee and its relatives; used to remove pollen from legs and body.
pollen comb. Pollen brush.
pollen compressor. Auricle; a concave space at the junction of hind tibia and metatarsus in the honeybee; used to compress masses of pollen before they are passed to the pollen basket.
pollen plate. Pollen basket.
pollex. Thumb of vertebrate pentadactyl forelimb.
pollination. Transfer of pollen from anther to stigma of flower(s); usually effected by gravity, wind, or insects.
pollinator. Agency which transfers pollen from anther to stigma of flower(s); usually an insect, bird, wind, or gravity.
polliwog. Tadpole.
polocyte. Polar body.
polyandry. Condition of one female mating with two or more males. Adj. polyandrous.
Polyarthra. Very common genus of rotifers; with 12, movable, sword- or blade-shaped lateral appendages.
polyaxon. 1. Neuron with several axons. 2. Sponge spicules with the parts extending in three or more directions.
Polycarpa. Genus of tunicates in the the Class Ascidiacea; in sand and mud bottoms along northern Pacific Coast.
Polycelis. Genus of fresh-water triclad Turbellaria; numerous small eyespots around the anterior margin.
Polycentropus. Genus of Trichoptera.
Polycera. Palio.
Polychaeta. One of the classes in the Phylum Annelida; segmentation well defined; most segments with a pair of lateral setaceous parapodia; head region with palps and tentacles; prostomium and peristomium present; often brightly colored; larva a trochophore; reproduction by budding in some spp.; many live in burrows or tubes in sand and mud bottoms; about 4,000 spp.
Polycheria. Genus of marine Amphipoda.
Polychoerus. Genus of small free-living flatworms in the Order Acoela; often found on eelgrass.
Polycirrus. Genus of marine Polychaeta; tentacles very numerous and extending in all directions.
Polycitor. Genus of sessile, gelatinous, colonial ascidians.
Polycladida. Order of large marine Turbellaria; pharynx opens into the main gastrovascular cavity which has many radiating diverticula; Notoplana common.
polyclimax hypothesis. Idea that within a given climatic area there may be several types of climax plant and animal communities, determined by variations in physiography, soils, temperature, etc. Cf. MONOCLIMAX.
Polycope. Genus of marine ostracods in the Order Cladocopa.
Polyctenidae. Bat bugs; family of Hemiptera ectoparasitic on bats; wingless, brownish, viviparous spp.; tropical and subtropical.
polycythemia. Abnormally large number of erythrocytes, sometimes as high as 15 million per cubic mm. of blood; usually the result of severe

diarrhea or other type of withdrawal of body fluids.

polydactyly. 1. Hereditary abnormality of man characterized by an extra digit on the hand or foot; a Mendelian dominant. 2. Occurrence of extra digits in any animal. (Hyperdactyly is a preferable term.)

Polydesmoidea. Order of millipedes; 19 to 22 body segments; male with one or two pairs of gonopods on seventh segment; Polydesmus found commonly in the U.S.

polydesmus. 1. Polydisc strobilation. 2. Budding off of two or more ephyrae simultaneously from a scyphistoma in certain Scyphozoa.

polydomous. Inhabiting several nests simultaneously, as an ant colony.

Polydora. Genus of tube-building colonial Polychaeta.

polyembryony. Production of two or more individuals from one zygote by the separation of the cells in the early stages of cleavage; occurs in identical twinning in man, regularly in the quadruplets of the armadillo, and in many parasitic Hymenoptera.

Polyergus. Genus of robber ants or Amazon ants; slave makers which usually use Formica ants to make their nests and rear their young.

polygamy. Having more than one mate at a time. Adj. polygamous.

polygene. Any gene which, singly, has little visible phenotypic effect, but which usually acts in conjunction with similar genes so that the complex may result in a pronounced phenotypic feature; e.g. multiple factors.

Polygonia. Genus of anglewing butterflies; lateral margins of wings irregular and angled; wings tawny or orange above with darker spots and borders, undersides mottled grays and browns; common in U.S.

Polygordius. Genus of annelids in the Class Archiannelida; threadlike, with numerous indistinct segments; to 1 in. long.

polygraph. 1. Lie detector; a device used to detect slight differences in blood pressure which occur when a subject's emotions and metabolism change when telling a lie. 2. Instrument for simultaneously recording the pulse from several different body areas.

polygyny. Mating of one male with two to many females; e.g. the fur seal.

Polygyra. Large N.A. genus of land snails; usually found in moist, more or less open woodlands.

polykaryocyte. Osteoclast.

Polykrikos. Genus of marine and brackish dinoflagellates; 2 to 16 individuals permanently joined in a chain.

Polymastia. Genus of marine sponges (Subclass Monaxonida); encrusting, with numerous flattened processes.

Polymastigina. Order of flagellates having three or more flagella; a few spp. are free-living, but the majority are parasitic.

Polymerurus. Genus of fresh-water gastrotrichs; caudal furca very long.

Polymesoda caroliniana. Fresh and brackish water bivalve mollusk found from S.C. to Tex.

Polymitarcys. Genus of mayflies.

polymorphism. Occurrence together of two or more distinct morphological forms of a sp.; the phenomenon is best exhibited by many hydroid coelenterates and the various castes of social insects. Adj. polymorphic.

polymorphonuclear leucocytes. Neutrophils.

Polymorphus. Genus of Acanthocephala parasitic in aquatic birds; larval stages in crustaceans.

Polymyaria. Taxon sometimes used to include those nematodes having many muscle cells in each quadrant when the body is seen in cross section.

Polynemidae. Threadfins; a family of fishes related to the mullets; pectoral fins divided, the lower portion having long threadlike rays; along sandy shores of warm seas; Polynemus common.

Polynemus. See POLYNEMIDAE.

Polynesian. Race of mongoloids found in Hawaii, Samoa, New Zealand, and other islands east of a line drawn between these three areas; skin yellow to light brown.

polyneuritis. Inflammation of many nerves at once.

Polynoë. Genus of peculiar marine Polychaeta; with a series of overlapping dorsal scale-like plates.

Polyodon spathula. American spoonbill or paddlefish; mostly in large rivers of the Mississippi Valley; to 6.5 ft. long; feeds on small organisms in mud with the use of the paddle-shaped snout; formerly abundant but now becoming uncommon.

Polyodontidae. Family of bony fishes which includes the spoonbills or paddlefish; to 6.5 ft. long; scales absent;

snout paddle-shaped and very long; teeth absent in adult; opercula rudimentary; Polyodon in Mississippi Valley and Psephurus of the Yangtse River.

Polyonyx. Genus of marine crabs which occur as commensals in the tube of Chaetopterus.

Polyophthalmus. Genus of Polychaeta.

Polyorchis. Genus of medusoid marine hydrozoan coelenterates; the hydroid stage is absent.

polyp. 1. Single attached individual of any colonial or solitary coelenterate; occasionally used to refer to an individual in a colony of Bryozoa (more properly known as a polypide). 2. A projecting mass of hypertrophied mucous membrane, especially in the nasal cavities.

Polypedates. See POLYPEDATIDAE.

Polypedatidae. Family of tree frogs occurring in Asia and Africa; Polypedates a common genus which usually oviposits in vegetation over water from which the tadpoles drop into the water.

Polypedilum. Genus of tendipedid midges.

polypeptide. Compound consisting of three or more linked amino acids.

Polyphaga. One of the two suborders in the insect Order Coleoptera; antennae variable, hind wings without cross veins. Cf. ADEPHAGA.

Polyphemus. Genus of small limnetic cladocerans; carapace absent, legs short and prehensile; see ONYCHOPODA.

polyphemus moth. See TELEA.

polyphyletic. Pertaining to a taxon derived from two or more ancestral sources. Cf. MONOPHYLETIC.

polypide. Soft parts of a single individual of a colony of Bryozoa.

Polyplacophora. True chitons; one of the two mollusk orders in the Class Amphineura; with a mid-dorsal row of eight calcareous plates; body elliptical and greatly flattened ventrally; usually on rocks in coastal shallows.

Polyplax spinulosa. Spiny rat louse in the Order Anoplura.

polyploidy. Condition in which the chromosome complement of the body cells is a multiple of the haploid (n) number greater than 2.

polypnea. Abnormally rapid rate of breathing.

Polypocephalus. See LECANICEPHALOIDEA.

polypod. Early embryological stage of some insects; appendage buds are present on all segments, including the abdominal segments.

Polypteridae. Small family of fresh-water fish; contains a single sp. of Polypterus.

Polypterus bichir. Bichir; a large cylindrical African fresh-water fish with well developed lungs and enameled scales; see CLADISTIA.

polysaccharide. Carbohydrate consisting of two or more disaccharides in a long chain; $(C_6H_{10}O_5)_{2000}$ is an approximate formula for cellulose; $(C_6H_{10}O_5)_{24-26}$ is a typical starch; $(C_6H_{10}O_5)_{12-18}$ is glycogen.

polysaprobic. 1. Pertaining to aquatic habitats marked by the absence or only small amounts of dissolved oxygen and excessive organic decomposition. 2. Pertaining to organisms in such habitats. Noun polysaprobe.

polyspermy. Penetration of a mature animal ovum by two or more spermatozoa; only one spermatozoan, however, functions in the actual fertilization process.

polyspondylic. Pertaining to a type of vertebral column in which the notochord is surrounded by many serially-arranged calcified rings; e.g. certain chimaeroid fishes.

Polystoma. Common genus of monogenetic trematodes parasitic on amphibians and turtles; with a large posterior holdfast.

polystomium. Any of numerous pores which collectively constitute the proctostome in certain jellyfishes. Pl. polystomia.

Polytoma. Genus of small biflagellate colorless Protozoa; stagnant fresh waters.

Polytomella. Genus of very small fresh-water Protozoa; four anterior flagella.

polytrochula. Type of free-swimming larva in some Polychaeta; similar to a trochophore except that there are several small posterior segments.

polytrophic. 1. In certain insects, the occurrence of ovarioles in which a nutritive cell is in close association with a developing egg. 2. Obtaining food from a variety of sources.

polytypic. Pertaining to a taxonomic category containing two or more immediately subordinate categories, such as an order containing two or more families and a genus containing two or more spp.

polytypic species. Any sp. showing conspicuous geographic variations in morphological characters (subspecies); subspecies adjacent to each other, however, commonly interbreed and show intergrades along the margins of their adjacent ranges; common in many land animals.

polyuria. Excessive urine formation.

Polyxenus. Common genus of millipedes in the Order Pselaphognatha; only 2 to 3 mm. long.

Polyzoa. Bryozoa.

Polyzonium. Genus of millipedes.

Pomacea. Apple snails; genus of large fresh-water snails; both gill and lung present; muddy substrates in Fla. and Ga.

pomace flies. See DROSOPHILIDAE.

Pomacentridae. Damselfishes; a large family of brightly colored marine percomorph fishes; mostly in tropical reef areas.

Pomatiopsis. Genus of amphibious fresh-water snails; common east of Great Plains.

Pomatomus saltatrix. Common bluefish; a large, oceanic, basslike sp.; common in warm and temperate seas; important commercial and game fish.

pomfret. 1. Any of several deep-bodied, blackish, pelagic, spiny-rayed fishes of the North Atlantic and Pacific; see BRAMIDAE. 2. Any of several edible East Indian marine fishes.

Pomoxis. Genus of N.A. fresh-water centrarchid fishes, including especially the white crappie (P. annularis) and the black crappie (P. nigromaculatus).

pompano. 1. Deep-bodied excellent food fish of the South Atlantic and Gulf Coast (Trachinotus). 2. Similar fish of the Calif. coast (Palometa).

pompano shell. Variable wedge shell; Donax variabilis, a small and extremely abundant bivalve; coloration highly variable; N.C. to Tex.

Pompholyx. Genus of plankton rotifers; lorica of dorsal and ventral plates; foot absent.

Pomphorhynchus. Genus of Acanthocephala parasitic especially in fishes.

Pompilidae. Family of spider wasps which stock their nests with paralyzed spiders and insects; see PEPSIS.

pond skaters. Common name of hemipterans in the Family Gerridae.

Ponera. Genus of primitive carnivorous ants which live in subterranean colonies of 20 to 30 individuals; tropical and Australian.

Pongidae. Family of the Order Primates; includes the anthropoid apes, such as the gibbon, siamang, orang utan, gorilla, and chimpanzee.

Pongo satyrus. Orang utan of swampy forests of Sumatra and Borneo; a large anthropoid ape; skull high, face flat, hair long and reddish; constructs crude nests in trees; herbivorous; formerly Simia satyrus.

pons Varolii. Broad mass of transverse nerve fibers on the ventral surface of the mammalian brain at the anterior end of the medulla oblongata.

Pontaster. Circumboreal genus of sea stars (Order Phanerozonia); disc small.

Pontella. Common genus of marine calanoid Copepoda.

Pontharpinia. Genus of marine Amphipoda.

pontine flexure. Slight dorsal bending of the brain in the region of the metencephalon.

Pontobdella. Genus of leeches parasitic on elasmobranch fishes.

Pontocypris. Genus of marine Ostracoda.

Pontodrilus. Genus of small oligochaetes found along seashores.

Pontonogeneia. Genus of marine Amphipoda.

Pontoporeia affinis. Fresh-water amphipod found in certain deep, cold lakes of northern N.A. and Europe.

Pontoporia. Genus which includes the La Plata dolphin, a slender cetacean to 6 ft. long; dirty white with a black dorsal line; bill long, armed with more than 200 sharp teeth.

Pooecetes gramineus. Vesper sparrow; a grayish sparrow with white outer tail feathers; open country in much of N.A.

poorwill. See PHALAENOPTILUS.

Popillia japonica. Pest beetle in the Family Scarabaeidae; Japanese beetle; grubs feed on roots of grasses, especially in lawns; adults feed on green vegetation, flowers, and fruits; northeastern states.

popliteal artery. Artery which supplies blood to the knee region.

popliteus. Tetrapod muscle which flexes the hind leg and rotates the flexed leg inward.

poppyfish. Pompano-like fish of the Calif. coast; Palometa simillima.

population cycles. Series of more or less oscillatory variations in the numbers of individuals in a sp., ranging from sparse to dense numbers; such oscillations for certain birds and mammals of the north temperate and subarctic areas commonly have peaks separated by intervals of about 4, 9, or 11 years.

population dynamics. 1. Study of the temporal changes in density of one or more spp. in a natural community. 2. Study of population phenomena in a community with special reference to: density, competition, available food, and physical and chemical changes in the environment.

population genetics. Study of the relative frequency of hereditary characters and alleles in large samples of whole populations of a sp.

population growth curve. See LOGISTIC CURVE.

Porania. Genus of sea stars (Order Phanerozonia); disc wide, arms short.

Poraniomorpha. Circumboreal genus of sea stars (Order Phanerozonia); disc wide, arms short.

porbeagle. See LAMNA.

Porcellana. Genus of small marine decapod crustaceans; abdomen folded under cephalothorax; chelae massive and flattened; often commensal with other marine invertebrates.

Porcellionides. Cosmopolitan genus of terrestrial Isopoda.

Porcellio scaber. Common sp. of terrestrial sowbug (Order Isopoda); about 12 mm. long; common in damp places, under bark of fallen trees, stones, etc.

porcupine. See ERETHIZONTIDAE.

porcupine fish. See DIODONTIDAE.

porcupine-rat. See ECHIMYIDAE.

Porella. Common genus of marine Bryozoa.

porgy. Any of several deep-bodied food fishes of the Mediterranean and Atlantic; see STENOTOMUS.

Porichthys. See HAPLODOCI.

Porifera. Phylum consisting of the sponges; the supporting skeleton is composed of calcium carbonate spicules, an organic material (spongin), silicon dioxide spicules, or spongin plus silicon dioxide spicules; the body is highly porous, and water passes continuously through its channels and cavities, there being no body cavity; sponges have no organs, only tissues; the most characteristic type of tissue consists of a layer of choanocytes; most of the 4500 spp. are marine, a few (the Spongillidae) being found in fresh waters.

Porina. Genus of marine Bryozoa.

Porites. Genus of colonial stony corals; cup with about 12 short septa.

porkfish. See GRUNT.

pork tapeworm. See TAENIA SOLIUM.

Porocephalida. Class of Linguatulida; mouth hooks with a basal arm; genital pore anterior in male, posterior in female; life cycle with intermediate host; Linguatula larvae in viscera of mammals, adults in nasal cavities of mammals.

Porocephalus. Genus of tongue worms parasitic in rodents and snakes.

porocyte. Barrel-shaped cell having a cavity running completely through from one end to the other; found in ascon sponges where they serve as minute channels from the external surface to the spongocoel; the actual opening through the porocyte is called a prosopyle.

Porohalacarus. Uncommon genus of fresh-water mites in the predominantly marine Family Halacaridae.

Porolohmannella. Uncommon genus of fresh-water mites in the predominantly marine Family Halacaridae.

Poromyia. Genus of marine bivalves in the Order Septibranchia.

Poronotus triacanthus. Butterfish or dollarfish; thin, deep-bodied marine sp. found along Atlantic Coast; scales very small; body to 10 in. long.

porphyrin. Organic compound in which four pyrrol nuclei are connected in a ring structure by -CH- groups; usually associated with metals, such as iron or magnesium; form important parts of the molecules of hemoglobin, cytochromes, and chlorophylls.

porphyropsin. Pigment in the retinal rods of fresh-water fishes and amphibians previous to metamorphosis.

Porphyrula martinica. Purple gallinule; head and under parts purple, wings light blue, back and tail olive green; fresh-water swamps, S.C. and Gulf states to tropical South America.

Porpita. Genus of marine hydrozoan coelenterates; an air-filled float keeps the colony at the surface, and the complex colonial system of polymorphic hydroids dangles downward in the water.

porpoise. Any of several small gregarious and carnivorous oceanic cetaceans in the Suborder Odontoceti; torpedo-shaped, to 8 ft. long; snout blunt; common in most seas; familiar American sp. is Phocaena phocaena, the harbor porpoise.

portal system. Series of large veins having a bed of capillaries at either end; e.g. hepatic portal and renal portal in vertebrates.

portal vein. See PORTAL SYSTEM.

Porthetria. Lymantria.

Portuguese man o' war. See PHYSALIA.

Portunidae. Family of marine swimming crabs; carapace broader than long, with a serrate anterior margin; last pereiopods paddle-like; Callinectes and Carcinides common American genera.

Portunus. Widely distributed genus of swimming crabs; body transversely oval; sometimes found far from shore in floating debris and brown algae.

Porzana carolina. Sora or sora rail; a small N.A. gray-brown rail with a chicken-like yellow bill; fresh-water marshes.

position effect. Phenotypic change produced by a chromosome aberration, presumably due to the interaction of genes which are not normally adjacent in the chromosome.

positive eugenics. Encouragement of marriages between individuals having desirable hereditary features, e.g superior mental ability.

possum. Opossum.

postabdomen. Posterior end of the body of Cladocera; bears a pair of terminal claws and is usually bent so as to extend ventrally or anteriorly.

postcardinal. Located posterior to the heart, e.g. postcardinal vein.

post cava. Posterior vena cava.

postcaval vein. Posterior vena cava.

postclimax. With reference to any particular ecological climax, the contiguous climax formation produced by a different climate, usually cooler and more moist; the deciduous forest, for example, has the northern coniferous forest as its postclimax. See PRECLIMAX.

posterior. Of or pertaining to the rear end; toward the rear end; caudal; in man, the back of the body is said to be posterior.

posterior adductor. Large but short muscle in the posterior portion of many bivalve mollusks; runs transversely from one valve to the other.

posterior chamber. Flat space in the eyeball; bounded anteriorly by the iris and posteriorly by the lens, suspensory ligament, and ciliary body; filled with aqueous humor and continuous axially with the anterior chamber by the diameter of the pupil

posterior choroid plexus. Thin membranous vascular area covering the fourth ventricle in the vertebrate brain.

posterior cricoarytenoid. One of two very small muscles in the posterior (dorsal) portion of the larynx.

posterior mesenteric artery. Small artery supplying the large intestine in certain vertebrates.

posterior pituitary. Posterior lobe of the pituitary gland of vertebrates; derived from the infundibulum of the brain during early embryology. See PITUITARY.

posterior retractor. One of a pair of muscles found in some bivalve mollusks; aids in retracting the foot.

posterior root. Dorsal root.

posterior vena cava. Large vein bringing blood to the heart from the kidneys, gonads, liver, and the general posterior body musculature in tetrapods.

postganglionic nerve fibers. See AUTONOMIC NERVOUS SYSTEM.

postlarva. Stage following the prolarva in certain fishes; marked by many different subjective features which separate them from typical juveniles; e.g. pigmentation, transparency, pelagic habit, flotation devices, external gills, etc.

postmentum. Most basal portion of an insect labium; a single median plate.

post mortem. Occurring or performed after death, especially an examination of the body.

postneural plate. One or two posterior median plates of a tortoise carapace.

postoral. Behind the mouth.

postorbital. Bone on the upper back orbital margin of many tetrapods.

postparietals. Pair of bones on either side of the dorsal midline at the back rim of the roof of the braincase in many tetrapods.

postzygapophysis. Inferior or posterior zygapophysis; an articulating vertebral process arising on the neural arch.

Potamanthus. Genus of Ephemeroptera; nymphs fossorial.

Potamilla. Genus of Polychaeta; in flexible tubes of sand.

potamobenthos. Collectively, the bottom organisms of a river.

Potamochoerus. Bush pigs; a genus of African wild hogs; large herds travel through dense forests rooting up the ground extensively.

Potamocypris. Common genus of fresh-water Ostracoda.

potamodromous. Pertaining to fishes that migrate freely up and down river.

Potamogale. See POTAMOGALIDAE.

Potamogalidae. Family which includes the giant water-shrew (Potamogale) of equatorial African forests; otter-shaped, to 2 ft. long; head shovel-shaped, ears and eyes reduced; nocturnal, burrow in river banks and feed on aquatic invertebrates.

potamology. Study of rivers, especially their biology, physics, and chemistry.

potamoplankton. Plankton found in running waters.

Potamopyrgus. Small genus of fresh-water snails; Fla. and Tex.

Potamyia. Genus of Trichoptera.

potato beetle. See LEPTINOTARSA.

potato worm. Large green and white larva of a hawkmoth; feeds on the leaves of the potato plant; see SPHINGIDAE.

potential energy. In the broad sense, energy in an inactive form owing to structure or position, and not manifest as motion; e.g. the energy possessed by sugar or any other fuel, or the energy possessed by any object by virtue of its position above a substrate.

Poteriodendron. Genus of colonial monoflagellate Protozoa; cell with a small collar and enveloping lorica.

Poterion. Genus of sponges in the Class Demospongiae; includes Neptune's goblet, a large goblet-shaped sponge.

potoo. See NYCTIBIIDAE.

potoroo. Small ratlike kangaroo found commonly only in Tasmania; genus Potorous.

Potorous. See POTOROO.

Potos caudivolvulus. Common kinkajou, a mammal in the Family Procyonidae; nocturnal and arboreal; head catlike, tongue very long, eyes large; body slender, tail long and prehensile; Mexico, Central and South America.

potter wasps. See EUMENES and EUMENIDAE.

potto. See LORISOIDEA and PERIODICTICUS.

pouched jerboa. See ANTECHINOMYS.

poulard. 1. Castrated hen. 2. Hen in which the oviduct has been severed.

pout. See AMEIURUS.

powder down. Specialized down feather which grows continuously and frays off at the tip; used for dressing the other feathers; present especially in herons and parrots.

powder post beetles. See LYCTIDAE.

poyou. Six-banded armadillo of Argentina.

P-P factor. Obsolete term for niacin.

praeputium. 1. Sheathlike mass which encloses the retracted penis in many snails. 2. Prepuce. 3. Fold of labia minora over glans clitoridis.

prairie. Humid to semi-arid grassland; in the U.S. extending from the Mississippi Valley westward to the Rockies, and from Canada to Tex.

prairie chicken. See TYMPANUCHUS.

prairie dog. See CYNOMYS.

prairie falcon. See FALCO.

prairie rattler. See CROTALUS.

prairie wolf. Coyote.

pratincole. Any of several small insectivorous Old World birds having long, pointed wings and a deeply forked tail; found in flocks; Glareola common.

Pratylenchus. Genus of microscopic, free-living nematodes.

prawn. Any of numerous shrimplike decapod crustaceans; claws weak or absent; abdomen compressed; many spp. edible.

Praxillella. Northern genus of Polychaeta.

praying mantids. Members of the insect Order Mantodea; same as preying mantids.

preabdomen. Anterior broad abdominal region of scorpions.

preadaptation. Specialization for a habitat which an organism does not actually occupy when such a specialization appears; term usually applied to an advantageous mutation which will permit the invasion of a new habitat.

preanal. Anterior to the anus.

preanal fin. In cephalochordates, the single median ventral fin extending from atriopore to anus.

preantennal segment. Most anterior embryonic segment of a typical arthropod; later incorporated into the head and has its identity lost; similar to the precheliceral segment of arachnids.

prearticular. Narrow dermal bone on the median surface of the lower jaw in primitive tetrapods.

precava. Anterior vena cava.

precaval vein. Anterior vena cava.

precentrum. In some lower vertebrates, an anterior portion of a vertebral centrum; commonly bears neural and hemal arches; in some fishes such centra alternate with centra having no arches.

precheliceral segment. See PREANTENNAL SEGMENT.

precipitin. Type of antibody which acts by coagulating or precipitating the antigen for which it is specific.

precipitin reaction. Formation of a visible precipitate at the interface when an antigen is carefully added to its corresponding antiserum.

precipitin tests. Biochemical tests between the blood sera or other body fluids of various animals in order to determine their relative phylogenetic interrelationships.

preclimax. With reference to any particular ecological climax, the contiguous climax formation produced by a different climate, usually warmer and drier; the deciduous forest, for example, may have grassland as its preclimax. See POSTCLIMAX.

precocial. 1. Pertaining to the young of ducks, shore birds, chickens, etc., which are covered with down and fully active at hatching. 2. Pertaining to the young of jackrabbits, deer, livestock, etc. which are fully haired at birth, have the eyes open, and are able to move about immediately after birth. Cf. ALTRICIAL.

precoracoid. Anterior ventral bony or cartilaginous part of the pectoral girdle; lies in front of the coracoid in many amphibians, reptiles, and monotremes.

precoxa. Small basal segment which precedes the coxopodite in certain crustacean appendages.

precystic. Of or pertaining to a protozoan intermediate between the trophozoite and mature cyst stages.

predaceous diving beetles. See DYTISCIDAE.

predation. Predatory behavior of any animal.

predator. Any animal that kills and consumes another animal.

preen. To trim, arrange, and dress feathers with the beak, especially with oil from the preen gland.

preen gland. Uropygial gland of birds; located on the back above the base of the tail; secretes an oil used in preening the plumage.

prefemur. Short leg segment found especially in certain terrestrial arthropods, e.g. centipedes; located basal to the true femur; sometimes called the second trochanter segment.

preferendum. Point or area in a gradient of an ecological factor which is "selected" by a motile organism.

Preformation theory. Early idea that the ovum contains a more or less perfect miniature of an adult animal, and that development consists of the growth and enlargement of this "homunculus" upon stimulation by the seminal fluid; some proponents maintained that it involved "encasement" of progressively more minute individuals, like a series of boxes one within another. Cf. SPERMATISTS, OVISTS.

prefrontal area. Foremost area of the frontal lobe of the brain; concerned chiefly with memory, conscience, obligation, etc.

prefrontal bone. One of two bones in the skull of many fishes, turtles, and some other tetrapods; one on each side of the middorsal line behind the nostrils.

preganglionic nerve fibers. Autonomic nervous system.

pregnancy. Condition or period during which the female carries developing young; term usually used with reference to vertebrates but may be used for any animal.

pregnancy test. Any of numerous biochemical tests used to determine early pregnancy, especially in the human female; e.g. when urine is injected into a virgin female rabbit, the ovaries will undergo certain morphological changes if the urine contains pregnancy hormones.

pregnant. 1. Carrying young in some stage of development. 2. Gravid.

prehallux. Calcar of the frog.

prehensile. Adapted for grasping or seizing, especially by curling or wrapping around, such as the long tail of American monkeys.

premaxilla. Paired bone at the anterior end of the upper jaw of certain vertebrates.

prementum. In typical insects, a transverse sclerite attached to the distal end of the postmentum; divided along the median line and bearing the labial palps and the ligula or glossae and paraglossae.

premolar. Certain tooth or pertaining to certain teeth in front of the true molars in mammals; variously modified; eight present in man (usually called bicuspids).

prenatal. Before birth.

Prenolepis. Large and common genus

of ants.

preopercle. L-shaped membrane bone lying in front of the operculum in fishes.

preoral. Situated in front of the mouth.

preoral somites. 1. First three embryonic segments in an arthropod embryo; in front of the mouth. 2. Those somites preceding the mouth in Annelida.

prepotency. Ability of one individual or strain to transmit a character to offspring, depending on relative character homozygosity and gene dominance.

prepubes. Pair of "marsupial bones" extending forward from the pubis and supporting the body wall in monotremes and marsupials. Sing. prepubis.

prepuce. Foreskin; fold of skin which may cover the tip of the penis (glans) in certain male mammals.

prepupa. Quiescent extra instar between the last larval instar and the true pupa in certain insects exhibiting hypermetamorphosis.

presbyopia. Defect of vision appearing in old age; a type of far sightedness produced by loss of lens elasticity.

Presbytis. Genus of Old World monkeys in the Superfamily Cercopithecoidea; langurs and hanuman; slender, long-tailed spp. with bushy eyebrows and a tuft of hair on the chin.

prescutum. In insects, the anterior portion of the mesonotum or metanotum.

presoma. Anterior muscular portion of the microscopic larva of a nematomorph larva.

presphenoid. Anterior ventral braincase bone in mammals.

presternum. 1. Anterior median element of the sternum in vertebrates, especially mammals. 2. In arthropods, a narrow anterior portion of a sternum; sometimes set off by a suture.

Prestwitchia. Genus of extinct horseshoe crabs ancestral to Limulus.

presumptive tissue. Embryonic tissue which, before differentiation, is known or presumed to develop into a definite adult tissue or organ.

prey. Any animal that is killed and consumed by another animal.

preying mantids. Members of the insect Order Mantodea; same as praying mantids.

prezygapophysis. Anterior or superior zygapophysis; an articulating vertebral process.

Priacanthidae. Catalufas, big eyes; a family of small, carnivorous, tropical and subtropical fishes; body short, scales rough, eyes large, mouth oblique, coloration bright; relatives of the sea basses.

Prianos. West Coast genus of marine sponges.

Priapulida. Small phylum of uncertain affinities; body cylindrical, annulated, to 3.5 in. long, and supplied with abundant spines and wartlike protuberances; mouth at the tip of a thick anterior proboscis; body wall includes cuticle, hypodermis, circular muscles, longitudinal muscles, and coelomic epithelium; nervous system consists of an anterior ring and ventral epidermal nerve cord; excretion by flame bulb system (solenocytes); growth by a series of molts; in soft mud of colder seas from the intertidal zone to depths of 500 m.; predaceous; only six spp. in two genera (Priapulus and Halicryptus).

Priapuloidea. Priapulida.

Priapulus. See PRIAPULIDA.

prickleback. See STICHAEIDAE.

prickly pen shell. Pinna rigida; a triangular marine bivalve 6 to 9 in. long; ribs beset with scales; tropical and subtropical.

primaquine. Antimalarial drug, especially effective against malaria organisms in the liver and spleen.

primaries. Outermost principal flight feathers of a bird; attached to digits III and IV.

primary embryo. Internal embryo formed in certain Bryozoa; by means of budding, small secondary embryos pinch off from it and become free-swimming larvae.

primary follicle. Undeveloped Graafian follicle in the ovary of higher vertebrates; it is small and consists of a developing ovum and a single layer of follicle cells, the follicular cavity not yet being developed.

primary mesentery. Mesentery in an anthozoan coelenterate; extends from the body wall to the stomodaeum.

primary oocyte. In oogenesis, the cell in which the prophase of the first meiotic division begins; a large cell derived by growth of an oogonium.

primary productivity. Total quantity of green plant (and other autotrophic) protoplasm produced per unit time in a specified habitat.

primary septum. Primary mesentery.

primary sex organs. Testes of the male and ovaries of the female.

primary sexual characters. Ovaries and testes.

primary spermatocyte. In spermatogenesis, the cell which undergoes the first meiotic division, forming two secondary spermatocytes; derived from a spermatogonium.

primary succession. Ecological succession which begins in a barren area and eventually may be expected to produce a climax; such areas are volcanic deposits, sand bars, newly-formed ponds, etc.

primiparous. Bearing young or producing eggs for the first time.

Primates. Order of mammals which includes lemurs, monkeys, tarsiers, marmosets, apes, and man.

primitive knot. Hensen's node.

primitive metamorphosis. See AMETABOLA.

primitive streak. Thickened dorsal longitudinal median strip of ectoderm and mesoderm formed in early bird and mammal embryos.

Primnoa. Genus of corals (Order Alcyonacea); finely branched; colony to 3 ft. high; on rocky bottom.

primordial germ cells. Early cells in an embryo which can be recognized as ancestral to the functional germ cells in the gonads of the mature animal.

primordium. Anlage; embryological cellular beginnings of a future tissue, organ, or other part before distinctive differentiation. Pl. primordia.

principle. See LAW.

principle of biogenesis. Biological principle which maintains that a living organism can originate only from a similar living organism; refutes the old concept of spontaneous generation.

Priodontes. See GIANT ARMADILLO.

Prionchulus. Genus of microscopic, free-living nematodes.

Prionocera. Genus of crane flies in the Family Tipulidae; larvae in wet organic mud.

Prionodesmacea. In some classifications, one of the three orders of mollusks in the Class Pelecypoda; two lobes of mantle separate and opening ventrally and posteriorly; siphons poorly developed or absent; fresh-water mussels and many familiar marine pelecypods belong in this order.

Prionodon. See LINSANG.

Prionotus. Genus of fishes which includes the sea robins; ugly bottom-living spp. with broad heads, slender bodies, and large fanlike pectoral fins.

Prionoxystus robiniae. See COSSIDAE.

Priscacara. Genus of fossil fishes in the Family Cichlidae.

prismatic layer. Middle lamella of a typical mollusk shell; composed of vertical, angular, close-fitting, prism-shaped units; located between the outer periostracum and the inner nacre.

Prismatolaimus. Common genus of microscopic, free-living nematodes.

Pristicephalus. Genus of fairy shrimps.

Pristina. Common and widely distributed genus of small fresh-water Oligochaeta.

Pristiophorus. Genus of saw sharks.

Pristiphora erichsoni. Larch sawfly; hymenopteran in the Family Tenthredinidae; often a pest on larch trees in Europe and N. A.

Pristis. Genus of sawfishes; sharklike rays to 18 ft. long; elongated snout bearing large toothlike projections along margins, for disabling prey.

Proales. Genus of fresh-water illoricate rotifers; P. wernecki forms microscopic galls on Vaucheria.

proamnion. Area of blastoderm in front of the early embryonic head in higher vertebrates; lacking in mesoderm.

Proanura. Extinct order of amphibians; trunk elongated, tail present, legs not modified for jumping.

proatlas. Small, extra neural arch anterior to the atlas in many reptiles.

Probezzia. Genus of ceratopogonid Diptera.

Probopyrus. Genus of marine Isopoda parasitic in gill chambers of decapods.

Proboscidactyla. See LIMNOMEDUSAE.

Proboscidea. Order of mammals which includes the extinct mastodonts (Mammut) and mammoths (Mammuthus) as well as the modern elephants (Loxodonta and Elephas); massive spp. with large head and ears, short neck, thick skin, few hairs, and the nose and upper lip modified into a muscular proboscis or trunk; two incisors in upper jaw of male modified into ivory tusks.

Proboscidiella. Genus of flagellates living in the intestine of termites.

proboscis. 1. Trunk of an elephant. 2. Long, more or less flexible snout, as in tapirs and shrews. 3. Tubular sucking or feeding organ with the mouth at the tip, as in certain leeches, planarians, and insects. 4. Cylindrical sensory, offensive, and defensive organ at the anterior end of certain invertebrates, e.g. nemertine worms. 5. Anterior armed protuberance in the Phylum Acanthocephala. 6. Narrow scoop-shaped anterior end of echiurid worms; used for feeding and as a sensory device. Pl. proboscides.

proboscis monkey. See NASALIS.

proboscis sheath. Muscular wall of the cavity containing the proboscis in the nemertine worms and some leeches.

proboscis worm. Any member of the Phylum Nemertea.

probuds. Unspecialized buds formed on the stolon of certain pelagic tunicates; they migrate over the surface of the body, become attached to the cadophore, and develop into definitive buds.

Procambarus. Large genus of fresh-water crayfishes occurring mostly in the southern and southeastern states; P. clarki is found in the Gulf states and has been introduced into Calif. and Nev.

Procavia. Dassies; an African genus of hyraxes; common in rocky areas; most active at dawn and dusk.

Procapra. Genus of goat-gazelles; the zeren is a sandy colored sp. of high Mongolian grasslands, and the goa is a long-haired form of Tibet.

procartilage. Early stage in cartilage development.

Procellariidae. Family of world-wide oceanic birds in the Order Procellariiformes; fulmars, shearwaters, and petrels; visit land only to nest; lay a single egg; Puffinus, Oceanodroma, and Fulmarus typical.

Procellariiformes. Order of oceanic birds which includes the albatrosses, shearwaters, fulmars, and petrels; nostrils more or less tubular; bill slightly hawklike at the tip; wings long, pointed; plumage compact and oily; usually nest on islands. See PROCELLARIIDAE, HYDROBATIDAE, DIOMEDEIDAE.

procephalon. That part of the head of an insect representing the fused premandibular segments.

procercoid. Solid elongated larval stage in the life history of certain tapeworms; usually retains the six oncosphere hooks at one end; develops in the body cavity of a fresh-water copepod; e.g. occurs in the life history of Dibothriocephalus latus.

procerebrum. In arthropods, the ganglia of the first true somite plus the more anterior ganglia. Cf. PROTOCEREBRUM.

Procerodes. Genus of small marine flatworms in the Order Tricladida; anterior end with two tentacle-like processes.

Prochordata. Taxonomic category which includes Tunicata and Cephalochordata, and sometimes Enteropneusta.

prochorion. 1. Albuminous coating surrounding a mammalian egg as it passes down the oviduct. 2. Zona pellucida of a fertilized mammalian ovum.

Procladius. Genus of tendipedid midges.

Proclymene. Circumboreal genus of Polychaeta.

Procoela. One of the five orders of amphibians in the Subclass Salientia; vertebrae procoelous; urostyle with two condyles; three modern families.

procoelous vertebra. Vertebra having a centrum which is concave on the anterior surface and convex on the posterior surface.

Proconsul. Genus of Miocene apes.

procoracohumeralis. Group of two dorsal pectoral girdle muscles in urodeles.

Procotyla. Genus of large pale triclad Turbellaria; ponds and streams.

proctodaeum. Ectodermal invagination of an embryo which later becomes the anus, or cloacal aperture and hindgut (in arthropods).

proctodeum. Proctodaeum.

proctology. Branch of medicine dealing with the anus and rectum and their diseases.

proctostome. Single opening of the gastrovascular cavity of the Coelenterata, Ctenophora, and Turbellaria; serves as both mouth and anus; usually referred to as the "mouth." (Term not in the literature of zoology generally, but a useful and definitive term employed regularly by this writer for many years.)

Procyonidae. Family of small to medium-sized carnivores; includes the raccoon, cacomistle, coati, kinkajou, and pandas; plantigrade, five toes on each foot; tail bushy and usually annulated.

Procyon lotor. Common N.A. raccoon, an omnivorous carnivore in the Family Procyonidae; upper parts grizzled, black, gray, and brown; black band across forehead and eyes; tail banded with six or seven dark rings; hibernates during cold weather; mostly nocturnal, common along streams.

Prodelphinus. Genus of small pelagic Atlantic dolphins; fish-eating, to 7 ft.

long.

Prodesmodora. Genus of microscopic, free-living nematodes.

Prodiamesa. Genus of tendipedid midges.

producer. Any organism able to synthesize organic compounds from simple inorganic substances, e.g. green plants.

production pyramid. See PYRAMID OF NUMBERS.

production rate. Total quantity of living protoplasm produced per population unit, per unit time in a specified habitat.

productivity. 1. Inherent capacity of an environmental unit to support organisms. 2. Rate of utilization of energy or formation of protoplasm by one or more organisms. See STANDING CROP, PRODUCTION RATE, PRIMARY PRODUCTIVITY, SECONDARY PRODUCTIVITY, and YIELD.

product rule of probability. If a particular double event consists of two independent events both of which must occur at the same time, then the probability of the double event is the product of the separate probabilities of the single event.

proecdysis. Especially in decapod crustaceans, a period of preparation for a molt, often accompanied by a high blood calcium level.

Proechidna. Zaglossus.

proenzyme. Precursor of an enzyme, e.g. pepsinogen is the proenzyme of pepsin.

proepimeron. 1. Insect prothoracic epimeron. 2. In Diptera, a posterior pronotal lobe.

proepipodite. Epipodite borne on the precoxa of certain crustaceans.

proerythroblasts. Large, early erythroblast.

proestrus. Period of increased follicular activity just preceding estrus in female mammals.

profundal. 1. Pertaining to that portion of the bottom of a lake which lies at a depth greater than that of the upper limit of the hypolimnion, or at a depth greater than the lakeward limit of rooted aquatic plants. 2. Generally referring to very deep portions of a body of water.

profundus nerve. Large branch of the fifth (trigeminal) cranial nerve; except in mammals, it frequently has a separate ganglion and emerges from the brain independently of the trigeminal.

progesterone. Ovarian hormone of mammals produced by the corpus luteum and which, along with estradiol, prepares the uterus for the reception of the ovum; also has a partial role in preparing the mammary glands for milk secretion; the "pregnancy hormone."

progestin. Progesterone.

proglottid. One of the body divisions of a tapeworm; each proglottid contains a full complement of both sets of sex organs.

proglottis. Proglottid.

prognathous. Having the jaws protruding beyond the general anterior facial margin.

Progne subis. Purple martin; largest U.S. swallow; male uniformly blue-black, female with whitish belly; breeds from Canada to Mexico.

prognosis. 1. Forecast of the course of a disease. 2. Outlook for recovery, judging from the severity and symptoms of the particular case.

Progomphus. Common genus of dragonflies.

Progoneata. Diplopoda.

prolactin. Lactogenic hormone.

prolamins. Simple proteins, soluble in 70 to 80% alcohol but insoluble in water and absolute alcohol.

prolan. Active principle of the anterior pituitary exclusive of the growth hormone. See CHORIONIC GONADOTROPHIN.

prolarva. Sac fry; newly hatched fish; obtains nutrition from the yolk sac.

prolegs. 1. Anterior pair of thoracic legs of insects. 2. Processes or appendages serving the purpose of legs in insects, especially on the abdominal segments of caterpillars, some fly larvae and sawfly larvae, etc.

proliferation. 1. Excessive or unusual multiplication of cells in a tissue. 2. Multiplication and spreading of individuals in a colonial sp.

proline. Common amino acid; $C_5H_9O_2N$.

Promenetus. Genus of small fresh-water snails; common in U.S.

promethea moth. See CALLOSAMIA.

Promicrops itaiara. Spotted jewfish; a large sluggish fish of Fla. reefs; to 8 ft. long.

promitosis. Simple type of mitosis, as in certain amoeboid protozoans.

promyelocyte. Early stage in the differentiation of a myelocyte.

pronation. Rotation of the palm downward. Cf. SUPINATION.

pronator profundus. Muscle in reptiles, birds, and amphibians which pronates and rotates the forefoot.

pronator quadratus. Mammalian muscle which pronates and rotates the hand.

pronator teres. Muscle of the forelimb of certain tetrapods.

Proneomenia. Genus of Arctic chitons (Amphineura).

pronephros. Primitive type of vertebrate kidney; persists only in adult cyclostomes, teleosts, and lungfishes; consists of 3 to 15 pairs of segmental tubules, each of which has a nephrostome opening into the body cavity; glomeruli project into the body cavity in the general vicinity of the nephrostome; the segmental tubules are all connected to a pronephric duct which carries excretions posteriorly to the cloaca. Adj. pronephric.

Proneurotes. Genus of marine Nemertea.

prongbuck. See ANTILOCAPRIDAE.

pronghorn. See ANTILOCAPRIDAE.

Pronocephalus. Genus of monostome trematodes in turtles.

Pronolagus. Genus of Asiatic rock-hares.

Pronorites. Genus of fossil ammonites.

pronotum. Dorsal sclerite(s) or dorsal surface of an insect prothorax.

Pronuba. Tegeticula.

pronucleus. 1. Sperm haploid nucleus after its penetration into an ovum but before its fusion with the ovum nucleus. 2. Ovum haploid nucleus after the completion of meiosis but before fusion with the sperm nucleus.

pronymph. 1. Newly hatched first nymph stage in some Odonata and Orthoptera; enclosed in a transient delicate sheath. 2. Stage in certain metabolous insects; larval tissues broken down and adult tissues just beginning to form.

pronymphal sheath. Thin, membranous, shroudlike covering enclosing e.g. newly hatched dragonflies and damselflies; shed in a few minutes.

prootic. 1. One of a pair of bones in the skull of certain vertebrates; each one forms the anterior surface of an ear capsule. 2. Located in front of the ear.

prophase. Initial stage of mitosis and meiosis during which the chromosomes and chromatids become discrete, the centriole (if present) divides, the nucleoli disappear, and the nuclear membrane disappears.

prophylaxis. Prevention of the occurrence of a disease by mechanical precautions or by taking drugs.

Propithecus. Sifakas; genus of silky lemurs of Madagascar.

propleuron. Lateral portion of the insect prothorax. Pl. propleura.

Propliopithecus. Genus of Oligocene fossil apes; featured by a reduced muzzle and enlarged brain.

propneustic. With only the prothoracic spiracles functional, e.g. certain Diptera larvae.

propodeum. In Hymenoptera, the posterior part of the thorax behind the wings and scutellum; overhangs or partly surrounds the insertion of the abdominal petiole.

propodite. Sixth segment of pereiopod in most malacostracan Crustacea.

propodium. Anterior part of the foot in certain mollusks.

propodus. Propodite.

propolis. Cement and varnish material derived from plant resins gathered by bees; used to cement and fill crevices in the hive as a protection against wind and weather.

proprietary medicine. Any preparation used in the treatment of diseases and protected from open competition as to name, composition, or manufacture by means of patent, secrecy, copyright, or trademark.

proprioceptors. Internal receptors which give information to the brain regarding movements, position of the body, and muscle stretch. Cf. INTEROCEPTOR.

Proptera. Common genus of unionid mollusks; eastern half of U.S.

propterygium. Heavy cartilaginous rod supporting the base of each of the pectoral and pelvic fins in elasmobranch fishes. Pl. propterygia.

propupa. 1. Prepupa. 2. In Thysanoptera, certain scale insects, etc., the third nymphal instar; an active stage featured by wing pads; followed by a quiescent fourth nymphal instar commonly called a pupa.

propygidium. 1. Tergite anterior to the pygidium, especially in certain beetles. 2. Seventh tergum in beetles.

prorennin. Renninogen.

Prorhynchella. Genus of rhabdocoel Turbellaria.

Prorhynchus. See ALLOEOCOELA.

Prorocentrum. Genus of yellow-brown marine biflagellates with a small anterior spine; may cause red water outbreaks.

Prorodon. Genus of ovoid to cylindrical ciliates with uniform ciliation; fresh and salt waters.

proscapula. 1. Clavicle of teleost fishes. 2. Vertebrate clavicle.

proscolex. Fluid-filled cyst stage in the early development of a tapeworm.

Proscorpius. Genus of primitive Silurian air-breathing arachnids.

prosencephalon. Forebrain.

Prosobranchia. Prosobranchiata.

Prosobranchiata. One of the three orders of the mollusk Class Gastropoda; shell almost always present, provided with an operculum; head with two tentacles and two eyes; ctenidia anterior to heart; marine and fresh-water spp., the former usually with trochophore and veliger larval stages.

prosocoel. Cavity of the embryonic prosencephalon.

prosoma. Anterior part of the body, especially the cephalothorax in certain arthropods.

Prosopium. Genus of whitefish in the Family Coregonidae; _P. cylindraceum_ is the round whitefish or pilotfish of cold waters of the Great Lakes area northward through much of Canada to the Arctic Ocean; _P. williamsoni_ is the Rocky Mountain whitefish, chiefly a stream sp. from Colo. to Vancouver Island.

Prosopygia. Phylum category used by a few specialists; includes Bryozoa and Brachiopoda.

prosopyle. 1. Minute opening through a porocyte cell in simple ascon sponges; see POROCYTE. 2. In sycon and leucon sponges, a small pore connecting incurrent canals with radial canals (flagellated chambers).

Prosostomata. Group of families of digenetic trematodes in which the proctostome is at the anterior end.

prostate gland. 1. Mass of muscle and glandular tissue surrounding the base of the urethra in male mammals; at the moment of sperm release it secretes an alkaline fluid which has a stimulating effect on the action of sperm. 2. Gland associated with the male reproductive system in many snails and a few other invertebrate groups.

prosternum. Ventral plate of an insect prothorax.

prosthetic group. Any non-protein substance which is chemically combined with a protein molecule.

prosthetics. Branch of medicine and surgery pertaining to artificial substitutes for parts of the body, e.g. false legs and false teeth.

Prosthiostomum. Genus of polyclad Turbellaria.

Prosthogonimus macrorchis. Oviduct fluke of poultry; intermediate hosts are fresh-water snails and dragonfly nymphs and adults.

Prostoma. Genus of small fresh-water nemertines in the Order Haplonemertea; occasionally common in vegetation and debris of lakes, ponds, and sluggish streams.

prostomial tentacles. Sensory tentacles located on the prostomium of typical polychaetes.

prostomium. Extreme anterior end of most annelids, especially Archiannelida, Polychaeta, and Oligochaeta; it is above and anterior to the mouth and has the form of a half segment, only the dorsal portion being present.

protamine. Any of a series of simple, basic proteins; especially abundant in fish spermatozoa.

protamine-zinc-insulin. Substance used by diabetics in place of insulin (alone) because of its slower and longer lasting effects.

protandry. 1. Condition in certain hermaphroditic animals where the mature male and female gametes are formed at different times, sometimes as much as days or weeks apart; in many worms, mollusks, and hagfishes. 2. Among hermaphroditic animals, the formation of functional sperm before the formation of mature ova. 3. Among a few insects, the appearance of males earlier in the year than females. Adj. protandrous.

protarsus. 1. Tarsus of an insect first leg. 2. In ticks and some other arachnoids, the leg segment between the tibia and tarsus.

protaspis. Minute ovoid or circular first larval instar of a trilobite.

protease. General term for any protein-digesting enzyme; proteolytic enzyme.

protective coloration. Coloration and color patterns of animals which tend to make them blend into their surroundings and thus be less easily found by animals preying upon them; examples are the ptarmigan, many moths, and the cottontail rabbit.

protective resemblance. Disguise.

protegulum. Semicircular or semielliptical embryonic brachiopod shell.

Proteida. One of the three orders of caudate amphibians; contains a single Family Proteidae; body slightly depressed; tail with a pronounced fin; permanent bushy gills; lungs present, but permanently aquatic.

Proteidae. Family of salamanders in the Order Proteida; includes the European olm (Proteus) and the American mud puppy or water dog (Necturus).

protein. One of a great number of complex organic compounds in protoplasm; all contain carbon, hydrogen, oxygen, nitrogen, and usually small amounts of sulfur, sometimes phosphorus, and occasionally iron, iodine, or other elements; the structural units of the protein molecule are the amino acids, of which more than 20 kinds are known; all amino acids contain the NH_2 radical arranged in the molecule as follows: NH_2-CHR-COOH, where R is hydrogen or a ring or chain of carbon atoms and -COOH is an acid group; examples of amino acids are: glycine, alanine, serine, threonine, valine, norleucine, leucine, isoleucine, aspartic acid, glutamic acid, arginine, lysine, citrulline, cystine, methionine, phenylalanine, tyrosine, thyroxine, iodogorgoic acid, tryptophane, histidine, proline, and hydroxyproline; by various combinations of these amino acids, an infinite variety of large-molecule proteins is possible. See CONJUGATED PROTEINS and SIMPLE PROTEINS.

proteinase. Any protein-digesting enzyme which acts on native protein molecules and converts them into polypeptides.

Proteles. Aardwolves; a genus of striped hyena-like mammals (Family Hyaenidae) of South Africa; feed on insects, carrion, and occasionally small mammals; smaller than a true hyena and lacking the strong jaws and teeth of hyenas.

Protenor. Genus of bugs.

Proteocephaloidea. Order of tapeworms in the Subclass Cestoda; scolex with four acetabula and an apical sucker or glandular device; adult in the intestine of cold-blooded vertebrates; Proteocephalus in fresh-water fishes, amphibians, and reptiles.

Proteocephalus ambloplitis. Tapeworm commonly found in the small- and largemouth black basses.

Proteolepas. See APODA.

proteolysin. Any substance producing proteolysis.

proteolysis. Enzymatic or other type of hydrolytic breakdown of proteins into proteoses, peptones, and other substances.

proteolytic enzyme. Protease.

Proteomyxa. Small order of naked amoeboid Protozoa with filopodia which often branch and anastomose; mostly parasitic in algae and higher plants in fresh and salt water.

proteose. Intermediate protein digestion product; formed in the stomach and small intestine of man.

Proteroglypha. Group of snakes including the families Hydrophiidae and Elapidae; all dangerously poisonous; venom-conducting fangs on anterior part of upper jaw.

Proteromonas. Genus of pyriform flagellate Protozoa; one flagellum directed anteriorly, another posteriorly; parasitic in intestine of lizards.

Proterosuchus. See EOSUCHIA.

Proterozoic era. One of the five major subdivisions of geological time; lasted from about 1 billion to 550 million years ago; an age of volcanic activity, erosion, and sedimentation; soft-bodied invertebrates and simple plants were present, but fossil evidence is scarce.

Proteus. Genus of gram-negative, motile, rod-shaped bacteria; soil and water saprophytes, common in decaying organic matter and in the human intestine; certain spp. cause infections in lower vertebrates.

Proteus anguineus. European olm, an unpigmented blind salamander found in caves and underground streams.

prothonotary warbler. See PROTONOTARIA.

prothoracic gland. Special endocrine gland in the prothorax of many larval insects; controls growth and imaginal differentiation, especially during the last larval instar.

prothorax. First segment of an insect thorax; bears a pair of legs but no wings. Adj. prothoracic.

prothrombin. Organic material produced from vitamin K in the liver; a normal constituent of blood; during blood clotting it combines with calcium to form thrombin, a catalyst which converts fibrinogen to fibrin; see CLOTTING.

Protista. General category or taxon which includes all unicellular algae, protozoans, bacteria, yeasts, etc.

Protobranchia. Order of pelecypod mollusks; gill with two divergent rows of short, flat filaments; foot with flat ventral surface; two adductor muscles; Nucula, Solemyia, and Yoldia typical.

Protoceratops. Genus of small, hornless, herbivorous dinosaurs, to 6 ft. long; fossils, including eggs, found in deserts of Mongolia.

protocerebrum. Collectively, the ganglia of the first true somite in arthropods. Cf. PROCEREBRUM.

protochlorophyll. Precursor of chlorophyll a and chlorophyll b in plants; $C_{55}H_{70}O_5N_4Mg$.

Protochordata. Prochordata.

Protociliata. Subclass of the Class Ciliata; consists of ciliates having two to several hundred nuclei of only one kind; no cytostome; almost all parasitic in the intestine of frogs and toads; some authorities consider this group a special category of flagellates; Opalina common.

protocneme. Primary mesentery of many anthozoan coelenterates; extends from body wall to stomodaeum.

protoconch. Larval shell of Gastropoda.

protocone. Chief cusp of a mammalian upper molar tooth.

protoconid. Chief cusp of a mammalian lower molar tooth.

protoconule. Minor cusp near the anterior margin of mammalian upper molar.

protocooperation. Association of two different spp. in which each normally gains an advantage, but in which each is able to survive when separated from the other; e.g. the case of certain marine crabs which place anemones on the carapace.

Protodonata. Order of fossil insects resembling modern dragonflies; many spp. attained a huge size, including Meganura with a wingspread of 27 in.

Protodrilus. Genus of small hermaphroditic annelids in the Class Archiannelida; segmentation marked by ciliated rings.

protogyny. In hermaphroditic spp., the formation of functional ova before the formation of functional sperm.

Protohippus. Genus of Miocene horses; milk and permanent teeth cemented and moderately long-crowned; all feet with three toes.

Protohydra. Genus of small, solitary hydroids found in tide pools, coastal swamps, and sandy beaches; tentacles absent; medusa stage lacking; total length to 3 mm.

protomerite. In certain gregarine protozoans, the anterior (usually smaller) of the two cells which make up the individual. Cf. DEUTOMERITE.

Protomonadina. Order of the Class Mastigophora; heterogeneous group with one or two flagella; mostly parasites; pellicle absent; holozoic or saprozoic; Oikomonas and Trypanosoma typical.

Protomonas. Genus of marine and freshwater amoeboid Protozoa.

Protonemertea. Palaeonemertea.

protonephridium. Unit of a flame bulb system; branched or unbranched ciliated osmoregulatory and excretory tubule ending blindly in the body cavity or tissues as flame bulbs or solenocytes; opens to the outside via a nephridiopore; see FLAME BULB SYSTEM. Pl. protonephridia.

protonephromixium. Type of nephromixium in which the coelomoduct is grafted on to the canal of the protonephridium; the external opening (nephridiopore) carries both genital and excretory products; in certain Polychaeta. Pl. protonephromixia.

Protonotaria citrea. Prothonotary warbler; wedge-shaped bill; upper parts yellowish olive and gray, head and under parts yellow; wooded areas in eastern half of U. S. as far north as Dela. and southern Minn.

protonymph. First instar of certain mites.

Protoopalina. Large genus of ciliates parasitic in Ambystoma guts; cell cylindrical, with two similar nuclei.

Protoparce. Important genus of pest moths in the Family Sphingidae; P. sexta is the tobacco worm moth whose larva feeds on leaves of the tobacco plant; P. quinquemaculata is the tomato worm whose larva feeds on leaves of the tomato plant.

protoplasm. The complicated self-perpetuating living material making up all organisms; it is at once a solution of water and many dissolved materials and also a complex colloidal system. See CHEMICAL COMPONENTS OF PROTOPLASM.

protoplast. 1. Currently, all of the living material inside the cellulose wall of plant cells; not applicable to animal cells. 2. Originally, the living unit, composed of cytosome and nucleus.

protopod. Protopodite; see BIRAMOUS APPENDAGE.

protopodite. See BIRAMOUS APPENDAGE.

protopod oligomero. Early embryological stage of some insects; head and thoracic segments and their appendage buds are present, but the abdomen is unsegmented.

protopod polymero. Early embryological stage of some insects; head, thorax, and abdominal segments are present, and limb buds are apparent on head and thorax.

Protopterus. See LEPIDOSIRENIDAE.

Protoptila. Genus of Trichoptera.

Protosauria. Order of extinct reptiles, especially abundant in Permian times; agile, lizard-like, land-living spp.

Protospondyli. Order of bony fishes which includes the carnivorous bowfin or dogfish, Amia calva, of sluggish rivers and lakes of the eastern half of the U.S.; dorsal fin long and continuous; tail short, heterocercal; scales cycloid; head bony; air bladder large.

Protospongia. Genus of uniflagellate fresh-water Protozoa; each cell with a gelatinous conical collar; colonies of 6 to 60 cells imbedded in a common gelatinous mass.

Protostomia. Group of higher phyla in which cleavage is determinate, mesoderm and coelom are formed by proliferation of mesodermal bands, and the blastopore becomes the mouth; includes the Annelida, Arthropoda, and Mollusca. Cf. DEUTEROSTOMIA.

Protothaca. Common genus of Pacific Coast clams; littleneck clams; thick-shelled spp. related to Venus but smaller.

Prototheria. Primitive subclass of mammals; includes the monotremes or egg-laying mammals; see MONOTREMATA.

Prototremata. Order of brachiopods in the Class Articulata.

prototroch. Pre-oral ring of cilia in a trochophore larva.

prototype. Primitive form regarded as ancestral to a particular taxon; archetype.

Protozoa. 1. Phylum consisting of unicellular animals; they are marine, fresh-water, or parasitic, and occur in all habitats wherever there is water or a certain amount of dampness; 25,000 described living spp. and 20,000 described fossil spp. 2. A corresponding subkingdom of the animal kingdom.

protozoea. Early larval stage of shrimps and prawns; seven pairs of appendages.

protozoology. Study of all aspects of the biology of Protozoa, or single-celled animals.

protractor muscle. 1. One of a pair of muscles found in some bivalve mollusks; aids in extending the foot. 2. Any muscle that protracts or extrudes a structure.

Protrichoniscus. Genus of terrestrial Isopoda.

protrochula. Free-swimming ciliated larva of certain polyclad Turbellaria; presumed to foreshadow the trochophore.

Protura. Order of minute primitive insects; no antennae, eyes, wings, or metamorphosis; in damp, dead vegetation on the soil; Acerentulus common.

proventriculus. 1. Anterior part of the bird stomach into which digestive enzymes are secreted, the posterior part being the gizzard. 2. In insects, the same as the gizzard. 3. In earthworms, the thin-walled crop just before the gizzard.

provitamin. Chemical precursor of a vitamin; e.g. carotene is provitamin A.

proximal. Situated toward or near a point of reference or attachment.

proximal convoluted tubule. Portion of a kidney tubule which begins immediately below Bowman's capsule and is coiled about the capsule.

Prunella. See PRUNELLIDAE.

Prunellidae. Small family of sparrow-like European and Asiatic birds; includes the accentors; Prunella common.

Prunum. Genus of marine gastropods; marginellas; small smooth spp. with large body whorl and low spire; Carolinas to West Indies.

Przhevalsky's horse. Tarpan.

psalterium. Omasum.

Psaltriparus minimus. Bush-tit; a small sp. in the Family Paridae; plain and gray-backed, with a stubby bill and long tail; western U.S. to Guatemala.

Psammobia. Genus of small, oblong marine bivalves.

Psammochares. See PSAMMOCHARIDAE.

Psammocharidae. Spider wasps, a family of small to large wasps; nest in burrows, tree holes, and crevices of all kinds, usually provisioned with spiders; Psammochares a large and widely distributed genus.

Psammodus. See BRADODONTI.

psammolittoral. Sandy shoreline habitat of a lake or stream; extends from below the water's edge to a few meters above the water's edge; occasionally applied to similar marine shorelines.

psammon. 1. Collectively, the microorganisms inhabiting the interstices of the psammolittoral. 2. Psammolittoral.

Psammophis. Genus of venomous African and Asiatic snakes.

Pselaphidae. Cosmopolitan family of minute yellowish and reddish beetles found in duff, under objects, and in ants' nests.

Pselaphochernes. Genus of Pseudoscorpionida.

Pselapognatha. Order of millipedes; a subclass in some classifications; 11 segments; integument poorly sclerotized; cosmopolitan; Polyxenus typical.

Psephenidae. Riffle beetles; a very small family of aquatic beetles; usually on rock or gravel bottoms of streams and wave-swept shores; larvae (water pennies) broadly oval, extremely flattened, and scale-like; Eubrianax and Psephenus common.

Psephenus. See PSEPHENIDAE.

Psephurus gladius. Paddlefish of the Yangtze River.

Pseudacris. Genus of small swamp cricket frogs; eastern and central states.

Pseudaeginella. Genus of caprellid Amphipoda.

pseudaposematic. Especially among insects, the color or form imitation by one sp. of another sp. having a disagreeable taste or other effective defense mechanism.

Pseudarchaster. Circumboreal genus of sea stars (Order Phanerozonia).

Pseudechiniscus. Large and widely distributed genus of Tardigrada.

Pseudechis. Genus of venomous Australian snakes.

Pseudemys. Sliders, cooters; genus of about ten spp. of turtles found in U.S. inland waters.

Pseudione. Genus of isopods parasitic on marine shrimps and anomurans.

pseudobranch. 1. False gill or accessory gill found in some fishes and mollusks; often does not have a respiratory function. 2. In some fishes, a gill-like structure on the inner surface of the operculum near the upper edge.

Pseudobranchus striatus. Mud siren, a small eel-shaped amphibian in the Order Meantes; swamps, ditches, and ponds in S. C. to Fla.; to 8 in. long.

pseudocardinal teeth. Set of internal interlocking teeth just below the umbones in many bivalve mollusks.

Pseudoceros. Genus of brightly colored marine polyclad Turbellaria.

Pseudochama. Genus of marine bivalves; right valve attached to rocks of intertidal zone.

Pseudochironomus. Genus of tendipedid midges.

Pseudochirops. Genus of diurnal phalangers of Papuan and Australian forests.

Pseudochlamys. Genus of fresh-water amoeboid Protozoa; with a discoid test.

Pseudochordodes. Genus of Nematomorpha.

Pseudocloeon. Common genus of Ephemeroptera.

Pseudococcidae. Large and destructive family of coccid insects; adults covered with mealy or cottony wax secretion; the mealybugs; Pseudococcus an important genus.

Pseudococcus. Large and destructive genus of insects in the homopteran Family Pseudococcidae; P. citri is the citrus mealybug which is a pest on citrus and other plants.

pseudocoel. Chief body cavity of certain invertebrate phyla (e.g. Rotatoria and Nematoda) which is not everywhere lined with an epithelium of mesodermal origin; usually the inner surface of the body wall is lined with mesoderm, but the endodermal digestive tract has no covering of mesoderm.

Pseudocoelomata. Group of phyla having a pseudocoel type of body cavity, including the Nematoda, Nematomorpha, Acanthocephala, Rotatoria, Gastrotricha, Kinorhyncha, and Entoprocta.

pseudocyesis. Striking psychosomatic condition in a woman who imagines herself pregnant; brought on by a fear of pregnancy or intense desire for it; the breasts and abdomen may enlarge, and pseudolabor pains may even occur; usually rectified by establishing the reason for the disturbed emotional state.

Pseudocytheretta. Genus of marine Ostracoda.

pseudoderm. Type of epidermis in certain compactly arranged sponges.

Pseudodiaptomus. Plankton genus of

marine calanoid copepods.

Pseudodifflugia. Genus of fresh-water amoeboid Protozoa; most of cell enclosed in a vase-shaped test covered with foreign particles.

pseudodominance. Expression of a recessive trait owing to the abnormal absence of the corresponding (normally) dominant allele; caused by a chromosome aberration or deficiency.

Pseudoechinoidea. Bothriocidaroida.

pseudogamy. Parthenogenetic development of an ovum as the result of stimulation by a male gamete whose nucleus does not actually fuse with that of the ovum; occurs in a few nematodes and in higher plants.

Pseudois. Bharal; genus of blue-gray sheep of the Himalayas.

Pseudoleon superbus. Dragonfly of southwestern states.

Pseudolimnophora. Genus of crane flies in the Family Tipulidae.

Pseudomelatoma. Genus of medium-sized marine snails.

Pseudopallene. Genus of Pycnogonida.

Pseudophryne australis. Australian toad which lays its eggs along the margins of temporary pools.

Pseudophyllidea. Order of tapeworms in the Subclass Cestoda; body segmented or unsegmented; scolex usually with two to six shallow bothria, occasionally absent; larval forms parasitic in a variety of invertebrates and invertebrates; adults in vertebrate intestines; Dibothriocephalus latus is the common fish tapeworm of man; larvae of Ligula occur in freshwater fishes, and the adults are in fish-eating birds.

pseudopod. Pseudopodium.

pseudopodium. More or less temporary protrusion of protoplasm extended out from certain cells; functions in locomotion and/or feeding; shape ranges from broad and blunt to very long, thin, and sometimes branching; common to all amoeboid protozoans, and present in certain flagellate protozoans, and in some of the epithelial cells lining the digestive tract of certain lower metazoans; common to all phagocyte cells throughout the animal kingdom. Pl. pseudopodia. See RHIZOPODIUM, FILOPODIUM, AXOPODIUM, and LOBOPODIUM.

pseudopregnancy. Pregnancy-like physiological and morphological changes occurring in a non-pregnant female mammal; develops after unsuccessful copulation in the rabbit and mouse.

Pseudoprorodon. Genus of ovoid to cylindrical holotrich Ciliata; marine and fresh water.

pseudopupa. Quiescent stage between any two larval instars in certain insects, especially some Coleoptera.

Pseudopythina. Genus of small, whitish marine bivalves; usually attached to various invertebrates.

Pseudorca crassidens. False killer whale; a small black whale found in all oceans; to 16 ft. long.

Pseudoscarus. Common genus of parrot fishes.

pseudoscolex. Modified anterior proglottid of certain tapeworms in which the scolex is absent or rudimentary; functions in attachment.

Pseudoscorpionida. Order of small flattened arachnids; pseudoscorpions, chelonethids, false scorpions; 1 to 10 mm. long; unsegmented cephalothorax broadly joined to 11- or 12 segmented abdomen; pedipalps large and chelate; usually two or four eyes; tracheae opening by four spiracles; silk spun with chelicerae and used for molting and hibernation nests; sluggish, carnivorous, and living under bark, debris, etc.; occasionally in houses; about 1900 spp.; Apocheiridium, Chthonius, Chelifer, and Garypus common.

Pseudosida bidentata. Cladoceran found especially in southern U.S.

Pseudosperchon. Sperchonopsis.

Pseudospora. Genus of small amoeboid Protozoa parasitic in algae and flagellate Protozoa.

Pseudosquilla. Genus of mantis shrimps (Stomatopoda).

pseudostigmatic organ. Prominent club-shaped sensory organ on each side of the cephalothorax in certain oribatid mites.

Pseudostylochus. Genus of polyclad Turbellaria.

Pseudosuccinea. Discarded generic name for some spp. now included in Lymnaea.

pseudotracheae. 1. False tracheae. 2. In some Diptera, the grooves on the labella with which they scrape their food.

Pseudotriton ruber. Red salamander; Penn. to Gulf of Mexico; near ponds and streams in hilly areas; to 5 in. long.

pseudovelum. Narrow shelflike flange projecting inward from the margin of certain jellyfish.

pseudovum. Egg which develops parthenogenetically.

pseudozoea. Free-swimming larval stage of Stomatopoda.

Psilaster. Circumboreal genus of sea stars (Order Phanerozonia).

Psilenchus. Common genus of microscopic, free-living nematodes.

Psilotreta. Genus of Trichoptera.

Psittacidae. Only family of birds in the Order Psittaciformes; includes the parrots.

Psittaciformes. Order of brightly-colored birds; 315 spp. include the parrots, parakeets, cockatoos, cockateels, keas, lovebirds, macaws, etc.; beak characteristically parrot-like, tongue fleshy, feet adapted for grasping, with two toes in front and two behind; forests of tropics and subtropics; only U.S. genera are Conuropsis and Rhynchopsitta.

psittacosis. Parrot fever; contagious virus disease of parrots and parakeets; communicable to man in whom it causes fever and pulmonary disorders.

psoas muscle. One of two paired muscles originating on thoracic and lumbar vertebrae and inserting on the base of the femur and junction of ilium and pubis.

psocid. See CORRODENTIA.

Psocidae. Widely distributed family of insects in the Order Corrodentia; the psocids.

Psocoptera. Corrodentia.

Psocus. Common genus of insects in the Order Corrodentia.

Psolus. Genus of sea cucumbers having a flat, well-defined "ventral" sole.

Psophiidae. Family of birds (3 spp.) in the Order Gruiformes; trumpeters; cranelike and rail-like; lowland jungles of northeastern South America.

Psorophora. Genus of blood-sucking mosquitoes found in North and South America; larvae voracious and predatory.

Psoroptes communis. Scab mite of sheep; parasitic in skin.

Psychidae. Large family of small to medium-sized hairy moths; females larviform and living in a loose bag or case; larvae in cases of silk and bits of vegetation and feeding on leaves of many trees and shrubs; bagworms; Thyridopteryx is especially important.

Psychoda. Common genus of insects in the dipteran Family Psychodidae; larvae and pupae usually found in foul waters such as filter beds and sewage.

Psychodidae. Family of minute insects in the Order Diptera; moth flies and sand flies; adults scaly and hairy; larvae terrestrial and aquatic, feed on all kinds of decaying vegetable matter; Psychoda and Phlebotomus are familiar genera.

psychology. Science of the mind, mental state, mental processes, and behavior of higher animals.

Psychomyia. Genus of Trichoptera.

psychoneurosis. Neurosis.

psychosexual isolation. Effective genetic isolation of two spp. which do not interbreed in the wild state even though there are no morphological or genetic barriers to such a mating.

psychosis. Serious and prolonged mental disorder, such as dementia precox, paranoia, manic-depressive, and persistent and grave despondency.

psychosomatic medicine. Branch of medicine dealing with effects of emotions and thoughts on physiology and physical condition; some such manifestations are asthma, high blood pressure, and hay fever.

Psychromaster. Genus of darters now usually includes in Etheostoma.

psychrometer. Instrument consisting essentially of two thermometers, one of which has a bulb kept wet with water; when moved through the air, the difference in temperature reading between the two is a measure of relative humidity.

Psylla. See PSYLLIDAE.

Psyllidae. Family of minute insects in the Order Homoptera; jumping plant lice; Psylla mali is a pest of apple trees, and P. pyricola occurs on pear trees.

ptarmigan. See LAGOPUS.

PTC. Phenylthiocarbamide.

pteralium. One of several articulating sclerites at the base of an insect wing. Pl. pteralia.

Pteranodon. Genus of Cretaceous flying reptiles in the Order Pterosauria; jaws greatly elongated into a toothless beak; back of skull projected posteriorly into a large crest; wingspread to 25 ft.

Pteraspis. See OSTRACODERMI.

Pteraster. Common genus of circumboreal sea stars; finlike membrane around edge of arms; cushion stars.

pterergate. Abnormal worker ant having vestigial wings but unmodified thorax.

Pterichthyodes. Genus of extinct placoderm fishes.

Pterichthys. Genus of Devonian ostracoderms.

Pteriidae. Family of bivalve mollusks which includes the pearl oysters and wing shells; valves quite unequal in size; mostly warm seas.

Pterobranchia. Small phylum containing about 20 spp. of minute, sessile, colonial creatures living on the sea bottom; except for the distal end, the animals are enclosed in branching tubes, often composed of series of rings; body roughly vase-shaped, with a short proboscis, collar, and trunk; collar bears two or several branched ciliated arms used in collecting food; digestive tract U-shaped with the mouth at the junction of the proboscis and collar; some spp. have anterior gill clefts; with a poorly defined dorsal mass of nervous tissue, and a rodlike mass of tissue just below it which has been considered a primitive notochord; reproduction by both budding and syngamy; larva sessile but reminiscent of certain echinoderm larvae; sometimes considered a section of the Phylum Chordata or of the Phylum Enteropneusta; Rhabdopleura and Cephalodiscus are the best known genera.

Pterocera. Genus of scorpion shells or finger shells; marine gastropods having long finger-like processes on outer lip of shell; Pacific and Indian oceans.

Pteroclidae. Old World family of birds in the Order Columbiformes; sandgrouse; chunky, pigeon-like game birds of sandy, open country.

pterodactyl. See PTEROSAURIA.

Pterodactyla. Pterosauria.

Pterodina. Testudinella.

Pterodrilus. Genus of small leechlike oligochaetes commensal and parasitic on eastern U.S. crayfishes; Family Branchiobdellidae.

Pterolepis. See ANASPIDA.

Pteromalidae. Large and cosmopolitan family of Hymenoptera which includes the jewel wasps; minute, metallic spp.; many are important as parasites of insect pests.

Pteromedusae. Suborder of the coelenterate Order Trachylina; body slender, bipyramidal, with four swimming lobes at equator; Tetraplatia in marine plankton.

Pteromonas. Genus of fresh-water biflagellate Protozoa; cell surrounded by gelatinous mass and thin lorica.

Pteronarcella. Common genus of insects in the Order Plecoptera; western U.S.

Pteronarcys. Common genus of insects in the Order Plecoptera.

Pteroneura. Saras; a genus of giant Brazilian otters; to 6 ft. long; body long, legs short, head flat, and tail with lateral flanges.

Pterophoridae. Small cosmopolitan family of moths; plume moths; small grayish to brownish spp.; each wing usually has two or three clefts reaching almost to the origin of the wing.

Pterophryne. Histrio.

Pteropoda. Suborder in the gastropod Order Opisthobranchiata; sea butterflies; small, pelagic, free-swimming gastropods with the sides of the foot modified to form fins; shell present or absent.

Pteropus. See MEGACHIROPTERA.

Pterosagitta draco. Cosmopolitan sp. of Chaetognatha; single pair of lateral fins; 10 mm. long.

Pterosauria. Order of Jurassic and and Cretaceous flying reptiles; pterodactyls; had membranous wings extending from the sides of the body, supported by the forelimbs and chiefly by the enormously elongated fourth finger; examples are Rhamphorhynchus and Pteranodon.

Pterostichus. Genus of ground beetles.

pterostigma. Thick, opaque cell along the anterior margin of the wings of certain insects.

Pterosyllis. Genus of Polychaeta.

pterotheca. That portion of a pupal case which covers the developing wings of an insect.

pterothorax. Collectively, the two wing-bearing thoracic segments in insects, especially when they are more or less fused.

pterotic. Bone in the outer dorsal portion of the otic capsule in many fishes.

Pterotrachea. Genus of pelagic marine snails; shell delicate and small, body elongated, translucent, and compressed; swim upside down, the foot being used as a fin; predaceous; tropics and subtropics.

pteroylglutamic acid. PGA, folic acid.

pterygial. 1. One of several small bones at the base of the pectoral fin rays in bony fishes. 2. Pertaining to a pterygium.

pterygiophore. One of many nodules of cartilage or bone which articulates at the base of certain fins in a typical fish.

pterygium. 1. Generalized vertebrate limb. 2. In certain weevils, a lobe at the tip or lateral portion of the snout. 3. Small basal wing lobe in certain Lepidoptera. 4. Cuticle at the base of a nail. 5. Abnormal small thickened mass of conjunctiva which covers part of the human cornea. Pl. pterygia.

pterygoid. One of two bones forming lateral basal portions of the cranium of lower vertebrates.

pterygoideus. Mammalian muscle between the base of the skull and the median surface of the mandibular ramus.

pterygoquadrate cartilage. Palatoquadrate cartilage.

Pterygota. Subclass in the Class Insecta which includes the more advanced orders; usually with wings.

Pterygotus. See EURYPTERIDA.

pteryla. One of the special areas of a bird skin which bears feathers; a feather tract. Pl. pterylae.

Ptilanthura. Genus of marine Isopoda.

ptilinum. In certain Diptera, an inflatable organ which is thrust out through a frontal suture above the base of the antenna at emergence from the pupa. Pl. ptilina.

Ptilogonatidae. Small family of passerine birds which includes the silky flycatchers or phainopeplas; Phainopepla nitens is a slim glossy black bird with a crest and white patches on the wing; southwestern U.S. and Mexico.

Ptilonorhynchidae. See BOWERBIRD.

Ptilosaurus. Genus of sea feathers found along Pacific Coast.

Ptilostomis. Common genus of Trichoptera.

Ptinidae. Widely distributed family of small oval or cylindrical beetles; spider beetles; head small; chiefly scavengers, some spp. being destructive to animal and vegetable foodstuffs.

ptomaine. Any of a group of bases formed by bacterial action on proteins; some are harmless, others are highly toxic.

ptosis. Drooping or one or both upper eyelids.

ptyalin. Digestive enzyme present in the saliva of some mammals; digests starches to maltose.

Ptychobranchus. Genus of fresh-water bivalve mollusks (Unionidae); central and southern U.S.

Ptychocheilus lucius. "White salmon" or squawfish of the Colorado River basin (Family Cyprinidae); largest N. A. true minnow, to 5 ft. long and 80 lbs.

Ptychodactiaria. Suborder of coelenterates in the Order Actiniaria; no ciliated areas on mesenteric filaments; Ptychodactis and Dactylanthus typical.

Ptychodactis. Genus of anemones found in cold seas (Suborder Ptychodactiaria).

Ptychodera. Genus of Enteropneusta; found among coral reefs.

Ptychoptera rufocincta. Common phantom crane fly in the Family Ptychopteridae; larvae in shallow ponds, with the tip of the breathing tube at the surface film.

Ptychopteridae. Phantom crane flies; small family of Diptera closely related to the true crane flies; larvae aquatic and characterized by a long posterior respiratory tube; Ptychoptera common.

Ptychoramphus aleuticus. Cassin's auklet; a small sea bird in the Family Alcidae; Pacific Coast of N.A.

Ptygura. Genus of sessile rotifers; most of body hidden by a tube constructed of fecal pellets, debris, etc.

puberty. Attainment of sexual maturity, especially as applied to human beings; manifestations usually appear at an age of 12 to 17 years; in boys it is marked by change of voice and involuntary seminal discharges, in girls by menstruation.

pubescent. Covered with feathers, setae, or fine, soft hairs.

pubic louse. See PHTHIRUS PUBIS.

pubic symphysis. Midventral plane of fusion between the two halves of the pelvic girdle in most mammals, many reptiles, and a very few birds.

pubis. Anterior ventral portion of the pelvic girdle.

puboischiadic fenestra. Large opening in the pelvic girdle, bounded by the pubis and ischium; in turtles, Sphenodon, and lizards; cordiform foramen.

puboischiofemoralis externus. Ventral muscle of the hind leg of reptiles and urodeles.

puboischiofemoralis internus. Dorsal muscle of the hind leg of reptiles and urodeles.

puboischiotibialis. Ventral muscle of the hind leg of reptiles and urodeles.

pubotibialis. Ventral muscle of the hind

leg of reptiles and urodeles.

pudendum. Vulva; female external genitalia.

Pudu. Genus of diminutive deer (pudu) found in South America, especially Chile and Ecuador.

puff adder. 1. See BITIS. 2. Hog-nosed snake; see HETERODON.

puffbird. Any of a group of arboreal and forest birds in the Family Bucconidae (Order Piciformes); New World tropics; bill curved and robust; head exceptionally large, coloration brownish and grayish; often perch with the head and neck feathers fluffed out; burrow in river banks.

puffer. See TETRAODONTIDAE.

puffin. See FRATERCULA.

Puffinus. Genus of oceanic birds in the Family Procellariidae; shearwaters; graceful in flight, often skimming over the surface of waves for long distances.

Pugettia. Genus of kelp crabs; small spp. found especially in kelp along the Pacific Coast.

puku. See ADENOTA.

Pulex irritans. Human flea in the Order Siphonaptera; also occurs on domestic and wild animals; cosmopolitan except in tropics; unimportant as a disease transmitter.

pulmocutaneous arches. Sixth pair of aortic arches in amphibians; supply arterial blood to the lungs and skin.

pulmonary. Of or pertaining to the lung.

pulmonary artery. Artery carrying blood to the lungs of tetrapods.

pulmonary vein. 1. One of the veins carrying oxygenated blood from the lungs to the left auricle in tetrapods. 2. One or more veins of invertebrates which bring blood from respiratory areas to the pericardial chamber.

Pulmonata. One of the three orders in the Class Gastropoda; portion of mantle and mantle cavity modified into an air-breathing lung; gills absent; fresh-water snails, land snails, and land slugs; shell a simple spiral, or absent; one or two pairs of tentacles; monoecious.

pulmonate. 1. Pertaining to the Pulmonata. 2. Having lungs or similar respiratory organs.

pulp. See DENTAL PULP.

pulp cavity. Pulp chamber; see DENTAL PULP.

pulp chamber. Central cavity of a tooth and its roots; see DENTAL PULP.

pulsating vesicle. Small bladder-like organ in rotifers; it receives the excretions of the flame bulb system and periodically discharges its contents to the outside via the cloaca.

pulse. Wave of increased blood pressure which passes rapidly from the heart along the arteries every time the ventricles contract and discharge additional blood into the aorta; the increase in pressure may be felt as a corresponding wave of dilation passing along the arteries; cannot ordinarily be detected in veins.

pulvillus. 1. Soft pad between the tarsal claws in many insects; variously modified. 2. Cushion of short hairs on the underside of tarsal segments in some insects. Pl. pulvilli.

Pulvinaria. Important genus of scale insects.

puma. See COUGAR.

pumpkinseed. See LEPOMIS.

Punctum. Genus of very small land snails; western states.

Puncturella princeps. Little chink; a very small keyhole limpet of the Atlantic Ocean.

Pungitius pungitius. Nine-spined stickleback; a circumpolar sp. of both fresh and salt waters; 9 to 11 dorsal spines.

punkie. Any of a wide variety of small biting or non-biting flies.

Punnett, Reginald Crundall. English geneticist (1875-); contributed to an understanding of gene linkage in sweet peas and sex linkage in poultry.

pupa. 1. Stage between the last larval instar and the adult in holometabolous insects; non-feeding, usually non-motile, and may be enclosed within the last larval skin, a cell, or cocoon; extensive anatomical changes take place within the pupa, and the adult emerges by a splitting of the pupal exoskeleton; 2. In certain of the Paurometabola, a quiescent fourth nymph instar. Pl. pupae.

pupa coarctata. Insect pupa in which the last larval exoskeleton is retained as a pupal covering.

pupa libera. Insect pupa with soft integument, functional mandibles, and appendages movable and free from the body.

pupa obtecta. Rigid insect pupa with hard integument and no mandibles; appendages sheathed and immovable against body.

puparium. In certain Diptera, a barrel-shaped thick larval exoskeleton in which the pupa is formed. Pl. puparia.

pupate. To metamorphose into a pupal stage.

pupation hormone. Hormone secreted by gland cells associated with insect cerebral ganglia; induces pupation.

pupfish. See CYPRINODON.

pupil. See IRIS.

Pupilla. Circumpolar genus of small land snails.

Pupillidae. Family of terrestrial Gastropoda; small to minute spp.; shell cylindrical or conical, blunt and multispiral; aperture small, toothed; common in N.A. in moist wooded areas.

Pupipara. One of the three tribes in the dipteran Suborder Cyclorrhapha; abdomen indistinctly segmented; ovoviviparous, one larva being produced at a time.

Pupoides. Common genus of small land snails; all continents except Europe.

pure line. Succession of generations of a sp. homozygous for all genes.

purine. Portion of the nucleic acid molecule; $C_5H_4N_4$.

purine base. Portion of a nucleoprotein molecule composed of a benzene ring and a five-membered ring; e.g. adenine, uric acid, and xanthine.

Purkinje, J. E. Czech physiologist, histologist, and embryologist (1787-1869); first used the term "protoplasm."

Purkinje cells. Large neurons with many branches in the middle layer of the cerebral cortex.

purple. See BASILARCHIA.

purple finch. See FINCH.

purple gallinule. See PORPHYRULA.

purple grackle. See GRACKLE.

purple martin. See PROGNE.

purple salamander. See GYRINOPHILUS.

purple sea urchin. Arbacia punctulata, a common sea urchin; spines long and stiff; test to 2 in. across; purplish coloration; on rock or shell bottoms of Atlantic Coast.

purple snail. See THAIS.

purple star. See ASTERIAS.

Purpura. Thais.

purse crab. See BIRGUS.

purse web spider. See ATYPIDAE.

pus. Yellowish to creamy-white liquid or semifluid inflammation exudate; consists of bacteria, leucocytes, disintegrating tissue, and tissue fluids.

puss moth. See NOTODONTIDAE.

Pusula. Genus of small, ribbed, cowry-like marine snails.

putrefaction. Decomposition of proteins by bacteria and yeasts; characterized by the production of foul-smelling substances such as hydrogen sulfide, ammonia, and mercaptans. Cf. FERMENTATION.

Pycnoclavella. Genus of ascidians.

Pycnogonida. Class in the Phylum Arthropoda; includes the sea spiders or Pantopoda; leg span 3 mm. to 50 cm.; compared with the legs, the body mass is insignificant; it consists of a segmented so-called "cephalothorax" and vestigial button-like abdomen; suctorial mouth at tip of cylindrical proboscis; appendages include a pair of chelicerae (rarely), palps (often absent), ovigerous legs (lacking in some females), and four pairs of very long thin walking legs (rarely five or six pairs); eggs carried by males; direct development, or with a four-legged larval stage; sluggish; on algae, hydroids, or bottom debris of sea from tidal zone to deep sea; 600 described spp.; Pycnogonum, Nymphon, Colossendeis, and Achelia representative.

Pycnogonum. See PYCNOGONIDA.

Pycnonotidae. See BULBUL.

Pycnophyes. Genus of Kinorhyncha; head and neck exceptionally spiny and retractable; two terminal bristles; usually in mud or sea bottom.

Pycnopodia. Genus of large Pacific Coast sea stars with a wide disc and 18 to 24 arms in the adult; sunflower stars; to 30 in. in diameter.

Pycnopsyche. Genus of caddis flies.

pycnosis. In dying and moribund cells, the phenomenon of contraction of nuclear material of the cells into a compact densely-staining mass. Adj. pycnotic.

pygal plate. Small median posterior dorsal plate of a tortoise carapace.

Pygathrix. Genus of langurs of southeastern Asia; coloration a bright pattern of yellow, brown, chestnut, white, black, and red; the doucs.

pygidium. 1. Posterior platelike part of the trilobite body; consists of 2 to 28 small segments, usually fused. 2. Distinct abdominal portion of certain scale insects; formed by fused segments II to VIII; secretes and molds the scale. 3. Caudal body region of certain crustaceans. 4. Terminal body segment or terminal tergum of certain insects. 5. Sensory plate on ninth abdominal segment of fleas. 6. Anal segment in certain annelids;

ganglion and coelom lacking.
pygmy anteater. See DIDACTYLUS.
pygmy armadillo. See ZAEDYUS.
pygmy backswimmers. See PLEIDAE.
pygmy flying possum. See ACROBATES.
pygmy hippopotamus. See CHOEROPSIS.
pygmy locusts. See TETRIGIDAE.
pygmy marmoset. See CEBUELLA.
pygmy mouse. See BAIOMYS.
pygmy owl. See GLAUCIDIUM.
pygmy rattlesnake. See SISTRURUS.
pygmy right whale. See NEOBALAENA.
pygmy sand cricket. See TRIDACTYLIDAE.
pygmy shrew. See MICROSOREX.
pygmy sperm whale. See KOGIA.
Pygopodidae. Family of lizards found in Australia, Tasmania, and New Guinea; body snakelike, forelimbs absent, hind limbs small and flaplike; Pygopus typical.
pygopods. Collectively, the tenth abdominal appendages of certain insects.
Pygopus. See PYGOPODIDAE.
Pygospio. Genus of small Polychaeta; live in sand tubes.
pygostyle. Terminal group of five or six fused vertebrae in most modern birds (Superorder Neognathae).
Pylodictis. See GOUJON.
pyloric caecum. 1. Tubular blind pouch attached at the junction of stomach and intestine, especially in certain fishes; secretory and absorptive. 2. One of two large digestive caeca in each arm of a sea star.
pyloric chamber. Posterior chamber of the stomach in decapod crustaceans and many other Malacostraca; contains the gastric mill.
pyloric gland. Type of gastric gland in the stomach wall; secretes pepsin.
pyloric sphincter. Sphincter muscle at the junction of the stomach and duodenum; pyloric valve.
pyloric stomach. Pentagonal aboral stomach of the sea star; receives the secretion of the ten hepatic caeca.
pyloric valve. Pyloric sphincter.
pylorus. Area of junction of stomach and small intestine of vertebrates.
pyorrhea. Inflammation of the sockets of the teeth, sometimes leading to loosening of the teeth.
Pyralidae. Pyralididae.
Pyralididae. Family of small moths whose larvae are plant feeders; snout moths; includes several important pest spp.
Pyrameis. Vanessa.
Pyramidella. Genus of small pyramid shells (Gastropoda).
Pyramidellidae. Pyramid shells; family of marine Gastropoda; shell small and conical, elongate, or ovate; operculum horny, spiral; eyes sessile and tentacles ear-shaped.
pyramid of biomass. Weight relationships between the various trophic levels involved in a particular food web; plants (at the base of the pyramid) have the greatest weight, herbivores are next, primary carnivores next, and so on to the dominant carnivore at the top of the pyramid (or end of the food web).
pyramid of energy. Energy relationships between the various trophic levels involved in a particular food web; plants (at the base of the pyramid) represent the greatest amount of energy utilization, herbivores next, primary carnivores next, and so on to the dominant carnivore at the top of the pyramid (or end of the food web).
pyramid of numbers. Numerical relationships between the various organisms involved in a particular food web; plants (at the base of the pyramid) are most abundant, herbivores are next most abundant, primary carnivores next, and so on to the least abundant dominant carnivores at the top of the pyramid (or end of the food web).
pyramid shell. See PYRAMIDELLIDAE.
Pyramimonas. Genus of green pyramidal or heart-shaped fresh-water Protozoa; four anterior flagella.
Pyrausta nubilalis. European cornborer; an important moth, especially east of the Mississippi; larva feeds on stalks and crowns of corn, sorghum, millet, hemp, etc.
Pyraustidae. Large family of moths containing many important pest spp.; webworms; Pyrausta (European cornborer) is most notable.
Pyrenaster. Widely distributed genus of sea stars (Order Phanerozonia).
Pyrenean. Ibero-insular.
pyrenoid. Small discrete mass of protein imbedded in a chromatophore of a photosynthetic protozoan; presumably a center for the formation of granules of starch and allied reserve foods.
Pyrgulopsis. Genus of carinate fresh-water snails; Great Lakes area, Ala., and Nev.
pyridoxal phosphate. See PYRIDOXINE.
pyridoxine. Vitamin of the B complex (B_6); present in cells as pyridoxal

phosphate which is a coenzyme that functions in the breakdown of amino acids in the liver; deficiency produces anemia in dogs and hogs, dermatitis in rats, and paralytic death in hogs, rats, and chicks; abundant in yeast, cereal grains, milk, and liver; $C_6H_{12}O_3N$.

pyriform. Pear-shaped.

pyriformis. Muscle arising on the vertebral column and inserting on the head of the femur in tetrapods; rotates the thigh outward.

pyriform lobe. Lateral exposed lobelike portion of the olfactory cerebral cortex in lower mammals.

pyrimidine base. Portion of a nucleoprotein molecule; a six-membered ring containing two nitrogen atoms.

pyrimidine desoxyriboside. Portion of a nucleic acid molecule.

Pyrocephalus rubinus. Vermilion flycatcher in the Family Tyrannidae; male with vermilion head and under parts; tail and upper parts blackish; southwestern U.S., Central and South America.

Pyromaia. Genus of attenuated spider crabs.

Pyrosoma. See PYROSOMIDA.

Pyrosomida. Order of free-swimming colonial tunicates in the Class Thaliacea; colonies compact, tubular, and with a common tunic; Pyrosoma chiefly in tropical seas.

Pyrotheria. Order of extinct South American ungulates; certain large spp. are similar to early elephants; early Tertiary.

Pyrrhocoridae. Red bugs, fire bugs, cotton stainers; family of medium-sized to large herbivorous Hemiptera; many spp. bright red; Dysdercus attacks cotton plants.

Pyrrhuloxia sinuata. Pyrrhuloxia; a slender gray and red finch with a crest; bill small and stubby; Ariz., N. M., Tex., and northern Mexico.

pyrrole ring. A cyclic part of the structure of certain amino acids; $(CH)_4NH$.

Pyrula. Ficus.

pyruvic acid. Intermediate substance in the metabolic utilization of many cellular fuels; simple organic acid, $CH_3COCOOH$; commonly results from the partial oxidation of glucose; may give rise to, or be derived from, the following (among others) in cell metabolism: glucose, alcohol, starch, amino acids, acetic acid, glycerol, glycogen, and lactic acid.

Python. Genus of large southeastern Asiatic snakes in the Family Boidae; P. reticulatus, the regal python, attains 30 ft. and inhabits southeastern Asia; P. molurus, the Indian python, attains 25 ft. and inhabits India, China, Java, and the Malay Peninsula.

Pyxicola. Genus of marine and fresh-water peritrich ciliates; in a lorica closed by a corneous operculum.

Pyxidicula. Genus of fresh-water amoeboid Protozoa; with a rigid patelliform test.

Pyxidium. Genus of fresh-water peritrich ciliates; stalk short, simple, and unbranched.

Q

Q disc. Anisotropic disc.

Q fever. Mild typhus-like disease of man, cattle, sheep, and goats; caused by a rickettsia (Coxiella burneti) transmitted through the bite of several ticks or via the air from tick excreta.

Q_{10} law. See VAN'T HOFF'S LAW.

quachil. Large Mexican and Central American pocket gopher (Heterogeomys).

Quadraspidiotus perniciosus. Sp. of scale insect; an important pest of fruit trees and many kinds of ornamentals and shrubs; San José scale.

quadrat. Sample random square of a substrate where the vegetation and/or fauna are carefully measured qualitatively and quantitatively.

quadrate. Small cartilage-bone at the posterior end of the vertebrate upper jaw; articulates with the lower jaw in bony fishes, amphibians, reptiles, and birds; becomes incus in mammals.

quadrato-jugal. Bone at the lateral posterior portion of the head on either side in some of the lower vertebrates.

quadratus femoris. Ventral muscle of the hind leg in certain amphibians, e. g. frog.

quadratus labii inferioris. Short muscle which originates on the lower anterior border of the mandible and inserts on the lower lip; depresses lower lip.

quadratus labii superioris. Group of three small muscles extending between the orbit area and the upper portion of the lip on each side.

quadriceps femoris. In man, the combination of the rectus femoris, vastus internus, vastus externus, and vastus intermedius muscles of the front of the thigh; their common tendon surrounds the patella and ends on the tuberosity of the tibia.

Quadrula. Genus of thick-shelled fresh-water mussels; especially useful in button manufacturing.

Quadrulella. Genus of fresh-water Protozoa; most of amoeboid cell enclosed in a vase-shaped test of quadrangular plates.

quadruped. Any four-footed animal, e. g. most mammals and many reptiles.

quagga. South African wild ass; became extinct in the late nineteenth century.

quahog. See VENUS and MERCENARIA.

quail. Any of various small gallinaceous game birds related to the pheasants; nest on ground, feed on vegetation and insects; Colinus and Lophortyx occur in the U.S.

quarantine. Enforced isolation of any person(s), other animals, or place infected with a contagious disease.

quarter deck. Common name for certain spp. of Crepidula.

Quaternary period. Present geological period, estimated to have begun about 1 million years ago. See RECENT and PLEISTOCENE epochs.

queen conch. Any of several large tropical conchs (Gastropoda), especially in Cassis and Strombus; shells used in cameo work.

queriman. Any of several West Indian marine mullets.

question mark. See ANGLE-WING.

quetzal. See PHAROMACRUS.

quill. 1. Hollow basal portion of the shaft of a typical feather. 2. Long, hollow, cylindrical structure mixed in with the coarse and fine hairs of a porcupine; finely pointed, barbed, and easily detached.

quillback. See CARPIODES.

Quincuncina. Genus of fresh-water bivalve mollusks (Unionidae); Gulf drainages.

quinine. Alkaloid extracted from the bark of cinchona trees and used extensively an as antimalarial; the sulfate or some other salt is usually taken by mouth as pills or capsules; a blood plasma level of 5 milligrams per liter for four to six days will interrupt the asexual cycle of Plasmodium organisms in the erythrocytes; used as both prophylactic and cure for malaria, although it has been largely replaced by other, more effective drugs.

quinnat salmon. *Oncorhynchus tschawytscha*, the chinook or king salmon; spawns in rivers from Calif. to the Bering Strait.

quinone. A hydrogen acceptor used in physiological experimentation.

quinsy. Suppurative inflammation of the tonsils.

Quiscalus. See GRACKLE.

R

raad. Local name for electric catfish.

rabbit. 1. In the broad sense, any of the numerous members of the mammal Order Lagomorpha. 2. In the strict sense, any member of the Order Lagomorpha which burrows and produces naked young, such as Sylvilagus.

rabbit-bandicoot. See MACROTIS.

rabbit fever. Tularemia.

rabies. Hydrophobia; acute, infectious, often fatal, virus disease of the central nervous system; occurs chiefly among dogs and wolves but may be communicated to man in the bite of these animals.

raccoon. See PROCYON.

raccoon-dog. See NYCTEREUTES.

race. 1. Subspecies or genetic variety. 2. General and arbitrary division of humanity based on hair texture, hair color, skin color, head shape, etc.

racemose gland. Compound gland with multiple branching ducts which end in individual acini, the whole having the appearance of a bunch of grapes; e.g. the pancreas.

racer. See COLUBER.

racerunner. See CNEMIDOPHORUS.

Rachiglossa. Subdivision of snails in the Suborder Pectinibranchia; predatory, siphon present, radula with not more than three teeth per transverse row; e.g. Thais, Buccinum.

rachion. Margin of a lake or ocean where maximum wave action and turbulence occur.

rachis. 1. Thickened longitudinal central part of the trunk of a trilobite; a flat pleuron is located laterally on each side; see GLABELLA. 2. Feather shaft.

rachitis. Rickets.

Rachycentridae. Sergeant fishes or crab-eaters; a family of swift, voracious shore fishes; mouth large; body long, cylindrical, and pikelike; Rachycentron canadus common in warm and temperate seas.

Rachycentron. See RACHYCENTRIDAE.

racial immunity. Resistance which members of a sp. or other natural group have against an infection normally occurring in close relatives.

radial canal. 1. One of the many internal canals of sponges; it may or may not be lined with choanocytes and leads toward the main spongocoel or an osculum. 2. One of the five canals projecting out from the ring canal in the water vascular system of echinoderms. 3. In Scyphozoa and hydrozoan medusae, any gastrovascular canal extending in a radial fashion.

radial cleavage. Type of early embryological cleavage; first two cleavages vertical, the third horizontal; each blastomere in the upper tier lies directly over the corresponding blastomere of the lower tier with reference to the polar axis.

radiale. 1. Carpal bone or cartilage which articulates with the radius. 2. Radial plate of a crinoid. 3. Small bone or cartilage which supports one of more rays in the fin of a fish.

radialia. Slender cartilaginous rays which fan out from the basalia and form the broader part of the pectoral and pelvic fins of elasmobranch fishes.

radial nerve. 1. One of five or more nerves radiating out from the central nerve ring of an echinoderm. 2. Large nerve arising from the brachial plexus; supplies the arm and hand.

radial sector. One of the veins in an insect wing.

radial symmetry. Condition obtaining in most Coelenterata, Ctenophora, and adult Echinodermata; parts arranged around one longitudinal axis; any plane drawn through this oral-aboral axis will divide the animal into two similar halves. Adj. radially symmetrical.

Radiata. Discarded term for a major branch of the animal kingdom which included all radially symmetrical animals.

radiating canals. Minute temporary canals associated with the contractile vacuoles of certain ciliate Protozoa; they have the function of gathering fluid from the surrounding cytoplasm and feeding it into the contractile vacuole proper.

Radicipes. Genus of corals (Order Alcyonaria); branched, rodlike colonies, to 3 ft. high.

radioautograph. Autoradiograph.

radiobiology. Study of the effects of radiant energy on living organisms.

radiocarbon. C^{14}, an isotope of carbon normally present in the air and all organic materials; used in dating fossil and subfossil remains; also used in experimental physiological work.

radioecology. Radiation ecology; study of ecology with special reference to effects and trophic community pathways of radioisotopes.

Radiolaria. One of the orders of the Class Sarcodina; these protozoans are mostly spherical with long radially-arranged filamentous pseudopodia; protoplasm separated into inner and outer portions by a porous central capsule of organic composition; usually also with an outer secreted spiny skeleton of silicon dioxide or strontium sulfate; marine and pelagic.

radiophosphorus. P^{32}, a radioactive isotope of phosphorus; much used in physiological and ecological research.

radio-ulna. Single bone, produced by the fused radius and ulna in the forearm of certain tetrapods.

radius. 1. One of five general areas on the body of an echinoderm; bears tube feet. 2. One of five small radiating ossicles in Aristotle's lantern. 3. Anterior of the two bones in the forearm of most tetrapods. 4. One of the veins in an insect wing. 5. One of the fine lines radiating from the central portion of a fish scale. Pl. radii.

Radix. Discarded generic name now included in Lymnaea.

radula. Flexible, tongue- or file-like structure in the anterior part of the digestive tract of all mollusks except the Pelecypoda; bears a series of transverse rows of teeth and by means of a special system of muscles, it is moved back and forth rapidly against the upper surface of the pharynx and thus macerates the ingested food; in some groups it can be protruded from the mouth for scraping or drilling.

Rafinesque, Constantine S. American naturalist (1783-1840); traveled extensively in the U.S. and wrote profusely on botany, fishes, and conchology.

Rafinesquina. Genus of brachiopods in the Order Prototremata.

ragfish. See MALACICHTHYES.

ragworm. Common name of the polychaete Nereis.

rail. See RALLUS.

railroad worm. Bioluminescent beetle larva found in Uruguay; row of green light spots along each side of the body and two red lights at the anterior end.

rainbow fish. Any of numerous brightly colored fishes, especially wrasses and parrot fishes associated with coral reefs.

rainbow snake. Long, slim, colubrid snake found in wooded swampy areas of southeastern U.S.; bluish black with three longitudinal stripes; in the genus Abastor.

rainbow trout. See SALMO.

rain forest. Dense evergreen forest in a region of heavy rainfall, high humidity, and no long dry season; found in parts of the tropics, subtropics, and a few temperate areas (e.g. Olympic Peninsula of Wash.). Cf. TROPICAL RAIN FOREST.

rainwater fish. See LUCANIA.

Raja. Common genus of skates (cartilaginous fish in the Suborder Batoidea).

Raji. Batoidea.

Rajida. Batoidea.

rale. Any abnormal respiratory sound detected by auscultation.

Rallidae. Cosmopolitan family of birds in the Order Gruiformes; includes the rails, coots, limpkin, etc.; chicken-like marsh birds of secretive habits; nest on ground; short wings, short tail, and long toes; Rallus, Fulica, and Gallinula typical.

Rallus. One of several genera of rails in the bird Family Rallidae; R. elegans is the king rail of eastern U.S.; R. longirostris is the clapper rail of Atlantic, Gulf, and Calif. salt marshes; R. limicola is the Virginia rail or marsh hen of N.A. generally.

ramate mastax. Rotifer mastax specialized for grinding bits of ingested food.

ramet. Any single individual of a clone.

Ramex. Genus of Polychaeta.

Rámon y Cajal, Santiago. Spanish histologist (1852-1934); professor at Madrid; best known for contributions to the fine structure of the neuron and nervous system; established the neuron as the unit of nervous tissue; shared Nobel prize in physiology and medicine with Golgi in 1906.

Ramphastidae. See TOUCAN.

ramus. 1. Branch, especially one of two branches of a projecting appendage or part, e.g. a ramus of a biramous appendage or a caudal ramus in a crustacean. 2. Branch of a nerve. 3. Vertebrate mandible. 4. Posterior flat part of the lower jaw of a vertebrate. 5. Branch or process of the pubis or ischium. 6. One of the two portions of the incus in the trophi of a rotifer mastax. Pl. rami.

ramus communicans. Short nerve connecting one of the spinal nerves with its corresponding sympathetic ganglion.

Rana. Large and very common genus of frogs found over most of the U.S.

Rana aurora. Common red-legged frog found in ponds from Vancouver Island to Lower Calif.; head and body to 3 in. long; dorsal surface brown, yellowish, or olive, with diffuse spots.

Rana capito. Gopher frog, a gray and black southeastern U.S. sp. found in crayfish and turtle burrows.

Rana catesbiana. Bullfrog, a large sp. occurring over much of the eastern two-thirds of N.A. and introduced farther west; in ponds and streams; head and body 6 to 8 in. long.

Rana clamitans. Common green frog or spring frog; southeastern Canada and eastern half of U.S.; dorsal surface bright green to brown; head and body 3 to 5 in. long.

Rana goliath. Giant frog of African Cameroons and Gabon; largest living frog; head and body 12 in. long, and 11 lbs. total weight.

Rana palustris. Common American pickerel frog; east Tex. to Hudson Bay and eastward; head and body to 2.5 in. long; ponds, meadows, and other wet places; dorsal surface brownish, with four rows of dark brown patches.

Rana pipiens. Common leopard frog or grass frog occurring over much of N. A. in ponds, marshes, and meadows, often a considerable distance from open water; head and body 3 to 5 in. long.

Rana sylvatica. Common American wood frog; northeastern states and in a broad band extending northwest to Alaska; in damp woods; head and body to 2 in. long; dorsal surface brown or tan, often with a yellowish or reddish cast.

Ranatra. Common genus of water scorpions in the insect Family Nepidae.

Randallia. Genus of marine crabs (Suborder Brachyura).

range. Geographic area in which a particular organism, sp., other taxon, or community occurs. Cf. HOME RANGE.

Rangia cuneata. Thick-shelled bivalve mollusk; brackish waters, Ala. to Tex.

Rangifer. Genus of ruminants in the Family Cervidae; includes the various spp. of caribou or northern N.A. (R. caribou, etc.) as well as the semidomesticated reindeer (R. tarandus) which has been recently introduced into N.A.; slightly palmated antlers in both sexes; neck maned; hoofs broad, flat, and deeply cleft; males to 300 lbs.

Ranidae. One of the three families of amphibians in the Order Diplasiocoela; true frogs; upper jaw with teeth; tongue forked posteriorly; world-wide except for Australian and Neotropical regions; Rana the most familiar genus and the only one represented in the U.S.

Ranodon. Genus of Asiatic land salamanders.

raphe. Ridge or furrow marking the midline of a bilaterally symmetrical structure, such as the line of union of the two halves of the medulla oblongata.

Raphicerus. Genus of small antelopes in the Family Bovidae; includes the grysbok and steinbok; southern and eastern Africa.

Raphidae. See DODO.

Raphidiodea. Small order of terrestrial insects which includes the snakeflies or serpentflies; elongated, fragile, with biting mouth parts, long antennae, and complete metamorphosis; prothorax long, slender, and necklike; two pairs of similar wings with many small veins; larvae mostly arboreal.

Raphidiophrys. Genus of spherical fresh-water amoeboid Protozoa; gelatinous envelope covered with siliceous scales.

Raphus. See DODO.

raptorial. 1. Predatory; preying upon other animals. 2. Adapted for seizing or tearing the prey, e.g. claws and beak. 3. Pertaining to predatory birds such as eagles, hawks, and owls.

Rasahus. Genus of nocturnal assassin

bugs in the hemipteran Family Reduviidae; corsairs; may inflict a painful bite when handled.

rasorial. Having feet adapted for scratching the ground, e.g. a chicken.

rasse. See VIVERRICULA.

Rassenkreis. A sp. composed of several to numerous intergrading subspecies inhabiting imperfectly separated adjacent biotopes.

rat. General term used to include a wide variety of medium-sized rodents, such as members of the following genera: Rattus, Sigmodon, and Dipodomys. See also CRICETIDAE and MURIDAE.

rat-chinchilla. See ABROCOMIDAE.

ratel. Honey badger (Mellivora); small mammal in the Family Mustelidae forming a line between badgers and weasels; the tropical African sp. is omnivorous, the southern Asiatic sp. feeds primarily on honey.

ratfish. See HYDROLAGUS.

rat flea. Any flea parasitizing rats, but especially Xenopsylla cheopis, which transmits bubonic plague.

Rathbunula. Genus of marine harpacticoid Copepoda; sandy beaches.

Rathkea. Genus of marine coelenterates in the Suborder Anthomedusae; mostly in colder waters.

Rathke's pouch. Embryological sac which forms as a dorsal evagination of the roof of the mouth of vertebrate embryos and develops into the anterior lobe of the pituitary.

Ratitae. Discarded taxonomic category which included the living flightless birds. Cf. CARINATAE.

rat-kangaroo. See BETTONGIA.

rat mite. See ECHINOLAELAPS.

rat snake. See ELAPHE.

rat-tailed maggot. See TUBIFERA.

rattlesnake. Typical venomous North and South American snakes in the genera Sistrurus and Crotalus; tail with a series of cornified button-like structures making up the rattle; when the snake is disturbed or coiled ready to strike, the tip of the tail is held upright and vibrated rapidly so as to produce a buzzing sound.

Rattus. Genus of omnivorous rats in the Family Muridae; Rattus norvegicus is the common and cosmopolitan rat variously called the Norway, house, barn, wharf, domestic, or brown rat; about 10 in. long, without the tail; R. rattus rattus (black rat) is a smaller sp. with longer tail and larger ears; it is much less common and appears to be driven out wherever it is in contact with R. norvegicus; R. rattus alexandrinus is the roof rat, also uncommon.

ratufa. Any of several Oriental palm-squirrels, especially in the genus Ratufa.

raven. Large black bird (Corvus corax) of northern Asia, Europe, and N.A.; larger than a crow, to 27 in. long; common scavenger.

ray. See STING RAY and DEVIL RAY.

Ray, John. English naturalist (1627-1705); summarized botanical knowledge of his time in a 2860-page work, Historia Plantarum Generalis, which did much to advance plant taxonomy; characterized each genus by one name and a short diagnosis; each of the sp. in a genus was also characterized by a short diagnosis; published a book on quadrupeds and another on insects.

rayed semele. Semele radiata, a small white orb-shaped bivalve; Ga. southward.

razor-billed auk. See ALCA.

razor clam. See ENSIS and SOLEN.

reaction. Ecological effects of organisms on their habitat.

recapitulation. Biogenetic law, first elucidated by Haeckel; embryological development of the individual traverses certain adult states of the phylogenetic series which have given rise to the particular group, thus compressing the racial history into a single lifetime; e.g. human embryos exhibit a transient gill slit (fish) stage during development; few zoologists adhere to the literal meaning of the biogenetic law.

Recent epoch. Holocene; present epoch and subdivision of the Quaternary period; estimated to have begun 15,000 to 20,000 years ago; chiefly postglacial and warm; characterized by the rise of modern man and mammals.

receptaculum seminis. 1. Seminal receptacle. 2. Spermatheca.

receptor. Any cell, tissue, or organ of an animal which, in cooperation with the nervous system, detects internal or external stimuli.

recessive. See RECESSIVE CHARACTER.

recessive character. 1. Character (feature) derived from one parent that is marked by an alternative (dominant) character (feature) from the other parent. 2. One of a pair of alternate characters determined by paired genes located on opposite members of a chromosome pair; its effect is excluded by the alternative (dominant) character. 3. A character that is evident in the phenotype only when the individual is homozygous for the gene(s) responsible for it.

reciprocal cross. 1. Mating of a male from source A with a female from source B as compared to taking the opposite sexes from the same two sources. 2. Mating of a male heterozygous for a Mendelian character with a homozygous recessive female, or a mating between a homozygous recessive male and a heterozygous female; the offspring of two such reciprocal matings will occur in the same ratios provided the genes in question are carried on the autosomes.

recognition marks. Special obvious marks possessed by certain animals; presumably used for identification by others of their sp. and to warn the latter of danger; the white rump patch of the American antelope and the white tail of the cottontail rabbit.

recombination. Occurrence of new gene combinations in an organism which were not present in either parent; produced by crossing over during gametogenesis, random segregation of chromatids, and chance recombination of gametes.

rectal caecum. One of two small caeca associated with the sea star rectum; function not definitely established, perhaps excretory.

rectal gills. Complex arrangement of gill-like structures forming the inner surface of the rectum of a dragonfly nymph.

rectal gland. Gland associated with cloaca or rectum of certain vertebrates; function usually uncertain.

rectrices. Tail feathers; sing. rectrix.

rectum. Terminal part of intestine in vertebrates and many higher invertebrates; opens to the outside by way of an anus.

rectus abdominis. Human ventral abdominal muscle which originates in the pubic symphysis area and inserts on the cartilages of the fifth, sixth, and seventh ribs; compresses the abdomen and flexes the trunk.

rectus anticus femoris. Narrow muscle on the anterior surface of the thigh in certain tetrapods.

rectus femoris. Mammalian muscle which originates on the ilium and acetabulum, and inserts on the patella; extends the leg.

rectus muscle. See EYE MUSCLES.

recurrent mutation. Any mutation which occurs relatively frequently within a population.

Recurvirostra americana. Common American avocet in the Family Recurvirostridae; large shore bird with long legs and slender, upturned bill; black and white coloration; common around ponds, lakes, and marshes west of the Mississippi; captures small aquatic animals by sweeping the bill through the water along the bottom.

Recurvirostridae. Widely distributed but small family of shore birds in the Order Charadriiformes; avocets and stilts; legs extremely long; _Recurvirostra_ and _Himantopus_ typical.

red abalone. _Haliotis rufescens_; a large brick-red abalone of the Calif. coast.

red admiral. See PYRAMEIS.

red ant. Any of many reddish spp. of ants; _Monomorium pharaonis_ is a small widely distributed sp. often infesting habitations.

red-backed salamander. See PLETHODON CINEREUS.

red-backed vole. See CLETHRIONOMYS.

red bat. See LASIURUS.

redbearded sponge. See MICROCIONA.

red-bellied snake. See STORERIA.

redbird. Any of several predominantly red-colored birds, e.g. cardinal and scarlet tanager.

red blood cell. Ervthrocyte.

red blood corpuscle. Erythrocyte.

red body. Capillary mass on the inner surface of the air bladder of some teleost fishes; presumably controls the gas content of the bladder.

redbreast. 1. American robin. 2. European robin. 3. Yellowbelly sunfish or red-breasted bream (_Lepomis auritus_) of southeastern states; belly orange-red.

red bugs. See TROMBIDIIDAE and PYRRHOCORIDAE.

red caltrop. Water chestnut.

redd. Spawning area of trout or salmon on the bottom of a lake or stream; usually a cleared circular depression in gravel.

red deer. 1. Common deer of temperate Europe and Asia (_Cervus elaphus_); smaller than the American wapiti but in the same genus; found wild in only a few areas. 2. American white-tailed deer in summer coat.

redear sunfish. Lepomis microlophus, a brightly colored small centrarchid somewhat similar to the bluegill; southeastern U.S.

red-eyed vireo. Vireo olivaceous, a vireo common in Canada and the U.S.; coloration gray and olive green.

redfin. See LUXILUS.

red grouse. Ptarmigan (Lagopus) related to the willow ptarmigan; does not have seasonal color changes; British Isles.

redhead. See NYROCA.

redheaded woodpecker. See MELANERPES.

red hind. Important Caribbean food fish, Epinephelus guttatus; a red-spotted grouper.

redhorse. See MOXOSTOMA.

Redi, Francesco. Italian naturalist (1626(?)-1698); by means of controlled experiments, he showed that maggots did not arise spontaneously from rotting meat; he was thus the first to show that spontaneous generation was a myth; also well known as an experimental entomologist.

redia. More or less cylindrical larval stage in the life history of many trematodes; usually has a small nonfunctional anterior gut, oral sucker, and rudimentary appendages; produced by a sporocyst and usually imbedded in the tissues of a host snail; a redia gives rise to daughter rediae or cercariae by asexual internal proliferation. Pl. rediae.

red-legged frog. See RANA AURORA.

red marrow. Bone marrow having a reddish cast, more abundant in immature vertebrates.

redox potential. Oxidation-reduction potential; in biology, the relative ability of protoplasm or a substance in protoplasm to be oxidized or reduced by another particular substance; also applies to certain aspects of aquatic ecology; can be measured electrically with great precision.

red phalarope. See PHALAROPUS.

redpoll. See ACANTHIS.

red salamander. See PSEUDOTRITON.

redside shiner. See RICHARDSONIUS.

red snow. Watermelon snow; summer snowbanks, especially in mountainous areas, which have a pinkish coloration owing to enormous numbers of red protozoans.

red spider. See TETRANYCHIDAE.

red squirrel. See TAMIASCIURUS.

redstart. 1. Small warbler, Setophaga ruticilla, especially common east of the Rocky Mountains; male black with bright orange patches on wings and tail, and white belly. 2. Small European songbird, Phoenicurus phoenicurus.

red-tailed hawk. See BUTEO.

red tide. Occasional red appearance of inshore waters along the south Atlantic and Gulf states (and elsewhere) owing to enormous numbers of red protozoan flagellates, especially Gymnodinium and Gonyaulax; produces death of fishes and other animals by means of toxic substances released into the water.

reduced hemoglobin. Hemoglobin which is carrying no respiratory oxygen; bluish red in color.

reduction division. Meiosis.

Redunca. Genus of ruminants in the the Family Bovidae; includes the reedbucks; small, fawn-colored spp. of South Africa; females hornless.

Reduviidae. Family of predaceous insects in the Order Hemiptera; assassin bugs, kissing bugs; with a narrow head and stout beak; Rasahus, Rhodnius, and Triatoma are important genera.

red water. 1. Red tide. 2. Any of several diseases of certain domestic animals; marked by reddish urine.

redwing. 1. European thrush having red under wing coverts. 2. Locally, the American red-winged blackbird.

red-winged blackbird. See BLACKBIRD.

reedbird. See DOLICHONYX.

reedbuck. See REDUNCA.

referred pain. Sensation of pain in one part of the body, even though the actual pain stimuli may be originating in some distant part of the body; e.g. a headache in cases of constipation.

reflex. Reflex action.

reflex act. Reflex action.

reflex action. Involuntary response to a stimulus in animals possessing nerves, a nerve cord, and central nervous system; may or may not involve a conscious sensation; the knee jerk, blinking of the eye when an object is thrust near it, withdrawal of the hand from a hot object, and secretion of tears when the cornea is irritated are all examples. See REFLEX ARC.

reflex arc. Specific pathway taken by a nerve impulse through various neurons and synapses involved in a reflex action; it always includes: (1) one or more afferent neurons transmitting the impulse away from the area of stimulation and toward the nerve cord and/or brain, (2) one or more adjustor neurons in the nerve cord and/or brain, and (3) one or more efferent (effector) neurons which transmit the impulse to the effector (muscle, gland, etc.); most vertebrate reflexes are complex. See REFLEX ACTION.

refractory period. Small fraction of a second following the point at which a nerve or muscle begins functional activity and during which the nerve or muscle will not respond to a second stimulus.

refugium. Isolated unmodified locality which is surrounded by an area drastically modified by geological, climatic, or other physical alteration; often a center for relict spp. Pl. refugia.

Regadrella. Genus of glass sponges (Hexactinellida).

Regalecus. Genus of greatly elongated compressed, ribbon-like marine fishes with dorsal fin extending the length of the body (Order Allotriognathi); oarfish, king of the herrings; to 20 ft. long; often the source of "sea serpent" stories.

regal moths. See CITHERONIIDAE.

regal python. See PYTHON.

regeneration. Capacity of plants and animals to replace tissues and larger parts lost by injury, fragmentation, or other processes; crustaceans and sea stars will regenerate lost appendages, and fragments of a turbellarian will often each regenerate the necessary parts to become a complete new individual; limited to tissue replacement and wound repair in mammals.

regressive evolution. Appearance of characters in a taxon which are usually considered more primitive than those normally present in such a taxon; e.g. the occurrence of a predominantly cartilaginous skeleton in the sturgeon which is descended from ancestral bony fishes.

Regularia. Subclass of echinoids; madreporite interambulacral, anus central and aboral, Aristotle's lantern present.

regulation. 1. Process whereby a damaged or partial embryo produces a complete, normal embryo. 2. Morphological, physiological, or behavioral adaptation to a change in environmental conditions.

Regulus satrapa. Golden-crowned kinglet, a very small warbler-like bird; upper parts olive-gray but with a yellow or orange crown; usually associated with evergreen trees.

Reighardia. Genus of tongue worms parasitic in gulls and terns.

reindeer. See RANGIFER.

Reissner's membrane. Membrane separating the scala vestibuli from the cochlear duct in the inner ear.

Reithrodontomys. Genus of small harvest mice in the Family Cricetidae; feed on vegetable material; common in southeastern states, central states, and Great Plains.

relapsing fever. Any of several acute infectious diseases marked by muscular and joint pains and a series of high fevers, each five to seven days long; produced by various spp. of the spirochaete Borrelia in the blood; transmitted by the bite of lice and ticks.

relaxin. Ovarian hormone which relaxes the pelvic ligament and facilitates parturition in certain mammals.

relic. Relict.

relict. 1. Any sp. surviving in a small local area and widely separated from closely related spp. or other individuals in the same sp.; usually indicative of a continuous population at a previous time and a subsequent separation from the main distribution area by climatic or other changes; e.g. the occurrence of the same sp. of butterfly on the high peaks of widely separated mountain ranges. 2. Sp. or other taxon which is a survivor of a nearly extinct group.

remiges. Row of large flight feathers along the back of the "forearm" and "hand" of a bird. Sing. remex.

Remora. See DISCOCEPHALI.

renal artery. 1. Short artery supplying the vertebrate kidney. 2. Similar artery in certain higher invertebrates.

renal capsule. Bowman's capsule; see MALPIGHIAN BODY.

renal corpuscle. Malpighian body.

renal portal system. System of blood vessels carrying blood from the hind legs and posterior trunk region to the kidneys in fishes, amphibians, reptiles, and birds.

renal threshold. That blood concentration level of a substance up to which it will not be extracted by the kidneys and appear in the urine.

renal vein. 1. Short vein draining the kidney in vertebrates. 2. Similar vein in certain higher invertebrates.

renette. Ventral excretory cell in certain microscopic nematodes.

Reniceps. See SHOVELHEAD.

Reniera. Genus of marine sponges (Subclass Monaxonida); cushion or incrustation growth form; temperate to Arctic seas.

Renifer. Genus of distome trematodes parasitic in reptiles.

Renilla. Genus of alcyonarian anthozoans in the Order Pennatulacea; colony disclike; the sea pansy.

renin. Proteolytic enzyme liberated by the kidneys; has the effect of increasing blood pressure by hydrolyzing a plasma globulin which in turn constricts the blood vessels.

rennet. Extract of the mucosa or contents of the stomach of an unweaned calf; contains rennin.

rennin. Enzyme contained in the gastric juice of mammals; causes the coagulation of casein in the presence of calcium.

renninogen. Proenzyme in the gastric glands; converted to rennin upon being secreted.

renopericardial pores. Two small openings in the pericardium in many Pelecypoda; each is an opening of a kidney.

replacement bone. Cartilage bone.

replete. See HONEY POT.

reproduction. Ability of all organisms to duplicate themselves in kind and thus perpetuate the sp.; fundamental property of protoplasm.

reproductive isolation. Prevention of interbreeding between two or more populations because of different breeding seasons or incompatibility of the reproductive organs in the two sexes.

reproductive system. Collectively, all organs and structures having to do with the processes of reproduction.

Reptantia. Section of crustaceans in the Order Decapoda; lobsters, crayfishes, crabs, etc.; body not compressed, more or less cylindrical or depressed; rostrum usually small or absent.

Reptilia. Class of poikilothermal vertebrates which includes all reptiles; snakes, lizards, turtles, tortoises, alligators, crocodiles, etc.; body covered with scales or scutes; two pairs of legs, each usually ending in five toes with horny claws; skeleton entirely bony; one occipital condyle; heart with two atria and an incompletely divided ventricle (divided in crocodiles); respiration by means of lungs; fertilization internal; eggs large and yolky; embryonic membranes present during development; more than 4000 modern spp.

reservoir. Small cavity near the anterior end of certain flagellates; it receives the products of adjacent contractile vacuole(s) and opens externally by way of the cytopharynx and cytostome; e.g. in Euglena.

reservoir host. Sp. which harbors a parasite which is pathogenic for some other sp.; e.g. cattle are the reservoir host for the human sleeping sickness trypanosome.

residual air. Volume of air remaining in the lungs after as complete an expiration as possible; about 1000 cc. in man.

resilium. Internal hidden portion of the ligament of a bivalve mollusk.

resolution. In biology, the relative ability of the optical system of a microscope to show fine details.

resonating cavities. Cavities which resonate the sounds produced by the vocal cords and give the quality of the sounds; e.g. lungs, trachea, larynx, pharynx, mouth, and nose.

resorption. Simplification of tissue structure or disappearance of certain structures when animals undergo physiological and morphological changes; e.g. the loss of reproductive organs in a turbellarian, and the disappearance of the tail in a developing tadpole.

respiration. 1. Physical processes of breathing (diffusion, inspiration, and expiration). 2. The sum total of all physical and chemical processes by which organisms utilize organic materials as sources of energy and heat; usually oxygen is also used, and carbon dioxide and water are the chief end products.

respiratory enzyme. Any of a series of enzymes which catalyze oxidation reactions by the removal of hydrogen from a substrate, e.g. dehydrogenase.

respiratory movement. Any obvious movement of a whole animal or a part of an animal and concerned with bringing oxygen to respiratory surfaces and taking carbon dioxide away; e.g. movement of the mammalian chest and diaphragm, opercular movements of a fish, and undulations of an aquatic oligochaete.

respiratory organ. Any specialized wet organ of the body associated with the exchange of oxygen and carbon dioxide, e.g. gills, lungs, body surface of an amphibian, etc.

respiratory pigment. Colored protein substance which is a transporter of oxygen in the blood or other body fluid of an animal; e. g. hemoglobin and hemocyanin.

respiratory quotient. RQ; ratio of volume of carbon dioxide expired to the volume of oxygen inspired per unit time with reference to type(s) of food being utilized; theoretical RQ for carbohydrates, fats, and proteins are 1.0, 0.7, and 0.8, respectively.

respiratory system. Collectively, all organs of an animal associated with the absorption of oxygen and the release of carbon dioxide.

respiratory tree. 1. An internal finely-branched double caecum of the cloaca in Holothuroidea; functions in respiration and is alternately filled and partially emptied with sea water via the cloacal aperture. 2. Collectively, the trachea, bronchi, and bronchioles.

response. Physiological or behavioral reaction of an organism to a stimulus; the reaction may be pronounced and rapid, or obscure and very slow.

resting cell. Any cell not actively undergoing mitosis.

resting egg. Egg characteristically produced by certain fresh-water invertebrates, especially in the autumn or at other times of the year when ecological conditions are changing markedly; usually thick-shelled, fertilized, diploid eggs which hatch more abundantly after a period of dormancy; same as winter eggs; e. g. many rotifers and cladocerans produce resting eggs.

resting nucleus. Cell nucleus which is not actively undergoing mitosis.

rete. Network, especially a network of blood vessels or nerve fibers.

rete cords. Ducts in the embryos of higher vertebrates; connect the undifferentiated gonad with some of the mesonephric tubules.

rete mirabile. Abrupt splitting of a small artery or vein into a knot of small vessels which then reunite into a similar vessel which proceeds on its course, as in a kidney glomerulus and the wall of a fish swim bladder.

rete ovarii. Female homologue of the rete testis; vestigial in the adult.

rete testis. Network of ducts, especially in the testis of a mammal; receives sperm from the seminiferous tubules and connects with the epididymis.

reticular connective tissue. Open network of cells found in lymph nodes, spleen, and bone marrow; the intercellular spaces are filled with other types of cells.

reticular fibers. Network of fine scleroproteinaceous connective tissue fibrils found in many vertebrate tissues.

reticulate. Having the structure or appearance of a network; often composed of crossing fibers, ridges, or veins.

reticulin fibers. Network of fine proteinaceous connective tissue fibrils found in many vertebrate tissues.

Reticulotermes. Genus of subterranean termites (Order Isoptera); common in western half of U. S.

reticulocyte. Immature erythrocyte showing a fine reticulum when vitally stained.

reticulo-endothelial system. Special phagocytic macrophages forming much of the lining of blood and lymph vessels in the bone marrow, spleen, lymph nodes, liver, etc.

reticulum. 1. Second of the four compartments of the stomach of ruminants; passes cuds to the mouth. 2. Any network or weblike structure, such as the meshlike structure of some cells.

retina. Innermost nervous tissue layer of the eyeball; contains special receptor rods and cones at the ends of the nerve fibers; all of the nerve fibers leave the eyeball by way of the optic nerve; an analogous structure occurs in certain invertebrate eyes.

retinaculum. 1. Hooked prominence which holds an egg mass in place in barnacles. 2. Device linking fore and hind wings of certain insects. 3. Device which holds the furcula in place on the ventral side of the abdomen of Collembola. 4. Proboscis retractor muscle in some marine worms. Pl. retinacula.

Retinella. Large genus of land snails; common in Europe and N. A.

retinene. Yellow carotinoid breakdown fraction formed when visual purple is stimulated by light, the other fraction being a protein (opsin).

retinula. Group of elongated pigmented cells (usually four to eight) in the basal portion of an arthropod compound eye; each retinula cell passes at its base into a nerve fiber which enters an optic ganglion; a long, cylindrical pigmented rhabdom runs down the center of a retinula. Pl. retinulae. See

OMMATIDIUM.

retractor muscle. Any muscle that retracts or withdraws a structure or part of the body.

retrocerebral organ. Small glandular organ of unknown function in rotifers; located anterior and dorsal, near the brain.

retrograde metamorphosis. Any metamorphosis involving loss or modification of fundamental phylogenetic features, e.g. the metamorphosis of the tunicate tadpole into the sessile adult.

retrogression. Degeneration; passing from a more advanced to a less advanced degree of specialization during the course of the embryological development of an animal or during the course of the phylogenetic development of any larger taxon.

retropharyngeal band. A pair of ciliated grooves located side by side where a tunicate branchial chamber joins the esophagus.

retrorse. Turned backward.

return migration. Type of migration shown by salmon, many birds, etc. which hatch in one area, migrate considerable distances to breed or mature, and then return to the general or specific place of origin.

Retusa. Genus of tiny cylindrical lathe shells (Gastropoda); widely distributed in the North Atlantic.

Rhabdammina. Genus of Foraminifera.

Rhabdias. Genus of rhabditid nematodes parasitic in the lungs of amphibians and snakes.

rhabdion. Any of the sclerotized pieces which are imbedded in the buccal wall of many nematodes.

rhabdite. Short rodlike type of rhabdoid.

Rhabditida. Order of nematodes in the Class Phasmidea; esophagus composed of: a cylindrical corpus which is swollen posteriorly, a narrow isthmus, and a swollen posterior bulbar region; mouth cavity without a stylet.

rhabditiform larva. Free-living larva of many soil nematodes and a few parasitic spp.; characterized by an esophagus having a bulb at about midlength as well as a large posterior bulb; rhabditiform larvae of human parasites are not infective.

Rhabditina. Suborder of rhabditid nematodes; mouth surrounded by two, three, six, or no lips; caudal alae, if present, with papillae; free-living in soil or parasitic in plants and animals; common genera are Rhabditis, Strongyloides, and Rhabdias.

Rhabditis. Very large genus of rhabditid nematodes; found in soil, decaying organic matter, and as parasites in many plants and animals.

Rhabdocoela. Order of small Turbellaria; gastrovascular cavity without diverticula; nervous system with only two longitudinal trunks; marine and fresh-water, and a few parasitic spp.; common genera are Dalyellia, Macrostomum, Stenostomum, and Microstomum.

Rhabdocoelida. Order of turbellarian flatworms; usually split into two orders: Rhabdocoela and Alloeocoela.

Rhabdodermella. Genus of marine sponges; usually urn-shaped, to 2 in. long.

rhabdoids. Minute rodlike structures formed in the epidermal or subepidermal tissues of many Turbellaria; function unknown but they are thought to form mucus upon disintegration; three common types are usually recognized: rhabdites, rhammites, and chondrocysts.

Rhabdolaimus. Common genus of microscopic, free-living nematodes.

rhabdom. Long, cylindrical, rodlike pigmented mass in the center of each retinula of an arthropod compound eye; composed of four rhabdomeres; receives the light rays, but the appropriate corresponding impulse is generated in the surrounding retinula cells which pass the impulses on to nerve fibers. See OMMATIDIUM.

rhabdome. Rhabdom.

rhabdomere. See RHABDOM.

Rhabdomonas. Genus of aquatic sulfur bacteria.

Rhabdopleura. Widely distributed genus in the Phylum Pterobranchia; with two tentacle-bearing arms.

Rhabdopleuridea. Class in the Phylum Pterobranchia; collar with two branched arms; colonial; one gonad, gill slits absent; Rhabdopleura typical.

Rhabdostyla. Genus of peritrich Ciliata; inverted bell-shaped, with a short rigid stalk; attached to invertebrates in fresh and salt waters.

Rhachianectes glaucus. Gray whale of Pacific Coast, now uncommon; a whalebone whale reaching 40 ft.

Rhacophorus. Polypedates.

Rhadinea. See YELLOW-LIPPED SNAKE.

Rhagionidae. Widely distributed family

of snipe flies; adults small to medium, head small; predaceous larvae terrestrial or fresh-water; adults of some spp. bite man; Atherix common.

Rhagodes. See SOLPUGIDA.

rhagon. Type of arrangement of internal chambers and canals in certain sponges; the spongocoel is bordered by flagellated chambers which open into it by wide apopyles.

Rhagovelia. Genus of broad-shouldered water striders in the Family Veliidae.

rhammite. Long slender type of rhabdoid.

Rhamphocorixa. Genus of water boatmen in the Family Corixidae.

Rhamphorhynchus. Genus of Jurassic flying reptiles in the Order Pterosauria; body 2 ft. long.

Rhamphostomella. Genus of marine Bryozoa.

rhamphotheca. Horny covering of the bill of a bird.

Rhantus. Genus of predaceous freshwater beetles in the Family Dytiscidae.

Rhea. See RHEIFORMES.

rhebok. See PELEA.

Rheiformes. Order of terrestrial birds which includes the rheas; three toes on each foot, head and neck partly feathered; Rhea to 5 ft. tall; open plains of South America.

rheobase. Minimal continuing electric current which will cause excitation, especially of nerve or muscle.

rheocrene. Spring in which the water emerges from the ground and runs off without forming a pool.

rheoplankton. Plankton of running waters.

rheotaxis. Behavioral response of an animal to a current of water; positive rheotaxis is heading into a current; negative rheotaxis is moving away from a region of current.

rhesus. See MACACA.

rhesus factor. Rh factor.

Rheumatobates. Genus of water striders in the Family Gerridae; males bizarre, with deformed hind legs.

Rh factor. Antigen determined by heredity and present in the erythrocytes of about 85% of the white population, such blood being Rh+; the remainder of the population is said to have the Rh- condition; if Rh+ blood is repeatedly transfused into an Rh- person, the latter develops an anti-Rh agglutinin which will hemolyze the corpuscles of Rh+ blood upon further transfusions; repeated reverse transufsions (an Rh+ individual receiving Rh- blood) produce the same results; an Rh- mother bearing an Rh+ fetus (which received the Rh+ gene from the father) may become immunized by the Rh+ fetal erythrocytes entering her circulatory system; in subsequent pregnancies her anti-Rh agglutinin will enter the fetal circulation through the placenta and hemolyze the fetal blood with possibly fatal results; this situation is known as erythroblastosis in the fetus or newborn child.

rhinencephalon. Most anterior olfactory portion of the vertebrate brain; olfactory lobes.

Rhineodon typicus. Very large tropical shark; whale shark; to 45 fet. long; the largest of all fishes; feeds on plankton by means of gill rakers.

Rhineura floridana. Thunder worm or worm lizard (reptile) of Fla.; lavender color and the general appearance of a large earthworm; burrows in soft soil; to 9 in. long.

Rhinichthys. Genus of small brightly colored stream fishes in the Family Cyprinidae; dace; common and widely distributed in U. S.

Rhinobatos. Genus of sharklike rays (Suborder Batoidea); guitar fish, fiddlerfish.

Rhinoceros. Genus of mammals in the Family Rhinocerotidae; rhinoceroses with a single horn on the snout; the large Indian rhinoceros is found in northern India south of the Himalayas; the Java rhinoceros is a smaller sp. found from India to Malaya, Sumatra, Java, and Borneo.

rhinoceros beetle. Any of several very large tropical beetles in which the male has a stout upright horn on the head.

Rhinocerotidae. Family of mammals in the Order Perissodactyla; includes the rhinoceroses; large spp. with thick armor-like hide and three hoofed toes on each foot; one or two heavy upright horns on snout; almost hairless; wooded and grassy areas of Africa and southeastern Asia; Rhinoceros, Dicerorhinus, Ceratotherium, and Diceros.

Rhinocheilus. See LONG-NOSED SNAKE.

Rhinocryptidae. Family of insectivorous birds in the Order Passeriformes; tapaculos; stout-bodied, weak fliers; brownish and grayish coloration; Central and South America; chiefly in dry plains and scrublands.

Rhinoderma. Genus of Chilean frogs;

male carries the developing eggs in his vocal sacs.

Rhinodrilus fafueri. Giant earthworm of Ecuador; attains a length of more than 6 ft. and has more than 600 segments.

Rhinoglena frontalis. Conical, free-swimming rotifer; large anterior dorsal proboscis.

Rhinolophidae. Family of Old World horseshoe-nosed bats.

rhinophore. One of two tentacle-like projections on the dorsal surface of the head of certain opisthobranch mollusks; presumably olfactory.

Rhinopithecus. Genus of snub-nosed monkeys; nose long and turned up; sturdy, omnivorous, long-haired spp. of midmountain areas of southeastern Asia where a snow cover is present during most of the year.

Rhinopomatidae. Family of mouse-tailed bats; primitive insectivorous spp. found throughout southern Asia, from Egypt to Burma and Sumatra.

Rhinoptera bonasus. Large ray (Order Batoidea) having a snout with a fancied resemblance to that of a cow; Cape Cod to Fla.

Rhipicephalus. Genus of ticks found on dogs, cattle, sheep, and horses; some spp. transmit protozoan and rickettsia diseases.

Rhipidistia. Order of extinct lobe-finned fishes; skull ossified; pineal eye present; tail heterocercal; Devonian to Carboniferous; Osteolepis typical.

Rhipidodendron. Genus of fresh-water biflagellate Protozoa; cells imbedded at tips of long tubular confluent gelatinous tubes.

Rhipidoglossa. Subdivision of the gastropod Suborder Aspidobranchia; radula with rows of numerous narrow teeth radiating out like the ribs of a fan.

Rhipidogorgia. Genus of tropical corals.

Rhithrogena. Common genus of Ephemeroptera.

Rhithropanopeus. Genus of marine mud crabs (Brachyura).

Rhizobium. Genus of nitrogen-fixing bacteria forming a symbiotic or commensal relationship in leguminous plant roots.

Rhizocephala. Order of barnacles (Subclass Cirripedia); adults highly modified and parasitic on crabs; body saclike, with absorptive rootlike structures penetrating and ramifying throughout the host body; no shell, appendages, or digestive tract; Sacculina is important because it parasitizes edible crabs.

Rhizochrysis. Fresh-water genus of naked amoeboid Protozoa; one or two chromatophores; placed in the Mastigophora although flagella are absent.

Rhizocrinus. Common deep-sea genus of sea lilies (Crinoidea).

Rhizogeton. Genus of coelenterates in the Suborder Anthomedusae.

Rhizoglyphus hyacinthi. Bulb mite (Order Acarina); tunnels in bulbs and roots.

Rhizomastigina. Order of flagellate Protozoa; colorless, one flagellum, amoeboid; fresh water and damp soil; e.g. Mastigamoeba and Mastigina.

Rhizomyidae. Family of root rats or bamboo rats; nocturnal spp. usually found in duff and trash of the soil in bamboo groves; Ethiopia and southern Asia.

rhizoplast. Intracellular fibril which connects the basal granule or blepharoplast with the nucleus (or parabasal body) in flagellate protozoans; when present in ciliate protozoans, it has no connection with the nucleus.

rhizopod. 1. Any member of the Class Sarcodina. 2. A rhizopodium.

Rhizopoda. Sarcodina.

rhizopodium. Filamentous, branching, and anastomosing pseudopodium; common in Foraminifera. Pl. rhizopodia.

Rhizorhina. Genus of marine copepods parasitic on crustaceans.

Rhizostoma. Genus of large scyphozoan jellyfish belonging to the Order Discomedusae; no peripheral tentacles; oral arms partially fused; usually tropical and subtropical.

Rhizostomae. Suborder of the Coelenterata Order Discomedusae; oral arms fused and with eight lobes; numerous small proctostomes; tentacles lacking; Rhizostoma and Cassiopeia typical.

Rhoda. Common genus of euphausid crustaceans; often very abundant in the seas.

Rhodacmea. Genus of fresh-water snails; Ala. to Ill.

Rhodesian man. See HOMO RHODESIENSIS.

Rhodeus amarus. Bitterling; a small European minnow recently introduced into N.Y.; female with a long ovipositor used to deposit eggs in the mantle cavity of fresh-water mussels.

Rhodine. Genus of tube-building Polychaeta.

Rhodites. Genus of cynipid wasps which form galls on rose plants.

Rhodnius. Genus of bugs which transmit Chagas' disease in Central and South America.

rhodopsin. Protein pigment contained in the rods of the retina; it is bleached by light and must be present for vision in dim light.

rhombencephalon. Hindbrain.

rhombocoel. Cavity of the embryonic rhombencephalon.

Rhombognathus. Genus of marine mites in the Family Halacaridae.

rhomboideus. Mammalian muscle originating on the spinous processes of thoracic vertebrae and inserting on the scapula.

rhombomere. Neuromere.

rhopalium. Tentaculocyst; marginal sense organ of certain jellyfishes; hollow, club-shaped organ borne in one of eight (usually) peripheral emarginations. Pl. rhopalia.

Rhopalocera. One of the three insect suborders in the Order Lepidoptera; skippers and butterflies; mouth parts modified into a siphon-like proboscis; antennae enlarged into knobs or hooks at tips; veins of forewings and hind wings unlike; wings of each side held together by a membrane near base of hind wing; wings held vertically above body when at rest; pupa not covered by a silken cocoon; day fliers.

Rhopalophrya. Genus of sapropelic marine and fresh-water holotrich Ciliata; cell cylindrical and furrowed.

Rhopalosiphum nymphaeae. Water lily aphid in the Order Hemiptera; an insect which feeds on a wide variety of aquatic plants as well as almond, apricot, and plum trees.

Rhopalura. Genus of Mesozoa parasitic in the tissues and body cavities of a wide variety of marine invertebrates; see ORTHONECTIDA.

Rhopilema. Uncommon genus of jellyfishes in the Order Discomedusae; diameter to 12 in.

Rhyacophila. Common genus of Trichoptera; larvae naked.

Rhyacotriton olympicus. Small salamander found in and near swift streams in western Wash. and Ore.; Olympic salamander; lungs reduced.

Rhynchelmis. Genus of fresh-water oligochaetes.

Rhynchiscus. Genus of small diurnal tropical American bats.

Rhynchobdellida. One of the three orders of leeches; mouth a small pore in the oral sucker through which a muscular proboscis can be protruded; jaws absent; three or six or more annuli per segment in middle of body; blood colorless; Placobdella and Glossiphonia common.

Rhynchobothrius. Genus of trypanorhynchid tapeworms; found especially in the spiral valve of sharks.

Rhynchocephalia. Order of primitive lizard-like reptiles; Sphenodon punctatum (tuatara) the only living sp., in New Zealand; mid-dorsal row of spines; vertebrae amphicoelous; with abdominal ribs; anal opening transverse; with a well developed median pineal eye; to 30 in. long; aquatic, terrestrial, and in burrows; carnivorous.

rhynchocoel. Dorsal tubular cavity in nemertines; contains the inverted proboscis and has no opening to the outside.

Rhynchocoela. Nemertea.

rhynchodaeum. 1. Anterior dorsal tubular cavity in nemertines; opens to the outside by an anterior proboscis pore and is obliterated when the proboscis is protruded. 2. A similar cavity present in some leeches, but here it is the extreme anterior end of the digestive tract.

Rhynchodemus. Large genus of land planarians.

Rhynchomesostoma. Genus of rhabdocoel Turbellaria.

Rhynchomonas. Genus of colorless marine and fresh-water Protozoa; one flagellum imbedded in an anterior projection, another trailing.

Rhynchophanes mccowni. McCown's longspur; a sparrow-like bird of the Great Plains.

Rhynchopidae. Family of birds in the Order Charadriiformes; skimmers; maritime spp. with lower bill longer than upper; feed on surface animals in oceans.

Rhynchophora. Suborder of Coleoptera; a group of vegetable feeders with a prolonged beak; often included in the Suborder Polyphaga.

Rhynchopsitta pachyrhyncha. Thick-billed parrot of northern Mexico and southern Ariz.; bright green, with heavy black bill and red forehead; to 17 in. long; the only parrot in the U.S.

Rhynchoscolex. Genus of small rhabdocoel Turbellaria; with an anterior proboscis-like projection.

Rhynchospio. Genus of Polychaeta.
Rhynchota. Hemiptera.
Rhynchotalona falcata. Small littoral cladoceran; northern states.
Rhynchothorax. Genus of Pycnogonida.
Rhynchotragus. Genus of African dik-diks.
Rhynchotremata. Genus of brachiopods in the Order Telotremata.
Rhynchotus. See TINAMIFORMES.
Rhynchozoon. Genus of marine Bryozoa.
Rhynochetidae. Family in the Order Gruiformes; the kagu; long-legged grayish bird of highland forests of New Caledonia.
Rhyssa. Genus of ichneumon flies which parasitize siricid wasps.
rib. One of a series of long, curved, paired bones more or less enclosing the trunk cavity and articulating with the vertebral column; present in most vertebrates.
ribbed mussel. See MODIOLUS.
ribbed pod shell. See SILIQUA.
ribbonfish. See TRACHYPTERUS.
ribbon snake. See THAMNOPHIS.
ribbon worm. Common name of any member of the Phylum Nemertea.
riboflavin. Vitamin B_2 of the B complex; essential to proper growth and formation of enzymes associated with intermediary metabolism; deficiency produces cheilosis (inflammation and cracking at corners of mouth); abundant in green leaves, milk, eggs, liver, and yeast; $C_{17}H_{20}O_6N_4$.
ribonuclease. Enzyme which hydrolyzes ribonucleic acid.
ribonucleic acid. RNA; constituent of nucleoproteins of cytoplasm and chromatin; composed of long chains of phosphate and sugar molecules (ribose) with several bases arranged as side groups: adenine, guanine, cytosine, and uracil; carrier and mediator of genetic information.
ribose. A pentose characteristic of nucleic acids and nucleotides, usually in the form of ribonucleic acid.
ribosome. Microsome.
ricebird. Any of several birds common in rice fields; see especially DOLICHONYX.
rice cup shell. Cylichna oryza; a tiny whitish marine gastropod shell resembling a grain of rice.
rice rat. See ORYZOMYS.
rice tenrec. See ORYZORICTES.
rice water weevil. See LISSORHOPTRUS.
rice weevil. See SITOPHILUS.
Richards, Alfred N. American physiologist (1876-); contributed much to our knowledge of the physiology of kidneys.
Richardson ground squirrel. See FLICKERTAIL.
Richardsonius. Small genus of redside shiners (Cyprinidae) found in Lahontan, Bonneville, and Columbia basins.
Richardson's owl. See AEGOLIUS.
Richmondena cardinalis. Common American cardinal or redbird in the Family Fringillidae; male all red, with crest; female brown; bill stout, conical.
ricin. Toxic albumin extracted from the castor bean seed.
Ricinulei. Small and poorly known order of tropical arachnids; body elongate-oval; abdomen nine- or 12-segmented, last four segments forming a retractile tubercle; cephalothorax with an anterior dorsal hood; chelicerae and pedipalps short; nine spp.; Cryptostemma and Cryptocellus examples.
rickets. Disease of children characterized by soft, malformed bones produced by improper calcium and phosphate utilization owing to a deficiency of sunshine and/or vitamin D in the diet; sometimes produced by insufficient amounts of calcium in the diet.
rickettsia. Any of numerous spp. in the genus Rickettsia; bacteria-like organisms which have certain protozoan characteristics; intracellular parasites commonly transmitted by arthropods; cause typhus fever, trench fever, and Rocky Mountain spotted fever.
rictus. 1. Posterior corner of the mouth. 2. Fissure or cleft.
ridged top shell. Margarites cinerea, a small marine snail of the Atlantic.
riffle beetles. See PSEPHENIDAE.
right atrium. Receiving chamber of the heart on the right side of the body; transmits blood to a ventricle; fishes have a single atrium but all tetrapods have two.
right ventricle. Right posterior chamber of the heart in birds, mammals, and a few higher reptiles; pumps blood to the lungs.
right whale. See BALAENA and EUBALAENA.
right whale dolphin. See LISSODELPHIS.
rigor mortis. Stiffening of body after death.
rinderpest. Infectious virus disease of cattle, sheep, goast, and wild herbivores; marked by inflammation of mucous membranes, especially in the

intestine.

ring-billed gull. Larus delawarensis; a black, white, and gray sp. with a black subterminal band on the bill; inland lakes of Alaska, Canada, and northcentral and northwestern U.S.

ring canal. 1. Circular canal in the region of the mouth in echinoderms; a part of the water vascular system. 2. Peripheral ring-shaped portion of the gastrovascular system of many free-swimming stages of the Coelenterata, especially in hydrozoan medusae and jellyfish.

ring dove. 1. European pigeon (Columba palumbus). 2. Small dove of southeastern Europe and Asia; often caged.

Ringer's fluid. Isotonic salt solution for human tissues; each 100 ml. contains 0.82 to 0.90 g. sodium chloride, 0.025 to 0.035 g. potassium chloride, and 0.030 to 0.036 g. calcium chloride; various modifications used for other vertebrates.

ringhals. South African cobra which sprays or spits venom, often at the eyes of the intended victim.

ring-necked duck. See NYROCA.

ring-necked snake. See DIADOPHIS.

ring-tailed cat. See BASSARISCUS.

ringworm. Any of several contagious fungus diseases of the skin of man and domestic animals; ringlike pigmented patches covered with vesicles or scales.

Rio Grande perch. See CICHLASOMA.

Riparia. See HIRUNDINIDAE.

riparian. Pertaining to an organism living on the shore of a lake or river.

Ripistes. Genus of tube-building fresh-water naidid oligochaetes.

ripple bugs. See VELIIDAE.

risorius. Muscle which extends from the angle of the jaw to the angle of the mouth; draws out corner of mouth and compresses cheek.

Rissa tridactyla. Kittiwake; a whitish gull-like sea bird of circumarctic and subarctic regions; winters south to Mediterranean and southern U.S.

Risso's dolphin. See GRAMPIDELPHIS.

river deer. See HYDROPOTES.

river dolphin. Any of several fresh-water dolphins (mammals), especially Inia, Platanista, and Lipotes.

RNA. Ribonucleic acid.

roach. 1. European fresh-water fish (Rutilis rutilis) in the Family Cyprinidae. 2. Any of several U.S. fresh-water fishes, especially Notemigonus crysoleucas, Hesperoleucas spp., and several spp. in the Family Centrarchidae.

roadrunner. See GEOCOCCYX.

roan antelope. See HIPPOTRAGUS.

robalo. Any of several pikelike marine fishes of the Caribbean area; see CENTROPOMUS.

robber ants. See POLYERGUS.

robber flies. See ASILIDAE.

robber frog. Either of two spp. of Eleutherodactylus (Leptodactylidae); the larger one (to 3.5 in. long) is often called the barking frog and occurs in limestone caves and ledges in southern Tex., N.M., and Ariz.; another (to 1.2 in. long) is found in southern Fla.; short spp. with a wide head; tadpole remains in the egg and a tiny frog hatches out.

robin. American sp. is Turdus migratorius; European robin is a warbler-like bird (Erithacus rubecula).

Roccus saxatilis. Striped bass; a fine food and game fish (Family Serranidae) of the Atlantic Coast which has also been planted along the Pacific Coast; a carnivorous inshore sp. which spawns in bays or streams; to 70 lbs.

Rocinela. Genus of marine Isopoda.

rock barnacles. Group of spp. of barnacles that have no stalk and are especially abundant on rocky shores; Balanus very common.

rock bass. See AMBLOPLITES.

rock borer. Any of numerous marine clams which bore holes into soft rock.

rock cockle. See VENERUPIS.

rock cod. 1. Any of numerous codlike marine fishes commonly associated with rocky bottoms. 2. A variety of the common codfish.

rock crab. See CANCER.

rock dove. Type of wild pigeon from which certain domestic pigeons are derived.

rock eel. See PHOLIS.

rockfish. Any of a great many fishes inhabiting rocky sea bottom; Sebastodes (Family Scorpaenidae) and Pholis (Family Pholidae) common.

rock hind. Any of several large groupers and related fishes found in West Indian waters; see EPINEPHELUS.

rock-kangaroos. See OSPHRANTER.

rock lobster. See PANULIRUS.

rock rabbit. Pika.

rock rat. See PETROMYIDAE.

rock shell. See MUREX.

rock squirrel. Citellus variegatus, a gray and black squirrel of rocky areas in Mexico, west Tex., Ariz., and N.

rock sturgeon. See ACIPENSER.
rock wallaby. See PETROGALE and PERADORCAS.
rock wren. See SALPINCTES.
Rocky Mountain sheep. See BIGHORN.
Rocky Mountain canary. Burro.
Rocky Mountain spotted fever. Human disease of the endothelium produced by the rickettsia Dermacentroxinus rickettsi transmitted by the bite of several spp. of wood ticks (Dermacentor); characterized by a rash, high fever, and high death rate, and has been reported from nearly all states; vaccination and certain antibiotics are of great value.
Rocky Mountain subregion. Zoogeographic subdivision of the Nearctic region; includes most of the U.S. Rocky Mountain area west of meridian 100 and extends through Baja California, northern Mexico, and the Mexican plateau. Cf. CALIFORNIA SUBREGION.
Rocky Mountain whitefish. See PROSOPIUM.
Rodentia. Very large order of mammals which includes the true rodents or gnawing mammals; squirrels, gophers, mice, rats, porcupines, and beavers; each limb with five toes and claws; two incisors in each jaw, rootless, chisel-like, and continuously growing; canine teeth absent; lower jaw moves forward and backward, as well as laterally; body usually less than 12 in. long; chiefly herbivorous; world-wide (except for Australian region) in many types of habitats; more than 6400 spp. and subspecies.
rods. Minute elongated rod-shaped cells in the retina of the vertebrate eye; in man, a rod is about 60 microns long and 2 microns thick; they are especially concerned with dim light vision and are about 20 times as abundant as cones; neurons synapse with the rods.
roe. 1. Roe deer (Capreolus) of the Old World. 2. Eggs of fishes. 3. Occasionally, the swollen ovaries or extruded eggs of decapod crustaceans.
roebuck. Male of the Old World roe deer (Capreolus).
roller. Any of certain Old World birds which turn over or tumble in flight; coloration bright; feed mostly by capturing insects on the wing.
Romalea microptera. One of the short-winged, heavy-bodied lubber grasshoppers of southeastern states.
Rondelet, Guillaume. French physician and naturalist (1507-1566); famous for his Die Piscibus Marinis, containing descriptions and illustrations of a great many aquatic animals, especially from the Mediterranean.
roof rat. Grayish brown variety of the common black rat (Rattus rattus); often nests in trees and roofs; cosmopolitan in warm regions.
rook. Common gregarious European bird (Corvus frugilegus) very similar to the American crow.
rooter skunk. See CONEPATUS.
root maggot flies. See ANTHOMYIIDAE.
root nodule bacteria. Bacteria which live symbiotically in small nodules on the roots of leguminous plants; able to use atmospheric nitrogen and convert it into compounds which can be used in the nitrogen metabolism of the plant bearing the bacteria and nodules. See NITROGEN CYCLE and NITROGEN-FIXING BACTERIA.
root rat. See RHIZOMYIDAE.
rootstock. Hydroid hydrorhiza or stolon; the horizontal attached portion of a colony.
rorqual. See BALAENOPTERA.
rose-breasted grosbeak. Pheucticus ludovicianus (Thraupidae); coloration black and white with throat, breast, and wing lining pink; eastern N.A.
rosebud. Common name for certain pelagic ctenophores.
rose chafer. See MACRODACTYLUS.
rose comb. Low papillated type of comb occurring in Wyandotte fowl.
rosefish. See SCORPAENIDAE.
rose slug. Larva of certain sawflies; feed on rose bushes.
rosette. See CESTODARIA.
rosorial. Pertaining to gnawing, as in rodents.
Ross, Sir Ronald. English physician (1857-1932); discovered in 1898 that malaria parasite is transmitted by mosquitoes; received Nobel prize in physiology and medicine in 1902.
Rossia. Genus of small squids with a short, globose body; to 3 in. long.
Ross's seal. See OMMATOPHOCA.
Rostanga. Genus of nudibranchs often found on red sponges.
rostellum. 1. Cone-shaped or cylindrical projection on the anterior end of the scolex in certain tapeworms; often retractile and armed with hooks. 2. Tubular mouth parts of certain insects.
Rostratulidae. Small family of birds in the Order Charadriiformes; painted snipes; South America, Africa, southern Asia, and Australia.

Rostrhamus. See KITE.

rostrum. 1. Median pointed process at the anterior end of the cephalothorax in many decapod crustaceans. 2. Beak of Hemiptera. 3. Rigid extension of the head in weevils. 4. Any snoutlike prolongation of the head. 5. Anterior end of some gregarine protozoans. 6. Small median ventral plate of certain barnacles. 7. Nonretractile gastropod snout. 8. Grooved extension of many gastropod shells.

rosy boa. Stout, heavy-bodied snake in the genus Lichanura (Family Boidae); abdomen red; to 3 ft. long; a constrictor of dry rocky hills of southern Calif., Ariz., and Mexico.

rosy synapta. Synapta roseola; a long, bright red, wormlike sea cucumber in the Order Apoda.

Rotalia. Genus of Foraminifera.

Rotaria neptunia. Common sp. of elongated rotifer which creeps about in a leechlike fashion.

rotator. Muscle that rotates a part of the body.

Rotatoria. Phylum which includes the rotifers or wheel animalcules; microscopic wormlike to spherical organisms found in great abundance and variety in all types of fresh-water habitats; a few spp. occur in salt and brackish waters; anterior end bears tufts of cilia and is surrounded by a complex ciliary band used chiefly in feeding and locomotion; body wall consists of a thin cuticle, syncitial hypodermis, and scattered longitudinal and circular muscle fibers; there is usually a posterior "foot" and two "toes"; the digestive tract has a muscular pharynx (mastax) containing minute sclerotized jaws (trophi); flame bulb system present; nervous system simple; males minute and degenerate, or unknown in many spp.; about 1500 spp.

rotenone. See DERRIS.

rotifer. Common name of any member of the Phylum Rotatoria.

Rotifera. Rotatoria.

rotula. 1. One of the small radial bones in the aboral portion of Aristotle's lantern. 2. Patella.

rotule. Wheel-like end of a sponge birotulate spicule.

Rouget cell. Cell having pseudopodial processes surrounding a capillary.

roughage. Indigestible residue in the large intestine; consists of tough fibers, cartilage, cellulose, etc.

rough file shell. Lima scabra; a yellow scalloped bivalve; N.C. to West Indies.

rough-legged hawk. See BUTEO.

rough periwinkle. Littorina rudis; a small, solid, sculptured periwinkle of variable coloration.

rough piddock. See ZIRFAEA.

rough-winged swallow. See STELGIDOPTERYX.

rouleaux. Clumps of erythrocytes in the shape of piles of coins.

round clam. See VENUS.

roundheaded apple borer. See SAPERDA.

round sting rays. Small roundish sting rays to 2 ft. long; common in shallows; Urolophis.

round whitefish. See PROSOPIUM.

round window. Fenestra rotunda.

roundworm. Any member of the Phylum Nematoda.

Rousettus. Genus of tailed fruit-bats; tropical Africa, and Asia from Burma to Palestine.

rove beetles. See STAPHYLINIDAE.

royal antelope. See NEOTRAGUS.

royal jelly. Highly nutritious food which is regurgitated by young adult worker bees and fed to all very young bee larvae; constitutes the sole food of larvae destined to become queens.

RQ. Respiratory quotient.

rubber boa. See CHARINA.

rubber snake. See CHARINA.

ruby-throated hummingbird. See ARCHILOCHUS.

rudd. Small European fresh-water fish in the minnow family.

ruddy duck. See OXYURA.

rudimentary. Essentially the same as vestigial, but with chief emphasis on small size.

ruffed grouse. See BONASA.

rugae. Ridges or folds, as the inner surface of the stomach.

rugose. 1. Roughened. 2. Full of wrinkles, especially irregular wrinkles.

Ruini, Carlo. Italian anatomist, known for his accurately illustrated book on horse anatomy (1598).

rumen. First and largest compartment of the four-part stomach of ruminants; food is ingested after little chewing and is passed to the rumen immediately and then to the reticulum; later, small masses, or cuds, are regurgitated, chewed thoroughly, and then reswallowed to the third compartment, or omasum; sometimes called the paunch.

ruminant. Any animal which chews a cud, e.g. camel, deer, giraffe, cattle, etc.

runner. 1. Small marine game fish of the Carolinas to the West Indies; *Caranx ruber*. 2. Any of several other active, leaping marine fishes of warm seas.

Rupicapra. Genus of ruminants in the Family Bovidae; includes the chamois, an antelope-like sp. about the size of a goat with a light brown coat marked with black; mountains of Europe and eastern Mediterranean countries, introduced into New Zealand; extremely agile, especially in rocky inaccessible places.

Rupicola. Genus of terrestrial birds which includes the cocks-of-the-rock; head with a large, elevated, compressed crest; coloration brilliant; total length to 13 in.; dense forests of tropical South America.

rusa. Small deer (*Cervus*) of the East Indies.

Russel's viper. Brightly marked venomous snake of southern and southeastern Asia.

Rustella. Genus of extinct brachiopods in the Order Palaeotremata.

rut. Same as estrus, except that rut is often used to designate the period of sexual desire in both sexes rather than the female along; usually applied to deer, cattle, and other ruminants, where it occurs once per year.

Ruvettus pretiosus. Escolar; a large, deep-water, mackerel-like fish of the Mediterranean and other warm seas; a good food sp.

Rynchopidae. Family of birds in the Order Charadriiformes; skimmers; bill bladelike, with very long mandible; *Rynchops* typical.

S

Sabella. Common genus of peacock worms (Polychaeta).

Sabellaria. Genus of polychaete annelids living in hard, twisted tubes made of sand and a secreted glutinous material; peristomium greatly enlarged; numerous gill filaments.

saber-toothed tiger. Any of numerous catlike mammals having greatly elongated upper canines; small to large spp.; Oligocene to Pleistocene.

Sabinea. Genus of marine shrimps.

sable. See MARTES.

sable antelope. See HIPPOTRAGUS.

sablefish. Gray to black Pacific food fish in the Family Scorpaenidae.

saccharase. Invertase.

saccharide. Any of a series of carbohydrates including the mono-, di-, tri-, and polysaccharides.

saccharin. Synthetic sweetening agent, much sweeter than sugar.

Saccocirrus. Genus of small marine worms in the Class Archiannelida.

Saccoglossus. Dolichoglossus.

Saccopharynx. See LYOMERI.

Sacculina. See RHIZOCEPHALA.

sacculus. Saclike portion of the membranous labyrinth of the inner ear; it originates from the base of the endolymphatic duct and gives rise to the cochlea; functions in equilibration and is relatively small in the higher vertebrates.

saccus vasculosus. Thin vesicle on the floor of the diencephalon of most fishes; function unknown but presumably sensory.

sacral. Of or pertaining to the sacrum.

sacral vertebra. Vertebra of tetrapods which articulates with the pelvic girdle; one is present in amphibians but there are two or more (fused) in other tetrapods; five in man.

Sacramento perch. See ARCHOPLITES.

sacred fish. See MORMYRUS.

sacroiliac. Collectively, pertaining to the sacrum and ilium, especially the associated joint and ligaments.

sacrospinalis. Mammalian lumbar muscle which helps maintain the arch of the spinal cord.

sacrum. Bony mass resulting from the fusion of two or more vertebrae between the lumbar and coccygeal regions in vertebrates; in man the sacrum is usually composed of five fused vertebrae and forms the posterior wall of the pelvis. Adj. sacral.

saddle-billed stork. See EPHIPPIORHYNCHUS.

Sagartia. Genus of sea anemones; commonly reproduce by longitudinal fission.

sagebrush vole. See LAGURUS.

sage grouse. See CENTROCERCUS.

sage hen. See CENTROCERCUS.

sage sparrow. See AMPHISPIZA.

sage thrasher. See OREOSCOPTES.

sagitta. Larger of the two otoliths in most fishes.

Sagitta. Common genus in the Phylum Chaetognatha.

sagittal plane. Vertical longitudinal plane drawn through the body of a bilaterally symmetrical animal so as to divide it into right and left halves.

Sagittariidae. See SECRETARY BIRD.

Sagittarius. See SECRETARY BIRD.

sagittocysts. Pointed epidermal vesicles in certain acoel Turbellaria; each contains a central explosively protrusile rod or needle-like structure.

Saiga. Genus of antelopes in the Family Bovidae; includes the saiga, a sp. noted for its swollen nose; steppes of southeastern Europe and western Asia.

sailfish. See ISTIOPHORUS.

sailor's choice. Any of several marine food fishes (porgies) of the Atlantic Coast of North and South America, especially Lagodon rhomboides.

Saimiri. Squirrel-monkeys; a genus of small, long-legged tropical South American monkeys; insect and fruit eaters.

Saint-Hilaire, Etienne Geoffroy. French zoologist (1772-1844); comparative anatomist; opponent of Cuvier.

Saissetia oleae. Insect in the homopteran Family Coccidae; the black scale; pest on citrus, apricot, oleander, pepper, olive, and other trees.

saki. Brazilian leaf-eating monkey of the Guianas and Amazon basin; see PITHECIA and CHIROPOTES.

sakiwinki. See PITHECIA.

salamander. General term for the tailed amphibians (Subclass Caudata).

Salamandra. Genus of European salamanders; S. salamandra is the European fire salamander, black with yellow spots; adults terrestrial; S. atra, the European Alpine salamander, has no aquatic stage and the females are viviparous.

Salamandridae. Family of salamanders; newts and efts; vomerine teeth in two diverging rows between internal nares; at breeding season the adults migrate to ponds and streams where the eggs are fertilized and deposited in water; Europe, eastern Asia, and N. A.; Notophthalmus and Taricha the only U. S. genera; Triturus and Salamandra also typical.

Salamandroidea. In some classifications, a suborder of amphibians which includes the Amphiumidae, Plethodontidae, Desmognathidae, and Salamandridae.

Saldidae. Shore bugs; a family of predaceous Hemiptera commonly found in marshes and bogs or along the margins of bodies of water; small, delicate spp. 3 to 7 mm. long; Saldula common.

Saldula. See SALDIDAE.

Salenia. Genus of deep-sea echinoids in the Order Stirodonta.

Salientia. Same as Anura; large subclass of amphibians; includes the frogs and toads; two pairs of well developed legs; hind legs enlarged for leaping, usually webbed between toes; head and trunk fused without an intervening neck; tail absent in adult; ribs reduced or absent; skin loose; fertilization and egg deposition external; larva an aquatic tadpole, with ovoid head and body and a long compressed tail; in metamorphosis, the gills and tail are absorbed, legs and lungs develop, and head and mouth change in relative size, so that the adult is essentially fitted for terrestrial life; nearly 2000 spp.

saline. Salty or containing salt.

Salinella. Genus of organisms (thus far unclassified) found in Argentina salt beds; wormlike, consisting of a single layer of cells enclosing a digestive cavity; anus and mouth present; digestive cavity and external surface of body ciliated; 2 mm. long; usually included with the Mesozoa.

saliva. Digestive and lubricative juice secreted into the oral cavity of many vertebrates and invertebrates.

salivary gland chromosomes. Unusually large chromosomes, especially in the salivary gland cells of certain Diptera; useful for cytological study of chromosome structure.

salivary glands. Glands in the head or oral cavity region which secrete a digestive juice; present in many vertebrates and invertebrates.

salivation. Discharge of saliva.

Salmacina. Genus of Polychaeta.

Salmincola. Genus of ectoparasitic copepods of trout and whitefish.

salmine. Simple protein found in salmon sperm.

Salmo. Genus of bony fishes which includes the Atlantic salmon and several spp. of trout; S. salar, the Atlantic salmon (to 50 lbs.), spawns in streams of eastern N. A. and western Europe; the Sebago salmon is a closely related landlocked form of New England fresh waters; unlike the Pacific salmon (Oncorhynchus), the Atlantic salmon may spawn two or more times during the life history; S. gairdneri, the rainbow trout, was originally native to Pacific Coast streams of N. A., but has now been introduced over much of the U. S. and other countries where ecological conditions are suitable; S. clarki, the cutthroat trout, is native to cold mountain streams and lakes of the western states; S. trutta, the brown trout, was originally native to Europe but has been planted extensively in the U. S. and elsewhere; S. agua-bonita is the small golden trout, originally restricted to mountainous areas of Calif.

salmon. See ONCORHYNCHUS, SALMO, and SALMONIDAE.

salmon fly. Common name of several large spp. of insects in the Order Plecoptera, especially West Coast spp.

Salmonella. Genus of gram-negative, non-sporeforming, motile, rod-shaped bacteria; many spp. are important intestinal parasites of man, domestic mammals, and birds; produce typhoid and paratyphoid infections through polluted water or food.

Salmonidae. Family of bony fishes which includes the salmon and trout;

native to cool waters of Northern Hemisphere; mouth relatively large; small dorsal adipose fin anterior to base of tail; scales small, dentition well developed; examples are Salmo, Oncorhynchus, Cristivomer, and Salvelinus.

Salmonoidei. Suborder taxon sometimes used to include the fish families Osmeridae, Thymallidae, Salmonidae, Coregonidae, etc.; adipose fine present, oviducts absent.

Salmopercae. Small order of bony fishes which includes the Percopsidae (troutperches) and Aphredoderidae (pirateperches); all small N.A. spp.

Salmopercida. Salmopercae.

salp. See THALIACEA.

Salpa. See SALPIDA.

Salpida. Order of tunicates in the Class Thaliacea; cylindrical or prism-shaped; tunic thick but transparent; muscle bands incomplete below; in chains or solitary; pelagic in all seas; Salpa common.

Salpina. Mytilina.

Salpinctes obsoletus. Rock wren; a gray sp. of rocky canyons and slopes; western Canada and U.S., south to Costa Rica.

Salpingoeca. Large genus of marine and and fresh-water monoflagellate Protozoa; cell with a collar and enveloped in a lorica.

salpingotomy. Sterilization by tying and cutting the oviducts.

salpinx. 1. Trumpet-shaped tube. 2. Eustachian tube. 3. Fallopian tube.

saltation. 1. Jumping or leaping. 2. Abrupt variation in a sp.; mutation.

Saltatoria. Group of families of insects in the Order Orthoptera which leap, e.g. grasshoppers and crickets.

saltatorial. Adapted for leaping.

saltatory conduction. Theory which maintains that action currents in neurons jump from one node of Ranvier to the next.

Salticidae. Very large family of jumping spiders; active daylight hunters, runners, and jumpers; snares are not built, but nests for the night, molting, and hibernating are usually constructed; Salticus and Phidippus common.

Salticus scenicus. Common zebra spider in the Family Salticidae; on fences and outside walls, occasionally indoors.

salt marsh. 1. Low-lying vegetation-covered land adjacent to the sea and regularly flooded by the high tide. 2. Similar inland areas associated with saline springs or lakes, though not regularly inundated.

Salvadora. Small genus of swift, slim, colubrid snakes to 3 ft. long; with a shield-like plate over the nose; yellow and brown stripes; southwestern deserts.

salvarsan. Arsphenamine.

salve bug. See AEGA PSORA.

Salvelinus. Small genus of fresh-water trouts; S. fontinalis is the brook trout, originally native to northeastern U.S. and southeastern Canada, but now introduced in many other suitable areas; usually smaller than other American trouts; S. malma is the Dolly Varden trout, in coastal streams from Japan to Calif.; trouts in this genus are often called char or charr; e.g. S. alpinus is the Arctic charr, found in the circumarctic area.

sambar. Sambur; any of a group of large shaggy deer (Cervus) of India, Ceylon, southeastern Asia, East Indies, and the Philippines.

Samia cecropia. Largest sp. of lepidopteran in the U.S.; wing expanse to 6.5 in.; cecropia moth, emperor moth.

Samytha. Widely distributed genus of tube-building Polychaeta.

sand badger. See ARCTONYX.

sand bug. See EMERITA.

sand clam. See MYA.

sand collar. Collar-shaped structure of agglutinated sand grains formed especially by Natica, a common genus of marine snails; holds extruded eggs.

sand crab. See EMERITA and OCYPODA.

sand cricket. Any of several large crickets of western and southern U.S.; in the Family Stenopelmatidae.

sand dab. See HETEROSOMATA.

sand dollars. Spp. of echinoderms in the Class Echinoidea; test very flat, thin, and circular; covered with a coat of fine spines; abundant on sandy bottoms; Clypeaster and Dendraster common.

sand eel. See AMMODYTES.

sanderling. See CROCETHIA.

sandfish. Any of a wide variety of marine fishes which frequent or burrow in sandy shores.

sand flea. 1. Amphipod found on damp marine beaches. 2. Chigoe.

sand flounder. See WINDOWPANE.

sand fly. Punkies, no-see-ums; any of numerous small biting or non-biting flies which breed in damp sand, especially along the sea shore, e.g. Phlebotomus.

sandfly fever. Mild virus disease of man; occurs in Mediterranean area and as far east as India; transmitted by the bite of Phlebotomus, the sandfly.

sand grouse. Any of numerous pigeon-like game birds of barren plains and deserts of southern Asia, Europe, and Africa; usually yellowish to brown coloration; in the Family Pteroclidae.

sandhill crane. See GRUS.

sand hopper. Common name of amphipods found on the sand of the intertidal zone; burrow and hop about vigorously.

sand lance. Sand launce; see AMMODYTES.

sand launce. See AMMODYTES.

sand lizard. Any of numerous active, medium-sized lizards inhabiting sandy desert areas of the southwest and Mexico; legs and toes long; with a skin fold across the throat.

sandpiper. Any of a group of small, widely distributed shorebirds included in several genera; Actitis and Pisobia typical.

sand puppies. See HETEROCEPHALUS.

sand roller. See PERCOPSIDAE.

sand shark. See CARCHARIAS.

sand skink. See NEOSEPS.

sand wasp. Any one of many large wasps which burrow in sandy places; Bembix common.

San José scale. See QUADRASPIDIOTUS.

Saperda candida. Sp. of long-horned beetle; roundheaded apple borer.

Sapphirina. Genus of brightly colored marine copepods (Suborder Harpacticoida).

Sappho. Genus of fork-tailed hummingbirds; mostly ruby and greenish coloration; Peru, Bolivia, and Argentina.

saprobe. Organism living in decaying matter. Adj. saprobic, saprobiotic.

Saprodinium. Genus of compressed ctenostome ciliates with posterior spines; sapropelic in fresh and salt waters.

sapropel. Type of organic bottom deposit of some lakes, rivers, and polluted areas; formed under continuing anaerobic conditions and marked by the production of methane and hydrogen sulfide. Adj. sapropelic.

saprophagous. Pertaining to animals which feed on other dead animals and/or dead plants.

Saprophilus. Genus of ovoid to pyriform holotrich Ciliata; usually in decomposing material in fresh waters.

saprozoic nutrition. 1. Nutrition by the absorption of dissolved salts and simple organic materials from the surrounding medium and synthesizing them into protoplasmic constituents; usually used with reference to fungi and certain protozoans. 2. Also occasionally refers to any organism feeding on decaying animal matter.

sapsucker. See SPHYRAPICUS.

Saratogan epoch. Last of three subdivisions of the Cambrian period.

Sarcocystis. Genus of sarcosporidian protozoans; in muscle tissues of higher vertebrates.

Sarcodina. One of the classes in the Phylum Protozoa; naked cells without a surrounding pellicle, although many spp. have a secreted skeleton or test; all spp. have temporary protoplasmic extensions, or pseudopodia, which are used chiefly in food getting and locomotion; free-living, parasitic, marine, and fresh-water; Amoeba most familiar.

sarcolemma. Thin membrane surrounding an individual voluntary muscle fiber; scattered nuclei are contained within.

sarcoma. Malignant tumor consisting of mesenchyme and connective tissue fibers.

Sarcophaga. Important genus of flies in the Family Sarcophagidae; larvae saprophagous, parasitic in invertebrates, or produce myiasis in vertebrates.

Sarcophagidae. Large family of medium to large flies; fleshflies, blowflies; many pest spp.; larvae saprophagous or parasitic; some cause myiasis in man and domestic animals; Sarcophaga the most important genus.

Sarcophilus. Genus of powerful, carnivorous, badger-like marsupials of Tasmania (Family Dasyuridae); Tasmanian devils.

sarcoplasm. Protoplasmic material filling the interstices between the fibrils of muscle tissue.

Sarcopterygii. Choanichthyes.

Sarcoptes scabiei. Common infectious itch mite of man and other mammals; female burrows about in skin and produces intense itching, redness, and tenderness; infection known as scabies or seven-year itch.

Sarcoptidae. Family of mites in Superfamily Acaroidea; itch mites, mange mites; skin parasites of birds and mammals; Sarcoptes, Psoroptes, and

Cnemidocoptes common.

Sarcoptiformes. Taxon (suborder) sometimes used to include certain families of mites; stigmata absent or on various parts of the body; chelicerae modified for chewing.

Sarcosporidia. One of the subclasses of Sporozoa; spores naked; muscle parasites of higher vertebrates; Sarcocystis common.

sarcostyle. 1. One of many discrete minute bundles of muscle fibrillae in a fibril; in cross section each sarcostyle is a Cohnheim's field. 2. Column of a dactylozooid in certain hydrozoan coelenterates.

sarcotheca. Sheath of a hydrozoan sarcostyle.

sardine. Any of several small marine fishes which are commercially canned, especially Sardinia pilchardus of France, Italy, Portugal, and Spain, and young sprat and herring of Scandinavia.

Sardinia. Genus of herring-like marine fishes; includes the commercial sardine of Italy, Spain, Portugal, and France.

Sardinops caerulea. Pilchard or California herring; West Coast fish averaging 9 in. long; usually caught in purse seines; canned for human food and processed as fish oil and fish meal.

sargassum fish. See ANTENNARIIDAE.

saro. Giant Brazilian otter; See PTERONEURA.

Sarsia. Genus of marine hydrozoans; the small medusa buds off daughter medusae from the bell or manubrium.

Sarsiella. Genus of marine Ostracoda.

sartorius. Large muscle of the thigh in certain tetrapods; flexes the hip and knee and rotates the leg.

sassaby. See DAMALISCUS.

Satan eurystomus. Widemouth blindcat; a small, pale, blind catfish found only in artesian wells at San Antonio, Tex.

satellite. Spherical or rodlike tip of a chromosome located at some distance from the main body of the chromosome and connected to it by a fine thread.

satin fin. Small minnow (Erogala whipplii) with whitish lower fins; clear streams of eastern U.S.

Saturniidae. Family of large brightly colored moths; giant silkworm moths; some spp. spin silk which is commercially important; larvae feed chiefly on leaves of deciduous and evergreen trees; familiar genera are Samia, Telea, Attacus, and Tropaea.

satyr. See NEONYMPHA.

Satyridae. Family of brownish or grayish butterflies; the satyrs or wood nymphs.

Satyrus. Genus of common U.S. butterflies; the nymphs; coloration dark brownish, with yellow or orange bands on the forewings.

sauger. See STIZOSTEDION.

Sauria. Suborder of the reptilian Order Squamata; lizards; body slender, usually with four limbs; mandibles fused anteriorly; eyelids usually movable; mostly oviparous; usually feed on small invertebrates but a few spp. are herbivorous; widely distributed in tropics and temperate zones; more than 2500 spp. in 26 families.

Saurischia. Order of Upper Triassic, Jurassic, and Cretaceous dinosaurs; reptile-like dinosaurs; commonly divided into two suborders, Theropoda and Sauropoda.

Sauromalus. Genus of large, herbivorous lizards with small scales; chuckwallas; southwestern U.S. and northwestern Mexico; body to 12 in. long, black or brownish coloration.

Sauropoda. Suborder of giant Jurassic and Cretaceous semi-aquatic, herbivorous, quadrupedal dinosaurs; examples are Brontosaurus, Diplodocus, and Brachiosaurus.

Sauropsida. Loose taxonomic category which includes all birds, all living reptiles, and extinct reptiles with the exception of the mammal-like forms.

Sauropterygia. Large, extinct order of marine and shore-dwelling reptiles; limbs paddle-like; Plesiosaurus and Elasmosaurus typical.

saury. Scomberesox saurus, a common oceanic needlefish of temperate Atlantic, Pacific, and Indian oceans.

savanna. 1. Plains area characterized by coarse grasses and sparsely scattered trees, especially in tropical areas where rainfall is seasonal, as in parts of Africa. 2. Grassland with scattered trees, grading into open plains on the one hand and woodland on the other hand; tropical or subtropical; rainfall not especially seasonal.

savannah. Savanna.

savanna sparrow. See SPARROW.

sawback. Any of several spp. of turtles in the genus Graptemys; southern states.

sawbelly. See ALOSA.

sawfish. See PRISTIS.
sawflies. See TENTHREDINIDAE and CIMBICIDAE.
saw sharks. Group of sharks, especially Pristiophorus and Pliotrema, which resemble the true sawfish (Pristis) but have lateral gill slits.
saw-whet owl. See AEGOLIUS.
Saxicava. Genus of marine bivalve mollusks; burrow in sand, mud, clay, and soft rock.
Saxidomus nuttalli. Washington clam; a large, edible, West Coast form; in sand and mud bottoms; sometimes called the butter clam or money shell.
Sayornis. Genus which includes the American phoebes in the Family Tyrannidae.
scabbard fish. Any of several long, thin, compressed, silvery marine fishes, especially Lepidopus of European coastal waters.
scabies. Mange type of disease in many mammals; see SARCOPTES.
scab mite. Psoroptes ovis, a small mite which causes scab on livestock.
scala media. 1. Central and smallest of the three chambers of the cochlea of higher vertebrates; triangular in cross section and filled with endolymph; its lower surface consists largely of the organ of Corti. 2. Cochlear duct.
scalare. Common angelfish, Pterophyllum scalare.
scala tympani. Most ventral of the three longitudinal chambers of the cochlea of higher vertebrates; lined with squamous epithelium, filled with perilymph, and connects with the scala vestibuli at the tip of the cochlea.
scala vestibuli. Uppermost of the three chambers of the cochlea of higher vertebrates; lined with squamous epithelium, filled with perilymph, and connects with the scala tympani at the tip of the cochlea.
scale. 1. Small flattened rigid dermal or epidermal plate forming an external covering of the body as in fishes, reptiles, the tails of a few mammals, the legs of certain birds, and parts of certain invertebrates. 2. Microscopic powdery plates covering the wings of certain insects, especially moths and butterflies. 3. Single secreted dorsal plate covering the body of a scale insect.
scaled quail. See CALLIPEPLA.
scale formula. Conventional way of expressing the scale pattern of a fish; 8+ 70 + 12, for example, indicates 8 scales in an oblique series above the lateral line, 70 scales in the lateral line, and 12 in an oblique series below the lateral line.
scale insects. See DIASPIDIDAE.
scalenus medius. Muscle which originates on the first rib and inserts on the transverse processes of the second to sixth cervical vertebrae; flexes neck laterally.
scale parasite. See APHELINIDAE.
scale-tail. See ANOMALURUS.
scale worm. Any of numerous polychaete worms with dorsal overlapping scales.
Scalibregma. Common genus of short, stout Polychaeta.
scallop. Any of a wide variety of marine bivalves which are able to swim in the manner of jet propulsion by rapidly clapping the valves together; diameter 1 to 6 in.; see especially PECTEN; other U.S. genera are Chlamys, Leptopecten, Placopecten, and Aequipecten.
Scalopus. See TALPIDAE.
scalpellum. Lancet-like piercing paired mouth part, as in some Diptera. Pl. scalpella.
Scalpellum. Large genus of barnacles; mostly in cold, deep waters.
scaly anteater. Pangolin; see PHOLIDOTA.
scaly-leg mite. See CNEMIDOCOPTES.
scaly-tailed squirrel. See ANOMALURUS.
scansorial. Specialized for climbing.
Scapanus. Genus of moles; several spp. in humid mountains of Wash., Ore., and Calif.
scape. 1. Basal segment or group of segments in the antennae of certain insects. 2. Feather shaft. 3. Peduncle of a dipteran halter.
Scaphander. Genus of very small marine snails; canoe shells; spire of shell sunken; aperture narrow above, very wide below.
Scaphiopus. N.A. genus of spadefoot toads (Family Pelobatidae); head and body to 70 mm.; nocturnal, in burrows during daylight; S. holbrookii, the most common spadefoot, in Mass. to Ark. and south to Fla. and Tex.; S. couchii in Okla. to Ariz.; S. hammondii in the western half of U.S. generally.
Scaphirhynchus platorhynchus. Common white sturgeon or shovel-nosed sturgeon of Mississippi drainage; snout

broad and flat.

scaphognathite. Bailer; modified second maxilla of the crayfish, lobster, etc.; creates a current of water forward over the gills in the branchial chamber.

Scapholeberis. Genus of small fresh-water Cladocera; usually among aquatic vegetation.

Scaphopoda. One of the classes in the Phylum Mollusca; tooth or tusk shells; shell tubular, slightly curved, and open at both ends; conical "foot" and small "head" with surrounding fringe of threadlike tentacles at the larger end of the shell; 0.5 to 6 in. long; gills absent; stand obliquely in sandy or muddy sea bottom with posterior end projecting slightly; radula present; larva an atypical trochophore; about 350 living spp.; Dentalium most common.

Scaptognathus. Genus of marine water mites.

scapula. In higher vertebrates, the dorsal portion of the pectoral girdle on each side; the shoulder blade in man.

scapulohumeralis posterior. Dorsal muscle of the pectoral girdle of reptiles and birds.

scarab. See SCARABAEIDAE and SCARABAEUS.

Scarabaeidae. Large family of small to very large, deep-bodied beetles; scarabs, June beetles, May beetles, cockchafers, dung beetles; many brilliantly colored spp., especially in the tropics; distal segments of antennae flat, expanded, and fitting together as the leaves of a book; some spp. feed on vegetation and carrion; adults of many feed upon dung and roll it into balls which they bury in soil; the female in such spp. deposits her eggs in these dung balls; familiar genera are Phyllophaga, Popillia, Macrodactylus, and Scarabaeus.

Scarabaeus sacer. Ancient Egyptian sacred dung beetle or scarab; see SCARABAEIDAE.

Scardafella inca. Inca dove in the Family Columbidae; a small grayish brown dove; Tex., N.M., and Ariz. to Costa Rica.

Scaridium. Genus of rotifers having a very long foot and toes.

scarlet snake. See CEMOPHORA.

scarlet tanager. See THRAUPIDAE.

Scatella stagnalis. Shore fly in the Family Ephydridae; larvae in cold and hot springs.

scatology. Study of excrement, especially of birds and mammals, in order to determine their food habits.

Scatophagidae. Dung flies; larvae in this family live in dung; adults tawny and hairy, predaceous on other small insects.

scats. Feces or droppings, especially of mammals and carnivorous birds.

scaup. See NYROCA.

scavenger. Any animal that feeds on dead animal material which it has not killed.

scavenger beetle. Any beetle in the aquatic Family Hydrophilidae.

scavenger flies. See SEPSIDAE.

Sceliphron. Common genus of mud-dauber wasps.

Sceloporus. Genus of small, active lizards found over the whole of N.A. from southern Canada to Panama, but most abundant in arid regions; includes the swifts, pine lizard, and fence lizard; scales keeled, pointed, and bristling.

Scepanotrocha. Genus of bdelloid rotifers.

Schaudinn's fluid. Excellent fixative for protozoans; consists of 66 ml. saturated mercuric chloride, 33 ml. absolute or 95% alcohol, and 1 ml. glacial acetic acid.

Schick test. Means of determining susceptibility to diphtheria; a very small amount of diluted diphtheria toxin is injected below the surface of the skin; the production of an inflammation in this area indicates an inadequate amount of the antitoxin in the body.

Schiff test. One of several chemical tests for the presence of carbohydrates, cholesterol, allantoin, urea, and uric acid in urine.

Schilbeodes. Genus of small fishes in the Family Ameiuridae; madtoms; head flat, mouth large, pectoral spine with a basal poison gland; common in streams of eastern half of U.S.

Schistocephalus. Genus of tapeworms; plerocercoid in the coelom of fresh-water fishes; mature form in water birds.

Schistocerca. Common genus of grasshoppers in the U.S.; S. americana is the American "locust."

Schistocomus. Genus of polychaetes usually found in the rocky intertidal zone.

Schistosoma. Genus of trematodes in the Order Digenea; parasites in the blood vessels of mammals; sexes separate and body greatly elongated;

three common spp. occur in man in the tropics; aquatic snails are intermediate hosts, and the definitive hosts become infected by the penetration of cercariae through the skin during wading or when drinking water; these three spp. are S. haematobium, S. japonicum, and S. mansoni.

Schistosomatium. Genus of distome trematodes found in the veins and liver of rats and mice.

Schizaster. Circumboreal genus of heart urchins (Order Spatangoida).

Schizobranchia. Genus of Polychaeta.

Schizocardium. Genus in the Phylum Enteropneusta.

Schizocoela. Group of phyla having schizocoelic mesoderm formation.

schizocoelic mesoderm formation. Embryonic formation of mesoderm characterized by the development of mesodermal bands growing forward from the blastopore as cords of cells between ectoderm and endoderm; these cords then split and the two resulting sheets of cells become associated with the ectoderm and endoderm, respectively, the large space between being the coelom; found in Bryozoa, Phoronidea, Mollusca, Sipunculoidea, Priapuloidea, Echiuroidea, Annelida, and Arthropoda.

Schizocoelomata. Schizocoela.

Schizocystis. Genus of sporozoans parasitic in Diptera, Annelida, and Sipunculoidea.

schizogamy. Reproduction resulting in the division of an animal into a sexual and an asexual individual, as in certain Polychaeta.

schizogenesis. Reproduction by fission, either in protozoans or in multicellular animals.

schizogony. Multiple asexual fission, especially as found in many sporozoans; e.g. the asexual reproductive cycle of Plasmodium in the human blood stream.

Schizomida. Uropygi; small order of the arachnid Subclass Arachnida; whip scorpions; cephalothorax divided into a head and two free thoracic segments; pedipalps robust, first legs sensory; nocturnal in tropics and subtropics; Mastigoproctus common.

Schizonopora. Common genus of marine Bryozoa.

schizont. Large intracellular stage in the life history of sporozoan Protozoa; formed by growth of the trophozoite and gives rise to merozoites by multiple fission.

Schizopera. One of three tribes in the dipteran Suborder Cyclorrhapha; with a suture around base of antenna.

schizophrenia. Type of psychosis typified by loss of contact with reality and by change and disintegration of personality.

Schizopoda. Euphausiacea.

schizopod larva. Mysis larva; one of the first three larval stages of the lobster and some other decapods; 13 pairs of appendages.

Schizoporella. Common genus of marine Bryozoa.

Schizothaerus. Common genus of West Coast gaper clams.

Schizotricha. Genus of marine coelenterates in the Suborder Leptomedusae.

schizozoites. Cells which are the products of multiple fission of a trophozoite in certain gregarine Sporozoa.

Schleiden, Matthias Jakob. German botanist (1804-1881); collaborated with Schwann; proved (1838) that plant tissues are composed of cells and that the nucleus is an essential part of the cell.

Schoinobates. Great gliders; a genus of large Australian phalangers; arboreal, feed on eucalyptus flowers, shoots, and leaves; accomplished gliders; total length to 3 ft.

schoolmaster. See LUTIANIDAE.

Schüffner's dots. Minute dots seen in erythrocytes infected with Plasmodium when stained with Romanowsky's stain.

Schultze, Max J. S. German biologist (1825-1874); well known for researches on the cell theory, structure of protoplasm, Protozoa, and nerve endings.

Schwann, Theodor. German physiologist and histologist (1810-1882); collaborated with Schleiden; proved that the cell is the basis of animal as well as plant structure; made contributions to the knowledge of nerve cell structure, pepsin, and yeasts.

Schwann's sheath. Neurilemma; thin structureless envelope surrounding nerve fibers of the vertebrate peripheral nervous system.

Sciaenidae. Marine and fresh-water family of percomorph fishes which produce a purring or drumming sound; fresh-water drum or sheepshead (Aplodinotus grunniens) common from Hudson Bay drainages south to Guatemala, of some commercial importance; edible marine weakfish and Pacific white sea bass (Cynoscion); edible marine drums (Pogonias and Sciaenops); common croaker (Micropogon). Also see MENTICIRRHUS.

Sciara. See SCIARIDAE.

Sciaridae. Small family of dipterans in the Tribe Nematocera; dark-winged fungus gnats; feed on fungi in damp wooded areas of the Holarctic region; Sciara common.

sciatic nerve. Large vertebrate nerve arising from the sciatic plexus on each side and passing into each thigh.

sciatic plexus. Interconnections of several spinal nerves just before they enter the hind limb of tetrapods as the sciatic nerve.

scientific name. Double (genus and species) name of any organism; sometimes a third (subspecies) name may also be used; a complete scientific name involves these two or three elements plus the name of the individual first describing that particular form plus the date of the published description; except for special taxonomic and publication purposes, however, the last two elements are customarily omitted.

Scincidae. Cosmopolitan family of lizards; includes the skinks; small, active, elongated spp. with smooth scales; some with diminutive legs and reduced toes; Eumeces common in U. S.

Scionopsis. Genus of brightly colored tube-building Polychaeta.

Scirtes. See HELODIDAE.

Scirtothrips citri. Important insect in the Order Thysanoptera; a pest on oranges and lemons.

scissiparity. Asexual fission of a metazoan into two new individuals.

scissor-tailed flycatcher. See MUSCIVORA.

Sciuridae. Family of rodents which includes the squirrels, marmots, chipmunks, etc.; usually arboreal, with bushy tail and strong hind legs; vegetarians, especially seeds and nuts; common U.S. genera are Sciurus, Citellus, Tamias, Eutamias, Marmota, and Glaucomys.

Sciuromorpha. Suborder of rodents; includes the sewellel, squirrels, pocket gophers, pocket mice, kangaroo rats, and beavers.

Sciuropterus. Genus of European flying squirrels.

Sciurus. Common genus of Old World and N.A. tree squirrels (Family Sciuridae); S. hudsonicus is the widely distributed red squirrel, S. fremonti and S. douglasi are the chickarees of western states, S. carolinensis is the eastern gray squirrel, S. aberti the tuft-eared squirrel, and S. niger includes most of the fox squirrels of the eastern half of the U.S.

Scleocrangon. Genus of marine shrimps.

sclera. Sclerotic coat.

scleral ring. Series of small bony plates imbedded in the anterior portion of the sclerotic coat of the eye in birds, reptiles, and actinopterygian fishes.

sclerite. Any one of the hard plates forming a portion of the exoskeleton of arthropods, especially insects; two adjacent sclerites are separated by a linear suture of more or less softer cuticle.

scleroblasts. Special amoebocytes in sponges; secrete spicules.

sclerocoel. In the cephalochordates, that portion of the coelom which is located dorsolateral to the pharynx and between it and the myotomes.

Scleroparei. Order of bony fishes which includes the mail-cheeked fishes; rockfishes, scorpion fishes, sculpins, and sticklebacks; with a bony plate across the cheek below the eye; body and head usually with extra spines; chiefly marine, few in fresh waters.

Scleropareida. Scleroparei.

Scleroplax. Genus of marine crabs (Suborder Brachyura); commensal in burrows of Urechis and Callianassa.

scleroprotein. Simple fibrous protein abundant in fibrous connective tissue.

Scleroptilum. Genus of long, slender sea pens (Order Pennatulacea); orange, to 12 in. high; deep sea.

sclerosis. Hardening of vertebrate tissues after injury or disease; produced chiefly by an increase in fibrous connective tissue.

sclerotic bones. Small bones of the eye capsule in bony fishes.

sclerotic coat. Outer, tough, whitish, connective tissue coating of the eyeball; forms the transparent cornea anteriorly.

sclerotic ring. Ring of bony plates surrounding the eye in birds.

sclerotium. Microscopic cystlike structure formed by slime molds; each contains 10 to 20 nuclei. Pl. sclerotia.

sclerotized. Pertaining to specially hardened or thickened areas of the arthropod exoskeleton, especially in insects.

sclerotome. Mass of mesenchyme in the ventral portion of a mesodermal somite of an early vertebrate embryo; destined to develop into vertebrae and ribs.

Scolecolepis. Genus of Polychaeta.

scolex. Anterior headlike structure of tapeworms; commonly bears suckers, lappet-like projections, spines, and/ or hooks. Pl. scolices or scoleces.

Scolia. See SCOLIIDAE.

Scoliidae. Family of black hairy wasps with bright spots or bands; hairy flower wasps; larvae ectoparasitic on beetle grubs; Scolia typical.

scoliosis. Abnormal lateral curvature of the spine.

Scolopacidae. Large cosmopolitan family of shore birds in the Order Charadriiformes; woodcock, snipe, dowitcher, sanderling, godwit, yellow-legs, sandpiper, curlew, etc.; found mostly around inland bodies of water; usually in the northern part of the Northern Hemisphere during the breeding season.

Scolopendra. Common genus of large centipedes.

Scolopendrella. Common genus of Symphyla; first legs imperfect, head not sharply separated from trunk; total length 1 to 3 mm.

Scolopendromorpha. Order of centipedes; 21 or 23 pairs of legs; body to 10 in. long; Scolopendra common and widely distributed.

scolophore. In insects, a spindle-shaped bundle of sensilla attached to the integument; presumably auditory.

Scoloplos. Genus of burrowing Polychaeta; northern seas.

Scolopoctyptops. Genus of large centipedes.

Scolytidae. Family of minute to small beetles with clubbed antennae; engraver beetles, bark beetles, ambrosia beetles, shot-hole borers; larvae and adults feed and burrow in many different shrubs and trees; especially important pests on conifers where they make extensive branched channels under the bark; Scolytus, Dendroctonus, and Xyleborus examples.

Scolytus rugulosus. Important pest sp. of beetle in the Family Scolytidae; shot-hole borer in many kinds of trees; another sp. in the genus Scolytus transmits Dutch elm disease.

Scomberesocidae. Family of marine bony fishes; includes the needlefishes, billfishes, and skippers; both jaws elongated to form a slender beak; teeth weak and small; body elongated; several posterior dorsal and ventral finlets; Scomberesox common.

Scomberesox. See SCOMBERESOCIDAE.

Scomberomorus. Genus of mackerels of the tropical and subtropical Atlantic; includes the Spanish mackerel, king mackerel, and cavallo (cero).

Scomber scombrus. Common Atlantic mackerel; fusiform, swift fish occurring in the North Atlantic in large schools; upper surface bluish with numerous transverse bands; lower parts whitish to coppery; to 22 in. and 4 lbs.; important food fish; several other important mackerels are in other genera.

Scombridae. Family of large marine percomorph fishes which includes the mackerels (Scomberomorus and Scomber), bonito or skipjack (Sarda), and tuna, horse mackerel, or albacore (Thunnus); numerous important commercial spp.

scopa. Ventral abdominal brush for collecting pollen; in certain Hymenoptera.

Scopes trial. Court trial (1925) of John T. Scopes, a high school biology teacher; brought to court because he taught Darwinism, contrary to Tennessee law; defended by Clarence Darrow and other able lawyers; prosecuted by William Jennings Bryan; convicted but later released by state supreme court; main effect of the trial was to discourage the passage of similar prohibitory legislation by other states.

Scopidae. Family in the bird Order Ciconiiformes; Scopus umbretta, the hammerhead, is a brownish heron-like bird of tropical Africa; bill flat, wide.

scopula. 1. Small dense tuft of hairs in arthropods, especially on the tarsus. 2. Comblike row of stiff hairs on the swollen endite of the first segment of the pedipalp in some spiders. 3. Brushlike adhesive organelle in certain peritrich Ciliata.

scorbutic. Pertaining to scurvy.

scorched mussel. Mytilus exustus, a small mussel found from N.C. to the Gulf of Mexico.

Scorpaena. See SCORPAENIDAE.

Scorpaenichthys marmoratus. Cabezon, a large Pacific sculpin found along rocky bottoms in shallow waters; skin soft, scales lacking, and with a large spine near each eye; to 30 in. long.

Scorpaenidae. Family of bony fishes which includes the common edible ocean perch or rosefish; Scorpaena the tropical and subtropical scorpion fishes and lionfishes; ugly spp. with many protuberances, frills, and spines; Sebastodes (rockfishes); these genera are abundant along rocky shores; some have poisonous spines.

Scorpididae. Family of Pacific percomorph fishes; body deep and compressed; dorsal fin single; several important food spp.

Scorpio. Old World genus of scorpions.

scorpion. Any member of the arachnid Order Scorpionida.

scorpion fish. See SCORPAENIDAE.

scorpion flies. See MECOPTERA.

Scorpionida. Arachnid order which includes the more than 600 spp. of scorpions; body elongated, chelicerae small, pedipalps large and chelate, second abdominal segment with a pair of ventral pectines; six-segmented postabdomen with a terminal bulbous sting; viviparous and nocturnal; four pairs of book lungs; mostly tropical and subtropical; Hadrurus, Centrurus, and Vejovis typical.

scorpion shells. See PTEROCERA.

scoter. See MELANITTA.

Scotiaptex nebulosa. Great gray owl, an uncommon sp. of Canadian forests, rare in U.S.

scotopic vision. Vision involving the retinal rods; twilight vision.

screamer. See AMHIMIDAE.

screech owl. See OTUS.

screw shell. Any of numerous small marine gastropods with long, slender spires, e.g. Turritella.

screwworm. See COCHLIOMYIA.

scrobicula. Smooth area surrounding a boss on the test of a sea urchin.

Scrobicularia. Genus of marine bivalves which burrow into sandy and muddy bottoms but secure food by searching the surface with a long intake siphon.

scrod. Any of several immature marine fishes, especially cod and haddock.

scropula. Hairy pad at the tip of each leg in many spiders.

scrotum. Bag or pouch of skin in the pelvic region of many male mammals containing the testicles.

scrub bird. See ATRICHORNITHIDAE.

scrub jay. See APHELOCOMA.

scrub typhus. Tsutsugamushi fever.

Scruparia. Genus of marine Bryozoa.

scud. Common name of amphipods, especially fresh-water spp. and marine beach spp.

Scudder, Samuel H. American entomologist (1837-1911); founder of American insect paleontology and specialist on Orthoptera and Lepidoptera.

sculpin. See COTTIDAE.

scup. See STENOTOMUS.

scurvy. Condition produced by insufficient vitamin C in the diet; marked by weakness, anemia, spongy gums, and subcutaneous hemorrhages.

scute. 1. Any external plate of horny or bony consistency; present in some fishes and many reptiles. 2. Especially thick chitinous or sclerotized plate of the arthropod exoskeleton. 3. Any scale-like structure.

scutella. Small dermal bone or thin bony plate. Pl. scutellae.

Scutella. Genus of fossil sand dollars.

scutellate. 1. Covered with scales, plates, scutella, or scutellae. 2. Having a scutellum, as in beetles.

Scutelleridae. Shield bugs; a small family of small to large Hemiptera with enormously enlarged scutellum; coloration bright; 4 to 25 mm. long; plant feeders; some pest spp.; Eurygaster typical.

scutellum. 1. A part of the dorsal thoracic exoskeleton in many insects. 2. In Coleoptera, the triangular piece between the elytra. 3. In Hemiptera, the triangular piece between the bases of the hemelytra. 4. Any plate or scale, such as those on the toes of a bird. Pl. scutella.

Scutigera. Cermatia.

Scutigerella. See SYMPHYLA.

Scutigeromorpha. Order of centipedes; 15 pairs of long legs; spiracles mid-dorsal; Cermatia common.

Scutiosorex. Genus of girder-backed shrews of the Congo basin; vertebral column greatly modified by a girder-like complex.

Scutopterus. Neoscutopterus.

Scutovortex. Genus of mites in the Family Oribatidae; usually under rocks and in debris between tide lines.

scutum. 1. One of a pair of large basal calcareous plates in the barnacle skeleton. 2. Chitinous plate of an arthropod exoskeleton. 3. Discrete integumentary plate. 4. Middle sclerite of an insect notum. 5. Dorsal shield of a tick. Pl. scuta.

Scyllaea. Genus of pelagic nudibranchs commonly found among sargassum weed; small branched gills on a median posterior crest and on two pairs of lateral cerata.

Scyllium canicula. Small "dogfish"

shark.

Scymnognathus. Genus of late Permian therapsids.

Scypha. Genus of small, vase-shaped marine sponges in the Class Calcarea; European and N.A. coasts; (often incorrectly called Grantia and Sycon).

Scyphacella. Genus of marine Isopoda.

Scyphidia. Genus of cylindrical peritrich Ciliata; attached to substrates in fresh and salt waters.

scyphistoma. Small fixed polyp stage of scyphozoan jellyfish; solitary or colonial; distal end lobed or bearing tentacles. See STROBILA.

Scyphomedusae. Scyphozoa.

Scyphozoa. One of the three classes of coelenterates; the true marine jellyfishes; bell- or umbrella-shaped free-swimming medusae; velum and stomodaeum absent; tentaculocysts present; polyp reduced or absent; about 200 spp.; examples are Aurelia and Cassiopeia.

scyphula. Scyphistoma.

Scyra. Genus of marine crabs (Suborder Brachyura).

Scytodidae. Family of spitting spiders; some spp. spit out a mucilaginous secretion which glues the prey to the substrate.

Scytomonas. Genus of monoflagellate holozoic Protozoa; stagnant fresh waters.

Scytonotus. Genus of small millipedes.

sea anemone. General inclusive term applied to all those anthozoan coelenterates which have no skeleton; includes especially the larger, brightly-colored spp. of the Order Actiniara which have a fancied resemblance to a flower.

sea ape. 1. Thresher shark. 2. Obsolete term for sea otter (Enhydra).

sea arrow. See OMMASTREPHES.

sea bass. Any of numerous marine fishes, especially those in the Family Serranidae; Centropristes striatus is the most important sea bass of the Atlantic Coast.

sea bat. See PATIRIA.

sea beef. 1. Chitons which have been prepared for use as human food, especially in the West Indies. 2. Flesh of a porpoise or whale.

sea blubber. Jellyfish.

sea bream. Any of numerous marine fishes in the Family Sparidae.

sea butterfly. Any member of the gastropod Order Pteropoda.

sea catfish. Any of numerous fishes in the Family Ariidae; most spp. incubate the eggs in the mouth.

sea cauliflower. See DUVA.

sea chrysanthemum. Common name of various sea anemones.

sea coot. See OIDEMIA.

sea cow. See TRICHECHUS and HYDRODAMALIS.

sea crayfish. See PANULIRUS.

sea cucumber. Any member of the echinoderm Class Holothuroidea.

sea eagle. Any of several eagles found primarily along the seashore; feed chiefly on fish.

sea elephant. See MIROUNGA.

sea fan. See GORGONACEA.

sea feather. See PENNATULACEA.

sea fir. Large bushy hydrozoan coelenterate (Abietinaria); common and widely distributed.

sea grapes. See MOLGULA.

sea gull. In the broad sense, any gull frequenting the sea or fresh waters.

sea hare. See APLYSIA.

sea horse. See HIPPOCAMPIDAE.

seal. See ARCTOCEPHALUS, CALLORHINUS, PHOCA, ERIGNATHUS, HALICHOERUS, CYSTOPHORA, and MIROUNGA.

sea lamprey. See PETROMYZON.

sea lemon. Anisodoris nobilis, a large lemon-yellow nudibranch; to 8 in. long.

sea leopard. 1. Leopard seal; either of two large spotted Antarctic seals in the genus Hydrurga (Phocidae). 2. Common U.S. harbor seal.

sea lily. Crinoid echinoderm which is sessile and stalked; see CRINOIDEA.

sea lion. Any of a Pacific group of marine mammals in the Family Otariidae; larger than true seals, underfur absent; genera are Zalophus, Eumetopias, and Otaria.

sea louse. Marine isopod, especially the smaller and parasitic spp.

sea mat. Flat encrusting mass of marine Bryozoa; the zooecia are packed closely together in a single layer.

sea mouse. See APHRODITE.

sea nettle. Dactylometra quinquecirrha, a large jellyfish.

sea otter. See ENHYDRA.

sea pansy. Renilla reniformis, a remarkable coelenterate (Order Pennatulacea) having a flower-like appearance; stalk soft, rachis platelike.

sea peach. Baglike, peach-colored tunicate.

sea pen. See PENNATULA and PINNA.

sea perch. Any of a family of viviparous perchlike fishes of the Pacific Coast, usually in the surf; body compressed and mostly less than 12 in. long.

sea pork. See AMAROUCIUM.

sea plume. Sea fan.

sea potato. Common name for several tunicates.

sea purse. Shield- or purse-shaped egg case of a skate and certain sharks; often found washed up on marine beaches.

sea raven. See HEMITRIPTERUS.

Searlesia. Genus of gray, robust, marine snails.

sea robin. See PRIONOTUS.

sea scorpion. See EURYPTERIDA.

sea serpent. See REGALECUS.

seaside sparrow. Ammospiza maritima; a sparrow of salt marshes from Mass. to Fla. and west to Tex.

sea slug. Any member of the Suborder Nudibranchia.

sea snail. See LIPARIDAE.

sea snake. Any of about 50 spp. of poisonous snakes found especially in tropical seas (but absent from the tropical Atlantic); related to the cobras; usually found within 20 miles of shore.

seasonal isolation. Prevention of interbreeding between two or more populations because of their breeding during different times of the year.

sea spider. See PYCNOGONIDA.

sea squirt. Common name of tunicates, especially the sessile spp. in the Class Ascidiacea; when irritated, the body contracts and squirts streams of water out of the mouth and cloacal (atrial) aperture.

sea star. Starfish; a member of the Class Asteroidea (Phylum Echinodermata); there are usually five radially disposed arms; sea stars are serious predators on oysters.

sea sturgeon. See ACIPENSER.

seatworm. Enterobius vermicularis.

sea urchins. Spp. of echinoderms in the Class Echinoidea; test globular, subglobular, or hemispherical.

sea walnut. Any member of the Phylum Ctenophora which is about the size and shape of a small walnut.

sea whip. Any of several marine hydrozoans having the form of long whiplike colonies.

sebaceous glands. Epidermal glands of mammals which project into the dermis and which secrete a fatty substance (sebum); each gland usually opens into a hair follicle.

Sebastes. See SCORPAENIDAE.

Sebastodes. Genus of fishes in the Family Scorpaenidae; includes the rockfishes; viviparous, with armored head; cool and temperate seas; S. paucispinis is the brown to red bocaccio of the Pacific Coast; S. goodei is the similar chilipepper of the same area; and S. mystinus is the black rockfish; important commercial and sport spp.

sebum. Fatty material secreted by sebaceous glands.

second antennae. Posterior pair of antennae on the head of Crustacea.

secondaries. Those flight feathers of a bird which are attached to the radius and ulna area.

secondary embryo. See PRIMARY EMBRYO.

secondary mesentery. Mesentery in an anthozoan coelenterate which extends from the body wall only part way to the stomodaeum; trinary mesenteries are still more narrow than secondary mesenteries.

secondary oocyte. In oogenesis, the larger of the two haploid cells resulting from the first meiotic division.

secondary productivity. Total quantity of animal (and other heterotrophic) protoplasm produced per unit time in a specified habitat.

secondary radial symmetry. Symmetry of adult echinoderms which follows an earlier primary bilateral symmetry of the larvae.

secondary septum. Secondary mesentery.

secondary sexual characters. External characters which distinguish the two sexes but which have no direct role in reproduction; the deeper voice, beard, and Adam's apple are examples in man.

secondary spermatocyte. In spermatogenesis, each of the two haploid cells resulting from the first meiotic division.

secondary succession. Ecological succession which has been interrupted and set back by some catastrophe such as a forest fire, flood, or plowing.

second form male. Male cambarine crayfish in which the first pleopods are soft and incapable of functioning in sperm transfer; such an instar follows the first form male instar.

second intermediate host. Host in which the more advanced larval stage(s) of a parasite occur; e.g. a fish is the second intermediate host of the human liver fluke, a snail being the first intermediate host.

second maturation division. Second meiotic division; see MEIOSIS.

second maxillae. 1. Fifth (last) pair of head appendages in typical Crustacea; variously modified for feeding. 2. Fourth pair of head appendages in centipedes; used for feeding.

second meiotic division. See MEIOSIS.

second polar body. Small nonfunctional haploid cell resulting from the second meiotic division of a secondary oocyte; the other cell resulting from this division (ootid) is larger and becomes a functional ovum.

second ventricle. Cavity of the right half of the vertebrate telencephalon.

secretagogue. One of a number of substances contained in food which are believed to stimulate glands at the intestinal end of the stomach; these glands, in turn, secrete gastrin into the blood stream, and gastrin stimulates the secretion of gastric juice by the stomach wall.

secretary bird. Long-legged African bird of prey, Sagittarius serpentarius, in the Family Sagittariidae; crest of black feathers suggesting a cluster of quill pens; 4 ft. high; grayish blue coloration with black on wings, tail, and upper legs; feeds largely on other vertebrates; hunts on foot.

secretin. Hormone secreted by the wall of the small intestine of certain vertebrates; it stimulates the pancreas to secrete water and salts.

secretion. 1. Product of any cell, gland, or tissue which is released through the cell membrane(s); in contrast to an excretion, it is usually useful to the animal. 2. The passage of a useful material elaborated by a cell to the outside of the cell membrane.

section. A very thin slice of an organism or part of an organism; usually stained and mounted on a slide for study under the microscope; such slices are customarily 6 to 15 microns thick.

Sedentaria. One of the two orders in the annelid Class Polychaeta; body composed of two or more regions with unlike segments and parapodia; head appendages modified or absent; pharynx without jaws; in burrows or tubes in the sea bottom; feed on detritus and plankton; Chaetopterus, Arenicola, and Amphitrite are familiar.

sedentary. Sessile; permanently attached to a substrate, such as a barnacle or sponge.

Sedgwick-Rafter counting chamber. Special microscope slide which encloses one-milliliter sample under a large coverslip within a space customarily 20 x 50 x 1 mm.; used in making quantitative plankton estimates.

seed oyster. Very small oyster used in cultivation and transplantation.

seed shells. Common name of small fresh-water pelecypods, especially those in the genus Pisidium.

seed shrimp. Any member of the crustacean Subclass Ostracoda.

seedsnipe. See THINOCORIDAE.

seed tick. Six-legged larval stage of a tick.

seed weevil. See LARIIDAE.

Segistra. Genus of primitive spiders which build simple webs.

segment. 1. Fundamental linearly arranged body subdivisions of the Annelida, Arthropoda, and Chordata; somite, metamere, arthromere (in Arthropoda only). 2. Subdivision of an arthropod appendage between areas of flexibility (joints); podite. 3. Incorrectly, a tapeworm proglottid.

segmental blood vessel. Blood vessel in the body wall of typical annelids; supplies blood to its tissues and to the nephridia.

segmental muscles. 1. Muscles of certain invertebrates and vertebrates; serially arranged according to the fundamental body segments. 2. Myotomes.

segmentation. 1. Metamerism. 2. Cleavage.

segmentation cavity. Blastocoel.

segregation. See MENDELISM.

seiche. Periodic oscillation of the surface (or subsurface strata) of a large lake; period ranges from a few minutes to several hours; appears after hard and continuous winds or as the result of rapid and pronounced changes in barometric pressure; amplitude usually several inches to several feet.

Seila. Genus of small marine gastropods with many whorls forming a long spire.

Seinura. Genus of microscopic, free-living nematodes.

Seisonidea. Order of marine rotifers in the Class Digononta; epizoic on the gills of the crustacean Nebalia; the body is curiously elongated and leech-like.

Seison nebaliae. Rotifer which lives as a commensal on Nebalia, a marine crustacean.

Seiurus. See OVENBIRD.

Selachii. Order of cartilaginous fishes

including sharks, skates, and rays; each gill in a separate cleft at sides of pharynx; spiracle behind each eye; cloaca present; almost exclusively marine.

Selaginopsis. Genus of marine hydrozoan coelenterates.

Selasphorus rufus. Common rufous hummingbird in the Family Trochilidae; western U. S. and Canada.

selective breeding. Selection and breeding of parents having desirable characters so that these characters will be perpetuated in the offspring.

selectively permeable membrane. Differentially permeable membrane; a membrane capable of regulating the rates at which different substances penetrate by diffusion or when moved by some other force; marked variations in permeabilities of membranes for various compounds have been demonstrated, with variations also noted in rates of penetration of the same substance through different membranes; all cell membranes are selectively permeable, most being freely permeable to water, molecules of dissolved gases, small organic molecules, and organic molecules soluble in lipids, but in most cases movements of ions through such membranes are restricted.

Selenarctos tibetanus. Himalayan bear, a common black sp. of Central Asia.

Selene vomer. Warm water marine fish of East and West coasts; greatly compressed, with the eyes high and the mouth low; dorsal and anal fins partly falcate; to 2 lbs.

selenodont. Type of dentition featured by crescentic molar cusps, as in ruminants.

self-fertilization. Union of egg and sperm produced by a single hermaphroditic animal; not a common process, since such animals usually copulate and cross fertilize; occurs in tapeworms readily.

sella turcica. 1. Depression of the sphenoid bone; contains the pituitary body. 2. Transverse bar formed by the fusion of apodemes in the posterior segments of certain decapod crustaceans.

Semaeostomae. Suborder of the coelenterate Order Discomedusae; manubrium with four, long, frilly oral lobes; tentacles present; Aurelia, Cyanea, and Pelagia typical.

Semele. Genus of small, rounded, marine bivalves.

semen. Product of male reproductive system including sperm and the secretions of various glands associated with the reproductive tract; the term is usually used with reference to mammals. Adj. seminal.

semicircular canals. Three (usually) minute tubular canals arranged in the three dimensions of space and forming a portion of the inner ear of most vertebrates; they function in equilibration and the detection of direction and turning of the body; only one canal in hagfish and two in lampreys.

semifossorial. Pertaining to an animal which may dig or burrow but does not do so regularly or as an essential part of its daily activities; e. g. skunk and raccoon.

semilunar ganglion. Gasserian ganglion.

semilunar valve. One of the crescent-shaped valves in the vertebrate heart at the exit of the aorta and pulmonary artery; prevents blood from returning to the ventricles from the arteries.

semimembranosus. Large muscle originating on the ischium and inserting on the femur and upper tibia in tetrapods; functions in flexing the leg and rotating it inward.

seminal receptacle. Duct, cavity, or sac specialized for receiving and storing sperm in the female reproductive system of many invertebrates.

seminal vesicle. Swollen portion of a male reproductive duct in which sperm are stored and/or which secretes a fluid useful in the transmission of sperm during copulation.

Seminatrix. See BLACK SWAMP SNAKE.

seminiferous tubules. Minute coiled tubules constituting much of a vertebrate testis; they consist mainly of germinal epithelium, and during active spermatogenesis all developmental stages may be found; mature sperm are formed by the meiotic divisions and maturation of the cells making up the walls of the tubules.

semipalmate. Having the anterior toes only partly webbed, as in some shore birds.

semipalmated plover. Charadrius semipalmatus, a small plover which breeds in Arctic America and winters in South America; feet semipalmate.

semipalmated sandpiper. Small common N. A. sandpiper, Ereunetes pusillus; feet semipalmate; winters from southern U. S. to South America.

semipermeable membrane. Selectively permeable membrane.

semiplacenta. Type of placenta in which the chorionic villi are closely associated but not fused with the uterine lining.

semipupa. Prepupa.

Semite. Arab.

semitendinosus. Posterior thigh muscle in certain vertebrates.

Semnopithecus. Presbytis.

Semotilus atromaculatus. Northern creek chub or horned dace; a small and widely distributed fish in the minnow family; male has a row of tubercles along the side of the head above the eye and nostril during the breeding season; streams and rivers of N. A.

Senecella calanoides. Limnetic copepod found in a few large cold lakes of the Great Lakes region.

sennet. Spet.

sense capsule. Group of bones more or less enclosing one of the sense organs.

sense organ. Simple to complex receptor.

sensilla. 1. Small sense organ, especially a simple epithelial sense organ. Pl. sensillae. 2. Pl. of sensillum.

sensillum. Receptor complex of an insect cuticle, composed of a sense cell or group of cells plus associated chitinous structures; may be merely an innervated hair, a flat sensory plate, or a complex sunken pit (sensory pit); thought to function mostly in olfaction; most commonly on the antennae. Pl. sensilla.

sensitization. 1. Physiological preparation of a tissue or organ by one hormone so that it will react when exposed later to another hormone. 2. Anaphylaxis. 3. Susceptibility of tissues to to the antigen action of a serum or other substance.

sensory corpuscle. 1. Cluster of special tactile cells encapsulated in connective tissue and supplied by sensory nerve fibers in the subcutaneous tissues of higher vertebrates; e.g. Grandry's corpuscle, Meissner's corpuscle. 2. A capsule of concentric layers of connective tissue containing free nerve endings; located in the subepidermal tissues of craniates; various types, including corpuscles of Pacini, Krause, and Golgi-Mazzoni.

sensory nerve. Peripheral nerve consisting of fibers which carry impulses only inward or toward the central nervous system, and away from a receptor.

sensory-neuro-motor mechanism. Collectively, the combination of receptors, neurons, and muscle fibers concerned with motor reactions to stimuli, especially as applied to lower invertebrates.

sensory neuron. Neuron carrying stimuli toward the central nervous system and away from a receptor.

sensory pit. See SENSILLUM.

sensory plate. See SENSILLUM.

sensory root. Short dorsal root of a spinal nerve as it enters the spinal cord; carries only sensory impulses.

sensory system. Collectively, all organs, tissues, and cells associated with the reception of stimuli.

Sepedon. See TETANOCERIDAE.

sepia. Secretion of a glandular sac near the anus in Sepia, the common European cuttlefish; the material is ejected into the water as a means of camouflage and escape from enemies; the ink is processed into rich brown pigment used especially in watercolor painting.

Sepia. Genus to which the common European and Mediterranean cuttlefish belongs; has much the same appearance as a squid but the body is shorter, heavier, and shield-shaped; there is an internal calcareous "cuttlebone" skeleton.

Sepioidea. Group of mollusks in the Suborder Decapoda; includes the cuttlefish; fourth pair of tentacles modified and retractable into pits; eyes with a cornea; fins not united posteriorly; Sepia and Sepiola typical.

Sepiola. Genus which includes a common European cuttlefish.

Sepsidae. Cosmopolitan family of small cylindrical flies; black scavenger flies, spiny-legged flies.

sepsis. Poisoning caused by putrefaction products.

septal filaments. Sinuous threads arranged along the free edges of septa in anthozoan coelenterates; they contain an abundance of gland cells and nematocysts; in the basal portion of the polyp the filaments become free of the septa, lie loosely in the gastrovascular cavity, and are known as acontia.

Septibranchia. Order of pelecypod mollusks; gills reduced to horizontal muscular septa dividing the mantle cavity; Cuspidaria and Poromyia typical.

septicemia. Bacterial infection in blood stream; serious, often fatal, form of blood poisoning.

Septifer. Genus of marine bivalves; shell subtriangular.

septomaxilla. Small bone on the back margin of the external naris in certain tetrapods.

septum. Any wall, membrane, or partition separating two cavities or distinct masses of tissue. Pl. septa.

septum pellucidum. Double membrane in the anterior part of each of the lateral ventricles of the brain.

sere. Series of communities which follow one another in slow but definite sequence, ending in a climax typical of a particular climate and geographic area; such series may be completed in a hundred years or up to thousands of years. See ECOLOGICAL SUCCESSION, HYDROSERE, XEROSERE.

sergeant fish. Cobia, crab eater; *Rachycentron canadus* (Rachycentridae); a marine fish found in all warm and temperate seas; to 5 ft. long.

sergeant major. Small striped fish common among coral reefs.

Sergiolus. Common genus of spiders in the Family Gnaphosidae.

Serialaria. Genus of marine Bryozoa in the Order Ctenostomata.

serial homology. Fundamental anatomical correspondence in repetitive or serial structures within single organisms, e.g. appendages of a crayfish, vertebrae.

Sericostoma. Uncommon genus of Trichoptera; usually in mountainous areas.

seriema. Large long-legged Brazilian bird with a crested head; also, a similar, smaller bird of northern Argentina; in the Family Cariamidae.

serine. Common amino acid; $C_3H_7O_3N$.

Serinus canarius. Common domesticated canary bird, originally derived from wild finch stock in the Canary Islands, Azores, etc.; extensively bred in Germany and Austria for the market.

Seriola. Common genus of mackerel-like carnivorous fishes; includes the amberjack of the Atlantic Coast and the California yellowtail of the Pacific Coast.

serology. Study of serums and the nature, production, and interactions of antigens and antibodies.

serosa. 1. Peritoneal membrane covering coelomic organs and lining the coelomic cavity in vertebrates; mesothelium. 2. Chorion of vertebrate embryos. 3. In insects, an outermost membrane of the developing embryo.

serotinal. Pertaining to the latter part of the summer season.

serotonin. Cyclic organic compound found in blood, the brain, certain tumors, nettles (plants), and many invertebrate tissues; raises blood pressure, contracts blood vessels and smooth muscle; action similar to that of adrenalin but more pronounced; now synthesized commercially.

serous membrane. See SEROSA (def. 1).

serow. See CAPRICORNIS.

serpent blenny. See LUMPENUS.

Serpentes. Suborder of the reptilian Order Squamata; the true snakes; limbs, ear openings, sternum, and urinary bladder absent; eyes without lids; tongue slender, bifid, and protrusible; scaly skin shed several times per year; oviparous or viviparous; more than 1700 spp.

serpentfly. See RAPHIDIOIDEA.

serpent star. Common name of members of the echinoderm Class Ophiuroidea.

Serphidae. Pointed-tail wasps; cosmopolitan family of minute to small spp.; ovipositor straight and retractile; prey on other small insects.

Serpula. Genus of polychaete annelids which live in secreted calcareous shells; anterior end with numerous gill filaments.

Serranidae. Large family of percomorph fishes; includes the sea basses (*Roccus*) and white perch (*Morone americana*).

Serranus. Genus of tropical and subtropical sea basses (Serranidae); *S. subligarius* is the only known hermaphroditic vertebrate.

serrate. Notched or toothed along the edge, like a saw.

serratus anterior. Serratus magnus.

serratus magnus. Large, flat, anterior trunk muscle in certain vertebrates.

Serripes. Genus of heavy-shelled marine bivalves.

serrula. 1. One of two comblike organs found on each chelicera in the Pseudoscorpionida; used for cleaning the palps. 2. Comblike row of stiff hairs on the swollen endite of each pedipalp in some spiders.

Sertoli cells. Large cells interspersed among the spermatogonia in vertebrate testes; spermatids become imbedded or attached to them and they are presumed to have nutritive and endocrine functions.

Sertularella. Common genus of marine coelenterates in the Suborder Leptomedusae.

Sertularia. Genus of sessile, colonial, marine hydrozoan coelenterates.
serum. Blood serum. Pl. sera.
serum albumin. Chief protein of human blood plasma.
serum globulin. Fraction of blood responsible for the production of certain antibodies.
serval. Long-legged African wildcat (Felis capensis) with spotted coat and no ear tufts.
Servetus, Michael. Spanish physician and theologian (1511-1553); established the general course of the blood from the right ventricle to the lungs and back to the left auricle.
sesamoid bone. 1. An ossification in a tendon, e.g. knee cap.
Sesarma. Common genus of marine crabs occurring in shallows; carapace rectangular; includes the wharf crab or wood crab (S. cinereum).
Sesiidae. Aegeriidae.
sessile. 1. Attached to a substrate; not motile. 2. Without a stalk.
Sessilia. Suborder of Ciliata in the Order Peritricha; usually attached to submerged objects; e.g. Vorticella.
sessoblast. Sessile statoblast; attached to zooecium wall; e.g. in some spp. of Plumatella.
seston. Collectively, all living and dead suspended microscopic particulate matter in aquatic habitats.
seta. One of many types of hairlike or needle-like projections of the exoskeleton in arthropods; their size and complexity vary greatly; comparable secreted structures occur in most annelids and their relatives, some gastrotrichs, tardigrades, and nematodes. Pl. setae.
setaceous. 1. Bearing bristles; setose. 2. Composed of bristles. 3. Having a bristle-like form or texture.
seta sac. Small muscular sac into which setae may be withdrawn, especially in the Phylum Echiuroidea and the Class Oligochaeta.
Setifer. See HEDGEHOG TENREC.
Setonyx. See PADEMELON.
Setophaga. See REDSTART.
setose. Setaceous.
seventeen-year locust. See CICADIDAE.
seven-year itch. See SARCOPTES.
Sewall Wright effect. Genetic drift.
sewellel. See APLODONTIDAE.
sex cell. Gamete.
sex chromosomes. Special sex-determining chromosomes, not occurring in identical number or shape in both sexes; in most animals the male has a pair of unlike sex chromosomes, the so-called X and Y chromosomes, while the female has a pair of similar X chromosomes; in other spp. the male has only a single X chromosome and no corresponding chromosome to make up the pair (XO condition); in certain Trichoptera, Lepidoptera, fishes, amphibians, and all birds, the female has the XY or XO conditions and the male the XX; sex chromosomes have also been determined for a few plants.
sex determinant. Establishment of the sex by means of sex chromosomes at the time of zygote formation, or by some other ecological or physiological means; examples of the latter are rotifers, honeybee, and sipunculid worms.
sex hormone. Any hormone having a morphological or physiological effect upon the reproductive organs, secondary sex characters, or sexual behavior.
sex-limited character. 1. Any factor restricted to one of the two sexes; may be morphological, physiological, or behavioristic. 2. A character determined by a gene present in both sexes but expressing itself in only one.
sex-linked characters. Genetic characters determined by genes located on one or more of the sex chromosomes; e.g. a colorblind gene in man may be carried on the X chromosome.
sex-linked genes. See SEX-LINKED CHARACTERS.
sex ratio. Ratio of average number of males per hundred females in a population; examples are man (105), cattle (107.3), dog (118.5), pigeon (115), horse (98.3), and chicken (94).
sexual reversal. Change from the functional male to the functional female condition, or vice versa, among certain animals; the phenomenon is very rare and has been most thoroughly studied in poultry; usually brought about by pathological conditions or malfunctioning of certain endocrine glands; no authenticated case of sex reversal is known for human beings.
sexton beetle. See SILPHIDAE.
sexual mosaic. Gynandromorph.
sexual reproduction. Any reproduction which involves gametes.
Seymouria. See COTYLOSAURIA.
shad. See ALOSA.
shaft. 1. Central exposed midrib of a typical feather; an extension of the basal hollow quill. 2. Long slender

portion of a long bone.

shag. European cormorant (widely distributed, common in U.S.) or the double-crested American cormorant (Phalacrocorax).

shagreen. Sharkskin tanned with the placoid scales intact, thus making a rough sandpaper-like surface.

shark. General term for all cartilaginous fishes in the selachian Suborder Squali.

shark sucker. See DISCOCEPHALI.

sharp bill. See OXYRUNCIDAE.

Sharpey's fibers. Connective tissue fibers running from periosteum inward through the substance of a bone.

sharp-shinned hawk. See ACCIPITER.

sharp-tailed grouse. See PEDIOECETES.

sharp-tailed snake. Contia tenuis; a small brown snake of the Pacific Coast area from Vancouver to central Calif.

shearwater. See PUFFINUS.

sheatfish. Large catfish (Silurus glanis) of central and eastern Europe; to 400 lbs.

sheathbill. White subantarctic sea bird having a horny sheath over the base of the upper mandible; in the Family Chionididae.

sheath of Schwann. Neurilemma.

sheep. See OVIS.

sheep frog. Small toad (1 in. long) of extreme southern Tex.; Microhyla areolata; head narrow, skin dark and loose.

sheep louse. Trichodectes ovis; a small louse in the Order Mallophaga; parasitic in sheep.

sheepshead. 1. Deep-bodied fish of Atlantic and Gulf coasts; with broad incisor teeth; see SCIAENIDAE. 2. See LABRIDAE and SPARIDAE.

sheep's wool sponge. See HIPPOSPONGIA.

sheep tapeworm. See MONIEZIA EXPANSA.

sheep tick. See HIPPOBOSCIDAE.

sheet-web weavers. See LINYPHIIDAE.

Shelford, Victor E. Pioneer American ecologist (1877-); professor at the University of Illinois; made important contributions to many fields of animal ecology, but especially to succession, aquatic ecology, tiger beetle biology, community structure, and bioecology.

shellac. See LAC INSECT.

shellcracker. See LEPOMIS.

shellfish. Common category which includes shelled mollusks and crustaceans, especially those used as human food.

shell gland. 1. Small paired gland in the head or anterior thoracic region of certain crustaceans; supposedly excretory. 2. Mehli's glands. 3. Nidamental gland; a specialized portion of the oviduct in elasmobranchs and chimaeras; secretes albumin and a horny case around the eggs.

shield bug. See SCUTELLARIDAE.

Shigella. Genus of short, non-motile, gram-negative, rod-shaped bacteria; cause mild diarrhea to very severe and fatal dysentery.

shiner. Any of numerous silvery marine and fresh-water fishes; see NOTROPIS.

shipworm. See TEREDO and BANKIA.

shoebill. Whalehead; a large storklike wading bird of the White Nile area; bill enormous and shoe-shaped; Balaeniceps rex; in the Family Balaenicipitidae.

shore bird. Any bird which frequents the shores of the oceans, lakes, and rivers, especially the snipes, plovers, turnstones, etc.

shore bug. See SALDIDAE.

shore crab. See CARCINIDES.

shore flies. See EPHYDRIDAE.

short clam. Mya truncata; a circumpolar rectangular sp. reaching nearly 3 in.

short-eared owl. See ASIO.

short-horned grasshopper. Grasshopper in the Family Acrididae; antennae shorter than body; medium to large spp.; most common locusts and grasshoppers.

short-tailed shrew. See BLARINA.

short-tailed snake. Stilosoma extenuatum; a small, aggressive, upland, burrowing constrictor of Fla.; 12 to 24 in. long.

shot-hole borer. See SCOLYTUS.

shoulder blade. Scapula.

shoulder girdle. Pectoral girdle.

shovelhead. 1. Shark related to the hammerhead, but with the head narrower; Reniceps tiburo; warm Atlantic and Pacific oceans. 2. Shovel-nosed sturgeon; see SCAPHIRHYNCHUS.

shoveller. See SPATULA.

shovel-nosed snake. Any of several small southwestern banded desert snakes in the genus Chionactis.

shovel-nosed sturgeon. See SCAPHIRHYNCHUS.

shrew. See SORICIDAE.

shrike. See LANIIDAE.

shrimp. Any of a large number of marine decapod crustaceans having long legs and compressed abdomen; many spp. edible; mostly small but a few to 8 in. long; not clearly distinguishable from prawn, sometimes smaller and larger individuals in the same sp. being known as shrimp and prawn, respectively.

Shull, A. Franklin. American zoologist (1881-); professor at University of Michigan; made numerous contributions to insect genetics.

shuttle shell. Any of certain spindle-shaped marine gastropod shells, especially Volva.

Sialia. Genus of birds in the Family Turdidae; includes the several spp. of American bluebirds.

Sialidae. Small family of insects in the Order Neuroptera; alderflies; wing expanse 20 to 40 mm.; aquatic larvae with seven pairs of lateral abdominal gill filaments; north and south temperate zones.

Sialis. See ALDERFLY and SIALIDAE.

siamang. See SYMPHALANGUS.

Siamese twins. Identical twins born joined together in varying degrees.

sib. 1. A brother or sister. 2. In the plural (sibs), two or more offspring of the same parents; siblings.

Sibbaldus musculus. Sulfur-bottom whale or blue whale; a widely distributed whalebone whale; to 103 ft. long, the most massive animal of all time; grayish above and under parts whitish to yellowish or grayish.

Sibekia. Genus of tongue worms parasitic in crocodiles.

siblings. See SIB (def. 2).

Sicista. Genus of small striped mice; northeastern to southeastern Europe and to central Asia.

sicklebill. Any of numerous birds having a sickle-shaped beak.

sickle cell anemia. Hereditary disease of Negros; marked by ulcers, anemia, and erythrocytes having a crescentic or sickle shape.

Sida crystallina. Common cladoceran among aquatic plants of lakes and ponds.

Siderastrea. Genus of reef-building corals of the tropical Pacific and Indian oceans.

sideswimmer. Common name of amphipods; so-called because of their habit of swimming along the bottom on the side of the body.

sidewinder. See CROTALUS.

Sidneyia. Genus of trilobites.

sierra. Any of certain tropical and subtropical mackerel-like marine fishes.

sieve plate. Porous platelike structure found just below an osculum of some sponges.

sifaka. See PROPITHECUS.

Sigalion. Genus of flattened burrowing Polychaeta.

Sigara. Common genus of water boatmen in the Family Corixidae.

Sigillinaria. Genus of colonial ascidians.

Sigmadocia. Genus of marine sponges.

Sigmodon. Genus of rodents in the Family Cricetidae; cotton rats; medium-sized herbivorous rats with rough coat; in grasslands and brushy areas; herbivorous, sometimes farm and ranch pests; southern states, Central America, and northern South America.

sigmoid curve. See LOGISTIC CURVE.

sigmoid flexure. S-shaped bend in the human colon just before its junction with the rectum.

Signalosa petenensis. Threadfin shad, a small forage fish in the Family Clupeidae; Gulf Coast, Mississippi Valley, and Central America.

sika. Any of several deer (Cervus) of medium size found in Japan and northern China.

silicoblast. Sponge cell which secretes all or part of a spicule.

Silicoflagellidae. Family of marine flagellate Protozoa; cell enveloped in a siliceous shell.

Siliqua costata. Ribbed pod shell; a fragile elongated marine bivalve of the East Coast.

silken fungus beetles. See CRYPTOPHAGIDAE.

silk glands. 1. Complex glands in the abdomen of spiders; secrete fluids which harden into thin threads on contacting air as they leave the body via the spinnerets. 2. In the silkworm and related moth larvae, glands which secrete the material which hardens into silk upon exposure to air. 3. Any of various glands in other arthropods which secrete silken threads; openings often near the mouth.

silkworm moth. See BOMBYX.

silky flycatcher. See PTILOGONATIDAE.

silky lemur. See INDRIDAE.

Silpha. Common genus of flattened beetles in the Family Silphidae; scavengers, predators, and herbivores.

Silphidae. Family of minute to large beetles; carrion beetles, sexton beetles; feed and breed on carrion and decaying plant material; Necrophorus and

Silpha common.

Silurian period. One of the six geological subdivisions of the Paleozoic era, between the Devonian and Ordovician periods; lasted from about 390 million to 350 million years ago; age of submergence and mountain building; first land plants; fauna characterized by many ostracoderms and the appearance of placoderms and air-breathing arthropods (scorpions).

Siluridae. Family of European and Asiatic catfishes; adipose fin absent, anal fin long and confluent with the caudal fin; includes Silurus.

Siluroidei. Suborder taxon sometimes used to include the Ariidae and Ameiuridae (catfishes).

Silurus glanis. Sheatfish, a large brown European catfish (Siluridae); a voracious river sp. weighing to 400 lbs.

silverfish. 1. The tarpon. 2. Any of numerous silvery fishes. 3. Silversides. 4. See THYSANURA.

silver fox. See VULPES.

silver-haired bat. See LASIONYCTERIS.

silverline system. Collectively, in a ciliate protozoan, all those pellicle and subpellicle structures which stain in silver impregnation methods; includes granules, fibrils, pits, etc.

silver perch. Any of numerous marine and fresh-water silvery perchlike fishes.

silversides. See ATHERINIDAE.

silverspot. Any of several spp. of fritillary butterflies having silvery spots on the under surface of the wings.

Simia. Pongo.

Simocephalus. Common genus of fresh-water Cladocera.

simple eye. General term to include eyespots and eyes which have a relatively simple structure and do not result in true image formation.

simple gland. Any gland contained within a single connective tissue covering.

simple goiter. See GOITER.

simple metamorphosis. See PAUROMETABOLA.

simple protein. Protein built up solely of amino acids; e.g. egg white, hair, and milk proteins. Cf. CONJUGATED PROTEINS.

simple sugar. See MONOSACCHARIDE.

Simpsoniconcha ambigua. Common fresh-water bivalve mollusk (Unionidae); usually under stones in streams; central U.S.

Simuliidae. Family of small humpbacked flies in the Order Diptera; buffalo gnats, black flies; adult females are vicious biters and pests familiar to the fisherman and the farmer whose stock is sometimes tormented by clouds of these insects; larvae and pupae occur only on rocks, etc. in swift streams, sometimes in enormous numbers; several spp. are important transmitters of the nematode Onchocerca volvulus which causes human onchocerciasis in Africa and Central America; Simulium is a common American genus.

Simulium. See SIMULIIDAE.

simung. See LUTROGALE.

Sinantherina. Genus of sessile, colonial rotifers; body with two to four opaque protuberances.

Sinanthropus pekinensis. Peking man; tooth, mandible, and skull material found in Pleistocene deposits near Peking, China; brain volume about 915 to 1220 cc.; probably able to talk; associated with crude stone implements.

single comb. Upright bladelike type of comb occurring in Leghorn fowl.

sinistral. 1. Left-handed. 2. Coiled or turned toward the left. 3. In gastropods, having the aperture to the left of the shell axis when the shell is held with the spire pointing upward and the aperture facing the observer.

sino-auricular node. Pacemaker.

Sinum. Genus of marine gastropods; baby's ear; body whorl very large.

sinus. 1. Any space in the tissues of an animal, especially a greatly expanded vein or other type of blood cavity. 2. Cavity in a bone of the skull which is normally air-filled and which communicates with the respiratory passages. 3. Channel for venous blood. 4. Dilation of a canal or vessel. 5. Long, narrow cavity through which pus discharges. Pl. sinuses.

sinus gland. Minute ductless gland in the eyestalk of various crustaceans; closely associated with the X organ; controls the physiology of chromatophores, regulates molting, and influences the deposition of calcium carbonate in the exoskeleton.

sinusoid. Terminal blood channel composed of an irregular anastomosing vessel, especially in the liver, pancreas, spleen, etc.

sinus venosus. Large, thin-walled chamber which receives blood from veins and transmits it to the (right) auricle in fishes, amphibians, and reptiles.

Siphateles. Small genus of chubs in the Family Cyprinidae; Calif., Ore., and Wash.; to 12 in. long.

Siphlonurus. Common genus of Ephemeroptera.

Siphloplecton. Genus of Ephemeroptera; eastern U.S.

siphon. 1. One of two specialized structures at the posterior end of most Pelecypoda; the edges of the mantle are more or less fused so as to form a short to long tubular incurrent siphon (ventral) and a corresponding excurrent siphon (dorsal); water is drawn into the mantle cavity through the former and is passed out of the suprabranchial chambers and adjacent mantle cavity through the latter. 2. Short, tubular structure in cephalopod mollusks; extends from the mantle cavity to the outside and used for locomotion on the jet propulsion principle. 3. Slender ciliated tube along one edge of much of the digestive tract of sea urchins, usually from the beginning of the small intestine to the beginning of the large intestine; thought to carry sea water directly to the intestine and aid in washing residues from the latter. 4. One of two wide tubes in tunicates; a current of water passes into the incurrent siphon (mouth) and thence to the branchial chamber, peribranchial chamber, and out of the body by way of the excurrent siphon. 5. Long, tubular, fleshy extension carried in the canal of the shell of certain large marine gastropods; contains both incurrent and excurrent tubes. 6. Feeding polyp of certain siphonophore coelenterates; has only a single, long, contractile tentacle originating basally.

Siphonactinia. Genus of elongated burrowing sea anemones in the Order Actiniaria.

Siphonaptera. Order of insects which includes the true fleas; minute to small; body laterally compressed, wingless; mouth parts piercing and sucking; head small; simple eyes present or absent; antennae small, in grooves; legs large, adapted for jumping; adults intermittent bloodsuckers of birds and mammals; metamorphosis complete; larvae apodous, feed on animal and vegetable debris; pupae in cocoons; 1000 spp.; important genera are Pulex, Ctenocephalides, Xenopsylla, Ceratophyllus, Echidnophaga, Tunga, and Histricopsylla.

Siphonodentalium. Common genus in the mollusk Class Scaphopoda.

siphonoglyph. Special, deep-walled, groovelike portion of the anthozoan stomodaeum lined with flagellated cells which create a current of water flowing from the region of the proctostome downward to the interior.

Siphonophora. Order of pelagic marine hydrozoan coelenterates; floating or swimming colonies composed of several kinds of polyps; colonies often massive; nematocysts abundant and powerful; medusae remain attached to the parent hydroid generation by a stem or disc; examples are Physalia and Velella.

siphonophore. Any member of the Order Siphonophora.

Siphonopoda. Cephalopoda.

Siphonosoma. Genus of large worms in Phylum Sipunculoidea.

siphonostomatous. 1. In certain marine snails, with the shell aperture elongated, notched, and channel-like for the protection and extension of the siphon. 2. Having tubular lips and mouth.

siphonozooids. Small modified polyps as found in hydrozoans in the Order Pennatulacea; tentacles rudimentary or absent, septa reduced, siphonoglyph strongly developed; function chiefly in driving a current of water through the fleshy colony.

siphuncle. Long strand of connective tissue which is connected to the posterior end of the body of a pearly nautilus and passes successively through all of the septa in the shell to the innermost chamber.

Siphunculata. Anoplura.

Sipunculoidea. Phylum including a small group of worms which live in tubes in the sea bottom or burrow about freely in the upper portions of mud and sand bottoms; anterior third of the animal (introvert) slender but the rest of the body more swollen and even bulbous; mouth surrounded by a dense cluster of small, hollow, ciliated tentacles; digestive tract U-shaped with the anus one-quarter of the way from the anterior end; coelom capacious; body wall muscular; one or two nephridia (metanephridia), simple circulatory system, simple nervous system, and sexes separate; larva typically a trochophore; to 18 in. long; about 250 spp., 20 along U.S. coasts; Phascolosoma common.

Sipunculus. Genus in the Phylum Sipunculoidea.

siren. See MEANTES.

Sirenia. Order of herbivorous aquatic mammals; includes the manatee, dugong, and Steller sea cow; large, spindle-shaped spp. with no hind limbs and paddle-shaped forelimbs; muzzle blunt, external ears absent, hairs few; modern spp. in tropical and subtropical rivers and seas; Trichechus, Dugong, and Hydrodamalis typical.

Sirenidae. Single family of eel-shaped amphibians in the Order Meantes; Siren and Pseudobranchus typical.

Siren lacertina. Mud eel, an eel-shaped amphibian in the Order Meantes; swamps, d ches, and ponds from Va. to Ind. and south to Fla. and Tex.; to 30 in. long.

Sirenoidei. Order of lungfishes which includes all modern spp.; see DIPNOI.

Sirex. Genus of hymenopterans in the Family Siricidae; larvae attack conifers.

Siricidae. Family of large metallic wasplike insects in the Order Hymenoptera; horn-tails, wood wasps; abdomen with apical spine; oviposition in trees and shrubs where larvae burrow about; Sirex and Tremex typical.

siskin. See SPINUS.

Sistrurus. Genus of venomous American rattlesnakes in the Family Crotalidae; ground rattlesnakes; tail with a series of cornified button-like structures making up the rattle; S. catenatus, the massasauga, to 3.5 ft. long, occurs from the Great Lakes area southwesterly into Ariz., Tex., and Mexico; S. miliarius, the pygmy rattler, to 2 ft. long, occurs from N.C. to Tex. and north to Ark., Okla., and Colo.; these spp. are the least dangerous of all the rattlers.

Sisyra. See SISYRIDAE.

Sisyridae. Spongilla flies; small family of Neuroptera in which the larvae live in the cavities and on the surface of fresh-water sponges; Climacia and Sisyra common.

Sitaris. Genus of beetles in the Family Meloidae; larvae in nests of solitary bees.

sitatunga. See STREPSICEROS.

Sitodrepa. Stegobium.

Sitophilus. Important pest genus of weevils; S. granarius is the granary weevil, and S. oryza the cosmopolitan rice weevil.

Sitotroga cerealella. Angoumois grain moth; larva feeds on kernels of grain; a widely distributed and very important pest sp.

Sitta. See SITTIDAE.

Sitticus palustris. Small brown or gray spider in the Family Salticidae; common east of Rockies.

Sittidae. Family of passerine birds which includes the nuthatches; in both Old and New worlds; small tree climbers; bill long and thin, tail short; habitually go down tree trunks head first; Sitta is the familiar genus of U.S. nuthatches.

skate. Any sp. in several genera of cartilaginous fishes in the Family Rajidae; body greatly depressed and disclike; tail short, without a spine; caudal fin vestigial; abundant on sea bottom; Raja common.

skatol. Putrefaction product of tryptophane found in feces; C_9H_9N.

skeletal muscle. Voluntary muscle.

skeletal system. Collectively, all of the supporting structures in an animal, such as bone and cartilage.

skeleton. Collectively, the hard supporting structures of an animal; may be external, as in a mollusk, arthropod, or nematode, or internal, as in an echinoderm or vertebrate.

skeleton shrimp. See CAPRELLA.

Skenea planorbis. Trumpet shell, orb shell; a minute, flat, marine gastropod found on both sides of the Atlantic.

Skiffia. Genus of small Mexican fresh-water fishes.

skimmer. See RHYNCHOPIDAE.

skin. External covering or integument of an animal, especially a vertebrate.

skin beetles. See DERMESTIDAE.

skin gill. Dermal branchia.

skink. See SCINCIDAE.

skipjack. Any of several marine fishes that swim near the surface and jump out or otherwise break the surface, especially the bonito (Sarda) and butterfish (Poronotus).

skippers. 1. Those members of the lepidopteran Suborder Rhopalocera in the Family Hesperiidae; diurnal, stout-bodied spp. which make short, swift flights. 2. See SCOMBERESOCIDAE.

skua. See CATHARACTA.

skull. That part of the bony and/or cartilaginous framework of a vertebrate head which encloses the brain and sense organs and to which the jaws are usually fastened.

skunk. See SPILOGALE, MEPHITIS,

and CONEPATUS.

skylark. European lark (Alauda arvensis) noted for its song and ability to fly upward vertically.

Slabberia. Genus of marine coelenterates in the Suborder Anthomedusae.

slave-making ant. Any sp. of ant which enslaves another sp. of ant; the latter (slave ants) perform essentially all of the chores involved in maintaining the slave-making colony and nest; a slave-making sp. may be either obligatory or facultative.

sleeper. See ELEOTRIDAE.

sleeping sickness. 1. Disease of human beings produced by the protozoan Trypanosoma gambiense (West Africa) and by T. rhodesiense (East Africa); the parasite is transmitted from one person to another through the bite of the tsetse fly. 2. Uncommon virus disease of man in the U.S.; encephalitis.

slider. See PSEUDEMYS.

slime eel. See MYXINOIDIA.

slime gland. Any gland which secretes slime or mucus; present in many higher invertebrates and some vertebrates, especially mollusks, annelids, fish, and amphibians.

slimy salamander. See PLETHODON.

slipper animalcule. Any member of the ciliate genus Paramecium; cell outline slipper-shaped.

slipper shell. See CREPIDULA.

sloth. See BRADYPODIDAE.

sloth bear. Coarse-haired bear of India and Ceylon; notable for its long mobile snout and tongue; see MELURUS.

slow loris. See NYCTICEBUS.

slowworm. See ANGUIS.

slug. 1. Type of terrestrial gastropod; shell usually reduced to a rudiment buried in the small mantle; foot elongated. 2. Occasionally, a similar type of marine gastropod. 3. Uncommonly, a smooth, soft, moth or sawfly larva which creeps about in a sluglike fashion.

small intestine. Principle organ of digestion and absorption in tetrapods; elongated (25 ft. in man), tubular, and situated between the stomach and large intestine; in man the small intestine consists of the duodenum, ileum, and jejunum.

small-mouth black bass. See MICROPTERUS.

smegma. Sebaceous, cheesy material which accumulates under the prepuce and around the clitoris area; composed mostly of castoff epithelial cells.

smell. Special sense, involving the ability to detect volatile substances in suspension in the air; the term is also often used to apply to the detection of dissolved substances by aquatic animals.

smelt. 1. Any of a group of small, slender, silvery marine fishes of the Northern Hemisphere; run up rivers to spawn; see OSMERUS. 2. Any of numerous other marine and fresh-water fishes which resemble the true smelts.

smew. Small merganser duck of northern Europe and Asia.

Smicridea. Genus of Trichoptera.

Sminthillus. Genus of small frogs found in Cuba, Peru, and Brazil; one sp. is known to deposit only one egg at a time.

Sminthopsis. Genus of marsupials about the size of mice (Family Dasyuridae); hind legs long and adapted for leaping; Australian region.

Sminthurides. Common genus of insects in the Order Collembola; terrestrial and semiaquatic, but a few spp. sometimes found on the surface of puddles and ponds.

Sminthurus hortensis. Common sp. of insect in the Order Collembola; sometimes a garden pest, especially on vegetables.

Smittina. Genus of encrusting marine Bryozoa.

smolt. A young silvery salmon as it descends to salt water.

smooth dogfish. See MUSTELUS.

smooth hound. Smooth dogfish; see MUSTELUS.

smooth jingle. Anomia simplex, the common jungle shell of the East Coast.

smooth muscle. Nonstriated muscle.

snail. Common name of typical representative of the molluscan Class Gastropoda.

snailfish. See LIPARIDAE.

snake. See SERPENTES.

snakebird. See ANHINGA.

snake blenny. See LUMPENUS.

snake doctor. See ODONATA and ANISOPTERA.

snake eel. See OMOCHELYS.

snakefly. See RAPHIDIODEA.

snake star. Common name of members of the echinoderm Class Ophiuroidea, especially spp. in the genus Amphipholis, which have exceptionally long arms.

snake weasel. See POECILOGALE.

snapper. See LUTIANIDAE.

snapping shrimp. See ALPHEIDAE.

snapping turtle. See CHELYDRIDAE.

Snellen's test. Eye test involving a chart with lines of letters of different sizes.

snipe. See CAPELLA.

snipe eel. Any of several long ribbonlike or threadlike eels in the Family Nemichthyidae; beak long and sometimes recurved; widely distributed at moderate and great depths; Nemichthys typical.

snipe flies. See RHAGIONIDAE.

Snodgrass, R. E. Outstanding insect anatomist (1875-1962); associated with U. S. Dept. of Agriculture for many years.

snook. 1. See CENTROPOMIDAE. 2. Any of several snooklike marine fishes in certain other families.

snout beetles. See CURCULIONIDAE.

snout mites. See BDELLA.

snout moths. See PYRALIDIDAE.

snowbird. Snow bunting; see BUNTING.

snow bunting. See BUNTING.

snowcap. Any of several white-crowned Central American hummingbirds.

snow cock. Any of several large gallinaceous birds of the tundra of central and western mountains of Asia.

snow flea. See ACHORUTES and BOREUS.

snow goose. See CHEN.

snow grouse. Ptarmigan.

snow scorpionflies. See BOREUS.

snowshoe hare. See LEPUS.

snow worm. Any of several small oligochaetes sometimes found in or on snow (near soil), especially in mountain areas.

snowy egret. See FLORIDA.

snowy owl. See NYCTEA.

snub-nosed monkey. See RHINOPITHECUS.

Snyderichthys copei. Leatherside chub in the Family Cyprinidae; Utah, Wyo., and Ida.

social hierarchy. In a local population of vertebrates, the existence of a system of dominance and subordination among the individuals; dependent upon such factors as strength, maturity, size, aggressiveness, sex, etc.

social hormones. Recognition secretions or ectohormones presumably found on the external surface of many social insects; secreted only by certain castes but found on all individuals as the result of grooming.

social insects. Societies or aggregations of different types (castes) of individuals among insects; e. g. ants, termites, and some bees and wasps; a hive of bees has only three castes in its social organization: queen, workers, and drones.

society. 1. Integrated, organized group of individuals such as a colony of wasps, ants, beetles, or termites, in which there is considerable specialization and division of labor. 2. Group of animals of the same sp. mutually bound together by behavior and physiological attractions, such as a colony of oysters, school of fish, flock of birds, and herd of antelope. 3. Community of subdominant plants and animals within an association, e. g. the community of shrubs or herbs within a forest.

society screw. Society thread; standard thread specifications used on microscope objectives; originated by the Royal Microscopical Society.

sociology. In the zoological sense, the study of the origin, development, organization, and interrelationships of groups of organisms comprising a single sp.; occasionally used to indicate these relationships between groups of different spp.

socket. In an articulation or joint, a concavity bearing the base of a segment, appendage, or organ.

sockeye salmon. Blueback salmon; see ONCORHYNCHUS.

sodium glychocholate. Bile salt which saponifies fats.

sodium taurocholate. Bile salt which saponifies fats.

soft coral. Any of a group of anthozoans (Order Alcyonacea) in which the polyps are imbedded in a gelatinous mass; skeleton of separate calcareous spicules.

soft palate. Especially in man, the posterior part of the palate which is not supported by bone but consists of fibrous connective tissue and mucous membrane.

soft-rayed fin. In fishes, a fin whose supporting skeletal elements are soft and pliable.

soft scales. See COCCIDAE.

soft-shelled clam. See MYA.

soft-shelled crab. Any marine crab that has just molted and has the new exoskeleton in an unhardened condition; the expression is usually used to refer to the common edible crab, Callinectes.

softshell turtle. See TRIONYX.

sokhor. See MYOSPALAX.

sol. 1. Fluid colloidal system. 2. Abbr. for solution.

solar plexus. Large nerve plexus located behind the stomach in man; it is formed from several autonomic nerves, contains several ganglia, and distributes nerves to abdominal viscera.

Solaster. Genus which includes the familiar large sun-star sea stars; 7 to 14 arms.

solation. Conversion of more solid, jellylike physical phase of protoplasm into the more liquid phase.

soldier. One of several castes in such social insects as ants and termites; specialized for guarding the nest and biting.

soldier beetles. See CANTHARIDAE.

soldier bug. Any of several bugs in the Family Pentatomidae; suck the blood of other insects.

soldier fish. Squirrel fish; any of numerous brightly colored spiny marine fishes in the Family Holocentridae; common around coral reefs; some are prized for food; Holocentrus common.

soldier fly. See STRATIOMYIIDAE.

sole. See HETEROSOMATA.

Soleidae. Family of Heterosomata which includes the soles.

Solemya. Genus of fragile marine clams in the Order Protobranchia; awning clams; yellowish to brown, usually 1 to 2 in. long; Atlantic Coast.

Solen. Pelecypod genus which includes the small razor clam; shell long, narrow, and with gaping ends; both coasts.

solenia. Gastrodermal tubes connecting adjacent polyps in certain colonial Anthozoa; usually enclosed in mats or stolons.

Solenichthyes. Order of marine bony fishes which includes the pipe fishes and sea horses; whole body elongated or head held at right angle to trunk; tubular mouth at end of snout; male usually with a ventral brood pouch.

Solenichthyida. Solenichthyes.

Solenidae. Family of marine Pelecypoda which includes the razor clams; shells long, narrow, and with gaping ends; siphons short; Solen, Ensis, and Siliqua common.

Solenius. Genus of thread-waisted wasps.

solenocyte. Special type of flame bulb containing a single long flagellum; in trochophores and related larvae, and in adult kinorhynchs, priapulids, some polychaetes, and cephalochordates.

solenodon. See SOLENODONTIDAE.

Solenodontidae. Solenodons; family of large vicious ratlike insectivores with long tubular snouts and heavy curved claws; Haiti and Cuba.

Solenogastres. Aplacophora.

Solenoglypha. Group of two families of venomous snakes, the Viperidae and Crotalidae; paired erectile fangs in front part of upper jaw are folded backward when not in use.

Solenophrya. Genus of fresh-water suctorians; stalk absent, cell enclosed in a lorica; tentacles in bundles.

Solenopsis. Genus which includes several fire ants; sting causes a burning sensation; a South American sp., S. saevissima, has gained a foothold in the southern U.S. where it is a serious pest.

soleus. Muscle present in many mammals; originates on the shafts of the tibia and fibula and inserts in the heel region; extends foot and rotates it inward.

sol-gel transformation. See COLLOIDAL SYSTEM.

Solifugae. Solpugida.

solitaire. 1. Flightless bird somewhat like the dodo but more slender; found on Rodriguez but extinct since about 1750. 2. See MYADESTES.

solitary wasps. See EUMENIDAE.

Sollasina. See OPHIOCISTIOIDEA.

Solmaris. Genus of marine coelenterates in the Suborder Narcomedusae.

Solmissus. Pacific genus of marine hydrozoan coelenterates.

soln. Abbr. for solution.

Solpugida. Order of arachnids which includes solpugids and sun spiders; 1 to 5 cm. long; cephalothorax with first pair of legs, followed by three free thoracic segments; thorax and abdomen broadly joined; abdomen ten-segmented; pedipalps leglike; large, pale, active, common in dry tropics and subtropics; 700 spp.; examples are Eremobates, Ammotreca, Rhagodes, and Galeodes.

soma. 1. Axial portion of a bilaterally symmetrical animal, not including the paired limbs. 2. Collectively, the whole body excepting the germ cells; see SOMATIC CELLS.

Somateria. Genus of sea ducks in the Family Anatidae; includes the eider ducks; female lines her nest with soft feathers (eider down) plucked from her body; northern part of Northern Hemisphere.

somatic cells. Collectively, those cells of the plant or animal body exclusive of the reproductive cells.

somatic crossing over. Reciprocal exchange between homologous chromosomes during somatic mitosis.

somatic mutation. Mutation in somatic tissues; results in a patch of tissue genetically distinct from the rest of the body; e.g. patches of feathers or hair differing from the surrounding normal coloration; may sometimes be induced by X-ray treatment.

Somatochlora. Common genus of dragonflies.

somatocoel. Early embryonic portion of the echinoderm coelom.

Somatogyrus. Genus of fresh-water snails; southeastern quarter of U.S.

somatoplasm. Collectively, the body cells of an organism as distinct from the germ cells or germ plasm.

somatopleure. In vertebrate embryos, a mass of tissue on each side consisting of ectoderm plus the upper or parietal of the two layers of mesoderm of the primary mesoblast; gives rise to much of the body wall.

somatotrophic hormone. Somatotrophin; growth-stimulating hormone of the anterior pituitary which acts on cellular protein synthesis; over- or undersecretion during early life produces gigantism or dwarfism, respectively; oversecretion later in life produces acromegaly.

somatotrophin. Somatotrophic hormone.

somesthetic area. Part of the parietal lobe of the brain which receives impulses from the touch, heat, cold, and kinesthetic receptors.

somite. 1. Segment or metamere. 2. One of a series of segmentally arranged masses of tissue (especially muscle) on either side of the spinal cord, notably in vertebrate embryos.

song sparrow. Any of numerous subspecies of Melospiza melodia, a common N.A. sparrow with a sweet song.

Sonora. See GROUND SNAKE.

Sonoran zone. See UPPER AUSTRAL and LOWER AUSTRAL.

sooty mangabey. Large arboreal monkey (Cercocebus) of forested Africa; dark fur, white upper eyelids.

sora. See PORZANA.

Sørensen, Søren P. L. Danish chemist who first proposed the numerical pH scale in 1909.

Sorex. See SORICIDAE.

Soricidae. Family of small, slender, mouselike mammals in the Order Insectivora; shrews; includes the smallest of mammals; eyes small; do not burrow but forage on the ground surface, feeding voraciously on small invertebrates and a few rodents; many American spp. of Sorex.

Sorocelis americana. Whitish planarian from Ozark caves and springs.

soupfin shark. Pacific shark prized for food by Asiatics; to 6 ft. long; Galeorhinus.

sow bug. Hog louse; see HAEMATOPINUS.

sowbug. Common name of terrestrial isopods, especially those that do not roll up into a ball.

sp. Abbr. for species (singular).

spadefish. 1. Local name for paddlefish. 2. Deep-bodied marine fish of tropical and subtropical Atlantic coastal areas; related to the angelfishes.

spadefoot toad. See PELOBATIDAE.

Spalacidae. Family which includes a single sp. of herbivorous mole-rats (Spalax); fur thick and woolly, limbs short and powerful, eyes minute; tunnel continuously; eastern Europe, Caucasus, and Asia Minor.

Spalax. See SPALACIDAE.

Spallanzani, Lazzaro. Italian biologist (1729-1799); early opponent of the spontaneous generation theory; studied vertebrate embryology and regeneration.

Spalocopus. Genus of South American ratlike mammals in the Family Octodontidae; the cururos; ears small; found mostly above ground.

Spaniotoma. Genus of tendipedid midges.

Spanish fly. 1. Lytta vesicatoris, a European beetle which is an important source of cantharidin. 2. Processed or unprocessed extract of L. vesicatoris.

Spanish mackerel. See SCOMBEROMORUS.

spanworm. Larva of a moth in the Family Geometridae.

Sparassidae. Family of giant crab spiders and banana spiders; mostly tropical; Heteropoda common.

Sparganophilus. Genus of earthworms.

sparganum. Plerocercoid, especially a large plerocercoid. Pl. spargana.

Sparidae. Family of percomorph fishes; scups, sea bream, porgies, sheepshead; deep-bodied spp. with manlike incisor teeth; valuable food spp.; mostly tropical and subtropical, but Stenotomus (scup, porgy) and Archosargus (sheepshead) found on Atlantic and Gulf coasts.

sparling. European smelt, Osmerus eperlanus.

sparrow. 1. General term which includes the true sparrows as well as other sparrow-like birds not in the Family Fringillidae. 2. In the narrow sense, any of a group of finches; common American spp. are savannah sparrow (Passerculus sandwichensis), grasshopper sparrow (Ammodramus savannarum), tree sparrow (Spizella arborea), chipping sparrow (S. passerina), field sparrow (S. pusilla), white-crowned sparrow (Zonotrichia leucophrys), and song sparrow (Melospiza melodia). See PLOCEIDAE and PASSER.

sparrow hawk. 1. Any of several small Old World hawks in the genus Accipiter. 2. See FALCO.

spasmoneme. Contractile fibril in the stalk of certain ciliates; composed of fused myonemes.

spasticity. Condition produced by an upper motor neuron injury; marked by stiff muscles and rigid limbs.

spat. Microscopic larval stage of an oyster.

Spatangoida. Order of echinoderms in the Class Echinoidea; heart urchins; test elongated, not flattened, and somewhat heart-shaped; mouth near the narrow end on oral surface; anus peripheral at broad end; Aristotle's lantern absent; Echinocardium a common American genus which burrows in sand.

Spatangus. Genus of heart urchins.

Spathidium. Large genus of marine and fresh-water holotrich Ciliata; flask- or sac-shaped.

Spatula clypeata. Common shoveller or broadbill duck of Northern Hemisphere; bill especially broad at tip.

spatulate. Spoon-shaped or spatula-shaped.

spawn. 1. In aquatic animals, to produce or deposit eggs or sperm. 2. To produce eggs or young. 3. Eggs of fishes and higher aquatic invertebrates.

spay. 1. Third-year male European red deer. 2. To remove ovaries from a female animal, especially a mammal.

spearfish. Name sometimes applied to marlins, swordfish, sailfish, and closely related spp.

special creation. Theory which accounts for the evolutionary creation of life by some supernatural power, either once or at successive intervals; also includes the concept that each sp. has been created separately and individually, without any evolution.

specialization. Any peculiarity of morphology, anatomy, physiology, or behavior which especially fits an organism or one of its parts for a particular function or habitat; adaptation.

speciation. Process of the formation of new spp. through the mechanisms of evolution.

species. Group of organisms which actually (or potentially) interbreed and which are reproductively isolated from all other such groups; the fundamental category of classification, intended to designate a single kind; each sp. is designated by a scientific name (q. v.). (The same spelling is used for both sing. and pl.; the term "specie" is not used in biology.)

speckled bass. Black crappie.

speckled trout. Brook trout; see SALVELINUS.

spectacled bear. See TREMARCTOS.

speculum. A lustrous, brightly colored patch on the wings of certain birds, especially ducks. Pl. specula.

Spelaeogriphacea. Order of primitive and poorly known peracarid crustaceans, somewhat similar to the Thermosbaenacea; elongated, cylindrical, up to 8.6 mm. long; only a few specimens of a single sp., Spelaeogriphus lepidops, known from a cave pool on Table Mountain, South Africa.

Spemann, Hans. German embryologist (1869-1941); studied growth and differentiation by means of transplants and tissue isolation; ascribed the fate of tissues in embryonic development to "organizers"; awarded the 1935 Nobel prize in physiology and medicine.

Speocarcinus. Genus of small marine burrowing crabs.

Speophila. Genus of cave planarians.

Speothus. Genus of bush dogs; elongated, fuzzy, short-legged, doglike, burrowing mammals of the savannahs of the Amazon drainage.

Speotyto cunicularia. Common burrowing owl of unforested areas in western states; to 9 in. long; brownish coloration; usually nests in deserted mammal burrows.

Sperchon. Genus of Hydracarina; mostly in cold waters.

Sperchonopsis verrucosa. Common sp. of Hydracarina.

Sperchopsis. Old World genus of hydrophilid beetles.

sperm. Spermatozoa.

spermaceti. 1. Whale oil. 2. White waxy solid obtained from whale oil and from the oil of certain dolphins.

spermary. An invertebrate testis, especially one of simple structure, as in Hydra. Pl. spermaries.

spermatheca. Small bulbous sac associated with the female reproductive system in many invertebrates, especially insects; receives sperm from the male during copulation; sometimes present in hermaphroditic animals.

spermatic cord. Collectively, a vas deferens, nerves, and blood vessels on each side as they pass from the scrotum through the inguinal canal into the body cavity of many mammals.

spermatid. Each of four haploid cells resulting from the two meiotic divisions of a primary spermatocyte; metamorphoses into a mature sperm cell.

Spermatists. Group of eighteenth century biologists who believed that a sperm cell contained a complete miniature individual. Cf. OVISTS.

spermatocyte. In spermatogenesis, a cell derived directly from a spermatogonium; known as a primary spermatocyte, it undergoes the first meiotic division and produces two secondary spermatocytes, each of which undergoes a second meiotic division to form two spermatids and subsequent spermatozoa.

spermatogenesis. Series of cell divisions and chromosome and cytoplasmic changes involved in the production of functional spermatozoa, beginning with the undifferentiated germinal epithelium; a single mature spermatogonium gives rise to four mature spermatozoa.

spermatogonium. Precursor of the mature male reproductive cells in animals; spermatogonia usually form much of testis tissue and multiply by ordinary mitosis; a mature spermatogonium gives rise to a spermatocyte directly.

spermatophore. Capsule or packet formed by certain types of male invertebrates and a few vertebrates; contains sperm and is usually transferred to the female to facilitate fertilization; found in such diverse groups as mollusks, leeches, copepods, newts, etc.

spermatozoa. Small haploid male gametes of animals; with the exception of nematodes, decapods, diplopods, and mites, they are flagellated. Sing. spermatozoan and spermatozoon.

spermatozoan. See SPERMATOZOA.

spermatozoon. See SPERMATOZOA.

Spermatozopsis. Genus of sickle-shaped protozoans; four (rarely two) anterior flagella; stagnant fresh waters.

sperm duct. Any duct more or less specialized for the conduction of sperm.

sperm funnel. Frilled funnel-shaped opening of a vas deferens in the earthworm, etc.; imbedded in seminal vesicle tissue and serves to transmit sperm from the seminal vesicle to the narrow vas deferens.

spermiducal glands. Small glands associated with the sperm ducts of oligochaetes; sometimes called prostate glands.

sperm oil. Pale yellow lubricating oil found in the blubber and head cavities of the sperm whale; contains spermaceti.

spermophile. See CITELLUS.

sperm jacket. Spermatophore.

sperm whale. Large whale, to 60 ft. long; the cachalot, Physeter catodon; feeds on fishes and large squids; head about one-third of body length, roughly square in section; sperm oil secreted into a large reservoir in the head; a valuable waxy substance, ambergris, formed in the stomach and sometimes cast up and found floating in the sea; used as a perfume fixative.

spet. Sennet; a small barracuda of southern Europe.

Sphaerella. Haematococcus.

sphaeridium. Minute glassy sphere attached to a boss by means of connective tissue; found in abundance on the tests of Echinoidea, especially in the ambulacral areas; presumably a balancing apparatus. Pl. sphaeridia.

Sphaerium. Genus of small fresh-water pelecypods; usually 7 to 15 mm. long, but a few spp. to 25 mm.; fingernail clams.

Sphaerodactylus. Genus of Cuban geckos; introduced into Fla.

Sphaeroma. Common genus of marine Isopoda; S. terebrans (a marine borer) occasionally enters fresh waters on the Gulf Coast.

Sphaeromyxa. Genus of sporozoan Protozoa (Order Myxosporidia) in marine fishes.

Sphaeronectes. Genus of coelenterates in the Order Siphonophora.

Sphaerophrya. Genus of spherical, unstalked suctorian Protozoa; fresh waters; sometimes parasitic in other Protozoa.

Sphaerosyllis. Genus of small Polychaeta.

Sphaerotilus. Genus of filamentous, sheathed bacteria common in sewage and grossly polluted waters.
Sphaerozoum. Genus of Radiolaria.
Sphaerularia. Genus of minute parasitic and free-living nematodes; one sp. parasitic in bees.
Sphalloplana. Genus of flatworms in the Order Tricladida; whitish, eyeless, in U.S. caves.
Sphecidae. Cosmopolitan family of long, slender wasps with long abdominal pedicel; mud-dauber wasps; adults build mud cells singly or in series; Sphex and Sceliphron common.
Spheciospongia. Genus of giant tropical sponges.
Sphecius. See CICADA KILLER.
Sphecodes. Common genus of bees in the Family Halictidae; shiny black and red coloration; social parasites in nests of Halictus (in this same family).
sphenethmoid. 1. One of a pair of bones forming a ring around each of the olfactory nerves in certain amphibians. 2. A chondral bone of lower vertebrates corresponding to the presphenoid and orbitosphenoid of mammals.
Sphenisciformes. Order of birds which includes the penguins; flightless spp. with the wings paddle-like and adapted for swimming; feathers small, scale-like, and covering entire body; feet webbed; excellent swimmers and divers; thick layer of fat beneath skin; coastal areas of Southern Hemisphere from Antarctica to the Galapagos Islands; six genera and about 20 spp. and varieties; Aptenodytes forsteri, emporer penguin, an inhabitant of Antarctica, is the largest (to 4 ft. tall); Spheniscus demersus is the Cape penguin.
Spheniscus. See SPHENISCIFORMES.
Sphenodon. See RHYNCHOCEPHALIA.
Sphenodontidae. Family of lizards in the Order Rhynchocephalia.
sphenoid. Irregular wedge-shaped bone at base of mammalian skull.
sphenoidal sinus. Cavity of variable size in the anterior part of the sphenoid bone, especially in man; opens into the nasal cavity.
sphenoid teeth. Small teeth on the sphenoid bones of certain lower vertebrates.
sphenopalatine foramen. Opening in the lateral surface of a palatine bone.
Sphenophorus. Cosmopolitan genus of weevils; many economic spp.
spherical aberration. Inherent defect in an optical system (including microscope objectives) whereby all light rays traversing it are not refracted to precisely the same degree, resulting in an indistinct image and some shape distortion; this defect is corrected to a large degree by cutting off the marginal rays, changing the shape of the lens, and combining several lenses (in an objective) having slightly different physical properties; achromatic objectives are corrected spherically for one color, and apochromatic objectives are corrected spherically for two colors.
spherical symmetry. Condition obtaining in certain Protozoa in which any plane drawn through the center point of the cell will divide the cell into two similar halves.
Sphex. Common genus of mud-dauber wasps.
sphincter. Muscle having its fibers in a circular arrangement around an opening so as to close that opening upon contraction.
sphincter ani. Either of the two anal sphincter muscles, especially in man.
sphincter colli. Thin, flat, ventral neck muscle in primitive tetrapods.
sphincter of choledocus. Sphincter muscle of the bile duct.
sphincter of Oddi. Common sphincter muscle for the bile duct and the pancreatic duct at their juncture with the duodenum.
Sphindidae. Family of small beetles of Holarctic distribution.
Sphingidae. Family of large, robust, attractive moths; hawkmoths, hummingbird moths, sphinx moths; wings large, narrow, and pointed; mouth parts very long; hover and feed from flowers at dusk; examples are Sphinx and Protoparce.
Sphinx chersis. Common sp. of hawkmoth; see SPHINGIDAE.
sphygomograph. Sphygmometer; any instrument used to measure the force and frequency of the pulse.
sphygmomanometer. Instrument used to measure arterial blood pressure; most frequently, pressure is applied around a part of the body, usually a limb, by means of a pneumatic bag enclosed by a non-extensible cuff or bandage, so that this pressure is exerted through the various tissues to the artery whose blood pressure is being determined; the condition of blood flow and pressure within the artery may then be established in rela-

tion to the externally applied pressure.

sphygmometer. Sphygmograph.

Sphyraenidae. Family of marine percomorph fishes which includes the barracudas; carnivorous, voracious, pike-like spp. of warm seas; to 6 ft. long; some edible.

Sphyranura osleri. Common monogenetic trematode parasitic on the gills of _Necturus._

Sphyrapicus. Common U.S. genus of sapsuckers in the Family Picidae; feed on insects and vegetable material; sometimes damage trees by drilling through the cambium in order to feed on sap.

Sphyrion. Genus of bizarre copepods parasitic on marine fishes; growth form having the appearance of a tumor.

Sphyrna. Genus of hammerhead sharks; head with a very large lateral lobe on each side with eye at tip; mostly tropical and subtropical; to 12 ft. long.

spicule. 1. Needle-like mass of bone, especially in actively forming bone tissue. 2. One of the minute calcium carbonate or silicon dioxide bodies which form the supporting skeleton of many sponges, radiolarians, sea cucumbers, etc.; spicules may be irregular, simple and needle-like, or multiradiate and complex.

spider. See ARANEAE.

spider beetles. See PTINIDAE.

spider bugs. See EMESA.

spider conch. See LAMBIS.

spider crab. Any of numerous marine crabs (Brachyura) having especially long legs; see especially LIBINIA.

spider mite. See TETRANYCHIDAE.

spider monkey. See ATELES.

spider wasp. Any of numerous spp. of wasps which kill or paralyze spiders and place them in their brood cells as a food source for the larval wasps; see PSAMMOCHARIDAE.

Spilogale. Genus of nocturnal carnivores in the Family Mustelidae; includes the several spp. of spotted skunks of the U.S. and Mexico; about as large as a half-grown housecat; dorsal color pattern in conspicuous stripes or connected black and white spots; black ventrally; feed on insects, small vertebrates, and fruit; the offensive odorous secretion is sprayed from two perianal glands.

spinal accessory nerve. One of the eleventh pair of cranial nerves in higher vertebrates; innervates the pharynx, larynx, and trapezius muscle; a motor nerve.

spinal ataxia. Degeneration of sensory nerve tracts; inherited as either a dominant or recessive.

spinal canal. Long cavity formed by the dorsal arches of the vertebrae; contains the spinal cord.

spinal column. Vertebral column.

spinal cord. Thick longitudinal bundle of nerve fibers extending from the brain posteriorly along the dorsal side in vertebrates; (the term is not correctly used for invertebrates).

spinal ganglion. Dorsal root ganglion.

spinal nerve. One of the paired peripheral nerves arising from the nerve cord in vertebrates; such nerves originate as a dorsal root and a ventral root which fuse to form the nerve proper in the immediate vicinity of the spinal cord.

spina sterni communis. Anterior projection of the sternum.

spindle. Metaphase arrangement of chromosomes, spindle fibers, and poles in a mitotic division.

spindle attachment. Centromere.

spindle cells. Spindle-shaped or pointed cells having large nuclei in frog blood; presumed to have a role in clotting.

spindle fibers. Minute fibrils making up a portion of the mitotic spindle during mitosis and meiosis.

spine. 1. Pointed projection on a bone. 2. A stiff, pointed, external process or appendage. 3. Vertebral column.

spinedace. See LEPIDOMEDA and MEDA.

spine-ribbed Murex. _Murex fulvescens,_ a handsome, spiny marine gastropod of N.C. to Tex.; to 6 in. long.

spinetail. See PRUNELLIDAE.

spiniferous. Spiny; bearing one or more spines.

spinner. Adult mayfly.

spinneret. 1. One of four to six blunt conical structures on the posterior ventral surface of typical female spiders; bear openings of the silk glands. 2. Minute pore at the posterior end of certain aquatic nematodes; serves as an exit for a sticky attachment secretion.

spinoblast. Free-floating, spiny type of statoblast; e.g. in _Pectinatella._

spinocerebellar tract. Nerve tract transmitting impulses from the spinal cord to the cerebellum.

spinodiencephalic tract. Nerve tract transmitting impulses from the spinal cord to the diencephalon.

spinomedullary tract. Nerve tract

transmitting impulses from the spinal cord to the medulla oblongata.

spinose. Bearing many spines.

Spinosphaera. Genus of Polychaeta.

spinous. 1. Bearing many spines. 2. Having the form of a spine.

Spintharus flavidus. Common yellowish spider of eastern U.S.; in the Family Theridiidae; to 4 mm. long.

spinule. Small spine.

Spinulosa. One of the orders in the the echinoderm Class Asteroidea; marginal plates small; dermal branchiae on both surfaces; pedicellariae almost lacking; Henricia and Solaster common.

Spinus pinus. Pine siskin, a small, heavily-streaked, brown finch with a little yellow on wings and tail; evergreen forests. See FINCH.

spiny anteater. Echidna.

spiny dogfish. See SQUALUS.

spiny dormouse. See PLATACANTHOMYIDAE.

spiny eel. See HETEROMI.

spiny-headed worm. Any member of the Phylum Acanthocephala.

spiny jingle. Anomia aculeata, a jingle covered with prickly scales on the upper valve; East Coast.

spiny-legged flies. See SEPSIDAE.

spiny lobster. See PANULIRUS.

spiny-rayed fin. Fin of a fish which is supported entirely or partially by sharp, stiff, non-pliable rays.

Spio. Common genus of Polychaeta.

Spiochaetopterus. Genus of small, tube-dwelling Polychaeta.

spiracle. 1. Stigma; external opening of a trachea in insects and most other terrestrial arthropods; usually located on the lateral surface of certain thoracic and abdominal segments. 2. One of a pair of dorsolateral openings on the head of elasmobranch and ganoid fishes for the passage of respiratory water; embryologically, derived from a pair of gill slits. 3. The excurrent aperture of a tadpole gill chamber. 4. Nasal aperture of Cetacea.

spiracular disc. Lobed disclike structure surrounding the spiracles at the posterior end of many aquatic Diptera larvae, especially Tipulidae.

spiral cleavage. Type of early embryonic cleavage in which the cleavage is diagonal to the polar axis, resulting in successive alternating tiers of cells; spindles of third cleavage, for example, are inclined diagonally, so that the resulting upper tier of four cells is displaced sideways, resting in the angles between the lower and larger cells; in polyclads, annelids, and mollusks (except cephalopods).

spiral organ. Organ of Corti.

spiral valve. 1. Flat, twisted valvelike structure in the truncus arteriosus of certain amphibians; functions in guiding a large portion of poorly-oxygenated blood from the ventricle to the pulmocutaneous arches. 2. Spiral partition in the intestine of elasmobranchs, ganoids, and lungfish; slows passage of food and facilitates absorption.

Spirastrella. Genus of marine sponges (Subclass Monaxonida); irregularly cylindrical; with many associated small invertebrates.

spire. Upper part of a coiled gastropod shell.

spireme. Tangle of long, discrete, linear chromosomes at the beginning of a mitotic prophase.

spire shell. Any of many small to minute marine snails with an umbilicus and horny operculum; cosmopolitan.

Spirifer. Genus of fossil Brachiopoda.

spirillum. 1. Any member of the bacterial genus Spirillum. 2. Any thread-like spiral-shaped microorganism.

Spirillum. Genus of motile bacteria consisting of rigid spiral-shaped spp. with 5 to 20 terminal flagella; to 50 microns long; common in polluted and stagnant waters; a few spp. are minor pathogens in animals.

Spirinchus. Small genus of smelts; enter Pacific coastal streams to spawn.

Spirobolus. Narceus.

Spirobranchus. Genus of Polychaeta.

Spirochaeta. Genus of large, slender, non-flagellated organisms usually classified with the bacteria; in sewage and stagnant waters.

spirochaete. Any of many kinds of long, nonflagellated, spiral, microscopic organisms; usually classified with bacteria; some are pathogenic, such as Treponema pallidum which produces syphilis.

Spirochona. Genus of chonotrich Ciliata; usually attached to marine invertebrates.

Spiromonas. Genus of small, elongated, spirally twisted, biflagellate Protozoa; stagnant fresh waters.

spironeme. Contractile coiling thread in the stalk of certain Ciliata.

Spirontocaris. Large genus of small marine shrimps; often placed in Hippolyte.

Spirorbis. Genus of minute marine Polychaeta; secrete an irregular calcareous tube.

Spirostomum. Genus of large, greatly elongated, free-living Ciliata occurring in both fresh and salt waters.

Spirotricha. Order of ciliate Protozoa with an adoral zone of membranelles; peristome not extending beyond the general cell surface; Stentor and Euplotes common.

Spirotrichonympha. Genus of complex pyriform flagellates inhabiting the digestive tract of termites; with many flagella arranged in spiral bands.

Spirozona. Genus of spindle-shaped holotrich Ciliata; sapropelic, fresh water.

Spirula. Genus of cephalopod mollusks in the Suborder Decapoda; shell partly internal and partly external, composed of a loose spiral of two or three coils; the common sp. is only 1 in. in diameter; deep waters in the tropics, but shells are occasionally washed ashore on the Atlantic Coast.

Spirurida. Order of phasmid nematodes; esophagus with only two parts, the anterior being more or less cylindrical and muscular and the posterior being long, cylindrical, and glandular; adults parasitic in tissues or digestive tract of vertebrates.

Spirurina. Suborder of spirurid nematodes; immature stages usually in insects; examples are Wuchereria, Onchocerca, and Loa.

Spisula. Genus of surf clams; large, heavy spp. to 7 in. long.

spitting spiders. See SCYTODIDAE.

spittle bug. See CERCOPIDAE.

Spiza americana. Dickcissel; a songbird of central U.S.; common in grassy fields and pastures.

Spizella. See SPARROW.

splake. Hybrid offspring between a lake trout and a brook trout; often fertile.

splanchnic. Pertaining to viscera, especially with reference to embryonic development.

splanchnic nerves. In man, three large nerves on each side of the body; originate from thoracic sympathetic ganglia numbers five to twelve; innervate visceral plexuses.

splanchnocoel. Coelomic cavity formed within the main mesodermal masses of a coelomate embryo.

splanchnocranium. That part of the skull comprising the jaws and their attachments; derived from gill arch skeleton. Cf. NEUROCRANIUM.

splanchnopleure. In vertebrate embryos, a mass of tissue on each side consisting of endoderm plus the inner or lower layer of mesoderm of the primary mesoblast; forms intestinal musculature and connective tissue.

spleen. Mass of lymphoid tissue imbedded in the abdominal mesentery of vertebrates; it is interposed in the blood circulatory system and is an important source of lymphocytes; also serves to remove foreign bodies and old erythrocytes from the blood stream.

splenius capitis. Muscle which originates on the anterior vertebrae and inserts on the back of the head on one side of the midline; rotates and extends head and neck and flexes sideways.

splenocyte. Monocyte found or originating in the spleen.

splittail. See POGONICHTHYS.

splitter. 1. Taxonomist who tends to divide his material to a greater extent than other specialists in the same field, such as dividing one genus into two or more genera, etc. 2. Taxonomist who recognizes small morphological variations as discrete taxonomic entities.

Spondylomorum. Genus of holophytic flagellates; tightly arranged colonies of four-flagellated cells; fresh water.

Spondylus. See THORNY OYSTER.

sponge. 1. Common name of any member of the Phylum Porifera. 2. Commercial product obtained by processing and cleaning away all but the spongin skeleton of some sponges.

sponge crab. See DROMIA.

Spongia. Genus of sponges in the Class Demospongiae; the important commercial sponge.

Spongilla. Common genus of fresh-water sponges.

spongilla flies. See SISYRIDAE.

Spongillidae. Family of sponges in the Class Demospongiae; includes all fresh-water sponges; grow as delicate encrusting or branching masses in clean lake waters and slow streams; some such masses may attain a volume of one cubic foot; Spongilla common and widely distributed.

spongin. Fibrous network forming part or all of the skeleton of many sponges; essentially the entire skeleton of commercial sponges is composed of this substance; a sulfur-containing protein.

spongioblast. Special mesenchyme cell of sponges which secretes spongin.

spongoblast. Spongioblast.

spongocoel. 1. Paragaster; main cavity in the center of a simple or complex sponge. 2. Collectively, those canals and cavities of a complex sponge which convey water from flagellated chambers to the outside.

Spongodes. Genus of anthozoan coelenterates; colony branching, dendritic, and without a central axis; polyps soft and leathery.

Spongomonas. Genus of fresh-water biflagellate Protozoa; cells imbedded in granulated gelatinous masses.

spongy bone. Type of bone occurring especially at the ends of long bones; consists of a framework of interlaced and anastomosing bars of bone arranged in such a way as to give maximum strength and rigidity.

spontaneous generation. Abiogenesis; old superstition which maintained that living animals arose spontaneously from dead materials; worms were thought to have originated from mud, flies and maggots from dead animal carcasses, etc.

spoonbill. 1. Any of several large wading birds having a broad spoonlike bill, especially the roseate spoonbill, Ajaia ajaja, a bright pink heron-like bird, found in the southern states and south well into South America; in the Family Threskiornithidae. 2. See POLYODONTIDAE.

spoon shell. See PERIPLOMA.

spore. In sporozoan Protozoa, a thin- or thick-walled caselike structure in which a few (often two, four, or eight) sporozoites are formed by multiple fission.

sporoblast. In the life history of many Sporozoa, a series of one, two, four, or many bodies which form in an oocyst; each sporoblast secretes a membrane and transforms into a spore.

sporocyst. 1. Saclike larval stage in the life history of many trematodes; usually imbedded in the viscera of a snail host; originates from a miracidium and usually produces rediae by asexual internal proliferation. 2. In certain Protozoa, a protective envelope in which spores are formed.

sporogony. Series of stages in the life history of sporozoan Protozoa; begins with the fusion of gametes and ends with the production of sporozoites, e. g. the asexual stages of Plasmodium in the stomach wall of the Anopheles mosquito.

Sporonchulus. Genus of microscopic, free-living nematodes.

sporont. 1. Zygote formed during protozoan life histories, especially in sporozoans. 2. Gamont; in certain sporozoans, a gametocyte which develops into one or more sporoblasts.

Sporozoa. One of the five classes of the Phylum Protozoa; locomotor organelles absent; all spp. found as parasites of members of all metazoan phyla; cell structure simple, although some of the reproductive processes are complex.

sporozoite. One of the stages in the life history of many sporozoan Protozoa; released from spores and usually penetrate new host cells.

sport. Organism showing a mutation, especially a pronounced mutation.

sporulation. Type of multiple fission in certain Protozoa; preceded by rapid succession of nuclear divisions; e. g. in Plasmodium.

spotted bat. See EUDERMA.

spotted cowry. Cypraea exanthema; a white and chestnut-colored cowry of the Caribbean; to 4 in. long.

spotted dolphin. See STENELLA.

spotted fever. See ROCKY MOUNTAIN SPOTTED FEVER.

spotted owl. See STRIX.

spotted sandpiper. See ACTITIS.

spotted toad. See BUFO.

spotted turtle. Small fresh-water turtle of the eastern states; see CLEMMYS.

spp. Abbreviation for species (plural).

sprain. Wrenching of a joint with more or less damage to muscle and/or tendon fibers.

sprat. Small herring-like marine fish of European waters; canned as sardines or anchovies in Norway; Clupea sprattus.

spring azure. Lycaena pseudoargiolus, a small blue butterfly; common in U. S.

springbok. See ANTIDORCAS.

springbuck. See ANTIDORCAS.

spring cankerworm. See PALEACRITA.

springfish. See EMPETRICHTHYS, CRENICHTHYS, and CHOLOGASTER.

spring frog. See RANA CLAMITANS.

spring overturn. Variable period during the spring when a lake is homothermous and is subject to complete top-to-bottom circulation by wind action.

spring peeper. See HYLA.

springtail. Common name of small insects in the Order Collembola; able to leap prodigiously by means of a triggered abdominal springing organ.

spring tide. Exceptionally high tide

which occurs twice per lunar month, when there is a new or full moon, and the earth, sun, and moon are in a straight line.

spruce beetle. 1. Generally, any beetle which attacks spruce trees. 2. Specifically, beetles of the genus Dendroctonus whose larvae attack the cambium and outer xylem.

spruce borer. Adult beetle or larva of of certain beetles which attack and bore into spruce trees.

spruce budworm. Harmologa fumiferana, a moth in the Family Tortricidae; larvae are serious pests of evergreens in the northeastern states and often defoliate trees.

spruce grouse. See CANACHITES.

sprue. Chronic tropical disease marked by diarrhea, ulcers of the digestive tract, and a smooth, shining tongue; produced by a dietary deficiency of folic acid.

spur. 1. Long, sharp spine on the distal end of the second tibia of a honey bee; used to dislodge wax from the ventral surface of the abdomen. 2. Any stiff, sharp spine.

SQH. Status quo hormone found in certain early insect larvae and nymphs; secreted by the corpora allata and inhibits premature metamorphosis.

squab. 1. Large nestling pigeon. 2. Any nestling or recently hatched bird.

Squali. Suborder of selachian fishes which includes the true sharks; body spindle-shaped; tail heterocercal; five to seven pairs of gill slits at least partly lateral; upper eyelids free; pectoral fins not especially large; familiar genera are Chlamydoselachus, Mustelus, Carcharias, Rhineodon, Squalus, Pristiophorus, and Pliotrema.

Squalida. Superfamily of sharks in the Suborder Squali; e.g. dogfish sharks.

Squalus acanthias. Common temperate and subarctic shark found chiefly in inshore waters; spiny dogfish, grayfish; usually less than 3 ft. long.

squama. 1. Broad scale-like exopod of the second antenna of certain crustaceans. 2. Squame, calypter. 3. Scale-like structure.

Squamata. Very large order of reptiles which includes lizards and snakes; skin with epidermal scales or plates; vertebrae usually procoelous; male copulatory organ double and eversible; anus transverse.

squamate. Scaly.

squamation. Arrangement of scales.

squame. Squama, calypter.

squamosal. Bone of the posterior side wall of the skull on either side in most vertebrates.

squamous. 1. Scale-like. 2. Composed of scales. 3. Covered with scales.

squamous epithelium. Epithelium composed of very flat thin cells, such as the layer of cells lining human blood vessels.

squamula. Calypter.

squash beetle. See DIABROTICA.

squash bug. See COREIDAE.

Squatarola squatarola. Black-bellied plover, a widely distributed sp. which breeds on tundra and winters in the U.S., South America, South Africa, India, and Australia.

Squatina. Genus of sharks which feed on animals of the sea bottom; angel sharks, monkfishes.

Squatinella. Genus of loricate rotifers; head with a wide transverse shield; lorica with one or two very long spines.

squawfish. See PTYCHOCHEILUS.

squeaker. 1. An Australian crow shrike. 2. Very young squab. 3. Stridulating aquatic beetle. 4. See BETTONGIA.

squid. Type of carnivorous cephalopod mollusks; body globose to elongated and cigar-shaped; eight arms and two tentacles originate from the head and surround the mouth; the former are shorter and bear an abundance of suckers, but the latter bear suckers only near the tips; with a pair of rounded or triangular posterior fins on the sides of the enveloping mantle; with a chitinoid quill, a pen-shaped skeletal structure imbedded in the mantle dorsally; Loligo, Illex, and Rossia common.

squill. Crustacean in the genus Squilla; see HOPLOCARIDA.

Squilla. See HOPLOCARIDA.

squirrel. See SCIURIDAE.

squirrel fish. Any of several brightly colored (usually red) marine fishes; see especially BERYCOMORPHI and HOLOCENTRIDAE.

squirrel-monkey. See SAIMIRI.

stable fly. See STOMOXYS.

stable tea. Culture solution made by boiling dried horse manure in water; commonly used for protozoans, rotifers, and microcrustaceans.

stadium. 1. Instar; interval between two ecdyses, especially in insects, but equally applicable to other arthropods, nematodes, etc. 2. Any one stage in the life history of an insect, e.g. pupa. Pl. stadia.

stag. Hart or mature male deer.

stag beetle. See LUCANIDAE.

staggard. Fourth-year male European red deer.

staghorn coral. See ACROPORA.

Stagmomantis. See MANTODEA.

staircase shell. See WENTLETRAP.

staircase phenomenon. Treppe.

stalk-eyed flies. See DIOPSIDAE.

standard deviation. Special form of expressing the average deviation from the mean for a group of values; the square root of the sum of the squared deviations from the mean divided by the sample size.

standard error of the mean. Standard deviation divided by the square root of the sample size.

standard length. Distance from the tip of the snout to the end of the last vertebra in fishes.

standing crop. Total number or weight of living organisms momentarily present in an environmental unit.

Stanley, Wendell Meredith. American biochemist (1904-); well known for his studies on the submicroscopic structure of proteins and viruses; shared the Nobel prize in chemistry in 1946 for his isolation and crystallization of the tobacco mosaic virus.

Stannius, Friedrich Hermann. German zoologist (1808-1883); made important contributions to comparative vertebrate anatomy and physiology.

stapedial artery. Artery in reptiles and amphibians; supplies the jaws and outer parts of the head.

stapedius. Very small muscle of the mammalian middle ear; connects the tympanum and stapes.

stapes. Innermost of three minute auditory ossicles in the middle ear of mammals; has roughly the shape of a stirrup and transmits vibrations to the vestibular window and endolymph of the cochlea; similar in origin to the columella of lower tetrapods and the hyomandibula of fishes.

Staphylinidae. Large family of elongated beetles; rove beetles; elytra short; larvae and adults predators or scavengers; Creophilus, Myrmedonia, and Termitomimus common.

Staphylococcus. Genus of spherical bacteria occurring singly or in clusters; common in dairy products and water; some forms are parasitic and produce skin infections such as boils and abscesses.

starch. White tasteless solid polysaccharide occurring as granules in many plant tissues; especially abundant in seeds, bulbs, tubers, etc. in such plants as rice, wheat, potatoes, corn, and beans; general formula for a typical starch is $(C_6H_{10}O_5)_{25}$.

star coral. See FAVIA.

starfish. Sea star; member of the Class Asteroidea (Phylum Echinodermata).

stargazer. Any of several ugly, rough-skinned, tropical marine fishes; mouth, nostrils, and eyes high on the head, making it possible for them to lie almost buried in the sand awaiting their prey; some have electric shocking organs in the head; see URANOSCOPIDAE.

star-nosed mole. See TALPIDAE and CONDYLURA.

stasis. 1. Lack of proper peristalsis of of the intestines, and feces retention. 2. Decrease in the velocity of blood through an organ or part. 3. Cessation of flow of any body fluid through a duct, organ, or vessel. 4. Cessation of growth of a part of the body.

stationary phase. See LOGISTIC CURVE.

statoblast. Resistant biconvex sclerotized capsule composed of two valves and containing an undifferentiated mass of germinative cells; the surface of the valves is sculptured and contains peripheral air cells and often spines or barbs; such structures are produced on the funiculus of most fresh-water Bryozoa by asexual budding; upon disintegration of the colony, they sink or float about freely, and when environmental conditions are favorable (usually in spring) a statoblast germinates to give rise to a new polypide and eventually a whole colony.

statocyst. Lithocyst, otocyst; small organ of equilibration found in many invertebrates; structure highly variable but often an open pit, vesicle, or closed chamber having one to many special cells (lithocytes) each of which contains a movable concretion (statolith) composed chiefly of calcium carbonate; each lithocyte is supplied with nerve fibers; other types of statocysts consist of capsules of ciliated cells containing a fluid in which one or more statoliths are suspended.

statolith. See STATOCYST.

statorhab. Short tentacular process bearing a statolith in the Trachymedusae.

Stauridia. Genus of marine coelenterates in the Suborder Anthomedusae;

mostly British seas.

Staurojoenina. Genus of complex flagellates occurring as symbionts in the digestive tract of termites; pyriform to cylindrical; four tufts of flagella at anterior end; digest cellulose and make it available for the host.

Stauromedusae. One of the orders of the coelenterate Class Scyphozoa; attached to the sea bottom by means of an adhesive organ at the end of a stalk extending from the top of the bell, the whole animal having a goblet shape; eight marginal lobes on the upward-facing bell, and each lobe bears a cluster of knobby tentacles; Lucernaria common.

Stauronereis. Genus of Polychaeta.

Staurophora. Genus of marine coelenterates in the Suborder Leptomedusae; medusae to 8 in. in diameter.

Staurophrya. Genus of fresh-water suctorian Protozoa; six processes; tentacles not knobbed.

steady state. In biological systems, a condition of maintenance of (relative) constancy of a system, where the system may be intracellular, cellular, or even up to the community in size and complexity; usually implies constancy of concentrations and organization of materials, achieved by various mechanisms, such as differential permeability of cellular membranes, electrical forces at membranes and interfaces, and active transport of substances into and out of cells or their parts, with the maintenance depending directly or indirectly on appropriate, readily available, metabolically released sources of energy; at the level of a whole organism or community of organisms it implies the maintenance of incoming and outgoing substances and energy sources; not comparable to a true chemical equilibrium.

steamer clam. See MYA.

steapsin. Lipase of pancreatic juice.

Steatoda borealis. Common orange-brown spider of northern half of U.S.; in the Family Theridiidae.

steatopygia. Having enlarged buttocks (a human anthropological trait).

Steatornis. See GUACHARO.

Steatornithidae. See GUACHARO.

steelhead trout. West Coast sea-run race of the rainbow trout.

steenbock. Steinbok; see RAPHICERUS.

Steenson, Nils. Nicolaus Steno; Danish scientist and Catholic priest (1638-1686); investigated glands and the muscular system; pioneer of paleontology and geological stratification.

Steganopus tricolor. Wilson's phalarope; breeds in marshes of northern Great Plains and Great Basin; winters in southern South America.

Steggoa. Genus of Polychaeta.

Stegobium panacea. Drugstore beetle in the Family Anobiidae; feeds on many dried animal and vegetable products.

Stegocephalia. One of the three subclasses of amphibians; cranium and cheeks covered with bony plates; fossil spp. often with ventral bony plates on trunk; most spp. extinct and primitive.

Stegocephalus. Genus of marine Amphipoda.

Stegodon. Genus of Pliocene and Pleistocene elephants.

Stegolophodon. Genus of Miocene and Pleistocene elephants.

Stegomyia. Old genus name for the yellow-fever mosquito (Aedes).

Stegophryxus. Genus of degenerate isopods; parasitic on marine shrimps and anomurans.

Stegosauria. Suborder of Jurassic and Lower Cretaceous dinosaurs in the Order Ornithischia; heavy, upland quadrupeds with very small head and large plates, spikes, and scutes forming clumsy protuberances and armor; Stegosaurus more than 20 ft. long.

Stegosaurus. See STEGOSAURIA.

Stegoselachii. Order of extinct placoderm fishes; body depressed and covered with small denticles; pectoral fins very large; marine; Gemuendina common.

steinbok. See RAPHICERUS.

Stelgidopteryx ruficollis. Rough-winged swallow; a brown-backed sp. found throughout the U.S.

stellate. Star-shaped.

Stellatta. West Coast genus of feltlike sponges with stinging spicules.

stellate ganglion. 1. In cephalopods, a large star-shaped posterior ganglion on the internal surface of the mantle on each side. 2. In vertebrates, a ganglion formed by the fused inferior cervical ganglion and the first sympathetic ganglion.

Steller sea cow. See HYDRODAMALIS.

Steller sea lion. See EUMETOPIAS.

Steller's jay. Cyanocitta stelleri; a jay of western coniferous forests; head and neck blackish, back dark brown, rest of body mostly blue.

stem mother. Female resulting from a fertilized egg and giving rise to a series of parthenogenetic female generations; e.g. some aphids and cladocerans.

Stenella plagiodon. Spotted dolphin of the Atlantic and Pacific oceans.

Stenelmis. Large genus of elmid beetles.

Steno. Genus of long-beaked dolphins; pelagic fish-eating spp. of tropical and temperate seas.

Steno, Nicolaus. Same as NILS STEENSON.

stenobath. Animal which is restricted to only a narrow depth range in the oceans, e.g. acorn barnacles, reef corals, commercial sponges, and deep-sea crinoids. Cf. EURYBATH.

stenobenthic. Living on the sea bottom over only a narrow depth range. Cf. EURYBENTHIC.

Steno bredanensis. Long-beaked dolphin; warmer parts of Atlantic and Indian oceans.

Stenocaris. Genus of marine Copepoda.

Stenochironomus. Genus of tendipedid midges.

Stenodus mackenzii. Inconnu; a large oily fresh-water fish in the Family Coregonidae; chiefly in the MacKenzie River system but also in Alaska and nearby Siberia.

stenogastric. Pertaining to a narrow abdomen, as in certain insects.

stenohaline. Able to tolerate only a certain narrow range of dissolved salts in the environment, e.g. certain jellyfish and many fishes. Cf. EURYHALINE.

stenohygric. Pertaining to an animal which can ordinarily maintain itself over only a narrow range of relative humidity, e.g. water buffalo (stenohygric wet) and camel (stenohygric dry).

stenoky. Relatively narrow range of ecological conditions tolerated by a particular sp.; reef corals and rain forest monkeys are examples. Adj. stenokous. Cf. EUROKY.

Stenolaemata. One of the three classes in the Phylum Bryozoa; lophophore circular, zooecium cylindrical and limy; marine; Crisia and Tubulipora typical.

Stenonema. Common genus of Ephemeroptera.

Stenopelmatidae. Family of Orthoptera which includes the camel crickets, sand crickets, and cave crickets; robust, brown or gray, usually wingless, and often with a high arched back; antennae long; widely distributed, mostly nocturnal.

stenophagous. Subsisting on only one or a few kinds of food.

Stenopidae. Family of shrimps which live in the paragaster of certain marine sponges.

Stenopodidea. Small suborder of crustaceans in the Section Natantia; banded shrimps; third legs especially long and robust, all other legs thin and weak; especially common in tropical tidepools; Stenopus typical.

stenopodium. Long, slender type of crustacean appendage, especially as contrasted with the broad, flat phyllopodium type of appendage. Pl. stenopodia.

Stenopus. Genus of tropical shrimps noted for their bizarre coloration.

Stenosaurus. Genus of Jurassic alligators.

stenosis. Failure of the heart valves to open fully.

Stenosmylus. Genus of neuropteran flies in the Family Osmylidae.

Stenostomum. See RHABDOCOELA.

stenotele nematocyst. Large spherical nematocyst; upon discharge, a coiled tubule penetrates the prey tissues and secretes a paralyzing fluid; the tubule has an enlarged base bearing spinules.

stenothermal. Pertaining to an organism which is able to maintain itself over only a relatively narrow range of temperature; e.g. tropical jellyfish and cut-throat trout. Cf. EURYTHERMAL.

Stenothoë. Genus of marine Amphipoda.

Stenotomus versicolor. Scup or porgy; common inshore deep-bodied food fish of the Atlantic Coast.

stenotopic. Having a narrow range of suitable ecological conditions; a sp. character.

stenozonal. Pertaining to an animal which is restricted to a narrow range of altitude in mountainous country; e. g. mountain goat, tuft-eared squirrel, and snowshoe rabbit. Cf EURYZONAL.

Stentor. Genus of large trumpet-shaped ciliate protozoans commonly found in stagnant fresh waters.

stentorin. Blue pigment in the protozoan Stentor.

Stephanasterias. Genus of small northern sea stars.

Stephanauge. Genus of small sea anemones in the Order Actiniaria; commonly attached to the stems of sea pens.

Stephanoceros. Genus of sessile fresh-water rotifers; corona drawn out into long pointed lobes bearing whorls of setae.

Stephanodrilus obscurus. Small leech-like oligochaete commensal and parasitic on western U.S. crayfishes; in the Family Branchiobdellidae.

Stephanolepis. See MONACANTHIDAE.

Stephanoprora. Genus of flukes found in the loon and gulls.

Stephanosella. Genus of marine Bryozoa.

Stephanosphaera. Genus of green, freshwater, biflagellate Protozoa; usually eight cells per colony.

Stephanurus dentatus. Swine kidney worm, a nematode parasite which matures in the kidneys of swine.

steppe. 1. In the narrow sense, the level treeless semiarid plains of south-central U.S.S.R.; formerly used for grazing but now mostly for grain crops. 2. In the broad sense, any treeless grassy plains of temperate to tropical latitudes, including prairies, pampas, veldt, and Russian steppes.

stercobilin. Brown fecal pigment resulting from the action of bacteria on bilirubin; $C_{33}H_{46}O_6N_4$.

stercóral pocket. Caecum which connects with the rectum of many arachnids; serves as storage place for feces.

Stercorariidae. Family of high-latitude maritime birds in the Order Charadriiformes; includes the skua and jaegers; see STERCORARIUS and CATHARACTA.

Stercorarius. Genus in the Family Stercorariidae; includes the jaegers, several spp. of large maritime birds of the Northern Hemisphere; often pursue gulls and terns, forcing them to disgorge or release their prey; nest on tundra and feed upon mammals and the young of other birds.

Stereobalanus. Genus in the Phylum Enteropneusta.

stereoblastula. Early blastula larva in which the blastocoel remains filled with cells, as in Acanthocephala, and certain annelids and mollusks.

Stereochilus marginatus. Many-lined salamander in the Family Plethodontidae; tongue attached at front; adults aquatic in spring; larvae aquatic; coastal plain of Va. to Ga.

stereogastrula. 1. Parenchymula; a larval type found in some sponges; modified gastrula larva in which the invaginated flagellated cells form a solid internal mass. 2. Any solid gastrula, as in many coelenterates.

Stereolepis gigas. Giant Pacific sea bass or Pacific jewfish; dark, heavily-built sp. of the Calif. coast and southward; to 7 ft. long.

stereoline glutinant. See NEMATOCYSTS.

stereoscopic microscope. See BINOCULAR MICROSCOPE.

stereoscopic vision. Binocular vision.

sterile. 1. Pertaining to barren or infertile organisms; incapable of reproducing or giving rise to gametes, owing to some morphological or physiological abnormality. 2. Pertaining to any tissue incapable of producing gametes. 3. Free of living organisms, especially microorganisms, e.g. a sterile fluid.

sterilization. 1. Process of killing all living cells, especially bacteria, by means of various physical and chemical agencies. 2. Rendering an organism incapable of reproducing; in male human beings it is usually done by cutting the two vasa deferentia through two small scrotal incisions, and in females the oviducts are usually cut and tied off.

sterilize. 1. To make sterile or deprive of the power of reproduction by chemical or physical agents, or by the surgical removal of essential reproductive tissues or organs. 2. To free of living organisms, especially by physical or chemical means.

sterlet. Small sturgeon of the Caspian Sea and its inlets; highly esteemed for its flesh and roe.

Sterna. Common and widely distributed genus of birds in the Family Laridae; includes some of the terns; slender, narrow-winged spp. with great powers of flight; bill slender and tapered, tail forked; smaller than gulls; especially abundant along seacoasts, but often far inland in large numbers; obtain food by plunging headfirst into water; S. albifrons is the least tern which breeds on Gulf, Atlantic, and Calif. coasts.

sternal artery. Artery in the crayfish and many other Malacostraca; brings blood from the base of the dorsal abdominal artery to the ventral thoracic and ventral abdominal arteries.

sternal gills. Thin-walled finger-like gills on the sternites of some of the thoracic segments in certain amphipods.

sternal rib. Distal cartilaginous portion of the thoracic ribs in many reptiles.

sternal sinus. Extensive cavity in the ventral portion of the thorax of the crayfish and many other Malacostraca; blood comes into this sinus and then passes into the gills.

Sternaspis. Genus of short, distended, grublike Polychaeta.

sternebra. Discrete segment of the sternum, especially early in life.

sternite. Ventral plate of an arthropod segment.

sternocleidomastoideus. Muscle which inserts on the mastoid and occipital area and originates on the sternum and clavicle; depresses and rotates head and flexes head and neck.

sternohyoid. Ventral muscle in the region of the sternum, larynx, and jaw symphysis of tetrapods.

sternomastoid. Small flat muscle in some tetrapods; derived from the trapezius; in man, located between the sternum and the mastoid region.

Steropteryx. Genus of small, bizarre, deep-bodied, luminescent marine fishes with very large eyes; hatchet fishes.

Sternorrhyncha. One of the two suborders in the insect Order Homoptera; plant lice, scale insects, mealybugs, etc.; beak arising between first coxae; antennae usually long.

Sternotherus. Common genus of musk turtles; in sluggish waters and ponds of eastern half of U.S.; carapace to 4 in. long; brownish coloration.

sternum. 1. Breastbone; a midventral chest bone of tetrapods to which the ventral ends of most of the ribs are attached; the pectoral girdles are attached at its anterior end in some forms. 2. The chief ventral plate of most of the body segments of arthropods.

steroid. Any of a large variety of chemical compounds consisting of four hydrocarbon rings, three of them with six carbons linked together, and one with five, to which various substituents are attached at particular locations; the rings are joined by the sharing of pairs of carbon atoms; examples are the vitamins D, bile acids, sex hormones, and adrenal cortical hormones.

sterol. Steroid which possesses an -OH group attached at some position, thus making the compound an alcohol, plus a long hydrocarbon side-chain; many types, widely distributed among plants and animals.

stethoscope. Any instrument used to convey sounds in the chest or other parts of the body to the ears of another person.

Stewart's organs. Coelomic pouches associated with Aristotle's lantern in some Echinoidea; serve as internal gills.

Sthenelais. Genus of flattened burrowing Polychaeta.

Stichaeidae. Pricklebacks; family of cold-water bottom-dwelling marine fishes; body long, slender; dorsal fin very long, spinous; head small; Lumpenus common.

Stichocotyle. Genus of slender trematodes in the Order Aspidocotylea; parasitic in spiral valve and bile ducts of skates.

Stichopus. Genus of Pacific Coast sea cucumbers; to 18 in. long.

Stichostemma. Prostoma.

Stichotricha. Genus of marine and freshwater hypotrich Ciliata; slender, ovoid to fusiform.

stick insect. See PHASMIDA.

stickleback. See GASTEROSTEIDAE.

sticktight flea. See ECHIDNOPHAGA.

sticky chromosomes. Chromosomes whose broken ends behave as though they were sticky and tend to adhere to one another; uncommon in animals.

Stictomys. Genus of mountain pacas; brown burrowing spp. of Ecuador and Colombia mountains; smaller than Cuniculus.

stigma. 1. Spiracle or breathing pore of many terrestrial arthropods. 2. Dense, discolored area of a wing in certain insects. 3. One of the many small slits in the pharyngeal wall of tunicates. 4. Any small spot, scar, or hole. 4. Simple eyespot in certain Protozoa and unicellular algae; usually composed of a clump of pigment granules and an adjacent light-sensitive portion of the protoplasm. Pl. stigmata.

stilbestrol. Diethylstilbestrol.

Stiles, Charles W. American medical zoologist (1867-1941); professor at Johns Hopkins University; contributed extensively to human parasitology, especially hookworm investigations.

stiletto flies. See THEREVIDAE.

Stilifer. Genus of gastropods parasitic on sea stars.

stillborn. Dead at birth.

Stilobezzia. Genus of ceratopogonid Diptera.

Stilosoma. See SHORT-TAILED SNAKE.

stilt. See HIMANTOPUS.

stimulus. Any change in the external or internal environment that is capable of producing a physiological or behavioral response in an organism; the response may be pronounced and rapid to obscure and very slow. Pl. stimuli.

stingaree. Sting ray.

sting rays. Group of cartilaginous fishes in the Suborder Batoidea; whip-tailed sting rays; body thin and discoid, with a long, thin tail bearing one to several saw-edged spines on upper surface; produce ugly wounds that heal slowly; Dasyatis common.

stink bug. Any of a large number of foul-smelling insects in the hemipteran Family Pentatomidae; body flat and shield-shaped; many are pest spp. on economic plants.

stink gland. Any gland which secretes an ill-smelling substance, especially in certain insects.

stinkpot. Common name of a small musk turtle of eastern and southern U.S.

stilt sandpiper. See MICROPALAMA.

stipes. 1. Large basal piece of a typical insect maxilla; bears the lacinia, galea, and palp. 2. Any stalk or peduncle of an animal or an organ. 3. First or second segment of a millipede mandible. Pl. stipites.

Stirodonta. Order of echinoids with rigid test, gills, and keeled teeth; Salenia typical.

stirrup. Stapes.

Stizostedion. Genus of large fresh-water game fishes in the Family Percidae; S. canadense (sauger), to 20 in., from Hudson Bay drainages as far south as Ala. and Ark.; S. vitreum (walleye pike, pikeperch), to 36 in., common over much of Canada and the eastern half of the U.S. as far south as Ala., Ark., and Nebr.

St. Martin, Alexis. See BEAUMONT.

stoat. See MUSTELA.

Stokesiella. Genus of small biflagellate fresh-water Protozoa; cell in a delicate, stalked, vaselike lorica.

stolon. 1. Cylindrical, horizontal, stemlike attachment structure of certain bryozoans and hydrozoans; individuals grow out or up from it at intervals. 2. In certain pelagic tunicates, a fingerlike ventral outgrowth which gives rise to probuds.

Stolonifera. Order of alcyonarian anthozoans; polyps grow upright from a mat or stolon-like base; skeleton of warty calcareous spicules or compact tubes and platforms; Clavularia and Tubipora are examples.

Stoloteuthis. Genus of squids; body short, globose; arms united by a broad web; commonly less than 30 mm. long.

stoma. 1. Mouthlike opening, especially in lower invertebrates. 2. Mouth cavity of nematodes. Pl. stomata.

stomach. One of the organs of the digestive tract in which considerable protein digestion usually occurs; the stomach has a relatively large diameter compared with other alimentary tract organs; usually located between the esophagus and intestine and present in many invertebrates as well as vertebrates.

stomach-intestine. Long, undifferentiated portion of the digestive canal of most oligochaetes and polychaetes; forms the posterior three-quarters or more of the digestive tract.

stomatogastric ganglion. A visceral ganglion closely associated with the stomach in certain invertebrates, e.g. crayfish.

Stomatopoda. See HOPLOCARIDA.

Stomatopora. Genus of marine Bryozoa; colony thin, crustlike.

stomochord. Anterior diverticulum of the gut in hemichordates; cells large and vacuolated; sometimes considered a primitive notochord.

Stomochorda. Hemichordata.

stomodaeum. 1. In embryology, an anterior depression lined with ectoderm; develops into the mouth area (and foregut in arthropods). 2. In anthozoan coelenterates and ctenophores, a pharynx-like tube which leads from the proctostome into the main gastrovascular cavity. 3. That portion of a rotifer digestive tract between the mouth opening and the stomach. 4. Buccal cavity of tunicates, just inside the mouth. 5. Mouth cavity of certain nematodes. 6. Foregut of arthropods.

stomodeum. Stomodaeum.

stomodoeum. Stomodaeum.

Stomolophus. Genus of scyphozoan jellyfish belonging to the Order Discomedusae.

Stomotoca. Genus of marine coelenterates in the Suborder Anthomedusae.

Stomoxyidae. Family of small flies similar to houseflies; stable flies, horn flies; adults of some spp. blood-sucking; larvae saprophagous; Stomoxys and Haematobia are important genera.

Stomoxys calcitrans. Stable fly or biting housefly in the Family Stomoxyidae; a blood-sucking sp. common wherever there are habitations in Europe and N. A.; transmits anthrax, tetanus, and infectious anemia of horses.
Stomphia. Genus of sea anemones in the Order Actiniaria; column soft, plastic.
stone canal. Short canal connecting madreporite and ring canal in echinoderms.
stonecat. See NOTURUS.
stone crab. See LITHODES and MENIPPE.
stone curlew. See BURHINIDAE.
stonefly. Common name of any adult insect in the Order Plecoptera.
stone roller. 1. Small fish in the minnow family; widely distributed in U. S. streams; internal anatomy peculiar in that the intestine is wound around the air bladder; Campostoma anomalum. 2. See HYPENTELIUM.
stony coral. Any true coral in the Order Madreporaria; with a dense calcareous exoskeleton secreted around the base of the polyp.
Storeria. Small genus of non-poisonous viviparous American snakes; DeKay's snake, a common secretive sp. found in eastern half of the U. S.; the red-bellied snake, of central and eastern states, has similar habits.
stork. Any of a group of Old World migratory wading birds; common European sp. (Ciconia ciconia) has white plumage with black wing quills and greater coverts; in the U. S. the wood ibis is the only true stork.
stormy petrel. See HYDROBATES.
strabismus. Cross eyes; eyeballs do not focus on the same point.
strand rat. See BATHYERGUS.
Strataegus antaeus. A large reddish-brown beetle in the Family Dynastidae; ox beetle; male with three large pronotal horns, female with one.
stratification. 1. Occurrence of superimposed communities of plants and animals at different distances above ground; e. g. a wooded area might contain ground, herb, shrub, and treetop communities. 2. Occurrence of sharply differentiated temperature gradients in bodies of water.
stratified epithelium. Epithelium composed of two to many layers of cells; usually the lowermost cells are columnar or cuboidal, while the outermost (exposed) cells are squamous and flat; the skin and lining of the oral cavity are composed of stratified epithelium.
Stratiomyiidae. Family of soldier flies; brightly colored spp. whose larvae are usually found in decaying vegetation, dung, or in the mud of shallow ponds; Stratiomyia common.
stratum compactum. Dense inner layer of dermis in the vertebrate integument; composed of white and yellow connective tissue fibers, a few smooth muscle cells, blood vessels, and nerves.
stratum corneum. Outermost layer of horny cells in the epidermis of terrestrial vertebrates.
stratum germinativum. Malpighian layer; lowermost layer of columnar or cuboidal cells of the epidermis from which all of the overlying epidermal cells are derived.
stratum granulosum. Layer of dark, granular cells lying above the stratum germinativum in thick skin.
stratum lucidum. Indistinct outermost refractive layer of the stratum germinativum; in certain few areas of the human skin, e. g. the finger tips; contains eleidin, a keratin-like substance derived from the dissolution of granules in the underlying stratum granulosum.
stratum reticulare. Tunica propria.
stratum spongiosum. Outer layer of connective tissue in the dermis of the skin.
strawberry bass. 1. Calico bass. 2. Black crappie; see POMOXIS.
strawberry crab. Small European marine spider crab with dorsal pink tubercles.
strawberry leaf roller. Any of several moths whose larvae feed on, and roll up in, strawberry leaves at pupation.
strawberry moth. Any of several moths whose larvae feed on leaves of strawberry plants.
straw worm. See EURYTOMIDAE.
Streblocerus. Uncommon but widely distributed genus of Cladocera; littoral.
Streblomastix. Genus of flagellates in the intestine of termites.
Streblospio. Genus of Polychaeta.
Strepsiceros. Genus of African ruminants in the Family Bovidae; includes the antelope-like kudu and sitatunga; the former is common in hilly, bushy country; the latter prefers marshy areas; other common representatives are the nyala and bushbucks; horns spiral and large.
strepsinema stage. Twisted chromosome threads during the prophase

stage of meiosis.

Strepsiptera. Small order of minute insects with complete metamorphosis and hypermetamorphosis; free-living or endoparasitic; atrophied biting mouth parts; females larva-like; males beetle-like, with forewings reduced to clublike processes, and hind wings large and fanlike; larvae and females parasitic in hemocoel of Hymenoptera, Homoptera, etc.; Stylops and Xenos most familiar.

streptasters. Short, spiny, monaxon sponge spicules.

Streptocephalus. Common genus of fairy shrimps.

Streptococcus. Large genus of bacteria in which the spherical cells occur in chains; some spp. are pathogenic and cause such infections as scarlet fever, puerperal fever, rheumatic heart disease, and septic sore throat.

streptoline glutinant. See NEMATOCYSTS.

streptomycin. Antibiotic elaborated by a soil fungus and now produced commercially in large quantities; effective against such infections as typhoid fever, tularemia, influenza, and tuberculosis.

Streptoneura. Prosobranchiata.

Streptopelia. Genus of Old World doves; two spp. introduced locally in Fla. and Calif.

Streptosyllis. Genus of small Polychaeta.

Striaria. Genus of millipedes in the Order Nematomorphoidea.

striated border. Fine lines on the exposed surface of epithelial cells, especially absorptive cells; extend at right angles to the cell surface.

striated muscle. Voluntary muscle.

Striatura. Genus of small land snails.

stria vascularis. Special vascular epithelium which lines the outer wall of the cochlear duct of the ear.

stridulation. Production of sounds in insects by rubbing two modified parts of the body together; common in crickets and grasshoppers.

stridulator. Stridulating organ; any divice for producing sounds by rubbing two parts together, especially among insects; usually one part is file-like and the other is scraper-like; often on the legs.

Strigeidae. Large family of holostome trematodes; adults in intestine of birds, mammals, and a few reptiles; constriction divides body into a flattened, concave, or spoonlike anterior end and cylindrical posterior end.

Strigidae. Family of birds in the Order Strigiformes; includes the typical owls (not including barn owls).

Strigiformes. Order of birds which includes the 133 spp. of owls; chiefly nocturnal predators; head and eyes large; beak short, curved, and powerful; claws sharp, feet modified for grasping; nest in banks, hollow trees, old buildings, etc.; Tyto, Bubo, Otus, Nyctea, and Speotyto typical.

strigil. 1. In many insects, a tibial comb, scraper, or antenna cleaner. 2. In Corixidae (Hemiptera), a currycomb-like structure on the dorsal surface of the abdomen.

Strigilla. Genus of small marine bivalves; warm waters.

striped bass. See ROCCUS.

striped dolphin. See LAGENORHYNCHUS.

striped muscle. Voluntary muscle.

striped swamp snake. Tropidoclonion lineatum, a small snake of the south-central states (Kan., Mo., and Tex.); olive colored with a dorsal and two lateral stripes; 9 to 21 in. long; also called lined snake.

Strix. Genus of owls; S. varia is the common large barred owl of the eastern half of the U.S.; S. occidentalis is the rare spotted owl of western states.

strobila. 1. Scyphistoma stage of scyphozoan jellyfish in process of forming a series of transverse constrictions resembling a pile of saucers distally; each such saucer-shaped body eventually is released to form a free-swimming immature jellyfish, the ephyra stage. 2. Main body of a tapeworm, not including the anterior scolex, and consisting of a linear string of proglottids; strobilus.

Strobilidium. Genus of pyriform or turnip-shaped spirotrich Ciliata; fresh and salt waters.

Strobilops. Genus of land snails; eastern and central U.S.

strobilus. Tapeworm strobila.

stroma. 1. Connective tissue matrix of an organ. 2. Fine filmy framework of an erythrocyte.

Stromateidae. Large family of marine fishes; harvestfishes, dollarfishes; body short and greatly compressed, scales smooth; Poronotus common on U.S. East Coast.

Stromatocystites. See EDRIOASTEROIDEA.

Strombidae. Gastropod family which includes the conchs; 80 large, showy spp.

of warm seas; Strombus and Lambis typical.

Strombidium. Genus of marine and fresh-water spirotrich ciliate Protozoa; ovoid to spherical; with several large anterior sickle-shaped membranelles.

Strombus gigas. Giant conch; a marine gastropod of southern Fla. and the West Indies; largest American shell, to 16 in. long; carrion feeder; shell used for cameos and ornaments; flesh sometimes eaten by man.

strong cockle. Cardium magnum, a large and common bivalve; to 5 in. long; yellowish brown, with dark spots and streaks.

strongyle. 1. Type of large sponge spicule which is rounded at both ends. 2. Any nematode in the Suborder Strongylina.

Strongylidium. Genus of fresh-water and marine hypotrich Ciliata.

strongyliform larva. Larval stage of many free-living and some parasitic nematodes; body elongated, and esophagus long and cylindrical, with a single posterior muscular bulb; the strongyliform larva of hookworms is infective to the host.

Strongylina. Suborder of rhabditid nematodes; mouth surrounded by three, six, or no lips; bursa with rays; parasitic in land vertebrates; some important genera are Ancylostoma, Necator, and Haemonchus.

Strongylocentrotus droehbachiensis. Common circumboreal green sea urchin; test to 4 in. in diameter; spines slender, short; in the Order Camarodonta.

Strongyloidea. Order of nematodes equivalent to the Suborder Strongylina.

Strongyloides stercoralis. Small nematode parasitic in the large and small intestines of man in the tropics and subtropics; the life history is very complicated, and there are free-living stages in the soil; man becomes infected when the skin comes in contact with soil containing filariform larvae which penetrate through the skin.

Strongylostoma. Genus of small rhabdocoel Turbellaria.

Strongylura. See BELONIDAE.

Strongylus vulgaris. Nematode parasite of the horse caecum and large intestine.

Strophitus. Common genus of fresh-water bivalve mollusk (Unionidae).

struggle for existence. See DARWINISM.

Struthio. See STRUTHIONIFORMES.

Struthioniformes. Order of birds which includes the ostrich; largest of living birds (Struthio camelus), to 8 ft. tall and 300 lbs. or more; flightless and terrestrial; head, neck, and legs sparsely feathered; two toes on each foot; omnivorous; flocks of 10 to 50 in arid Africa and Arabia.

Strymon melinus. Butterfly in the Family Lycaenidae; bean lycaenid; larvae feed on legumes and other plants.

studfish. One of several bluish or greenish killifish (Fundulus) of southern and central U.S.

stumpknocker. See LEPOMIS.

sturgeon. See ACIPENSERIDAE.

Sturnella. Genus of birds in the Family Icteridae; includes the meadowlarks; breast yellow, upper parts brown and buff.

Sturnidae. Large family of dark-colored Old World passerine birds which includes the common European starling (Sturnus vulgaris), now common over most of U.S.; gregarious, open country spp., but sometimes a nuisance in cities.

Sturnus. See STURNIDAE.

Sturtevant, Alfred Henry. American geneticist (1891-); noted for his work on the linear arrangement of genes; made the first chromosome map (for Drosophila) which located many genes.

Styela. Common genus of solitary tunicates; body simple and irregularly oval.

Stygobromus. Genus of eyeless amphipods found in wells, seeps, springs, and caves in central U.S.

Stygonectes. Genus of eyeless amphipods; caves and wells in Tex.

Stylactis hooperi. Coelenterate in the Suborder Anthomedusae; hydroids in dense colonies on shells of mud snail (Nassa).

Stylantheca. See STYLASTERINA.

Stylaria. Genus of microscopic, fresh-water oligochaetes; prostomium a long tentacular process.

Stylarioides. Genus of tube-building Polychaeta.

Stylasterina. Suborder of colonial hydrozoan coelenterates; polyps embedded in upright branching calcareous growths; a regular component of coral reefs; tropical and temperate oceans; Stylantheca in Calif.

Stylatula. Genus of long, slender sea pens.

style. Any macroscopic, slender, bristle-like process.

stylet. Spear- or needle-shaped piercing structure in certain invertebrates; usually associated with the mouth and used in feeding and/or defense; found at the tip of the proboscis in nemertines and in the anterior end of the digestive tract of many nematodes.

Stylifer. Cosmopolitan genus of marine snails; shell thin, elongated; parasitic on the integument of echinoderms.

Stylochaeta. Gastrotrich genus usually included in Dasydytes.

Stylochoplana. Genus of polyclad Turbellaria; usually on algae.

Stylochus. Genus of brightly colored marine polyclad Turbellaria; often feed on oysters.

Stylocometes. Genus of fresh-water suctorians; cell irregular, with unbranched arms and no stalk.

Stylocordyla. Genus of small deep-water marine sponges in the Subclass Monaxonida; stalk slender, long.

stylomastoid foramen. Small opening behind the bulla region of mammals; carries the seventh cranial nerve.

Stylommatophora. Suborder of pulmonate snails which includes the terrestrial pulmonates; two pairs of tentacles; mantle cavity transformed into a lung.

Stylonychia. Genus of flat, oval to reniform, ciliate Protozoa; body rigid, with ventral cirri, adoral membranelles, and bristles; no simple cilia.

Stylopidae. Widely distributed family in the insect Order Strepsiptera; parasitic on bees.

Stylops. Genus in the insect Order Strepsiptera; larvae and females parasitic on bees.

Stylotella. Common genus of marine sponges (Subclass Monaxonida); encrusting or lobed; shallow water.

subalpine. Pertaining to an area just below timberline in mountainous regions.

subarcual. One of several gill arch muscles in bony fishes.

subarticular tubercle. In amphibians, a small projection on the ventral surface of a digit, at the junction of two phalanges.

subcartilaginous. Partially composed of cartilage.

subchela. Prehensile appendage of certain marine crustaceans (e.g. Squilla); the terminal segment folds back against the more robust penultimate segment. Adj. subchelate.

subclavian artery. Large artery arising in the upper chest region and supplying the arm (some lower vertebrates), or having branches supplying the neck, thorax, spinal cord, arm, and brain (some higher vertebrates).

subclavian vein. Vein draining the forelimb or pectoral fin area in vertebrates.

subclimax. Stage immediately preceding a climax formation; usually persists for a long time and is replaced by the climax spp. only slowly; e.g. in portions of eastern U.S. pine forests are often a subclimax to the hardwood stage, or climax.

subclinical. Pertaining to an early stage of an infection when the presence of the infecting organisms or its pathological effects cannot be detected.

subcoracoscapularis. Dorsal pectoral girdle muscle in reptiles.

subcosta. One of the veins in an insect wing; immediately posterior to the costa.

subcostal. One of a series of deep thoracic muscles in amniotes.

subcutaneous tissue. A combination of loose fibrous connective tissue and fat lying just below the dermis of vertebrate skin.

subdermal lymph space. One of several large flat lymph-filled cavities just under the skin, especially in amphibians.

subdominant. A sp. which exerts some degree of control and dominance in a community but which is still of less importance than the dominant(s); e.g. trout in a small lake characterized by a large-mouth black bass population.

subendostylar coelom. In cephalochordates, a small portion of the coelomic cavity located within the midventral wall of the pharynx.

Suberites. Genus of sponges in the Order Monaxonida; commonly grow on snail shells.

subesophageal ganglion. Subpharyngeal ganglion.

subfamily. Taxonomic category intermediate between a family and a tribe; in the animal kingdom subfamily names customarily have the suffix inae.

subgalea. Maxillary sclerite, attached to the stipes in many insects.

subgenital pit. One of four concavities on the oral surface of certain scyphozoan jellyfishes.

subgenus. Optional taxonomic category between the genus and sp.; used mostly for large genera where certain spp. groups are apparent; the subgenus name is capitalized and inserted within parentheses between the genus and sp. names; the second element in the following is the subgeneric name: Palaemon (Palaeander) northropi.

subimago. See EPHEMEROPTERA.

subintestinal vessel. Blood vessel located below the intestine in certain invertebrates, e.g. earthworm.

subitaneous egg. See SUMMER EGG.

sublingual gland. Small gland beneath the tongue on either side.

sublittoral zone. 1. In lake biology, that portion of the bottom which extends from the lower limit of the eulittoral zone to the lakeward limit of rooted aquatic vegetation, or through a comparable depth range in those areas which contain no rooted aquatics. 2. Also in lake biology, that portion of the bottom between the lakeward limit of rooted aquatics and the upper limit of the hypolimnion. 3. In oceanography, that portion of the bottom between 50 and 200 meters deep.

sublobular vein. One of the large veins of the liver.

subluxation. Incomplete dislocation of bones of a joint.

submarginal. Near the margin.

submaxilla. Lower jaw, lower maxillary bone, or mandible.

submaxillary gland. Gland on either side near the angle of the lower jaw.

submedian. Situated near the midline.

submentum. Basal sclerite of an insect labium by which it articulates with the head.

submucosa. Layer of fibrous connective tissue lying just beneath the mucosa in the respiratory, digestive, and reproductive tracts of vertebrates and certain higher invertebrates.

subneural blood vessel. Blood vessel on the ventral surface of the annelid nerve cord; often broken up into several smaller vessels.

subneural gland. Small gland just below the brain in tunicates; function not definitely known.

subnotochordal rod. Hypochord.

subopercle. In fishes, that opercular bone which lies just posterior and ventral to the large opercle; the most ventral and posterior opercular bone.

suborbital. 1. One of several small bones below the eye of fishes. 2. Located below the eye.

suborder. One of two or more taxonomic subdivisions of an order, the members of each suborder having certain taxonomic characters in common.

subpharyngeal ganglion. Invertebrate ganglion (usually bilobed) located on the ventral surface of the pharynx; the first in the serially-arranged ventral ganglia in arthropods and annelids.

subphylum. Major taxonomic subdivision of a phylum, each such subdivision having certain striking features; e.g. the separation of the Phylum Chordata into three subphyla: Tunicata, Cephalochordata, and Vertebrata.

subplantigrade. Indistinctly plantigrade.

subquadrate. Almost square in shape.

subscapularis. Dorsal pectoral girdle muscle in mammals.

subscapular vein. Vein which drains the shoulder in the frog and certain other vertebrates; joins the innominate vein.

subsegment. In arthropods, a secondary division of an appendage segment into two or more portions.

subspecies. Well defined geographic or physiological aggregate of local populations which differs from other such subdivisions of the same sp.; the subspecies name is customarily a third (uncapitalized) element in a scientific name; e.g. the common eastern white-foot mouse is known as Peromyscus leucopus leucopus, while the northern white-footed mouse is Peromyscus leucopus noveboracensis.

substrate. 1. Material which is changed as the result of the presence of an associated enzyme. 2. Chemical or biological medium. 3. The ground or any other solid object to which an animal may be attached, on which it moves about, or with which it is otherwise associated.

substratum. Substrate.

subtilin. Antibiotic elaborated by Bacillus subtilis.

subumbrella. Concave surface of coelenterate medusae.

subuncus. One of two paired accessory portions of the trophi in certain rotifers.

subunguis. 1. Cornified under surface of a claw. 2. A small cornified transverse area just under the tip of a nail or hoof.

subvertebral lymph space. Large lymph-filled space containing the kidneys and bounded by the dorsal body wall and the peritoneum; especially evident in the Anura.

succession. See ECOLOGICAL SUCCESSION.

succus entericus. Intestinal juice.

sucker. See CATOSTOMUS and CATOSTOMIDAE.

suckerfish. Remora.

suckermouth minnow. See PHENACOBIUS.

sucking stomach. Specialized muscular stomach in certain invertebrates, e.g. the spider foregut; used in sucking up liquid food.

suckle. In mammals, to nourish the young at the breast.

sucrase. Invertase; a sucrose-digesting enzyme in the intestinal juice of certain vertebrates; sucrose is converted to glucose and fructose.

sucrose. Crystalline disaccharide, $C_{12}H_{22}O_{11}$, obtained from sugar cane, sugar beets, sorghum, maple trees, etc.

Suctoria. One of the classes of the Phylum Protozoa; motile immature forms have cilia, and the sessile mature stages have long piercing and/or suctorial tentacles used in catching the prey and feeding; with a macronucleus and micronucleus; a few are parasitic, but most spp. are commensal or free-living in fresh and salt waters; Tríchophrya, Podophrya, and Acineta common.

suctorial pad. Adhesive pad.

Sudan III. Red biological dye having a strong affinity for lipids.

Sudan IV. Scarlet biological dye having a strong affinity for lipids.

sugar. Sweet, crystalline carbohydrate with 6, 12, or 18 carbon atoms in the molecule (monosaccharide, disaccharide, and trisaccharide, respectively); many play important roles in plant and animal metabolism; common naturally occurring sugars are cane sugar, beet sugar, maple sugar, etc.

sugar beet curly top. Virus disease of sugar beets transmitted by a beet leafhopper (Circulifer).

sugar bird. Any of numerous small tropical birds which suck up the nectar of flowers.

sugarcane borer. See CRAMBIDAE.

sugar mite. Small mite sometimes found in unrefined sugar.

Suidae. Family of mammals in the Order Artiodactyla; includes the European wild boar, domestic pigs, and African wart hog; Sus, Phacochoerus, and Babirussa.

Suiformes. Suborder in the Order Artiodactyla; includes the pigs and hippos.

Sula leucogastra. Booby; a large, gregarious, fish-eating, oceanic bird of the tropics and subtropics (Family Sulidae); occasional along the Gulf and south Atlantic Coast of U.S.

sulcate. Grooved or furrowed.

sulcus. Furrow or groove, such as those on the surface of the brain. Pl. sulci.

sulfa drugs. Group of sulfanilamide and related organic compounds especially effective against streptococcus infections.

sulfur-bottom whale. See SIBBALDUS.

sulfur cycle. World-wide circulation and reutilization of sulfur atoms, chiefly due to metabolic processes of plants and animals; plants take up inorganic sulfur, mostly as sulfates, and convert much of it into organic compounds; such sulfur atoms may then be assimilated into the bodies of one or more successive animals; excretion, burning, and bacterial action on dead organisms return the sulfur atoms to an inorganic state.

sulfurs. See PIERIDAE.

Sulidae. Family of gregarious birds in the Order Pelecaniformes; gannets, boobies; large sea birds which feed on fish which they capture by diving from the wing; Sula and Morus common.

summation. Additive effects of rapidly repeated stimuli applied to a muscle or nerve.

summer egg. Subitaneous egg; type of egg produced especially during the warm season by certain fresh-water invertebrates, particularly rotifers, turbellarians, and many fresh-water crustaceans; such eggs are usually thin-shelled, parthenogenetic, and hatch very soon after deposition.

summer flounder. Paralichthys dentatus; a common flounder of the East Coast south of Cape Cod; coloration variable, depending on the background; to 25 lbs.; a desirable food fish.

summer kill. Complete or partial kill of a fish population in ponds or lakes during the warm months; variously produced by excessively warm water, by a depletion of dissolved oxygen, and by the release of toxic substances from a decaying algal bloom, or by a combination of these factors.

Sumner, James. American biochemist (1887-); first isolated a crystalline enzyme (urease) in 1926; shared the Nobel prize in chemistry in 1946.

sun animalcules. Heliozoa.

Sunapee trout. *Salvelinus aureolus*, a small trout of N. H., Vt., and Me. lakes.

sun bear. See HELARCTOS.

sunbird. Any of numerous small brilliantly colored passerine songbirds (Family Nectariniidae) of warmer forested parts of the Old World; nectar feeders; bill down-curved; nest hanging and purselike.

sun bittern. See EURYPYGIDAE.

Suncus. Genus of fat-tailed shrews found in Europe, Asia, and Africa; 2 in. long.

sundew. Insectivorous plant commonly found in boggy areas; certain modified leaves have sticky glandular structures to which insects become stuck.

sundial shell. See ARCHITECTONICA.

sunfish. 1. Any of numerous N. A. freshwater centrarchid fishes, usually weighing less than 1 lb.; body deep and compressed; see LEPOMIS. 2. See MOLA.

sunflower star. See PYCNOPODIA.

sun fly. 1. Any member of a small family of Diptera, the Helomyzidae. 2. Also, see APHELINIDAE.

sun gem. Brightly colored Brazilian hummingbird.

suni. See NESOTRAGUS.

sunset shell. Common name of marine bivalves in the genus *Tellina*.

sun spider. See SOLPUGIDA.

sun star. Any of numerous sea stars with a large disc and more than five arms.

Suomina. Genus of small rhabdocoel Turbellaria.

superciliary. Pertaining to the region of the eyebrow.

superclass. Taxonomic category below a phylum which includes a number of classes having certain features in common.

superfamily. Taxonomic category below an order and including a number of families which have certain features in common; superfamily name usually ends in *oidea*.

superfecundation. Especially in lower mammals, the fertilization by successive acts of coitus of two or more eggs liberated at the same ovulation.

superfemale. Exaggerated (or otherwise abnormal) female characters and ovaries in certain animals owing to an abnormal chromosome complement in the body cells; in *Drosophila* a (sterile) condition is usually produced by two sets of autosomes and three X chromosomes per cell.

superficial cleavage. Type of cleavage in the early embryology of insects; yolk is concentrated in the center of the egg, and the nuclei, cytoplasm, and separating cell membranes are restricted to a thin outermost layer of the embryo.

superior. In human anatomy, toward or pertaining to the head end of the body.

superior mesenteric vein. Vein which drains the small intestine and mesentery in certain vertebrates.

superior vena cava. Precava; anterior vena cava.

supermale. Exaggerated (or reduced) male characters and testes in certain animals owing to an abnormal chromosome complement in the body cells; in *Drosophila* such a (sterile) condition is usually produced by three sets of autosomes and one X chromosome per cell.

superorganism. Epiorganism; colony of closely interdependent castes and individuals, such as a colony of ants; the interdependence is often so pronounced that the whole colony is considered as a type of higher organism.

superposition image. See COMPOUND EYE.

superspecies. Artenkreis; group of closely related spp. which are all or mostly allopatric.

Suphisellus. Genus of small dytiscid beetles.

supination. Rotation of the palm upward. Cf. PRONATION.

supinator. Any muscle whose action turns the ventral side up, e.g. the rotation of the palm upward.

supplemental air. Extra amount of air which can be forcibly exhaled after a normal exhalation; about 1000 cc.

supplementary factors. See INTERACTION OF SUPPLEMENTARY FACTORS and MUTUALLY SUPPLEMENTARY FACTORS.

supportive tissues. Connective tissues.

suppressor genes. Inhibitor genes.

suppuration. Formation of pus, or discharging pus.

supra-anal opening. Opening of the excurrent canal in many bivalve mollusks.

suprabranchial chamber. Epibranchial chamber; one of two or more tubular chambers running longitudinally just dorsal to the gills in most bivalve mollusks; they pass a current of water posteriorly and out of the excurrent siphon.

supracardinal vein. Either of two transient veins in the mammalian embryo; located in the dorsolateral thoracic and abdominal regions.

supracoracoideus. Large muscle arising on the coracoid plate and inserting on the under surface of the humerus; not present in mammals.

supraesophageal ganglion. Suprapharyngeal ganglion.

supramaxillary. Small lateral bone of the upper jaw at the angle of the mouth in bony fishes.

supraneuston. Collectively, those animals living on the upper surface of the surface film in an aquatic habitat; e.g. water striders, whirligig beetles, springtails, etc.

supraoccipital. Cranial bone just above the foramen magnum.

suprapharyngeal ganglion. Invertebrate ganglion located on the dorsal surface of the pharynx; usually two or more such ganglia form the brain or major portion of the brain; especially prominent in annelids and arthropods.

suprarenal gland. Adrenal gland.

suprarenin. Synthetic epinephrine.

suprascapula. Cartilaginous dorsal extension of the scapula in certain vertebrates.

supraspinous fossa. That portion of the scapula just above the scapular ridge.

supratemporal. Small dorsal posterior bone in the skull of primitive amphibians, many reptiles, and fishes.

supratidal. Above the high-tide line.

surf bird. See APHRIZA.

surf clam. See MACTRA and SPISULA.

surf duck. Scoter; see OIDEMIA.

surf fish. See EMBIOTOCIDAE.

surgeon fish. The tangs; any of numerous compressed, brightly colored, tropical fishes; with one or more movable lancelike spines on each side of the body near the base of the tail.

Suricata. Meerkats or suricates; genus of small mammals of the sandy South African veldt (Family Viverridae); related to the mongooses but look somewhat like a cross between a lemur, dog, and raccoon; dry and rocky places.

Surinam toad. See PIPIDAE.

Surnia ulula. Hawk owl; a sp. of Holarctic coniferous forests; occasional in northern U.S.

surra. Disease of stock animals and dogs in Africa and southern and southeastern Asia; caused by Trypanosoma evansi which is transmitted by fleas and biting flies.

survival curve. Survivorship curve; curve showing the relationship between time and the number of surviving individuals in a particular population, beginning with the date of birth, hatching, etc.

survival of the fittest. See NATURAL SELECTION and DARWINISM.

survivorship curve. Survival curve.

Sus. Genus of omnivorous mammals in the Family Suidae; four wild spp. occur in Europe and Asia, including the common wild boar (S. scrofa) in which the male has large upturned tusks in both jaws; also includes the domestic hogs.

suslik. Common ground squirrel, Citellus citellus, of Europe and Asia.

suspensorium. In most fishes, a chain of three bones from the hyomandibular to the palatines; in birds and reptiles it consists of a single quadrate bone.

suspensory ligament. Minute ring of fibrous tissue extending between the periphery of the lens of the eye and the ciliary muscle; maintains the position of the lens and by varied tension exerted on it, the shape of the lens is flattened or thickened.

sustentacular cells. Special supporting cells found particularly as a framework for olfactory cells and in the organ of Corti.

susu. See PLATANISTIDAE.

Sutton, W. S. American cytologist (1876-1916); discovered meiotic pairing (synapsis) of chromosomes; noted the close parallel between chromosome behavior and Mendelian segregation.

suture. 1. Irregular line of junction at the interlocking edges of certain skull bones of vertebrates; the suture areas remain fibrous or cartilaginous in the lower vertebrates, permitting continuing growth, but in birds and mammals the suture eventually becomes bony and rigid. 2. In arthropods and certain echinoderms and mollusks, the junction between two skeletal plates.

Swainson's warbler. See LIMNOTHLYPIS.

swallow. See HIRUNDINIDAE.

swallow-tailed kite. See KITE.

swallowtails. See PAPILIONIDAE.

Swammerdam, Jan. Dutch microscopist (1637-1680); published detailed studies of the anatomy of insects and other small invertebrates; work on the anatomy of the honeybee is a classic.

swamp deer. See BARASHINGA.

swamp eel. See SYMBRANCHII.

swampfish. See CHOLOGASTER.

swamp rabbit. Sylvilagus aquaticus; a large, dark-colored rabbit found in marshes and swamps; S. C. to Ill. and Tex. and southward.

swamp sparrow. Melospiza georgiana; a dark-colored sparrow of eastern N. A.

swan. See CYGNUS.

swarm spore. Zoospore.

Swartkrans man-ape. Large Pleistocene gorilla-like primate; remains found in South Africa in 1949.

sweat. See SWEAT GLANDS.

sweat bee. See HALICTIDAE.

sweat glands. Long tubular epidermal glands of mammals which project into the dermis and secrete sweat or perspiration to the surface of the epidermis; sweat is a very dilute watery solution containing small amounts of inorganic salts and certain nitrogenous excretory products; the chief function of these glands is the regulation of body temperature by evaporation; the number of glands present varies greatly from one sp. to another; the action of the sweat glands is under the control of the autonomic nervous system.

sweetbread. Popular designation for the thymus of young animals, the pancreas, and the testes, when these are used for food.

swift. 1. Kit fox, Vulpes velox; a small fox of the plains states and southern Canada. 2. See SCELOPORUS and UTA (lizards). 3. See MICROPODIDAE (birds).

swim bladder. Elongated gas-filled sac dorsal to the digestive tract in most bony fishes; chiefly a hydrostatic organ, but vascularized and used as an accessory lunglike structure in lungfishes and some primitive Palaeopterygii; occasionally acts as a sense organ or serves in sound production; the bladder may be connected to the anterior part of the digestive tract by a duct, or it may be completely closed off; occasionally the swim bladder is double.

swimmeret. Pleopod.

swimmer's itch. Temporary rash produced on the skin of man by the superficial penetration of trematode cercariae that normally enter the bodies of water birds; common among bathers in fresh-water lakes in many areas of the world; uncommon on marine shores.

swimming bell. 1. In certain siphonophore coelenterates, a large, contracting, umbrella-shaped structure at the top of the colony; propels the colony through the water. 2. In the broad sense, any umbrella-shaped or bell-shaped coelenterate which moves through the water by successive contractions.

swimming crab. See PORTUNUS.

swine. See SUS.

swordfish. See XIPHIIDAE.

sword-razor shell. Ensis directus, the common razor shell of the Atlantic Coast.

swordtail. 1. Local name sometimes applied to any of a wide variety of insects having long, sword-shaped ovipositors. 2. See XIPHOPHORUS.

Sycandra. Genus of calcareous marine sponges.

sycon. Type of canal system found in sponges of intermediate complexity; consists of two sets of cylindrical overlapping radial canals, only the inner radial canals being lined with choanocytes; occurs in Scypha.

Sycon. Scypha.

Syconosa. Heterocoela.

Syllidae. Family of small marine Polychaeta; reproduction by asexual budding.

Sylvicapra. See DUIKER.

Sylviidae. Large family of small active passerine birds; includes the (1) Old World warblers, (2) New World gnatcatchers, (3) kinglets of the Northern Hemisphere, and (4) Australian warblers; bill small; coloration olive, green, brown, or gray above; examples are Polioptila and Corthylio.

Sylvilagus. Genus of which includes the common and widely distributed American cottontail rabbits; much smaller than the jackrabbit and snowshoe hare.

Sylvisorex. Genus of equatorial forest shrews of Africa and Asia.

Sylvius. Jacques Dubois.

Symbiocladius. Small genus of tendipedid midges; larvae under wing covers of mayfly or stonefly nymphs.

symbiont. Any organism involved in a symbiosis.

symbiosis. 1. Broad term used for all types of close association between different spp., including parasitism, commensalism, and mutualism. 2. Also, a synonym for mutualism.

symbiote. Symbiont.

Symbranchii. Order of eel-like fishes; dorsal and anal fins lacking rays; pectorals absent; gill openings fused into a single transverse ventral slit; tropical rivers and swamps; swamp eels; Fluta and Synbranchus typical.

Symbranchiida. Symbranchii.

Symmetrodonta. One of the two orders in the mammal Infraclass Pantotheria; Jurassic spp. having molar teeth with three molar teeth with three main cusps arranged in a triangle.

symmetry. Basic body arrangement and disposition of parts in an organism. See RADIAL, BILATERAL, and SPHERICAL SYMMETRY, and ASYMMETRY.

sympathetic ganglion. See AUTONOMIC NERVOUS SYSTEM.

sympathetic nervous system. 1. Autonomic nervous system. 2. That portion of the invertebrate nervous system (especially in insects) which innervates the viscera.

sympathetic trunk. Longitudinal strand of nervous tissue on either side of the spinal column; part of the autonomic nervous system; has ganglia and many fibers connecting with the spinal nerves, brain, viscera, glands, and smooth muscle throughout the body.

sympathin. Neurohumor facilitating the passage of impulses across the synapses of the sympathetic nervous system to the effector; physiologically closely similar to epinephrine (adrenalin).

sympatric. Pertaining to two or more populations of closely related spp. which occupy identical or broadly overlapping geographical areas.

Sympetrum. Common genus of dragonflies.

symphalangism. Stiff fingers; a Mendelian dominant.

Symphalangus. Genus of anthropoid apes of the Malay Archipelago; the siamangs; black, arboreal spp. with slender body and exceptionally long limbs.

Symphurus. See HETEROSOMATA.

Symphyla. Small class of small to minute, whitish, cylindrical, terrestrial arthropods; to 8 mm. long; head with one pair of antennae, one pair of mandibles, one pair paragnaths, and two pairs maxillae; trunk of 15 to 22 segments, with 12 pairs of legs; under stones and debris in damp places; about 100 spp.; Scutigerella immaculata (garden centipede) sometimes damages seeds and young shoots; Scolopendrella also common.

symphysis. Line of junction and fusion between two bones originally separate and distinct; usually such areas are connected and covered with fibrous cartilage.

symplectic. In certain fishes, a small bone located at the junction of the hyomandibular, quadrate, and metapterygoid.

Symplegma. Genus of colonial encrusting tunicates.

Synalpheus. Genus of snapping shrimps, especially along the Atlantic Coast.

synapse. Junction of the processes of one neuron with those of the next; there is no direct protoplasmic connection but perhaps at most a simple physical contact between two plasma membranes; the transmission of impulses from one neuron to another occurs solely through the synapse, and in one direction only for a particular synapse; a synapse most commonly occurs between the terminal endings of the axon of one neuron and the dendrites of the nerve cell body of another neuron; each neuron customarily has synaptic associations with several or numerous other neurons (sometimes as many as several hundred).

Synapseudes. Genus of marine Isopoda.

Synapsida. One of the subclasses of reptiles; extinct mammal-like reptiles; Carboniferous, Permian, and Triassic; contains three orders: Pelycosauria, Therapsida, and Ictidosauria.

synapsis. See MEIOSIS.

Synapta. Genus of long, wormlike, translucent sea sucumbers; usually lie buried in mud or sand with the tentacles projecting into the water; one sp. attains 6 ft.

synaptic knobs. End-feet; knoblike terminal swellings of an unmyelinated axon; make contact with the dendrites or nerve cell body of another neuron.

Synaptida. Same as Apoda (Holothuroidea).

Synaptomys. Genus of small, robust, herbivorous rodents in the Family Cricetidae; lemming mice; usually in bogs and swamp areas; northern N.A.

Synaptosauria. One of the six subclasses of reptiles; all spp. extinct; large marine reptiles with paddle-like appendages, although early primitive types lived on land; includes two orders:

Protorosauria and Sauropterygia.

synarthrosis. Joint allowing little or no movement, e.g. suture of skull.

Synbranchus. See SYMBRANCHII.

Syncarida. One of the divisions of the Malacostraca; small group of small spp. with generalized structure; carapace absent, uropods present, thoracic appendages biramous and not chelate; Anaspides (in the Order Anaspidacea) is 20 to 40 mm. long and occurs in streams and lakes of Tasmania; it has seven free thoracic segments and well developed pleopods; Bathynella and Parabathynella (in the Order Bathynellacea) occur mostly in European and Japanese wells and subterranean waters; they are wormlike, 1 mm. long, and have eight free thoracic segments and one pair of pleopods (or pleopods absent).

Syncaris. Small genus of fresh-water prawns found in a few streams of the Calif. coast.

Syncera. Genus of small snails found near the high tide line; common in southern Calif.

syncerebrum. Arthropod brain formed by union of true brain with one or more of the ventral ganglia.

Syncerus. Genus of ruminants in the Family Bovidae; includes the African and Cape buffaloes; large, heavy spp., with little hair.

Synchaeta. Common genus of limnetic rotifers; pear-shaped or pyramidal.

Syncoelidium. Genus of triclad Turbellaria; commensal on gill lamellae of Limulus.

Syncoryne. Genus of marine coelenterates in the Suborder Anthomedusae; hydroid colony to 1 in. high; medusa to 0.5 in. in diameter.

Syncrypta. Genus of colonial freshwater biflagellate Protozoa.

syncytium. 1. Single cell with two or more nuclei. 2. Mass of protoplasm containing scattered nuclei which are not separated from each other by intervening cell membranes (animals) or cell walls (plants). Adj. syncytial. Pl. syncytia.

syndactyly. 1. Zygodactyly; human abnormality characterized by the fusion of two of the fingers; the so-called "webbed" condition; a Mendelian dominant. 2. Type of foot in which the anterior toes are more or less fused and have a common sole; e.g. kangaroos and birds.

Snyderella. Genus of flagellates found in the intestine of termites.

Syndesmis. Genus of marine rhabdocoel Turbellaria; in the digestive tract of echinoderms.

syndesmosis. Synarthrosis in which the adjacent bones are held together by fibrous connective tissue.

Syndiamesa. Genus of tendipedid midges.

syndrome. Group of symptoms or indications which collectively characterize a disease or infection.

synecology. 1. Study of the ecology of communities. 2. Study of environmental conditions and adaptations of spp. taken in association. Cf. AUTECOLOGY.

Synemosyna. Genus of small antlike spiders in the Family Salticidae.

Synentognathi. Order of marine bony fishes which includes the needlefish, garfish, halfbeaks, and flying fishes; dorsal fin posterior; pectorals high on body.

syneresis. Shrinkage of a gel so that fluid is forced out at the surface of the mass.

synergism. Joint action of substances is greater than the sum of the action of each of the individual substances; e.g. action of certain combinations of drugs and hormones.

Syngamus trachea. Rhabditid nematode parasitic in the trachea of poultry; the gapeworm.

syngamy. 1. Union of male and female gametes to form a zygote. 2. Fertilization. 3. Conjugation. Adj. syngamic.

Syngnathidae. Family of bony fishes in the Order Solenichthyes; pipefishes; body extremely elongated and head snoutlike; skin armed with rings of bony plates; fins small and soft rayed; abundant in warm seas, especially in littoral; Syngnathus common.

Syngnathiformes. Order of fishes which corresponds to the Solenichthyes.

Syngnathus. See SYNGNATHIDAE.

Synidotea. Genus of marine Isopoda.

synizesis. Clumping of chromatin during synapsis.

synkaryon. Fertilization nucleus.

Synodontidae. Lizard fishes; a family of elongated voracious marine fishes in the Order Iniomi; mouth very wide, supplied with numerous sharp teeth; Synodus common.

Synodus. See SYNODONTIDAE.

synonym. In biological nomenclature, one of two or more different names applied to the same taxon.

synonomy. Chronological list of all

scientific names that have been applied, correctly or incorrectly, to a given taxon; such a list properly includes dates of publication, author, and the journal citation.

synovia. Synovial fluid.

synovial fluid. Viscous transparent fluid secreted by a synovial membrane and filling the joint cavity surrounded by the membrane.

synovial membrane. Connective tissue membrane which completely encloses freely movable skeletal joints in a bag-like fashion; e.g. as in the knee joint.

synovial plicae. Folds on the inner surface of a ligamentous capsule.

Synpleonia. Genus of eyeless amphipods found in wells, seeps, caves, and springs in central U.S.

synsacrum. Fusion of some thoracic, all lumbar, all sacral, and some caudal vertebrae into a single mass in birds; attached to pelvis ventrally.

syntype. Any one of a number of specimens of the same taxon which formed the material studied by the original author in those cases where he did not designate a single type specimen.

Synura. Genus of yellow-brown, colonial, fresh-water biflagellates; 2 to 50 ovoid cells radially arranged; when present in large numbers, these protozoans impart a disagreeable odor and taste to water; S. uvella common.

Synurella. Genus of amphipods; ponds, streams, and springs of central U.S.

synxenic culture. Culture consisting of one or more individuals of a certain sp. with one or more associated spp.

Syphacia. Genus of nematode parasites of the rat and mouse intestine.

syphilis. Contagious venereal disease caused by a spirochaete, Treponema pallidum; an ulcer first appears at the site of infection (usually the external genitalia), followed by skin, mucosa, and constitutional disturbances, and finally characterized by infections in the viscera, muscles, bones, and nervous system.

syringium. 1. In most Hemiptera, a tubular muscular organ associated with the mouth parts; used for ejecting salivary secretions. 2. A similar tubular organ in certain insect larvae, especially for ejecting disagreeable fluids.

syrinx. Sound-producing organ of birds; situated at the bifurcation of the trachea and contains small, thin muscles which vibrate and whose tension can be varied to produce notes of different pitch.

Syrphidae. Family of small, brightly colored flies; flower flies, hover flies; many spp. mimic wasps or bees; hover about and feed on nectar of flowers; larvae saprophagous, phytophagous, or predaceous; familiar genera are Syrphus, Merodon, and Eumerus.

Syrphus. Genus of hover flies; larvae feed on aphids; see SYRPHIDAE.

Syrrhophus. Tropical and subtropical genus of frogs in the Family Leptodactylidae; two spp. in southern Tex., others as far south as Peru.

Systema Naturae. See LINNAEUS.

systematics. The study of the kinds and diversity of organisms and of their relationships.

systemic arch. Fourth aortic arch in tetrapods, supplying the trunk and limbs with blood; in adult amphibians and reptiles both right and left arches persist, in birds only the right, and in mammals only the left.

systemic heart. Special asymmetrical heart in most cephalopod mollusks; receives blood from the gills and pumps it to all parts of the body.

systole. 1. Contraction stage of a heart beat, especially contraction of the ventricles. 2. The contraction stage of a contractile vacuole in Protozoa. 3. Contraction of channels in plasmodia of slime molds.

systolic blood pressure. Blood pressure at the peak of arterial pressure as determined from a radial artery pulse with the sphygmomanometer and stethoscope; about 120 mm. mercury for normal adults at rest.

syzygy. 1. End to end attachment of certain gregarine Protozoa. 2. Fusion of organs without loss of their identity. 3. Reunion of chromosome fragments during meiosis. 4. Close union of skeletal elements, as in certain Crinoidea.

T

Tabanidae. Family of robust flies in the Order Diptera; horse flies, deer flies; large lateral eyes; males feed on plant juices, females blood-sucking; larvae usually aquatic or subaquatic; Tabanus and Chrysops are important genera.

Tabanus. Common cosmopolitan genus of horseflies in the Family Tabanidae; female a vicious biter on warm-blooded animals.

tabula. Horizontal partition of vertical canals of certain corals and hydrocorallines. Pl. tabulae.

tabular. Small dorsolateral paired bone in the primitive tetrapod skull.

Tachardia lacca. Southeastern Asiatic insect in the homopteran Family Lacciferidae; used in making commercial shellac; see LAC INSECT.

tachina flies. See TACHINIDAE.

Tachinidae. Large family of flies; tachina flies; larvae of most spp. parasitize larvae and pupae of insects, especially Lepidoptera; Gonia parasitizes cutworms, and Compsilura concinnata (tachina fly) is important as a parasite of the gypsy and brown-tail moths.

Tachopteryx thoreyi. Eastern U.S. dragonfly.

tachyauxesis. Relative growth of a part more rapidly than the growth of the whole organism.

tachyblastic egg. Thin-shelled gastrotrich egg which hatches in 12 to 70 hours after deposition. Cf. OPSIBLASTIC EGG.

tachycardia. Unusually rapid heart beat; either pathological or the result of muscular effort or emotional disturbances.

Tachycineta thalassina. Violet-green swallow; Alaska to Mexico, east of the Rockies.

tachygenesis. Development which skips certain normal embryonic or larval stages, e.g. as in certain crustaceans and insects.

Tachyglossus. Common genus of echidnas or spiny anteaters in the mammalian Order Monotremata; terrestrial and nocturnal spp. with coarse quill-like hairs and spines, cylindrical beak, and extensile tongue; legs short; feed on worms, ants, and other insects; Australian region.

Tachysphex. See LARRIDAE.

tachysterol. Isomer of ergosterol produced by ultraviolet radiation; upon further irradiation it is converted to calciferol.

Tachypleus. See XIPHOSURA.

tachytelic. Evolving at an unusually rapid rate, resulting in rapid speciation.

tactile. Pertaining to touch.

tactile corpuscle. Any one of many kinds of minute epidermal structures supplied with nerve endings and sensitive to touch; especially in vertebrates.

Tadarida. Common cosmopolitan genus of bats; free-tailed bats; general brownish coloration; to 4 in. long; T. mexicana, of the southwest, roosts in Carlsbad Caverns, N. M.

tadpole. Immature aquatic form of frogs and toads, especially before the appearance of the forelimbs.

tadpole larva. Characteristic minute free-swimming larval stage of the Tunicata; superficially tadpole-shaped; with the chief characters of the Chordata: gill slits, notochord, and hollow dorsal nerve cord; by complex metamorphosis and sometimes attachment to a substrate, it develops into the adult tunicate.

tadpole shrimps. Common name of members of the crustacean Order Notostraca.

Taenia. Common genus of tapeworms in the Order Cyclophyllidea; man becomes infected with T. saginata, the beef tapeworm, by eating poorly cooked beef containing the cysticerci; he becomes infected with T. solium, the pork tapeworm, by eating poorly cooked pork containing the cysticerci;

T. pisiformis is a common cat and dog tapeworm, the cysticerci being in the mesentery of a rabbit.

Taeniactis. See HEMIZONIDA.

taenidium. Spiral or circular chitinous band which stiffens the wall of an insect trachea. Pl. taenidia.

Taeniodonta. Order of Upper Paleocene to Middle Eocene mammals; large, powerful spp.; teeth in the form of long, rootless pegs.

Taeniogaster. Genus of dragonflies; eastern U.S.

Taenioglossa. Subdivision of the gastropod Suborder Pectinibranchia; radula with seven teeth per row; Littorina and Strombus typical.

Taeniopteryx pacifica. Salmon fly, an insect in the Order Plecoptera; found especially in Wash.; feeds on foliage and buds.

Taeniothrips. Common genus of thrips; one sp. is an important pest on pears.

Taenoidea. Cyclophyllidea.

Tagelus. Genus of elongated marine burrowing clams; jackknife clams.

tagma. General body division of an arthropod, consisting of two or more segments, such as head, thorax, and abdomen. Pl. tagmata.

tagmosis. Union of body segments into functional groups, e.g. the cephalothorax of certain arthropods.

taguan. See PETAURISTA.

tahr. See HEMITRAGUS.

taiga. Continuous coniferous forests of subarctic areas, such as those of northern Europe, Asia, and Canada.

tail bud. Simple embryonic protuberance in vertebrate embryos which either degenerates or develops into a functional tail.

tail coverts. Small feathers covering the dorsal and ventral basal parts of the rectrices.

tail fan. Posterior fanlike structure in the crayfish and lobster; consists of the telson and two uropods; the animal darts swiftly backward by means of strong ventral sweeping movements.

tailorbird. Any of several small passerine Asiatic birds which stitch leaves together to form and camouflage their nests.

takin. See BUDORCAS.

Talitrus. Genus of marine Amphipoda.

talon. Claw of a bird of prey or predatory mammal.

talonid. Accessory crown area on the posterior portion of certain mammalian lower molar teeth; used for crushing.

Talorchestia longicornis. Beach flea, a common amphipod of the high tide line of sandy marine beaches.

Talpa. Genus of European moles.

Talpidae. Family of subterranean mammals in the Order Insectivora; moles; forelimbs short and highly specialized for burrowing; eyes minute, fur soft and short; feed chiefly on insects and worms as they burrow extensively through the soil; Scalopus aquaticus is the common American mole east of the Rockies; Scapanus includes the several forms of the West Coast states; Condylura cristata is the star-nosed mole which has a snout with a terminal disc bearing a fringe of 22 fleshy processes.

talus. Astragalus.

tamandua. See TETRADACTYLUS.

tamarao. See ANOA.

tamarin. Any of several small tropical South American marmoset-like mammals; Mico, Tamarin, and Tamarinus.

Tamarinus. See TAMARIN.

tambour. Shallow metal or plastic cup or drum covered with a thin rubber membrane; two of these may be connected with a rubber tube, and if a delicate lever is attached to the center of the membrane of one tambour, the system may be used for making kymograph records of pulse, peristalsis, etc.

Tamiasciurus. Genus which is sometimes set off from Sciurus and thus includes the common red squirrel (T. hudsonicus) of eastern U.S. and the Douglas squirrel or chickaree (T. douglasi) of West Coast states.

Tamias striatus. Common chipmunk of eastern N.A. in the Family Sciuridae; a small terrestrial squirrel with conspicuous stripes above and below the eye and on the dorsal part of the trunk; well developed cheek pouches; alert, nervous behavior; several subspecies.

Tamoya. Genus of scyphozoan jellyfish in the Order Cubomedusae.

tanager. See THRAUPIDAE.

Tanaidacea. Order of malacostracan Crustacea; carapace small, first two thoracic segments fused with head; total length usually 1 to 5 mm.; intermediate between Cumacea and Isopoda; found on the sea bottom, usually in burrows or tubes; Apseudes and Tanais common.

Tanais. See TANAIDACEA.

tandan. Australian fresh-water catfish.

tang. See SURGEON FISH.

tangoreceptors. Touch receptors in the skin of vertebrates.

Tantilla. Genus of small, slender, burrowing snakes in the Family Colubridae; southern half of U.S. southward into South America; venomous but not harmful to man. See BLACK-HEADED SNAKE.

Tanypteryx. Genus of dragonflies.

Tanypus. Genus of tendipedid midges.

Tanysphyrus lemnae. Aquatic weevil which feeds on duckweed; Europe and U.S.

Tanystylum. Genus of pycnogonids.

Tanytarsus. Genus of non-biting midges in the dipteran Family Tendipedidae; larvae and pupae in fresh water.

tapaculo. See RHINOCRYPTIDAE.

tapetum. 1. Group of fibers of the corpus callosum. 2. Any of certain areas in the choroid coat or retina of many vertebrates. 3. See OMMATIDIUM.

tapetum lucidum. Iridescent pigmented choroid coat in the eyes of certain mammals; causes reflected night eyeshine.

tapeworm. Member of the Class Cestoidea (Phylum Platyhelminthes).

Taphozous. Tomb bats; genus of sheath-tailed bats common in Egypt; frequent old ruins.

Taphrocampa. Genus of creeping rotifers; body annulated; foot rudimentary.

Tapinoma. Genus of small ants which spurt or exude a malodorous secretion.

tapir. See TAPIRIDAE.

Tapirella. Genus of tapirs in the Family Tapiridae; black or dark brown spp. of Central and northern South American forests.

Tapiridae. Small family of mammals in the Order Perissodactyla; tapirs; heavy, piglike, hoofed spp. related to the horse and rhinoceros; four toes on front feet, three on hind feet; shy, nocturnal; forests of Central America, northern South America, and the Malay area; Tapirella and Tapirus.

Tapirus. Genus of tapirs in the Family Tapiridae; dark-colored, but with a large white patch in center of body; forests of Malay Peninsula, Borneo, and Sumatra.

tarantula. See THERAPHOSIDAE.

Tarantula. Common genus of whipscorpions in the arachnid Order Pedipalpi; southern states (not a spider).

tarantula hawk. See PEPSIS.

Tarantulidae. Family in the arachnid Order Pedipalpi; cephalothorax broader than abdomen; terminal filament lacking; pedipalps clawed; Tarantula common in southern states.

Tardigrada. Phylum which includes the water bears or bear animalcules; microscopic, cylindrical, mitelike creatures with four pairs of unsegmented, stout legs having terminal claws; body cavity possibly a hemocoel; muscular pharynx with a pair of sharp stylets used in piercing plant cells and small organisms so that the fluids may be sucked out for food; sexes separate but males rare; on substrates in fresh and salt water but most common in damp mosses; most spp. resist desiccation by forming a viable cyst stage; affinities uncertain, often considered a group of arachnids near the mites; about 300 spp.; Macrobiotus and Echiniscus common.

Tarentola. Genus of geckos.

Taricha. Triturus.

Tarnetrum. Genus of dragonflies.

tarpan. Przhevalsky's horse; a small, dun-colored wild horse of Asiatic steppes; Equus przhevalskii.

Tarpon. See MEGALOPIDAE.

tarsal. Pertaining to a tarsus.

tarsale. One of the cartilages or bones of the tarsus, especially one that articulates with the metatarsus.

tarsal glands. Glands of Meibom.

tarsier. See TARSIOIDEA.

Tarsioidea. Suborder of nocturnal mammals in the Order Primates; includes the tarsiers; small, agile, arboreal spp., about as large as a small squirrel; tail long, thin, tufted, and not prehensile; tips of digits padded; eyes enormous; mostly carnivorous, especially on insects and lizards; East Indian area; Tarsius typical.

Tarsius. See TARSIOIDEA.

tarsometatarsus. Fused tarsus and metatarsus bones in birds.

Tarsonemidae. Family of mites in the Superfamily Acaroidea; Pediculoides ventricosus normally feeds on insects in stored grain and straw but may attack skin of human beings and produce "grain itch."

tarsus. 1. Ankle; series of bones in the ankle region of the hind limbs of tetrapods. 2. The fifth segment from the base of an insect leg (often subdivided into two or more portions). 3. Distal segment(s) in the legs of certain other arthropods. Pl. tarsi.

tartar emetic. Sodium antimony tartrate; useful in the treatment of kala azar, schistosomiasis, and trypanosomiasis.

tartaruga. See PODOCNEMIS.

Tascobia. Genus of Trichoptera; larval case fibrous and purselike.

Tasmanian devil. See SARCOPHILUS.

Tasmanian wolf. See THYLACINUS.

taste. Special sense, involving the ability to detect substances in solution, either in the mouth or on some portion of the surface of the body; human taste buds are able to distinguish only sweet, sour, salty, and bitter sensations.

taste bud. Cluster of elongated cells located on a papilla of the tongue or in other parts of the oral cavity; specialized for the reception of taste stimuli (sweet, bitter, sour, salty); also on fins of some fishes.

tatou. Giant armadillo.

tattler. See HETEROSCELUS.

tatu. Common name of the nine-banded armadillo (Dasypus) of Tex. to Central America.

tatuasu. Giant armadillo.

Tatum, Edward Lawrie. American bacteriologist and biochemist (1909-); along with Beadle, discovered that the formation of a single enzyme is often governed by a single gene.

Taurotragus. Genus of oxlike African ruminants in the Family Bovidae; includes the eland.

tautog. Blackfish; edible fish (Tautoga) of the wrasse family; common along the U.S. Atlantic Coast; see LABRIDAE.

Tautoga. See TAUTOG.

tautonym. Scientific name in which the genus and sp. designations are the same, e.g. Phocaena phocaena (a sp. of porpoise).

tawny owl. Strix aluco; common owl of Europe and northern Africa; upper parts reddish brown with black markings; under parts buff with brown bars.

taxa. Plural of taxon.

Taxidea taxus. Common American badger in the Family Mustelidae; short, heavy, squat body with shaggy gray fur; legs short, powerful; claws of forefeet heavy and long; semifossorial; feeds mostly on small mammals (especially rodents), a few birds, and occasionally insects; often dig prey out of their burrows; prairies, plains, and open forests of central and western N.A.

taxis. Type of response by which an animal moves and orients itself toward or away from a particular stimulus.

Taxodonta. Pelecypod order designation used especially by paleontologists; includes shells having numerous similar teeth along hinge area and two equal adductor muscles; e.g. Arca and Nucula.

taxon. Any formal taxonomic unit or category of organisms; e.g. sp., genus, family, order, etc. Pl. taxa.

taxonomy. 1. Scientific naming of organisms and their classification with reference to their precise position in the animal or plant kingdom. 2. The theoretical study of classification, including principles, procedures, and rules. Adj. taxonomic.

Tayassu. See TAYASSUIDAE.

Tayassuidae. Small family of mammals in the Order Artiodactyla; includes peccaries or javelinas; small, dark-colored, piglike animals traveling in herds in deserts, brushy areas, and forests; Tayassu angulatus, with an indistinct white collar, is the collared peccary of Tex. to Patagonia; T. peccari is the white-lipped peccary of British Honduras to Paraguay.

Tayra. Common genus of long-legged marten-like animals of Mexico to Argentina; found in a wide variety of habitats.

teacher bird. 1. Ovenbird. 2. Red-eyed vireo.

teal. See ANAS.

teat. Nipple.

Tectibranchia. Suborder of gastropods in the Order Opisthobranchiata; head large, shell absent or thin and imbedded in folds of mantle and foot; sea slugs, sea hares, pteropods, etc.

Tectipes. Genus of marine Isopoda.

tectorial membrane. Narrow membrane extending longitudinally down much of the length of the scala media chamber of the inner ear of higher vertebrates; a part of the organ of Corti; function not definitely known.

tectum. Rooflike structure, especially the corpora quadrigemina over the mesencephalon.

Tedania. Widely distributed genus of marine sponges (Subclass Monaxonida); encrusting to massive.

Tedanione. Genus of marine sponges.

Tegenaria. Common genus of spiders in the Family Agelenidae; house spiders; build funnel-like webs in buildings.

Tegeticula alba. Small white moth having a symbiotic relationship with the western Yucca plant; the plant is absolutely dependent upon the moth for pollination; the moth derives benefit because the female oviposits in the ovary of the flower, and the developing larvae feed upon seeds which appear later.

tegmen. 1. Hardened leathery or parchment-like forewing in the insect orders Orthoptera and some Homoptera; term also sometimes used for the hemelytra of the Heteroptera. 2. General term for a coating, covering, or investment. 3. Ninth abdominal tergite of male insects. 4. Beetle elytron. 5. Covering of the oral disc of crinoid and blastoid echinoderms. Pl. tegmina.

tegmentum. 1. Gray matter covering the crura cerebri. 2. Upper layer of the calcareous plates of certain Amphineura.

tegula. 1. Small overhanging mesothoracic sclerite in Lepidoptera and Hymenoptera. 2. Calypter. 3. Patagium of a moth or butterfly.

Tegula. Large genus of small tropical marine gastropods; top shells; smooth iridescent shell partly covered by a dark periostracum.

tegumen. 1. Tegmen. 2. Tergum of Lepidoptera. 3. Dorsal hoodlike structure in the genitalia of male Lepidoptera.

Teidae. Family of New World lizards; body slender, limbs reduced in a few spp., tongue deeply bifid; Cnemidophorus, Ameiva, and Tupinambus typical.

teju. Large, edible, South American lizard.

tela choroidea. Choroid plexus.

tela subcutanea. Subcutaneous connective tissue; merges gradually into the deep portion of the corium.

Teleallagma daecki. Common damselfly of East Coast states.

Telea polyphemus. Large sp. of N. A. moth having ocherous, buff, or pink coloration; wingspread to 5 in.; the polyphemus moth; caterpillar feeds on leaves of deciduous trees and shrubs.

teledu. See MYDAUS.

telegony. Erroneous supposition that characteristics of a male involved in one mating are transferred through the female to the offspring of a mating with another male.

teleiophan. Quiescent pupa-like stage between the active nymph and active adult stage in typical Hydracarina.

telencephalon. Most anterior part of the forebrain of vertebrates; includes olfactory lobes and cerebral hemispheres.

Telenomus. Genus of minute to small, shiny, dark hymenopterans; parasitic on many harmful bugs and therefore of economic importance.

Teleodesmacea. Large order of marine Pelecypoda; gills reticulate, siphons well developed, ligament behind umbo, hinge teeth well developed.

teleology. With respect to animals, the doctrine of design and anticipatory purposeful behavior.

teleost. Any bony fish in the Neopterygii.

Teleostei. Neopterygii.

Teleosteica. Subclass of bony fishes; along with the Subclass Holosteica, usually combined into the Subclass Neopterygii.

Teleostomi. Obsolete taxon of fishes; included the Crossopterygii, Palaeopterygii, and Neopterygii.

telepod. Modified leg serving as a copulatory appendage on one of the posterior segments of male Diplopoda.

telescope fish. Unusual breed of goldfishes having protuberant eyes, short, thick body, and a large double tail fin.

Telestacea. Order of alcyonarian anthozoans; colony composed of a proliferating base having upright stems bearing lateral polyps; skeleton composed of more or less fused spicules; common genus is Telesto.

Telesto. See TELESTACEA.

Tellidora. Genus of small, triangular, Caribbean bivalves; growth lines heavy.

Tellina. Widely distributed genus of marine bivalves; tellin shells; several hundred spp.; thin, compressed, and delicately colored; hinge short and weak.

tellin shells. See TELLINA.

Telmatogeton. Genus of midges in the Family Tendipedidae.

teloblast. Special large cell of an annelid embryo; divides rapidly and produces many small cells.

telocoel. Cavity of a telencephalon, especially in an embryo.

Telodesmacea. In some classifications, one of the three orders in the Class Pelecypoda; mantle margins more or connected ventrally and posteriorly; siphons well developed; chiefly marine; includes the cockles, razor clams, quahogs, etc.

telolecithal egg. Ovum which has a large amount of yolk concentrated at the vegetal pole and the early embryo concentrated at the animal pole, as in the hen's egg.

telophase. Last stage in mitosis; characterized by: the formation of a new cell wall between the two nuclei (in plants) or the pinching-in of the cell membrane to form two cells (in animals), the disappearance of discrete chromosomes, the formation of nuclear membranes and nucleoli, and the disappearance of the spindle.

telophragma. Krause's membrane.

telopodite. 1. Long, segmented appendage of a trilobite trunk; originates from the basal coxopodite segment. 2. Collectively, those segments of an insect limb distal to the coxa.

Telosporidia. One of the subclasses of Sporozoa; sporozoites elongated; spores without polar capsules.

Telotremata. Brachiopod order in the Class Articulata; includes many fossil and most modern forms.

telotroch. 1. Preanal tuft of cilia in a trochophore, pilidium, or tornaria. 2. Motile stage of a vorticellid protozoan; with anterior and posterior rings of cilia; stalk absent.

telson. 1. Broad median posterior projection of the last body segment in many decapod crustaceans, Eubranchiopoda, Isopoda, and Limulus (spine-like); not a true body segment. 2. Twelfth abdominal segment in certain primitive insects and insect embryos. 3. The terminal stinging segment of a scorpion.

Temnocephala. Genus of aberrant flatworms in the Order Rhabdocoela; 5 to 12 anterior finger-like processes; ectocommensals on crustaceans, snails, and turtles; tropical and subtropical.

Temnospondyli. Extinct order of labyrinthodont amphibians; Devonian to Triassic.

Temora. Common genus of marine plankton copepods in the Suborder Calanoida.

temperate rain forest. Lush, heavily wooded area with long growing season in temperate latitudes; characterized by a dominant gymnosperm or angiosperm tree sp. with heavy undergrowth; rainfall heavy and distributed through most of the year; Olympic Peninsula, southeastern Australia, New Zealand, Japan, southern Chile and Argentina, etc.

template-antitemplate hypothesis. A physical hypothesis cited to explain gene duplication; the gene (template) serves as a mold from which a negative (antitemplate) is formed, and the antitemplate then serves as a mold for a positive (duplicate gene).

temporal. 1. Pertaining to the temples or areas of the head just behind the orbits. 2. Compound bone on the side of the human skull.

temporalis. Adductor muscle of the mandible in vertebrates.

temporal lobe. Lateral ventral portion of a cerebral hemisphere, especially in higher vertebrates.

temporary immunity. Any acquired immunity which is effective for only a relatively short time.

tenaculum. 1. In Collembola, a ventral catchlike process on the third abdominal segment; holds furcula in place. 2. Clasper of a shark. Pl. tenacula.

tench. Hardy fresh-water European fish in the carp family; introduced elsewhere; Tinca tinca.

Tendipedidae. Very large family of minute, delicate, non-biting, mosquito-like insects; midges; adults often in large swarms, especially in the evening; larvae and pupae mostly in fresh water and damp places; Tendipes most common.

Tendipes. Large genus of non-biting midges in the dipteran Family Tendipedidae; larvae and pupae common in all fresh waters; larvae of some spp. known as bloodworms.

tendon. Cord or band of fibrous connective tissue which attaches muscle to bone.

tendon of Achilles. Large tendon connecting the muscles of the calf of the leg with the bone of the heel.

Tenebrio molitor. Common beetle in the Family Tenebrionidae; larvae (mealworms) highly destructive to cereals and cereal products.

Tenebrionidae. Large family of sluggish, nocturnal beetles; darkling beetles; wings usually vestigial or absent; mostly phytophagous scavengers; many pest spp.; common genera are Tenebrio and Tribolium.

teneral. Callow.

Tenodera. Genus of insects which includes some of the praying mantids.

tenpounder. See ELOPIDAE.

Tenrec. Genus of common nocturnal tenrecs of Madagascar; snout long and pointed, legs short, claws sharp, tail absent; body spiny.

Tenrecidae. Family of mammals in the Order Insectivora which includes the tenrecs of Madagascar; some spp. spiny, some tail-less; snout long and pointed; size variable.

tensor. Muscle which stretches or makes rigid.

tensor fasciae latae. Lateral thigh muscle in certain vertebrates.

tentacle. Long, unsegmented, cylindrical or threadlike protuberance from an animal; variously specialized as sensory or food-getting devices; present in some Protozoa, most Coelenterata, Polychaeta, Bryozoa, etc.

tentacular bulb. Swollen base of a tentacle in many coelenterate medusae; the site of nematocyst formation and also digestion in the gastrovascular cavity.

tentacular crown. Lophophore of Bryozoa.

Tentacularia. Genus of tapeworms in the Order Trypanorhyncha; adults in bony fishes, larval stages in marine invertebrates and elasmobranchs.

tentacular pouch. One of two long pouchlike structures found in comb jellies; each contains the base of one of the long trailing tentacles.

Tentaculata. One of the two classes of the Phylum Ctenophora; long tentacles present.

tentaculocyst. Rhopalium; one of several (usually eight) small club-shaped sense organs along the margin of certain jellyfish.

tentaculozooid. In some hydrozoans, a polyp type which is little more than a single long tentacle.

tent caterpillar. See MALACOSOMA.

Tenthredinidae. Family of sawflies in the Order Hymenoptera; female with large concealed sawlike ovipositor of four parts; larvae feed mostly on foliage; Caliroa common.

tentillum. Branch of a tentacle.

tentorium. 1. Chitinous internal framework of the head capsule in insects; supports the brain. 2. Transverse fold of dura mater between the cerebellum and occipital lobes of the mammalian brain; sometimes ossified.

Tentorium. Genus of marine sponges (Subclass Monaxonida); size of an olive; cold seas.

Teratocephalus. Common genus of microscopic, free-living nematodes.

teratology. Study of abnormal development and congenital malformations, especially in man.

teratoma. Tumor composed of a disorderly arrangement of tissues and organs; usually produced by faulty embryology.

Teratornis merriami. Extinct giant condor.

tercel. Male falcon.

Terebella. Genus of marine Polychaeta; in tubes constructed of sand and debris; anterior cluster of tentacular gills.

Terebellides. Genus of tube-building Polychaeta; threadlike tentacles very numerous.

terebra. 1. Boring, piercing, or cutting ovipositor of an insect. 2. Mandibular sclerite of an insect.

Terebra. See TEREBRIDAE.

Terebrantia. One of the two suborders of the insect Order Thysanoptera; females with a sawlike ovipositor for depositing eggs in plant tissues.

Terebratalia. Genus of West Coast brachiopods.

Terebratella. Common genus of small brachiopods.

Terebratulina. Very common brachiopod genus in the Order Telotremata.

Terebridae. Auger shells; family of carnivorous gastropods with a very long tapered spire and small aperture; mostly tropical; Terebra common.

Teredo navalis. The common shipworm or teredo; a long wormlike pelecypod with a small body and united siphons; to 24 in. long; the small valves are used for excavating burrows in wood and clay by rasping movements; such burrows are lined with a calcareous coating and the worm remains in its original burrow for life, tunneling all the while; extremely destructive to wooden piling and ship hulls.

teres major. Muscle originating on the scapula and intermuscular septum and inserting on the lesser tubercle of the humerus; draws raised arm downward and backward and rotates it inward.

terete. Cylindrical and tapered.

Tergipes. Genus of nudibranchs.

tergite. Dorsal sclerite of an arthropod body segment, especially where there is only one dorsal sclerite in a segment.

tergosternal muscles. Special dorsoventral thoracic muscles in insects; cause raising of the wings by contracting; wings are lowered chiefly by the contraction of a set of dorsolongitudinal muscles; together these two sets of muscles produce flight.

tergum. 1. Transverse dorsal surface of a typical body segment of most arthropods; may consist of one or more sclerites. 2. One of a pair of distal calcareous plates in the barnacle skeleton. Pl. terga.

termen. Outer margin of an insect wing, between apex and anal or hind angle.

Termes. Common tropical genus of termites.

terminal arborizations. Finely divided ends of an axon.

terminal bars. Short masses of rodlike intercellular cement material, found especially between epithelial cells just below the exposed surface.

terminalis. First paired cranial nerve found in vertebrates.

termitarium. Natural or artificial nest or colony of termites; above ground termitaria are constructed of bits of earth or vegetable matter cemented together; in ground-dwelling spp. they are sometimes massive and as much as 25 ft. tall.

termite. Any member of the insect Order Isoptera.

Termitidae. Large family of tropical and subtropical termites; Termes and Nasutitermes typical.

Termitomimus. Genus of rove beetles living in termite nests.

Termitonicus. Genus of minute beetles living on the dorsal side of the head of certain termites; take bits of food as it is passed from one worker to another.

Termopsis. Genus of termites.

tern. See STERNA.

terramycin. Antibiotic extracted from Streptomyces, a soil fungus; effective against a wide variety of disease-producing organisms.

Terrapene. Genus of box turtles; generally distributed east of the Rocky Mountains but extending into Ariz. and not found in Minn. and the Dakotas; land spp., occasionally near water; plastron hinged; carapace highly arched.

terrapin. See MALACLEMYS.

terrarium. Indoor terrestrial habitat or garden and associated animals; it may consist of a whole room, or it may be only a small aquarium-sized glass enclosure.

Terricola. Taxon which includes the terrestrial spp. of triclad Turbellaria.

terricolous. Living in or on the ground.

territory. In the ecological sense, a specific area over which an animal (or pair of animals) establishes jurisdiction; it is vigorously defended and no other individual of the same sp. is allowed to come inside its boundaries. usually established for breeding purposes; frequently exhibited by certain mammals, reptiles, birds, and fishes.

tertiaries. Those flight feathers of a bird which are attached to the humerus area.

Tertiary period. Geological period lasting about 70 million years and ending about 1 million years ago; a subdivision of the Cenozoic era; characterized by modern plants and the rapid evolution and dominance of mammals; divided into Pliocene, Miocene, Oligocene, Eocene, and Paleocene epochs.

test. 1. Protective shell which loosely covers certain protozoans; it may be secreted or made from sand grains or bits of debris. 2. Rigid calcareous skeleton of the echinoderm class Echinoidea. 3. Shed exoskeleton of Limulus.

Testacea. Order of amoeboid Protozoa with a single-chambered chitinoid test which is often thickened or covered with sand grains and other foreign bodies; Arcella and Difflugia are examples.

Testacella haliotidea. Greenhouse "slug"; found especially in greenhouses, where it feeds on worms, snails, etc.; not a true slug but rather a snail with a very small ear-shaped shell which is carried on the dorsal surface near the posterior end.

test cross. Genetic mating of an animal with a parent or parent stock; mating between a homozygous recessive individual and a dominant phenotype; such a cross is used in determining whether the dominant is heterozygous or homozygous; a mating between a white guinea pig and a heterozygous black guinea pig results in 50% black and 50% white offspring; a mating between a white guinea pig and a homozygous black results in all black offspring.

testes. Plural of testis.

Testicardines. Group of brachiopods having a shell hinge and internal skeleton; anus absent.

testicles. Testes, especially in those mammals possessing a scrotum.

testis. Male gonad. Pl. testes.

testis cords. Branching anastomosing strands of cells in an embryonic testis, especially in mammals; later become the seminiferous tubules.

testosterone. Steroid hormone secreted by the testes, especially in higher vertebrates; responsible for the development and maintenance of sexual characteristics and the normal production of sperm; thought to be secreted by the connective tissue, or interstitial cells, between the seminiferous tubules; also known as androsterone (incorrectly) and androtin.

Testudinata. Chelonia.

Testudinella. Genus of flattened rotifers with a long annulated foot and no toes.

Testudinidae. Family of reptiles which includes the common turtles, terrapins, and land tortoises; plastron covered with 12 plates; world wide except for Australian region; about 80 U.S. spp.; Chrysemys, Emys, Clemmys, Terrapene, Testudo, and Gopherus typical.

Testudo. Genus of giant land tortoises of the Galapagos Islands and (rarely) islands of the Indian Ocean; strictly vegetarians; clumsy and sluggish; to 400 lbs. and 4 ft. long.

Tetanocera. Genus of marsh flies in the Family Tetanoceridae.

Tetanoceridae. Marsh flies; family of sluggish flies found at edges of ponds, streams, and swamps; larvae aquatic, float at surface among aquatic plants; Sepedon and Dictya most common U.S. genera.

tetanus. 1. Continuous state of contraction of a muscle produced by a series of very rapidly repeated stimuli; normal expression of skeletal muscle function. 2. Infectious disease marked by tonic contraction of voluntary muscles and caused by the toxin of a bacterium, Clostridium tetani, which gets into the tissues through cuts and bruises; when the disease affects the jaw muscles, it is known as lockjaw.

tetany. 1. Neuromuscular hyperexcitability produced by disturbance of blood calcium and by parathyroid insufficiency. 2. Tetanus (def. 1).

Tethya. Genus of sponges in the Subclass Monaxonida; tuberculate surface.

Tethymyia. Genus of midges in the Family Tendipedidae; wings vestigial.

Tethys. Aplysia.

Tetilla. Genus of very small marine sponges in the Class Demospongiae; often attached to muddy sand substrates.

tetra. Any of many small brightly colored fishes in the Family Characinidae; commonly used in tropical aquariums; native to South and Central America.

Tetrabelodon. Gomphotherium.

Tetrabranchia. One of the two orders in the mollusk Class Cephalopoda; includes the nautilus and its relatives; coiled shell external and symmetrical, divided by many transverse internal septa; two pairs of gills and nephridia; with many tentacles having no suckers.

Tetraceros. Chousingha; genus of tiny shy deerlike mammals of India; female hornless, male with four small horns; reddish brown; glades and open woodlands.

tetrachloroethylene. Drug used extensively for treatment of hookworms; also useful for intestinal trematode infections.

Tetraclita. Genus of barnacles; to 40 mm. in diameter.

Tetractinellida. One of the subclasses of sponges in the Class Demospongiae; spicules tetraxon or absent, spongin negligible or absent; body rounded or flattened, without branches; Geodia typical.

tetrad. Group of four chromatids formed by synapsis during meiosis, two chromatids having been derived from each of a pair of homologous chromosomes; a cell whose diploid number is 10 chromosomes will therefore have 5 sets of chromatids (5 tetrads), or a total of 20 chromatids.

Tetradactylus. Genus of South American anteaters; tamanduas; tail almost naked and prehensile, fur dense and close, ears large, and muzzle shorter than in other anteaters; claws extremely powerful.

Tetragoneuria. Common genus of dragonflies.

Tetragraptus. See GRAPTOLOIDEA.

Tetrahymena. Common genus of holotrich Ciliata.

Tetrameres. Genus of spirurid nematodes parasitic in the digestive tract of birds.

tetramerous. Having parts of the body arranged in fours, especially in a radially symmetrical animal.

Tetramitus. Genus of colorless flagellates having four flagella; fresh-water and marine.

Tetranychidae. Family of mites in the Superfamily Trombidoidea; spider mites, "red spiders"; feed on soft green parts of plants; some spp. spin silk; Tetranychus is the common "red spider," 0.4 mm. long.

Tetranychus. See TETRANYCHIDAE.

Tetrao. See CAPERCAILLIE.

Tetraodontidae. Family of marine bony fishes; puffers or swellfishes; similar to Diodontidae in habits and structure but skin covered with abundant prickles.

Tetraonidae. Family of birds in the Order Galliformes; grouse, ptarmigan, etc.; Northern Hemisphere only; examples are Bonasa, Dendragapus, Lagopus, Tympanuchus, and Centrocercus.

Tetraphyllidea. Order of tapeworms in the Subclass Cestoda; scolex with four bothridia; parasites in the intestine of elasmobranch fishes; probably two intermediate hosts in the life history; Phyllobothrium common.

Tetraplatia. See PTEROMEDUSAE.

tetraploid. Having a 4n or four haploid chromosome complement.

Tetrapoda. Superclass of vertebrates which includes all vertebrates normally having four appendages, namely the classes Amphibia, Reptilia, Aves, and Mammalia.

Tetrarhynchidea. Trypanorhyncha.

Tetrarhynchobothrium. Genus of trypanorhynchid tapeworms; found especially in the stomach and intestine of skates.

Tetrarhynchoidea. Trypanorhyncha.

Tetrarhynchus. Tentacularia.

tetrasomic. 1. Organism having one chromosome in quadruplicate and all others in duplicate, the somatic complement being 2n + 2 chromosomes. 2. Pertaining to such an organism.

Tetrastemma. Genus of marine nemertines in the Order Hoplonemertea.

Tetrastichidae. See TETRASTICHUS.

Tetrastichus. Cosmopolitan genus of minute hymenopterans which are parasitic on the eggs of moths and beetles and on the larvae of other parasitic hymenopterans; the most important genus in the Family Tetrastichidae.

tetraxon spicules. Four-rayed spicules occurring in some sponges.

Tetrigidae. Family of orthopteran insects; includes the pygmy locusts or grouse locusts; small spp. with a narrow head and a prothorax which extends over the wings to the tip of the abdomen and forward ventrally to shield the mouth parts; Tetrix common.

Tetrix. See TETRIGIDAE.

Tettigoniidae. Family in the insect Order Orthoptera; long-horned grasshoppers, katydids, and Mormon cricket; delicate greenish insects with long antennae.

Teuthoidea. Taxon which includes the squids.

Teuthophrys trisula. Fresh-water holotrich ciliate with three, large, anterior, spirally curved arms.

Texas cattle fever. See BABESIA.

Thais. Cosmopolitan genus of carnivorous marine snails; dogwinkles, dog whelks, purple snails; striated, thick, nodular shells, usually about 1 in. long.

thalamencephalon. Diencephalon.

thalamus. Gray matter forming the lateral walls of the diencephalon of the vertebrate brain; an area of sensory coordination.

Thalarctos maritimus. Polar bear (Family Ursidae); creamy white sp. found along coasts or at sea in Arctic America; average male weights 800 lbs.; omnivorous but feeds mostly on fish and seals.

Thalassacarus. Genus of marine mites in the Family Halacaridae.

Thalassarachna. Genus of marine mites in the Family Halacaridae.

Thalassema. Genus of marine worms in the phylum Echiuroidea; Atlantic and Pacific coasts.

Thalasseus. Small genus of terns found in warm and temperate seas; three spp. along south Atlantic, Gulf, and Calif. coasts of U.S.

Thalassicolla. Genus of radiolarians.

Thalassochelys. Genus of sea turtles.

Thaleichthys. See CANDLEFISH.

Thalessa. Genus of large ichneumon flies; adult female 3 to 6 in. long.

Thalestris. Common genus of marine harpacticoid copepods.

Thaliacea. Small class of pelagic tunicates; salps; body more or less cylindrical and supplied with bands of circular muscle which produce locomotion in some spp.; incurrent siphon anterior, excurrent siphon posterior; to 4 in. long; most spp. frequently produce long chains of many individuals by budding; Doliolum and Salpa common.

Thallasotrechus. Genus of carabid beetles.

Thamnocephalus platyurus. Common fairy shrimp in central and southwestern states; to 2 in. long.

Thamnophilus. Genus of small tropical Central and South American ant birds; bill deep, compresed, and hooked and toothed at the tip; in dense trees and bushes near the ground.

Thamnophis. Genus of non-poisonous viviparous snakes in the Family Colubridae; garter snakes and ribbon snakes;

common over N.A.; usually medium-sized but some individuals attain 44 in.; feed chiefly on amphibians, insects, and earthworms; occasionally feed on fish; about 21 spp. in U.S.

thanatosis. 1. Necrosis, gangrene, or death of a part. 2. Act of feigning death.

Tharyx. Genus of Polychaeta.

Theatops. Common genus of centipedes.

theca. Sheath or case, such as the covering of a tendon, the chitinous covering of an insect pupa, and the test of a protozoan. Pl. thecae.

theca granulosa. Layer of small cells surrounding an oocyte in a vertebrate ovary.

Thecamoebae. General taxonomic term used to include those amoeboid protozoans which are covered with a secreted or otherwise fabricated shell, e.g. Arcella. Cf. GYMNAMOEBAE.

thecated embryo. Type of early embryo formed by fresh-water hydras; during late cleavage the attached embryo secretes a resistant sclerotized yellowish shell or theca; the whole structure drops off the parent.

Thecla. Common genus of hairstreak butterflies in the Family Lycaenidae.

Thecodontia. Order of primitive reptiles; phytosaurs; examples are Ornithosuchus and Mystriosuchus.

Thecophora. Large suborder of the reptilian Order Chelonia; includes most turtles, tortoises, and terrapins; thoracic vertebrae and ribs fused to carapace.

Thecosomata. Group of genera of Pteropoda in which a shell is present.

Thecurus. Genus of small Sumatran porcupines.

theelin. See ESTROGENS.

Theileria. Genus of Sporozoa occurring in endothelium and erythrocytes of mammals; T. parva is parasitic in African cattle, transmitted by a tick.

Thelastoma. Genus of rhabditid nematodes parasitizing cockroaches and beetle larvae.

Thelazia. Genus of spirurid nematodes parasitic in the eye and tear ducts of mammals and under the nictitating membrane of birds.

Thelepus. Common genus of tube-building Polychaeta.

Thelotornis. Genus which includes the African twig snake, an arborial sp. having a remarkable resemblance to a twig.

Thelphusidae. Family of fresh-water crabs found along rivers of tropics and subtropics.

Thelyphonida. Pedipalpi.

thelytoky. Production of females by a parthenogenetic female. Adj. thelytokous.

Thenea. Genus of small, conical sponges in the Order Tetractinellida.

Theobaldia. Genus of widely distributed mosquitoes of temperate latitudes.

theory. Generalization based on many observations, measurements, and experiments conducted to test the validity of a hypothesis and found to support the hypothesis.

theory of the gene. All characters of an organism are referable to genes located on chromosomes.

therapeutics. 1. Science and art of healing; adj. therapeutic. 2. A technical account of the treatment of a disease.

Theraphosa. Genus of large spiders, to 3.5 in. long.

Theraphosidae. Family of spiders which includes the tarantulas or bird spiders; large, hairy burrowers or in vegetation; vicious biters though these bites are usually no more serious than a bee sting; Theraphosa, Eurypelma, and Dugesiella common.

Therapsida. Order of Permian and Triassic reptiles in the Subclass Synapsida; small, heavy-bodied spp., supposedly ancestral to primitive mammals; Cynognathus typical.

therapy. Treatment of disease.

Therevidae. Stiletto flies; a family of medium-sized to large hairy predaceous flies somewhat similar to robber flies; common in temperate meadows and woodlands.

Theria. One of the three subclasses of mammals; includes marsupials and placental mammals; cloaca usually absent; testes in a scrotal sac; female with Fallopian tubes, uterus, and vagina; mammary glands with nipples; viviparous; teeth usually specialized in form.

Theridiidae. Very large family of spiders which usually build irregular snare webs; viscid silk is thrown over the prey when it becomes caught in the snare; Latrodectus mactans (black widow) and Theridion tepidariorum (house spider) are familiar.

Theridion tepidariorum. Common house spider often found in homes and other buildings; female 5 to 6 mm. long; web in corners, more or less flat; found in all seasons.

Theridula. Common genus of spiders in the Family Therididae.

Therion. Genus of ichneumon flies.

Thermacarus. Hot-spring genus of Hydracarina.

Thermobia domestica. Insect in the Order Thysanura; firebrat; often found around fire places and ovens.

thermobiotic. Able to live at exceptionally high temperatures; applies especially to certain bacteria, blue-green algae, protozoans, spores, cysts, and resting stages.

thermocline. 1. Mesolimnion, discontinuity layer, metalimnion. 2. A single plane in the metalimnion where temperature drop is most rapid. See EPILIMNION, HYPOLIMNION.

thermolabile. Easily changed or decomposed by heat.

Thermonectus. Genus of dytiscid beetles; southwestern states.

thermophile. Any organism which thrives at unusually high temperatures, e.g. protozoans, algae, and bacteria of hot springs.

Thermosbaena. See THERMOSBAENACEA.

Thermosbaenacea. Order of curious malacostracan Crustacea; contains a single sp. in the genus Thermosbaena from a hot spring in Tunisia, three spp. of Monodella from subterranean waters in Jugoslavia and Italy, and one sp. of Monodella from a salt spring on the shore of the Dead Sea; body cylindrical and stubby, 2 to 3 mm. long.

thermostable. Not easily modified by moderate heat.

Theromorpha. Order of extinct mammal-like reptiles.

Theromyzon. Genus of fresh-water leeches; common in Europe, uncommon in U.S.

Theropithecus. Genus of large, baboonlike primates of southern Abyssinia.

Theropoda. Suborder of the dinosaur Order Saurischia; bipedal carnivores; examples are Compsognathus and Tyrannosaurus.

Thescelosaurus. Genus of birdlike dinosaurs.

thiaminase. Thiamine-digesting enzyme in the tissues of many fresh-water fishes.

thiamine. Vitamin B_1, a member of the vitamin B complex; essential for normal carbohydrate metabolism; deficiency produces poor growth, loss of appetite, loss of muscle tone, and beriberi (nervous and muscular atrophy and paralysis); abundant in yeast, cereal germ, eggs, liver, and lean pork; $C_{12}H_{17}ON_4S$.

thiamine pyrophosphate. Coenzyme derived from thiamine; aids the catalysis of pyruvic acid to acetaldyhyde and carbon dioxide.

Thiara granifera. Fresh-water snail native to East Indies and southwestern Pacific islands but introduced into Fla.

thick-headed flies. See CONOPIDAE.

thick-knees. See BURHINIDAE.

thigmocytes. 1. Special cells in the blood of certain arthropods; upon exposure to air, a cut, or broken surface, they disintegrate and aid in rapid coagulation of the blood. 2. Name occasionally given to blood platelets of vertebrates.

thigmotaxis. Movement of an organism toward or away from an object which provides a touch stimulus.

Thigmotricha. Suborder of the ciliate Order Holotricha; most spp. commensal in the mantle cavity of marine and fresh-water bivalve mollusks.

thigmotropism. Turning of an organism or part of an organism toward or away from an object which provides a touch stimulus.

Thinocoridae. Small family of birds in the Order Charadriiformes; seedsnipes; small, plump vegetarians of South America.

Thiobacillus. Genus of gram-positive, non-sporeforming, rod-shaped, sulfur bacteria; oxidize sulfur and its compounds; common in aquatic and damp habitats.

third eyelid. Nictitating membrane.

third ventricle. Cavity of the diencephalon and the posterior part of the telencephalon in the vertebrate brain.

Thoburnia. Genus of suckers found especially in Va. streams.

Thomisidae. Family of crab spiders; short, crablike appearance; do not spin silk but secure prey by stealth; Misumena, Philodromus, and Xysticus common.

Thomomys. Genus of pocket gophers in the Family Geomyidae; numerous spp. and subspecies in western N.A.; upper incisors not grooved.

Thompsonia. Genus of Cirripedia parasitic on crabs, hermit crabs, etc.

thoracic. Pertaining to the thorax.

Thoracica. One of the orders of barnacles in the Subclass Cirripedia; mantle and six pairs of thoracic appendages; common genera are Lepas and

Balanus.

thoracic basket. Collectively, the ribs and sternum of tetrapods.

thoracic cavity. Chief body cavity in the thoracic (chest) region of vertebrates and arthropods; in mammals it contains the heart and lungs and is separated from the abdomen by the diaphragm.

thoracic duct. Large mammalian lymph vessel which drains the trunk and hind legs and discharges lymph into the blood stream in the jugular-subclavian region.

thoracic vertebra. Vertebra in the upper trunk portion of the spinal column; 12 in man.

Thoracophelia. Genus of Polychaeta.

Thoracosteida. Gasterosteidae.

thorax. 1. Chest region of tetrapods. 2. In insects, the three leg-bearing segments between head and abdomen. 3. In crustaceans and arachnids, the central portion of the body between head and abdomen; number of segments variable; frequently one or more segments are fused with the head to form a cephalothorax.

thornback. 1. Local name for several rays (fishes) with dorsal spines. 2. Local name for a spiny European spider crab.

thorny-headed worm. Member of the Phylum Acanthocephala.

thorny oyster. Chrysanthemum shell, Spondylus americanus; a ribbed and brightly colored tropical and subtropical bivalve.

Thos. Genus sometimes used to include the jackals (Canis).

Thracia. Duck-bill shells; genus of marine clams which burrow into mud or sand; shell thin.

thrasher. See TOXOSTOMA.

Thraupidae. Large New World family of passerine birds which includes the tanagers; mostly tropical and subtropical; males brightly colored, often more or less red, or yellow and black; western, scarlet, and summer tanagers of the U.S. are all in the genus Piranga.

thread capsule. Nematocyst.

thread cell. See NEMATOCYST.

thread fin. See ALECTIS and POLYNEMIDAE.

threadfin shad. See SIGNALOSA.

thread-legged bugs. See EMESA.

thread tube. See NEMATOCYSTS.

threadworm. Member of the Phylum Nematoda and Nematomorpha.

three-day fever. Sandfly fever.

three-toed sloth. See BRADYPODIDAE.

three-toed woodpecker. See PICOIDES.

threonine. Common amino acid; $C_4H_9O_3N$.

thresher shark. See ALOPIAS.

threshold. Specific stimulus intensity below which a given irritable tissue exhibits no response.

Threskiornis. See IBIS.

Threskiornithidae. Family of birds which includes the ibises and spoonbills; in the Order Ciconiiformes; large, gregarious, wading birds with face bare of feathers.

thrips. Any member of the insect Order Thysanoptera.

Thrips. Common insect genus in the Order Thysanoptera; T. tabaci (the onion thrips) is an important pest of tobacco, beans, onions, cabbage, and many other plants.

thrombin. Catalyst formed from calcium and prothrombin in clotting blood; converts fibrinogen into fibrin; see CLOTTING.

thrombocyte. Small, spindle-shaped, nucleated cell in the blood of most nonmammalian vertebrates; functions in blood clotting.

thrombogen. Prothrombin.

thrombokinase. Thromboplastin.

thromboplastin. Material released from blood platelets and injured tissues; essential to blood clotting; same as thrombokinase. See CLOTTING.

thrombosis. Formation of a clot (thrombus) in a blood or lymph vessel.

thrombus. Plug or clot in a lymph duct, blood vessel, or heart; may variously consist of coagulated blood, fibrin, globulin, masses of bacteria or protozoan parasites, platelets, leucocytes, or cellular growth.

Throscinus. Genus of dryopid beetles.

throwback. Individual or character in an individual which has reverted to the wild, more primitive, or original phenotype.

thrush. Any of about 700 spp. of songbirds in the Family Turdidae; usually solidly colored dorsally with a spotted breast; common American spp. are the hermit thrush (Hylocichla guttata) and wood thrush (H. mustelina).

Thryomanes bewicki. Bewick's wren; upper parts brown, under parts dull white; southern two-thirds of U.S. south to central Mexico.

Thryothorus ludovicianus. Carolina wren; coloration brown, buff, and dusky; southeastern quarter of U.S. but also as far as Ontario and Tex.

Thuiaria. Genus of marine coelenterates in the Suborder Leptomedusae.

thunderfish. 1. Electric catfish. 2. See MISGURNIS.

thunder worm. See RHINEURA.

Thunnidae. Family of tunas, now included in the Scombridae.

Thunnus thynnus. Common tunafish, horse mackerel, or albacore; a large Atlantic oceanic fish with a very close relative in the Pacific.

Thuricola. Genus of long, cylindrical peritrich ciliates; in a lorica having a valvelike door; marine and fresh waters.

Thyas. Genus of Hydracarina; mostly northern.

Thylacinus. Genus of wolflike marsupials in Tasmania (Family Dasyuridae); commonly known as the Tasmanian wolf, now nearly extinct.

Thylacis. Genus of stocky Australian short-nosed bandicoots.

Thylogale. See PADEMELON.

Thymallidae. Holarctic family of freshwater bony fishes; includes the graylings; with a flaglike dorsal fin; in cold streams and rivers; Thymallus the only N.A. genus; Arctic grayling from Mackenzie River to Alaska and the Arctic Ocean; Montana grayling in the headwaters of the Missouri River; Michigan grayling, now extinct, formerly in streams of northern Mich.

Thymallus. See THYMALLIDAE.

thymine. Pyrimidine base; component of deoxyribonucleic acid; involved in cellular metabolism; $C_5H_6ON_2$.

thymocytes. Lymphocyte-like cells in the cortex of the thymus gland.

thymus. Spongy gland in the chest, neck, or pharyngeal region of vertebrates; in man it is in the upper chest and lower neck and usually disappears during youth and puberty; functions not clearly established.

Thyone. Very common genus of sea cucumbers.

Thyonepsolus. Genus of small sea cucumbers.

Thyreocoridae. Negro bugs; a family of small blackish hemipterans which gather in large numbers on early spring flowers.

thyridium. Small whitish or hyaline spot in a wing of certain Neuroptera, Hymenoptera, and Trichoptera.

Thyridopteryx ephemeraeformis. Common U.S. sp. of bagworm moth; larvae feed on leaves of shrubs and trees and construct bags from silk and bits of vegetation.

thyroglobulin. Protein formed by the thyroid gland; composed partly of thyroxin.

thyroid cartilage. Large ventral cartilage of the mammalian larynx.

thyroid fenestra. Puboischiadic fenestra.

thyroid gland. Endocrine gland peculiar to chordates; in the pharyngeal region; in man it is an H- or U-shaped brownish organ just below the larynx; secretes thyroglobulin which contains the active fraction, thyroxin; the chief role of this hormone is the regulation of energy metabolism. See CRETINISM, MYXEDEMA, and GOITER.

thyrotropic hormone. Hormone secreted by the anterior pituitary; stimulates the thyroid gland to secrete thyroxin; thyrotropin.

thyrotropin. Thyrotropic hormone.

thyroxin. Thyroxine; derivative of tyrosine, abundant in the thyroid gland; $C_{15}H_{11}O_4I_4N$; see THYROID GLAND.

Thysanoëssa. Genus of pelagic euphausid crustaceans.

Thysanoptera. Order which includes about 5000 spp. of thrips; minute, slender insects, with simple metamorphosis and mouth parts modified for rasping and sucking; wings absent, vestigial, or well developed, fringed with hairs or bristles; parthenogenesis common, males unknown in some spp.; nymphs and adults scrape epidermis on flowers, leaves, and fruits and suck juices; many pest spp.; familiar genera are Thrips, Herco-thrips, Leptothrips, and Scirtothrips.

Thysanosoma actinioides. Fringed tapeworm of sheep in the western half of the U.S.; each proglottid has a fringe on the posterior border.

Thysanozoon. Genus of polyclad Turbellaria.

Thysanura. Order of primitive insects (bristletails, silverfish); body scaly, tapered and depressed, and whitish to grayish or silvery in color; no metamorphosis; wings absent; antennae long; with two or three posterior cerci; secretive and photonegative; in buildings and in debris of the floor of forest and woodland; Lepisma and Thermobia common.

tiang. See DAMALISCUS.

Tiaroga cobitis. Loach minnow; a brightly colored minnow of the Gila

River basin; mostly in riffle areas.

Tiaropsis. Genus of marine coelenterates in the Suborder Leptomedusae.

Tibellus oblongus. Common spider in the Family Thomisidae; elongated, light gray or yellow with darker longitudinal stripes; throughout U.S.

tibia. 1. Anterior of the two bones between the knee and ankle in hind limbs of tetrapods; the shin bone. 2. The fourth segment from the base of an insect leg; also, one of the segments in the legs of certain other arthropods.

tibiale. Embryonic tarsal bone, partly represented in the adult astragalus.

tibialis anticus. Muscle on the front of the shank which flexes the tarsus and elevates the inner border of the foot; in certain tetrapods.

tibialis posticus. Long but small muscle on the posterior surface of the shank in certain tetrapods; extends the tarsus and turns the foot in.

Tibicina. Same as Magicicada; see CICADIDAE.

tibiofibula. Fused tibia and fibula in certain tetrapods.

tibiotarsus. Fused tibia and tarsal bones in birds.

tick. See IXODOIDEA.

tick fever. 1. Rocky Mountain spotted fever. 2. Colorado tick fever.

tidal air. Amount of air inhaled and exhaled when a person is at rest and breathing quietly.

tidal flats. Flat sea bottom, usually muddy and wide, which is exposed at low tide.

tidal zone. Intertidal zone.

tide. Alternate periodic rise and fall of water level in oceans and seas produced by the gravitational pull of the sun and moon; typically, high tide at any one point occurs every 12 hours and 26 minutes; tidal amplitude varies from a few inches to more than 50 ft., depending on shoreline configuration, location, barometric pressure, wind, and relative positions of sun, moon, and earth.

Tiedemann's bodies. Small caeca or pouches of the ring canal in many echinoderms; produce amoebocytes which are released into the coelomic fluid.

tiger. 1. Large Asiatic carnivore in the cat family (Panthera tigris); tawny coloration with transverse black stripes; some specimens to 500 lbs.; occasionally included in the genus Felis. 2. Occasionally "tiger" is used locally for the jaguar or leopard.

tiger beetles. See CICINDELIDAE.

tiger cat. See DASYURUS and OCELOT.

tiger Lucina. Lucina tigrina, a large, heavy, sculptured bivalve; Fla. to Tex.

tiger moths. See ARCTIIDAE.

tiger salamander. See AMBYSTOMA.

tiger shark. A large viviparous shark of warm seas; to 15 ft. long; Galeocerdo cuvier.

tiger shell. Large cowrie (Cypraea tigris) covered with brown spots.

tiger snake. 1. Black-fanged snake of southern and eastern Africa. 2. See NOTECHIS.

Tigriopus. Genus of marine copepods; usually in tide pools.

tigroid bodies. Nissl granules.

tilefish. See BRANCHIOSTEGIDAE.

Tillina. Common genus of kidney-shaped holotrich ciliates; salt and fresh waters.

Tillodontia. Order of Upper Paleocene to Middle Eocene mammals; large bearlike animals with rodent-like skull and dentition.

Tima. Genus of marine coelenterates in the Suborder Leptomedusae; hydroid minute.

timber line. That altitude or latitude above which trees do not grow, usually because of the adverse climatic and soil conditions.

timber rattler. See CROTALUS.

Timeliidae. Chamaeidae.

Tinamiformes. Order of birds in the Superorder Palaeognathae; tinamous; partridge-like game birds to 18 in. long; sternum keeled; wings short, able to fly, but usually run; southern Mexico to southern South America; about 45 spp.; examples are Tinamus and Rhynchotus.

Tinamus. See TINAMIFORMES.

Tinca tinca. See TENCH.

tine. Prong of an antler.

tinea. Ringworm.

Tinea pellionella. Case-making clothes moth; widely distributed lepidopteran whose larvae live in portable cases and feed on wool, hair, and other dried animal products; important pest of woolen clothes, tapestries, upholstery, etc.

Tineidae. Important family of Lepidoptera which includes the clothes moths; small, somber spp.; larvae feed on dried animal and vegetable matter and fungi; Tinea an important genus.

Tineola biselliella. Webbing clothes moth, the larva of which feeds on wool, fur, and feathers.

Tingidae. Lace bugs; a family of minute to small hemipterans with the dorsal part of the head, prothorax, and hemelytra raised into a lacelike embossed pattern; plant feeders, some being economic pests.

Tintinnidium. Genus of sessile fresh-water and marine spirotrich Ciliata; cell conical or trumpet-shaped but surrounded by a hollow cylindrical lorica.

Tintinnopsis. Genus of marine and fresh-water spirotrich ciliate Protozoa; with a vase-shaped chitinoid lorica.

Tiphia. See TIPHIIDAE.

Tiphiidae. Family of small hairy wasps; larvae ectoparasitic on beetle grubs; some spp. important in the control of pest beetles; Tiphia common.

Tipula. Common genus of dipterans in the Family Tipulidae; larvae mostly aquatic or semi-aquatic.

Tipulidae. Family of insects in the Order Diptera; crane flies; slender, long-legged insects with narrow wings, having the general appearance of large, clumsy mosquitos; larvae in damp habitats, or aquatic; Tipula common.

Tisbe. Marine genus of copepods in the Suborder Harpacticoida.

tissue. Aggregation of similar cells having the same function(s), e.g. blood, connective, epithelial, nerve, lymph, and muscle.

tissue culture. Technique for maintaining bits of tissue growing in a sterile liquid or solid culture medium.

Titanosaurus. Genus of large, widely distributed, amphibious, herbivorous dinosaurs of Cretaceous times.

titi. Brazilian leaf-eating monkey in the Family Colobidae.

titmouse. See BAEOLOPHUS.

Tivela stultorum. Large, heavy-shelled, light brown bivalve found on sandy shores of the coast of Calif. and Mexico; pismo clam; to 6 in. long.

Tmarus. Common genus of spiders in the Family Thomisidae.

toad. Any of a large number of tail-less amphibians; usually have rough or warty skin and are customarily found in cool, moist terrestrial habitats; plump and more sluggish than frogs; parotid gland forms a conspicuous raised area behind and above the tympanum; eggs laid in water; Bufo common in U.S. Cf. FROG.

toad bug. See GELASTOCORIDAE.

toad crab. See HYAS.

toadfish. See HAPLODOCI.

tobacco worm moth. See PROTOPARCE.

tocopherol. Vitamin E; essential to reproduction; deficiency produces sterility, death of embryos, and muscular dystrophy in young animals; abundant in green leaves and vegetable fats; actually a complex of four closely related compounds of which the one whose formula is given (alpha tocopherol) is the most potent; $C_{29}H_{50}O_2$.

Todidae. Small West Indian family of birds in the Order Coraciiformes; todies; bill flattened, serrated; insectivorous and burrowers; Todus the only genus.

tody. See TODIDAE.

toe biter. Electric-light bug.

Tokophrya. Common genus of stalked, pyramidal, fresh-water Suctoria.

Tolypeutes. See APAR.

tomato horn worm. Larva of the tomato worm moth; see PROTOPARCE.

tomato worm moth. See PROTOPARCE.

tomb bat. See TAPHOZOUS.

tomcod. See MICROGADUS and GADIDAE.

tomentose. Covered with a dense coat of short matted hairs or setae.

Tomista schlegeli. Sp. of gavial (Family Gavialidae) in Borneo and Sumatra rivers; to 15 ft. long; general habits and structure similar to those of alligators and crocodiles, but the snout is extremely long and slender, like the handle of a frying pan.

tomium. Cutting edge of a bird bill. Pl. tomia.

Tomocerus. Common genus of Collembola.

Tomopteris. Genus of pelagic Polychaeta; body flattened and parapodia long and delicate.

tomtit. Local name for the titmouse or wren.

tone. Tonus.

tongue. 1. Muscular organ, usually freely movable and often protrusile, on the floor of the oral cavity of most vertebrates. 2. Mollusk radula. 3. Proboscis of certain insects, especially some bees and Lepidoptera.

tongue papillae. Elevations on the upper surface of the tongue; in the cat and certain other mammals the papillae are cornified and form a rasping surface. See FILIFORM, FUNGIFORM, and CIRCUMVALLATE PAPILLAE.

tonguetied chub. See PARAEXOGLOSSUM.

tongue worm. 1. Member of the Phylum Linguatulida. 2. Also occasionally certain members of the Phylum Enteropneusta.

Tonicella. Lepidochiton.

Tonna. Common genus of tropical marine gastropods which includes certain tun or cask shells; large, thin, globular or triangular shells of deep waters; to 10 in. long.

tonofibrils. Fine fibrils in the cytoplasm of certain cells, especially epithelium.

tonoreceptors. Receptor cells of the inner ear of most terrestrial vertebrates.

tonsil. 1. Any mass of lymphoid tissue in the oral cavity or pharynx of tetrapods. 2. In man, a small almond-shaped mass of lymphoid tissue on either side of the posterior surface of the lower pharynx; the palatine tonsils. See ADENOIDS.

tonus. Continuous but moderate activity of a tissue or organ; especially true of muscles associated with keeping the limbs and body in proper position when standing, sitting, etc.; produced by a type of tetanic contraction of muscle which resists stretch.

tooth. 1. In most vertebrates, one of the hard structures on the jaws, used in mastication, attack, or defense; typically composed of hard dentine surrounding a soft pulp and covered with enamel on the crown; type and number of teeth vary greatly from one sp. to another; in the permanent dentition of man there are two incisors, one canine, two premolars, and three molars on each side of each jaw, totaling 32 teeth. 2. In invertebrates, any analogous structure occurring in association with the digestive tract. 3. Any toothlike prominence, especially on the shells of mollusks or exoskeleton of arthropods.

toothed herring. See HIODON.

tooth shell. Common name of a member of the mollusk Class Scaphopoda.

tope. Any of several kinds of sharks.

topi. See DAMALISCUS.

topknot. Any of several small European flounders having one to several elongated rays at the anterior end of the dorsal fin.

top minnow. See POECILIIDAE.

topotype. Specimen taken at the type locality of a taxon.

top shell. Any of a wide variety of marine gastropods whose shell has a very definite pyramidal or toplike shape with a flat base; Calliostoma common on the West Coast; Trochus cosmopolitan.

tora. See ALCELAPHUS.

tornaria. Ciliated free-swimming larva of some Enteropneusta; structurally similar to larvae of many echinoderms.

Tornatina canaliculata. Channeled lathe shell; a small cylindrical gastropod found along Atlantic and Gulf coasts.

Torpedo. Common genus in the cartilaginous fish Family Torpedinidae; torpedo rays, electric rays; body disc-shaped and highly flattened; two large electric organs in front part of disc are used to produce shocks and stun prey; tail fin present.

torpedo ray. See TORPEDO.

torpid. Inactive and sluggish, as a hibernating animal.

torpor. Dormancy, aestivation, sluggishness, or suspended animation.

torques. Any distinct cervical collar of hair, feathers, special integument, or coloration.

torsion. In typical gastropods, a developmental twisting phenomenon which brings the mantle cavity to the front of the body, and the visceral and pallial organs altered in position through 180° in relation to head and foot; the visceral mass may undergo further coiling or spiraling but this is associated with shell growth and is independent of the original torsion process.

tortoise. Terrestrial turtle, common in arid and semiarid regions.

tortoise beetle. Common name for members of the beetle Family Chrysomelidae.

tortoise scales. See COCCIDAE.

tortoise shell. Mottled horny material forming overlapping dorsal plates on the hawksbill turtle; used for inlaying and various ornaments.

tortoise-shell butterfly. See NYMPHALIS.

Tortricidae. Family of small, cryptically colored moths whose larvae feed on the leaves of deciduous and evergreen trees; some larvae nest in rolled leaves; leaf rollers. See OLETHREUTIDAE.

torulus. Universal joint socket for the articulation of an insect antenna.

torus. Swelling or smoothly bulging projection or ridge. Pl. tori.

Totanus. Genus which includes the greater and lesser yellow-legs (Family Scolopacidae); large, slim, gray shore birds with yellow legs; shores of salt and fresh waters.

totipalmate. In birds, having all four toes joined by a web.

toucan. Any of a group of brightly colored, fruit-eating birds of Tropical America; in the Family Ramphastidae (Order Piciformes); beak tremendous but of light construction; nest in tree cavities.

touch. A special sense, variously developed in different animals; in man, "touch" often includes the sensations of contact, pressure, heat, and cold.

touraco. Plantain eaters; African forest birds in the Family Musophagidae, related to the cuckoos; tail long, crest erectile, bill short and stout, and coloration brilliant.

tower shell. Any of numerous small marine gastropods with long, slender spire, e.g. Turritella.

towhee. Any of certain American birds in the Family Fringillidae; front and upper parts dark colored; under parts variously white, gray, brown, chestnut, yellow, or black; common American spp. are the green-tailed towhee (Oberholseria chlorura), brown towhee (Pipilo fuscus), Abert's towhee (P. aberti), and eastern towhee or chewink (P. erythrophthalmus).

Townsend's solitaire. See MYADESTES.

toxa. Simple curved sponge spicule.

Toxascaris. Common genus of nematodes; adult in small intestine of dog, cat, and related wild mammals.

toxemia. Pathological conditions produced by the presence of bacterial toxins in the blood stream.

toxic. Poisonous or pertaining to a poison.

Toxiglossa. Division of the gastropod Suborder Pectinibranchia; radula with two long teeth in each row; Conus typical.

toxin. Material (exotoxin) secreted by a living organism; capable of inducing the production of antitoxins when injected into the tissues of an animal (especially man); true toxins are formed by certain bacteria, a few higher plants, and by such venomous animals as poisonous snakes. See ENDOTOXIN.

toxin-antitoxin. Nearly-neutralized mixture of diphtheria toxin and antitoxin used in vaccination.

Toxocara. Genus of rhabditid nematodes parasitic in the digestive tract of carnivores; T. canis common in the dog.

Toxoglossa. Group of families of marine carnivorous gastropods (especially Conidae and Terebridae); radula with few large teeth serving as poison fangs to carry the poison secreted by a large proboscis gland.

toxoid. Toxin which has been inactivated (usually by heat or chemicals) but which is still able to induce antibody formation when introduced into the tissues.

toxophore group. That portion of a toxin molecule responsible for its injurious action. Cf. HAPTOPHORE.

Toxopneustes. Common tropical genus of sea urchins.

Toxostoma. Common American genus of thrashers in the Family Mimidae; thrushlike spp.; upper parts brown or gray, with or without whitish wing bands; tail long.

Tozeuma. Genus of small transparent marine shrimps.

TPN. Triphosphopyridine nucleotide.

trabecula. 1. Septum or other inward extension of an enveloping substance, forming an essential part of the stroma of an organ or tissue; usually used with reference to connective tissue; e.g. the individual nodules of a lymph node are separated by trabeculae. 2. Small bar, rod, or bundle of fibers.

trace element. 1. Any element essential for normal metabolism but required in only minute amounts, e.g. cobalt and iodine. 2. Any element which is present in only minute amounts in an organism, in soil, water, or some other substrate.

trachea. 1. One of a complex system of branched air-filled respiratory tubules in the bodies of insects, a few terrestrial arachnids, and myriapods; they communicate to the exterior by means of small paired lateral openings, or spiracles; at their internal distal ends the finest tracheal branches are terminated by minute tracheoles which supply individual cells of the body; oxygen and carbon dioxide diffuse inward and outward to and from the tissues through the tracheal system. 2. In most terrestrial vertebrates, the "wind pipe" or tube which leads from the larynx to the point of its bifurcation into two bronchi. Pl. tracheae.

tracheal commissure. In terrestrial arthropods, a large transverse (usually) connecting tube of the tracheal system.

tracheal gill. Gill of aquatic naiad

stages of certain insects; contains tracheae, as in the Odonata and Ephemeroptera; occasionally the term is extended to include gills of certain aquatic Diptera and Coleoptera larvae.

tracheal system. Collectively, all of the branched air-filled tubules in insects and certain other terrestrial arthropods; see TRACHEA.

Trachelas tranquilla. Common hunting spider in the Family Clubionidae.

Trachelipus. Terrestrial genus of Isopoda.

Trachelius. Genus of large, fresh-water holotrich Ciliata; oval to spherical; anterior end with a finger-like process.

Trachelobdella. Genus of leeches parasitic on marine shore fishes and occasionally on fresh-water fishes; body segments considerably subdivided.

Trachelocerca. Marine genus of greatly elongated holotrich ciliate Protozoa.

Trachelomonas. Genus of fresh-water, ovoid or spheroidal flagellates having a single flagellum and an enveloping lorica which is often spinous; green, yellow, or brownish coloration.

Trachelophyllum. Genus of marine and fresh-water holotrich Ciliata; cell elongated and flexible, with a long necklike portion.

tracheoles. Fine, intracellular or intercellular, thin-walled, tapering, respiratory tubules in the tissues of insects, a few terrestrial arachnids, myriapods, and Onychophora; the tracheoles have blind terminal endings and originate from the ends of the smallest tracheae; they are partly filled with fluid, and the amount of absorptive surface for the exchange of carbon dioxide and oxygen is governed by the relative length of this capillary water column in the end of the tracheole, which, in turn, is determined by the changing physiological and osmotic conditions in the tissues containing the tracheoles.

Trachinidae. See WEEVER.

Trachinotus. See POMPANO.

Trachinus. See WEEVER.

Trachodon. Genus of Cretaceous dinosaurs in the Order Ornithischia; duckbills; massive, bipedal, herbivorous spp. to 40 ft. long; jaws broad and flat, bearing large numbers of flat teeth arranged in a pavement fashion; same as Anatosaurus.

trachoma. Contagious virus disease of the conjunctiva and cornea.

Trachycardium. See COCKLE.

Trachydemus. Genus of Kinorhyncha.

Trachylina. Order of hydrozoan coelenterates; medusoid generation dominant and craspedote, often large; hydroid generation reduced or absent; mostly marine; a common example is Liriope.

Trachymedusae. Suborder of the coelenterate Order Trachylina; periphery of medusa smooth; gonads on radial canals; Liriope and Aglantha typical.

Trachypterus. Genus of marine ribbonfishes or dealfishes (Order Allotriognathi); body compressed and deep anteriorly but tapering toward the tail.

tract. 1. Series of tubular organs concerned with the passage of materials necessary in animal metabolism, e.g. digestive tract, respiratory tract. 2. In the central nervous system, a bundle of nerve fibers having the same origin, termination, and function.

tractellum. Anterior flagellum of Mastigophora; used mostly for pulling. Pl. tractella.

Tragelaphus. Strepsiceros.

tragopan. Any of several spp. of Tragopan, a genus of brightly colored Asiatic pheasants having wattles and two erectile horns on the head.

Tragulidae. Family of mammals in the Suborder Ruminantia; includes the chevrotains, mouse deer, and napu; small, hornless, deerlike spp. of southeastern Asia and western Africa; males with large canine teeth; Tragulus typical.

Tragulus. See NAPU and TRAGULIDAE.

tragus. 1. Small flap which can be pulled down over the ear drum in certain mammals, e.g. bats. 2. Cartilaginous projection in front of the external auditory meatus in man.

transamination. Transfer of amino nitrogen from one amino acid to another.

Transennella. Genus of marine bivalve mollusks; in coarse sand substrates.

transformer gene. Recessive gene in Drosophila which, when homozygous, will transform diploid females into sterile males.

transient immunity. Temporary immunity.

Transition zone. One of Merriam's North American life zones, located between the Canadian and Upper Austral zones; found as an east-west strip through most of the northern states and extending south through the Appalachians and in many large areas in the West; east of longitude 100° this zone is often called the Alleghenian zone.

translocation. 1. In genetics, the transfer of a chromosome part to a different or non-homologous chromosome. 2. Also, the two chromosomes resulting from such transfer.

transmethylation. Physiological transfer of methyl radicals from choline, etc. to other substances in the body.

transparent Lyonsia. Lyonsia arenosa, a small clam of the North Atlantic.

transplantation. Removal of part of an organism from its normal location to an abnormal location on the same or another individual.

transverse ark shell. Arca transversus, a strongly-ribbed marine bivalve; New England to Fla.

transverse colon. That portion of the large intestine which extends transversely across the abdominal cavity; lies between the ascending and the descending colon.

transverse fission. Asexual reproduction by division of an animal body at right angles to the long axis; common in ciliate protozoans, many flatworms, etc.

transverse process. Lateral projecting process on each side of a vertebra in tetrapods.

transversospinalis system. Especially in reptiles, the series of small muscles connecting the vertebral spines and transverse processes.

transversus abdominalis. Broad flat muscle lying beneath the obliquus internus and forming the innermost muscle layer of the body wall of certain vertebrates.

trapdoor spiders. See CTENIZIDAE.

trapezium. Small carpal bone in typical mammals.

trapezius. Muscle between the scapula and vertebral column; draws head backward or sideways and rotates scapula.

trapezoid bone. Small carpal bone in typical mammals.

Trapezostigma. Genus of dragonflies.

trauma. Any wound or injury, especially one of major importance.

Travisia. Genus of spindle-shaped Polychaeta.

tree-badger. See HELICTIS.

tree crickets. Group of slender, mostly light-colored, grasshopper-like orthopterans in the Family Gryllidae; usually in trees, bushes, and other vegetation; the males stridulate, especially on warm summer evenings.

tree duck. See DENDROCYGNA.

tree frog. See HYLIDAE.

treehopper. See MEMBRACIDAE.

tree kangaroo. Any kangaroo of the arboreal genus Dendrolagus; New Guinea and Australia.

tree line. Timber line.

tree mouse. See DEOMYS.

tree oyster. See ISOGNOMON.

tree salamander. Aneides lugubris, a brownish salamander of western Calif.; on the ground and in trees; 5 in. long; dorsal surface brownish.

tree shrews. Small, squirrel-like primates of the Oriental Region; Tupaia a common genus. See TUPAIOIDEA.

tree snail. Any of many kinds of tropical snails which live chiefly in trees and shrubbery.

tree swallow. See IRIDOPROCNE.

tree toad. See HYLA.

tree-toad chigger. See HANNEMANIA.

Tremarctos. Genus of spectacled bears; small and shaggy, with yellow rings around the eyes and a creamy muzzle, throat, and chest; Andes.

Trematoda. One of the three classes of the Phylum Platyhelminthes; the flukes; all parasitic; body covered with a resistant cuticle; one or more muscular suckers on the external surface; proctostome anterior; gastrovascular cavity usually with two main branches; several thousand spp.; familiar genera are Gyrodactylus, Fasciola, Clonorchis, and Schistosoma.

Trembley, Abraham. Swiss naturalist (1700-1784); noted for his detailed work on the natural history of freshwater hydroids.

Tremex. Genus of hymenopterans in the Family Siricidae; larvae infest deciduous trees.

trench fever. Relapsing or five-day nonfatal fever produced by a rickettsia transmitted in the bite of the body louse.

trepang. Eviscerated, dried sea cucumbers prepared as a human food, especially in the Orient; bêche de mer.

trephine. 1. Special saw for removing a disk of bone from the skull. 2. To operate upon a skull with a trephine.

trephocyte. Type of lymph or blood cell which transports substances from one cell to another; especially in invertebrates.

Trepobates. Genus of small water striders in the Family Gerridae.

Trepomonas. Genus of fresh-water, parasitic, and coprozoic Protozoa; cell flattened, often lobed, and with four flagella on each side.

Treponema. Genus of slender spirochaetes parasitic in birds and mammals; motile, slender, spiral rods; cause syphilis and yaws in man.

treppe. Staircase phenomenon; gradual increase in the extent of muscular contraction (up to a maximum level) produced by rapidly repeated stimulations.

Treptostomata. Order of fossil Bryozoa; zooecium limy, of two parts; operculum present; Ordovician to Permian; Amplexopora and Hallopora typical.

Triacanthagyna. Genus of dragonflies.

triacetin. Glyceryl trinitrate (nitroglycerin); a poisonous and explosive compound which is a vasodilator and cellular narcotic.

Triactinomyxon. Genus of cnidosporidian in the intestinal epithelium of oligochaetes.

Triaenodes. Genus of Trichoptera.

Triaenophorus. Genus of tapeworms in the Order Pseudophyllidea; adults in fresh-water fishes, larvae in copepods.

Trialeurodes vaporariorum. Common and widely distributed pest insect in the homopteran Family Aleyrodidae; the greenhouse white fly; a general feeder on many plants.

triangle spider. See HYPTIOTES.

triangularis. Muscle which originates on the lower edge of the mandible and inserts on the lower lip near the corner of the mouth; pulls down corner of mouth.

Triannulata. Genus of small leechlike oligochaetes commensal and parasitic on western U.S. crayfishes; in the Family Branchiobdellidae.

Triarthra. Filinia.

Triarthrus. Common genus of trilobites.

Triassic period. Earliest of three geological subdivisions of the Mesozoic era; lasted from 190 million to 155 million years ago; an age of rivers and flood plains, temperate to subtropical climates, and conifers and cycads; characteristic animals were bony fishes, first mammals, and giant reptiles.

Triatoma. Genus of bugs which transmit Chagas' disease in Central and South America; cone-nosed bugs (Family Reduviidae).

Triaxonida. Hexactinellida.

triaxon spicules. Three-rayed spicules in some sponges.

tribe. Taxonomic category inserted between the genus and subfamily; in zoology, tribe names customarily end in ini.

Tribolium confusum. Common beetle in the Family Tenebrionidae; confused flour beetle; larvae very destructive to flour and other cereal products; extensively used for the experimental study of population dynamics.

tricarboxylic acid cycle. Citric acid cycle.

Tricellaria. Genus of marine Bryozoa.

triceps brachii. Extensor muscle along the back of the upper arm in tetrapods; usually has three points of origin on the pectoral girdle and humerus.

triceps femoris. Muscle on the anterior surface of the thigh in certain tetrapods.

Triceratops. See CERATOPSIA.

Trichechidae. Family in the Order Sirenia; includes the manatees.

Trichechus latirostris. Common American manatee or sea cow (Order Sirenia) found in coastal rivers and estuaries from Fla. to Central America and in the West Indies; two other spp. in northeastern South America and West Africa; sluggish, ponderous animal with no hind limbs and tail rounded and flattened; to 15 ft. long; feed on aquatic vegetation. Cf. DUGONG.

trichina. Common name of Trichinella spiralis.

Trichinella spiralis. Nematode which causes trichinosis in man, hog, and rats; man becomes infected by eating poorly-cooked pork containing encysted larvae; these develop into adults in the small intestine, and the larvae which are then produced get into the circulatory system and become lodged in the body everywhere and encyst in the voluntary muscle tissues; causes diarrhea, muscular pain, fever, and death in severe cases.

Trichinelloidea. Order of nematodes usually included as a portion of the Suborder Dorylaimina.

trichinosis. Trichiniasis; see TRICHINELLA SPIRALIS.

trichites. 1. Minute rods associated with the cytostome and cytopharynx wall of many ciliates and a few flagellates; project into endoplasm and facilitate the intake of food particles. 2. Hairlike siliceous spicules in certain sponges; often arranged in fascicles.

Trichiuridae. See CUTLASS FISH.

Trichiurus. See CUTLASS FISH.

Trichloris. Genus of triflagellate fresh-water Protozoa; cell bean-shaped, chromatophore large.

trichobothrium. 1. Invertebrate sensory hair or cluster of hairs, especially in arthropods. 2. Hair-bearing spot in certain insects. Pl. trichobothria.

Trichobranchus. Genus of tube-building Polychaeta.

Trichocephalus. Trichuris.

Trichocerca. Large and common genus of creeping or free-swimming rotifers; lorica a single cylindrical piece; toes spinelike, with several small basal spinules; occasional in marine habitats.

Trichocorixa. Common genus of water boatmen (aquatic bugs in the Family Corixidae).

trichocysts. Abundant, minute, rodlike organelles imbedded in the ectoplasm of many ciliates; each is a small mass of fluid; by means of mechanical or chemical stimulation they are extruded through minute surface pores; the extruded trichocysts harden into gelatimous needle-like structures immediately on contact with water; thought to function in attachment and defense.

Trichodectes. Genus of "bird lice" (Order Mallophaga) parasitic on mammals; T. bovis is common on cattle; T. canis is common on dogs and may serve as the intermediate host for the tapeworm Dipylidium caninum.

Trichodina. Genus of peritrich ciliate Protozoa; low barrel-shaped, with a large, toothed horny attachment disc; commensal on fresh-water animals.

trichogen. Cell which produces a hair or bristle in insects.

Trichogramma evanescens. Minute hymenopteran in the Family Trichogrammatidae; female oviposits in eggs of other insects, including those of many pest spp.

Trichogrammatidae. Family of minute insects in the Order Hymenoptera; oviposit in eggs of other insects and the resulting larvae parasitize the enclosing egg; Trichogramma the most important genus.

Trichomonas. Large genus of pyriform flagellates in invertebrate and vertebrate digestive and reproductive tracts; four anterior flagella and another along the edge of an undulatory membrane; T. tenax is a commensal in the human mouth, T. hominis is a commensal in the human large intestine, and T. vaginalis is a mild parasite of the human vagina.

Trichoniscus. Terrestrial genus of Isopoda.

Trichonympha. Genus of complex flagellates living in the digestive tract of termites and roaches; symbiotic, necessary for digestion of the cellulose eaten by the host; many flagella arranged in longitudinal rows.

Trichopelma. Genus of compressed fresh-water holotrich ciliates; surface longitudinally furrowed.

trichophore. 1. Saclike structure which produces an annelid seta. 2. Trichopore.

Trichophrya. Genus of fresh-water and marine suctorian Protozoa; rounded or elongated, unstalked, and with groups of tentacles.

trichopore. Trichophore; opening in an insect cuticle below the base of the trichogen.

Trichoptera. Order of insects which includes the caddis flies; adults soft-bodied, with two pairs of wings coated with hairs and scales; complete metamorphosis; mouth parts feebly biting; antennae and legs long; larvae aquatic, most spp. living in distinctive movable or fixed elongated cases in running or standing water; cases constructed of sand grains, gravel, or bits of vegetation cemented together with a salivary secretion; pupal stage usually passed in the modified larval case; common genera are Hydropsyche and Limnephilus; about 5000 spp.

Trichopterygidae. Family of minute beetles usually found in foliage; many spp. less than 0.25 mm. long.

Trichosomoides. Genus of long, slender nematodes; parasites in birds and mammals.

Trichospira. Genus of cylindrical, sapropelic, fresh-water holotrich ciliates.

Trichostemma. Genus of small oval marine sponges (Subclass Monaxonida).

Trichostomata. Large suborder of Ciliata in the Order Holotricha; peristome lined with rows of free cilia and containing cytostome.

Trichostrongylus. Genus of rhabditid nematodes parasitic in the stomach and small intestine of sheep, goats, and other ruminants.

Trichosurus. Common Australian genus of brush-tailed phalangers; body sturdy and compact; face foxlike or kangaroo-like; nocturnal and arboreal.

Trichotria. Genus of rotifers; lorica one heavy boxlike piece.

Trichotropis borealis. Northern hairy-keel; a small, top-shaped, Atlantic snail with "hairy" periostracum.

Trichurata. Chromadorida.

Trichuris trichiura. Whipworm, a nematode parasitizing the caecum and large intestine of man; infection results from the ingestion of eggs in contaminated food and water; the anterior portion of the worm is long and filamentous.

Trichys. Genus of rat-porcupines inhabiting mountains of Borneo, Sumatra, and Malacca.

triclad. Any member of the flatworm Order Tricladida.

Tricladida. Order of large Turbellaria; gastrovascular cavity with three highly diverticulated branches; marine, fresh-water, and terrestrial; Dugesia, Procotyla, and Bdelloura are familiar genera.

Triconodonta. Fossil order of Jurassic mammals of uncertain phylogenetic relationships; jaws long, teeth relatively undifferentiated.

Tricorythodes. Common genus of Ephemeroptera.

tricuspid. Having three points or cusps.

tricuspid valve. Valve separating the right atrium and the right ventricle in the mammalian heart; consists of three membranous flaps and prevents backflow when the ventricle contracts.

Tridacna. Genus of giant marine clams of coral reefs of the tropical Pacific and Indian oceans; to 4.5 ft. long and 550 lbs.

Tridactylidae. Family of orthopteran insects which includes the pygmy sand crickets; very small burrowing spp. in damp areas near water; Tridactylus typical.

Tridactylus. See TRIDACTYLIDAE.

trifurcate. Forked into three parts.

trigeminal nerve. One of the fifth pair of cranial nerves of vertebrates; innervates nasal, maxillary, and mandibular areas; a mixed nerve.

triggerfish. See BALISTIDAE and PLECTOGNATHI.

Triglidae. Family of marine fishes which includes the sea robins; see PRIONOTUS.

Triglopsis thompsoni. Deep-water sculpin (Cottidae) found in deep lakes from Arctic Canada to the Great Lakes.

Trigona. Genus of minute to small tropical stingless honeybees.

trigonid. High, well-developed triangular area of a primitive mammalian lower molar.

Trigonoporus. Genus of polyclad Turbellaria.

triiodothyronine. Substance formed from thyroxin in the general tissues; more active than thyroxin.

Trilobita. One of the classes in the the Phylum Arthropoda; the trilobites; now extinct, but abundant in Cambrian to Permian times; 1 to 70 cm. long and composed of head, trunk, and pygidium; body with two longitudinal furrows dividing it into a thick median rachis and two flat lateral pleura; trunk of 2 to 29 segments; head with five pairs of appendages, and all trunk segments except last with a pair of elongated sppendages, each bearing an epipodite; all marine, mostly bottom dwellers; more than 10,000 described spp.

Trilobus. Large and common genus of enoplid nematodes; mostly terrestrial and fresh-water soils.

Trilophodon. Miocene genus of ancestral elephants.

Trimerorhachis. Genus of primitive labyrinthodonts.

trimorphic. Animal having three fundamental forms or castes, such as the honeybee. Cf. DIMORPHIC and POLYMORPHIC.

Trimorphodon. See LYRE SNAKE.

Trimusculus. Genus of marine limpets.

Trimyema. Genus of ovoid, flattened, holotrich ciliates; long caudal cilium; main ciliation in several spiral rows; fresh and salt waters.

Trinema enchelys. Amoeboid protozoan found in fresh water and damp soils; cell mostly invested in a hyaline ovoid test, compressed anteriorly, and built of circular siliceous scales.

Tringa solitaria. Solitary sandpiper; an olive-brown sp. of Alaska to Wash. and Ore.; winters in Tropical America.

trinomial. Designation of a plant or animal by a scientific name composed of three parts, the last being a subspecies designation. See SUBSPECIES.

Triodontophorus. Genus of parasitic nematodes; one important sp. occurs in the horse digestive tract.

triolein. Common animal and vegetable fat.

Trionychidae. Family which includes the soft-shelled turtles of N.A., Asia, and Africa; Trionyx the common American genus; see TRIONYCHOIDEA.

Trionychoidea. Small superfamily of turtles; body depressed, oval, and with a rubbery or leathery carapace; head long, with a tubular snout; feet well webbed; includes the soft-shelled turtles (Trionychidae).

Trionyx. Common genus of American soft-shelled turtles; to 35 lbs.; excellent for the table; common except in north and middle Atlantic drainages, mountain states, and Pacific slope; see TRIONYCHIDAE and TRIONYCHOIDEA.

Triopha. Genus of tuberculous nudibranchs.

Triops. Genus of Anostraca; similar or identical to Apus.

Triphora. Genus of minute marine snails with long spire.

triphosphopyridine nucleotide. TPN; important substance in cellular carbohydrate metabolism and in citric acid oxidation; same as coenzyme II.

triple-banded basket shell. Nassa bivittata; a small dog whelk of Atlantic Coast sand flats.

tripletail. See LOBOTES.

triploblastic. Of or pertaining to the fundamental body structure of metazoans where embryonic ectoderm, mesoderm, and endoderm layers eventually give rise to all adult tissues. See DIPLOBLASTIC.

triploidy. Abnormal 3n chromosome number in an animal.

Tripneustes. Genus of large tropical sea urchins; test to 6 in. in diameter.

tripton. Collectively, all of the dead suspended particulate matter in aquatic habitats; abioseston. Cf. PLANKTON and SESTON.

Tripyla. Common genus of microscopic, free-living nematodes.

Trischistoma. Genus of microscopic,

trisomic. Aberrant chromosome condition in which one extra chromosome is present, the cell complement totaling 2n + 1 chromosomes.

Tritella. Genus of caprellid Amphipoda.

Triticella. Genus of marine Bryozoa; usually on legs or in branchial chambers of crabs.

tritocerebral commissure. Narrow transverse band of nervous tissue connecting the circumesophageal connectives immediately anterior to the subesophageal ganglion in certain arthropods.

tritocerebrum. Collectively, the ganglia of the third segment of arthropods; chiefly an embryological term.

Tritogonia verrucosa. Fresh-water bivalve mollusk (Unionidae); Ala. and Miss.

Triton. Common genus of European salamanders.

triton shell. Any of several very large marine gastropods; shell long, heavy, and wrinkled and ridged.

Tritonalia. Ocenebra.

tritonymph. Third instar of many Acarina.

Triturus. Common genus of salamanders (newts, efts) in the Family Salamandridae; the American spp., formerly included in this genus, are now placed in Notophthalmus and Taricha; Triturus cristatus is the European crested newt; Diemyctylus an alternate.

triungulin. Active first larval stage of Strepsiptera and certain hypermetamorphic beetles.

Trivia. Genus of small, ovoid, tropical marine gastropods; coffee-bean shells.

trivial name. 1. Common or vernacular name, as distinct from the scientific name. 2. The sp. name, as distinct from the generic name in a scientific name.

trivium. Collectively, those three arms of a sea star which are farthest from the madreporite.

troch. Band of cilia found in a trochophore and related larval types.

trochal disc. Coronal disc.

trochanter. 1. Second segment of an insect leg; also the second segment of the legs of certain other arthropods. 2. Prominence on the upper part of the femur in many vertebrates.

trochantin. 1. Small sclerite at base of insect coxa. 2. Articular sclerite on Orthoptera mandible. 3. Smaller trochanter of the femur in vertebrates.

Trochelminthes. Discarded phylum name used for the Rotatoria and Gastrotricha, or the Rotatoria alone.

Trochilia. Genus of marine and freshwater holotrich ciliates; ovate, with posterior spine.

Trochilidae. Large family of birds (319 spp.) in the Order Micropodiformes (Apodiformes); hummingbirds; small to very small birds restricted to North and South America; mostly tropical and subtropical; 14 U.S. spp.; bright iridescent coloration; beak awl-shaped; tongue very long and tubular; wings beat very rapidly, often with a humming sound; feed on insects and nectar; able to hover at flowers while nectar is sucked up; nest delicate, cup-shaped, and attached to a twig; Archilocus and Selasphorus common in U.S.

trochilus. Any of a wide variety of birds, especially certain hummingbirds and Old World warblers.

Trochilus. Genus of Jamaican hummingbirds.

trochlear nerve. One of the fourth pair of cranial nerves of vertebrates; innervates the superior oblique muscle of the eye; a motor nerve.

Trochocystites. See CARPOIDEA.

trochophore. Generalized type of minute, translucent, free-swimming larva found in several invertebrate groups, including many marine turbellarians, nemerteans, brachiopods, Phoronidea, Bryozoa, mollusks, sipunculids, and some annelids; the body is more or less pear-shaped with the larger end uppermost; externally, it is provided with a prominent circlet of cilia and sometimes one or two accessory ciliary circlets; commonly there are also anterior and posterior ciliary tufts; internally, there is a complete digestive tract, usually a pair of excretory tubules, muscle and nerve fibers, sense organs, and a band of mesoderm.

Trochosphaera. Genus of rare globular rotifers having a superficial resemblance to a trochophore.

trochosphere. Trochophore.

Trochospongilla. Small genus of freshwater sponges.

Trochostoma. Genus of burrowing sea cucumbers; tentacles unbranched; without tube feet on the trunk.

trochozoon. Hypothetical ciliated free-swimming ancestral trochophore-like form which is thought by some zoologists to be the phylogenetic forerunner to the Mollusca, Bryozoa, Annelida, and Rotatoria.

trochus. Inner ciliated ring or zone at the anterior end of a rotifer.

Trochus. Very large cosmopolitan genus of marine snails; shell conical, with a flat base and flattened whorls; top shells.

Troctes divinatorius. Minute insect (1 mm.) in the Order Corrodentia; book louse or cereal psocid; often a pest because it feeds on prepared cereals, book bindings, herbarium specimens, and dried zoological specimens.

troglobiont. Animal restricted to caves and/or underground waters.

Troglocambarus maclanei. Florida cave crayfish.

Troglodytes. Genus which includes several common spp. of wrens, e.g. the house wren (T. troglodytes) and winter wren (T. aedon); the latter winters in southern U.S. and Mexico.

Troglodytidae. Widely distributed family of insectivorous passerine birds which includes the wrens; small, energetic, brown-backed spp., usually smaller than sparrows, with slender bills; tail usually held more or less vertically; several American genera.

Trogloglanis pattersoni. Toothless blindcat; a small, blind, pale catfish known only from artesian wells at San Antonio, Tex.

Troglotrema salmincola. Intestinal fluke of dogs and wild fish-eating mammals, especially in northwestern U.S.; the first intermediate host is a snail, and the second intermediate host is a salmon or trout.

Trogon elegans. Common coppery-tailed trogon of southern Ariz., and Mexico (bird Order Trogoniformes); profile slightly parrot-like; head and upper parts green; under parts bright red.

Trogonidae. Only family of birds in the Order Trogoniformes; includes the trogons.

Trogoniformes. Order of birds in tropics of both hemispheres; trogons and quetzal; non-migratory forest spp. which feed on fruits and insects; bill short and broad, edges toothed, and with stout bristles at the base; legs small and weak; plumage brilliantly metallic in color; Trogon and Pharomacrus typical.

troilus butterfly. A common swallowtail butterfly; upper surface of wings a clouded greenish color.

Trombicula. Large genus of blood-sucking mites; certain spp. transmit the rickettsia of tsutsugamushi fever.

Trombidiformes. Taxon (suborder) sometimes designated to include certain families of mites; one pair of stigmata at mouth, or stigmata absent; chelicerae piercing.

Trombidiidae. Family of robust mites covered with short hairs; harvest mites, chiggers, red bugs; larvae parasitic on insects or vertebrates; Eutrombicula alfreddugesi the most common of several American spp. attacking the skin of man and causing severe dermatitis.

Trombidoidea. Superfamily of mites; usually free-living.

Tropaea luna. Large light green moth

(luna moth) with purple-brown markings and long tail-like projections on the hind wings; wing expanse to 6 in.; especially common in the Ohio and Mississippi river valleys.

trophallaxis. 1 Exchange of food between an adult and larval insect; e.g. in certain Hymenoptera, a worker taking food to a larva receives a drop of saliva from the larva. 2. Mutual licking, shampooing, and exchange of food and exudates between adult insects.

trophectoderm. Trophoblast.

trophi. 1. Collectively, a set of minute, complicated, sclerotized jaws imbedded in the mastax of rotifers. 2. Collectively, the insect mouth parts.

trophic. 1. Pertaining to nutrition. 2. Pertaining to productivity.

trophic hormone. A stimulator hormone, e.g. those of the anterior pituitary.

trophic level. One of the several successive levels of nourishment in a pyramid of numbers, food web, or food chain; plant producers constitute the first (lowest) trophic level, and dominant carnivores constitute the last (highest) trophic level.

trophoblast. Layer of extra-embryonic nutritive ectoderm which forms an outermost layer of a blastodermic vesicle; in mammals it is part of the chorion and attaches to the uterine wall.

trophocytes. 1. Cells which form nutritive material, especially fat. 2. Cells of an insect fat body. 3. Nutritive cells of an insect gonad. 4. Fat cells used as a food source, especially in insects.

trophoderm. Trophoblast.

trophodisc. 1. Female gonophore in certain Hydrozoa. 2. Endodermal mass in certain hydrozoan gonophores; nourishes developing eggs or sperm.

Trophon. Genus of small North Atlantic snails.

trophosome. Collectively, all of the asexual structures in a polypoid hydrozoan coelenterate colony.

trophothylax. Pocket in the first abdominal segment of the larvae of certain ants.

trophozoite. Protozoan during active feeding stage in its life history; term used especially with reference to parasitic spp.

tropical rain forest. Dense evergreen dicot forest characteristic of lowland tropical areas with more than 60 or 80 in. of rain per year, high temperatures, and high humidity; a heterogeneous mixture of many spp. of trees, rather than continuous stands of one, two, or three spp.; occurs in Central America, Amazon drainage, Orinoco drainage, central and western Africa, Madagascar, Indo-Malay area, Borneo, New Guinea, etc.

Tropical zone. One of Merriam's North American life zones; in the U.S. it consists of the lowermost one-third of the Florida peninsula, the region of Brownsville, Tex., and the valley of the Colorado River along the California-Arizona border.

tropic bird. See PHAËTHON.

Tropicorbis. Genus of fresh-water snails; Tex., La., Central and South America.

Tropidoclonion. See STRIPED SWAMP SNAKE.

Tropidonotus. Genus of reptiles which includes the European ring snake.

tropin. Opsinin.

tropism. Response of plants and sessile animals to a stimulus, especially light and gravity; growth curvature or turning reaction, such as the downward growth and proliferation of the basal part of a colonial hydrozoan and the upward growth of the functional polyps.

Tropisternus. Common genus of hydrophilid beetles.

trout. See SALMO, SALVELINUS, and SALMONIDAE.

troutperch. See PERCOPSIDAE.

trumpeter. 1. See PSOPHIIDAE. 2. Trumpeter swan. 3. Breed of domestic pigeons. 4. Any of numerous kinds of marine fishes which emit a trumpeting sound when taken out of water.

trumpeter swan. See CYGNUS.

trumpet fishes. See AULOSTOMUS.

trumpet shell. See SKENEA.

trumpet worms. See PECTINARIA.

truncate. 1. Having the extremity cut off, flattened, square, or even. 2. Lacking a normal apex, as in certain snail shells.

truncated borer. Barnea truncata, a common marine bivalve to 3 in. long; bores in clay or rocks.

Truncilla. Genus of unionid mollusks; Mississippi drainage system.

truncus arteriosus. Heavy muscular base of the aorta just as it leaves the ventricle in lower vertebrates.

trunkfish. See OSTRACIIDAE.

Tryngites subruficollis. Buff-breasted

sandpiper; breeds in Arctic America; winters in Argentina.

Tryonia clathrata. Small fresh-water snail of certain Great Basin streams; shell elongated and ribbed.

Trypanoplasma. Cryptobia.

Trypanorhyncha. Order of tapeworms in the Subclass Cestoda; scolex with four bothria and four spiny proboscides; adults in the intestine of sharks and rays; intermediate hosts include a wide variety of marine invertebrates and fishes; Tetrarhynchus, Tentacularia, and Haplobothrium common.

Trypanosoma. Protozoan genus in the Family Trypanosomidae; parasitic in the vertebrate circulatory system; usually transmitted by blood-sucking arthropods; T. lewisi in rats, T. evansi in domestic animals in the Orient, T. brucei in African game mammals; T. gambiense and T. rhodesiense cause two kinds of human sleeping sickness in Africa, and T. cruzi causes Chagas' disease in Central and South America.

Trypanosomidae. Family of flagellates parasitic in some plants and many metazoans; cell flat and elongated; usually the flagellum forms the outer margin of the undulating membrane which extends along one side of the body.

Trypanosyllis. Genus of Polychaeta.

Trypetesa. Genus of barnacles in the Order Acrothoracica.

Trypetidae. Family of small to medium-sized flies; fruit flies; adults visit flowers, fruit, and foliage; larvae feed on fruits and other parts of plants in the Family Compositae; many pest spp.; Ceratitis and Dacus important genera.

trypsin. Enzyme which hydrolyzes proteins, in neutral or slightly alkaline solution, to smaller polypeptides; secreted by the vertebrate pancreas into the small intestine but also occurs in many invertebrates; an active form of trypsinogen, formed as the result of the activator action of enterokinase.

trypsinogen. Inactive form of trypsin secreted by the pancreas; see TRYPSIN and ENTEROKINASE.

tryptophane. Abundant essential amino acid; $C_{11}H_{12}O_2N_2$.

tsetse fly. One of a group of Diptera (Glossina) which transmit African sleeping sickness from one human being to another; slightly larger than a housefly.

tsutsugamushi fever. Scrub typhus; a human rickettsial disease of the southwestern Pacific and southeastern Asia; characterized by a skin ulcer at the site of inoculation, damage to small blood vessels all over the body, chills, dizziness, fever, and skin eruptions; sometimes fatal; transmitted by the bite of mites in the genus Trombicula.

tuatara. See RHYNCHOCEPHALIA.

tube feet. Numerous small fluid-filled closed muscular tubes in echinoderms; project outside the body and have a muscular bulblike ampulla at the internal end; used in clinging, food handling, locomotion, and respiration; see WATER VASCULAR SYSTEM.

Tubella pennsylvanica. Fresh-water sponge; thin, slimy growth form; eastern U.S.

tube-nosed bat. See HARPIOCEPHALUS.

tuber cinereum. Small conical process of the subthalamus on the ventral surface of the mammalian brain; lies between the mammillary bodies and the infundibulum.

tuberculin. Sterile extract of tuberculosis bacteria; used for diagnostic skin tests, especially in children.

tuberculosis. Contagious, infectious disease caused by the bacterium Mycobacterium tuberculosis; although chiefly a human disease, strains also occur in cattle, birds, and other vertebrates; usually transmitted through the sputum of a person having pulmonary tuberculosis, less frequently through milk from tuberculous cows; bacteria form characteristic nodules in infected tissues, especially the lungs, bones, intestines, lymph tissues, and skin; infection spreads easily through the blood and lymph.

tuberous sclerosis. Epiloia.

tube shell. See CAECUM.

tube-web spider. See ATYPIDAE.

tubiculous. 1. Living in a constructed tube. 2. Pertaining to a spider which spins a tubular web.

Tubifera tenax. Sp. of drone fly; larva is a common rat-tailed maggot found in shallow stagnant puddles and has a long posterior respiratory tube; occasionally produces intestinal myiasis.

Tubifex. Common genus of small fresh-water oligochaetes; anterior end buried in the substrate, and posterior end waves and undulates rapidly for aeration and respiration; sometimes in short vertical chimney-like tubes; often present in enormous numbers, especially in polluted waters.

Tubipora. Genus of alcyonarian anthozoans; the organ-pipe coral; a colony consists of long parallel upright polyps supported by skeletal tubes composed of fused spicules; the tubes are united at various levels by transverse connections; usually associated with coral reefs.

Tubulanus. Genus of nemertines in the Order Palaeonemertea; mostly in marine littoral.

tubular gland. Any gland composed of a simple tubular cavity or a series of connected tubular cavities.

Tubularia. Genus of sessile, colonial, marine hydrozoan coelenterates.

Tubulidentata. Order of African mammals which includes the aardvarks (Orycteropus); body piglike, with few hairs and long ears and snout; tongue slender, sticky, and protrusible, used for picking up ants and termites, the chief food; legs and claws powerful, used in burrowing and for tearing up ant and termit nests; mostly nocturnal; Africa.

Tubulifera. One of the two suborders in the insect Order Thysanoptera; females without a special ovipositor.

Tubulipora. Genus of Bryozoa in the Order Cyclostomata.

tucotuco. See CTENOMYIDAE.

tufted titmouse. See BAEOLOPHUS.

tui. New Zealand bird noted for its ability to mimic speech; glossy black with white patches on neck and shoulders; Prosthemadera novaseelandiae.

tularemia. Disease of rodents transmitted by the bites of flies, fleas, ticks, and lice; may be acquired by man through handling of infected animals; caused by the bacterium Pasteurella tularensis which produces an ulcer at the point of inoculation and inflammation of the lymph glands.

tulip mussel. Modiolus tulipa, a small striped mussel found from N. C. to the West Indies.

tulip shell. See FASCIOLARIA.

tullibee. See LEUCICHTHYS.

Tulotoma. Genus of nodulous fresh-water snails; Alabama River system.

tumblebug. Dung beetle; see SCARABAEIDAE.

tumbler. Variety of pigeons which turn somersaults while in flight.

tumbling flower beetles. See MORDELLIDAE.

tumor. 1. Swelling or enlargement. 2. Mass of new tissue or neoplasm which persists and grows independently of its surrounding tissues; malignant tumors continue growing and eventually kill the patient; benign or non-malignant tumors grow slowly, sometimes stop growing, and do not recur after surgical removal.

tuna. See THUNNUS, EUTHYNNUS, and NEOTHUNNUS.

tundra. Treeless areas of high latitudes and altitudes; characterized by growths of low herbs, lichens, mosses, and grasses.

Tunga penetrans. Sp. of burrowing flea (Order Siphonaptera) common in tropical South America and Africa; chigoe, chigger, jigger; pregnant females burrow into the skin of domestic animals and man, especially under the toenails.

tunic. Cuticular layer covering body surface of all members of the Subphylum Tunicata; secreted by underlying thin body wall.

tunica adventitia. Outermost fibro-elastic layer of various tubular organs, such as arteries, vas deferens, esophagus, uterus, ureter, etc.

tunica albuginea. Fibrous membrane enclosing an organ or other part.

tunica intima. Innermost endothelial layer of an artery or vein.

tunica media. Middle layer of arteries and veins.

tunica mucosae. Mucous membrane, especially of the digestive tract.

tunica muscularis. Double layer of muscle tissue (circular and longitudinal) in the wall of the digestive tract.

Tunicata. Subphylum in the Phylum Chordata; includes the tunicates, ascidians, salps, and sea squirts; highly modified marine chordates having a more or less cylindrical or globular shape; body wall covered with a secreted cuticular tunic; sessile, free-swimming, or suspended at surface; solitary or colonial; with two prominent distal openings (mouth and cloacal aperture); coelom small; large perforated food-gathering pharynx suspended in an extensive peribranchial chamber; circulatory and nervous systems simple; hermaphroditic, but reproduction by budding or sexually; larva a tadpole-shaped creature with notochord, gill slits, and dorsal nerve cord; about 1600 spp.

tunica vaginalis. Peritoneum lining of the mammalian scrotum.

tunicin. Polysaccharide resembling cellulose, found in the tunic of the Tunicata.

tunny. Tuna.

tun shell. See TONNA.

Tupaioidea. Suborder of primates which includes the tree shrews of southeastern Asia; insectivorous and arboreal.

Tupinambis. Genus of tropical American lizards in the Family Teiidae; large, carnivorous, active spp.

tur. One of several wild goats (Capra) of the Caucasus and Pyrenees.

turban shells. See TURBINIDAE.

Turbatrix. See ANGUILLULA.

Turbellaria. One of the three classes of the Phylum Platyhelminthes; free-living flatworms; epidermis ciliated; usually hermaphroditic but reproduction by transverse fission in some spp.; proctostome usually ventral; no suckers or hooks; marine, fresh-water, and terrestrial; Microstomum and Dugesia typical.

turbinal. In the shape of a scroll-like whorl or spiral, e.g. the conchae or whorls of thin bone in the nasal passages of certain higher vertebrates.

turbinate. 1. Spiral shaped but with the spirals decreasing rapidly in diameter from base to apex. 2. Turbinal.

Turbinidae. Large family of marine snails; turban shells; up to 8 in. long; mostly in warm seas; Turbo common.

Turbo. See TURBINIDAE.

Turbonilla. Common and cosmopolitan genus of marine snails.

turbot. Common name applied to several European flatfishes, especially halibuts and flounders.

Turdidae. Large family of cosmopolitan passerine songbirds; includes the thrushes, robins, and bluebirds.

Turdus. See ROBIN.

turkey. See MELEAGRIDAE.

turkey blackhead. An important disease of turkeys; transmitted in the eggs of a nematode (Heterakis gallinae) from one turkey to another; a protozoan infection.

turkey buzzard. See CATHARTES.

turkey vulture. See CATHARTES.

Turko-Tatar. Race of mongoloids found from Turkey through central Asia; skin light yellow, head short and high, face long.

Turner, William. English naturalist (1510-1568); wrote a book on natural history of birds, but much of the material was borrowed from Aristotle and Pliny.

Turnicidae. Small family of birds in the Order Gruiformes; the hemipodes; quail-like, but seldom fly; to 8 in. long; southern part of Eastern Hemisphere.

Turnix. Button quails; genus of small three-toed quail-like birds of grassy plains in southern Eurasia and northern Africa; mostly brownish with blackish markings.

turnover number. Number of molecules of a substrate acted upon per minute by one molecule of an enzyme; a specific characteristic of the enzyme.

turnover rate. In ecology and physiology, that fraction of a component in an animal or population which is released (or enters) per unit time; e.g. the amount of phosphorus which is taken into an animal per day with reference to the amount present in the whole body.

turnover time. In ecology and physiology, the time necessary to replace the total quantity of a component in an animal or population of animals.

turnstone. See ARENARIA.

Turris. Genus of marine coelenterates in the Suborder Anthomedusae; common and widely distributed.

Turritella. Cosmopolitan genus of long, slender-spired, sculptured, marine snails; mostly in deep waters.

Turritopsis. Genus of marine coelenterates in the Suborder Anthomedusae.

Tursiops. See DOLPHIN.

turtle. 1. In the broad sense, any member of the reptilian Order Chelonia; see CHELONIA, TERRAPENE, TESTUDO, KINOSTERNON, CHELYDRA, MACROCHELYS, DERMOCHELYS. 2. In the narrow sense, any aquatic chelonian.

turtledove. 1. Any of several small Old World wild doves with a long slender tail; noted for plaintive call and tameness. 2. U.S. mourning dove.

Turtonia. Circumboreal genus of very small marine bivalves.

tusk shell. Common name of a member of the mollusk Class Scaphopoda.

tussock moth. See LYMANTRIIDAE.

twig snake. See THELOTORNIS.

two-lined salamander. One of several small (3 in.) salamanders in the genus Eurycea found east of the Mississippi River; swim freely; dorsal color yellowish to brownish, limited by a black dorsolateral stripe on each side.

two-toed sloth. See BRADYPODIDAE.

tyee salmon. See ONCORHYNCHUS.

Tylenchida. Order of phasmid nematodes; esophagus composed of a cylindrical corpus which is swollen posteriorly, a narrow isthmus, and a swollen posterior bulbar region; mouth surrounded by eight, six, or no lips; mouth cavity armed with a protrusile spear or stylet; feed on fluids of living cells, or in the hemocoel of insects; examples are Tylenchus and Heterodera.

Tylencholaimus. Genus of microscopic, free-living nematodes.

Tylenchus. Genus of tylenchid nematodes found chiefly in damp soils; feed mostly on fluids of plant cells; the meadow nematodes.

Tylobolus. Genus of millipedes.

Tylodina. Soft-shelled limpet-like gastropod living on certain sponges, especially Verongia.

Tylopoda. Suborder of the mammal Order Artiodactyla which includes the Family Camelidae (camel, llama, alpaca, vicuna, and guanaco); stomach in four parts; feet soft and broad; Lama and Camelus.

Tylos. Genus of terrestrial Isopoda.

tylus. In Hemiptera, the distal part of the clypeus.

tympanic canal. Lowermost of the three compartments of the cochlea of the inner ear.

tympanic cavity. Cavity of middle ear, between ear drum and auditory capsule.

tympanic membrane. Thin, membranous eardrum which receives external vibrations of the air and transmits them to the middle ear.

Tympanuchus. Genus which includes the prairie chickens (Family Tetraontidae); large, henlike birds of prairies and brushy grasslands; at least one sp. was formerly abundant east of the Mississippi River; the extinct heath hen of the New England area is also in this genus.

tympanum. 1. Tympanic membrane of vertebrates. 2. Membrane on either side of the basal abdominal segment of the Orthoptera; part of the auditory apparatus; similar membranes occur on the legs of certain insects. 3. Complex organ on each side of the metathorax or base of the abdomen in many Lepidoptera; thought to be an auditory organ. 4. In the prairie chicken and other grouse, a bare area on the neck which marks the position of an inflatable sac during mating dances.

type. Single designated plant or animal specimen that serves as the basis for the original name and description of any taxon, such as a type sp., type genus, etc. See HOLOTYPE.

type locality. Specific geographic area in which the type specimens of a particular taxon were first collected.

type species. Particular sp. in a genus that is first designated as having the characteristics of the genus in which it occurs.

type specimen. The single specimen, or one of a group of similar specimens, which serves as the basis for the description and name of a new sp.

Typhlichthys. See AMBLYOPSIDAE.

Typhlocoelum. Genus of monostome trematodes; in ducks.

Typhlogobius. See PINKFISH.

Typhlomolge rathbuni. Colorless blind salamander; permanently aquatic and neotenic; underground waters near San Marcos, Tex.

Typhlonectes. Genus of viviparous caecilian amphibians; aquatic throughout life in northern South America; T. compressicauda attains a length of 3 ft.

Typhlopidae. Family of small, shining, burrowing snakes of the tropics and subtropics; worm snakes; teeth in both jaws; feed mostly on soil insects; Typhlops typical.

Typhloplana. Genus of rhabdocoel Turbellaria; zoochlorellae abundant in parenchyma.

Typhlops. See TYPHLOPIDAE.

typhlosolar blood vessel. One of two median blood vessels in most segments of typical annelids; receives blood from the dorsal vessel and delivers it to the typhlosole.

typhlosole. Longitudinal thickening or fold which projects downward into a digestive cavity thereby increasing the absorptive surface; e.g. in the stomach-intestine of the earthworm and in the intestine of many bivalve mollusks.

Typhlotriton. Genus of whitish blind salamanders found in caves of Mo. and nearby Kan.

typhoid fever. Infectious bacterial disease obtained from contaminated water or food; marked by fever, aches, prostration, and abdominal rash; bacteria especially abundant in kidney and small intestine; complications frequent; vaccination useful.

typhus fever. Human disease produced by Rickettsia prowazeki; transmitted through the feces and into the incision made by the biting of human head and body lice (Pediculus humanus); the rickettsia attacks the endothelial cells, and the mortality rate is 5 to 60% in untreated cases; the disease breaks out under excessively crowded, unsanitary conditions.

Typosyllis. Genus of Polychaeta.

Tyrannidae. Large family of passerine birds which includes the New World flycatchers, peewees, kingbirds, and phoebes.

Tyrannosaurus. Genus of large, bipedal, carnivorous, Cretaceous dinosaurs; to 47 ft. long; head 4 ft. long, bearing teeth 3 to 6 in. long; hind legs massive; digits with large powerful claws.

Tyrannus. Genus of kingbirds in the Family Tyrannidae; North and South America.

Tyroglyphus. Common genus of minute mites which live on dead or decaying animal and plant material; includes the cheese mite, often found on unpackaged cheeses.

tyrosine. Common amino acid; $C_9H_{11}O_3N$.

Tyto alba. Common barn owl; face white or buff and heart-shaped; coloration buff above, vermiculated and spotted with gray and white; essentially cosmopolitan except for Asia north of the Himalayas.

Tytonidae. Family of birds in the Order Strigiformes; includes the barn owls; facial disc heart-shaped; 13 to 18 in. long; non-migratory but world-wide.

U

uacari. See CACAJAO.

ubiquinone. Coenzyme Q; cellular coenzyme of the electron transmitter system.

Uca. Genus of marine decapod crustaceans; fiddler crabs; body wide and short; front of carapace expanded; one of the two chelae in the male is greatly enlarged and held horizontally across the front of the body; abundant along muddy beaches, mud flats, and salt marshes, where they make their burrows, often 8 to 24 in. deep.

udder. Large pendant mammary gland provided with two or more nipples, as in cows.

Ugrian. Race of mongoloids in the Arctic of Europe and Asia; skin light yellow to brown, hair black to red, cheekbones prominent; includes Lapps, Tungus, and Chukchi.

Uintatherium. Genus of Eocene mammals almost as large as modern elephants but with three pairs of large protuberances and horns on the dorsal surface of the head; canine teeth well developed; to 7 ft. tall.

ulcer. Open sore which heals slowly and with difficulty; usually on skin, exposed mucous membranes, or internal mucous membranes; develops from injuries, poor tissue nourishment, and parasitic infections.

ulna. Posterior of the two bones in the forearm of most tetrapods.

ulnare. Small wrist bone at the end of the ulna in tetrapods.

ulnocarpalis. Ventral muscle of the hand of amphibians.

Uloboridae. Family of hackle-banded weaver spiders; spin geometric orb webs or sectors of orbs; includes the feather-legged spiders which have prominent tufts of setae on some of the leg segments; Uloborus common.

Uloborus. See ULOBORIDAE.

Ulocentra. Genus of darters now usually included in Etheostoma.

ultimobranchial bodies. Small epithelial bodies which appear on the last gill pouch of vertebrate embryos; fate variable and obscure; function unknown.

ultracentrifuge. Any centrifuge having an extremely high rate of rotation; separates and sediments large organic molecules.

ultramarine jay. See APHELOCOMA.

ultramicroscope. Microscope utilizing lateral or scattered light rather than a vertical parallel beam, thus rendering exceptionally small objects visible; the substage is illuminated from the side, and objects then appear light against a dark field.

ultramicroscopic. 1. Visible with an ultramicroscope but not with an ordinary microscope. 2. Too small to be seen with any type of optical microscope. 3. Too small to be detected with any type of device, including the electron microscope.

ultraviolet. Just beyond the visible violet end of the spectrum; wave lengths between 2000 and 4000 Å; such rays have pronounced actinic and chemical properties.

Uma notata. Fringe-toed lizard of the Colorado and Mojave deserts; gray or white coloration; to 5 in. long.

Umbellula. Genus of deep-sea Pennatulacea; clusters of polyps at the distal end of a long slender stem which has a bulbous base imbedded in the mud.

umbilical cord. Tough, stalklike projection from the ventral surface of the embryo of placental mammals; contains blood vessels and connects the fetus to the placenta; severed at birth.

umbilicate. 1. With a small depression. 2. Navel-shaped. 3. Having an umbilicus.

umbilicus. 1. Umbilical cord. 2. One of two small openings into the quill of a typical feather; the superior umbilicus is at the upper end of the quill, just below the origin of the first barbs; the inferior umbilicus is at the tip of the quill. 3. In gastropod shells, the small chink or slit between the reflected inner lip and the body whorl.

umbo. 1. One of the prominences on either side of the hinge region in a bivalve mollusk; marks the point of origin of growth in the juvenile bivalve. 2. Posterior lower "beak" of a brachiopod shell. 3. Boss or rounded eminence. Pl. umbones.

Umbra. See UMBRIDAE.

umbrella ant. Any of several tropical leaf-cutting ants which carry a bit of leaf over the body in an umbrella-like manner.

umbrella bird. South and Central American bird having an umbrella-like crest on the head.

umbrette. Hammer-head; a brown, storklike, African wading bird; body 2 ft. long, bill large, head crest well developed; nest a dome-shaped mass of twigs and mud with a small side entrance.

Umbridae. Family of carnivorous bony fishes which includes the mud minnows; two hardy spp. of Umbra in the U. S. and one in southeastern Europe; 4 in. usual length; dorsal fin more or less posterior; tail fin rounded; scales large; common in ponds and stagnant waters with muddy bottoms.

Ummidia. Common genus of trapdoor spiders of southeastern states; cylindrical burrow in the ground closed with a corklike lid.

unau. Two-toed sloth, Choloepus hoffmanni, of Tropical America.

unciform bone. Distal carpal bone in mammals.

uncinate. Hooked at the tip.

uncinate mastax. Rotifer mastax in which the trophi are specialized for laceration of ingested food.

uncinus. 1. Small hooked or hooklike structure. 2. Small hooked seta of certain Polychaeta. 3. Marginal tooth of the radula in certain Gastropoda.

Unciola. Genus of marine Amphipoda.

uncus. One of two lateral toothlike pieces in the trophi of a rotifer; simple to complex structure. Pl. unci.

underfur. Fine, dense, inner coat of fur as contrasted with an outer, coarser, protective coat of guard hairs.

underwing moth. See CATOCALA.

undulant fever. Brucellosis.

undulating membrane. 1. Thin, living membrane extending out and along one side of certain parasitic flagellates; a flagellum forms the outer margin of the undulating membrane. 2. Thin, living membrane in certain ciliates; composed of one or two linear rows of cilia, which are more or less fused; most commonly in or near the cytopharynx; longer than a membranelle.

Unguiculata. Taxon of mammals which have claws or nails; includes the following living orders: Insectivora, Dermoptera, Chiroptera, Primates, Edentata, and Pholidota.

unguiculate. 1. Having nails or claws. 2. Any mammal with nails or claws.

unguis. Fingernail, toenail, anterior surface of a claw, or the anterior surface of a hoof.

ungulate. 1. Having hoofs. 2. Any hoofed mammal.

unguligrade. Pertaining to animals that walk on the enlarged nail or hoof, such as the horse.

unicellular. Consisting of one cell.

unicorn. 1. Mythical horselike animal having a single, large, straight, twisted horn. 2. The narwhal (Monodon).

unicorn shell. Any of several marine snails having a finger-like spine on the aperture of the shell.

Unio. Elliptio.

Uniomerus. Genus of fresh-water bivalve mollusks (Unionidae); central and southern U. S.

Unionicola. Common genus of Hydracarina; mostly parasitic or commensal in gills or mantle cavity of fresh-water mussels.

Unionidae. Large, cosmopolitan family of fresh-water mussels; inner surface of shell pearly, outer surface tan to black and often sculptured; glochidium present; especially abundant in central U. S.; includes nearly all U. S. fresh-water mussels.

uniovular twins. Monozygotic twins.

uniparous. Pertaining to an animal that produces only one egg or young at a time.

unipennate muscle. Any muscle in which the fibers are arranged obliquely and attach to a tendon along one side.

unipolar. Having a single process, as a nerve cell.

uniramous appendage. Arthropod appendage with a simple longitudinal disposition of segments. Cf. BIRAMOUS APPENDAGE.

unisexual. Of or pertaining to an animal which produces either eggs or sperm, but not both; dioecious.

unisexual reproduction. Any type of

reproduction involving a parent of only one sex (female). See PARTHENOGENESIS.

univalent. Chromosome which does not posess, or does not become associated with, its corresponding homologous chromosome during synapsis.

univalve. 1. Any mollusk having a one-piece shell. 2. Pertaining to a one-piece shell or carapace.

universal donor. Person having blood type O and whose blood cells contain neither antigen A nor antigen B.

universal symmetry. Spherical symmetry.

univoltine. Pertaining to organisms which have only one generation during a growing season; usually applied to birds and mammals.

upland plover. See BARTRAMIA.

Upogebia. Common genus of burrowing shrimplike marine decapod crustaceans (Suborder Anomura).

Upper Austral zone. One of Merriam's life zones, located between the Transition and Lower Austral zones; east of the Rockies found as a broad, irregular, east-west strip through the central one-third of the U. S.; west of 105° found as large areas at mid- and lower elevations; east of 100° this zone is often called the Carolinian zone, and west of 100° it is often called the Upper Sonoran zone.

Upper Sonoran zone. See UPPER AUSTRAL ZONE.

Upupidae. See HOOPOE.

ur. Aurochs.

urachus. Canal in the fetus of certain mammals; connects the bladder with the allantois.

uracil. A pyrimidine base; component of ribonucleic acid; $C_4H_4O_2N_2$.

Urania. Genus of diurnal, brightly colored West Indian and South American moths.

Uranoscopidae. Stargazers; a family of tropical marine fishes; head cuboid, mouth almost vertical, eyes on upper surface of head; often nearly buried in sand bottom.

Uranotaenia. Genus of mosquitoes.

Urceolaria. Genus of aberrant peritrich ciliates; peristome oblique; basal end a horny attachment disc; commensal on planarians.

Urceolus. Genus of colorless marine and fresh-water monoflagellate Protozoa; cell flask-shaped and plastic.

urchin. Sea urchin.

urea. One of the chief products of protein metabolism in animals; in vertebrates it is formed in the liver and excreted by the kidneys; $CO(NH_2)_2$.

urease. Enzyme which converts urea into ammonia and carbon dioxide; in many seeds, some invertebrates, blood, gastric mucosa, bacteria, and other fungi.

Urechis caupo. Sp. of echiurid worm especially common in shallows along Calif. coast; to 18 in. long; builds U-shaped burrows and secretes a mucous net which entraps food particles swept through the burrow by peristaltic contractions of the worm.

uremia. Presence of urinary compounds in the blood owing to faulty renal function.

ureotelic. Pertaining to those animals whose chief break-down and excretory product of amino acids is urea by way of the ornithine cycle; characteristic of fish, amphibians, turtles, and mammals.

ureter. Duct connecting the metanephric kidney with the urinary bladder in higher vertebrates.

urethra. Duct which carries urine from the urinary bladder to the exterior in mammals; it is joined by the vas deferens (one or two) in the male.

Uria. Genus of birds in the Family Alcidae; includes the murres; widely distributed, gregarious, oceanic birds which breed along rocky shores, especially at high latitudes; head, neck, and upper parts dark colored; under parts light colored; excellent swimmers and divers.

uric acid. Main excretory product of proteins and nucleic acids in uricotelic animals; although man is not uricotelic, the urine does contain some uric acid derived from metabolism of purine bases; $C_5H_4O_3N_4$.

uricotelic. Pertaining to those animals whose chief breakdown and excretory product of amino acids is uric acid; chiefly in lizards, snakes, birds, insects (except Diptera), and terrestrial gastropods.

urinalysis. Chemical, physical, and biological examination of the urine.

urinary bladder. Sac used primarily for the temporary retention of fluid or semifluid excretory wastes; present in some invertebrates as well as in vertebrates; in the latter it stores urine which originates in the kidneys; absent in birds.

urinary tubule. Long convoluted tubule

leading away from Bowman's capsule in the kidney of reptiles, birds, and mammals; urine is formed by differential reabsorption and secretion in these tubules which ultimately unite and empty their contents into the ureter. See MALPIGHIAN BODY and NEPHRON.

urine. Excretory product of vertebrates and many invertebrates; usually formed in kidneys or other excretory structures; in birds and reptiles urine is solid or semisolid; in man the urine is 96% water and contains certain salts, acids, urea, uric acid, and other dissolved organic materials.

uriniferous tubule. Urinary tubule.

urinogenital. Urogenital.

Urnatella gracilis. Only known fresh-water sp. of Entoprocta (Class Urnatellida); rare in running waters of eastern U.S.

Urnatellida. Class of the Phylum Entoprocta; small colonies formed from a basal plate; uncommon in fresh water; Urnatella gracilis typical.

urn bodies. Minute, more or less ovoid, vase-shaped, multicellular structures found in the coelom of certain Sipunculoidea; move about by ciliary action and accumulate excretory granules; some eventually are voided to the outside through the nephridia.

Urobatis. Genus of sting rays.

Urocentrum. Genus of short, ovoid, fresh-water holotrich Ciliata; constricted behind middle; tuft of fused cilia at posterior end.

Urochordata. Tunicata.

urochrome. Yellow pigment in the urine; a breakdown product of hemoglobin.

Urocoptis. Large genus of West Indian and Fla. land snails; associated with limestone outcrops.

Urocyon cinereoargenteus. Common N. A. gray fox; essentially omnivorous; found in a variety of habitats, but not so common in cold areas as the red fox; numerous subspecies.

Urodela. Caudata.

urodeum. Ventral portion of the monotreme cloaca with which the bladder connects.

urogenital duct. In the broad sense, any duct which carries both urine and reproductive cells, as in the male frog.

urogenital papilla. Small protuberance bearing a common reproductive and excretory pore; common in certain lower vertebrates, such as fishes.

urogenital system. Collectively, the organs of reproduction and excretion in vertebrates.

Uroglena. Genus of yellowish-brown fresh-water flagellates; ovoid biflagellate cells arranged in the periphery of a spherical gelatinous mass; all cells connected by means of fibrils which run inward and meet in the center of the mass.

Uroglenopsis. Genus of yellowish-brown fresh-water flagellates; ovoid biflagellate cells arranged in the periphery of a spherical gelatinous mass; when extremely abundant it imparts a fishy odor to water.

Uroleptus. Genus of marine and fresh-water hypotrich Ciliata; elongated, posterior end tail-like.

Urolophis. Genus of round sting rays.

Uronema. Genus of marine and fresh-water holotrich ciliate Protozoa; cell ovoid, anterior end not ciliated.

Uronychia. Genus of marine hypotrich ciliate Protozoa; posterior cirri very large.

Uropeltidae. Family of small burrowing snakes inhabiting India and Ceylon; Uropeltis typical.

Uropeltis. See UROPELTIDAE.

Urophycis. See GADIDAE.

uropod. Appendage of the posterior abdominal segment(s) in certain Malacostraca; three stout pairs used for jumping in Amphipoda, one tactile pair in Isopoda, and one pair making up much of the tail fan in the crayfish, etc.

Uropygi. Schizomida.

uropygial gland. Large gland which opens on the dorsal side of the posterior end of the body in birds; secretes an oily fluid used in preening feathers.

uropygium. Prominence at the posterior end of the body of a bird; attachment of the tail feathers.

Urosalpinx. Genus of small marine snails with a shell having a ridged surface and a scalloped lip; U. cinerea is the common oyster drill or borer of the Atlantic Coast; it is 1 in. long and destroys oysters by drilling a small hole in the shell and sucking out the juices by means of a long proboscis.

Urosoma. Genus of fresh-water hypotrich Ciliata.

urosome. 1. General posterior or abdominal region of an arthropod. 2. Tail region of a fish.

urostege. One of the large ventral scales posterior to the anus in most snakes.

Urostyla. Genus of marine and fresh-water hypotrich Ciliata; ellipsoid, flexible.

urostyle. Long, unsegmented posterior part of the vertebral column in the Anura.

Urotricha. Genus of small, oval to conical, fresh-water holotrich Ciliata; one or more long caudal cilia and several short tentacles around mouth.

Urozona. Genus of short, ovoid, fresh-water holotrich Ciliata; constricted in middle; with a posterior tuft of fused cilia.

Ursala butterfly. Common butterfly in the genus Basilarchia; upper side black and pale blue or green.

Ursidae. Family of mammals which includes the bears; large spp. with short, heavy legs and plantigrade feet with five digits; tail very short; molars of the crushing type with broad, flat crowns; American genera are Euarctos, Ursus, and Thalarctos.

Ursus. Genus of omnivorous bears (Family Ursidae); includes the various spp. of grizzly bears and Alaskan brown bears; coloration brownish to yellowish; U. horribilis (grizzly) the best known sp., originally common in western N. A. but exterminated in most areas; U. gyas, U. middendorffi (Kodiak bear), and their relatives are the giant brown bears (to 1500 lbs.) of the Alaskan Peninsula and associated islands.

urticaria. Sudden rash accompanied by severe itching; usually lasts only a few days; arises from irritation of internal mucous membranes, emotions, or menstruation.

Urticina. Genus of large, stout sea anemones (Order Actiniaria).

urus. Large, extinct blackish-brown wild ox of Europe (Bos primigenius); same as aurochs and ur; thought to be ancestral to domestic cattle; by crossing several primitive breeds of Mediterranean cattle, a Berlin zoologist recently obtained offspring which are markedly similar to the original urus.

Uta. Genus of lizards which includes the swifts and utas; small, flattened, active lizards with small scales; western states and Mexico, especially in deserts.

uterine bell. Anterior portion of the female reproductive tract in Acanthocephala; by peristalsis it takes up eggs from the pseudocoel and passes them posteriorly.

uterus. 1. In female mammals (excluding monotremes), a muscular expansion of the reproductive tract in which the embryo and fetus develop; usually paired, but single in primates; opens externally by way of the vagina. 2. Womb. 3. Portion of the female reproductive tract in many invertebrates.

uterus bicornis. Bicornate uterus.

uterus bipartitus. Uterus which is almost completely divided along the median line; with a single opening into the vagina; carnivores, ruminants, pigs, and horses and their relatives.

uterus duplex. Double uterus, each with a separate opening into the vagina; rodents.

uterus simplex. Single median uterus, as in primates.

utriculus. Saclike portion of the membranous labyrinth of the inner ear; the three semicircular canals and sacculus originate from it, and its chief function is in equilibration.

uvea. Middle coat of the eyeball; composed of the choroid, ciliary muscle, and iris.

uvula. 1. Small pendant fleshy lobe at the posterior edge of the soft palate. 2. Lobe on the lower surface of the cerebellum. 3. Prominence on the inner lower anterior surface of the bladder.

V

vaccinate. To inoculate with any vaccine or antigen, the purpose being to instigate the production of corresponding antibodies which will have a protective effect against future natural exposures to the antigen.

vaccination. Act of inoculating with a vaccine or antigen.

vaccine. Suspension of dead or weakened disease-producing organisms (antigen) introduced into animal tissues which will induce the formation of antibodies specifically antagonistic toward that antigen; often consists of dead bacteria, e.g. the use of dead typhoid bacteria for vaccination against typhoid fever.

Vacuolaria. Genus of fresh-water biflagellate Protozoa; with green chromatophores; cell plastic, anterior end narrow.

vacuole. Any clear, fluid-filled internal cavity bounded by a plasmalemma in a cell; some vacuoles are quite small and abundant, others are large. See CONTRACTILE VACUOLE and FOOD VACUOLE.

vacuome. System of cell vacuoles staining with neutral red.

vagility. Relative power of dispersal or ability to overcome ecological barriers; e.g. the housefly and English sparrow have high vagilities.

vagina. Part of the female reproductive duct; receives the male penis or cirrus during copulation; present in mammals and certain invertebrates.

Vaginicola. Genus of fresh-water and marine peritrich ciliates; cell surrounded by a vase-shaped lorica.

vagus nerve. One of a pair of tenth cranial nerves of vertebrates; arises on the medulla oblongata and innervates (where present) the larynx, trachea, lungs, esophagus, stomach, intestine, gall bladder, lateral line organs, heart, gastric glands, and pancreas; a mixed nerve.

Vahlkampfia. Genus of very small amoebae; fresh-water or parasitic.

Valentinia. See ACORN MOTH.

valine. Common amino acid; can be converted to glucose in the body; $C_5H_{11}O_2N$.

Valkeria. Genus of marine Bryozoa.

vallate papillae. Group of 8 to 12 large papillae arranged in the form of a V at the back of the tongue.

Vallentinia. Genus of marine hydrozoan coelenterates with well developed medusae.

valley perch. See HYSTEROCARPUS.

Vallonia. Common genus of small land snails.

Valvata. See VALVATIDAE.

Valvatidae. Family of small, hermaphroditic fresh-water snails; shell depressed, sometimes almost discoidal; operculum multispiral; Valvata common.

valve. 1. One of the two shells in a typical bivalve mollusk or brachiopod. 2. One of the sheath pieces of an ovipositor in certain insects. 3. Flap- or cuplike structure which permits flow of a fluid in only one direction through a vessel, and prevents backflow.

valve of Heister. Mucosa folds at the junction of the neck of the gall bladder and the cystic duct.

valves of Kerkring. Large transverse folds in the small intestine of mammals.

vampire bat. Any bat which bites and laps up vertebrate blood or is erroneously believed to feed on blood; few spp. are true vampires; see DESMODUS.

Vampyrella. Genus of amoeboid Protozoa (Order Proteomyxa) which penetrate algal cells and feed on the contained protoplasm.

Vampyrops. Genus of striped leaf-nosed bats of Tropical America.

Vampyrum. Genus of large leaf-nosed bats of Amazon forests.

Van Beneden, Edouard. Belgian zoologist (1846-1910); demonstrated constancy in chromosome number in body cells; discovered reduction in chromosomes during maturation and restoration as the result of fertilization.

Van Beneden, Pierre-Joseph. Belgian parasitologist (1809-1894); author of several important parasitology texts.

Van Cleave, Harley J. American zoologist (1886-1953); professor at the University of Illinois; active researcher in several groups of invertebrates but best known for his contributions on Acanthocephala.

vane. Collectively, the many closely-spaced parallel barbs coming out on either side of the central shaft of a typical bird feather.

Vanessa. Genus of common N. A. butterflies; upper surface of wings usually various patterns of purplish black, tawny, and brownish, often with white spots; includes Hunter's butterfly, red admiral, and painted lady (V. cardui).

vanga-shrikes. See VANGIDAE.

Vangidae. Small family of arboreal passerine birds; vanga-shrikes; bill stout, hooked; Madagascar.

Van Leeuwenhoek, Antony. Dutch microscopist (1632-1723); made early simple microscopes with magnifications as high as X270; examined and drew a wide variety of organisms, objects, and materials with these instruments and reported his observations in a long series of letters to the Royal Society of London.

Van Slyke apparatus. Apparatus used for precise measurement of gas absorbed or produced, originally designed for analysis of blood gases, but adapted for many different chemical analyses involving gases; two basic types are available, volumetric and manometric, the latter being more precise.

Van't Hoff's law. Q_{10} law; doubling of the rate of a chemical reaction for every ten-degree C. increase in temperature; not a precise generalization; as indicated by many biological processes and activities, the measured rate is usually 1.5 to 4.0.

Varanidae. Family of very large carnivorous and carrion-feeding lizards occurring in southern Asia, Africa, and Australia; monitors; tongue extremely long and cleft; neck and tail long; body exceptionally muscular; Varanus salvator, to 8 ft. long, in Ceylon and the Malay Archipelago; V. komodoensis (dragon lizard), to 12 ft. long and the largest of all living lizards, found especially on the island of Komodo.

Varanus. See VARANIDAE.

variable coloration. Coloration which varies from one season to another (ptarmigan), or from one short time to another, depending on the background coloration (flatfishes, chameleon).

variable wedge shell. See POMPANO SHELL.

varicose veins. Permanently distended and tortuous veins, especially in the legs.

varied thrush. See IXOREUS.

variety. Term often used loosely to designate various infraspecific forms; more precisely, the term is restricted to discontinuous variants within a single interbreeding population.

variola virus. One of several related viruses; cause smallpox, chickenpox, etc.

varves. Thin layers of sediments laid down on the bottoms of certain lakes; usually one such layer is formed each year, especially during the spring runoff near river inlets; the deposits contain microfossils whose deposition may be accurately dated.

vas. General term for a duct, vessel, or tube. Pl. vasa.

vasa deferentia. Pl. of vas deferens.

vasa efferentia. Pl. of vas efferens.

vasa vasorum. Small blood vessels in the walls of large blood vessels.

vascular. 1. Full of vessels, especially blood vessels. 2. Pertaining to vessels, especially circulatory vessels.

vas deferens. 1. Tube carrying sperm from the testis or epididymis to the cloaca or urethra in reptiles, birds, and mammals. 2. General term applied to a tubule carrying sperm in many invertebrates. Pl. vasa deferentia.

vasectomy. Vasotomy.

vas efferens. 1. One of the many tubules carrying sperm from the testis on each side to the mesonephros or epididymis in vertebrates. 2. In invertebrates, the small collecting ducts which bring sperm to the vas deferens. Pl. vasa efferentia.

vase sponge. Any of several marine sponges having the general shape of a vase; Callispongia common.

Vasicola. Genus of marine and freshwater holotrich Ciliata; cell annulated, stalked, and in a vase-shaped lorica.

vasoconstriction. Decrease in diameter

(constriction) of small blood vessels, especially arterioles, resulting in a decrease in blood supply to an organ or tissue.

vasodilation. Increase in diameter of small blood vessels, especially arterioles, resulting in an increase in blood supply to an organ or tissue.

vasomotor. Pertaining to constriction and dilation of blood vessels.

vasomotor nerves. Vertebrate sympathetic nerves which control the dilation and constriction of arterioles.

vasopressin. Neurohormone secreted by the posterior lobe of the pituitary; it constricts arterioles, thereby raising blood pressure, and promotes resorption of water by the kidney tubules.

vasotomy. Surgical cutting of the one or more vasa deferentia so as to render a male animal incapable of fertilizing the female, owing to the absence of sperm in the seminal fluid.

vastus externus. Prominent muscle on the anterolateral surface of the thigh in certain tetrapods; straightens the knee.

vastus internus. Muscle of the anterior surface of the thigh in certain tetrapods; extends the leg.

vastus medialis. Median thigh muscle of man; originates on the femur and inserts on a tendon in the knee region; extends the leg and rotates the patella inward.

vector. Animal which transmits parasites from one host to another; e.g. the female Anopheles mosquito is the vector of human malaria.

veery. Thrush common in overgrown woodlands of eastern U.S.; upper parts cinnamon brown, under parts white to creamy; Hylocichla fuscescens.

vegetable pole. Misnomer for vegetal pole.

vegetal hemisphere. Egg and early embryo hemisphere containing the vegetal pole; opposite animal hemisphere.

vegetal pole. 1. One of two general regions in the early embryo of many animals; it consists of large cells containing an abundance of yolk and is opposite the animal pole. 2. That portion of the ovum or zygote which is rich in yolk granules.

vein. 1. Blood vessel carrying blood toward the heart and away from the tissues and organs; relatively thin-walled. 2. Thickened, rodlike portion of an insect wing; supports and stiffens the wing.

Vejovis. Common western and southwestern genus of scorpions.

velarium. Narrow shelflike peripheral extension in a few jellyfish; the peripheral tentacles are attached at the upper basal surface of the velarium.

velar tentacles. Tentacles associated with the mouth and velum in cephalochordates; serve to strain out large particles.

Velella. Genus of marine hydrozoan coelenterates; a projecting air-filled float keeps the colony at the surface, and above it is a flat, broad "sail"; the complex colonial system of polymorphic hydroids dangles downward in the water; by-the-wind sailor.

Velia. Widely distributed genus of broad-shouldered water striders (Family Veliidae).

veliger. Free-swimming larva of most marine Gastropoda, Scaphopoda, and Pelecypoda; develops from the trochophore and has the beginning of a foot, mantle, shell, etc

Veliidae. Broad-shouldered water striders, ripple bugs; family of small, gregarious, plump-bodied surface-dwelling hemipterans; common in small streams; Microvelia typical.

Velocitermes. Genus of termites.

velum. 1. Thin, shelflike projection extending inward from the periphery of hydrozoan medusae. 2. Band of cilia in front of the mouth in marine veliger larvae (Gastropoda). 3. Membranous appendage or spine at the distal end of the anterior tibia in many insects. 4. Special membranelle bordering the cytostome in certain ciliates. 5. Membrane surrounding the mouth of cephalochordates; located within the atrium. 6. Flexible transverse plate guarding the entrance to the respiratory tube in the lamprey; also in the oral cavity of Amphioxus. 7. Ciliated ridge or lobed ciliated swimming device of many marine invertebrate larvae. 8. Membrane. Pl. vela.

velum interpositum. Roof of the third or fourth ventricle of the mammalian brain.

velum transversum. Faint transverse depression on the dorsal surface of an early embryonic vertebrate brain; divides the prosencephalon into the telencephalon and diencephalon.

Velutina laevigata. Velvet shell; gastropod with a thin, transparent shell having a large aperture; Atlantic Ocean.

velvet. Soft skin covering the growing antlers of deer or elk.

velvet ants. See MUTILLIDAE.

velvet shell. See VELUTINA.

velvet water bugs. See HEBRIDAE.

vena cava inferior. Posterior vena cava, postcaval vein, postcava.

vena cava superior. Anterior vena cava, precaval vein, precava.

venation. 1. System and arrangement of veins in an insect wing. 2. Arrangement of veins (blood vessels) in any part of an animal body.

venereal disease. Infectious disease usually acquired through sexual intercourse, especially syphilis, gonorrhea, chancroid, etc.

Venericardia. Genus of small marine bivalves; cold waters.

Veneridae. Family of widely distributed marine bivalves; shell heavy, ligament external; tropical spp. often brightly colored.

Venerupis staminea. Rock cockle; a common Pacific Coast bivalve; to 3.5 in. long; an important food sp.

venin. Any of the specific toxic substances found in animal venoms.

venipuncture. Surgical puncture of a vein.

venom. Poisonous secretion of various reptiles, scorpions, bees, etc.; chemical composition highly variable.

venous. Pertaining to vein(s).

venous blood. That blood which is returning to the heart from the tissues and capillaries; low in oxygen and high in carbon dioxide except when returning from respiratory capillaries.

vent. Cloacal aperture; posterior opening of the digestive tract in vertebrates; the term is used correctly only when it functions in the elimination of reproductive products and/or excretory substances in addition to the feces.

venter. Abdomen, belly, or ventral surface.

ventral. Pertaining to the under or lower surface; in human beings, the ventral side (belly) is directed forward and is usually called the anterior surface of the body.

ventral abdominal artery. Artery in the crayfish and many other Malacostraca; carries blood along the ventral portion of the abdomen and supplies the abdominal musculature and posterior appendages.

ventral aorta. Aortic trunk.

ventral fins. Pelvic fins; in fishes; the pair of fins corresponding to the hind limbs of higher vertebrates; usually located on the abdomen, but sometimes as far forward as the throat.

ventral line. Faint longitudinal line visible on the ventral surface of some nematodes; produced by a thickening of the hypodermis which contains the ventral nerve strand.

ventral plate. In many arthropod embryos, a thickened ventral mass of blastoderm.

ventral root. Ventral of the two roots by which spinal nerves originate from the nerve cord in vertebrates; carries only motor neurons; same as motor root.

ventral thoracic artery. Artery in the crayfish and many other Malacostraca; carries blood along the ventral surface of the thorax and supplies the mouth region, esophagus, and anterior appendages.

ventricle. 1. One of the main pumping chambers of the heart; thick-walled and muscular; present in certain higher invertebrates as well as vertebrates; fishes and amphibians have a single ventricle, and reptiles, birds, and mammals have two. 2. One of the several cavities in the chordate brain.

ventricose. Inflated; swollen asymmetrically.

ventriculus. In insects, the same as the stomach.

Ventridens. Common genus of land snails.

ventro-intestinal blood vessel. One of a pair of prominent blood vessels located in the gut wall in most segments of typical annelids; receives blood from the subintestinal vessel and delivers it to the gut wall.

venule. Small vein.

Venus' flower basket. Common name of a sponge (Euplectella) having a complex, delicate, skeletal network of silicon dioxide spicules; when dried and cleaned the skeleton has the shape of a long tubular vase.

Venus flytrap. Small insectivorous bog plant found especially in the Carolinas; leaves consist of two lobes which are edged with spines and hinged in a clamlike fashion; when an insect touches the inner sticky surfaces, the lobes quickly close and entrap the prey; after digestion has progressed, the two lobes reopen.

Venus' girdle. See CESTIDA.

Venus mercenaria. Variously known as hardshell clam, round clam, quahog, cherrystone clam (when small), and little neck clam; on sand or mud bottoms along Atlantic Coast; shell ovate to heart-shaped; to 4.5 in. long; extensively used as food; the shells were the chief source of Indian wampum.

verdin. Small gray titmouse with yellowish head; southwestern states; Auriparus flaviceps.

verge. Male copulatory organ, especially in certain snails.

Vermes. Obsolete taxonomic term which included all more or less wormlike phyla.

Vermetus. Widely distributed genus of sessile marine snails; shell tubular, with the lower whorls free and extended in a linear fashion.

vermicide. Any drug used for treating an infection of parasitic worms.

vermicular. Wormlike in shape, configuration, or motion.

Vermicularia. Vermetus.

vermiform. Worm-shaped.

vermiform appendix. Appendix.

vermifuge. Any drug taken to eliminate parasitic worms from the body; anthelminthic.

vermilion flycatcher. See PYROCEPHALUS.

Vermiliopsis. Genus of Polychaeta.

vermin. Term in common usage which includes all noxious, parasitic, and nuisance animals, e.g. lice, rats, mice, flies, bedbugs, crows, snakes, prairie dogs, etc.

vermis. Median dorsal connecting lobe of the cerebellum.

Vermivora. See MNIOTILTIDAE.

vernacular name. Common or local name for any animal.

Verongia. Genus of West Coast sulfur sponges; coarse-textured clustered growth form.

verrucose. Covered with wartlike projections.

verruga peruviana. Eruption of skin nodules in late stages of Oroya fever.

vertebra. One of the segments of the spinal column; in certain lower vertebrates each vertebra consists of several pieces which never become united, but in the higher vertebrates it is a single fused bony piece bearing various processes and covering the spinal cord; according to their location, vertebrae may be cervical, thoracic, lumbar, sacral, or caudal.

vertebral column. Series of bones or cartilages along the dorsal side of vertebrates surrounding the spinal cord and supporting the body.

Vertebrata. Subphylum of the Phylum Chordata; includes all fishes, amphibians, reptiles, birds, and mammals; enlarged brain enclosed in a cranium, or brain case; segmented vertebral column supports body; head, neck, trunk, and (usually) tail present.

vertex. Top of head of insects, between the eyes.

Vertigo. Large circumboreal genus of small land snails.

vervet. African monkey, Cercopithecus pygerythrus.

Vesalius, Andreas. Flemish anatomist (1514-1564); made many discoveries in human anatomy, most of which are contained in his chief work De Humani Corporis Fabrica; severely criticized because many of his discoveries were not in keeping with the ancient teachings of Galen.

vesicant. Any substance or drug, such as cantharidin, which produces blistering when applied to the skin.

vesicle. 1. Small fluid-filled sac or bladder. 2. Small hollow prominence on the surface of a gastropod shell or coral. 3. One of the three primary brain cavities.

vesicular connective tissue. Large, fluid-filled, polygonal cells forming the notochord of chordates during early development.

Vesicularia. Genus of marine Bryozoa.

vesicular vein. Short paired vein in the posterior abdominal region of certain lower vertebrates.

vesicula seminalis. Seminal vesicle.

Vespa diabolica. Yellow jacket (hymenopteran Family Vespidae); nests in or above ground.

vesper sparrow. See POOECETES.

Vespertilionidae. World-wide family of bats; ten genera in the U.S.

vespiary. Nest or colony of social wasps.

Vespidae. Family of medium to large colonial wasps; yellow and red with black or brown markings; paper wasps, yellowjackets, hornets; nest of papery consistency (chewed wood), often large; colonies arise from single overwintering females; powerful stingers; Vespula, Vespa, and Dolichovespula common.

vespoid. Wasplike.

Vespula. Common genus of hornets.

vestibular canal. Upper compartment of the cochlea of the inner ear.

vestibular glands. Small lubricating glands located at the vaginal opening in certain female mammals.

vestibular membrane. Reissner's membrane.

vestibular nerve. Branch of the auditory nerve; supplies the utriculus and ampullae of the membranous labyrinth.

vestibule. Outer cavity forming an entryway into a deeper cavity; found in the ear, entrance to the mammalian vagina, entrance to the pharyngeal region of certain prochordates, etc.

vestibulum vaginae. Especially in the human female, the triangular space between the clitoris and the labia minora.

vestigial. Small, degenerate, imperfectly developed, and incapable of normal functions, as applied to a part, organ, or tissue; e.g. vestigial wings in certain insects and vestigial muscles associated with body hair in man.

viable. Capable of becoming normally active, especially after being exposed to unfavorable environmental conditions.

vibraculum. Curiously modified zooid found scattered about on the surface of certain marine bryozoan colonies; has the shape of a long whiplike filament and sweeps back and forth across the surface at intervals and thus aids in keeping the colony free of debris. Pl. vibracula.

Vibrio. Genus of short, curved, rodlike bacteria with one to three flagella; widely distributed in polluted water, soils, and many animals; few pathogenic spp., e.g. those that cause Asiatic cholera and abortion in stock animals.

vibrissa. 1. One of the stiff hairs just inside or at the nostril. 2. One of the long whisker-like hairs on the face of the cat and its relatives; also on flanks, elbows, above eyes, etc. 3. One of the hairlike feathers around the mouth of insectivorous birds. 4. One of the curved bristles on each side of the mouth of certain Diptera. Pl. vibrissae.

Viceroy butterfly. A common N. A. nymphalid butterfly with rich brown wings having black and white markings; its color pattern is much different from those of its close relatives, and the fact that it resembles the Monarch butterfly (distasteful to birds) is supposedly an indication of Batesian mimicry.

vicuña. Wild South American herbivorous mammal (Lama vicugna); related to the camel, guanaco, llama, and alpaca; smaller than the guanaco; its silky fawn-colored wool is woven into cloth; mountain areas of Peru, Bolivia, and Ecuador.

villiform. Having the shape of fingerlike processes.

villikinin. Hormone produced by the action of stomach acid on the duodenal mucosa; accelerates movements of intestinal villi.

Villora. Genus of darters now usually included in Etheostoma.

villus. See INTESTINAL VILLI and CHORIONIC VILLI.

vinegar eel. See ANGUILLULA.

vinegar fly. Drosophila melanogaster.

vinegarroon. See WHIP SCORPIONS.

vinegar worm. See ANGUILLULA.

vine snake. Oxybelis aeneus; a long, slender, arboreal and bush-dwelling sp. of Ariz. to Yucatan; drab to gray above.

violet-green swallow. See TACHYCINETA.

violet snail. See IANTHINA.

violet tip. Common anglewing butterfly in the genus Polygonia; posterior part of hind wings violet.

viper. 1. In the general sense, any venomous snake. 2. In the technical sense, any member of the reptile families Viperidae and Crotalidae.

Vipera. Genus of venomous snakes in the Family Viperidae; vipers; Europe, Asia, and Africa; the common European spp. attain 2 ft.

Viperidae. Family of venomous snakes which includes the Old World vipers; paired erectile fangs in front part of upper jaw which are folded backward when not in use; no pit between nostril and eye; common genera are Vipera, Cerastes, and Bitis.

Virchow, Rudolf. German pathologist (1821-1902); founder of cellular pathology.

Vireo. See VIREONIDAE.

Vireonidae. Family of temperate and tropical American arboreal passerine birds; vireos; small olive- or gray-backed spp., slightly smaller than sparrows; feed on insects and fruits; Vireo the typical American genus; several common spp.

virgate mastax. Rotifer mastax used for sucking up fluid contents of plants and animals; trophi of the mastax also used for biting and nibbling.

virgin. 1. Human female who has never had sexual intercourse. 2. In the broad sense, any female animal which has never been impregnated by a male.

virgin forest. Any original forest stand of mature trees which has not been subjected to logging, grazing, or recent severe fire.

Virginia deer. See ODOCOILEUS.

virulent. Exceedingly harmful, infective, or toxic.

virus. Non-cellular living material which is self-perpetuating when present in the cells of a host plant or animal; cannot be seen with an ordinary microscope, but images can be projected with an electron microscope; chiefly protein, but some viruses contain carbohydrates and lipids also; smallpox, measles, and yellow fever are produced by viruses.

viscacha. See LAGOSTOMUS.

viscera. Collectively, all of the soft internal organs of an animal. Sing. viscus (rare).

visceral. Of or pertaining to the viscera.

visceral arch. One of the cartilaginous or bony arches developed in the walls of the pharynx and between gill slits to strengthen and support the gills and pharyngeal region in fishes. See VISCERAL SKELETON.

visceral ganglion. In bivalve mollusks, a ganglion located in the posterior part of the visceral mass, especially near the posterior adductor muscle.

visceral leishmaniasis. Kala azar.

visceral muscle. Nonstriated muscle.

visceral peritoneum. That portion of the peritoneum which covers the visceral organs located in the coelom of higher animals.

visceral pleura. See PLEURA.

visceral skeleton. Series of cartilage or bony bars and plates forming much of the support for the gills and lateral and ventral portions of the pharyngeal and throat region in fishes but reduced to form other structures in higher vertebrates; some cyclostomes have 8 to 15 archlike structures, but the condition, as found in the shark, consists of seven visceral arches and a number of accessory ventral pieces; the first arch (mandibular) forms the upper and lower jaws, the dorsal part of the second arch usually joins the lower jaw to the skull, and the remaining five (branchial) arches serve to support the gill arches; in tetrapods the jaws and malleus and incus, if present, are derived from the first arch, the columella and hyoid bone are derived from the second arch, and cartilages of the laryngeal and pharyngeal region are derived from the remaining arches.

vision. Special sense involving the ability to detect light; in higher animals it includes the perception of images, light intensity, and colors; in many simple animals it involves merely the ability to detect the presence of light with respect to intensity and source.

visual cortex. Posterior part of the cerebral cortex; concerned with vision.

visual purple. Rhodopsin.

vital capacity. Total maximum air capacity of the lungs in excess of the residual volume; about 4000 cc. in men and 3000 cc. in women.

vitalistic concept. Idea that the present organic world is the result of vital forces rather than the ecological, chemical, and physical forces at work.

vital staining. Staining of living cells with non-toxic dyes.

vitamin. One of many organic substances necessary for proper health, growth, reproduction, and metabolism in animals; the following vitamins are probably essential to human beings: A, thiamine, riboflavin, nicotinic acid, folic acid, pyridoxine, pantothenic acid, biotin, B_{12}, ascorbic acid, D, tocopherol, and K.

vitamin A. Vitamin derived from carotene of plants and converted to $C_{20}H_{30}O$ in the intestinal tract; essential to growth, development, and to normal skin and the regeneration of visual purple in the retina; deficiency produces night blindness and xerophthalmia (dry cornea and improper tear secretion); abundant in fish liver oils, eggs, and milk; stored in the liver.

vitamin B_1. Thiamine; aneurin.

vitamin B_2. Riboflavin.

vitamin B_6. Pyridoxine.

vitamin B_{12}. Cobalamin.

vitamin B complex. Series of at least eight water-soluble vitamins, some closely related, some distantly related, and most of which occur abundantly in one or more of yeast, liver, cereal germ, milk, liver, eggs, green leaves, and fruits. See THIAMINE, RIBOFLAVIN, NICOTINIC ACID, PARA-AMINOBENZOIC ACID, CHOLINE, INOSITOL, FOLIC ACID, PYRIDOXINE, PANTOTHENIC ACID, BIOTIN, and COBALAMIN.

vitamin C. Ascorbic acid.

vitamin D. Calciferol or activated 7-

dehydrocholesterol; a vitamin having the formula $C_{27}H_{44}O$ or $C_{28}H_{44}O$; essential to proper calcium and phosphorus metabolism for bone formation; deficiency produces rickets in children and osteomalacia (soft bones), especially in Oriental women; abundant in fish-liver oils, and produced by exposure of skin to ultraviolet radiation when certain precursors are converted to vitamin D; stored in liver.

vitamin E. Tocopherol.

vitamin K. Vitamin complex having the formula $C_{31}H_{46}O_2$; essential for production of prothrombin in liver, necessary for blood clotting; deficiency inhibits blood clotting; abundant in green leaves, and produced by certain intestinal bacteria.

vitamin P. Factor in lemon juice and certain red peppers; deficiency increases capillary permeability; probably a mixture of several organic compounds.

vitamin P-P. Niacin or nicotinic acid.

vitellarium. Yolk gland found in various invertebrates such as rotifers and flatworms. Pl. vitellaria.

vitelline. Pertaining to yolk.

vitelline artery. Embryonic artery supplying blood to the yolk sac area of vertebrates.

vitelline glands. Yolk glands.

vitelline membrane. Plasma membrane of an ovum.

vitelline vein. Embryonic vein draining blood from the yolk sac area of vertebrates.

vitrella. Bundle of two to five clear cells grouped around the refractive crystalline cone in the distal portion of a typical arthropod ommatidium. Pl. vitrellae.

vitreous body. 1. Mass of vitreous humor filling the eyeball behind the lens in vertebrates. 2. Clear portion of the retinal cells in the axial area of an insect eye.

vitreous humor. Jelly-like substance forming the vitreous body and filling the eyeball behind the lens; maintains the shape of the eyeball.

Vitrina. Cosmopolitan genus of very small terrestrial snails.

vitrodentine. Material similar to dentine but harder; covers teeth in sharks and some bony fishes.

vivarium. Enclosure, room, laboratory, or other area used for maintaining colonies of terrestrial animals indoors. Pl. vivaria.

Viverra. See CIVET.

Viverricula. Rasses; genus of small civets of the Oriental region; climb trees readily, feed on fruits and small animals.

Viverridae. Family of small, slender, carnivorous mammals which includes the mongooses and civets; originally in Africa and southern Asia.

viviparous. Pertaining to spp. in which the female produces eggs that are retained and nourished in the uterus or other part of the reproductive system until the young are mature enough to be released to the outside; birth of living young instead of eggs; the situation among mammals. Noun viviparity.

Viviparus. Genus of small fresh-water snails occurring especially in the eastern half of the U.S.

vivisection. Dissection or cutting operation upon a living animal, especially without the use of an anesthetic.

vixen. Female fox.

vocal cord. One of a pair of membranes stretched across the larynx and concerned with sound production; present in most mammals and many other tetrapods.

vocal sac. Inflatable sac located on either side of the base of the head in certain male Anura; acts as a resonator and amplifier for croaking sounds produced by the vocal cords.

volant. Volitant.

vole. Any of a large group of small mouselike rodents, especially in the genus Microtus.

volitant. 1. Specialized for flying or capable of flying. 2. Capable of moving about at a rapid rate.

voluntary muscle. Type of syncitial muscle tissue under voluntary control and characteristic of vertebrates, arthropods, and a few members of other groups; it consists of greatly enlarged cylindrical fibers with nuclei scattered around the edges; the cytoplasm has alternate light and dark banded fibers, and on high magnification the fibers are seen to consist of bundles of minute longitudinal fibrils; by means of connective tissue the fibers are held together in the form of large sheets, bands, or spindle-shaped masses (muscles); voluntary muscle tissue is capable of rapid contraction and relaxation.

voluntary striated muscle. Voluntary muscle.

Voluta. Genus of tropical marine gastropods; vase-shaped shell with tapering spire; many brightly patterned spp.

volvent. See NEMATOCYSTS.

Volvocales. See PHYTOMONADINA.

Volvocidae. Family of green biflagellates forming spherical colonies; fresh waters.

Volvox. Genus of Mastigophora in which the cells are formed into large spherical or subspherical colonies; certain cells become specialized as gametes; Volvox globator has colonies about 0.5 mm. in diameter; holophytic and fresh-water.

vomer. Bone in the skull of most vertebrates; in some vertebrates it is a double bone; in man it is a single bone forming part of the nasal septum; in fishes it is a median bone lying near the front of the roof of the oral cavity.

vomerine teeth. Two patches of small teeth located on the vomerine bones; in the anterior part of the roof of the mouth, especially in many reptiles and amphibians.

vomeronasal organ. Organ of Jacobson; one of two grooves or blind sacs in the roof of the mouth or posterior nasal channel in many tetrapods; presumed to pick up olfactory sensations from the mouth.

Vomer setapinnis. Dollarfish; body very compressed and deep, to 12 in. long; warmer areas of Atlantic Ocean.

Von Baer, Karl Ernst. Esthonian biologist (1792-1876); founder of modern embryology, discoverer of the mammalian ovum, and proponent of the germ layer concept.

Von Behr trout. Brown trout; see SALMO.

Von Frisch, Karl. Austrian zoologist (1886-); famous for his discoveries of message communication among bees by means of their "wagging" and "circling" dances.

Von Haller, Albrecht. Swiss author, botanist, and physiologist (1708-1777); made basic discoveries on muscle irritability; wrote an eight-volume work on human physiology.

Von Humboldt, Alexander. German scientist and explorer (1769-1859); noted especially for his extensive explorations and natural history observations in tropical Central and South America between 1779 and 1804; with Bonplant, he published the results of these expeditions in 23 volumes.

Von Mohl, Hugo. German botanist (1805-1872); suggested (along with Purkinje) the use of "protoplasm" for living cells and the contents of cells (in plants).

Von Siebold, Karl Theodor Ernst. German zoologist (1804-1885); made important contributions to comparative invertebrate anatomy and parasitology.

Vormela. Genus of mottled polecats; southeastern Europe to India and China.

Vorticella. Genus of inverted bell-shaped ciliate Protozoa; with a long contractile stalk; attached to submerged objects in fresh and salt waters.

Vulpes. Genus of carnivorous mammals in the Family Canidae; includes about ten spp. of N.A. red foxes and the several small kit foxes, or swifts, of the western states; silver or black foxes are merely color phases of the red fox which are raised commercially for their pelts; the cross fox is intermediate between the red and black.

vulture. Any of certain large carrion-feeding birds of tropics and temperate areas; American genera are Cathartes, Coragyps, and Gymnogyps.

Vultur gryphus. Andean condor of western South America; largest bird of prey, wingspread sometimes more than 10 ft.; male glossy black with an ashy white bar across each wing.

vulva. 1. Pudendum; collectively, the external portions of the female reproductive organs, especially in mammals; includes the labia majora, labia minora, mons veneris, clitoris, perineum, and vestibulum vaginae. 2. In certain invertebrates, the female genital pore.

wading rat. See DEOMYS.

wagtail. See MOTACILLIDAE.

wahoo. See ACANTHOCYBIUM.

Waksman, Selman A. American microbiologist (1888-); made important contributions to human microbiology, the decomposition of organic matter by microorganisms, and the production of antibiotics; discovered streptomycin.

Wala. Common genus of spiders in the Family Salticidae.

Waldeyer, Otto. German biologist (1863-1921); first named chromosomes.

walking fish. Any of various Asiatic and South American fishes which are able to leave the water and move about on land, or even in vegetation.

walking stick. See PHASMIDA.

wallaby. Any of a group of Australian marsupials in the Family Macropodidae; generally similar to kangaroos but smaller; see PETROGALE and LAGORCHESTES.

Wallace, Alfred Russell. English naturalist (1823-1913); evolved a concept of evolution similar to that of Charles Darwin; through the intercession of Lyell and Hooker, Wallace's essay and a summary of Darwin's conclusions were published together in 1858; most famous for his work on the zoogeographic realms of the world and on the peculiarities of island life; *The Geographical Distribution of Animals* appeared in 1876, *Island Life* in 1880.

Wallace's line. Imaginary line separating the Australian from the Oriental faunas; between Bali and Lombok, between Borneo and Celebes, and then continues east of the Philippines.

wallaroo. 1. In local usage, any of several large kangaroos. 2. Rock kangaroo; see OSPHRANTER.

walleye pike. See STIZOSTEDION.

walnut comb. Type of comb (shaped like a half walnut meat) resulting from a cross between homozygous fowl having a rose comb and a pea comb; example of interaction of two different dominant factors.

walrus. See ODOBAENIDAE.

waltzing mice. Mendelian double recessive character in mice which affects the development of the inner ears; such mice are unable to move in a straight line and run about in small circles.

wapiti. American elk; largest N.A. deer; antlers large and widely branching, neck maned, tail short, rump with large, light-colored patch; males to 600 lbs.; formerly over much of U.S. and southern Canada except for desert regions, but now scattered in the Rocky Mountain areas of the West; *Cervus canadensis* is the common American sp., but there are also several Asiatic spp.

warble flies. See HYPODERMA.

warbler. See MNIOTILTIDAE and SYLVIIDAE.

Warburg apparatus. System of a flask and calibrated tubes used in experimental work involving the manometric determination of respiration and gaseous exchanges.

Ward, Henry B. American zoologist (1865-1945); professor at the University of Nebraska and Illinois; with wide interests but best known as a parasitologist.

Wardius. Genus of amphistome trematodes of the muskrat intestine.

warm blooded. Having a constant body temperature, independent of the environmental temperature; typical only of birds and mammals; e.g. man has a normal body temperature of 98.6^{o} F., and many birds have temperatures of about 104.0^{o}.

warmouth. See CHAENOBRYTTUS.

warning coloration. Aposematic coloration; bright and conspicuous coloration possessed by certain animals which are distasteful or dangerous when attacked by other animals; such devices are presumed to warn away attackers; e.g.

the red frills of certain lizards, the white interior of the mouth of some snakes, and the bright color pattern of the harlequin grasshopper.

Warsaw. See GARRUPA.

wart hog. See PHACOCHOERUS.

warty sea star. See ECHINASTER.

Washington clam. See SAXIDOMUS.

wasp. Any of a great many winged insects in the Order Hymenoptera, all having a slender body, abdomen narrowly attached to the thorax, biting mouth parts, and a sting in the females and workers which can be used repeatedly; social or solitary; adults feed chiefly on insects and nectar; nests are burrows or constructed of "paper" (masticated wood fiber), mud, etc.

Wasserman test. Complement-fixation test used in the diagnosis of syphilis.

Watasenia. Genus of luminescent squids.

water bamboo. Small Oriental sp. of bamboo grown in ponds and used as human food; sometimes bears the encysted metacercariae of the human intestinal fluke, Fasciolopsis buski.

water bear. See TARDIGRADA.

water beetle. Any of a great many spp. of beetles found in fresh waters.

water boa. See EUNECTES.

water boatman. See CORIXIDAE.

waterbuck. See KOBUS.

water buffalo. See BUBALUS.

water bug. Any of numerous hemipterans inhabiting fresh-water ponds and streams.

water caltrop. Water chestnut.

water chestnut. Water caltrop, red caltrop; a rooted and floating aquatic plant; introduced and locally abundant in the Atlantic Coast states; its fruits are used extensively as food in the Orient and are important because they may bear the metacercariae of the human intestinal fluke, Fasciolopsis buski.

water-civet. See OSBORNICTIS.

water dog. See NECTURUS.

water flea. Any member of the crustacean Order Cladocera.

water hen. Florida gallinule (Gallinula).

water hyacinth. Large aquatic plant which forms dense floating mats in sluggish rivers and lakes in the southeastern states; such growths are sufficient to stop all river traffic and make conditions unfavorable for fishes; in the Orient it is important because parts of the plant used as human food may bear encysted cercariae of the human intestinal fluke, Fasciolopsis buski.

water measurer. See HYDROMETRIDAE.

watermelon snow. Red snow.

water mite. See HYDRACHNOIDEA.

water moccasin. Cotton mouth; see AGKISTRODON.

water opossum. See CHIRONECTES.

water penny. See PSEPHENIDAE.

water scavenger beetle. See HYDROPHILIDAE.

water scorpion. Any insect member of the hemipteran Family Nepidae.

water shrew. Any of many semiaquatic shrews living along the shores of running waters; hind feet usually with fringes of stiff hairs, and sometimes partially webbed.

water snake. Any of numerous marine and fresh-water snakes commonly found in water; see Natrix.

water spider. 1. European aquatic spider, Argyroneta aquatica, which constructs an underwater air-filled silken dome. 2. Any spider commonly found in marshy places; often run about on the surface film; Dolomedes and Pirata common in U.S.

water striders. Common name of insects in the hemipteran Family Gerridae.

water-tenrec. See LIMNOGALE and GEOGALE.

water tigers. Predaceous larvae of beetles in the Order Dytiscidae.

water turkey. See ANHINGA.

water vascular system. System of fluid-filled closed tubes and ducts (chiefly internal) peculiar to echinoderms; when well developed, it consists of (1) a sievelike madreporite usually on the surface of the body, (2) a stone canal leading from the madreporite and connecting with, (3) a ring canal around the mouth, from which (4) five long radial canals extend; each radial canal has (5) many short lateral canals, each of which has a tube foot at its tip; each tube foot is a muscular closed cylinder which has a sucker at its outside free end and a bulblike ampulla at its inner end in the coelom; when an ampulla contracts, its contained fluid causes the tube foot to extend; contraction of the tube foot causes the fluid to be forced back and distend the ampulla; tube feet may act independently or in a coordinated fashion; the water vascular system is filled with sea water containing some protein; depending on the sp., the water vascular system functions variously in clinging, food handling, locomotion, and respiration.

water weevil. See LISSORHOPTRUS.

Watson-Crick model. Model of the structure of the DNA molecule first suggested by two British investigators, J. D. Watson (at Harvard University) and F. H. C. Crick (at Cambridge University), in 1953; a helix consisting of two spiral intertwining molecular chains, with adenine, thymine, cytosine, and guanine arranged in pairs like rungs of a ladder between the chains; each half of the helix contains all the genetic "information" needed to construct the other half; Watson and Crick published their theory largely on the basis of previous X-ray diffraction studies of viscous DNA strands by M. H. F. Wilkins (University of London); all three men shared the 1962 Nobel prize in physiology and medicine.

wattle. 1. In certain fowl, one of a pair of wrinkled, fleshy, highly colored folds of skin hanging from the throat region. 2. In certain lizards, a median, wrinkled, fleshy, colored fold of skin hanging from the chin or throat region.

wattle bird. See CALLAEIDAE.

Waucoban epoch. Earliest of the three subdivisions of the Cambrian period.

waved whelk. Buccinum undatum, a common whelk with a gray ridged shell to 3 in. long; the edible whelk of northern Europe.

wave of depolarization. See MEMBRANE THEORY.

wavy Astarte. Astarte undata; a dark brown, ridged bivalve of the East Coast.

wavy carpet shell. See LIOCYMA.

wavy top shell. See MARGARITES.

waxbill. Any of numerous Old World birds in the weaverbird family; bill a waxlike white, pink, or red color; some spp. are caged as pets.

wax glands. Small ventral abdominal glands in worker bees; secrete the wax which is used in comb construction.

wax insect. Any of various insects in the Order Homoptera, especially scale insects, which secrete a waxlike substance.

wax pick. Spurlike projection on the distal end of the second tibia of the honeybee; used for removing bits of wax secreted by the ventral abdominal wax glands.

waxwing. See BOMBYCILLIDAE.

waxworm. See GALLERIA.

W chromosome. A sex chromosome as designated in female moths and birds; comparable to Y chromosome of mammals.

weakfish. See SCIAENIDAE.

weasel. See MUSTELA.

weasel-headed armadillo. See EUPHRACTUS.

weasel-lemur. See LEPILEMUR.

weatherfish. Any of several Old World loaches; burrow in bottom mud of ponds and streams but are reputed to swim about actively in the water during rains.

weaver bird. Any of numerous African and Asiatic finchlike birds which build elaborate nests of interwoven grass and other vegetation.

Weber-Fechner law. In order for a sensation to increase arithmetically, the stimulus must increase geometrically.

Weberian organ. Series of small ossicles and ligaments connecting the swim bladder with the ear in the fish Order Ostariophysi; a mechanical device for registering swim bladder air pressure changes in the ear.

Weberian ossicles. Series of four small bones forming the Weberian organ in certain fishes; convey pressure changes to inner ear; derived from four vertebrae.

Weber's line. Imaginary north-south line separating faunas which are mainly Asiatic from those which are mainly Australian; located west of Halmahera and Boeroe, and east of Timor. Cf. WALLACE'S LINE.

web-footed shrew. See NECTOGALE.

webworm. Any of a large variety of caterpillars which are more or less gregarious and spin irregular webs or masses of a silken material; see HYPHANTRIA.

Weddell's seal. See LEPTONYCHOTES.

wedge shell. See DONAX.

weever. Any of several edible marine fishes in the Family Trachinidae; head broad, eyes dorsal; dorsal fin with many sharp venomous spines; Trachinus common.

weevil. See CURCULIONIDAE.

Weismann, August. German biologist (1834-1914); most famous as the originator of the germ-plasm theory, which stresses the unbroken continuity of germ plasm and the nonheritability of acquired characters.

Wenrich, D. H. American zoologist

(1885-); professor at University of Pennsylvania; noted for his numerous contributions in protozoology, especially parasitic flagellates.

wentletrap. Any of a wide variety of deep-water marine gastropods having a whitish, tight, high-spiral whorl; staircase shell, ladder shells; usually to 1.5 in. long; widely distributed; Epitonium common in U.S.

wether. Castrated ram.

whale. General name applied to any large cetacean, as distinct from the smaller porpoises and dolphins; true mammals, although modified for a marine existence; spouting or blowing is the forcible exhalation of air from the lungs; see ODONTOCETI and MYSTICETI.

whalebone. See MYSTICETI.

whalehead. See SHOEBILL.

whale louse. Highly modified amphipod parasite of whales; see PARACYMUS.

whale shark. See RHINEODON.

wharf crab. See SESARMA.

wheatear. Small gray, black, and whitish land bird of northeastern Arctic America and Europe.

wheat midge. Minute dipteran, Thecodiplosis mosellana, destructive to growing wheat in Europe and N.A.

wheat sawfly. Any of several hymenopterans, especially Dolerus and Pachynematus, whose larvae feed on wheat stems and heads.

wheatworm. Nematode (Anguina tritici) which feeds on leaves and inflorescences of wheat, oats, and other grasses.

wheel animalcule. Common name sometimes used for members of the Phylum Rotatoria.

wheel bug. See ARILUS.

Wheeler, William Morton. American entomologist (1865-1937); for many years on faculty of Harvard University and curator of invertebrates at the American Museum of Natural History; noted for research on ants and other social insects.

whelk. See BUCCINUM and BUSYCON.

wherryman. Water strider.

whimbrel. See NUMENIUS.

whip-poor-will. See ANTROSTOMUS.

whip scorpion. Arachnid in the Order Pedipalpi; tip of abdomen with a slender terminal whip; Mastigoproctus giganteus common in Fla. to Ariz.; to 5.5 in. long; sometimes called the vinegarroon because it excretes a substance having the odor of vinegar.

whip snake. See MASTICOPHIS.

whiptail lizard. See CNEMIDOPHORUS.

whipworm. See TRICHURIS.

whirligig beetle. See GYRINIDAE.

whistler. See GLAUCIONETTA.

whistling frog. Hyla avivoca; a small frog with a birdlike voice; La. to Fla. and southern Ill.

whistling swan. See CYGNUS.

white admiral. Banded purple; a common butterfly in the genus Basilarchia.

white ant. Any member of the insect Order Isoptera.

whitebait. 1. See ALOSMERUS. 2. Any of a wide variety of small silvery-colored marine and fresh-water fishes.

white bass. See MORONE CHRYSOPS.

white blood cell. Leucocyte.

white cat. See ICTALURUS.

white corpuscle. Leucocyte.

white crab. See OCYPODE.

white-crowned sparrow. See SPARROW.

white cup shell. Cylichna alba; a tiny circumboreal marine gastropod having the superficial appearance of a grain of rice.

white-eye. Any of numerous small Old World passerine arboreal songbirds which usually have a ring of white around the eye; in the Family Zosteropidae; Japan to southern Asia, southern Africa, East Indies, and Australian region.

white fibrous connective tissue. Type of tissue comprising tendons; consists of a matrix of fine parallel fibrils which are pale, crossed, interlaced, often wavy in outline, and with scattered living cells; this tissue also occurs in the vertebrate dermis and intestinal wall.

whitefish. 1. See COREGONUS and PROSOPIUM. 2. White whale, or beluga.

white flies. See ALEYRODIDAE.

white-footed mouse. See PEROMYSCUS.

white forelock. Lock of unpigmented hair growing from one small patch of the anterior part of the scalp; a dominant mutant in man.

white ibis. 1. Asiatic ibis having white plumage and blue-black neck. 2. See EUDOCIMUS.

white-lipped frog. Leptodactylus labialis; a medium-sized, dark-colored frog with cream-colored lip; lays eggs in a frothy mass; in the Family Leptodactylidae; Mexico and southern Tex.

white matter. Vertebrate nerve tissue consisting chiefly of medullated fibers;

forms the outermost mass of the spinal cord and parts of the brain and peripheral nerves.

White Mountain butterfly. *Oeneis semidea*; a delicate, marbled, dark brown and gray butterfly of the White Mountains and Rocky Mountains.

white perch. 1. See MORONE. 2. Freshwater drum. 3. Crappie.

white rat. Albino form of the Norway rat, *Rattus norvegicus*; important laboratory animal.

whites. See PIERIDAE.

white salmon. See PTYCHOCHEILUS.

white sea jelly. *Aurelia aurita*.

white shark. See CARCHARODON.

white-sided dolphin. See LAGENORHYNCHUS.

white sturgeon. See SCAPHIRHYNCHUS and ACIPENSER.

white sucker. 1. Common sucker, *Catostomus commersonii*. 2. Any of several redhorses.

white-tailed deer. See ODOCOILEUS.

white-tailed mongoose. See ICHNEUMIA.

white-throated sparrow. *Zonotrichia albicollis*, a common N.A. sparrow which breeds in northern coniferous forests.

white-throated swift. See AERONAUTES.

white whale. See DELPHINAPTERUS.

white-winged dove. See ZENAIDA.

whiting. Any of numerous European, Australian, and N.A. marine food fishes more or less related to the cod. See MERLUCCIIDAE, MERLANGUS, and MENTICIRRHUS.

whole mount. Microscope slide having a whole organism or discrete part of an organism mounted under a coverslip.

whorled lizard. See ZONURIDAE.

widemouth blindcat. See SATAN.

widgeon. See MARECA.

Wiedersheim, Robert. German zoologist (1848-1923); well known for his comprehensive work on vertebrate anatomy.

wild ass. Any of several wild horselike animals of Asia and northeastern Africa; similar to the domesticated ass; e.g. kiang and onager.

wild boar. See SUS.

wild canary. American goldfinch.

wildcat. See BOBCAT.

wildebeest. See GORGON.

willet. See CATOPTROPHORUS.

Williamsonia. Genus of dragonflies.

willow sawfly. Any of several sawflies infesting willow trees.

Wilson, Edmund Beecher. American zoologist (1856-1939); especially well known for work in cytology, embryology, heredity, and experimental morphology.

Wilsonema. Common genus of microscopic, free-living nematodes.

Wilsonia. Genus of warblers in the Family Parulidae; includes three U.S. spp.: hooded warbler, Wilson's warbler, and Canada warbler.

Wilson's petrel. See OCEANITES.

Wilson's phalarope. See STEGANOPUS.

Wilson's warbler. See WILSONIA.

windowpane. Sand flounder; small, thin, translucent, American flounder, *Lophopsetta maculata*; first 10 or 12 dorsal fin rays free and branched at tips; Atlantic Coast.

windpipe. Trachea.

wing bud. 1. Simple thickening of body wall in bird and bat embryos; develops into a wing. 2. Embryonic protuberances in an immature insect; destined to become a wing.

wing covers. Elytra.

winged Strombus. *Strombus pugilis alatus*; a yellowish to buff conch of the southeastern U.S. coast; to 4 in. long.

wing pads. Short, flat, saclike growths encasing the undeveloped wings; common in stonefly nymphs and other Hemimetabola.

wing shell. See AVICULA.

winkle. 1. Periwinkle. 2. Any of numerous large marine snails, e.g. *Busycon*.

Winkler method. Chemical method of determining the quantity of dissolved oxygen in water.

winter egg. Resting egg; type of egg produced especially in the autumn or during radically changing ecological conditions by certain fresh-water invertebrates, especially rotifers and cladocerans; such eggs are diploid, thick-shelled, and are capable of remaining viable through drying and freezing.

winter kill. Complete or partial kill of the fish population in a lake or pond during a prolonged period of winter ice and snow cover; caused by oxygen exhaustion and lack of photosynthesis.

winter wren. See TROGLODYTES.

wireworm. 1. Larva of a click beetle; see ELATERIDAE. 2. See HAEMONCHUS.

Wirsung's duct. One of the pancreatic ducts.

wisdom teeth. Four molar teeth (third molars) which are last to erupt in man.

wisent. European bison (Bos bonasus), a sp. now represented by only about 100 animals in captivity.

wishbone. Fused clavicles and interclavicle in birds; see FURCULA.

witch milk. Mammary secretion produced in small quantities within a few days after birth in both sexes.

Wohlfahrtia. Genus of flies in the Family Sarcophagidae; the larvae of an Old World sp. produce human myiasis when the female oviposits in sores, cuts, and at any natural body opening; larvae of a N.A. sp. are able to penetrate unbroken skin.

wolf. See CANIS.

Wolff, Caspar Friedrich. German biologist (1733-1794); founder of observational embryology; elaborated the theory of epigenesis to replace the preformation concept; emphasized developmental differences between plants and animals.

Wolffian duct. Kidney duct occurring on each side in vertebrates; it is formed in all vertebrate embryos in the region of the pronephros but becomes the functional duct of the mesonephros in adult anamniotes where it also carries sperm in the male; in adult amniotes the functional kidney has its own special duct (ureter), and the Wolffian duct persists in the male as the epididymis, duct of epididymis, ejaculatory duct, and vas deferens.

wolf fish. See ANARHICHADIDAE.

wolf spider. See LYCOSIDAE.

wolverine. See GULO.

womb. Uterus, especially in mammals.

wombat. See PHASCOLOMYIDAE.

wood ant. 1. Any of several ants in the genus Formica which live in wooded areas and build large nests. 2. Carpenter ant. 3. Termite.

wood bison. See BISON.

woodborer. 1. Larva of any of many beetles, lepidopterans, or hymenopterans which tunnel in wood. 2. Any of several marine bivalve mollusks and crustaceans which burrow in submerged wood.

woodchuck. See MARMOTA.

woodcock. See PHILOHELIA.

wood crab. See SESARMA.

woodcreeper. See DENDROCOLAPTES.

wood duck. See AIX.

wood frog. See RANA SYLVATICA.

Woodhouse toad. See BUFO.

wood louse. Terrestrial isopod.

wood nymph. Common dark brown U.S. butterfly.

woodpecker. See PICIDAE.

wood peewee. See CONTOPUS.

wood piddock. See MARTESIA.

wood rat. See NEOTOMA.

wood rat flea. See HISTRICOPSYLLA.

Woodruff, Lorande L. American biologist (1879-1947); noted for his biology textbooks and extensive research in protozoology.

wood satyr. See NEONYMPHA.

Woodsholia lilliei. Small rhabdocoel flatworm especially abundant in the Woods Hole, Mass. area.

wood swallow. See ARTAMIDAE.

wood thrush. See THRUSH.

wood tick. Any of numerous ticks in the Family Ixodidae; cling to vegetation and attach to animals passing by.

wood tortoise. See CLEMMYS.

wood turtle. See CLEMMYS.

wood wasps. See SIRICIDAE.

woolly apple aphid. See ERIOSOMATIDAE.

woolly bear. See ARCTIIDAE.

woolly monkey. See LAGOTHRIX.

woolly rhinoceros. An extinct two-horned rhinoceros of Arctic regions in the Pleistocene epoch; hair dense and woolly.

wool sponge. A durable soft-fibered commercial sponge; Hippospongia canaliculata, found in the Caribbean area.

worker. Sterile female in a colony of social insects.

worm. 1. Common name applied to the members of several different phyla having elongated cylindrical legless members, including Platyhelminthes, Nematoda, Annelida, Nemertea, etc. 2. Sometimes used in a broader sense to include any elongated creeping animal, such as insect larvae, rotifers, and a few mollusks, in addition to the phyla listed above.

Wormaldia. Genus of Trichoptera.

worm lizard. See AMPHISBAENIDAE.

worm salamander. Batrachoseps attenuatus attenuatus, a thin wormlike salamander, 4 in. long, of western Calif. and southwestern Ore.

worm snake. See TYPHLOPIDAE and LEPTOTYPHLOPIDAE.

Wotton, Edward. English naturalist (1492-1555); wrote De differentiis Animalium; an exponent of Aristotle and his ideas.

woundfin. See PLAGOPTERUS.

wrasse. See LABRIDAE.

wren. See TROGLODYTIDAE.

wren-tit. See CHAMAEIDAE.

Wright's stain. Standard biological stain which is especially useful for blood cells and malaria parasites.

wryneck. Any bird of the genus Jynx; related to woodpeckers but with soft tail feathers and a curious way of twisting the neck and head; Eurasia and Africa.

Wuchereria. Genus of spirurid nematodes; two important spp., W. bancrofti and W. malayi, are parasites of the human lymphatic system in the tropics; persistent cases result in filariasis and elephantiasis; transmitted by mosquitoes.

Wyeomyia. Genus of mosquitoes found in eastern and southern states.

Wyulda. Genus of phalangers which includes the elangurra, a small phalanger living among rocks in northwestern Australia; face ratlike, tail hairless and scaly.

X

xanthine. White organic base found in most body tissues, urine, and some plants; may be oxidized to uric acid; $C_5H_4O_2N_4$.

xanthine oxidase. Dehydrogenase which catalyzes the oxidation of xanthine and hypoxanthine into uric acid.

Xanthocephalus. See BLACKBIRD.

xanthophore. Yellow pigment cell in the dermis of many vertebrates.

xanthoproteic reaction. Test for the presence of the benzene ring in proteins; addition of strong nitric acid to proteins produces a yellow color which changes to orange when an alkali is added; indicates tyrosine, tryptophane, and phenylalanine.

Xanthoura. Genus of green and yellow jays; Central and South America.

Xantusia. See XANTUSIIDAE

Xantusiidae. Small family of desert lizards; night lizards; small cylindrical spp. with scales of dorsal surface granular; southwestern U.S. to Central America; Xantusia common.

X chromosome. See SEX CHROMOSOME.

Xenarthra. Edentata.

Xenia. Genus of soft corals belonging to the Order Alcyonacea.

Xenochironomus. Genus of tendipedid midges.

Xenogale. Genus of black mongooses of the Congo region.

xenogenesis. 1. Spontaneous generation. 2. Production of offspring unlike either parent. 3. Alternation of generations.

xenoparasite. Parasite which infects an abnormal host through a wound.

Xenopeltidae. Family of harmless snakes containing a single sp., Xenopeltis unicolor; to 3 ft. long; southeastern Asia.

Xenopeltis. See XENOPELTIDAE.

Xenophanes. Greek philosopher (570-480 B.C.); recognized fossils as animal remains and realized that the presence of fossils on mountains indicated that they had once been beneath the sea.

Xenophora. Genus of marine Gastropoda; carrier shells; exterior covered with dead clam and snail shells and bits of coral; Carolinas to Caribbean.

Xenopidae. Family of African aquatic toads in the Order Opisthocoela; clawed toads; tongue absent, three hind toes with horny claws; Xenopus a common genus, used in human pregnancy tests.

Xenopsylla cheopis. Common rat flea; transmits bubonic plague.

Xenopterygii. Order of bony fishes which includes the clingfishes; scales absent; large adhesive disc on abdomen used for clinging to stones, shells, etc.; littoral of warm oceans; Gobiesox typical.

Xenopus. Genus of African clawed toads; used in pregnancy diagnosis, since the female quickly produces eggs when injected with the urine of a pregnant woman; see XENOPIDAE.

Xenos. Genus in the insect Order Strepsiptera; larvae and females parasitic in wasps.

xerarch succession. Complicated series of animal and plant community changes occurring during the slow conversion of barren land surface to the climax formation typical of the climate and geography of a particular region; such changes may involve only a hundred years or up to thousands of years.

xeric. Arid, lacking in moisture.

xerophthalmia. See VITAMIN D.

xerosere. Sere which begins on a relatively dry site and ends in a terrestrial community featured by the climax vegetation characteristic of the climate and geography of the particular region.

Xerus. Genus of African ground squirrels.

Xestobium rufovillosum. Common deathwatch beetle of Europe (Family Anobiidae); male sounds a mating call in spring by tapping its head against the wall of its burrows in hardwood timbers and furniture.

Xestospongia. Pacific Coast genus of hard, encrusting, whitish sponges.

Xiphias. See XIPHIIDAE.

Xiphiidae. Family of fishes which includes the swordfish (Xiphias gladius), a large and widely distributed warm-water oceanic sp. in which the upper jaw is prolonged into a flat, sharp-edged, swordlike structure; first dorsal fin short and high; adult without teeth and scales; total length to 16 ft.; an important game and food fish.

Xiphinema. Large genus of soil nematodes specialized for feeding on the fluids of plant cells, especially those of roots.

xiphiplastron. Posterior ventral plate of a tortoise plastron.

xiphisternum. One of the elements of the tetrapod sternum.

xiphoid cartilage. Ensiform cartilage.

Xiphophorus. Genus of fishes which includes the swordtails; small, brightly colored, fresh-water spp. of Central America; used extensively in tropical aquariums and in genetics research.

Xiphosura. Only living order of arthropods in the arachnoid Subclass Merostomata; king crabs, horseshoe crabs; body with a long terminal telson; cephalothorax massive, arched, and horseshoe-shaped; mouth in center of cephalothorax; wide abdomen with six pairs of appendages, the first modified as opercula, the second through sixth modified as book gills; in muddy and sandy shallows of marine shores; to 15 in. long, not including telson; one sp. (Limulus polyphemus) from Maine to Yucatan, three spp. of Tachypleus) on the east coast of Asia, and one sp. of Carcinoscorpius off southeastern Asia.

Xiphydriidae. Small family of Hymenoptera; larvae bore into deciduous trees.

Xironodrilus. Genus of small leechlike oligochaetes commensal and parasitic on eastern U.S. crayfishes; in the Family Branchiobdellidae.

Xironogiton. Genus of small leechlike oligochaetes commensal and parasitic on crayfishes; in the Family Branchiobdellidae.

X organ. Small ductless gland in certain crustaceans; closely associated with the sinus gland in the eyestalk; its secretion prevents or delays molting.

Xyleborus. See AMBROSIA BEETLES.

Xylocopa. See XYLOCOPIDAE.

Xylocopidae. Family of robust bees; carpenter bees; tunnels and nests in dry, solid wood, mostly in tropics; Xylocopa common.

xylophagous. Eating, boring in, or destroying wood, e.g. the shipworm and some marine annelids.

Xylotrya. Common genus of wood-boring marine bivalve mollusks; similar to Teredo; Mass. to Tex.

Xyrauchen texanus. Humpback sucker of the Colorado and Gila river systems.

Xysticus. Common genus of brown and gray crab spiders usually in loose bark or forest duff.

Y

yak. Wild or domesticated shaggy ox (Bos grunniens) of Tibet and adjacent Asiatic areas; large, stocky, and brownish to blackish; horns large and curved.

yapok. See CHIRONECTES.

yaws. Tropical skin disease caused by Treponema pertenue; featured by raspberry-like skin tumors which may run together and ulcerate, especially on the face, arms, feet, and genitals; the parasite is transmitted by flies and enters the tissues through any break in the skin.

Y chromosome. See SEX CHROMOSOME.

yellowback. Lampsilis anodontoides, a buff-colored mussel of central and southeastern states.

yellow bass. See MORONE INTERRUPTA.

yellow bat. See DASYPTERUS.

yellowbird. Any of many birds having a predominantly yellow color, e.g. the goldfinch and yellow warbler.

yellow cartilage. Cartilage containing an abundance of yellow connective tissue fibers.

yellow cat. Any of several yellowish catfishes of eastern and southern U.S.

yellow cells. Chlorogogue.

yellow fever. Acute, non-contagious, human virus disease transmitted by the yellow fever mosquito, Aedes aegypti; now restricted to central Africa and inland tropical Central and South America; one attack confers permanent immunity; also called yellow jack.

yellow fibrous connective tissue. Tissue consisting of scattered cells in a secreted matrix of elastic yellowish fibrils which may be straight, bent, or branched; attaches organs and tissues to each other.

yellowhammer. 1. Flicker. 2. Common yellowish European finch.

yellow-headed blackbird. Xanthocephalus xanthocephalus, a common bird of western marshes; male black with yellow head and neck.

yellow jack. Yellow fever.

yellow jacket. See VESPIDAE.

yellow-legs. See TOTANUS.

yellow-lipped snake. Rhadinaea flavilata, a small snake of swampy coastal areas of the southeastern states; upper lip yellow, back reddish brown; 8 to 16 in. long.

yellow marrow. Bone marrow having a yellowish cast owing to the presence of considerable fat; abundant in adult vertebrates.

yellow perch. Perca flavescens.

yellows. See PIERIDAE.

yellow spot. Macula lutea.

yellowtail. Any of various fishes having a yellowish tail. See SERIOLA and LUTIANIDAE.

yellowthroat. Any of several warblers having a yellow breast and throat; thickets and bushes of N.A.; Geothlypis trichas is the Maryland yellowthroat.

yellow warbler. Common N.A. warbler, Dendroica aestiva; small and predominantly yellow; in gardens, orchards, and shrubbery.

yield. Amount of protoplasm removed from a specified habitat per unit time.

Yoldia. Genus of widely distributed marine pelecypods in the Order Protobranchia; shell elongated and somewhat pointed posteriorly.

yolk. Protein and fat material in the ova of animals; depending on the sp., the amount in an ovum is highly variable; serves as food for the developing embryo.

yolk gland. Any gland having as its chief function the production of yolk material for the unfertilized egg or zygote; especially common among invertebrates, e.g. Platyhelminthes and Mollusca.

yolk plug. Small pluglike mass of yolky endoderm during the final stages of gastrulation, and just before obliteration by the advancing ectoderm, in certain vertebrate embryos, e.g. frog.

yolk sac. Sac containing yolk which is used for the nourishment of embryonic and immature fishes, reptiles, and birds; outgrowth from the ventral surface of the body and has a lumen in common with that of the digestive tract; in most mammalian embryos there is a homologous organ, but it contains no yolk.

Yucca moth. See TEGETICULA.

Z

Zaedyus. Genus of South American pygmy armadillos; to 8 in. long; a coastal sp. that browses in the jetsam of beaches.

Zaglossus. Long-billed anteater of New Guinea (Order Monotremata); an echidna with a long snout and long, stiff legs.

zagouti. See PLAGIODONTIA.

Zaitzevia parvula. Small elmid beetle of western U.S.

Zalophus californianus. Common California sea lion of the Pacific Coast of N.A.; to 600 lbs.; feeds on fish, crustaceans, and squids; when captured young, it is easily trained for circus work.

Zamenis. Large genus of blackish snakes of Eurasia.

Zanclea. Genus of marine coelenterates in the Suborder Anthomedusae; hydranth long, cylindrical.

Zanclidae. Family of fishes which includes only the Moorish idol (Zanclus canescens); body short, deep, and greatly compressed; dorsal and ventral fin rays very long; mouth prolonged and tubular; coloration yellow with black bars; warm Pacific Ocean.

Zanclus. See ZANCLIDAE.

Zapodidae. Family of rodents which includes the jumping mice or kangaroo mice; hind legs and tail greatly elongated; cheek pouches present; over much of U.S. and Canada as well as Asia; Zapus and Napaeozapus typical.

Zapus. See ZAPODIDAE.

Zavreliella. Genus of tendipedid midges.

Z chromosome. See SEX CHROMOSOME.

zebra. See EQUIDAE.

zebra caterpillar. Yellow and black striped larva of an American moth, Ceramica picta (Noctuidae); feeds on cultivated plants.

zebra spider. See SALTICUS.

zebra swallowtail. Common whitish swallowtail butterfly in the genus Papilio; eastern U.S.

zebra-tailed lizard. See CALLISAURUS.

zebu. Asiatic domestic cattle, Bos indicus; short horns, large ears, large dewlap, and a very prominent hump extending above the shoulders.

Zeidae. See ZEOMORPHI.

Zelinkiella synaptae. Curious sp. of marine bdelloid rotifer; corona expanded, foot modified into a sucker; commensal on the body surface or in the digestive tract of the small sea cucumber, Synapta.

Zelotes. Very common genus of spiders in the Family Gnaphosidae.

Zenaida asiatica. White-winged dove in the Family Columbidae; southwestern U.S. to Chile.

Zenaidura macroura. Common wild dove of N.A. (Family Columbidae); mourning dove; tail pointed, not fan-shaped; herbivorous.

Zenker's fluid. Common histological killing and fixing fluid; consists of 2.5 g. potassium bichromate, 5.0 g. mercuric chloride, 1.0 g. sodium sulfate, and 5 ml. glacial acetic acid in 100 ml. water.

Zenopsis. See ZEOMORPHI.

Zeomorphi. Small order of marine bony fishes; includes the John dories (Family Zeidae); body greatly compressed; skin silvery and naked except for dorsal and ventral series of bony bucklers; tail very small; dorsal fin with about ten very long anterior spines; to 2 ft. long; Zenopsis ocellata the American John dory; Zeus faber the European John dory.

Zeomorphida. Zeomorphi.

zeren. See PROCAPRA.

Zernike, Frits. Dutch physicist (1888-); invented phase contrast microscope; Nobel prize winner in physics.

zeuglodont. See ARCHAEOCETI.

Zeus. See ZEOMORPHI.

Zeuzera pyrina. Important pest sp. of moth whose larva burrows into the wood of many deciduous trees; leopard moth.

Zinjanthropus boisei. Fossil tool-making manlike creature found in 1959 in Tanganyika; bones determined at about 1,750,000 years old.

Ziphiidae. Family of mammals; includes the beaked whales, e.g. Mesoplodon and Ziphius; only two or four teeth in lower jaw.

Ziphius cavirostris. Goosebeak whale; a sp. of beaked whale found in all oceans; to 25 ft. long; color usually blackish above, pale below.

Zirfaea crispata. Rough piddock; a marine bivalve with anterior radiating wrinkles terminating in rasplike teeth; Atlantic Coast.

Z line. Krause's membrane.

zoaea. Zoea.

Zoantharia. One of the two subclasses of anthozoan coelenterates; sea anemones and corals; solitary or colonial polyps with a solid exoskeleton, or skeleton absent; with few to many tentacles and one, two, or no siphonoglyphs.

Zoanthidea. Order of zoantharian coelenterates; skeleton and pedal disc absent; several polyps commonly united basally; growing on shells of hermit crabs, sponges, hydroids, and stones; Epizoanthus common on hermit crabs.

Zoanthus. Common genus of sea anemones; polyps elongated, arising from a network of stolons.

Zoarces. Genus of European eelpouts.

zoarium. Colony of Bryozoa. Pl. zoaria.

zobo. Dzo.

zoea. One of several in a series of early larval stages of various marine crabs; marked especially by a long anterior spine and a long dorsal spine.

zoecium. Zooecium.

zona fasciculata. Thick middle layer of the adrenal cortex; cells arranged in parallel columns; bounded by zona glomerulosa and zona reticularis.

zona glomerulosa. Narrow outermost layer in the adrenal cortex; cells arranged in spheres or loops.

zona granulosa. Membrane-like mass of follicular cells surrounding an ovum in a Graafian follicle.

zona pellucida. 1. Delicate but thick membrane surrounding the mammalian egg; it disappears before implantation in the uterus. 2. Similar membrane in the egg of certain other higher vertebrates, e.g. hen's egg.

zona radiata. Layer of fine radial cytoplasmic striations lying between a primary oocyte and its enveloping theca granulosa in many vertebrate ovaries.

zona reticularis. Innermost layer of the adrenal cortex; consists of cords of cells only one cell thick.

zonary placenta. Placenta whose chorionic villi form a girdle-like band around its middle; notably present in carnivores.

Zoniagrion exclamationis. Californian damselfly.

zonite. Millipede body segment.

Zonites. Genus of thin-shelled terrestrial snails; in damp places.

Zonitoides. Common genus of land snails; shell depressed, with wide umbilicus; aperture long, without teeth.

Zonotrichia. See SPARROW.

Zonuridae. Family of lizards found in central and south Africa and Madagascar; whorled lizards; legs sometimes absent; scales often keeled.

zoobenthos. Collectively, the animals of the benthic region of a lake or sea.

Zoobotryon. Genus of marine Bryozoa; miniature treelike growth form.

zoochlorellae. Small green algae (often Chlorella) which occur as intracellular symbionts in certain protozoans, sponges, coelenterates, etc.

zooea. Larval stage of shrimps and prawns; distinct cephalothorax and abdomen and eight pairs of appendages.

zooecium. General secreted covering of the individuals making up a colony of Bryozoa; gelatinous, chitinoid, or calcareous. Pl. zooecia.

zoogamy. Sexual reproduction.

zoogeographic realms. Major divisions of the land masses of the world, according to their distinctive faunas; the division of the world into six such realms in accordance with the ideas of Alfred Russell Wallace is the most widely accepted plan. See ORIENTAL, AUSTRALIAN, ETHIOPEAN, NEOTROPICAL, NEARCTIC, and PALEARCTIC realm; see also MARINE ZOOGEOGRAPHIC REALMS.

zoogeography. Study of the geographic distribution and dispersal of animals over the surface of the globe.

zooglea. Bacteria imbedded in a gelatinous matrix.

Zooglossus seychellensis. Small frog of the Seychelles; eggs, laid under fallen foliage, hatch into gill-less and limbless tadpoles which work their way onto the back of the male and remain attached there by their sucker-like mouth until metamorphosis is completed.

zooid. 1. One individual of a colony of

animals; term used especially with reference to the Entoprocta, Bryozoa, and Hydrozoa. 2. Object which resembles an animal.

zoology. 1. Science dealing with all matters relating to animals. 2. A treatise on this subject. 3. Animal life of a particular region.

Zoomastigina. One of the two subclasses of the Mastigophora; chromoplasts lacking; body organization simple to complex; sexual reproduction unknown.

zoon. Any individual in a colony of animals; term seldom used.

zoophagous. Pertaining to animals which feed on animal material.

Zoophyta. Old taxonomic category which included a wide variety of sessile animals, especially coelenterates, sponges, and tunicates.

zoophyte. 1. Specifically, any member of the Phylum Bryozoa. 2. In the broad sense, any non-motile plantlike animal, e.g. sponges, sea anemones, hydroids, bryozoans, etc.

zooplankton. Collectively, all those animals suspended in the water of an aquatic habitat which are not independent of currents and water movements; most such organisms are microscopic and commonly include protozoans, rotifers, and small crustaceans; technically, certain large marine coelenterates and the basking shark are also members of the zooplankton. See PHYTOPLANKTON and PLANKTON.

zoospore. 1. Naked, flagellated cell formed within a sporangium; found in the life cycle of some green flagellates as well as in many algae; usually four are produced from one zygote. 2. One of the minute cells produced by a sporocyst.

Zootermopsis. Genus of termites (Order Isoptera) living in damp wood; western Canada to Va. and southward.

Zoothamnium. Genus of colonial ciliate Protozoa; each cell has the shape of an inverted bell with a basal stalk; the stalks arise by dichotomous branching, and there are interconnecting myonemes, so that the whole colony contracts when stimulated; fresh and salt waters.

zootomy. 1. Anatomy of animals. 2. Dissection of animals.

zootoxin. Any toxin of animal origin, e.g. snake venom.

zooxanthellae. Minute yellow or brown flagellates which often occur as intracellular symbionts in certain Foraminifera and Radiolaria.

Zoraptera. Very small order of minute rare tropical and subtropical insects; metamorphosis simple; biting mouth parts; wings present or absent; in decaying vegetation and debris; Zorotypus common.

zorille. Zoril, a weasel-like animal which produces a fetid odor; locally called a polecat; remarkably similar to the American skunk in color and habits; drier parts of Africa and the eastern Mediterranean area.

Zorotypus. Genus in the insect Order Zoraptera; under bark and in vegetable debris in southern states.

Zosteropidae. See WHITE-EYE.

Zygaenidae. Family of small, brightly colored or metallic moths.

zygapophysis. Articular process of a vertebra.

Zygeupolia. Genus of marine Nemertea.

Zygherpe. Genus of marine sponges.

Zygoballus bettini. Common dark brown spider in the Family Salticidae.

Zygobranchiata. Subdivision of gastropod mollusks; shell and animal with marginal or apical slit or marginal holes; ventricle pierced by rectum; marine abalones and keyhole limpets.

Zygocotyle. Genus of amphistome parasites of ducks and ruminants.

zygodactyl. With two toes projecting forward and two behind, as in a parrot.

Zygodactyla groenlandica. Marine coelenterate in the Suborder Leptomedusae; medusa to 5 in. in diameter, with 32 tentacles.

zygodactyly. Syndactyly; "webbed" toes or fingers, a human Mendelian dominant.

zygomatic arch. Arch of bone below the orbit on the front or side of the skull in certain vertebrates, as the cat and dog.

zygomaticus. Facial muscle which extends from the malar bone to the angle of the mouth; draws upper lip outward and upward.

Zygonemertes. Genus of slender marine nemertine worms.

Zygoptera. Suborder of the insect Order Odonata; includes the damselflies; abdomen very long, wings held vertically when at rest, hind wings not thickened near base; aquatic naiads elongated and with three posterior leaflike tracheal gills; common genera are Ischnura, Hetaerina, Lestes, and Argia.

zygote. Diploid cell resulting from the union of the male and female gametes. Adj. zygotic.

zygotene stage. Longitudinal pairing of homologous chromosomes during the prophase of the first meiotic division.

zymase. Enzyme complex produced by yeasts; catalyzes sugars to alcohol and carbon dioxide.

zymogen. Inactive form of an enzyme which becomes functional in the presence of an activator; e.g. trypsinogen is the inactive form which is activated into trypsin by enterokinase in the small intestine.

APPENDIX

This condensed taxonomic outline of the animal kingdom is intended as a general orientation reference and supplement to the foregoing portion of this volume. It is a scheme which happens to be generally acceptable to this particular author; other zoologists will prefer slight to major modifications of this plan. All of the taxa listed here are defined, but, in addition, many other taxa are defined but are not included in this taxonomic section.

In general, the most detailed classifications are given for the vertebrate groups, but it should be emphasized that where the classification is carried as far as the family level, only certain typical and important families are ordinarily cited, and the list is far from complete. Extinct taxa are indicated by asterisks.

PHYLUM PROTOZOA
- Subphylum Plasmodroma
  - Class Mastigophora
    - Subclass Phytomastigina
      - Order Chrysomonadina
        - Chloromonadina
        - Cryptomonadina
        - Dinoflagellata
        - Euglenoidina
        - Phytomonadina
    - Subclass Zoomastigina
      - Order Hypermastigina
        - Polymastigina
        - Protomonadina
        - Rhizomastigina
  - Class Sarcodina
    - Order Amoebina
      - Foraminifera
      - Heliozoa
      - Mycetozoa
      - Proteomyxa
      - Radiolaria
      - Testacea
  - Class Sporozoa
    - Subclass Cnidosporidia
      - Order Actinomyxidia
        - Helicosporidia
        - Microsporidia
        - Myxosporidia
    - Subclass Haplosporidia
    - Subclass Sarcosporidia
    - Subclass Telosporidia
      - Order Coccidia
        - Gregarinida
        - Haemosporidia
- Subphylum Ciliophora
  - Class Ciliata
    - Subclass Protociliata

Subclass Euciliata
- Order Chonotricha
- Order Holotricha
  - Suborder Apostomea
    - Astomata
    - Gymnostomata
    - Hymenostomata
    - Thigmotricha
    - Trichostomata
- Order Peritricha
  - Suborder Sessilia
    - Mobilia
- Order Spirotricha
  - Suborder Ctenostomata
    - Heterotricha
    - Hypotricha
    - Oligotricha

Class Suctoria

PHYLUM PORIFERA
- Class Calcarea
  - Order Homocoela
    - Heterocoela
- Class Hexactinellida
  - Order Hexasterophora
    - Amphidiscophora
- Class Demospongiae
  - Subclass Tetractinellida
    - Monaxonida
    - Keratosa

PHYLUM COELENTERATA
- Class Hydrozoa
  - Order Hydroida
    - Suborder Anthomedusae
      - Leptomedusae
      - Limnomedusae
  - Order Hydrocorallina
    - Suborder Milleporina
      - Stylasterina
  - Order Trachylina
    - Suborder Trachymedusae
      - Narcomedusae
      - Pteromedusae
  - Order Siphonophora
- Class Scyphozoa
  - Order Stauromedusae
  - Order Cubomedusae
  - Order Coronatae
  - Order Discomedusae
    - Suborder Semaeostomae
      - Rhizostomae
- Class Anthozoa
  - Subclass Alcyonaria
    - Order Alcyonacea
      - Coenothecalia
      - Gorgonacea
      - Pennatulacea
      - Stolonifera
      - Telestacea
  - Subclass Zoantharia
    - Order Actiniaria
      - Suborder Actinaria

Ptychodactiaria
Corallimorpharia
Order Madreporaria
Order Zonanthidea
Order Antipatharia
Order Ceriantharia
PHYLUM CTENOPHORA
Class Tentaculata
Order Cestida
Cydippida
Lobata
Platyctenea
Class Nuda
PHYLUM PLATYHELMINTHES
Class Turbellaria
Order Acoela
Order Rhabdocoela
Order Alloeocoela
Order Tricladida
Order Polycladida
Suborder Acotylea
Cotylea
Class Trematoda
Order Monogenea
Digenea
Aspidocotylea
Class Cestoidea
Subclass Cestodaria
Subclass Cestoda
Order Aporidea
Cyclophyllidea
Diphyllidea
Lecanicephaloidea
Nippotaeniidea
Proteocephaloidea
Tetraphyllidea
Trypanorhyncha
Pseudophyllidea
PHYLUM MESOZOA
Class Dicyema
Class Orthonectida
PHYLUM NEMERTEA
Class Anopla
Order Palaeonemertea
Heteronemertea
Class Enopla
Order Hoplonemertea
Bdellonemertea
PHYLUM NEMATODA
Class Aphasmidea
Order Chromadorida
Suborder Monhysterina
Chromadorina
Order Enoplida
Suborder Enoplina
Dorylaimina
Dioctophymatina
Class Phasmidea
Order Tylenchida
Order Rhabditida
Suborder Rhabditina

Strongylina
Ascaridina
Order Spirurida
Suborder Spirurina
Camallanina
PHYLUM NEMATOMORPHA
Class Gordioidea
Class Nectonematoidea
PHYLUM ACANTHOCEPHALA
Class Archiacanthocephala
Class Palaeacanthocephala
Class Eoacanthocephala
PHYLUM ROTATORIA
Class Digononta
Order Seisonidea
Bdelloidea
Class Monogononta
Order Flosculariacea
Collothecacea
Ploima
PHYLUM GASTROTRICHA
Class Macrodasyoidea
Class Chaetonotoidea
PHYLUM KINORHYNCHA
PHYLUM PRIAPULIDA
PHYLUM ENTOPROCTA
Class Loxosomatida
Class Pedicellinida
Class Urnatellida
PHYLUM CHAETOGNATHA
PHYLUM BRYOZOA
Class Phylactolaemata
Class Gymnolaemata
Order Ctenostomata
Cheilostomata
Cryptostomata*
Class Stenolaemata
Order Cyclostomata
Treptostomata
PHYLUM BRACHIOPODA
Class Inarticulata
Order Atremata
Neotremata
Class Articulata
Order Prototremata
Telotremata
Palaeotremata*
PHYLUM PHORONIDEA
PHYLUM ANNELIDA
Class Archiannelida
Class Polychaeta
Order Errantia
Sedentaria
Class Oligochaeta
Class Hirudinea
Order Acanthobdellida
Rhynchobdellida
Arhynchobdellida
Class Myzostoma
PHYLUM SIPUNCULOIDEA
PHYLUM ECHIUROIDEA

- PHYLUM ONYCHOPHORA
- PHYLUM ARTHROPODA
  - Class Trilobita*
  - Class Crustacea
    - Subclass Branchiopoda
      - Division Eubranchiopoda
        - Order Lipostraca*
        - Anostraca
        - Notostraca
        - Conchostraca
      - Division Oligobranchiopoda
        - Order Cladocera
          - Suborder Calyptomera
          - Gymnomera
    - Subclass Copepoda
      - Order Eucopepoda
        - Suborder Calanoida
        - Harpacticoida
        - Cyclopoida
        - Notodelphyoida
        - Monstrilloida
        - Caligoida
        - Lernaeopodoida
      - Order Branchiura
    - Subclass Mystacocarida
    - Subclass Cephalocarida
    - Subclass Ostracoda
      - Order Podocopa
      - Myodocopa
      - Cladocopa
      - Platycopa
    - Subclass Cirripedia
      - Order Thoracica
      - Acrothoracica
      - Apoda
      - Rhizocephala
      - Ascothoracica
    - Subclass Malacostraca
      - Division Leptostraca
        - Order Nebaliacea
      - Division Eumalacostraca
        - Subdivision Syncarida
          - Order Anaspidacea
          - Bathynellacea
        - Subdivision Peracarida
          - Order Mysidacea
          - Thermosbaenacea
          - Spelaeogriphacea
          - Cumacea
          - Tanaidacea
          - Isopoda
          - Amphipoda
        - Subdivision Hoplocarida
          - Order Stomatopoda
        - Subdivision Eucarida
          - Order Euphausiacea
          - Order Decapoda
            - Section Natantia
              - Suborder Penaeidea
              - Caridea
              - Stenopodidea

Section Reptantia
Suborder Macrura
Anomura
Brachyura

Class Arachnoidea
Subclass Merostomata
Order Xiphosura
Eurypterida*
Subclass Arachnida
Order Scorpionida
Order Amblypygi
Order Schizomida
Order Palpigrada
Order Araneae
Families - Agelenidae, Araneidae, Atypidae, Clubionidae, Ctenidae, Ctenizidae, Dictynidae, Gnaphosidae, Linyphiidae, Lycosidae, Oxyopidae, Pholcidae, Pisauridae, Salticidae, Scytodidae, Sparassidae, Theraphosidae, Theridiidae, Thomisidae, Uloboridae
Order Solpugida
Order Pseudoscorpionida
Order Phalngida
Order Ricinulei
Order Acarina
Suborder Trombidoidea
Families - Anystidae, Tetranychidae, Trombidiidae
Suborder Hydrachnoidea
Families - Halacaridae, Hydrachnidae, Pionidae
Suborder Ixodoidea
Families - Argasidae, Ixodidae
Suborder Parasitoidea
Families - Dermanyssidae, Macrochelidae, Parasitidae
Suborder Oribatoidea
Family Oribatidae
Suborder Acaroidea
Families - Analgesidae, Sarcoptidae, Tarsonemidae
Suborder Demodicoidea
Families - Demodicidae, Eriophyidae
Class Pycnogonida
Class Chilopoda
Order Scutigeromorpha
Lithobiomorpha
Scolopendromorpha
Geophilomorpha
Class Diplopoda
Order Pselapognatha
Limacomorpha
Oniscomorpha
Polydesmoidea
Nematomorphoidea
Juliformia
Colobognatha
Class Pauropoda
Class Symphyla
Class Insecta
Subclass Apterygota
Order Protura
Order Thysanura
Order Collembola

Order Aptera
Family Iapygidae
Subclass Pterygota
Order Orthoptera
Families - Acrididae, Gryllidae, Gryllotalpidae, Stenopelmatidae, Tetrigidae, Tettigoniidae, Tridactylidae
Order Blattaria
Order Grylloblattodea
Order Phasmida
Order Mantodea
Order Dermaptera
Order Plecoptera
Order Isoptera
Order Zoraptera
Order Embioptera
Order Corrodentia
Order Mallophaga
Order Anoplura
Order Ephemeroptera
Order Odonata
Suborder Anisoptera
Suborder Zygoptera
Order Thysanoptera
Suborder Terebrantia
Suborder Tubulifera
Order Hemiptera
Suborder Cryptocerata
Families - Belostomatidae, Corixidae, Gelastocoridae, Naucoridae, Nepidae, Notonectidae, Pleidae
Suborder Gymnocerata
Families - Anthocoridae, Cimicidae, Coreidae, Gerridae, Hydrometridae, Lygaeidae, Macroveliidae, Miridae, Nabidae, Pentatomidae, Phymatidae, Polyctenidae, Pyrrhocoridae, Reduviidae, Saldidae, Scutelleridae, Thyreocoridae, Tingidae, Veliidae
Order Homoptera
Suborder Auchenorrhyncha
Families - Cercopsidae, Cicadellidae, Cicadidae, Fulgoridae, Jassidae, Membracidae
Suborder Sternorrhyncha
Families - Adelgidae, Aleyrodidae, Aphididae, Coccidae, Dactylopiidae, Diaspididae, Eriosomatidae, Kermidae, Lacciferidae, Margarodidae, Phylloxeridae, Pseudococcidae, Psyllidae
Order Mecoptera
Order Megaloptera
Order Neuroptera
Order Trichoptera
Order Raphidiodea
Order Lepidoptera
Suborder Jugatae
Families - Hepialidae, Micropterygidae
Suborder Frenatae
Families - Aegeriidae, Arctiidae, Bombycidae, Citheroniidae, Coleophoridae, Cossidae, Crambidae, Dioptidae, Galleriidae, Gelechiidae, Geometridae, Lasiocampidae, Ly-

mantriidae, Lyonetiidae, Megalopygidae, Noctuidae, Notodontidae, Olethreutidae, Phycitidae, Plusiidae, Psychidae, Pterophoridae, Pyralididae, Pyraustidae, Saturniidae, Sphingidae, Tineidae, Tortricidae, Zygaenidae

Suborder Rhopalocera

Families - Danaidae, Hesperiidae, Lycaenidae, Morphoidae, Nymphalidae, Papilionidae, Parnassidae, Pieridae, Satyridae

Order Diptera

Suborder Orthorrhapha

Tribe Nematocera

Families - Blepharoceridae, Cecidomyiidae, Ceratopogonidae, Culicidae, Deuterophlebiidae, Dixidae, Mycetophilidae, Psychodidae, Ptychopteridae, Sciaridae, Simuliidae, Tendipedidae, Tipulidae

Tribe Brachycera

Families - Acroceratidae, Asilidae, Bombyliidae, Rhagionidae, Stratiomyidae, Tabanidae, Therevidae

Tribe Prosechomorpha

Tribe Gephyroneura

Families - Dolichopodidae, Empididae, Lonchopteridae

Suborder Cyclorrhapha

Tribe Aschiza

Families - Braulidae, Phoridae, Pipunculidae, Syrphidae

Tribe Schizopera

Families - Agromyzidae, Anthomyiidae, Calliphoridae, Chloropidae, Conopidae, Cuterebridae, Diopsidae, Drosophilidae, Gasterophilidae, Glossinidae, Hypodermatidae, Muscidae, Oestridae, Sarcophagidae, Scatophagidae, Sepsidae, Stomoxyidae, Tachinidae, Trypetidae

Tribe Pupipara

Families - Hippoboscidae, Nycteribiidae

Order Siphonaptera

Order Coleoptera

Suborder Adephaga

Families - Amphizoidae, Carabidae, Cicindelidae, Dytiscidae, Gyrinidae, Haliplidae, Paussidae

Suborder Polyphaga

Families - Anobiidae, Bostrichidae, Buprestidae, Cantharidae, Cerambycidae, Cetoniidae, Chrysomelidae, Coccinellidae, Cryptophagidae, Cucujidae, Curculionidae, Dermestidae, Dryopidae, Dynastidae, Elateridae, Elmidae, Erotylidae, Helodidae, Histeridae, Hydraenidae, Hydrophilidae, Hydroscaphidae, Lampyridae, Lariidae, Lucanidae, Lycidae, Lyctidae, Meloidae, Oedemeridae, Passalidae, Platypsyllidae, Pselaphidae, Psephenidae, Ptinidae, Scarabeidae, Scolytidae, Silphidae, Staphylinidae,

Tenebrionidae

Order Strepsiptera
Order Hymenoptera
Suborder Chalastogastra
Families - Cephidae, Cimbicidae, Oryssidae, Siricidae, Tenthredinidae, Xiphydriidae
Suborder Clistrogastra
Families - Agaontidae, Andrenidae, Anthophoridae, Aphelinidae, Apidae, Bembicidae, Bombidae, Braconidae, Chalcididae, Chrysididae, Cynipidae, Encyrtidae, Eumenidae, Eurytomidae, Figitidae, Formicidae, Halictidae, Ichneumonidae, Larridae, Megachilidae, Meliponidae, Mutillidae, Mymaridae, Nomadidae, Polistidae, Pompilidae, Psammocharidae, Pteromalidae, Scoliidae, Serphidae, Sphecidae, Tetrastichidae, Tiphiidae, Trichogrammatidae, Vespidae, Xylocopidae

PHYLUM LINGUATULIDA
Class Cephalobaenida
Class Porocephalida
PHYLUM TARDIGRADA
PHYLUM MOLLUSCA
Class Amphineura
Order Aplacophora
Order Polyplacophora
Class Monoplacophora
Class Scaphopoda
Class Gastropoda
Order Prosobranchiata
Suborder Aspidobranchia
Pectinibranchia
Order Opisthobranchiata
Suborder Tectibranchia
Nudibranchia
Order Pulmonata
Suborder Stylommatophora
Basommatophora
Class Pelecypoda
Order Protobranchia
Order Filibranchia
Order Eulamellibranchia
Order Septibranchia
Class Cephalopoda
Order Tetrabranchia
Suborder Ammonidea*
Nautiloidea
Order Dibranchia
Suborder Decapoda
Octopoda
PHYLUM ECHINODERMATA
Subphylum Pelmatozoa
Class Carpoidea*
Class Cystoidea*
Class Blastoidea*
Class Edrioasteroidea*
Class Crinoidea
Subphylum Eleutherozoa
Class Holothuroidea
Order Aspidochirota

Elasipoda
Dendrochirota
Molpadonia
Apoda
Class Echinoidea
Subclass Bothriocidaroida*
Subclass Regularia
Order Lepidocentroida
Melonechinoida*
Cidaroida
Aulodonta
Stirodonta
Camarodonta
Subclass Irregularia
Order Holectypoida
Cassiduloida
Clypeasteroida
Spatangoida
Class Asteroidea
Order Platyasterida*
Hemizonida*
Phanerozonia
Spinulosa
Forcipulata
Class Ophiuroidea
Order Ophiurae
Euryalae
Class Ophiocistioidea*
PHYLUM POGONOPHORA
PHYLUM PTEROBRANCHIA
Class Rhabdopleuridea
Class Cephalodiscidea
PHYLUM GRAPTOZOA*
PHYLUM ENTEROPNEUSTA
PHYLUM CHORDATA
Subphylum Tunicata
Class Ascidiacea
Order Enterogona
Pleurogona
Class Thaliacea
Order Pyrosomida
Salpida
Doliolida
Class Larvacea
Subphylum Cephalochordata
Subphylum Vertebrata
Class Agnatha
Subclass Ostracodermi*
Order Anaspida
Heterostraci
Osteostraci
Subclass Cyclostomata
Order Petromyzontia
Myxinoidia
Class Placodermi*
Order Acanthodii
Arthrodira
Antiarchii
Stegoselachii
Macropetalichthyida
Class Chondrichthyes

Order Cladoselachii*
Order Pleuracanthodii*
Order Selachii
Suborder Squali
Superfamily Heterodontida
Hexanchida
Lamnida
Squalida
Suborder Batoidea
Order Bradyodonti*
Order Holocephali
Class Osteichthyes
Subclass Palaeopterygii
Order Archistia*
Order Cladistia
Order Chondrostei
Families - Acipenseridae, Polyodontidae
Order Belonorhynchii*
Subclass Neopterygii
Order Protospondyli
Family Amiidae
Order Ginglymodi
Order Halecostomi*
Order Isospondyli
Families - Albulidae, Argentinidae, Clupeidae, Coregonidae, Elopidae, Engraulidae, Hiodontidae, Megalopidae, Osmeridae, Salmonidae, Thymallidae
Order Haplomi
Families - Esocidae, Umbridae
Order Bathyclupei
Order Iniomi
Families - Myctophidae, Synodontidae
Order Ateleopii
Order Giganturoidea
Order Lyomeri
Order Mormyrii
Order Ostariophysi
Families - Ameiuridae, Ariidae, Catostomidae, Characinidae, Cyprinidae, Electrophoridae, Gymnotidae, Siluridae
Order Apodes
Families - Anguillidae, Congridae, Nemichthyidae, Ophichthyidae
Order Colocephali
Family Muraenidae
Order Heteromi
Family Notacanthidae
Order Synentognathi
Families - Belonidae, Exocoetidae, Hemiramphidae, Scomberesocidae
Order Cyprinodontes
Families - Amblyopsidae, Anablepidae, Cyprinodontidae, Poeciliidae
Order Salmopercae
Families - Aphredoderidae, Percopsidae
Order Solenichthyes
Families - Hippocampidae, Syngnathidae
Order Anacanthini
Families - Gadidae, Macrouridae, Merluciidae
Order Allotriognathi

Order Berycomorphi
Order Zeomorphi
Family Zeidae
Order Percomorphi
Families - Anarhichadidae, Atherinidae, Aulostomidae, Blennidae, Branchiostegidae, Centrarchidae, Centropomidae, Cichlidae, Coryphaenidae, Eleotridae, Embiotocidae, Fistulariidae, Gobiidae, Haemulidae, Holocentridae, Istiophoridae, Labridae, Lumpenidae, Lutianidae, Mugilidae, Mullidae, Ophidiidae, Percidae, Pholidae, Polynemidae, Pomacentridae, Priacanthidae, Rachycentridae, Sciaenidae, Scombridae, Scorpaenidae, Serranidae, Sparidae, Sphryraenidae, Stromateidae, Trachinidae, Trichiuridae, Triglidae, Uranoscopidae, Xiphiidae, Zanclidae
Order Scleroparei
Families - Agonidae, Cottidae, Cyclopteridae, Dactylopteridae, Gasterosteidae, Liparidae, Scorpaenidae
Order Hypostomides
Order Heterosomata
Families - Pleuronectidae, Soleidae
Order Discocephali
Order Plectognathi
Families - Balistidae, Diodontidae, Molidae, Ostraciidae, Tetraodontidae
Order Malacichthyes
Order Chaudhurei
Order Xenopterygii
Order Haplodoci
Order Pediculati
Families - Antennariidae, Ceratiidae, Lophiidae, Melanocetidae
Order Opisthomi
Order Symbranchii
Subclass Choanichthyes
Superorder Crossopterygii
Order Rhipidistia*
Order Coelacanthini
Superorder Dipnoi
Order Ctenodipterini*
Order Sirenoidei
Families - Ceratodontidae, Lepidosirenidae
Class Amphibia
Subclass Stegocephalia
Order Leptospondyli*
Order Phyllospondyli*
Order Labyrinthodonti*
Order Gymnophiona
Family Caeciliidae
Subclass Caudata
Order Proteida
Family Proteidae
Order Mutabilia
Families - Ambystomidae, Amphiumidae, Cryptobranchidae, Hynobiidae, Plethodontidae, Salamandridae
Order Meantes

Family Sirenidae
Subclass Salientia
Order Amphicoela
Family Ascaphidae
Order Opisthocoela
Families - Discoglossidae, Pipidae, Xenopidae
Order Anomocoela
Family Pelobatidae
Order Procoela
Families - Brachycephalidae, Bufonidae, Hylidae, Leptodactylidae, Palaeobatrachidae*
Order Diplasiocoela
Families - Microhylidae, Polypedatidae, Ranidae
Class Reptilia
Subclass Anapsida
Order Cotylosauria*
Order Chelonia
Suborder Atheca
Family Dermochelidae
Suborder Thecophora
Superfamily Cryptodira
Families - Chelonidae, Chelydridae, Dermatemyidae, Kinosternidae, Platysternidae, Testudinidae
Superfamily Pleurodira
Families - Carettochelyidae, Chelydidae, Pelomedusidae
Superfamily Trionychoidea
Family Trionychidae
Subclass Ichthyopterygia*
Order Ichthyosauria
Subclass Synaptosauria*
Order Protorosauria
Order Sauropterygia
Subclass Lepidosauria
Order Eosuchia*
Order Rhynchocephalia
Order Squamata
Suborder Sauria
Division Ascalabota
Families - Agamidae, Chamaeleontidae, Gekkonidae, Iguanidae
Division Autarchoglossa
Families - Amphisbaenidae, Anguidae, Anniellidae, Helodermatidae, Lacertidae, Pygopodidae, Scincidae, Teiidae, Varanidae, Xantusiidae, Zonuridae
Suborder Serpentes
Families - Anilidae, Boidae, Colubridae, Crotalidae, Elapidae, Hydrophiidae, Leptotyphlopidae, Typhlopidae, Uropeltidae, Viperidae, Xenopeltidae
Order Loricata
Families - Alligatoridae, Crocodylidae, Gavialidae
Order Thecodontia*
Order Pterosauria*
Order Saurischia*
Suborder Theropoda
Suborder Sauropoda
Order Ornithischia*

- Suborder Ornithopoda
- Suborder Stegosauria
- Suborder Ankylosauria
- Suborder Ceratopsia

Subclass Synapsida*
- Order Pelycosauria
- Order Therapsida
- Order Ictidosauria

Class Aves
- Subclass Archaeornithes*
- Subclass Neornithes
  - Superorder Odontognathae*
    - Order Hesperornithiformes
    - Order Ichthyornithiformes
  - Superorder Palaeognathae
    - Order Aepyornithiformes*
    - Order Apterygiformes
    - Order Caenognithiformes*
    - Order Casuariiformes
      - Families - Casuariidae, Dromiceidae
    - Order Dinornithiformes*
    - Order Rheiformes
    - Order Struthioniformes
    - Order Tinamiformes
  - Superorder Neognathae
    - Order Sphenisciformes
    - Order Gaviiformes
      - Family Gaviidae
    - Order Podicipitiformes
      - Family Podicipitidae
    - Order Procellariiformes
      - Families - Diomedeidae, Hydrobatidae, Procellariidae, Pelecanoididae
    - Order Pelecaniformes
      - Families - Anhingidae, Fregatidae, Pelecanidae, Phaëthontidae, Phalacrocoracidae, Sulidae
    - Order Ciconiiformes
      - Families - Ardeidae, Balaenicipitidae, Ciconiidae, Cochleariidae, Phoenicopteridae, Scopidae, Threskiornithidae
    - Order Anseriformes
      - Families - Anatidae, Anhimidae
    - Order Falconiformes
      - Families - Accipitridae, Cathartidae, Falconidae, Pandionidae, Sagittariidae
    - Order Galliformes
      - Families - Cracidae, Megapodiidae, Meleagrididae, Opisthocomidae, Numididae, Phasianidae, Tetraontidae
    - Order Gruiformes
      - Families - Aramidae, Cariamidae, Eurypygidae, Gruidae, Heliornithidae, Mesitornithidae, Otididae, Psophiidae, Rallidae, Rhynochetidae, Turnicidae
    - Order Diatrymiformes*
    - Order Charadriiformes
      - Families - Alcidae, Aphrizidae, Burhinidae, Charadriidae, Chionididae, Dromadidae, Glareolidae, Haematopodidae, Jacanidae, Laridae, Phalaropodidae,

Recurvirostridae, Rostratulidae, Rynchopidae, Scolopacidae, Stercorariidae, Chionididae

Order Columbiformes
- Families - Columbidae, Pteroclidae, Raphidae,

Order Cuculiformes
- Families - Cuculidae, Musophagidae

Order Psittaciformes
- Family Psittacidae

Order Strigiformes
- Families - Strigidae, Tytonidae

Order Caprimulgiformes
- Families - Aegothelidae, Caprimulgidae, Nyctibiidae, Podargidae, Steatornithidae

Order Micropodiformes
- Families - Hemiprocnidae, Micropodidae, Trochilidae

Order Coliiformes
- Family Coliidae

Order Trogoniformes
- Family Trogonidae

Order Coraciiformes
- Families - Alcedinidae, Bucerotidae, Coraciidae, Meropidae, Motmotidae, Todidae, Upupidae

Order Piciformes
- Families - Bucconidae, Capitonidae, Galbulidae, Indicatoridae, Picidae, Ramphastidae

Order Passeriformes
- Families - Acanthisittidae, Alaudidae, Artamidae, Atrichornithidae, Bombycillidae, Callaeidae, Campephagidae, Certhiidae, Chamaeidae, Chloropseidae, Cinclidae, Conopophagidae, Corvidae, Cotingidae, Cracticidae, Dendrocolaptidae, Dicaeidae, Dicruridae, Drepanididae, Eurylaimidae, Formicariidae, Fringillidae, Funariidae, Grallinidae, Hirundinidae, Icteridae, Laniidae, Meliphagidae, Menuridae, Mimidae, Mniotiltidae, Motacillidae, Muscicapidae, Nectariniidae, Oriolidae, Oxyruncidae, Paradisaeidae, Paridae, Philepittidae, Phytotomidae, Pipridae, Pittidae, Ploceidae, Prunellidae, Ptilogonatidae, Ptilonorhynchidae, Pycnonotidae, Rhinocryptidae, Sittidae, Sturnidae, Sylviidae, Thraupidae, Timaliidae, Troglodytidae, Turdidae, Tyrranidae, Vangidae, Vireonidae, Zosteropidae

Class Mammalia
- Subclass Prototheria
  - Order Monotremata
- Subclass Allotheria*
  - Order Multituberculata
- Subclass Theria
  - Infraclass Pantotheria*
    - Order Pantotheria
    - Order Symmetrodonta

Infraclass Metatheria
- Order Marsupialia
  - Families - Caenolestidae, Dasyuridae, Didelphidae, Macropodidae, Notoryctidae, Peramelidae, Phalangeridae, Phascolomidae, Polydolopidae

Infraclass Eutheria
- Cohort Unguiculata
  - Order Insectivora
    - Families - Chrysochloridae, Erinaceidae, Macroscelididae, Potamogalidae, Solenodontidae, Soricidae, Talpidae, Tenrecidae
  - Order Dermoptera
  - Order Chiroptera
    - Suborder Megachiroptera
    - Suborder Microchiroptera
  - Order Primates
    - Suborder Tupaioidea
    - Suborder Lemuroidea
      - Families - Indridae, Lemuridae
    - Suborder Daubentonioidea
    - Suborder Lorisoidea
      - Families - Galagidae, Lorisidae
    - Suborder Tarsioidea
    - Suborder Anthropoidea
      - Superfamily Ceboidea
        - Families - Callithricidae, Cebidae
      - Superfamily Cercopithecoidea
        - Families - Cercopithecidae, Colobidae
      - Superfamily Hominoidea
        - Families - Hominidae, Pongidae
  - Order Tillodontia*
  - Order Taeniodonta*
  - Order Edentata
    - Families - Bradypodidae, Dasypodidae, Myrmecophagidae
  - Order Pholidota
    - Family Manidae
- Cohort Glires
  - Order Lagomorpha
    - Families - Leporidae, Ochotonidae
  - Order Rodentia
    - Families - Anomaluridae, Aplodontidae, Bathyergidae, Capromyidae, Castoridae, Caviidae, Chinchillidae, Cricetidae, Ctenodactylidae, Dasyproctidae, Dipodidae, Echimyidae, Erethizontidae, Geomyidae, Gliridae, Heteromyidae, Histricidae, Hydrochoeridae, Muridae, Octodontidae, Pedetidae, Platacanthomyidae, Rhizomyidae, Sciuridae, Spalacidae, Zapodidae
- Cohort Mutica
  - Order Cetacea

Suborder Archaeoceti
Suborder Odontoceti
Suborder Mysticeti

Cohort Ferungulata
- Order Carnivora
  - Suborder Creodonta*
  - Suborder Fissipedia
    - Families - Canidae, Felidae, Hyaenidae, Mustelidae, Procyonidae, Ursidae, Viverridae
  - Suborder Pinnipedia
    - Families - Odobaenidae, Otariidae, Phocidae
- Order Condylarthra*
- Order Litopterna*
- Order Notoungulata*
- Order Astrapotheria*
- Order Tubulidentata
- Order Pantodonta*
- Order Dinocerata*
- Order Pyrotheria*
- Order Proboscidea
  - Family Elephantidae
- Order Embrithopoda*
- Order Hyracoidea
- Order Sirenia
- Order Perissodactyla
  - Families - Equidae, Rhinocerotidae, Tapiridae
- Order Artiodactyla
  - Suborder Suiformes
    - Families - Hippopotamidae, Suidae, Tayassuidae
  - Suborder Tylopoda
    - Family Camelidae
  - Suborder Ruminantia
    - Families - Antilocapridae, Bovidae, Cervidae, Giraffidae, Tragulidae